T.J. BRALOWER

THE GEOLOGICAL ASSOCIATION OF CANADA SPECIAL PAPER 27

JURASSIC-CRETACEOUS BIOCHRONOLOGY AND PALEOGEOGRAPHY OF NORTH AMERICA

The Proceedings of a Symposium held
in honour of Ralph W. Imlay and George W. Jeletzky
at the 3rd North American Paleontological Convention
in Montreal, Quebec, August 6th, 1982

Edited by
Gerd E. G. Westermann

1984

G.E.G. Westermann
Department of Geology
McMaster University
Hamilton, Ontario L8S 4M1

Canadian Cataloguing in Publication Data

North American Paleontological Convention (3rd: 1982: McGill University)
Jurassic-Cretaceous biochronology and paleogeography of North America : the proceedings of a symposium held in honour of Ralph W. Imlay and George W. Jeletzky at the 3rd North American Paleontological Convention in Montreal, Quebec, August 6th, 1982

(The Geological Association of Canada special paper; 27)
Includes summaries in English and French.
ISBN 0-919216-23-4

1. Geology, Stratigraphic–Jurassic–Congresses. 2. Geology, Stratigraphic–Cretaceous–Congresses. 3. Paleontology–Jurassic–Congresses. 4. Paleontology–Cretaceous–Congresses. 5. Paleontology–North America–Congresses. I. Imlay, Ralph W. (Ralph Willard), 1908- II. Jeletzky, George W. III. Westermann, G.E.G. (Gerd Ernst Gerold), 1927- IV. Geological Association of Canada. V. Title. VI. Series.

QE701.N67 1982 551.7′66′097 C85-090004-2

International Standard Book Number: 0-919216-23-4

Geological Association of Canada
Department of Geology
Memorial University of Newfoundland
St. John's, Newfoundland
Canada A1B 3X5

Printed by: University of Toronto Press

CONTENTS

PREFACE

This volume is dedicated to two great North American paleontologists, Ralph W. Imlay of the United States Geological Survey and George W. Jeletzky of the Geological Survey of Canada, each of whom has contributed uniquely to Jurassic and Cretaceous research. Special tributes are given by their close associates Earle Kauffman and Tim Tozer.

The research of the two honourees extends over four decades and through the Jurassic and Cretaceous systems of the entire North American subcontinent from Mexico to the Arctic Islands and from the Atlantic to the Pacific, not to mention their work on other continents. Although their main concerns are the classification and distribution of ammonoids, bivalves and belemnites, they have used these molluscs extensively to unravel regional and global problems of: 1) paleogeography, including the past distributions of the inorganic and organic environments; and 2) biochronology in the construction of super-regional timescales.

In the early planning of this symposium, papers from scientists whose work is at the forefront of research in this area were invited. The topics included not only North American Jurassic and Cretaceous invertebrate megafossils and environments of deposition (the special interests of the honourees) but also the Jurassic origin and evolution of the North Atlantic Basin in the Proto-Atlantic, based on evidence from microfossils and deep-sea drilling, and the interrelation of ammonite distribution with allochthonous terrane reconstructions.

The invited specialists met with Drs. Imlay and Jeletzky to present Symposium #7 of the Third North American Paleontological Convention in Montreal on August 6, 1982. Two papers presented at the symposium were not available for publication; however, two additional papers (Longoria, and Westermann *et al.*) are included. Further, the papers by Callomon and by Jeletzky have been extended to encompass a wealth of original data and implications beyond that given in their earlier presentations.

Rather than variations on a single theme, the articles in this symposium volume are like the different movements of a large symphony on different modern themes from contemporary Jurassic and Cretaceous research of North America. Dissonances occur here and there, but the movements on biochronology, paleogeography (including paleogeology and paleobiogeography) and plate tectonics are logically interconnected. Continent-wide and, even more so, global understanding is possible only with strict time frames in the development of "four-dimensional" biomes. Biogeographic provinces change with time and can often be understood only when plotted on pre-drift globes, but the rapid frame changes required for good resolution need superior chronostratigraphy.

The *Cretaceous System of the Western Interior of North America* has been the topic of a previous publication in this Special Paper series (No. 13, edited by W.G.E. Caldwell, 1975), therefore, this volume concentrates on the Jurassic and peripheral North America, i.e., the ancestral North Atlantic, Mexico, the Cordillera and the North Slope.

The papers have been arranged geographically, i.e., clockwise from the eastern trailing edge of the subcontinent and the North American Basin of the Proto-Atlantic, to the more complicated Gulf region and Mexico, to the largely active Pacific leading edge, the Arctic, and finally, the cratonic interior. This leads to an inevitable mixture of taxa and of research methods applied but this cannot be helped.

The editor and the Geological Association of Canada gratefully acknowledge the generous support of the National Science and Engineering Research Council, the Geological Survey of Canada, and the Canadian Geological Foundation in providing grants towards the publication of this volume.

G.E.G. Westermann

RALPH W. IMLAY

It is my honor and privilege to present to you the world's most distinguished authority on Jurassic and Cretaceous biotas, stratigraphy and paleogeography, Dr. Ralph W. Imlay, Senior Paleontologist of the United States Geological Survey and Honorary Research Associate of the Department of Paleobiology at the Smithsonian Institution in Washington, D.C.

For Mesozoic specialists of my generation, and all that will follow, it seems that everywhere we go in the Western Hemisphere in search of interesting research problems, Ralph Imlay has been there before us and his pioneering geological investigations are already legend. In my own career, I followed Ralph to the University of Michigan, taught courses he had taught, became interested in Mesozoic mollusks using his collections, launched my professional career at the United States National Museum under his tutorage, and have followed his paleontological footprints all over the world – in Mexico, Alaska, the Caribbean, Saudi Arabia and the American Rockies. Ralph is the recognized leader in our science, whose meticulous research and interpretive skills have set a standard for the profession, and inspired many of us to seek similar levels of excellence. I am proud to count Ralph as my close friend, colleague, and mentor. He is a remarkable human being whose quiet, friendly, and unassuming manner belies the great magnitude of his contributions to paleontology, stratigraphy, and regional geology, and his widely acclaimed stature as the best in the world.

Ralph was born on February 21, 1908 in Hampton, Iowa and shortly afterward moved with his family to Reed Point, Montana, where he received his primary and secondary education. He attended the University of Montana, receiving his B.Sc. in Geology in 1930, and then moved to the University of Michigan where he worked with the great E.C. Case and L.B. Kellum. Here Ralph became interested in Kellum's pioneer geological and biological studies of interior Mexico. Through his Doctoral thesis work and subsequent contributions, Ralph became an integral part of the University of Michigan's Mexican Research team, with primary responsibilities for Jurassic and Lower Cretaceous regional geology, stratigraphy, and paleontology. His research in Mexico was to become the first comprehensive study of the Middle Mesozoic in this region, and had such a profound impact on the geological development of this nation that, in 1968, Ralph was made an Honorary Professor of Geology at the University of Mexico – only the fourth such honorary professorship to be granted in the history of this, the oldest University in the Western Hemisphere, since it's establishment in the 1500s.

Ralph received his Ph.D. from the University of Michigan in 1933, and after two years of teaching at Rutgers, and four years at Michigan between 1934 and 1940, he joined the Paleontology and Stratigraphy Branch of the U.S. Geological Survey where he continued his pioneering work in Mexico through 1945 as part of the United States Governments' Point 4 Program in that country. In 1940, Ralph began entirely new research on the Jurassic and Lower Cretaceous of the United States Gulf Coastal Plain, and adjacent Caribbean Islands. This pioneering study integrates surface and subsurface data to document the highly complex Mesozoic geological history of this important petroleum province. This work remains a modern standard. He was subsequently a leading figure in the formulation of the Geological Society of America-sponsored correlation charts for this region, as well as for those covering the Jurassic of North America. In 1945, Ralph began a 38 year investigation of the Jurassic stratigraphy, paleontology and paleogeography of the Western Interior, Pacific Coast, and Alaskan regions of North America which recently culminated in his world-known synthesis – *The Jurassic Paleogeography Of The Conterminous United States And Its Continental Setting,* published as U.S. Geological Survey Professional Paper 1062.

Ralph's international reputation as a Mesozoic stratigrapher and paleontologist grew steadily from his early work in the Caribbean and Mexican regions, through his studies in the United States, and led to a series of justly deserved awards, including an honorary Ph.D. from the University of Montana in 1959, his election as President of the Paleontological Society in 1964, his Honorary Professorship at the University of Mexico in 1968, and his receipt of the United States Government's Distinguished Service Award in 1974. He is an elected Fellow of the Geological Society of America and the Society of Economic Paleontologists and Mineralogists; he was an early moving force in the JOIDES Deep Sea Drilling Program, and an original member of its Paleontology and Stratigraphy Panel.

Ralph Imlay's contribution to our science has been immense, and especially in the fields of systematic paleontology, biostratigraphy, paleogeography, and regional geology. His impact is far greater than the more than 100 major publications, 2500 pages, and innumerable maps and charts that he has published. Ralph has, in addition, been a one-man service organization for field geologists working in Jurassic and Lower Cretaceous rocks throughout North America – identifying fossils, interpreting stratigraphy and paleoenvironments, providing dates and precise correlations, and generally counseling the North American Jurassic community for over 30 years. The main evidence of his impact on our science lies among reams of unpublished service reports in the open files of the United States Geological Survey which exceed, in magnitude, his published works. In addition to his important contributions to American Mesozoic studies Ralph has done definitive work on the Jurassic ammonites of Saudi Arabia and the Antarctic continent. Ralph has named and described more molluscan species, and has had more named after him, than almost any single living Mesozoic paleontologist. Yet to see him at work in Washington, and to hear his enthusiasm at this conference it seems as if he is just getting started.

Upon his official retirement from the Paleontology and Stratigraphy Branch of the U.S. Geological Survey, Ralph quietly smiled and confided in me that now he could really get down to work and do some paleontology – and so he has, leading us through the Jurassic, stage by stage, and ammonite by ammonite. Let us hope that he never changes his mind or his style; because one wonders in this day of machine-generated theory, models, formulas, and imaginary fossils, if there will ever be another Ralph Imlay or an equivalent set of real paleontologic and stratigraphic data, so lovingly and meticulously collected, as that which he has generated and left to us as his heritage. We hope so, and we thank you Ralph, for being among us through all of those prolific years, and for being with us today to receive our admiration.

E.G. Kauffman,
Department of Geological Sciences,
University of Colorado,
Boulder, Colorado 80302

J.A. JELETZKY

It is both appropriate and gratifying that the symposium "Jurassic and Cretaceous Biochronology and Biogeography of North America", held at the Third NAPC, Montreal, August 1982, was also a festival honouring Dr. Ralph W. Imlay and Dr. J.A. Jeletzky.

Dr. Jeletzky was born Jurij Alexandrovich Romanov at Pensa, Russia, in 1915. In 1930 he took the name Jeletzky, upon adoption by his stepfather. When he came to Canada he exchanged Jurij for George and it is by that name he is known to a host of friends and colleagues throughout the world.

His professional training was in Kiev, graduating as a geologist in 1938 and completing his graduate work in 1941. For more than 40 years he has been actively engaged in research in stratigraphy and paleontology. His early years of professional activity were in Russia, Poland and Germany. He was in Kiev when that city fell to the German armies in September 1941. With the return of the Soviet Forces in 1943 he moved west, first to Poland, then to Germany. At times he worked as a librarian but throughout these difficult times he managed to keep not only his family together, but also his collection of Cretaceous belemnites, which might be described as his paleontological first love. Several publications on the Cretaceous stratigraphy and belemnite faunas of Europe resulted from this early work. George's achievements during this period testify to the fortitude and determination that all his colleagues know and respect.

In 1948 he came to Canada where employment with the Geological Survey of Canada had been arranged. He is still associated with this organization, for, although he officially retired in 1981, he continues to work energetically. He himself describes his energy as somewhat diminished these days but as the occupant of a neighbouring office, I would say that this brings his level of activity to about 1.5 instead of 2.0 of the ordinary mortal.

His main contribution to the work of the Geological Survey has been to deal with Cretaceous stratigraphy and molluscan faunas, not only those that he has collected himself, but also large collections submitted for his assessment by colleagues in the Geological Survey, Provincial Surveys and oil and mining companies. His international reputation has also led to his working on collections submitted by the Deep Sea Drilling Project.

The breadth of his interests and activities has made a major impact in about half a dozen areas of the science.

His first field work in Canada was on Vancouver Island and over the years he has maintained an interest, and made further contributions there. His original task was to sort out the Cretaceous but ended up elucidating the whole Mesozoic and Cenozoic history to bring the state of knowledge on the rock and faunal succession to a state unmatched in the whole of the western Canadian Cordillera, indeed perhaps in the whole of the Cordillera. I write with some authority on this matter because it has fallen my lot to study the outstandingly important collections of Triassic fossils which he obtained, with important data on sequence, derived from his study of an area of great structural complexity. George's work on Vancouver Island led him to make further studies on the correlative rocks on the mainland of southern British Columbia.

The second main area of his field activity has been the northern Yukon, where he made many new finds, for example, the first discovery of marine Hauterivian, Barremian and Aptian.

Always one who has enjoyed challenge and controversy, George's work, which has generally been in areas of structural complexity, has always whetted his interest in tectonic problems. Similarly his work on correlating with fossils has led him to take a great interest in paleogeography and paleozoogeography. In consequence he has written extensively on topics such as tectonics, paleogeography and eustasy.

Another major field of activity has been his work on *Buchia*, the bivalve so important in Upper Jurassic-lowermost Cretaceous bichronology. George fought like a tiger for the retention of the name *Aucella* for these shells but eventually deferred to ruling of the International Commission on Zoological Nomenclature! For this work he has drawn not only on his own collections but on material from all over the world and has devised a biochronological scheme more refined than hitherto available.

Lower Cretaceous ammonoids, are a group that he has actively studied and on which he has published extensively. The faunas critical to defining and correlating the Jurassic-Cretaceous boundary have been a particular object of his studies.

Another field in which he has made a notable mark is in the study of the Coleoidea, which include the belemnites, his first love. Here, as with the buchias, he has not restricted his studies to his own collections. Recognized early on as the principal authority, he was invited to be coordinator of the Coleoidea Volume of the "Treatise on Invertebrate Paleontology". Although this volume has not yet appeared, George has published substantially on coleoids of all the Mesozoic systems from all over the world towards clearing the air so that the definitive summary can be prepared.

No account of Dr. Jeletzky's activities would be complete without mention of his writings on the methodology and philosophy of biochronology in which he has aggressively and effectively championed the unmatched, and he would probably say unmatchable, role that fossils play in the establishment of the Phanerozoic time scale.

For all this he has been rightly honoured: Fellow and Miller Medalist of the Royal Society of Canada; Billings Medalist of the Geological Association of Canada. That this volume is in his honour is a fully justified compliment that is coupled with the hope that his activity may long continue.

E.T. Tozer,
Geological Survey of Canada,
601 Booth Street,
Ottawa, Ontario K1A 0E8

Jurassic-Cretaceous Biochronology and Paleogeography of North America,
edited by G.E.G. Westermann,
Geological Association of Canada Special Paper 27, 1984

JURASSIC AMMONITE SUCCESSIONS IN NORTH AMERICA AND BIOGEOGRAPHIC IMPLICATIONS

R.W. Imlay
United States Geological Survey, E-501, United States National Museum, Washington, D.C. 20560

ABSTRACT

During the Early Jurassic, marine waters covered areas now represented by southern Mexico, Sonora, Nevada, Idaho, British Columbia, southern Alaska, northern Alaska, northernmost Canada, and East Greenland. The ammonites that lived in those waters are very similar to those of the same ages in western Europe and, likewise, during late Pliensbachian time, changed southward from a faunule dominated by Amaltheus *to a faunule dominated by Hildoceratid genera.*

During the Middle Jurassic, marine waters covered most of the same areas as during the Early Jurassic as well as a large area in the Western Interior Region. The ammonite succession, as in Europe, changed southward from being mostly boreal in Alaska to being mostly Tethyan in Mexico but from Oregon southward it included some genera that are indigenous to the eastern part of the Pacific Ocean.

During the Late Jurassic, marine waters covered most of the same areas during the Middle Jurassic. They differed by withdrawing from most of the Western Interior region of the United States by the end of the middle Oxfordian and by extending over a large area in the Gulf of Mexico region during and after the late middle Oxfordian.

During the entire Jurassic those marine waters received a variety of sediments of which most have been fairly accurately dated by ammonites. Such dating has made possible reasonable interpretations of the major events in the Jurassic history of the North American continent.

RÉSUMÉ

Pendant le Jurassique Inférieur, des eaux marines ont recouverte des régions qui sont présentement repésentées par: le Mexique méridional, le Sonora, le Névada, l'Idaho, la Colombie-Britanique, l'Alaska méridionale et septentrionale, les régions les plus au nord du Canada et l'est du Groendland. Les ammonites qui ont vécu dans ces eaux sont très similaires à celles du même âge, en Europe de l'ouest, et en outre, pendant la fin du Pliensbachien, varient méridionalement d'un faunule dominé par Amaltheus *à un faunule dominé par* Hildoceratid *genra.*

Pendant le Jurassique Moyen, des eaux marines recouvraient encore la plupart de ces mêmes régions de la période Jurassique Inférieure ainsi qu'une grande partie de la région de l'ouest intérieure. La sucession des ammonites, comme en Europe, varie méridionalement d'un assemblage principalement boréal en Alaska, à un assemblage Tethyen au Mexique. Cependant à partir de l'Oregon et plus au sud, la succession ammonitique inclus quelques genres qui sont indigènes à l'est de l'Océan Pacifique.

Pendant le Jurassique Supérieur, des eaux marines occupaient toujours la plupart des mêmes régions que précédemment. Elles diffèrent cependant à causedu recul des eaux de la région interne de l'Ouest des Etats-Unis, à la fin de l'Oxfordien moyen, et à cause du recouvrement d'une grande partie de la région du Golfe du Mexique durant et après la fin de l'Oxfordien moyen.

Durant la totalité du Jurassique, ces eaux marines ont reçu une variété de sédiments pour la plupart datés assez précisément par les ammonites. Ces dates ont permis une interprétation raisonnable des évènements majeurs survenus pendant le Jurassique sur le continent Nord-Américain.

INTRODUCTION

The interpretation and ideas presented herein are based mainly on data concerning the Jurassic ammonite successions and geographic occurrences in conterminous United States as presented by the writer in 1976 and published in 1980 in U.S. Geological Survey Professional Paper 1062. Since 1976, however, much additional data have been presented by the writer in

IUGS UNESCO
**Project 171:
Circum-Pacific Jurassic**

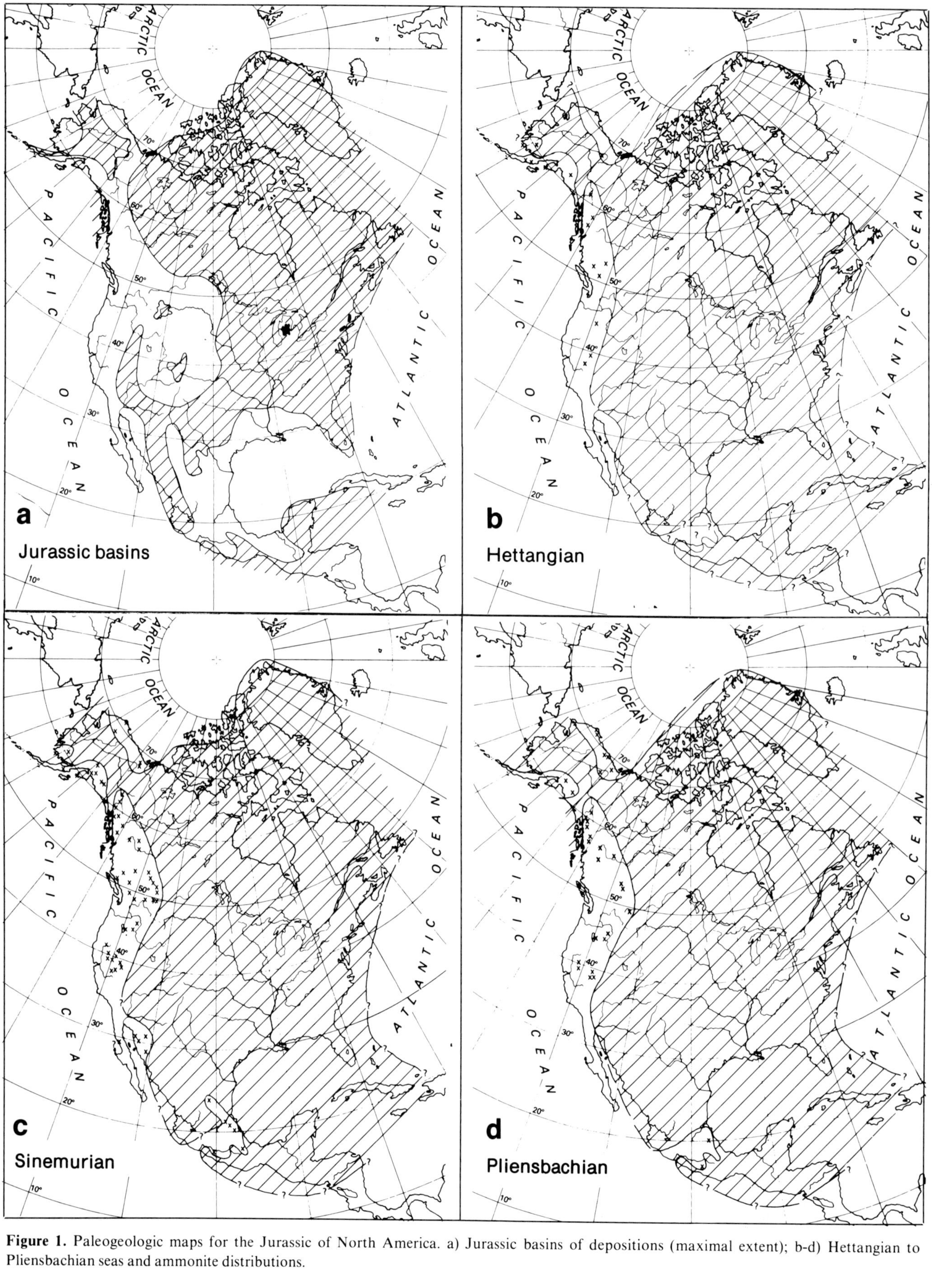

Figure 1. Paleogeologic maps for the Jurassic of North America. a) Jurassic basins of depositions (maximal extent); b-d) Hettangian to Pliensbachian seas and ammonite distributions.

seven additional Professional Papers. Of these papers, five deal with ammonites of Early to Late Jurassic ages obtained from various parts of Alaska, one with ammonites of Bathonian and Callovian ages from eastern Oregon and westernmost Idaho, and one with ammonites of late Callovian to middle Oxfordian ages from the Western Interior Region. Those Professional Papers that have been published are listed in the References.

Detailed data showing the probable distribution of Jurassic seas and the Jurassic ammonite succession in North America have been presented previously (Imlay, 1980a, Figs. 1-15) and are readily accessible to readers. The locations of the Jurassic basins of deposition in North America are shown on Figure 1a.

EARLY JURASSIC AMMONITE SUCCESSIONS AND BIOGEOGRAPHIC IMPLICATIONS

Hettangian ammonites have been found in western Nevada, eastern Oregon, southern Yukon, and in southern and northern Alaska (Fig. 1b). The ammonite succession in those areas appears to be essentially the same as in western Europe (Fig. 2). The earliest Hettangian is represented by a fairly smooth species of *Psiloceras* and by the ribbed subgenus *P.* (*Franziceras*); the middle Hettangian by *Waehneroceras* and *Discamphiceras*; and the upper Hettangian by *Schlotheimia.*

Sinemurian ammonites occur in the same area as Hettangian ammonites as well as in east-central, southern, and northwestern Mexico (Fig. 1c). Their succession is, likewise, similar to that in western Europe (standard zones in parentheses) except possibly during the middle Sinemurian, which is not represented in most areas in North America. Thus the basal European Sinemurian zone of *Arietites bucklandi* (Bucklandi Zone) is well represented by *Coroniceras* and in some areas by *Arietites, Charmasseiceras,* and *Paracaloceras.* The next higher zone of *Arnioceras semicostatum* (Semicostatum Zone) is represented in most areas by *Arnicoeras* and *Coroniceras* (*Paracoroniceras*). The next younger zone of *Caenisites turneri* (Turneri Zone) has not been identified in North America. The succeeding zone of *Asteroce-*

Stages	E. Ctr. Oregon (Imlay 1968)	Canadian Rocky Mountains (Frebold 1969; Frebold & Tipper 1970)	S. British Columbia (Frebold 1964a; Frebold & Tipper 1970)	S. Yukon & N. British Columbia (Frebold & Tipper 1970)	S. Alaska (Imlay 1968, 1981a; Imlay & Detterman 1973)	N. Alaska (Imlay 1955; Imlay & Detterman 1973)
Toarcian U.	*Catulloceras Dumortieria?*	*Phylseogrammoceras Grammoceras*	*Phylseogrammoceras Grammoceras*	*Catulloceras Grammoceras*	 *Grammoceras*	 *Pseudolioceras*
Toarcian	*Haugia Polyplectus*	*Phymatoceras Peronoceras*		*Phymatoceras Peronoceras*	*Haugia Phymatoceras*	
Toarcian L.	— ? —	*Harpoceras* cf. *exaratum Dactylioceras*	*Harpoceras* cf. *exaratum Dactylioceras*		*Dactylioceras* cf. *commune* — ? —	*Dactylioceras* cf. *commune* *D.* cf. *semicelatum*
Pliensbachian U.	*Arieticeras, Reynesoceras, Paltarpites Prodactylioceras Leptaleoceras* — ? —	*Pleuroceras Dactylioceras* *Amaltheus*		*Pleuroceras, Amaltheus, Paltarpites, Arieticeras Leptaleoceras*	*Pleuroceras, Amaltheus, Paltarpites*	*Pseudoamaltheus* *Amaltheus margaritatus* *A. stokesi*
Pliensbachian L.	— ? —	— ? — *Phricodoceras*	— ? — *Fanninoceras* *Acanthopleuroceras Uptonia, Tropidoceras* — ? —	*Prodactylioceras davoei, Becheiceras* — ? — *Platypleuroceras?*	*Uptonia, Apoderoceras Acanthopleuroceras*	
Sinemurian U.	*Eoderoceras Crucilobiceras* — ? —	*Eoderoceras Gleviceras*	*Echioceras* *Asteroceras*		*Crucilobiceras* *Microderoceras*	*Crucilobiceras*
Sinemurian L.	— ? — *Arnioceras* — ? —	*Arietites Arnioceras*	*Coroniceras bisulcatum Arnioceras*	*Paracoroniceras Arniotites*	— ? — *Arnioceras* — ? — *Coroniceras*	*Arnioceras* *Coroniceras*
Hettangian	*Schlotheimia Alsatites* — ? —	(? absent)	*Psiloceras canadense Paracaloceras Charmasseiceras*	*Psiloceras canadense*		— ? —
Hettangian	*Waehneroceras Psiloceras*		*Psiloceras* aff. *planorbis*	*Psiloceras erugatum*	*Waehneroceras Psiloceras*	*Psiloceras*

Figure 2. Succession and correlation of Early Jurassic ammonite faunas in North America.

ras obtusum (Obtusum Zone) may be represented by reported occurrences of *Asteroceras* in southern British Columbia (Frebold and Tipper, 1970) and in Shasta County, California (Sanborn, 1960, p. 11-14). The succeeding zone of *Oxynoticeras oxynotum* (Oxynotum Zone) is represented by occurrences of *Oxynoticeras* in Mexico (Erben, 1956, Pl. 37, Fig. 1; Hallam, 1965) western Nevada (Hallam, 1965) northwestern Canada (Frebold, 1964b), Table 1) and the Canadian Arctic islands. The highest Sinemurian zone of *Echioceras raricostatum* (Raricostatum Zone) is characterized by *Echioceras* and by the lowest occurences of *Crucilobiceras* at many places from southern Mexico to northern Alaska and to northeastern Canada (Imlay, 1980a, Fig. 13).

Pliensbachian ammonites have been found in North America in the same general areas as the Sinemurian ammonites, but at fewer localities in Mexico, Alaska (Imlay, 1981a) northwestern Canada and Arctic Canada (Fig. 1d). The basal Pliensbachian zone of *Uptonia jamesoni* (Jamesoni Zone) is possibly represented in southern Mexico and in western Nevada by ammonites that are assigned questionably to *Uptonia* (Hallam, 1965). It is probably represented by *Uptonia* in southern British Columbia (Frebold and Tipper, 1970), in the Wrangell Mountains in southern Alaska (Imlay, 1981a, Pl. 9, Figs. 1-4, 8, 12-16), and on Lisburne Ridge between the Ipnavik and Anaktuvik Rivers in north-central Alaska (Imlay, 1981a, Pl. Fig. 17). This zone is also represented in the Talkeetna Mountains in south-central Alaska by one occurrence of *Apoderoceras* (Imlay, 1981a, Pl. 8, Figs. 14, 18-23). The zone of *Tragophylloceras ibex* (Ibex Zone) is represented by the Wrangell Mountains at U.S. Mesozoic Locality 28534 by *Tropidoceras actaeon* (d'Orbigny) (Imlay, 1981a, Pl. 8, Figs. 1-9) in association with *Crucilobiceras* (Imlay, 1981a, Pl. 8, Figs. 10-12, 15-17). The succeeding zone of *Prodactylioceras davoei* (Davoei Zone) is possibly represented in the southern Yukon and in northern British Columbia by *Prodactylioceras* (Frebold, 1964b, Pl. 3, Fig. 2) in association with *Becheiceras* (Frebold, 1964b, Pl. 3, Fig. 1; Pl. 4, Fig. 1; Pl. 5, Fig. 1). It is possibly represented, also, in the Wrangell Mountains by *Prodactylioceras* at the top of its range above *Tropidoceras*.

The highest European Pliensbachian zones have been dated, by the ammonite *Arieticeras* in areas between southern Mexico (Erben, 1957; Hallam, 1965; Cameron and Tipper, 1981) and southern Alaska (Imlay, 1981a). That genus in eastern Orgeon, western and northern British Columbia, southern Yukon and southern Alaska is associated with specimens of *Protogrammoceras, Fontanelliceras* and *Leptaleoceras* that are identical, or nearly identical with species in the Mediterranean region. Those genera are likewise associated rarely with *Amaltheus* in the Talkeetna Mountains of southern Alaska (Imlay and Detterman, 1973) but are fairly common in the Tulsequah area of northwestern British Columbia and in the Whitehorse area of southernmost Yukon territory (Frebold and Tipper, 1970).

A very different ammonite faunule occurs in beds of late Pliensbachian age in northern and east-central Alaska, northwestern Canada, and the Canadian Arctic islands (Tipper, 1981; Imlay, 1981a). That faunule is dominated by the genus *Amaltheus*, which is represented by the same species in the same succession as in northwestern Europe and is not associated with the Hildoceratid genera *Arieticeras, Protogrammoceras, Fontanelliceras*, or *Leptaloceras*. These faunal differences between northern and southern Alaska are identical with those between northwestern Canada and areas to the south in the southern Yukon and northern British Columbia. They are, likewise, essentially identical with southward changes in ammonite assemblages between northwestern Europe and the Mediterranean Region (Donovan, 1957; Hallam, 1969). Evidently the southern boundary of abundant Amaltheidae in Europe is hundreds of kilometers farther south than the southern boundary in Alaska and the southern Yukon. This difference could be explained: 1) by faulting in which major blocks containing Hildoceratid ammonites were moved far north of their original positions (Jones, *et al.*, 1977, Fig. 1; Coney *et al.*, 1980, Fig. 1; Tipper, 1981), 2) by present latitudes being somewhat different from those of Jurassic time, or 3) by variations in the latitudinal position of faunal provinces due to changes in ocean currents.

Toarcian ammonites have been found in westernmost North America north of Mexico in the same areas as the Pliensbachian ammonites (Fig. 3a). The succession of Toarcian ammonites is similar to that in western Europe. In western Nevada the sequence from the base upward includes: 1) *Nodicoeloceras* of early Toarcian age, 2) *Catacoeloceras* of early to late Toarcian age, and 3) *Grammoceras* and *Pseudolioceras* of latest Toarcian age. In eastern Oregon the late Toarcian is represented (1) by *Haugia* and *Polyplectus* that correlate with the European zone of *Haugia variabilis*, and (2) by *Catulloceras* of latest Toarcian age. In the Talkeeta Mountains of southern Alaska the lowest lower Toarcian is represented by *Dactylioceras* and *D.* (*Orthodactylites*) *kanense;* the upper lower Toarcian by *Harpoceras* cf. *H. exaratum* and *Eleganticeras;* the lower upper Toarcian by *Haugia, Brodeia* and *Pseudolioceras;* and the uppermost Toarcian by *Grammoceras* and *Pseudolioceras* (Imlay, 1981a). In the DeLong Mountains of northwestern Alaska the lower part of the lower Toarcian is represented by *Dactylioceras* (*orthodactylites*) and the upper part of the lower Toarcian by *Harpoceras* and *Eleganticeras*.

These ammonite occurrences show that during Early Jurassic time marine waters from the Pacific Ocean extended as much as 800 kms east of the present western margin of the North American continent (Figs. 1a-c, 3a). In southern Mexico the sea extended northeastward from the Pacific Ocean nearly as far as the Tampico area during Sinemurian and possibly during Pliensbachian times. In extreme northwestern Mexico the sea extended eastward several hundred kilometers beyond the present western margin of Sonora. Farther north the sea extended eastward into areas that now include the western half of Nevada, westernmost Idaho and southwestern British Columbia where the sea attained its maximum spread. In southern Alaska the sea extended northward and eastward at least 320 kms.

In the Arctic region during the Early Jurassic marine waters transgressed southward nearly 300 kms into areas now represented by northern Alaska and extreme northwestern Canada. The Arctic sea also extended southward into the area of northernmost Arctic islands during Sinemurian to Toarcian times and into marginal areas in East Greenland during Pliensbachian and Toarcian times.

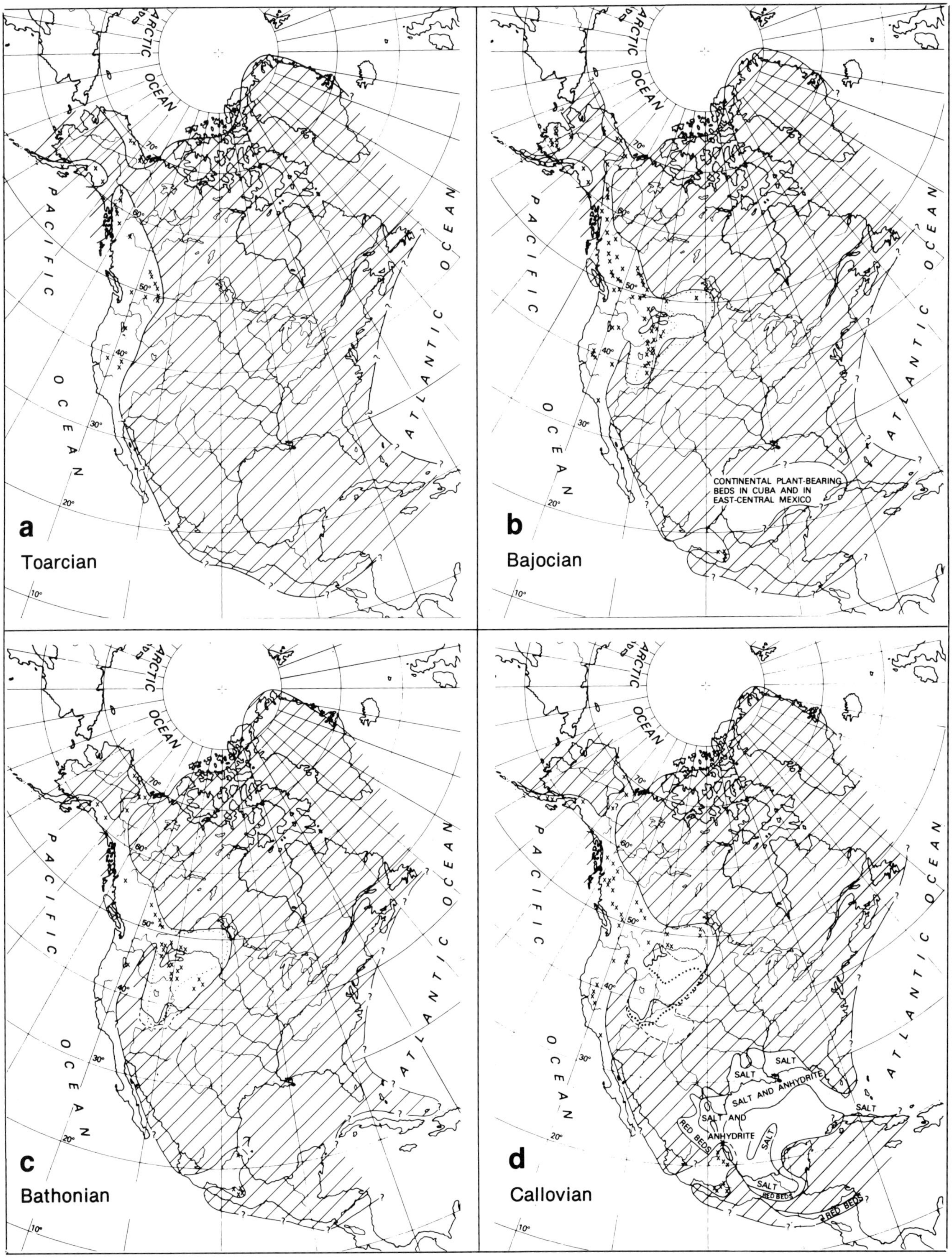

Figure 3. Toarcian to Callovian seas and ammonite distribution in North America.

In the Atlantic region sometime between latest Triassic and Middle Jurassic (Bathonian) times, marine waters penetrated into an area comprising the Scotian Shelf and the Grand Banks of Newfoundland. In those waters were deposited at least 6,100 meters of sediments ranging from bedded salt and anhydrite basally to limestone, mudstone, siltstone and sandstone at higher levels (Bartlett and Smith, 1971; McIver, 1972; Upshaw *et al.*, 1974). The exact age of the oldest sediments needs confirmation.

MIDDLE JURASSIC AMMONITE SUCCESSIONS AND BIOGEOGRAPHY

Aalenian and Bajocian ammonites have been found in North America mostly in the same areas as Early Jurassic ammonites (Fig. 3b). The major exception includes a large area in the Western Interior Region extending from northwestern Montana southward to central Utah and north-eastward to southern Saskatchewan.

The Aalenian is used herein as a stage at the base of the Middle Jurassic rather than as a basal substage of the Bajocian in order to be consistent with other papers that have been published or that will be published soon. Its use has been discussed by Westermann (1979). The matter was essentially settled at a recent international Meeting of the Stratigraphic Commission in Paris.

The ammonite genera present in the Aalenian of North America, as listed herein on Figure 4 and as listed previously by Imlay (1980a, F. 14), include *Pseudolioceras*, throughout the stage and *Tmetoceras, Erycites, Erycitoides, Eudmetoceras* and *Praestrigites* in the upper part of the stage.

The Bajocian ammonite succession in North America as discussed in detail by Westermann (1981) is similar to that in Europe (Imlay, 1973). Exceptions include *Alaskinia, Pseudotoites, Parabigotites, Pseudocidoceras, Arkelloceras* and *Zemistephanus* of early Bajocian age and *Megasphaeroceras* of early late Bajocian age (Imlay, 1965; Hall and Stronach, 1981). Those ammonites are associated with other genera that correlate with the European standard zones of *Hyperlioceras discites* (Discites Zone), *Sonninia ovalis, Wichellia laeviuscula* (Laeviuscula Zone), *Otoites sauzei* (Sauzei Zone), *Stephanoceras humphriesanum* (Humphriesanum Zone), and *Strenoceras subfurcatum* (Subfurcatum Zone), (Mouterde *et al.*, 1971; Parson, 1974; Hall and Westermann, 1980). An exception also includes *Lupherites* (Imlay, 1973) which is dated as early late Bajocian (1) because in eastern Oregon it directly overlies beds containing *Megasphaeroceras* (Imaly, 1973) in association with *Spiroceras, Stemmatoceras?* and *Stephanoceras*, and (2) because it is asso-

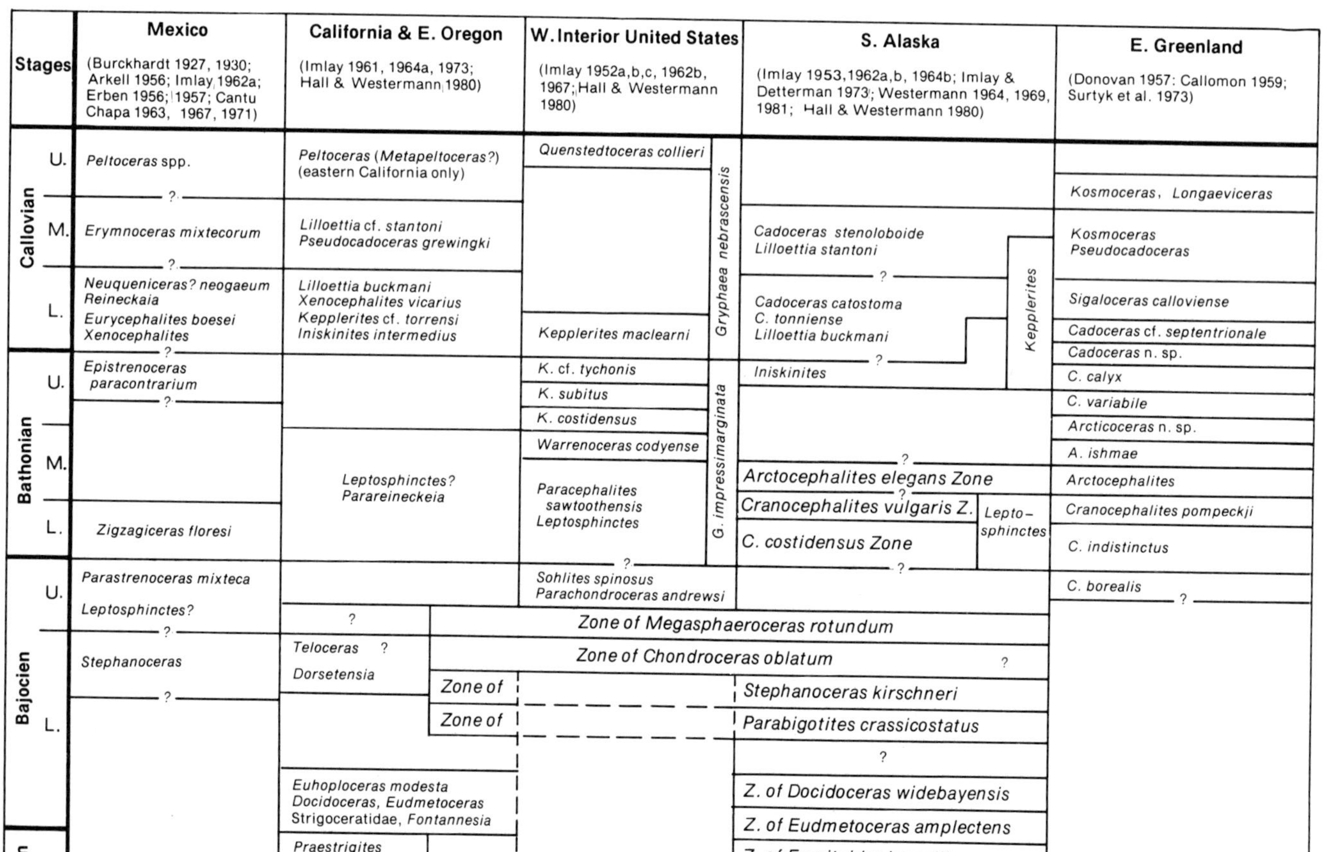

Figure 4. Succession and correlation of Middle Jurassic ammonite faunas in North America.

ciated with *Leptosphinctes* at Luphers locality 30 (Imlay, 1973).

Bathonian ammonites have been found in North America in most of the same areas at Late Aalenian and Bajocian ammonites but have not yet been recorded from California, or from the Alaska Peninsula. They are poorly represented in western British Columbia (Tipper and Cameron, 1980), but are fairly well represented in northern Alaska, between Cook Inlet and the Wrangell Mountains in southern Alaska, and in western Montana. The assignment of certain beds to the Bathonian rather than to the Callovian is uncertain because many of the ammonite genera are Boreal or East Pacific origin and hence may have little value in making correlations with Europe.

The Bathonian ammonite succession in North America differs considerably latitudinally (Imlay, 1980a). In northern Alaska (Imlay, 1976; Imlay and Detterman, 1973), northwestern Canada, and the Canadian Arctic islands the Bathonian ammonites from the oldest to youngest include the Boreal genera *Cranocephalites, Arctocephalites* and *Arcticoceras*. These genera are identical, or closely similar specifically to ammonites in East Greenland (Callomon, 1959). In southern Alaska the ammonite succession is dominated by the same genera plus the Boreal genus *Kepplerites* (Imlay, 1980b). It, also, includes: 1) the genera *Xenocephalites, Iniskinites,* and *Parareineckeia* and *Cobbanites?* which are characteristic of the East Pacific Realm, 2) the genera *Tuxednites, Talkeetnites* and *Chinitnites* which are known only from southern Alaska, 3) *Cadomites* which occurs nearly world-wide except in the Boreal Realm, and 4) in its lower and middle parts include species that were once identified as *Cobbanites* and that are now largely assigned to *Leptosphinctes*.

The Bathonian ammonite succession in the Western Interior Region of the United States and in the Canadian Rocky Mountains (Imlay, 1967, 1980a) is characterized from the base upward: 1) by *Paracephalites* and '*Cobbanites*' of early to early middle Bathonian age, 2) by *Warrenoceras* of probable late middle Bathonian age, and 3) by three successively younger species of *Kepplerites* that are similar to species in Greenland and are of late Bathonian age. *Warrenoceras* is so dated because of its close resemblance to *Arcticoceras*. *Paracephalites* is so dated because of its stratigraphic position and because of its resemblance to *Arctocephalites*. All of these genera except 'Cobbanites' are of Boreal origin, or are closely related to ammonites of Boreal origin.

In eastern Oregon the late Bathonian has been identified by an association of *Bullatimorphites, Xenocephalites, Kepplerites, Torricelliceras, Iniskinites, Cobbanites?, Choffatia,* and *Parareineckeia* (Imlay, 1981b). Of these genera *Kepplerites* and *Torricelliceras* are Boreal in origin, Bullatimorphites and *Choffatia* occur nearly worldwide except in the Boreal Realm, and the remainder represent the East Pacific Realm.

In southern Mexico the earliest and latest Bathonian are represented by *Zigzagiceras* and *Epistrenoceras* which are respectively European and Mediterranean in origin (but see Callomon, this volume; Westermann *et al.*, this volume).

Callovian ammonites have been found in North America in the same areas as Bajocian ammonites except in the Kuskokwim region of southwestern Alaska. In comparisons with Bajocian ammonite occurrences, Callovian ammonites have been found at more localities in southern Mexico, at fewer localities in Alaska and in asnorthwestern Canada, and have not been found in the Western Interior Region south of north-central and northwestern Montana.

In southern Alaska the lower Callovian is represented (Imlay, 1975, 1980a) by *Cadoceras, C. (Paracadoceras), Lilloettia,* and basally by *Iniskinites* and *Xenocephalites*. The middle Callovian is represented by *C. (Stenocadoceras)* and *Pseudocadoceras*. Most of the upper Callovian is represented by an unconformity. The Callovian exposed in British Columbia and in the southern Yukon bears a sequence of ammonites identical to that in southern Alaska except for the presence of *Quenstedtoceras* at the very top of one locality in British Columbia (Frebold and Tipper, 1975). In eastern Oregon the Callovian faunal sequence is similar to that in British Columbia except for the absence of *Cadoceras*. The lower Callovian is represented in the Trowbridge Shale by *Lilloettia, Kepplerites* and *Xenocephalites*. The middle Callovian is represented in the Lonesome Formation basally by *Lilloettia* and throughout by *Xenocephalites* and *Pseudocadoceras* (Imlay, 1981b).

These occurrences of ammonites of Middle Jurassic age show that marine waters from the Pacific Ocean extended into nearly the same area in western North America as during the Early Jurassic. In addition the sea that covered the present area of Oregon, Washington and southern British Columbia extended eastward at least 2000 kms from the present eastern margin of the Pacific Ocean and formed a large inland sea. That sea extended southward as far as southern Utah, eastward across most of the area comprising Wyoming and North Dakota, and northward across at least 250 kms of southern Saskatchewan. The resulting inland sea was surrounded on its north, east, south and southwest sides by land masses from which it received appreciable amounts of sand, silt and clay. It also surrounded a fairly large eastward-trending island (Belt Island) in central and north-central Montana. The western part of that island apparently extended southwestward and may have been an extension of an uplift along the southwest side of the inland sea. That inland sea withdrew completely from the Western Interior Region during middle Bajocian, late Callovian, and late Oxfordian to Tithonian times.

The latitudinal differentiation of the Middle Jurassic ammonite faunules in North America is similar to that in Europe and likewise may be related to differences in temperatures of the sea waters from north to south, to restriction of the sea in the Boreal Region at certain times, to northward movement of the North American continent, or perhaps by extensive movements of fault blocks in the Pacific Coast Region (Coney *et al.*, 1980).

LATE JURASSIC AMMONITE SUCCESSIONS AND BIOGEOGRAPHY

Oxfordian ammonites have been found in North America in the same areas as Callovian ammonites except for their presence in the Gulf of Mexico Region (Fig. 5a, b). Nonetheless, south of British Columbia a marked difference in ammonite distribution occurs near the middle of that stage. Thus ammonites of early to early middle Oxfordian age have not been found anywhere in Oregon, California, Mexico, or in the land areas bordering

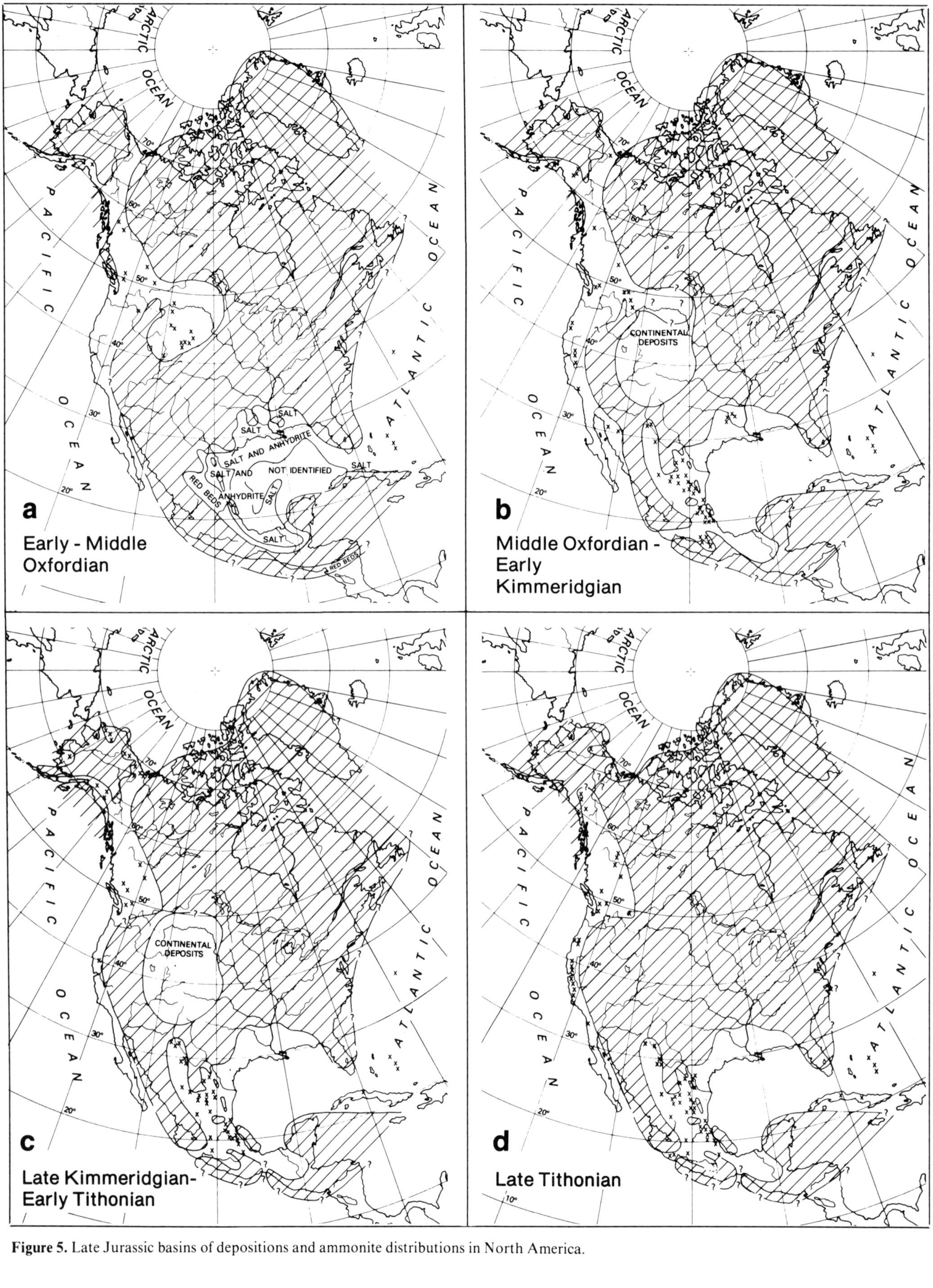

Figure 5. Late Jurassic basins of depositions and ammonite distributions in North America.

the Gulf of Mexico. Ammonites of that age, however, are common in the highest marine Jurassic beds in the Western Interior region, are fairly common from southern British Columbia northward and have not been found south of an occurrence of *Cardioceras (Scarburgiceras)* on the east side of the Snake River in westernmost Idaho about 6.4 kms south of the northeast corner of Oregon (Imlay, 1964a).

In contrast, during late middle to late Oxfordian time (Fig. 5b), ammonites became common: 1) in the Pacific Coast Region; 2) in areas now represented by southern, central, northcentral and northeastern Mexico; 3) in an area bordering the Gulf of Mexico on the west and north; and 4) in western Cuba. Ammonites of that age have never been found in the Western Interior Region of the United States where continental beds were deposited nearly as far north as the Canadian border. They do occur, however, in Canada a little northwest of Glacier Park near the boundary between Alberta and British Columbia.

The Oxfordian ammonite succession in North America (Fig. 6) likewise differs markedly latitudinally. From Arctic Canada and Alaska southward as far as southeastern Washington the early to middle Oxfordian is represented mostly by the Boreal ammonite *Cardioceras (Scarburgiceras)* which is succeeded by *C. (Subvertebriceras)* or by *C. (Cardioceras)*. The next younger late middle to late Oxfordian is represented by *Amoeboceras (Prionodoceras)* in association with the Boreal bivalve *Buchia concentrica* (Sowerby) (Imlay, 1980a). In addition, in southern Alaska the Tethyan ammonite *Dichotomosphinctes* of late middle to late Oxfordian age occurs in association with *Buchia concentrica* on the Alaska Peninsula (Imlay, 1961) and in the Iniskin Bay area west of Cook Inlet (U.S. Geol. Survey Mesozoic loc. 22449) (Imlay and Detterman, 1973). Similarly in the Western Interior Region of the United States the earliest early Oxfordian is represented by *C. (Scarburgiceras), C. (Maltoniceras)* and *Quenstedtoceras (Pavloviceras)* and the latest early Oxfordian by *Cardioceras cordiforme* (Meek and Hayden). The presence of *Buchia concentrica* (Sowerby) in northwestern Montana (Imlay, 1956) on the south edge of Glacier National Park shows that beds of Middle Oxfordian to Late Kimmeridgean age occur there at least locally. Its association with *Cardioceras (Scoticardioceras)* at USGS Mesozoic locality 27641 (Imlay, 1982) at the south edge of the park is good evidence of an age not later than early Middle Oxfordian.

In contrast to this ammonite sequence that is mostly of Boreal origin, the late middle to late Oxfordian in Oregon, California, Mexico, the Gulf Region of the United States, and Cuba includes ammonites of Tethyan origin (Imlay, 1980a) such as *Discosphinctes, Dichotomosphinctes, Orthosphinctes, Ochetoceras, Euaspidoceras* and *Aspidoceras*. These genera in California, Oregon and Mexico are associated with the bivalve *Buchia concentrica* (Sowerby) of Boreal origin but not with ammonites of Boreal origin except possibly for one occurrence of *Amoeboceras* reported by Burckhardt (1930) from a locality near San Pedro del Gallo in north-central Mexico.

Kimmeridgian ammonities have been found in North America in the same areas as late Oxfordian ammonites (Fig. 5c). During late Kimmeridgian time, however, they became more restricted westward in their distribution along the Pacific Coast from northwestern Mexico to southernmost British Columbia. This change apparently coincided with an enlargement of the area undergoing continental deposition in the Western Interior Region and with a western expansion of the land area lying west of that region.

The Kimmeridgian ammonite succession in North America differs latitudinally from Boreal in the Arctic Regions to Tethyan in Mexico. From the Arctic regions southward to southern British Columbia the beds of that age are characterized by the Boreal ammonites *Amoeboceras, A. (Amoebites),* and *A. (Prionodoceras)* along with the bivalve *Buchia concentrica* (Sowerby) at the top of its range (Imlay, 1980a). With these taxa in southern Alaska from the Talkeetna Mountains southwestward to the middle of the Alaska Peninsula occurs an abundance of *Phylloceras,* which genus is characteristic of the Tethyan and Pacific Realms. Much farther south the ammonite *Amoeboceras (Amoebites)* has been found in association with *Buchia concentrica* in east-central California near the Cosumnes River (Imlay, 1961). Still farther south in northwest Mexico, *Amoeboceras* occurs with the Tethyan ammonite *Idoceras* throughout 152 meters of beds and at the top of its range is associated with the bivalve *Aulacomyella* of middle to late Kimmeridgian age (Imlay, 1980a).

The Kimmeridgian ammonite sequence in southern, eastern and northcentral Mexico from the base upward is characterized by: 1) *Ataxioceras, Rasenia* and *Sutneria;* 2) *Idoceras;* and 3) *Glochiceras* in association with *Buchia concentrica* (Sowerby) and *B. mosquensis* (von Buch) (Cantu Chapa, 1971; Imlay, 1980a). These ammonites represent the early and most of the late Kimmeridgian. The association of *B. concentrica* with *B. mosquensis* indicates a middle Late Kimmeridgian age.

A similar ammonite sequence obtained from the subsurface and from outcrops in the southeastern and southcentral United States also includes *Haploceras, Metahaploceras, Nebrodites* and *Physodoceras* (Imlay, 1945). All of these genera plus many others occur in Kimmeridgian beds in Mexico (Burckhardt, 1930) and represent the Tethyan Realm.

Tithonian ammonites have been found in North America in the same areas as late Kimmeridgian ammonites (Fig. 5c, d). The major difference in faunal distribution occurs in the Pacific Coast Region from Baja California northward to southwestern Oregon where ammonites of early Tithonian age have not been found but where ammonites of late Tithonian age occur at many localities (Imlay and Jones, 1970).

The Tithonian ammonite succession in North America likewise differs markedly from north to south (Fig. 6). In most places in Alaska and Canada, Tithonian ammonites are rare but the bivalve Buchia is common. Consequently, the molluscan succession is characterized by species of *Buchia,* which from bottom to top, are represented by 1) *Buchia mosquensis* (von Buch) in association with *B. rugosa* (Fischer) of late Kimmeridgian to early late Tithonian age; 2) *Buchia piochii* in the Canadian Rocky Mountains occurs in beds similar to those that have yielded the ammonite "*Titanites*" of about early late Tithonian age (Frebold, 1964a) and 3) *Buchia fischeriana* and *B. unschensis* which in Arctic Canada are associated respectively with *Dorsoplanites* and *Craspedites* (Jeletzky, 1966; Frebold, 1961).

In addition, the early Tithonian on the Alaska Peninsula is

represented by the ammonite *Aulacosphinctoides* in association with *Buchia mosquensis* (von Buch) and *B. rugosa* (Fischer) (U.S. Geol. Survey Mesozoic loc. 29254) and is probably represented by one ammonite that resembles *Subplanites* (U.S. Geol. Survey Mesozoic loc. 29255). The late Tithonian on the peninsula is represented by *Buchia piochi* (Gabb), *B. fischeriana* (d'Orbigny), *B. unschensis* (Pavlov), and possibly by an ammonite resembling *Phylloceras knoxvillensis* Stanton (Stanton, 1895; Imlay and Jones, 1970).

The Tithonian sequence in western California and southwestern Oregon is entirely of late Tithonian age. It is characterized basally by the ammonite *Kossmatia* in association with *Buchia piochii* (Gabb). Its upper part is characterized by *Substeueroceras, Proniceras, Spiticeras* and *Parodontoceras.* This sequence rests unconformably on beds of Kimmeridgian age (Imlay and Jones, 1970).

The early Tithonian in eastern and northern Mexico is represented basally by *Hybonoticeras* of latest Kimmeridgian and earliest Tithonian age. Above follows a much thicker sequence of late early Tithonian age that is characterized by such ammonites as *Virgatosphinctes, Aulacosphinctoides, Subplanites* and *Mazapilites.* Above follow beds containing the same ammonites as occur in late Tithonian beds in California plus many other genera (Imlay, 1980a).

The Tithonian ammonite sequence in the Gulf Region of the United States appears to be nearly identical with that in Mexico. The early Tithonian is probably represented by a single specimen of *Virgatosphinctes;* the late Tithonian by *Parodontoceras, Hildoglochiceras,* and *Micracanthoceras;* the early late Tithonian by *Durangites;* and the late late Tithonian by *Substeueroceras, Proniceras* and *Aulacosphinctes.*

EXTENT OF LATE JURASSIC SEAS IN NORTH AMERICA

The occurrences of these ammonites of Late Jurassic age show that marine waters covered nearly the same areas in Canada, Greenland, and Alaska as during the Middle Jurassic (Fig. 3, 5). Farther south, however, several major changes occurred in the extent of the seas. First, in the Western Interior

Stages		E. & N. Mexico (Imlay 1943, 1952a; Erben 1957; Cantu Chapa 1963, 1967, 1971; Bridges 1965; Burckhardt 1906, 1912, 1919, 1927, 1930; Verma & Westermann 1973)		California, Oregon, W. Idaho (Imlay 1961, 1964a; Imlay & Jones 1970)		Canadian Rocky Mountains (Frebold 1964a; Frebold et al. 1959; Frebold & Tipper 1970)	S. Alaska (Imlay 1953; Imlay & Detterman 1973)		Arctic Canada (Frebold 1964a; Jeletzky 1965, 1966)
Tithonian	U.	*Berriasella, Substeueroceras, Aulacosphinctes, Protacanthodiscus, Protancycloceras*	*Parodontoceras, Hildoglochiceras, Micracanthoceras, Corongoceras*	*Substeueroceras, Proniceras, Spiticeras, Parodontoceras*	*Buchia* aff. *okensis*	*Buchia* cf. *fischeriana*	*Buchia fischeriana, B. piochii*		*Craspedites canadensis, Buchia unschensis*
				Spiticeras, Parodontoceras	*B. fischeriana*	?			*Buchia fischeriana, Dorsoplanites*; *Buchia richardsonensis*
		Kossmatia victoris, Durangites vulgaris, Parodontoceras butti, Dickersonia, Simoceras, Virgatosimoceras, Lytohoplites, Pseudolissoceras zitteli, Buchia mosquensis		*Kossmatia*	*B. piochii*	*Titanites occidentalis, Buchia piochii*			*Buchia piochii*
		?		?		?	?		?
	L.	*Virgatosphinctes, Aulacosphinctoides, Torquatisphinctes, Subplanites, Pseudolissoceras, Mazapilites*		(Angular unconformity)		*Buchia mosquensis*	*Buchia rugosa, B. mosquensis*		*Buchia mosquensis*
		Hybonoticeras							
Kimmeridgian	U.	*Glochiceras fialar, Buchia mosquensis, concentrica*							
		?					?		
	L.	*Idoceras* cf. *balderum*		*Idoceras, Amoeboceras dubium*	*Buchia concentrica*			*Buchia concentrica*	
		?							
		Ataxioceras, Rasenia, Sutneria cf. *platynota*							
Oxfordian	U.	*Discosphinctes carribeanus, Ochetoceras canaliculatum, Dichotomosphinctes plicatiloides, Euaspidoceras*		*Discosphinctes virgulatiformis, Dichotomosphinctes muhlbachi*		*Amoeboceras, Buchia concentrica*	*Amoeboceras spiniferum, Dichotomosphinctes*		*Amoeboceras, Buchia concentrica*
	M.	*Dichotomosphinctes durangensis*							
				?			?		
	L.			?		*Cardioceras, Goliathiceras*	*Cardioceras distans*		
				Cardioceras martini (in Idaho)			*C. martini*		*C.* [*Scarburgiceras*]

Figure 6. Succession and correlation of Late Jurassic ammonite faunas in North America.

region, the inland sea extended a little farther eastward during early Oxfordian time than it had previously, but then withdrew into the Montana area during middle Oxfordian time and as far west as westernmost Alberta during middle late Oxfordian time. Its withdrawal was followed by deposition of continental sediments (Morrison Formation), whose basal beds range in age from late middle to middle late Oxfordian and whose deposition continued during most, or perhaps all of the remainder of the Jurassic.

Second, along the Pacific Coast Region of the United States and northwestern Mexico the Late Jurassic sea did not extend as far inland as the Middle Jurassic sea and apparently withdrew west of the present land area from latest Kimmeridgian to earliest Tithonian time.

Third, in the Gulf of Mexico region occurred deposition of thick masses of salt, anhydrite and some red beds that are probably of late Callovian and early Oxfordian ages (Imlay, 1943, 1980a; Viniegra, 1971). The marine waters that furnished the salt could have come from the west through southern Mexico, or from the east in the area between Florida and Cuba, or from both directions.

Fourth, the Gulf of Mexico was formed during late middle Oxfordian to late Tithonian times as shown by the presence of thick marine sediments of those ages throughout much of southern, central and northern Mexico and much of the southern United States from Texas to Florida.

REFERENCES

Arkell, W.J., 1956, Jurassic Geology of the World: London, Oliver and Boyd, 806 p.

Bartlett, B.A., and Smith, Leigh, 1971, Mesozoic and Cenozoic history of the Grand Banks of Newfoundland: Canadian Journal of Earth Sciences, v. 8, p. 65-84.

Bridges, L.W. 2d, 1965, Estudios Geõlogicos en el Estado de Chihuahua – Pt. 1, Geologia del area de Plomosas, Chihuahua: México Univ. Nac. Antõnoma Instituto Geologia Boletin 74, p. 1-134.

Burckhardt, Carlos, 1906, La Faune Jurassique de Mazapil avec un Appendice sur les Fossiles du Cretacique Inferieur: Instituto de Geologia de México. Boletin 23, 216 p.

______, 1912, Faunas Jurassiques et Cretaciques de San Pedro del Gallo: Instituto de Geologia de México Boletin 29, 264 p.

______, 1919, 1921, Faunas Jurasicas de Symon (Zacatecas): Instituto de Geologia de México Boletin 33, 135 p.

______, 1927, Cefalopodos del Jurasico Medio de Oaxaca y Guerrero: Instituto de Geologia de México Boletin 47, 108 p.

______, 1930-31, Etude Synthetique sur le Mesozoique Mexicain: Mémoires Societé Paléontologique Suisse, v. 49, p. 1-23 (1930); v. 50, p. 123-280 (1931).

Callomon, J.W., 1959, The Ammonite Zones of the Middle Jurassic Beds of East Greenland: Geological Magazine, v. 96, no. 6, p. 505-513.

Cameron, B.E.B., and Tipper, H.W., 1981, Jurassic Biostratigraphy, Stratigraphy and Related Hydrocarbon Occurrences of Queen Charlotte Islands, British Columbia: *in* Current Research, Part A, Geological Survey of Canada Paper 81-1A, p. 209-212.

Cantu Chapa, Abelardo, 1963, Etude Biostratigraphique des Ammonites du Superieur et Cretace: Societé Géologique de France Mémoire 99, Nouvelle Serie, Tome 42, pt. 4, 102 p.

______, 1967, El Limite Jurasico-Cretacico en Mazatepec, Puebla (Mexico): Instituto Mexicano del Petrõleo Techologia de la Exploration Seccion Geologia, Monografia No. 1, p. 3-24.

______, La Serie Huasteca (Jurãsico Medio-Superior) del Centro Este de México: Instituto Mexicano del Petroleo Revista, v. 3, no. 2, p. 17-40.

Coney, P.J., Jones, D.L., and Monger, J.W.H., 1980, Cordilleran Suspect Terranes: Nature, v. 288, no. 5789, p. 329-333.

Donovan, D.T., 1957, The Jurassic and Cretaceous Systems in East Greenland: Meddelelser om Gronland, v. 155, no. 4, 214 p.

Erben, H.K., 1956, El Jurasico Inferior de México y sus Amonitas: México, D.F., Congreso Geologico International, 20th, 1956, 393 p.

______, 1957, New Biostratrigraphic Correlations in the Jurassic of Eastern and South-Central Mexico: *in* El Mesozoico del Hemisferio Occidental y Sus Correlations Mundiales: International Geological Congress 20th, Mexico, D.F., 1956, sec. 2, p. 43-52.

Frebold, Hans, 1961, The Jurassic Faunas of the Canadian Arctic – Middle and Upper Jurassic Ammonites: Geological Survey of Canada Bulletin 74, 43 p.

______, 1964a, Illustrations of Canadian Fossils – Jurassic of Western and Arctic Canada: Geological Survey of Canada Paper 63-4, 106 p.

______, 1964b, Lower Jurassic and Bajocian Ammonoid Faunas of Northwestern British Columbia and Southern Yukon: Geological Survey of Canada Bulletin 116, 31 p.

______, 1969, Subdivision and Facies of Lower Jurassic Rocks in the Southern Rocky Mountains and Foothills: Geological Association of Canada Proceedings, v. 20, p. 76-87.

Frebold, Hans, Mountjoy, E.W., and Reed, R.A., 1959, The Oxfordian Beds of the Jurassic Fernie Group, Alberta and British Columbia: Geological Survey of Canada Bulletin 53, 47 p.

Frebold, Hans and Tipper, H.W., 1970, Status of the Jurassic in the Canadian Cordillera of British Columbia, Alberta, and southern Yukon: Canadian Journal Earth Sciences, v. 7, p. 1-21.

______, 1975, Upper Callovian and Lower Oxfordian Ammonites from Southeastern Bowser Basin, British Columbia: Canadian Journal of Earth Sciences, v. 12, p. 145-157.

Hall, R.L., and Stronach, N.J., 1981, First Record of Late Bajocian (Jurassic) Ammonites in the Fernie Formation, Alberta: Canadian Journal Earth Sciences, v. 18, p. 919-925.

Hall, R.L. and Westermann, G.E.G., 1980, Lower Bajocian (Jurassic) Cephalopod Faunas of Western Canada and North American Assemblage Zones: Palaeontographical Americana, v. 9, no. 52, p. 1-93.

Hallam, Anthony, 1965, Observations on Marine Lower Jurassic Stratigraphy of North America, with Special Reference to United States: American Association of Petroleum Geologists Bulletin, v. 49, p. 1485-1501.

______, 1969, Faunal Realms and Facies in the Jurassic: Palaeontology, v. 12, pt. 1, pl. 18.

Imlay, R.W., 1943, Jurassic Formations of Gulf Region: American Association of Petroleum Geologists Bulletin, v. 27, no. 11, p. 1407-1533.

______, 1945, Jurassic Fossils from the Southern States, No. 2: Journal of Paleontology, v. 19, no. 3, p. 253-276.

______, 1952a, Correlation of the Jurassic Formations of North America Exclusive of Canada: Geological Society of America Bulletin, v. 63, no. 9, p. 953-992.

______, 1952b, Marine Origin of Preuss Sandstone of Idaho, Wyoming, and Utah: American Association of Petroleum Geologists Bulletin, v. 36, no. 9, p. 1735-1753.

______, 1952c, Summary of Jurassic History in the Western Interior of the United States: Billings Geological Society, Guidebook, 3d Annual Field Conference, Sept. 1952, p. 79-85.

______, 1953, Callovian (Jurassic) Ammonites from the United States and Alaska, Part 2, Alaska Peninsula and Cook Inlet Regions: United States Geological Survey Professional Paper 249-B, p. 41-108.

______, 1955, Characteristic Jurassic Mollusks from Northern Alaska: United States Geological Survey Professional Paper 274-D, p. 69-96.

______, 1956, Marine Jurassic Exposed in the Bighorn Basin, Pryor Mountains, and Northern Big Horn Mountains, Wyoming and Montana: American Association of Petroleum Geologists Bulletin, v. 40, no. 4, p. 562-599.

______, 1961, Late Jurassic Ammonites from the Western Sierra Nevada, California: United States Geological Survey Professional Paper 374-D, 30 p.

______, 1962a, Late Bajocian Ammonites from the Cook Inlet Region, Alaska: United States Geological Survey Professional Paper 418-A, 15 p.

______, 1962b, Jurassic (Bathonian or Early Callovian) Ammonites from Alaska and Montana: United States Geological Survey Professional Paper 374-C, p. C1-C32.

______, 1964a, Upper Jurassic Mollusks from Eastern Oregon and Western Idaho: United States Geological Survey Professional Paper 483-D, p. 1-21.

______, 1964b, Middle Bajocian Ammonites from the Cook Inlet Region, Alaska: United States Geological Survey Professional Paper 418-B, 61 p.

______, 1965, Jurassic Marine Faunal Differentiation in North America: Journal of Paleontology, v. 39, no. 5, p. 1023-1038.

______, 1967, Twin Creek Limestone (Jurassic) in the Western Interior of the United States: United States Geological Survey Professional Paper 540, 105 p.

______, 1968, Lower Jurassic (Pliensbachian and Toarcian) Ammonites from Eastern Oregon and California: United States Geological Survey Professional Paper 593-C, p. C1-C51.

______, 1973, Middle Jurassic (Bajocian) Ammonites from Eastern Oregon. United States Geological Survey Professional Paper 756, 100 p.

______, 1975, Stratigraphic Distribution and Zonation of Jurassic (Callovian) Ammonites in Southern Alaska: United States Geological Survey Professional paper 836, 28 p.

______, 1976, Middle Jurassic (Bajocian and Bathonian) Ammonites from Northern Alaska: United States Survey Professional Paper 854, 22 p.

______, 1980a, Jurassic Paleobiogeography of the Conterminous United States in its Continental Setting: United States Geological Survey Professional paper 1062, 134 p.

______, 1980b, Middle Jurassic (Bathonian) Ammonites from Southern Alaska: United States Geological Survey Professional Paper 1091, 42 p.

______, 1981a, Early Jurassic Ammonites from Alaska: United States Geological Sruvey Professional Paper 1148, 49 p.

______, 1981b, Jurassic (Bathonian and Callovian) Ammonites in Eastern Oregon and Western Idaho: United States Geological Survey Professional Paper 1142, 24 p.

______, 1982, Jurassic (Oxfordian and Late Callovian) Ammonites from the Western Interior Region of the United States: United States Geological Survey Professional Paper 1232, p. 1-44.

Imlay, R. W., and Jones, D.L., 1970, Ammonites from the *Buchia* Zones in northwestern California and Southwestern Oregon: United States Geological Survey Professional Paper 647-B, 59 p.

Imlay, R.W., and Detterman, R.L., 1973, Jurassic Paleobiogeography of Alaska: United States Geological Survey Professional Paper 801, 34 p.

Jeletzky, J.A., 1965, Late Upper Jurassic and Early Lower Cretaceous Fossil Zones of the Canadian Western Cordillera, British Columbia: Geological Survey of Canada Bulletin 103, 70 p.

______, 1966, Upper Volgian (Latest Jurassic) Ammonites and Buchias of Arctic Canada: Geological Survey of Canada Bulletin 128, 51 p.

Jones, D.L. Silberling, N.J., and Hillhouse, J., 1977, Wrangellia – A Displaced Terrane in Northwestern North America: Canadian Journal of Earth Sciences, v. 14, p. 265-2577.

McIver, N.L., 1972, Cenozoic and Mesozoic Stratigraphy of the Nova Scotia Shelf: Canadian Journal of Earth Sciences, v. 9, p. 54-70.

Mouterde, R., Enay, R., Carieu, E., Contini, D., Elmi, S., Gabilly, J., Mangold, Ch., Mattei, J., Rioult, M., Thierry, J., and Tintont, H., 1971, Les Zones du Jurassque en France: Société Géologique de France., C.R. No. 2, p. 76-102.

Parsons, C.F., 1974. The Suazei and "So-Called" Sowerby Zones of the Lower Bajocian: Newsletter on Stratigraphy, v. 3, p. 153-179.

Sanborn, A.F., 1960, Geology and Paleontology of the Southwest Quarter of the Big Bend Quadrangle, Shasta County, California: California Division of Mines Special Report 63, p. 3-26.

Stanton, T. W., 1895, Contributions to the Cretaceous Paleontology of the Pacific Coast – The Fauna of the Knoxville Beds: United States Geological Survey Bulletin 133, 132 p. (Issued Feb. 3, 1896).

Surlyk, F., Callomon, J.H., Bromley, R.G., and Birkelund, Tove, 1973, Stratigraphy of the Jurassic-Lower Cretaceous Sediments of Jameson Land and Scoresby Land, East Greenland: Greenland Geological Undersogelse Bulletin 105, p. 5-76.

Tipper, H.H., 1981, Offset of an Upper Pleisbachian Geographic Zonation in the North American Cordillera by Transcurrent Movement: Canadian Journal of Earth Sciences, v. 18, p. 1988-1992.

Tipper, H.W., and Cameron, B.E.B., 1980, Stratigraphy and Paleontology of the Upper Yakoun Formation (Jurassic) in Alliford Bay Syncline, Queen Charlotte Islands, British Columbia: *in* Current Research, Part C, Geological Survey of Canada, Paper 80-1C, p. 37-44.

Upshaw, C.G., Armstrong, W.E., Creath, W.B., Kidson, E.J., and Sanderson, G.A., 1974, Biostratigraphic Framework of Grand Banks: American Association of Petroleum Geologists Bulletin, v. 58, no. 6, pt. 2, p. 1124-1132.

Verma, H. and Westermann, G.E.G., 1973, The Tithonian (Jurassic) Ammonite Fauna and Stratigraphy of Sierra Catorce, San Luis Potosi, Mexico: Bulletins of American Paleontology, v. 63, no. 277, p. 103-320.

Viniegra, O. 1971, Age and Evolution of Salt Basins of Southeastern Mexico: American Association of Petroleum Geologists Bulletin, v. 55, no. 3, p. 478-494.

Westermann, G.E.G., 1964, The Ammonite Fauna of the Kialagvik Formation at Wide Bay, Alaska Peninsula, Part 1, Lower Bajocian (Aalenian): Bulletins American Paleontology, v. 47, p. 329-503.

______, 1969, The Ammonite Fauna of the Kialagvik Formation at Wide Bay, Alaska Peninsula, Part 2, *Sonninia sowerbyi* Zone (Bajocian): Bulletins American Paleontogy, v. 57, no. 255, 226 p.

______, 1979, Troublesome Definition of the Lower/Middle Jurassic Boundary: Canadian Journal of Earth Sciences, v. 16, p. 2060-2062.

______, 1981, Ammonite Biochronology and Biogeography of the Circum-Pacific Middle Jurassic: in The *Ammonoidea:* Systematics Association, Special Volume 18, p. 459-498.

Jurassic-Cretaceous Biochronology and Paleogeography of North America,
edited by G.E.G. Westermann,
Geological Association of Canada Special Paper 27, 1984

EARLY JURASSIC STRATIGRAPHY AND MICROPALEONTOLOGY OF THE GRAND BANKS AND PORTUGAL

J. Exton
Petro-Canada Exploration Inc., P.O. Box 2844, Calgary, Alberta T2P 3E3
F.M. Gradstein
Geological Survey of Canada, Bedford Institute of Oceanography, Box 1006, Dartmouth, Nova Scotia
B2Y 4A2

ABSTRACT

The Triassic-Lower Jurassic sedimentary sequence in the Grand Banks subsurface and in outcrop sections in the Lusitanian Basin of western Portugal are closely similar. Major vertical changes of lithology in both areas, from red beds and evaporites through shallow marine dolostones and limestones to calcareous shales, are synchronous within the limits of biostratigraphic resolution. Portuguese latest Pliensbachian beds are more calcareous and less shaly than below and above and indicate a slightly regressive cycle, as in France and England. The Early Toarcian bituminous shale of northwestern Europe did not extend to Portugal and the Grand Banks.

Foraminifer and ostracode assemblages of Portugal and the Grand Banks have many taxa in common and allow for the recognition of three zones in each area; the Zone of Involutina liassica *recognized in the Grand Banks is considered equivalent to the Zone of* Frondicularia terquemi *of the Lusitanian Basin which corresponds to the ammonite standard zones of Raricostatum through Tenuicostatum (latest Sinemurian-earliest Toarcian). The Zones of* Bairdiacypris *sp. and* Lenticulina dorbignyi *are common to both basins with age ranges of Early through Middle Toarcian and Late Toarcian-Aalenian respectively. The Grand Banks and Portuguese foraminiferal assemblages form part of a cosmopolitan Early Jurassic microfauna which spread throughout the world's epicontinental seas before continental fragmentation.*

RÉSUMÉ

Il y a une ressemblance remarquable entre les séries sédimentaires triassiques et jurassiques inférieures des Grands Bancs de Terre Neuve et les affleurements du Bassin Lusitanien du Portugal. Dans les deux régions, les changements verticaux majeurs de la lithologie, des grès et argiles rouges et évaporites à travers les dolomites marins d'eau peu profonde aux argiles calcaires, sont synchrones dans les limites de la résolution biostratigraphique. Les dépots lusitaniens du Pliensbachien supérieur sont plus calcaires et moins schisteux que les dépots au dessous et au dessus et indiquent un cycle légèrement régressif tel qu'en France et en Angleterre. Les argiles bitumineux du Toarcien inférieur du nord-ouest de l'Europe ne sont pas étendus au Portugal et aux Grands Bancs.

*Les assemblages de foraminifères et d'ostracodes du Portugal et des Grands Bancs sont très comparables et permettent de reconnaitre trois zones dans les deux régions. La Zone d'*Involutina liassica *des Grands Bancs est considerée comme étant équivelante à la Zone de* Frondicularia terquemi *du Bassin Lusitanien qui correspond aux zones d'ammonites de Raricostatum jusque'au Tenuicostatum (Sinémurien supérieur au Toarcien inferieur). Les Zones de* Bairdiacypris *et de* Lenticulina dorbignyi *sont communes aux deux bassins et s'étendent respectivement du Toarcien inférieur à l'Aalenien. Les assemblages foraminifères du Portugal et des Grands Bancs sont partie d'une microfaune cosmopolitaine du Jurassique inférieur qui s'est étendue à travers les mers épicontinentales du monde avant la séparation des continents.*

INTRODUCTION

Pre-drift reconstructions of the North Atlantic continental margins indicate that before the onset of sea-floor spreading the Lusitanian Basin of central Portugal was in close proximity to the Grand Banks of Newfoundland basin(s). Just how close these two areas were is still uncertain (Fig. 1). Kristoffersen (1978) shows a gap of several hundreds of kilometres between the southern Grand Banks and Iberia, while Le Pichon *et al.* (1977) and the Groupe Galice (1979) present somewhat tighter reconstructions in which Iberia has been moved westwards along the North Pyrenean fault. Laughton (1975) and Wilson (1975) have produced the closest reassembly by arbitrarily shuffling Orphan Knoll, Galicia Bank and Flemish Cap into the gap between Porcupine Bank and the Labrador Shelf.

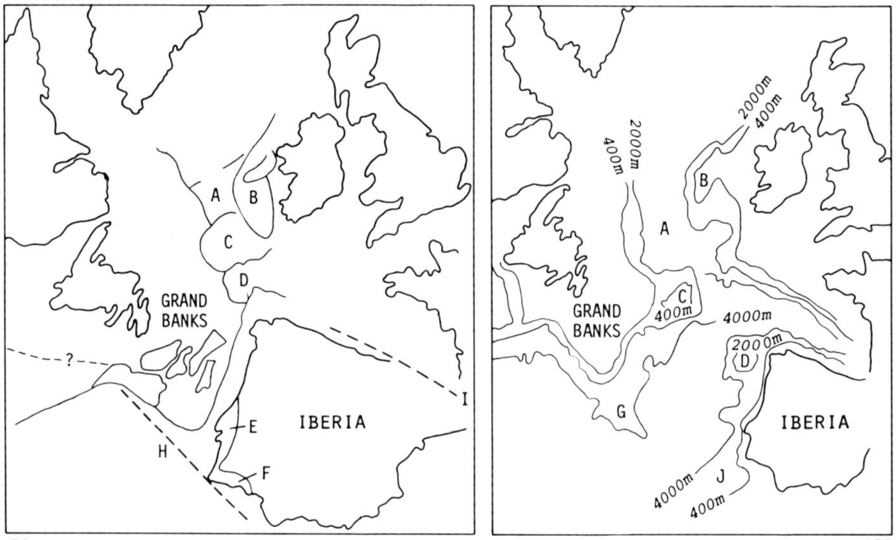

FG GSC

Figure 1. Two reconstructions of the relative pre-drift positions of Iberia and Grand Banks in Jurassic time (after Wilson, 1975 – left, and Kristoffersen, 1978 – right). A – Orphan Knoll, B – Porcupine Bank, C – Flemish Cap, D – Galicia Bank. In letters are indicated the Jurassic basins of Portugal and the 'subbasins' of the Grand Banks; E – Lusitanian Basin, Jurassic, F – Algarve; Grand Banks sub-basin terminology is in Figure 2b. G – Newfoundland Ridge, H – Azores-Gilbrator Ridge (Fracture Zone), I – North Pyrenean Fault, J – Gorringe Bank.

Whatever the case there is no marine magnetic evidence of oceanic crust of pre-Cretaceous age between Portugal and the Grand Banks, and it has been argued from geophysical evidence that the pre-Cretaceous sedimentary province of the southern Grand Banks extends eastwards beneath the Newfoundland Ridge, thus partially filling the gap in most reconstructions (Grant, 1977).

Oceanic conditions are thought to have existed between the Scotian Shelf and Moroccan basins by Early Jurassic time (Hallam, 1971b; Manspeizer *et al.*, 1978), although a Middle Jurassic age has also been suggested (Bhat *et al.*, 1975; Given, 1977). Deep Sea Drilling Project Site 534, east of Florida, U.S.A. and north of the Bahamas, indicates that the North Atlantic Jurassic ocean is not older than Middle Jurassic (Sheridan *et al.*, 1983). North of the Azores-Gilbraltar Ridge, however, the onset of continental separation and sea-floor formation was much later. Williams (1975) suggested an age of 130 Ma for crustal separation between Newfoundland and Iberia based upon extrapolation of the spreading rate at anomaly 32. This agrees well with the isotopic age of 135 $\pm$ 3 Ma for oceanic crust beneath the Gorringe Bank (Fig. 1) (Prichard and Mitchell, 1979). Keen *et al.* (1977) concluded a somewhat younger date of 115 Ma for the initiation of active spreading between the Grand Banks and Iberia. If these Early Cretaceous dates for the opening of this segment of the Atlantic are correct, and if we assume that before this event the area was largely occupied by continental crust, then it is probable that the basins on either side of the present-day ocean will show a broadly similar sequence of events in their tectonic, sedimentary and faunal development during the Jurassic.

In the first part of this account we compare the lithostratigraphy of the Grand Banks and Lusitanian Basins in the Triassic and Early Jurassic. The second part compares the Early Jurassic biostratigraphy of the Lusitanian Basin with that established for the Grand Banks. Calibrations are made to the standardized ammonite zonation and to the comprehensive foraminiferal zonation of Copestake and Johnson (in press). The third section of this study deals with Early Jurassic foraminiferal biogeography in the light of the known distribution of land and seas.

The Triassic and Jurassic rocks of Portugal have been well known since the late years of the nineteenth century (e.g., Choffat, 1880), but although sediments of Triassic age occur along parts of the eastern seaboard of Canada it is only within the past fifteen years that thick sequences of Cretaceous and Jurassic sediments have been discovered in offshore exploratory wells drilled on the Grand Banks and Scotian Shelf. In the Grand Banks subsurface over 3050 m of predominantly marine sediments have been assigned to the Jurassic on the basis of biostratigraphic analysis of sidewall cores and cuttings (Gradstein *et al.*, 1975). All of the standard European Jurassic stages have been recognized, and it is believed that the Grand Banks marine Jurassic sequence is one of the most complete stratigraphically in North America.

Both the Grand Banks and Portuguese Jurassic sequences have yielded diversified foraminiferal and ostracode faunas (Zbyszewski *et al.*, 1965; Ruget and Sigal, 1970; Ramalho, 1971; Gradstein, 1977, 1978; Exton, 1979). An examination of the above literature suggests substantial differences between coeval microfaunal assemblages from the Grand Banks and the Lusitanian Basin, especially in the Middle and Late Jurassic. For this reason a study is underway to determine if this apparent provincialism is a function of bias in (a) sampling, (b) taxonomy, or reflects (c) pronounced differences in depositional history of the two regions, or (d) results from the presence of physical barriers.

The results of studies such as this may contribute towards a better understanding of the development of the Grand Banks and Lusitanian basins, and help to define more precisely the timing of seafloor spreading in this segment of the North Atlantic Ocean.

LITHOSTRATIGRAPHY

Portugal and Grand Banks

The lithostratigraphic framework of the Grand Banks has been described previously by Amoco Canada and Imperial Staffs (1973) and Jansa and Wade (1975); detailed accounts of the Murre G-67 and Osprey H-84 wells (Fig. 2b), both of which penetrated sediments of Early Jurassic age, have also been published (Jansa *et al.*, 1976, 1977). A summary account of the Lusitanian Basin stratigraphy is found in Livret Guide G-14 (1980); data on Triassic exposures have been presented by Palain (1976) and the Lower Jurassic facies distribution has been summarized by Hallam (1971a) and Mouterde and Ruget (1975). Localities referred to below are shown in Figures 2a, 2b, and 3.

Subsidence and sediment accumulation in the Lusitanian Basin commenced in the Triassic as a result of normal movements along reactivated late Hercynian strike-slip faults. At least two major fault trends influenced basin geometry in the Triassic and Early Jurassic; a north-northeast trend parallel to the basin axis (and zone of maximum subsidence), and a northeast trend following the Nazaré line, to the north of which sediment accumulation was apparently greater. Intersection of these two trends by a less pronounced NW trending fault set may have resulted in the formation of several small sub-basins by local fault-block subsidence. To the northwest of Portugal in the vicinity of Galicia Bank, where the basin lies offshore, seismic studies suggest that the northwest-trending fault set had a more important control on basin subsidence (Mauffret *et al.*, 1978; Groupe Galice, 1979).

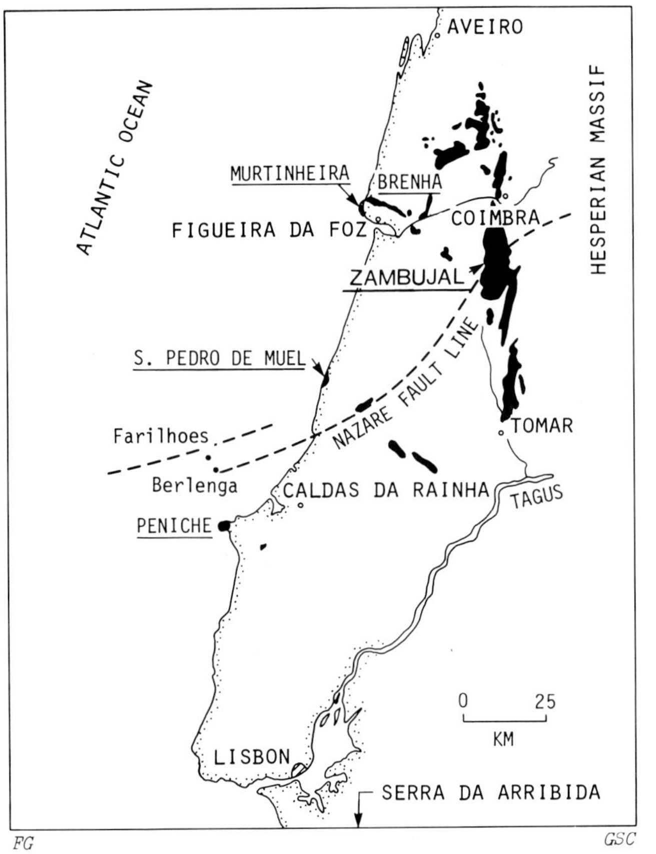

Figure 2a. Distribution of Lower Jurassic outcrops in west-central Portugal and location of sections studied.

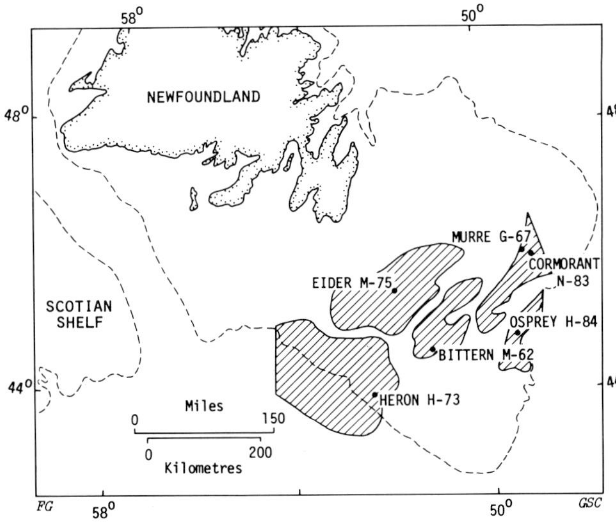

Figure 2b. Location of exploratory wells penetrating Lower Jurassic strata, Grand Banks of Newfoundland. Grand Banks sub-basins are from east to west: Jeanne d'Arc, Carson, Horseshoe, Whale, South Whale.

Little has been published concerning the early tectonic development of the Grand Banks area, but it has been suggested that faulting possibly began in the Late Triassic (Keen *et al.*, 1977) and that Jurassic sediments accumulated in structural basins created by block-faulting (Amoco Canada and Imperial Staffs, 1973). This seems all the more likely in view of the widespread Triassic and Early Jurassic faulting and graben formation in other areas bordering the North Atlantic (see Kent, 1977, for discussion).

Triassic sediments in both the Grand Banks and Portuguese basins rest unconformably upon a basement of igneous, metamorphic and sedimentary rocks of Precambrian and Paleozoic ages (Da Costa, 1950; Howie and Barss, 1975). The oldest dated (by palynology) sediments above the unconformity in the Lusitanian Basin occur near the base of the Silves Sandstone Formation, a fluviatile red-bed sequence exposed along the eastern margin of the basin, and are of Karnian-Norian age (Doubinger *et al.*, 1970; Adloff *et al.*, 1974). The same age range has been assigned to the upper part of the lithologically similar Kettle Red Beds (mudstones, sandstones and conglomerate) in the Osprey H-84 well in the Carson Basin (Jansa *et al.*, 1980), although pre-Mesozoic basement was not reached and this may not be the true age of earliest sediment accumulation. Middle Triassic sediments have been reported from the Bittern M-62 well in the Horseshoe Basin (W.A.M. Jenkins, as cited in Jansa and Wade, 1975). Triassic red beds (shales and sandstones) have been found overlying metamorphic basement in offshore wells in the Lusitanian Basin, but no information about their precise age is available.

Overlying the red-beds in both areas are thick evaporite sequences consisting predominantly of halite, but with minor anhydrite, dolostone and shale. The Dagorda evaporites occur over a large part of the Lusitanian Basin (Zbyszewski and de Faria, 1971) and are thought to pass eastwards into the Silves Sandstone, although the transition between the two is not exposed. The presence of the lower part of the Silves Sandstone beneath the Dagorda has been proved in one deep well onshore (Sao Mamede 1 drilled in 1958) and in several offshore wells. Palynological studies of shales in the upper part of the Dagorda sequence indicate an Hettangian age (Doubinger *et al.*, 1970); similarly the upper part of the Silves Sandstone Formation, which includes the dolomitic Pereiros Beds, has also been assigned to the Hettangian (see Palain, 1976). The base of the Dagorda evaporites has not been dated, but dolostones and anhydritic shales intercalated in the Late Triassic part of the Silves Sandstone Formation in the east (Palain, 1968, 1976) may represent the earliest marginal facies of the evaporite basin. In the Carson basin of the eastern Grand Banks, the Osprey Evaporite is of latest Triassic to Hettangian age (Jansa *et al.*, 1980). Evaporites, presumably of similar age, are seen as diapiric structures on seismic profiles from the South Whale, Whale, Horseshoe and Jeanne d'Arc basins. These have been correlated previously with the Argo salt of the Scotian Shelf (Jansa and Wade, 1975), but the latter is now considered to be of Hettangian-Sinemurian age (Barss *et al.*, 1979).

Towards the top of the Portuguese Dagorda sequence, and in the Pereiros Beds at the top of the Silves Sandstone Formation, dolostones form an important part of the succession. Near the hamlet of Dagorda, in the type area of the Dagorda evaporites (Choffat, 1882), crystalline, vuggy dolostones and fine-grained, laminated dolostones occur as large 'rafts' within red shales forming the core of an eroded diapiric structure (part of the large Caldas da Rainha diapir (Fig. 2a). At a similar stratigraphic level in the Grand Banks subsurface the lower part of the 'Murre Carbonate', which overlies anhydritic red shales of presumed Triassic age in the Murre G-67 well and the Osprey Evaporite in Osprey H-84, consists of fine crystalline and laminated dolostones with interbedded halite and anhydrite.

Over the whole of the Lusitanian basin the Lower Sinemurian is represented by the dolomitic Coimbra Beds, although the absence of ammonites makes it impossible to define accurately the boundary between the Hettangian and Sinemurian stages. Lithologically, these beds consist of thickly bedded dolostones, dolomitic limestones and limestones, and a variety of facies including crystalline dolostone, dolomitised oolitic, peletoidal and bioclastic limestones, and finely laminated dolostones have been recognised (Hallam, 1971a). The upper boundary of the Coimbra Beds is strongly diachronous, ranging in age from Late Sinemurian (Obtusum Zone) to Early Pliensbachian (Jamesoni Zone) (Mouterde *et al.*, 1971). In the western part of the onshore basin, to the north of the Nazaré fault (the area of maximum subsidence), argillaceous limestones of Late Sinemurian age overlie Hettangian and Early Sinemurian dolostones. The oldest ammonite faunas found in the basin occur in these limestones at São Pedro de Muel and in the well Verride 2 to the north (Mouterde and Ruget, 1975) and are indicative of the Obtusum Zone (Late Sinemurian). In the latest Sinemurian (Raricostatum Zone) of the western sections (Brenha, São Pedro de Muel, Peniche), soft marls and shales containing abundant ammonites form an important part of the succession for the first time. Along the north-eastern margin of the basin bordering the Hesperian Massif, dolostones and dolomitic limestones occur throughout the Sinemurian (e.g., Zambujal) and the facies change to interbedded limestones and marls occurs at the base of the Pliensbachian. Near Tomar, dolostones persist into the Pliensbachian (Jamesoni Zone) (Mouterde, Rocha and Ruget, 1971), and to the south of the River Tagus, dolostones and limestones predominate throughout the Lias (Seifert, 1963; Hallam, 1971a; Mouterde *et al.*, 1971).

The name 'Coimbra Beds' has acquired strong chronostratigraphic connotations in the literature of the Lusitanian Basin, possibly because in the type area around Coimbra the beds encompass almost the whole of the Sinemurian Stage. As a consequence the name has not been applied to the dolomitic successions to the south of the River Tagus (in the Serra da Arrabida) which range in age from latest Hettangian to Bajocian (Seifert, 1963; Zbyszewski *et al.*, 1965; Tintant, 1967). These outcrops together with those of Santiago da Cacem and the Algarve, because of their position peripheral to the subsiding basin and also because of their predominantly dolomitic and calcareous sediments, may be considered as an extension of the western Tethys carbonate platform.

Comparison with offshore Eastern Canada reveals striking similarities between the Murre Carbonate of the Grand Banks and the Sinemurian succession of western Portugal. In the Grand Banks subsurface a threefold division of the Murre Carbonate has been recognised (Jansa *et al.*, 1976): a lower dolostone-halite unit which occupies a similar stratigraphic position to the topmost Dagorda and Pereiros beds of the Lusitanian Basin (see above), a middle part lacking halite and consisting of crystalline dolostones with minor dolomitized oolitic and peloidal limestone and occasional anhydrite nodules (*cf.* the Coimbra Beds), and lastly an upper part which consists of micritic, skeletal and oolitic limestones (*cf.* the Upper Sinemurian of western Portugal). From palynological studies the lower and middle parts of the formation in the Murre G-67 well are of Hettangian-Sinemurian age, whereas the upper unit is Pliensbachian. Similar successions have been recorded from Bittern M-62 and Osprey H-84.

By analogy with the Coimbra Beds and overlying limestones, the upper boundary of the Murre Carbonate may be strongly diachronous. It is probable that on the northern and eastern Grand Banks the unit is of Sinemurian and possibly earliest Pliensbachian age, and that the youngest beds occur in the proximity of the inter-basin highs. The Murre Carbonate has previously been correlated with the Iroquois Formation of the Scotian Shelf which apparently extends throughout the Early Jurassic (although few dates are available) and is overlain by Middle Jurassic clastics of the Mohican Formation (Jansa and Wade, 1975; Given, 1977).

Overlying the Murre Carbonate is a thick shale and limestone sequence informally named the Whale unit (Jansa and Wade, 1975). This has been divided into three members (Jansa *et al.*, 1976) of which the lower shale member is Pliensbachian through Bajocian in age. The contact with the Murre Carbonate is conformable and gradational; in the Murre G-67 well the transition beds of interbedded limestone, marlstone and shale are approximately 30 m thick. Above this, dark- to medium-grey, calcareous and silty shales predominate, with occasional thin beds of siltstone or micritic limestone. Similar lithologies are also encountered in Cormorant N-83, Heron H-75 and Bittern M-62. The Whale unit is not present in the Osprey H-84 well where the Murre Carbonate is unconformably overlain by fine terrigenous clastics of Cenomanian through Coniacian age.

Throughout the Lusitanian Basin the Pliensbachian and Toarcian stages (defined by ammonite standard zones: Mouterde *et al.*, 1971) are represented in an unnamed sequence of shales, marlstones and limestones. In western sections this predominantly argillaceous facies first appeared in the latest Sinemurian (Raricostatum Zone), as discussed above. Towards the southeastern and southern margins of the basin, sedimentation and subsidence rates were lower than in the north and west, and the succession is more calcareous. At Brenha, over 300 m of sediment accumulated during the Pliensbachian and Toarcian, whereas in the same interval near Tomar, only 100 m accumulated.

In the interval of the Jamesoni through Margaritatus Zones sediments became progressively more argillaceous; the latter zone occurs in dark-grey shales which are in part bituminous. An abrupt change of lithology occurs throughout the basin at the top of the Margaritatus Zone, and the overlying Spinatum Zone is represented by interbedded hard limestones and softer blue-grey marls. Argillaceous sediments predominate at the base of the Toarcian Stage, and shale-marl-limestone sequences continue into the Aalenian (Middle Jurassic), although the proportion of limestones to marls and shales increases gradually from the Middle Toarcian onwards.

The Toarcian succession at Peniche is unique in the basin in that it contains coarse clastic deposits and massive limestones. Overlying the limestones and marls of the Spinatum Zone are olive-green, silty and micaceous shales. These continue throughout the Tenuicostatum and Falciferum Zones, the latter zone containing several lenses and thin beds of coarse sandstone. The Bifrons Zone is largely developed in a 'normal' interbedded shale and limestone facies, but above this are thickly bedded oolitic and bioclastic limestones containing variable amounts of quartz and feldspathic sand representing the remainder of the Toarcian Stage and the Aalenian Stage. This local influx of coarse clastic sediment and the occurrence of high-energy carbonates may have resulted from the uplift and emergence of a basement block to the west (Hallam, 1971a), evidence of which is seen in the isolated outcrop of granitic and metamorphic basement in the vicinity of the nearby Berlenga and Farilhoe Islands.

Summary

In the Early Jurassic, the Lusitanian region and the Grand Banks lay immediately to the north of the western Tethys carbonate platform which is represented in sediments of the Scotian Shelf, Serra da Arrabida and Algarve. Within the limits of resolution of Grand Banks biostratigraphy (see section 2), lithofacies changes in the Grand Banks basins are synchronous with those of the Lusitanian Basin. The lithological succession is also similar to those of northern Spain and the Aquitaine Basin, but differs from those of southern England and the Paris Basin where marine shales and marls with abundant ammonites occur throughout the Lower Jurassic and the dolomitic phase of sedimentation is absent. Another important difference between the Grand Banks-Lusitanian Basin and north-western Europe is the absence of a Lower Toarcian bituminous shale facies which is well developed and widespread in the latter area, e.g., the Jet Rock of England, Posidonienschiefer of Germany and the Schistes-Carton of northern France.

The Spinatum Zone of the Lusitanian Basin is developed in calcareous sediments of apparently shallower-water origin than the shales above and below. In Grand Banks wells, notably Murre and Cormorant, a slightly more calcareous or sandy interval occurs at the approximate level of the Pliensbachian/Toarcian boundary as determined by microfaunal analysis. Over much of north-western Europe the Spinatum Zone is developed in similar apparently regressive facies, e.g., the Marlstone of England, and the Banc de Roc and Grès médioliasiques of France.

FORAMINIFER AND OSTRACODE BIOSTRATIGRAPHY

Lusitanian Basin

Microfaunas have been examined from four Lower Jurassic sections at Zambujal, Brenha, São Pedro de Muel and Peniche and from the Middle Jurassic (Aalenian) of Murtinheira, Cap Mondego (Fig. 2a). The Zambujal faunal succession has been studied in detail (Exton, 1977, 1979). Foraminifera from the São Pedro de Muel section have been described previously by Ruget and Sigal (1970).

The Early Juriassic biostratigraphy outlined below is based upon data from Zambujal; preliminary results from the other three sections have demonstrated its regional significance. Both foraminifer and ostracode faunas have been used to define one local range zone containing two subzones, and two assemblage zones.

Where possible, the stratigraphic position of each zone is defined in terms of the standard European ammonite zones and stages. The ammonite zonal scheme applied to the Portuguese Early Jurassic (Mouterde, 1967a, 1967b; Mouterde *et al.*, 1971) is very similar to that used in northwestern Europe (Dean *et al.*, 1961), and the latter, more familiar scheme has been adopted in the following account (see Figs. 3 and 4). The ammonite zonation and lithostratigraphy of the sections discussed have been described in the following works: Zambujal (Mouterde *et al.*, 1964), Peniche (Mouterde, 1955), São Pedro de Muel (Mouterde, 1967a, 1967b), sections in the vicinity of Brenha (Schott and Staesche, 1957; Mouterde *et al.*, 1978; Gomes, 1963), Murtinheira at Cap Mondego (Mouterde *et al.*, 1978).

Frondicularia Terquemi Assemblage Zone.

Definition: The zone is defined by the known local range of *Frondicularia terquemi* d'Orbigny, *Marginulina prima* d'Orbigny, *Lingulina tenera* Bornemann and the ostracode genus *Ogmoconcha.*

Age: Late Sinemurian (Raricostatum Zone) through Early Toarcian (Tenuicostatum Zone, in part).

The base of the zone is diachronous, ranging in age from latest Sinemurian in the western part of the basin (Brenha, São Pedro de Muel and Peniche) to earliest Pliensbachian in the east (Zambujal), and corresponds to the lithological boundary between the Coimbra and Brenha beds. Because the dolostones and limestones of the Coimbra Beds are unsuited to microfaunal analysis, the lower range of the indicator species has not been determined. In other parts of Europe the ranges of *Frondicularia terquemi, Lingulina tenera,* and possibly also of *Marginulina prima* extend down into the Sinemurian and Hettangian stages (Bartenstein and Brand, 1937; Norling, 1972; Copestake and Johnson, in press).

Subspecies of the zonal indicators occur throughout the zone although their ranges may be locally restricted: these are *Lingulina tenera tenera* Bornemann, *L. tenera pupa* (Terquem), *Frondicularia terquemi bicostata* D'Orbigny, *F. terquemi sulcata* Bornemann, *F. terquemi dubia* Bornemann and *F. terquemi terquemi* d'Orbigny. Other common species include *Dentalina pseudocommunis* Franke, *Lenticulina gottingensis* Bornemann, *Astacolus varians* (Terquem) and *Eoguttulina liassica* (Strickland) which are present throughout the zone and range high into the Toarcian Stage.

Within the *F. terquemi* Zone two subzones, which have a basin-wide distribution, may be recognized.

1) Subzone of *Brizalina liasica.* This subzone corresponds to the upper part of the Margaritatus Zone (Late Pliensbachian) and is readily distinguished by the abundance of *Brizalina liasica* particularly in the western sections of Peniche, São Pedro de Muel and Brenha. Although *B. liasica* has a local stratigraphic range of Ibex through Margaritatus Zones it is only in the Margaritatus Zone that the species is common.

2) Subzone of *Ogmoconcha* cf. *ambo.* This species has been found only in the upper part of the Spinatum Zone and the lower part of the Tenuicostatum Zone. It occurs commonly with a robust variety of *Ogmoconcha amalthei. Ogmoconcha* cf. *ambo* is similar to *Ogmoconcha amalthei circumvallata* Dreyer as interpreted by Michelsen (1975), but the latter does not compare well to the type description of Dreyer. *O. ambo* Lord and Moorley was originally described from the Spinatum Zone of southern Germany. It has also been recorded from the Tethyan area, occurring with Group A (of Gordon, 1970) foraminiferal assemblages in the Spinatum Zone of Algeria (Maupin, 1977).

A third subzone, that of *Gammacythere ubiquita* Malz and Lord, may be recognizable in some areas of the basin, but this species has been found so far only in the Jamesoni and Ibex Zones of Zambujal. Malz and Lord (1976) named the species '*ubiquita*' because of its widespread occurrence in the Lower Pliensbachian of northwestern Europe.

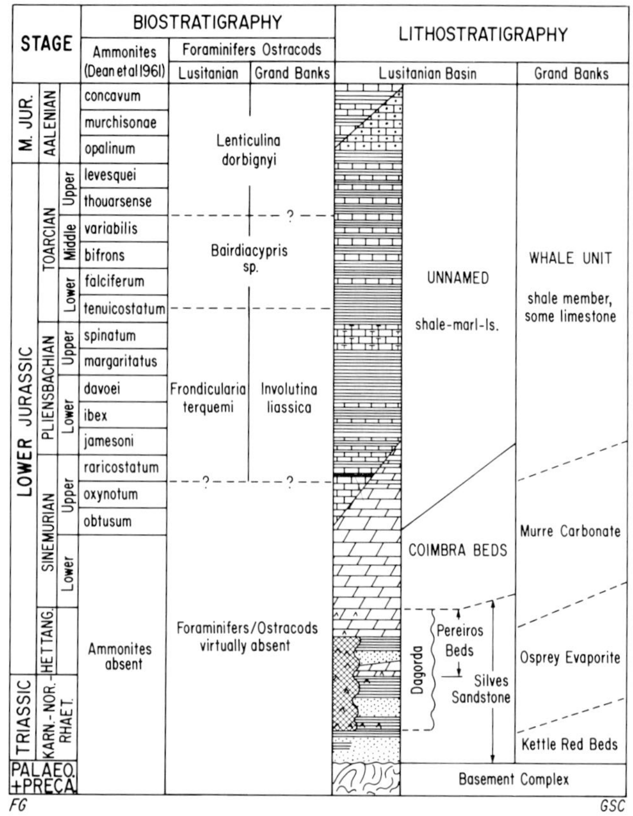

Figure 3. Litho- and biostratigraphic framework of the Triassic and Early (-Middle) Jurassic strata in the Lusitanian Basin, Portugal and the Grand Banks of Newfoundland.

Bairdiacypris sp. Assemblage Zone.

Definition: The zone is characterised by the abundant occurrence of an unnamed species of *Bairdiacypris,* which resembles closely the *Bairdiacypris* sp. recorded by Bate and Coleman (1975) from the Early Toarcian (Falciferum Zone) of Britain. Also of common occurrence within the zone are *Ektyphocythere debilis* Bate and Coleman which ranges into the overlying *Lenticulina dorbignyi* Zone, and *Kinkelinella sermoisensis* (Apostolescu) which has been found only in the Early and Middle Toarcian (Falciferum and Bifrons Zones) of Portugal, although its range in other areas of Europe is Early Toarcian through Aalenian.

Age: Early and Middle Toarcian. In the Zambujal section this zone ranges from Early Toarcian (Tenuicostatum Zone) through Middle Toarcian (Variabilis Zone). In the Brenha section its range cannot be determined precisely due to uncertainties in the ammonite stratigraphy of the Middle and Upper Toarcian section, but it is approximately the same age as at Zambujal. At Peniche the characteristic assemblage has been recovered only from the Bifrons Zone and its truncated range may be attributed to facies (see section 1 and discussion of *F. terquemi* Zone).

As a result of the mass extinction of foraminifer and ostracode species in the basal Toarcian, assemblages assigned to this zone are generally poor both in numbers of specimens and numbers of species. In the Tenuicostatum and Falciferum Zones holdover species from the underlying *F. terquemi* Zone dominate foraminiferal assemblages, notably *Lenticulina gottingensis, Dentalina pseudocommunis* and *Eoguttulina liassica.* Ostracode assemblages on the other hand are dominated by *Bairdiacypris* sp., which first appears in the Tenuicostatum Zone and is abundant through to the Variabilis Zone. The only common ostracode left over from the underlying zone is *Poly-*

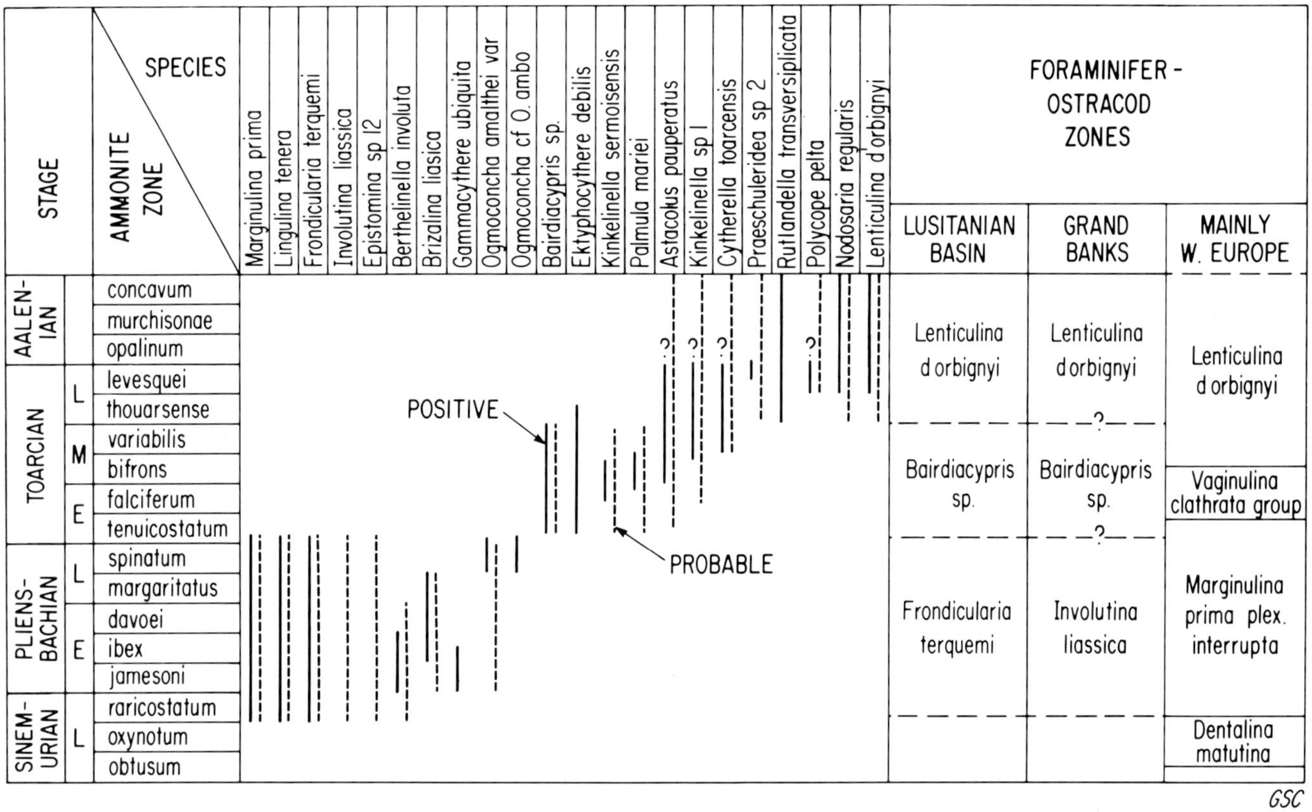

Figure 4. Stratigraphic distribution of foraminifer and ostracod taxa used in the biostratigraphic zonations for the Lusitanian and Grand Banks Sinemurian through Aalenian strata. To the right is shown the foraminifer zonation of Copestake and Johnson (in press) mainly for Northwestern Europe.

cope sp. I (of Exton, 1979). Faunal diversity increases steadily throughout the zone with the appearance of *Lenticulina toarcencis* Payard, *Astacolus pseudocrepidulatus* (Adams), *Citharina clausa* (Terquem), *Citharina colliezi* (Terquem), *C. clathrata* (Terquem), *Nodosaria globulata* Barnard, *Palmula mariei* and *P. chicheryi* (Payard). At some horizons in the Falciferum and Bifrons Zones either *Eoguttulina liassica* or *Ophthalmidium liasicum* dominates the foraminiferal faunas.

Towards the top of this zone several species occur which are more characteristic of assemblages of the overlying *L. dorbignyi* Zone. Most notable is *Kinkelinella sp.* I in the upper part of the Bifrons Zone.

Lenticulina Dorbignyi Assemblage Zone.

Definition: The zone is characterised by the occurrence of *Lenticulina dorbignyi* (Roemer), *Cytherella toarcensis* Bizon and *Kinkelinella sp.* I (Apostolescu). Whereas *L. dorbignyi* is restricted to this zone, both *C. toarcensis* and *Kinkelinella* sp. I also occur in small numbers towards the top of the preceding zone.

Age: Late Toarcian (Thouarsense and Levesquei Zones) to Aalenian (Concavum Zone). The age of this assemblage has been well determined in the Zambujal and Murtinheira sections, but the zone is also well represented in the Brenha section. In other areas of Europe the species listed above also range into the Aalenian Stage of the Middle Jurassic (Simon and Bartenstein, 1962; Apostolescu, 1959). Copestake and Johnson (in press) report that *Lenticulina dorbignyi* in Europe locally extends from the Bifrons Zone (Middle Toarcian) through the Concavum Zone (Aalenian), or slightly higher.

Assemblages are numerically larger and more diverse than those of the preceding zone. Common species include *Ammobaculites fontinensis* (Terquem), *Ammodiscus tenuissimus* (Gümbel), *Citharinella deslongchampsi* (Terquem), *Astacolus pauperatus* (Jones and Parker), *Nodosaria regularis* Terquem, *Nodosaria fontinensis* Terquem, *Lenticulina toarcense* Payard, and *Palmula tenuistriata* (Franke); ostracodes include *Cytheropteron alafastigatum* Fischer, *Monoceratina scrobiculata* Triebel and Bartenstein, *Polycope pelta* Fischer, *Praeschuleridea* species 1 and 2, and *Rutlandella transversiplicata* Bate and Coleman.

This zone has not been recorded from the Peniche section because of unfavourable facies (oolitic carbonates) for this microfauna.

Grand Banks

Gradstein (1977, 1978) recognized two foraminiferal zones in the Lower Jurassic sedimentary sequence of the Grand Banks penetrated by the wells Heron H-73, Murre G-67 and Cormorant N-83. The lower of the two zones, *Involutina liassica,* was assigned approximately a Pliensbachian age and the overlying zone of *Lenticulina dorbignyi* was used to define a Toarcian to Aalenian interval. The upper limit of each zone was determined by the extinction levels (down-well tops) of indicator species.

Further research (by J. Exton) has strengthened the biostratigraphic framework. More zonal indicator species are recognized, and the presence of both zones has been established in the Bittern G-62 well. In addition the *Bairdiactypris* sp. Zone of Portugal (Early-Middle Toarcian) is probably also present in the region but due to inadequate sampling is not well separated from the *Lenticulina dorbignyi* Zone.

Involutina Liassica Interval Zone. This zone has been defined (Gradstein, 1977) by the stratigraphically highest occurrence of the foraminifera *Involutina liassica* (Jones), *Brizalina liasica* (Terquem), *Lingulina tenera* Bornemann and *Epistomina* sp. 12. In addition *Marginulina prima* d'Orbigny, *Frondicularia terquemi* d'Orbigny and the ostracode *Ogmoconcha,* which are also present in the four wells under examination, became extinct at the same stratigraphic level and are equally valid as zonal indicator species.

Based largely on the presence of *M. prima, F. terquemi, L. tenera* and *Ogmoconcha* spp., first-order correlation can be made between the Grand Banks Zone of *I. liassica* and the Zone of *F. terquemi* of the Lusitanian Basin. Establishing the time-stratigraphic equivalence of these two zones and second-order correlation with ammonite zones are more problematic. Certainly the faunal extinction which defines the top of both zones is a widespread, almost sychronous event recognizable throughout Europe in the earliest Toarcian (within the Tenuicostatum Zone). The base of the *I. liassica* Zone, however, coincides approximately with the top of the Murre Carbonate, which consists of lithologies unsuited to microfaunal analysis, as are the Coimbra Beds of the Lusitanian Basin. None of the diagnostic species of this zone, with the possible exception of *Epistomina* sp. 12 which is known only from the Grand Banks, has a stratigraphic range restricted to the Pliensbachian Stage. Palynological study (Barss *et al.*, 1979) supports a Pliensbachian age for the zone but is not able to determine how much of the Sinemurian Stage may also be represented.

The following species of foraminifera encountered in the Grand Banks wells also occur in the *F. terquemi* Zone of Portugal: *Lenticulina gottingensis, Lenticulina vetusta, Lingulina tenera tenera, Lingulina tenera pupa, Marginulina prima prima, Frondicularia terquemi bicostata, Berthelinella involuta, Brizalina liasica* and *Eoguttulina liassica;* also present are species of *Ophthalmidium, Ammobaculties* and *Trochammina.* The ostracode fauna of the *I. liassica* Zone has yet to be studied in detail but it is dominated by smooth species of *Ogmoconcha;* several species of *Polycope,* including *Polycope* sp. 1, *P. decorata* and *P. cincinnata,* also occur. Despite these faunal similarities two of the zonal indicators, namely *I. liassica* and *Epistomina* sp. 12, have not been recorded from the Lower Jurassic of Portugal, and the latter species is known only from the Grand Banks. Epistominids are rare in the Lusitanian Basin; a few pyrite casts have been found in the Margaritatus Zone at Zambujal, and a small, smooth form occurs in the Ibex and Margaritus Zones at Peniche. *Brizalina liasica,* which occurs abundantly only in the upper part of the Margaritatus Zone in Portugal, appears to be abundant in Grand Banks assemblages particularly in the Murre G-67 and Cormorant N-83 wells. The possibility that the specimens are all derived from a single, thin bed where the species occurs in abundance cannot be ruled out.

Differences in the ostracode fauna between the Grand Banks and the Lusitanian Basin are also apparent, namely the absence of *Gammacythere ubiquita* and *Ogmoconcha ambo* in the Grand Banks assemblage. However, in view of the small numbers of specimens of these species recovered from surface samples from Portugal, and their short stratigraphic ranges, it is perhaps not surprising that so far they have not been recovered from the Grand Banks subsurface.

Inferred Age: Late Sinemurian? through earliest Toarcian (lower Tenuicostatum Zone).

Occurrence: Murre G-67 (7840 to 8454 ft.), Cormorant N-83 (6390 to 7690 ft.), Heron H-73 (10060 to 10560 ft.) and Bittern M-62 (9930 to 10500 ft.).

Lenticulina Dorbignyi Interval Zone. This zone has been defined (Gradstein, 1977) by the highest stratigraphic occurrence of *Lenticulina dorbignyi* and *Kinkelinella* sp. I (previously recorded as ostracod no. 84 of Klingler); its lower limit corresponds to the top of the *I. liassica* Zone. The stratigraphic range of *L. dorbignyi* is Middle Toarcian (Bifrons Zone) through Aalenian (Bartenstein and Brand, 1937; Simon and Bartenstein, 1962; Copestake and Johnson, in press), although it is common in most areas of Europe only in the Late Toarcian and Aalenian. *Kinkelinella* sp. I ranges from Early Toarcian (Falciferum Zone) through Aalenian (Apostolescu, 1959; Lord, 1974).

Of the foraminifera and ostracodes occurring in this zone the majority are found in abundance in Portugal only in the *L. dorbignyi* Zone. These include *L. dorbignyi, Nodosaria regularis, Citharinella deslongchampsi, Ammodiscus tenuissimus, Thurammina jurensis, Kinkelinella* sp. I, *Praeschuleridea* sp. 2, *Polycope pelta, Cytherella toarcensis* and *Monoceratina ungulina.* Towards the top of the zone, *Garantella* aff. *rudia* is common and may be indicative of an Aalenian age (Gradstein, 1977). This species has not been found in the Toarcian and Aalenian of Portugal. *Lenticulina toarcense, Astacolus pauperatus* and *Polycope* sp. 1 also occur, but these species range throughout much of the Toarcian Stage in Portugal.

The predominance of Late Toarcian species in the Grand Banks assemblages probably indicates that the Early and Middle Toarcian assemblages are numerically small and of low diversity, just as they are in Portugal. The Zone of *Bairdiacypris* sp. has so far been recognised in only two of the Grand Banks wells: Cormorant N-83 (±5280 to ±6210 ft.) and Bittern M-62 (±9200 to ±9730 ft.). In Bittern M-62 *Bairdiacypris* sp. is common, and *Astacolus mariei,* a species restricted to this zone in Portugal, also occurs. *Kinkelinella sermoisensis* is common in the Cormorant well, but *E. debilis* has not been recorded so far from the Grand Banks. A zone of *Bairdiacypris* sp. may be widespread on the Grand Banks below the Zone of *L. dorbignyi* but lack of detailed sampling prevents stratigraphic resolution.

Occurrence: Cormorant N-83 (5070 to 6210 ft.), Murre G-67 (6830 to 7800 ft.), Bittern M-62 (9200 to 9730 ft.). The zone is also tentatively identified in the Eider M-75 well (10,600 to T.D. at 11,582 ft.).

Age: Toarcian and Aalenian.

Comparison to Northwestern Europe

Copestake and Johnson (in press) furnish a detailed foraminiferal zonation, keyed to the same standard ammonite scheme used in this study. Calibrated ranges are reported for over 50 taxa, including many forms of subspecific rank. The following main differences with the Portuguese/Grand Banks zonation are noted.

1) The subspecies (plexus) *Marginulina prima spinata* Terquem is reported as high as the Falciferum Zone rather than the underlying Tenuicostatum Zone. The species s.l. in Portugal disappears in the latter zone, and as a result the upper boundary of the Portuguese *Frondicularia terquemi* Zone, which includes the range of *Marginulina prima* s.l., is drawn within the Tenuicostatum Zone rather than at the top (see Fig. 4).
2) *Frondicularia terquemi* 'plexus' *muelensis* (Ruget and Sigal) is reported as high as the Tenuicostatum Zone, which agrees with the Portuguese data, and also possibly several zones higher.
3) The stratigraphic range of *Lenticulina dorbignyi* in Portugal is Thouarsense through Concavum Zones, but Copestake and Johnson (in press) report this easily recognizable species in Britain also from the older Bifrons Zone (in the Mochras borehole). In Germany, Switzerland and France the species ranges into earliest Bajocian strata (see Wernli and Septfontaine, 1971). As a result, the *L. dorbignyi* Zone of Copestake and Johnson (in press) is of longer duration than the one used in this report.
4) A form called *Lenticulina quenstedti* (Gümbel) is reported in Britain and West Germany from Early Jurassic strata. This species has not been found in Lower Jurassic rocks in Portugal or the Grand Banks, and was not reported from the Early Jurassic in Central Europe by Wernli and Septfontaine (1971). *L. quenstedti* is generally thought to be a Middle and Late Jurassic taxon.

Summary

At least twenty-two foraminifera and ostracod species, all but two of which are present in Portugal and in the Grand Banks, provide a basis for recognizing three zones in the Lower Jurassic. The zones are: 1) *Frondicularia terquemi* in Portugal, which correlates to the Raricostatum to Tenuicostatum (in part) ammonite standard zones of the latest Sinemurian to earliest Toarcian; 2) *Involutina liassica* on the Grand Banks which stratigraphically equates the *Frondicularia terquemi* Zone; 3) *Bairdiacypris* sp. of the Toarcian Tenuicostatum to Variabilis ammonite standard zones which occurs in Portugal and, probably, in the two Grand Bank wells where it is not separated from the overlying zone of 4) *Lenticulina dorbignyi*, which in Portugal is of Late Toarcian-Aalenian age (Thouarsense to Concavum ammonite standard zones) and in the Grand Banks probably also Toarcian-Aalenian.

FORAMINIFER BIOGEOGRAPHY

Gordon (1970) has produced a relatively detailed account of Jurassic foraminiferal biogeography; in it he proposed the classification of assemblages shown here in Table I.

TABLE I

CLASSIFICATION OF JURASSIC FORAMANINIFERAL ASSEMBLAGES (AFTER GORDON, 1970)

Group A. Shelf Assemblages
1) Nodosariid and nodosariid-mixed assemblages (nodosariids not less than one-fifth of all specimens; no other calcareous species conspicuously present)
2) Dominantly simple arenaceous assemblages. (Arenaceous species at least four-fifths of all specimens)
3) Calcareous benthic species other than nodosariids conspicuous (amount to at least one-quarter of all specimens)
a) Epistominidae conspicuous
b) Opthalmidiidae conspicuous
c) Buliminidae conspicuous
d) etc. ....
Group B. Tethyan Assemblages
1) Complex arenaceous species dominant
2) Planktonic assemblages. (Occur at times on "shelf" regions also)

Although Gordon suggested percentage abundances for major components of each asemblage, it seems likely that the reality is a continuous compositional gradation between each assemblage type. The geographic distribution of assemblages was probably also as variable and dependent upon local and regional environmental changes as the distribution patterns found in modern shelf seas. Nevertheless, Gordon's classification adequately demonstrates the extremes of compositional variation to be found in Jurassic assemblages. Although no provision is made for deep-water assemblages (e.g., Luterbacher, 1972) the significance of his twofold division into 'Tethyan' and 'Shelf' assemblage groups is worthy of further consideration. Gordon admitted that the recognition of foraminiferal faunal provinces, such as those recognised for ammonites, was not possible owing to the sparsity of data outside of Europe, but he suggested that the Tethyan region (most of Alpine Europe and areas surrounding the Mediterranean) had its own distinctive assemblages of larger foraminifera, occasionally occurring together with involutinids, trochamminids and verneulinids. Furthermore, he stated that, in the Early Jurassic, "there are no published records of Tethyan faunas dominated by nodosariids" (Gordon, 1970, p. 1691) and implied that nodosariid-dominated assemblages were absent from the Tethys. Scheibnerova (1971), Luterbacher (1972) and Gradstein (1977) all contradict this, remarking that such assemblages (i.e., of the A type) do occur within the Tethys, although little has yet been published about them.

The literature examined for the compilation of the following biogeographic section is by no means exhaustive but is sufficient to demonstrate that Gordon's type-B1 assemblages are indeed geographically restricted to the Tethyan region in the Early Jurassic, and that group-A assemblages, typically known from northwestern Europe, had a more cosmopolitan distribution. Undoubtedly planktonic (B2-type) assemblages have not been reported from Early Jurassic strata.

Group A Assemblages. The vast majority of Early Jurassic

foraminiferal assemblages described to date fall into group A of Gordon's classification (Table I), hardly surprising in view of the broad range of assemblage composition allowed for this group (several present-day shelf faunas could be classified as type A3). Although numerous families occur in these faunas most of them are represented by only a small number of genera and species. A single family, the Nodosariidae, is usually the predominant component; the relative diversity of this family is indicated by the percentage data compiled in Table II.

TABLE II

PERCENTAGES OF NODOSARIIDS IN PUBLISHED SPECIES LISTS

Area	Stages	% Nodosariid Species	Source
Sicily	S.P.T.	67	Barbieri, 1964
Portugal	P.T.	76	Exton, 1979
Alaska	H.S.P.T.	71	Tappan, 1955
Sweden	H.S.P.T.	87	Norling, 1972
Germany	P.	80	Welze, 1968
Germany	H.S.P.T.	75	Pietrzenuk, 1961

The percentages given above are calculated from the total number of species recorded, not the total number of individuals, and do not indicate percentage composition of assemblages; H – Hettangian, S – Sinemurian, P – Pliensbachian, T – Toarcian

Because of the absence of detailed quantitative data, recognition of assemblage types A1 to 3 is seldom possible from published accounts. However, from the authors' experience and from comments on relative abundance that have been made in the literature (e.g., Pietrzenuk, 1961), assemblage type A1 seems to be the most common and the most widely distributed whereas type A3 is of sporadic occurrence, and the arenaceous type-A2 assemblages are the least common. The Zambujal assemblages fall within types A1 and A3; Pliensbachian assemblages have a nodosariid component ranging from 66% to 98% (of specimens) and those from the Toarcian between 23% and 92% (Exton, 1977). Type-A3 assemblages, in which *Ophthalmidium liasicum* (Kübler and Zwingli) is abundant, occur rarely within the Zone of *Bairdiacypris* sp. In the *Brizalina liasica* Subzone at Peniche, São Pedro de Muel and Brenha the index species accounts for over 25% of specimens in some assemblages. A similar classification may apply to the Grand Banks assemblages; certainly the scarcity of arenaceous specimens indicates the absence of type-A2 assemblages.

Early Jurassic group-A assemblages are now known in both the northern and southern hemispheres, from Alaska to Australia, (Fig. 5; Table III), and although this widespread distribution is not considered to be unusual because of the coarseness of the classification mentioned above, the compositional similarity of these assemblages at the level of species is quite remarkable. Of particular significance in the context of the zonal scheme presented earlier is the widespread distribution of the zonal indicator species.

At present, data from the southern hemisphere are sparse; we are aware of only three recorded occurrences of Early Jurassic foraminifera. Scheibnerova (1971) listed a number of species

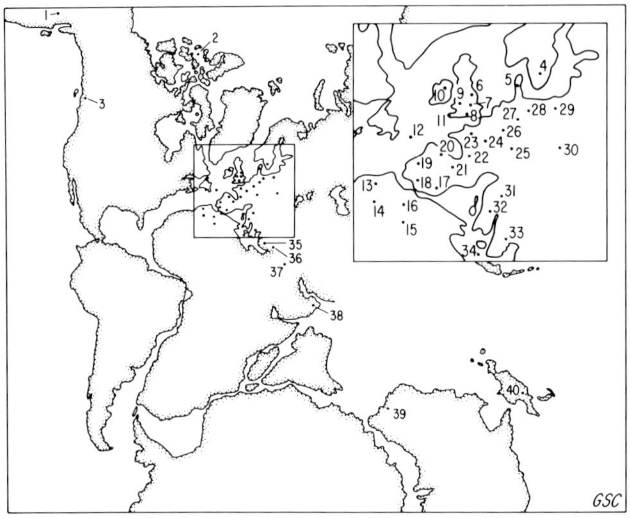

Figure 5. Early Jurassic paleogeography (following Smith and Briden, 1977) and locations from which Early Jurassic Groups A and B foraminiferal assemblages (Table I) have been reported. References to these assemblages are tabulated in Table III.

from western Australia including *Frondicularia terquemi*, and Hornibrook (1953) recorded a few species from New Zealand, of which *Lingulina evansi* Hornibrook is here considered to fall within the range of morphological variation exhibited by *Lingulina tenera*. The most complete account is that of Haig (1979) who provided a well illustrated microfaunal record from the Sinemurian/Pliensbachian of Papua New Guinea. The assemblage includes *Lingulina tenera*, *Frondicularia terquemi bicostata*, *Marginulina prima* and *Involutina liassica*, together with a number of other species common in the northern hemisphere. No foraminifera of definite Toarcian age have been recorded to date from the southern hemisphere.

From Alaska, Tappan (1955) has recorded a large number of nodosariids, many of which were described as new species. Both *Marginulina prima* and *Frondicularia terquemi* occur (*F. lustrata* Tappan must also be included in this species group) and exhibit the wide range of morphological variation found in their European counterparts; *Lingulina tenera* is surprisingly absent. The large number of arenaceous species recorded by Tappan suggests that some assemblages may be of A2 type. This is undoubtedly the case with the small arenaceous assemblage described by Souaya (1976) from Lower Jurassic sediments penetrated by the Linckens Island well in the Canadian Arctic. Late Toarcian assemblages have not yet been described from these areas, which probably accounts for the absence of *Lenticulina dorbignyi*.

Early Jurassic foraminiferal assemblages from northwestern Europe need little introduction. They have been the subject of numerous taxonomic and biostratigraphic studies over the past hundred years and more, and their uniformity of composition is well established. All of the data points in this area on Figure 5 refer to group A faunas. Space does not permit a detailed summary; the reader must refer to the works listed in Table III together with the useful summaries of stratigraphic and geographic distribution given by Brouwer (1969) and Norling (1972). Data from some areas of southern Europe are sparse and we have had to rely upon unillustrated species listed from Spain (Assens, 1971; Ramirez del Pozo, 1971) and identifications in thin section from southern Poland (Gazdzicki, 1975).

Our knowledge of group-A assemblages from the Tethyan area of southern Europe and North Africa is still pityfully small. Many records give only a brief list of species or are to be found in articles of a more general stratigraphic nature. Such are the records from northern Italy (Cousin and Neumann, 1971), Turkey (Bronnimann *et al.*, 1970) and southern Portugal (Bordalo da Rocha, 1976). This latter record is from the basal Toarcian Stage (Tenuicostatum Zone) and the faunal list consists almost entirely of nodosariid species. These include the three indicator species of the *F. terquemi* Zone as defined from the Lusitanian Basin of northwestern Portugal, although the sedimentary facies (predominantly limestones and dolostones with some argillaceous beds in the Toarcian) is more typical of shallow, basinal areas within the Tethyan carbonate platform (section 1).

The most detailed records of group-A assemblages from the Tethys are those from Sicily (Barbieri, 1964) and Algeria (Maupin and Vila, 1976; Maupin, 1977). Of the ninety-six species recorded from the Sinemurian, Pliensbachian and Toarcian stages of Sicily, 67% are nodosariids; the Algerian faunas are smaller, and again nodosariids form the predominant component. Two or more of the *F. terquemi* Zone indicators are present in both of these areas and *L. dorbignyi* occurs in the Late Toarcian. We have also studied a small fauna from samples collected by L.F. Jansa in Morocco, south of Ksar-es-Souk (outcrop 3 km east of Tunnel du Legionnaire, High Atlas; Geological Survey of Canada No. JN 73-X1B, samples 3 and 8), containing *Lingulina tenera, Citharina inaequistriata* (Terquem), *Eoguttulina liassica, Verneuilinoides mauritii* (Terquem) and a smooth species of the ostracode *Ogmoconcha* (see Plate I).

Because so few Early Jurassic type-A assemblages have been described from the Tethys, any detailed comparison with the better-known faunas of northwestern Europe would be premature. It is sufficient to note that wherever Early Jurassic type-A assemblages have been recorded, either from the Tethyan margins or from the epeiric seas to the north and south of the Tethys, they all show close compositional similarities even at the level of species.

TABLE III

LITERATURE REFERENCES TO THE EARLY JURASSIC GROUPS A AND B ASSEMBLAGES OF FORAMINIFERA (TABLE I), SHOWN IN THE PALEOGEOGRAPHIC MAP OF FIGURE 5

Group A Assemblages	
1. Tappan, 1955	Alaska
2. Souaya, 1976	Canadian Arctic
3. Cameron (in prep.)	Western Canada
4. Norling, 1972	Southern Sweden
5. Norvang, 1957	Denmark
6. Tate and Blake (in Blake), 1876	Northern England
7. Barnard, 1950b	Midlands of England
Horton and Coleman, 1977	
8. Barnard, 1950a	Southern England
9. Copestake and Johnson (in press)	North Wales
11. Lloyd *et al.*, 1973	Southwestern England
10. McGugan, 1965	Northern Ireland
23. Rioult and Bizon, 1961	Northern France (Normandy)
24. Champeau, 1961	Paris Basin
Cousin *et al.*, 1961	Paris Basin
26. Ruget and Sigal, 1967	NE France
22. Séronie - Vivien *et al.*, 1961	Aquitaine Basin
28. Bartenstein and Brand, 1937	N Germany
29. Pietrzenuk, 1961	NE Germany
Dreyer, 1967	NE Germany
27. Hoffmann and Martin, 1960	NW & SW Germany
25. Welze, 1968	S Germany
30. Gazdzicki, 1975	S Poland
21. Assens, 1971	NE Spain
20. Ramirez del Pozo, 1971	N Central Spain
19. Ruget and Sigal, 1970	NW Portugal
Exton, 1979	NW Portugal
18. Bordalo da Rocha, 1976	S Portugal
34. Barbieri, 1964	Sicily
31. Cousin and Neumann, 1971	N Italy
35. Bronnimann *et al.*, 1970	Turkey
16. Maupin and Vila, 1976	Algeria
Maupin, 1977	Algeria
13. This Paper	Morocco
12. Gradstein, 1977, 1978	Grand Banks
40. Haig, 1979	Paupa, New Guinea
39. Scheibnerova, 1971	W. Australia
Group B Assemblages	
14. Hottinger, 1967	Morocco
15. Bassoullet and Fares, 1969	Algeria
17. Gonzalez-Donoso, 1971	Southern Spain
32. Crescenti, 1973	Italy
37. Maync, 1966	Israel
Derin and Reiss, 1966	Israel
36. Bassoullet and Poisson, 1975	Turkey
38. Glennie *et al.*, 1974	Oman.
33. Bassoullet and Guernet, 1970	Greece

Group B Assemblages. Only assemblage type B1 (see Table I) has been recorded from the Early Jurassic; the pelagic type-B2 assemblage is unknown, although planktonic foraminifera may have evolved in the Triassic (Fuchs, 1967) and they were certainly present in the Middle Jurassic (Pazdrowa, 1969; Gradstein, 1977).

The larger foraminifera (with complex wall structures) had a geographic distribution restricted to the area of carbonate platform deposits surrounding the Early Jurassic Tethyan seaway, and they are particularly well known from areas around the present Mediterranean, from Morocco in the west (Hottinger, 1967) to Turkey (Bassoullet and Poison, 1975), Israel (Maync, 1966) and the Oman (Glennie *et al.*, 1974) in the east (see also Fig. 5).

Gordon (1970) concluded that differentiation into "Tethyan" and "Shelf" foraminiferal assemblages resulted in part from "the two different geotectonic-sedimentary environments involved" (*op. cit.*, p. 1701). A discussion along similar lines has been presented by Hallam (1968) who noted the coincidence between macrofaunal distribution patterns and sedimentary facies change within the Jurassic epeiric and Tethyan seas. The paleoecology of the larger foraminifera has seldom been discussed except for the often repeated generalization that these forms inhabited warm, shallow waters in nearshore environments (e.g., Scheibnerova, 1971; Bassoullet and Farès, 1969). In many instances they occur as part of a faunal association containing large megalodontid and lithiotid bivalves and dasycladacean algae; distribution patterns of *Lithiotis* (Broglio Loriga and Neri, 1976) and the foraminifer *Orbitopsella* (Maync, 1960) show close similarities. The B1 assemblage is quite diversified and includes genera in common with group-A assemblages, such as *Ophthalmidium, Textularia* and *Ammobaculites,* as well as less widespread siphonate valvulinids. However, as a generalization it may be stated that group-A and group-B2 assemblages had different ecologic requirements and are not found as mixed assemblages. Larger foraminifera are usually found in shallow-water, high-energy carbonate sediments devoid of ammonites, whereas nodosariid assemblages occur in basinal argillaceous sediments where ammonites are common. *Involutina,* which Gordon grouped in his B1 assemblage, shows a preference for this latter environment in the Early Jurassic although in the Triassic it occupied an ecologic niche similar to that of the large foraminifera (Zaninetti, 1976).

Because the shallow-water sediments in which B1 assemblages occur are generally devoid of ammonites, the larger foraminifera have been intensively studied to provide an alternative means of biostratigraphic subdivision and correlation (Hottinger, 1967, 1973; Bassoulett and Fourcade, 1979). The first abundant and diversified faunas appear in the Pliensbachian Stage (Hottinger, 1967) and are charcterized by species of *Orbitopsella.* The top of the Pliensbachian Stage is marked by the extinction of *Orbitopsella,* and as in northwestern Europe, Early Toarcian faunas are impoverished; *Lucasella cayeuxi* and *Lituosepta compressa* appear to be characteristic of the Late Toarcian.

SUMMARY AND CONCLUSIONS

The Triassic-Lower Jurassic lithological succession recorded from wells drilled on the Grand Banks is very similar to that observed in onshore exposures in the Lusitanian Basin of west-central Portugal. Both basins developed during a Late Triassic rifting event which affected most North Atlantic marginal basins. Major vertical changes of lithology occur within similar time intervals in both areas. The occurrence in the Grand Banks sub-surface of thick Triassic-Hettangian evaporites overlain by Hettangian-Sinemurian dolostones and limestones demonstrates a close similarity to three sedimentary basins of south-west Europe – the Aquitaine and Cantabrian basins and the Lusitanian Basin. This similarity is further emphasized by the absence of Late Pliensbachian sandstone or ironstone facies and thick Early Toarcian bituminous shales which occur in the more northerly European basins. Both the Grand Banks and Lusitanian basins lay immediately to the north of the Early Jurassic Tethyan carbonate platform, the northern boundary of which is probably represented by the dolomitic successions of the Serra da Arrabida (the southern margin of the Lusitanian Basin) and the Algarve (south coast of Portugal) and by a change to a more calcareous Lower Jurassic on the Scotian Shelf. This gradual north-south facies change through western Europe (more calcareous southwards) places some constraints upon the position of Portugal in continental reconstructions.

First-order correlation between foraminifer/ostracode zones of the Grand Banks and Lusitanian basins allow second-order correlation of ammonite zones to be carried to the Grand Banks subsurface. A Zone of *Frondicularia terquemi* in the Lusitanian Basin occurs in the ammonite zones Raricostatum through Tenuicostatum (latest Sinemurian through earliest Toarcian); this microfaunal zone correlates with the Zone of *Involutina liassica* of the Grand Banks subsurface. The top of both zones is marked by a major microfaunal extinction event which in Portugal occurs within the Tenuicostatum Zone. Zones of *Bairdiacypris* sp. and *Lenticulina dorbignyi* are common to both areas, the former occupying a range of Early to Middle Toarican and the latter Late Toarcian through Aalenian in the Lusitanian Basin.

Early Jurassic foraminiferal assemblages of the Grand Banks and Portugal form part of a cosmopolitan microfauna which occupied epicontinental seas before fragmentation of the continents in Middle to Late Jurassic time. A distinctive and geographically restricted, larger foraminiferal fauna occupied the marginal Tethyan carbonate platforms during the Early Jurassic; this fauna has not been found either in Portugal or the Grand Banks subsurface.

EXPLANATION OF PLATE I

Figure 1. *Verneuilinoides mauritii* (Terquem) Sample JN73-X1B
High Atlas. Morocco; x 180.

Figure 2. *Lingulina tenera* Bornemann var. Sample JN73-X1B
High Atlas, Morocco; x 180.

Figure 3. *Citharina inaequistriata* (Terquem) Sample JN73-X1B
High Atlas, Morocco; x 30.

Figure 4. *Marginulina prima* d'Orbigny
Jamesoni Zone, Zambujal, Portugal; x 70.

Figure 5. *Lingulina tenera* Bornemann var.
Heron H-73, 10260-10290 ft., Grand Banks; x 100.

Figure 6. *Frondicularia terquemi* d'Orbigny var.
Cormorant N-83, 6670 to 6700 ft., Grand Banks; x 85.

Figure 7. *Involutina liassica* (Jones)
Cormorant N-83, 6510-20 ft., Grand Banks; x 70.

Figure 8. *Epistomina* sp. 12 Gradstein
Cormorant N-83, 6630-60 ft., Grand Banks; x 55.

Figure 9. *Brizalina liasica* (Terquem)
Margaritatus Zone, São Pedro de Muel, Portugal; x 85.

Figure 10. *Palmula mariei* (Payard)
Bifrons Zone, Peniche, Portugal; x 70.

Figure 11. *Astacolus pauperatus* (Jones and Parker)
Cormorant N-83, 5910-5940 ft., Grand Banks; x 125.

Figure 12. *Lenticulina dorbignyi* (Roemer)
Levesquei Zone, Zambujal, Portugal; x 95.

Figure 13. *Berthelinella involuta* (Terquem)
Heron H-73, 10460 to 1,0490 ft., Grand Banks; x 125.

PLATE I

PLATE II

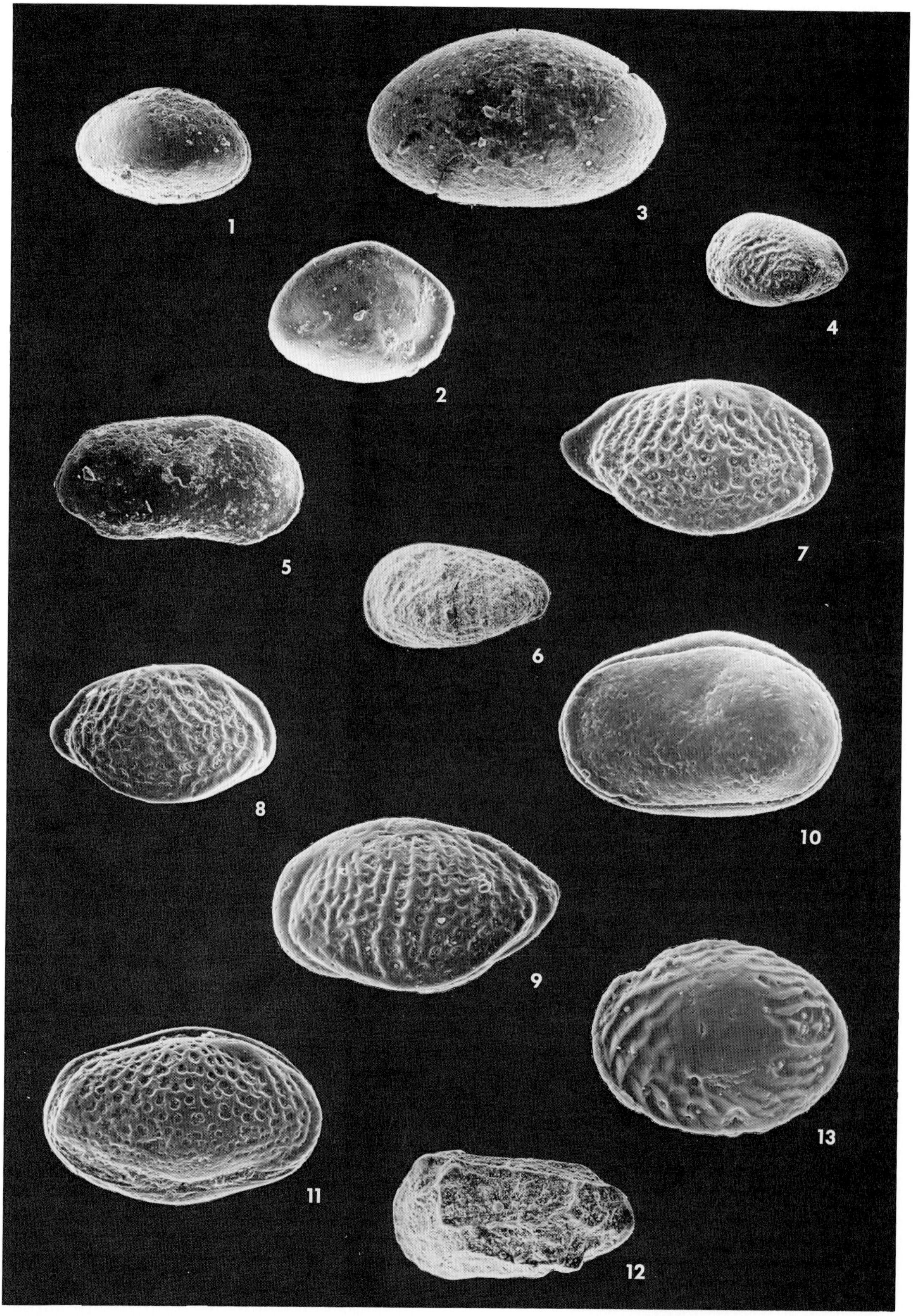

ACKNOWLEDGEMENTS

We are obliged to Drs. R.B. da Rocha and H.M. Ramalho (Lisbon), R. Mouterde and C. Ruget-Perrot (Lyon), R.C.L. Wilson (U.K.), and L.F. Jansa (Dartmouth, N.S) for cordial advice and field assistance in Portugal. The scanning electron microscope illustrations were prepared by F.C.Thomas (Dartmouth, N.S.).

REFERENCES

Adloff, M.C., Doubinger, J. and Palain, C., 1974, Contribution à la Palynologie du Trias et du Lias inférieur du Portugal. "Grès de Silves" au nord du Tage: Communicaçoes dos Serviços Geologicos de Portugal, v. 58, p. 91-144.

Amoco Canada Petroleum Company Ltd. and Imperial Oil Ltd., 1973, Regional Geology of the Grand Banks: Bulletin of Canadian Petroleum Geology, v. 21, p. 479-503.

Apostolescu, V., 1959, Ostracodes du Lias du Basin de Paris: Revue de l'Institute français de Pétrole, v. 14, p. 795-826, pls. 1-4.

Assens, J., 1971, Notas sobre el Jurasico de la Zona de Cameros: Cuadernos Geologia Ibérica, v. 2, p. 637-646.

Barbieri, F., 1964, Micropaleontologia del Lias e Dogger del Pozzo Ragusa I (Sicilia): Rivista Italiana Paleontologia e Stratigrafia, v. 70, no. 4, p. 709-831.

Barnard, T., 1950a, Foraminifera from the Lower Lias of the Dorset Coast: Quarterly Journal of the Geological Society of London, v. 105, p. 347-391.

______, 1950b, Foraminifera from the Upper Lias of Byfield, Northamptonshire: Quarterly Journal of the Geological Society of London, v. 106. p. 1-36.

Barss, M.S., Bujak, J.P. and Williams, G.L., 1979, Palynological Zonation and Correlation of Sixty-seven Wells, Eastern Canada. Geological Survey of Canada Paper 78-24, p. 1-119.

Bartenstein, H., and Brand, E., 1937, Mikro-paleontologische untersuchungen zur Stratigraphie des Nordwest-Deutschen Lias und Doggers: Abhandlungen Senckenbergischen Naturforschenden Gesellschaft, v. 439, p. 1-224, pls. 1-20.

Bassoullet, J.P., and Farès, F., 1969, Les Orbitopselles du Lias du Djebel Hafid (Algérie): Revue de Micropaléontologie, v. 12, p. 171-176.

Bassoullet, J.P., and Fourcade, E., 1979, Essai de Synthèse de Répartition de Foraminifères Benthiques du Jurassique Carbonaté Mésogéen: Comptes rendus des séances de la Société Géologique de France, v. 2, p. 69-71.

Bassoullet, J.P., and Guernet, C.L., 1970, Les Trias et le Jurassique de la Région des Lacs de Thèbes (Béotie et Locride, Grèce): Revue de Micropaléontologie, v. 12, p. 209-217.

Bassoullet, J.P., and Poisson, A., 1975, Microfacies du Jurassique de la Région d'Antalya (secteurs N. et N.W.) Taurus Lycien (Turquie): Revue de Micropaléontologie, v. 18, p. 3-14.

Bate, R.H., 1977, Jurassic Ostracods of the Atlantic Basin: *in* Swain, F.M., ed. Stratigraphic Micropaleontology of the Atlantic Basins and Borderlands: Developments in Paleontology and Stratigraphy No. 6: Elsevier, Amsterdam, p. 231-242.

Bate, R.J. and Coleman, B.E., 1975, Upper Lias Ostracoda from Rutland and Huntingdonshire: Bulletin of the Geological Survey of Great Britain, v. 55, p. 1-42.

Bernoulli, D. and Jenkyns, H.C., 1974, Alpine, Mediterranean, and Central Atlantic Mesozoic Facies in Relation to the Early Evolution of the Tethys: *in* Dott, R.H., and Shaver, R.H., eds., Modern and Ancient Geosynclinal Sedimentation: Society of Economic Paleontologists and Mineralogists Special Publication 19, p. 129-160.

Bhat, H., McMillan, N.J., Aubert, J., Porthault, B. and Surin, M., 1975, North American and African Drift – the Record in Mesozoic and Coastal Plain Rocks, Nova Scotia and Morocco: *in* Yorath, C.J., Parker, E.R., and Glass, D.J., eds., Canada's Continental Margins and Offshore Petroleum Exploration: Canadian Society of Petroleum Geologists Memoir 4, p. 375-389.

Blake, J.F., 1876, *in* Tate, R and Blake, J.F., Class Rhizopoda: The Yorkshire Lias, pt. 2, London p. 449-473.

Bordalo da Rocha, R., 1976, Estudo Estratigrafico e Paleontologico do Jurassico do Algarve Ocidental: Universidade Nova do Lisboa, Ciências da Terra, v. 2, p. 1-178.

Broglio Loriga, C. and C. Neri, 1976, Aspetti Paleobiologici e Paleogeografici della Facies a «Lithiotis» (Giurense Inf): Rivista Italiana Paleontoliga e Stratigraifa, v. 82, p. 651-706.

Bronnimann, P., Poisson, A. and Zaninetti, L., 1970, Facies et Foraminifères du Trias et du Lias Domuz Dag (Taurus lycien – Turquie): Rivista Italiana Paleontologia, v. 76, p. 1-36.

Brouwer, J., 1969, Foraminiferal Assemblages from the Lias of North-western Europe: Verhandelingen der Koninklijke Nederlandse Akademie van Wetenschappen, Afd. Natuurkunde I Reeks, Deel. 25, p. 1-48.

Champeau, H., 1961, Contributions a l'étude micropaléontologique du Lias du Bassin de Paris: Troisième Partie: Etude de la microfaune des niveaux nameax du Lias dans le sud-est du Bassin de Paris: *in* Colloque sur le Lias Français: Mémoire Bureau Recherche Géologiques Minières, v. 5, p. 437-443.

EXPLANATION OF PLATE II

Figure 1. *Ogmoconcha amalthei rotunda* Dreyer
Spinatum Zone, Peniche, Portugal; x 70.

Figure 2. *Ogmoconcha* cf. *ambo* Lord and Moorley
Spinatum Zone, Zambujal, Portugal; x 70.

Figure 3. *Ogmoconcha* sp. Sample JN73-X1B
High Atlas, Morocco; x 70.

Figure 4. *Gammacythere ubiquita* Malz and Lord
Jamesoni Zone, Zambujal, Portugal; x 60.

Figure 5. *Bairdiacrypris* sp.
Falciferum Zone, Zambujal, Portugal; x 130.

Figure 6. *Ektyphocythere debilis* Bate and Coleman
Bifrons Zone, Zambujal, Portugal; x 95.

Figure 7. *Kinkelinella sermoisensis* (Apostolescu)
Cormorant N-83, 6000 to 6030 ft., Grand Banks; x 95.

Figure 8. *Kinkelinella sermoisensis* (Apostolescu)
Bifrons Zone, Zambujal, Portugal; x 95.

Figure 9. *Kinkelinella* sp. I (Apostolescu)
Cormorant N-83, caved specimen from 6770 to 6800 ft.
Grand Banks; x 85.

Figure 10. *Cytherella toarcensis* Bizon
Cormorant N-83, 5820 to 5850 ft., Grand Banks; x 85.

Figure 11. *Praeschuleridea* sp. 2 Exton
Cormorant N-83, 6180-6210, Grand Banks; x 70.

Figure 12. *Rutlandella transversiplicata* Bate and Coleman
Levesquei Zone, Zambujal, Portugal; x 210.

Figure 13. *Polycope pelta* Fischer
Cormorant N-83, 6180-6210 ft., Grand Banks; x 145.

Choffat, P., 1880, Etude Stratigraphique et Paléontologique des Terrains Jurassiques du Portugal: Le Lias et le Dogger au Nord du Tage: Memorias de Secçao dos Trabalhos Geologicos de Portugal, v. 12, 73 p.

———, 1882, Note Préliminaire, Sur les Vallées Tiphoniques et les Eruptions d'Ophite et de Teschenite au Portugal: Bulletin de la Société Géologique de France, v. 10, p. 267-288.

Copestake, P. and Johnson, B., in press, 1984, Lower Jurassic (Hettangian-Toarcian) Foraminifera from the Mochras borehole, North Wales; Systematic and Biostratigraphical Micropaleontology: Institute of Geological Sciences, Report, p. 1-162.

Cousin, M., and Neumann, M., 1971, Microfaciès du Lias dans les Préalpes Juliennes Occidentales (Frioul, Italie): Revue de Micropaléontologie, v. 14, p. 35-49.

Cousin, N., Esptitalie, J. and Sigal, J., 1961, Contributions à l'étude Micropaléontologique du Lias du Bassin de Paris: Quatrième Partie: Sud du Bassin, Région d'Argenton-sur-Creuse et de La Chatre (Départements du Cher et de l'Indre): *in* Colloque sur le Lias Français: Mémoir Bureau Recherche Géologiques Minières, v. 5, p. 445-449.

Crescenti, U., 1973, Biostratigrapic Correlations in the Jurassic Facies of Central Italy by means of the Microfossils. Annales Instituti Geologici Publici Hungarici, v. 54, p. 209-213.

Da Costa, 1950, Quelques Remarques sur la Tectonique du Portugal: Boletim da Sociedade Geologica de Portugal, v. 8, p. 193-206.

Dean, W.T., Donovan, D.T. and Howarth, M.K., 1961, The Liassic Ammonite Zones and Subzones of the North-west European Province: Bulletin of the British Museum (Natural History), Geology, v. 4, p. 438-505.

Derin, B. and Reiss, Z., 1966, Jurassic Microfacies of Israel: Israel: Institute of Petroleum Special Publication, Tel-Aviv.

Doubinger, J., Adloff, M. and Palain, C., 1970, Nouvelles Précisions Stratigraphiques sur la Série de Base du Mésozoique Portugais: Comptes Rendus des Séances de l'Académie des Sciences de Paris, v. 270, p. 1770-1772.

Dreyer, E., 1967, Mikrofossilien des Rät und Lias von SW-Brandenburg: Jahrbuch für Geologie, v. 1, p. 491-531.

Exton, J., 1977, Benthonic Microfaunal Associations from the Liassic (Lower Jurassic) of Zambujal, West Central Portugal: M.Sc. Thesis, Carleton University, Ottawa, Canada, p. 1-321.

———, 1979, Pliensbachian and Toarcian Microfauna of Zambujal, Portugal: Systematic Paleontology: Carleton University Geological Paper 79-1, p. 1-104.

Fuchs, W., 1967, Über Ursprung und Phylogenie der Trias-"Globigerinen" und die Bedeutung dieses Formenkreises für das echte Plankton. Jahrbuch der Geologischen Bundesanstalt, v. 110, p. 135-177.

Gazdzicki, A., 1975, Lower Liassic ("Gresten Beds") Microfacies and Foraminifers from the Tatra Mts: Acta Geologica Polonica, v. 25, p. 385-398.

Given, M.M., 1977, Mesozoic and Early Cenozoic Geology of Offshore Nova Scotia: Bulletin of Canadian Petroleum Geology, v. 25, p. 63-91.

Glennie, K.W., Boeuf, M.G.A., Hughes Clarke, M.W., Moody-Stuart, M., Pilaar, W.F.H. and Reinhardt, B.M., 1974, Geology of the Oman Mountains: Verhandelingen van het Koninklijk Nederlands geologisch mijnbouwkundig Genootschap, v. 31, p. 1-423.

Gomes, J.N., 1963, Estudo Micrografico do Corte de Brenha PO13: Unpublished report of Companhia Portuguesa dos Petroleos.

Gonzales-Donoso, J.M., Rivas, P. and Vera, J.A., 1971, La Serie Liasica de Sierra Pelada (Zona Subbética, N. de Granada): Cuadernos Geologia Ibérica, v. 2, p. 205-226, Madrid.

Gordon, W.A., 1970, Biogeography of Jurassic Foraminifera: Geological Society of America Bulletin, v. 81, p. 1689-1704.

Gradstein, F.M., 1977, Biostratigraphy and Biogeography of Jurassic Grand Banks Foraminifera: *in* Proceedings of 'Benthonics 75' Halifax (N.S.): Maritime Sediments Special Publication, 1 Part B, p. 557-583.

———, 1978, Jurassic Grand Banks Foraminifera: Journal of Foraminiferal Research, v. 8, p. 97-109.

Gradstein, F.M., Williams, G.L., Jenkins, W.A.M. and Ascoli, P., 1975, Mesozoic and Cenozoic Stratigraphy of the Atlantic Continental Margin, Eastern Canada: *in* Yorath, C.J., Parker, E.R., and Glass, D.J., eds., Canada's Continental Margins and Offshore Petroleum Exploration: Canadian Society of Petroleum Geologists Memoir, v. 4, p. 103-131.

Grant, A.C., 1977, Multichannel Seismic Reflection Profiles of the Continental Crust Beneath the Newfoundland Ridge: Nature, v. 270, p. 22-25.

Groupe Galice, 1979, The Continental Margin off Galicia and Portugal: Acoustical Stratigraphy, Dredge Stratigraphy, and Structural Evolution: Initial Reports of the Deep Sea Drilling Project, v. XLIII. United States Government Printing Office, Washington, D.C., p. 633-662.

Haig, D.W., 1979, Early Jurassic Foraminiferids from the Western Highlands of Papua New Guinea: Neues Jahrbuch für Geologie und Paläeontologie Monatshefte, v. 4, p. 208-215.

Hallam, A., 1971a, Facies Analysis of the Lias in West Central Portugal. Neues Jahrbuch fü Geologie und Paläontologie Abhandlung, v. 139, p. 226-265.

Hallam, A., 1971b, Mesozoic Geology and the Opening of the North Atlantic: Journal of Geology, v. 79, p. 129-157.

Hoffmann, K. and Martin, G.P.R., 1960, Die Zone des *Dactylioceras tenui costatum* (Toarcian, Lias) in NW and SW Deutschland: Palaontologisches Zeitschrift, v. 34, p. 103-149.

Hornibrook, N de B, 1953, Jurassic Foraminifera from New Zealand: Transactions of the Royal Society of New Zealand, v. 81, p. 375-378.

Horton, A. and Coleman, B.E., 1977, The Lithostratigraphy and Micropaleontology of the Upper Lias at Empingham, Rutland: Bulletin of the Geological Survey of Great Britain, v. 62, p. 1-12.

Hottinger, L., 1967, Foraminifères Imperforés du Mésozoïque Marocain: Notes et Mémoires du Service Géologique du Maroc (Rabat), v. 54, p. 1-168.

———, 1973, Larger Foraminifera of the Mediterranean Jurassic and their Stratigraphic Use: Annales Instituti Geologici Publici Hungarici, v. 54, p. 497-504.

Howie, R.D., and Barss, M.S., 1975, Upper Paleozoic Rocks of the Atlantic Provinces, Gulf of St. Lawrence, and Adjacent Continental Shelf: *in* Offshore Geology of Eastern Canada: Geological Survey of Canada Paper 74-30, v. 2, p. 35-50.

Jansa, L.F., and Wade, J.A., 1975, Geology of the Continental Margin off Nova Scotia and Newfoundland: *in* Offshore Geology of Eastern Canada: Geological Survey of Canada Paper 74-30, v. 2, p. 51-105.

Jansa, L.F., Bujak, J.P., and Williams, G.L., 1980, Upper Triassic Salt Deposits of the Western North Atlantic: Canadian Journal of Earth Sciences, v. 17, p. 547-559.

Jansa, L.F., Gradstein, F.M., Harris, I.M., Jenkins, W.A.M., and Williams, G.L., 1976, Stratigraphy of the Amoco-IOE Murre G-67 Well, Grand Banks of Newfoundland: Geological Survey of Canada Paper 75-30, p. 1-14.

Jansa, L.F., Gradstein, F.M., Williams, G.L., and Jenkins, W.A.M., 1977, Geology of the Amoco Imp Skelley A-1 Osprey H-84 Well,

Grand Banks, Newfoundland: Geological Survey of Canada Paper 77-21, p. 1-17.

Keen, C.E., Hall, B.R., and Sullivan, K.D., 1977, Mesozoic Evolution of the Newfoundland Basin: Earth and Planetary Science Letters, v. 37, p. 307-320.

Kent, P.E., 1977, The Mesozoic Development of Aseismic Continental Margins: Journal of Geological Society of London, v. 154, p. 1-18.

Kristoffersen, Y., 1978, Sea-Floor Spreading and the Early Opening of the North Atlantic: Earth and Planetary Science Letters, v. 38, p. 273-290.

Laughton, A.S., 1975, Tectonic Evolution of the Northeast Atlantic Ocean – A Review: Norges Geologiske Undersokelse, v. 316, p. 169-193.

Le Pichon, X., Sibuet, J.C., and Francheteau, J., 1977, The Fit of the Continents Around The North Atlantic Ocean: Tectonophysics, v. 38, p. 169-209.

Livret-Guide G-14, 1980, Portugal: Introduction à la Géologie Générale – Excursions: 016A-045A, 26th Congrès Géologique International, Paris, 1980, p. 1-115.

Lloyd, A.J., Savage, R.J.G., Stride, A.H., and Donovan, D.T., 1973, The Geology of the Bristol Channel Floor: Philosophical Transactions of the Royal Society of London, v. 274, p. 595-626.

Lord, A., 1974, Ostracods from the Domerian and Toarcian of England: Palaeontology, v. 17, p. 599-622.

Luterbacher, H.P., 1972, Foraminifera from the Lower Cretaceous and Upper Jurassic of the Northwestern Atlantic: *in* Hollister, C.D., Ewing, J.I., *et al.*, 1972; Initial Report of the Deep Sea Drilling Project, v. 11, Washington, D.C., p. 561-593.

Malz, H., and Lord, A., 1976, *Gammacythere* n.g. (Ostracoda) and its Occurrence in the Lower Jurassic of N.W. Europe: Senckenbergiana Lethaea, v. 57, p. 249-263.

Manspeizer, W., Puffer, J.H., and Cousminer, H.L., 1978, Separation of Morocco and Eastern North America: A Triassic-Liassic Stratigraphic Record: Geological Society of America Bulletin, v. 89, p. 901-920.

Mauffret, A., Boillot, G., Auxiètre, J. and Dunand, J., 1978, Evolution Structurale de la Marge Continental au Nord-ouest de la Péninsule Iberique: Bulletin de la Société Géologique de France, v. 7, no. 20, 375-388.

Maupin, C., 1977, Données Micropaléontologiques Nouvelles et Précisions Stratigraphiques sur le Lias du Kef Ben Chike Bou Rouhou et du Kef Toumiette Nord (Chaine Calcaire Kabyle – Nord du Constantinois – Algérie): Revue de Micropaléontologie, v. 20, p. 91-99.

Maupin, C., and Vila, J.M., 1976, Microfossiles du Lias Supérieur au Djebel Youssef (Hautes plaines sétifiennes, Algérie): Revue de Mciropaléontologie, v. 19, p. 162-165.

Maync, W., 1960, Morphology and Occurrence of the Foraminiferal Genus *Orbitopsella:* Rivista Italiana Paleontologia e Stratigrafia, v. 66, p. 491-524.

______, 1966, Microbiostratigraphy of the Jurassic of Israel: Geological Survey of Israel Bulletin, v. 40, p. 1-56.

McGugan, A., 1965, Liassic Foraminifera from Whitepark Bay, County Antrim: Irish Naturalists Journal, v. 15, p. 85-87.

Michelsen, O., 1975, Lower Jurassic Biostratigraphy and Ostracods of the Danish Embayment: Danmarks Geologiske Undersogelse (2), v. 104, p. 1-287.

Mouterde, R., 1955, Le Lias de Peniche: Communicaçoes dos Serviços Geologicos de Portugal, v. 36, p. 87-115.

______, 1967a, Le Lias Moyen de São Pedro de Muel (Portugal): Communicaçoes dos Serviços Geologicos de Portugal, v. 52, p. 185-208.

______, 1967b, Le Lias du Portugal: Vue d'Ensemble et Division en Zones: Comunicaçoes dos Serviços Geologicos de Portugal, v. 52, p. 209-226.

Mouterde, R., and Ruget, C., 1975, Esquisee de la Paléogéographie du Jurassique Inférieur et Moyen au Portugal: Bulletin de la Société Géologique de France, ser 7, v. 18, p. 779-786.

Mouterde, R., Rocha, R.B. and Ruget, C., 1971, Le Lias Moyen et Supérieur de la Région de Tomar: Comunicaçoes dos Serviços Geologicos de Portugal, v. 55, p. 50-80.

______, 1978, Stratigraphie et Faune du Lias et de la Base du Dogger au Nord du Mondego (Quiaios e Brenha): Communicaçoes dos Serviços Geologicos de Portugal, v. 63, p. 83-104.

Mouterde, R., Ruget, C., and Moitinho de Almeida, F., 1964, Coupe du Lias au Sud de Condeixa: Comunicaçoes dos Serviçcos Geologicos de Portugal, v. 48, p. 61-91.

Mouterde, R., Ramalho, M., Rocha, R.B., Ruget, C., and Tintant, H., 1971, Le Jurassique du Portugal: Esquisse Stratigraphique et Zonale: Boletin da Sociedade Geologica de Portugal, v. 18, p. 73-104.

Norling, E., 1972, Jurassic Stratigraphy and Foraminifera of Western Scania, Southern Sweden: Sveriges Geologiska Undersökning ser. Ca., v. 47, p. 1-120.

Norvang, A., 1957, The Foraminifera of the Lias Series in Jutland, Denmark: Meddeleser fra Dansk Geologisk Forening, v. 13, p. 1-135.

Palain, C., 1968, Analyse Séquentielle et Lithostratigraphique de la Série de Base du Mésozoïque Portugais au Nord du Tage: Comptes Rendus des Séances de l'Académie des Sciences de Paris, ser. D., v. 267, p. 480-482.

______, 1976, Une Série Détritique Terrigène les Gres de Silves: Trias et Lias Inférieur du Portugal: Serviços Geologicos de Portugal Memoria, no. 25 p. 363.

Pazdrowa, O., 1969, Bathonian Globigerina of Poland: Annales de la Société Géologique de Pologne, v. 39, p. 41-56.

Pietrzenuk, E., 1961, Zur Mikrofauna einiger Liasvorkommen in der Deutchen Demokratischen Republic: Freiberger Forschungshefte, H. C113, p. 1-129.

Prichard, H.M. and Mitchell, J.G., 1979, K-Ar data for the Age and Evolution of Gettysburg Bank, North Atlantic Ocean: Earth and Planetary Science Letters, v. 44, p. 261-268.

Ramalho, M.M., 1971, Contribution a l'Étude Nicropaléontologique et Stratigraphique du Jurassique Supérieur et du Crétacé Inférieur des Environs de Lisbonne (Portugal): Serviços Geologicos de Portugal, Memoria no. 19, 212 p.

Ramirez del Pozo, J., 1971, Algunas Observaciones sobre el Jurasico de Alava, Burgos y Santander: Cuadernos Geologia Ibéria, v. 2, p. 491-508.

Rioult, M. and Bizon, J.-J., 1961, Contributions à L'étude Micropaléontologique du Lias du Bassin de Paris: Cinquièeme Partie: Basse-Normandie (Régions d'Isigny et sud de Bayeux): Colloque sur le Lias Francais; Mémoire Bureau Recherches Géologiques Minieres, v. 5, p. 451-458.

Ruget, C., and Sigal, J., 1967, Les Foraminifères du Sondage de Laneuville-devant-Nanc (Lotharingien de La Region Type): Sciences de la Terre, v. 12, p. 33-70.

______, 1970, Le Lias Moyen de São Pedro de Muel: Deuxieme Partie: Paleontologie 2: Les Foraminifères: Comunicaçoes dos Serviços Geologicos Portugal, v. 54, p. 76-108.

Sartoni, S. and Cresenti, U., 1962, Richerche Biostratigraphiche nel Mesozoico dell'Appennino Meridionale: Giornale di Geologia, Bologna, v. 27, p. 115-139 and p. 159-302.

Scheibnerova, V., 1971, Foraminifera and Their Mesozoic Biogeoprovinces: Records of the Geological Survey of New South Wales, v. 13, pt. 3, p. 135-174.

Schott, W. and Staesche, K., 1957, Über den Jura der Weiteren Umgebung von Coimbra (Portugal): Geologisches Jahrbuch, v. 74, p. 531-558.

Seifert, H., 1963, Beitrage zur Geologie der Serra da Arrabida in Portugal: Geologisches Jahrbuch, v. 81, p. 277-344.

Seronié-Vivien, R.M., Magné, J., and Malmoustier, J., 1961, Le Lias des Bordures Septentrionale et Orientale du Bassin d'Aquitaine: *in* Colloque sur le Lias Francias: Mémoire Bureau Recherches Géologiques Minières, v. 5, p. 757-791.

Sheridan, R.E., Gradstein, F.M., *et al.*, 1983, Early History of the Atlantic Ocean and Gas Hydrates on the Blake Outer Ridge: Initial Report of the Deep Sea Drilling Project, v. 76: United States Government Printing Office, Washington, D.C., 947 p.

Simon, W., and Bartenstein, H., eds., 1962, Leitfossilien der Mikropalaeontologie: Gebrüder Borntraeger, Berlin.

Smith, A.G., and Briden, J.C., 1977, Mesozoic and Cenozoic Paleocontinental Maps: Cambridge University Press, England.

Souaya, F.J., 1976, Foraminifera of Sun-Gulf-Global Linckens Island Well P-46, Arctic Archipelego: Micropaleontology, v. 22, p. 249-306.

Tappan, H., 1955, Foraminifera from the Arctic Slope of Alaska: Part 2, Jurassic Foraminifera: United States Geological Survey Professional Paper 236B, p. 19-90.

Tate, R., and Blake, F.J., 1876, The Yorkshire Lias: London, p. 449-473.

Tintant, H., 1967, Remarques sur le Jurassique de l'Arrabida: Comunicaçoes dos Serviços Geologicos de Portugal, v. 51, p. 5-19.

Welze, E., 1968, Foraminiferen und Fazies des Fränkischen Domeriums. Erlanger Geologische Abhandlungen, v. 69, p. 1-79.

Wernli, R., and Septfontaine, M., 1971, Micropaléontologie Compareé du Dogger du Jura Méridional (France) et des préalpes Médianes Plastiques Romandes (Suisse): Ecologae Geologicae Helvetia, v. 64, p. 437-458.

Williams, C.A., 1975, Sea-floor Spreading in the Bay of Biscay and its Relationship to the North Atlantic: Earth and Planetary Science Letters, v. 24, p. 440-456.

Wilson, R.C.L., 1975, Atlantic Opening and Mesozic Continental Margin Basins of Iberia: Earth and Planetary Science Letters, v. 25, p. 33-43.

Zaninetti, L., 1976, Les Foraminifères du Trias: Essai de synthèse et Corrélation Entre les domaines Mésogéens Européen et Asiatique: Rivista Italiana Paleontologia e Stratigrafia, v. 82, p. 1-258.

Zbyszewski, G. and de Faria, Barreto 1971, O Sal-gema em Portugal Metropolitano: Suas Jazidas, Caracteristicas e Aproveitamento: Estudos, Notas e Trabalhos do Serviço de Fomento Mineiro, v. 20, p. 5-105.

Zbyszewski, G., Da Viega Ferreira, O., Manmupella, G., and Torres de Assuncao, C., 1965, Noticia Explicativa de Folha 38-B, Setubal: Serviços Geologicos de Portugal, Lisboa, p. 1-134.

Jurassic-Cretaceous Biochronology and Paleogeography of North America,
edited by G.E.G. Westermann,
Geological Association of Canada Special Paper 27, 1984

MICROFOSSIL ZONATION ACROSS THE JURASSIC-CRETACEOUS BOUNDARY ON THE ATLANTIC MARGIN OF NORTH AMERICA

P. Ascoli
Geological Survey of Canada, Atlantic Geoscience Centre, Dartmouth, Nova Scotia B2Y 4A2
C.W. Poag
United States Geological Survey, Woods Hole, Mass. 02543
J. Remane
Institute de Géologie, Université de Neuchâtel, 2000 Neuchâtel 7, Switzerland

ABSTRACT

A biostratigraphic study of the COST G-2 well (Georges Bank Basin, New England Shelf) has shown that its Late Jurassic and Early Cretaceous foraminiferal and ostracod assemblages are almost identical to coeval assemblages from the adjacent Scotian Basin, and are very similar to those from the Mobil-Gulf Bonnition H-32 well in the East Newfoundland Basin. The biozonation developed for the Late Jurassic and Early Cretaceous of the Scotian Basin, therefore, can be applied in the Bonnition H-32 and COST G-2 wells. In addition, abundant calpionellid assemblages are present across the Jurassic-Cretaceous boundary in the COST G-2 and Bonnition H-32 wells and are identical to those of the Mediterranean area which are calibrated with ammonite zones. It is possible, therefore, to recognize on the North American margin the standard calpionellid zones "A" (with 3 subzones; Late Tithonian), "B" (latest Tithonian-earliest Berriasian), possibly "C" (Middle Berriasian), and "D" (Late Berriasian-Early Valanginian). The calpionellid biozonation may be used, in turn, to calibrate the stratigraphic ranges of foraminiferal and ostracod zone markers. This has provided a comprehensive, integrated, fourfold system of microfossil zonation and correlation for the Late Jurassic-Early Cretaceous strata which can be used from the East Newfoundland Basin to the Georges Bank Basin. In addition, the foraminiferal and ostracod zonations can be extended as far south as the Baltimore Canyon Trough. The presence in this region of calpionellids, Foraminifera and Ostracoda originally described from Europe and North Africa is evidence of a westward extension of the Tethyan domain during Tithonian-Valanginian time. The abundance of calpionellids confirms that the North American margin, the Iberian Peninsula, and northwest Africa were separated by open ocean during this time.

RÉSUMÉ

Une étude biostratigraphique du sondage COST G-2 (Georges Bank Basin, New England Shelf) a montré que les associations de foraminifères et d'ostracodes du Jurassique supérieur et du Crétacé inférieur y sont presque identiques aux associations du même âge rencontrées dans le Scotian Basin adjacent et qu'elles y sont très similaires à celles du sondage Mobil-Gulf Bonnition H-32 dans le East Newfoundland Basin. Par conséquent on peut appliquer à l'étude des sondages Bonnition H-32 et COST G-2 la biozonation développée pour le Jurassique supérieur et le Crétacé inférieur du Scotian Basin. De plus, on trouve de nombreuses associations à calpionelles subjacentes et sous-jacentes à la limite Jurassique-Crétacé des sondages COST G-2 et Bonnition H-32. Ces associations sont identiques à celles de la région méditerranéenne qui sont calibrées avec les zones d'ammonites. On a ainsi la possibilité d'y reconnaître les zones standard à calpionelles suivantes: zone "A" (avec trois sous-zones, d'âge Tithonique supérieur), zone "B" (partie supérieure du Tithonique supérieur-partie inférieure du Berriasien inférieur), probablement zone "C" (Berriasien moyen) et aussi zone "D" (Berriasien supérieur-Valanginien inférieur). A son tour, cette biozonation des calpionelles peut être employée pour calibrer les distributions stratigraphiques des foraminifères et ostracodes marqueurs de zones. Grâce à cela on a pu établir un système quadruple, complet et bien documenté, de zonation par microfossiles et de correlation pour les couches du Jurassique supérieur et du Crétacé inférieur allant du East Newfoundland Basin au Georges Bank Basin. D'autre part, la biozonation à foraminifères et ostracodes peut être étendue au sud, aussi loin que le Baltimore Canyon Trough. La présence dans cette région de calpionelles, foraminifères et ostracodes, à l'origine décrites en Europe et en Afrique du Nord, indique l'existence d'une extension occidentale du domaine de la Tethys pendant la période Tithonique-Valanginien. L'abondance de calpionelles confirme que, à cette période, la côte Nord-américaine, la péninsule Ibérique et l'Afrique du nord-ouest étaient separées par un océan profond.

INTRODUCTION

Several published stratigraphic investigations of the North American continental margin have pointed out that accurate recognition of the top of the Jurassic has been difficult. Gradstein (1976), for instance, established an *Anchispirocyclina lusitanica* Zone of Tithonian age on the Grand Banks, on the basis of "few specimens" of this species observed in only one well (Amoco-Imp.-Skelly Egret K-36). He stressed that "The species (*A. lusitanica*) so far has not been recorded in other Grand Banks wells, but occurs in Scotian Shelf wells, probably in a comparable litho- and biostratigraphic position". We, therefore, have focused our attention upon documenting assemblages of four different groups of microfossils (calcareous and arenaceous benthic Foraminifera, Ostracoda, and calpionellids) that indicate proximity to the Jurassic-Cretaceous boundary. Herein, we summarize our investigation of 28 wells, ranging geographically from the Baltimore Canyon Trough to the East Newfoundland Basin (a distance of 2300 km) (Fig. 1).

The wells are: COST B-3 and COST G-2 (Baltimore Canyon Trough and Georges Bank Basin respectively), Shell Mohawk B-93, Shell Mohican I-100, Shell Naskapi N-30, Shell Moheida P-15, Chevron Acadia K-62, Shell Oneida 0-25, Shell Demascota G-32, Shell Cree E-35, Mobil-Tetco Cohasset D-42, Shell Onondaga E-84, Mobil-Tetco-Pex Migrant N-20 and Mobil-Tetco-Pex Thebaud I-94 (southwest Scotian Basin); Texaco-Shell Intrepid L-80, Mobil Sable Island C-67, Shell Penobscot L-30, Shell Abenaki J-56, Mobil-Texaco-Pex Venutre B-13, Mobil-Tetco-Texaco Citnalta I-59, Shell Wyandot E-53, Shell Sauk A-57, Union *et al.* Jason C-20 and Mobil Dauntless D-35 (northeast Scotian Basin); Amoco IOE Puffin B-90 (southwest Grand Banks) and Chevron Hibernia P-15 and Mobil-Gulf Bonnition H-32 (East Newfoundland Basin). (We have used English or U.S. customary units for measuring depths in these wells because that is the system used during drilling and reference to original samples must be made on this basis.) A multizone analysis of these wells clearly establishes the approximate stratigraphic position of the Jurassic-Cretaceous boundary and permits its recognition even in facies where only one of the four fossil groups is present.

The micropaleontological stratigraphy of the Canadian Atlantic Shelf basins (Labrador Shelf excluded) has been published chiefly in the following papers: 1) Gradstein *et al.* (1975) described Pliensbachian to Pleistocene microfossil assemblages from 17 deep exploratory wells on the Scotian Shelf and Grand Banks of Newfoundland; 2) Ascoli (1976) established a Bathonian to Miocene foraminiferal-ostracod zonation from 10 wells on the Scotian Shelf and SW Grand Banks; 3) Gradstein (1976, 1978) established a foraminiferal zonation for the Jurassic from eight wells on the Grand Banks; 4) Jansa *et al.* (1980) studied in detail the calpionellid and foraminiferal-ostracod stratigraphy across the Jurassic-Cretaceous boundary on the Grand Banks and Scotian Shelf; 5) Ascoli (1981), published informally an

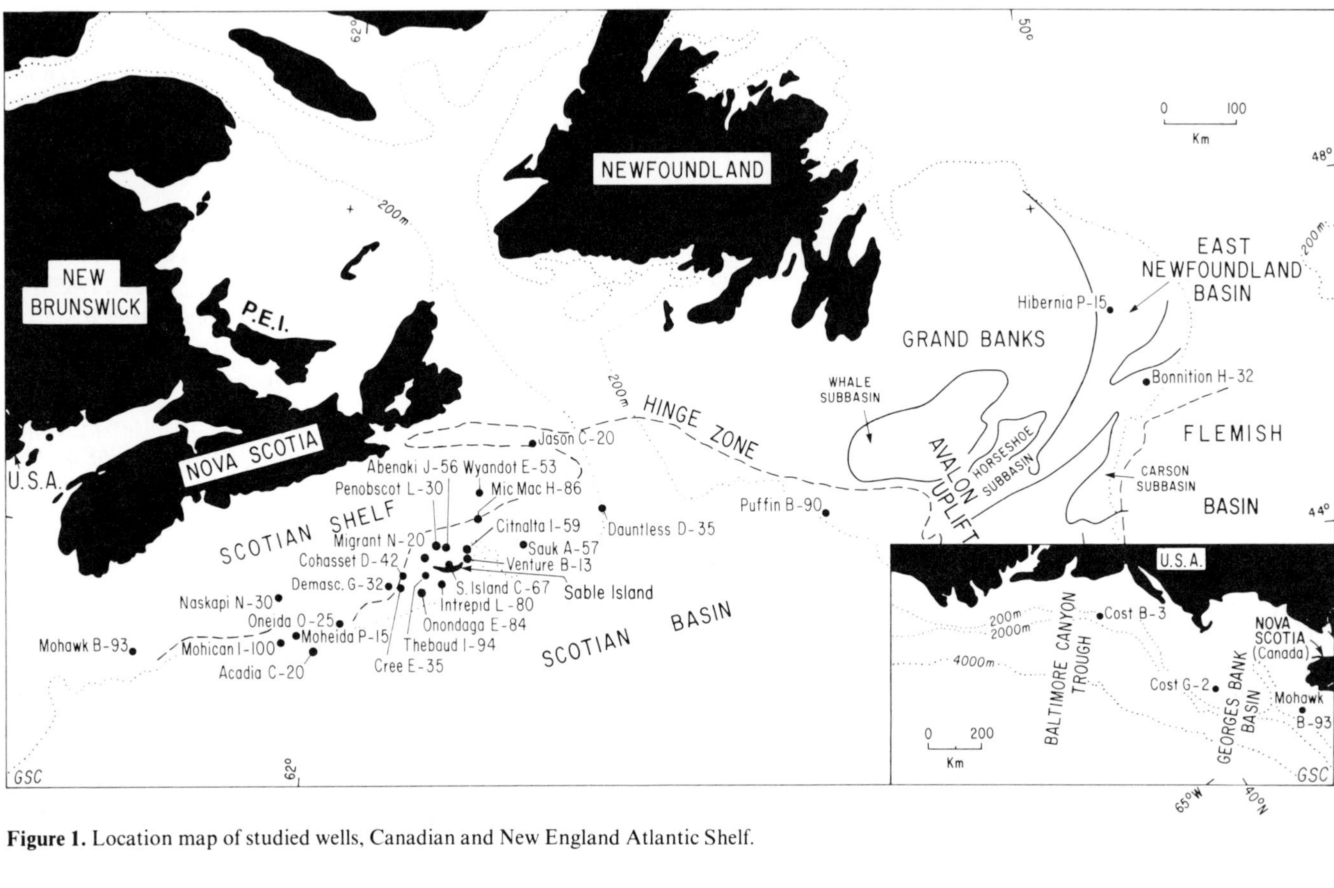

Figure 1. Location map of studied wells, Canadian and New England Atlantic Shelf.

updated version of his 1976 Jurassic foraminiferal-ostracod zonation, using 19 wells on the Scotian Shelf and Grand Banks; 6) Bujak and Williams (1977, 1978) and Barss *et al.* (1979) have published palynological zonations and correlations for 67 wells in the offshore region of eastern Canada.

The micropaleontological stratigraphy of the adjacent United States Atlantic Shelf basins has been published chiefly by Poag (1977, 1979, 1980, 1982a, 1982b) and Valentine (1977, 1980, 1982), who based their studies primarily on five deep stratigraphic test wells. Informal open-file reports have been prepared by Bebout (1980) and Steinkraus (1979, 1980). Poag showed that important aspects of Ascoli's (1976) biozonation could be easily extrapolated and applied in the Upper Jurassic-Lower Cretaceous sediments of the Georges Bank Basin and Baltimore Canyon Trough.

GEOLOGIC SETTING AND STRATIGRAPHIC FRAMEWORK

The large sedimentary basins of the submerged Atlantic margin of North America began to form as down-dropped basement blocks during the Triassic-Jurassic rifting that initiated the separation of North America from Europe and Africa (Emery and Uchupi, 1965; Mattick *et al.*, 1974; Ballard and Uchupi, 1975; James and Wade, 1975; Schlee *et al.*, 1976; Sheridan, 1974, 1976; Given, 1977; Klitgord and Behrendt, 1979; Schlee *et al.*, 1979; Austin *et al.*, 1980; Klitgord and Grow, 1980; Jansa and Wiedmann, 1982; Klitgord *et al.*, 1982; Poag, 1982a, 1982b). The continental margin subsequently subsided due to crustal cooling associated with sea floor spreading and to the accumulation of thick prisms of sedimentary rocks (5 to 8 km thick; Steckler and Watts, 1978; Watts and Steckler, 1979; Heller *et al.*, 1982; Watts, 1982).

The structural development and the temporal sequences of lithofacies and biofacies are quite similar from the Baltimore Canyon Trough (off New Jersey) to the East Newfoundland Basin, although some local and latitudinal differences have been noted. Rifting began in the Triassic under arid continental conditions. Early deposits in the young pull-apart basins are chiefly arkosic red beds, lignitic shales, evaporitic lenses, and basaltic flows and dikes (Jansa and Wade, 1975; Manspeizer *et al.*, 1978; Jansa *et al.*, 1980; and Manspeizer, 1982). Marine deposition began in the Late Triassic and Early Jurassic with the widespread accumulation of halite, followed by dolomitic limestone and anydrite in middle to outer shelf locations. Carbonate facies dominated the outer shelf from the Early Jurassic to the Early Cretaceous and a prominent discontinuous carbonate bank fringed the shelf edge from Florida to Newfoundland (Emery and Uchupi, 1972; Schlee *et al.*, 1976, 1979; Given, 1977; Eliuk, 1978; Poag, 1978, 1982a, 1982b; Jansa, 1981). Terrigenous clastics buried the carbonate bank during the Early Cretaceous and a chiefly siliciclastic regime was maintained until the Paleogene. Shelf carbonates were again widespread during the Paleogene, but since the Oligocene, a terrigenous regime has been dominant (Poag, 1978; 1982a, 1982b).

A formal lithostratigraphic terminology has been introduced for the Scotian Basin (McIver, 1972; Jansa and Wade, 1975; Eliuk, 1978), and as drilling has progressed in other Canadian basins, the initial stratigraphic framework has been extended and modified. A recent study by Poag (1982a, 1982b) has extended the Scotian Basin formations into the adjacent Georges Bank Basin. During this study it has become clear that the major biostratigraphic events and the microfossil assemblages are also closely similar among these basins. The Mesozoic biostratigraphy, in particular, is nearly coincident from New Jersey to Newfoundland.

BIOZONATION

We use calcareous benthic foraminifera, arenaceous benthic foraminifera, ostracods and calpionellids to zone and correlate the Kimmeridgian to Valanginian sediments of the studied area. Foraminiferal and ostracod zones are chiefly informal "assemblage zones" whose upper boundaries are defined by the highest stratigraphic occurences of "zonal markers" – after which all zones are named (Ascoli, 1976; Ascoli, 1981). In addition to "zonal markers", we also use "zone diagnostic species" to help identify zone boundaries. "Zone diagnostic species" are species other than "zonal markers" whose highest stratigraphic occurences are also in that particular zone. Calpionellid zones were originally defined chiefly by the lowest stratigraphic occurences of characteristic forms (Remane, 1964; Allemann *et al.*, 1971). In order to be consistent with foraminiferal and ostracod zones, we have redescribed the calpionellid zones in terms of the highest stratigraphic occurrences of their characteristic forms (Figs. 2, 6, 7). All zones are discussed in sequence from oldest to youngest.

The arenaceous benthic foraminiferal and ostracod zones are best recognized in littoral and inner neritic coarse-grained deposits; the calpionellid and calcareous benthic foraminiferal zones are best recognized in finer grained middle and outer neritic, and bathyal deposits. There is, in many cases, of course, a certain "environmental overlap" between coeval zones, in the sense that arenaceous Foraminifera and/or ostracod assemblages are present in calcareous Foraminifera zones and vice versa.

The relationships between the biostratigraphic zones and lithostratigraphic units on the Scotian Shelf and Grand Banks, have been described by Ascoli (1976). Poag (1982a, 1982b) has described similar relationships in the Georges Bank Basin.

Calcareous Benthic Foraminiferal Zones

Planularia Tricarinella-Epistomina Mosquensis Assemblage Zone. The top of this zone is defined by the highest stratigraphic occurrence of *Planularia tricarinella* (Reuss) and *Espistomina mosquensis* Uhlig. Other diagnostic species having their highest stratigraphic occurence at the top of this zone are *Epistomina* gr. *mosquensis* Uhlig, *Conicospirillina* gr. *basiliensis* Mohler, *Lenticulina brueckmanni* Mjatliuk, *L. vistulae* Bielecka and Pozaryski, *Conorboides marginata* Lloyd, *?Conorboides* aff. *paraspis* (Schwager) and possibly *Epistomina* aff. *uhligi* Mjatliuk. The base of the zone corresponds to the top of the *Epistomina soldanii-Conorboides paraspis* Zone of Ascoli (1981). Diagnostic species having their highest stratigraphic occurrence in the lower part of this zone include *Eoguttulina inovroclaviensis* (Bielecka and Pozaryski), whereas *E. liassica* (Strickland) occurs commonly in it (Fig. 3).

AGE	CALCAREOUS BENTHIC FORAMINIFERA ZONES
VALANGINIAN-BERRIASIAN	Lenticulina saxonica bifurcilla Lenticulina busnardoi
TITHONIAN	Epistomina uhligi Epistomina stellicostata
KIMMERIDGIAN	Epistomina mosquensis Planularia tricarinella

AGE	OSTRACOD ZONES
VALANGINIAN-BERRIASIAN	Schuleridea aff. praethoerenensis Asciocythere sp.1
TITHONIAN	Schuleridea sp.1 (2) Galliaecytheridea postrotunda (2) Hutsonia gr. collinsensis (1)
KIMMERIDGIAN	Schuleridea triebeli (2) Cytherelloidea weberi (2) Eocytheropteron decoratum (1)

(2)=highest stratigraphic occurrence at top of zone
(1)=highest stratigraphic occurrence in lower part of zone

AGE		ARENACEOUS BENTHIC FORAMINIFERA ZONES
VALANGINIAN-BERRIASIAN		Everticyclammina virguliana
TITHONIAN		Anchispirocyclina lusitanica (2) Ammobaculites coprolithiformis (1)
KIMMERIDGIAN	L	
	E	Alveosepta jaccardi
OXFORDIAN		

AGE			CALPIONELLID ZONES
VALANGINIAN	L		(no calpionellids)
	E	E	Calpionellites
BERRIASIAN	L	D	Calpionellopsis
		C	Calpionella
	E	B	(Calpionella alpina "acme zone")
TITHONIAN	L	A	Crassicollaria
	E		(no calpionellids)

GSC

Figure 2. Foraminiferal, ostracod and calpionellid assemblage zones in the Kimmeridgian-Valanginian interval of the Canadian and New England Atlantic Shelf. Where no number follows the zonal species name, the highest stratigraphic occurrence of that species is at the top of the nominate zone.

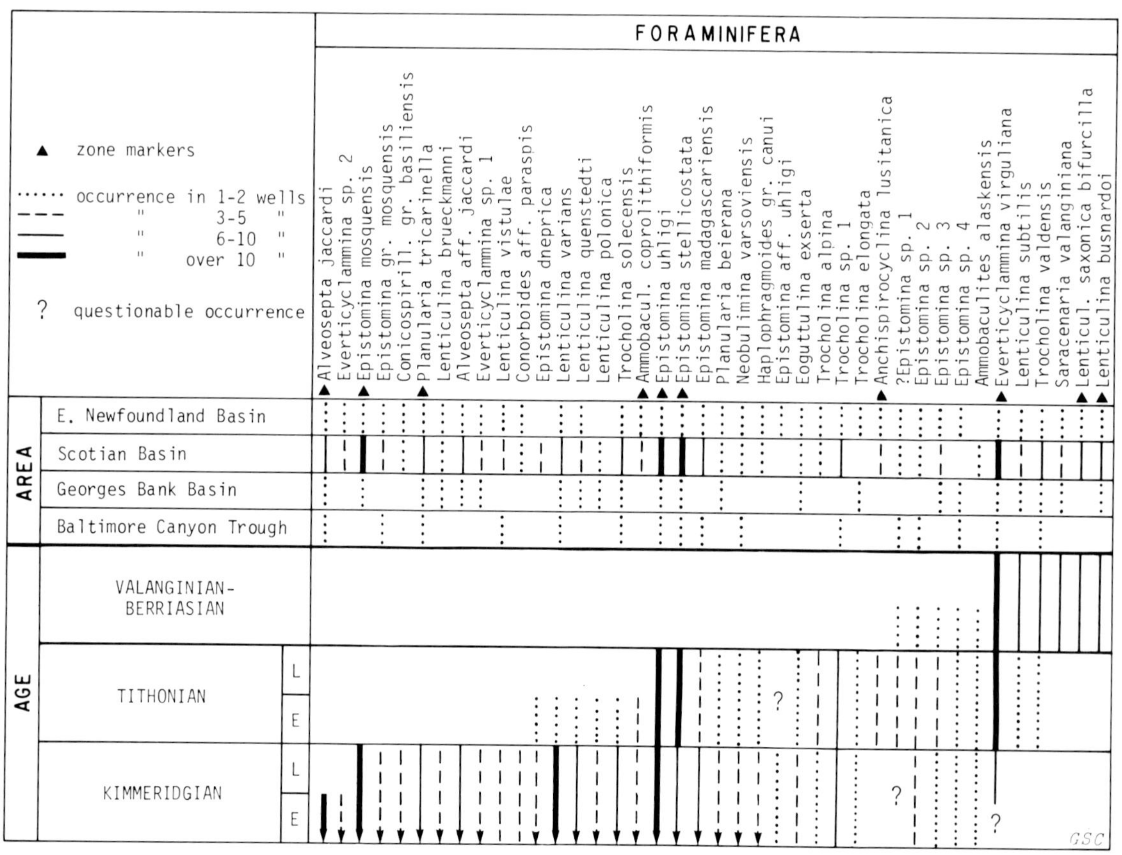

Figure 3. Stratigraphic ranges and areal distribution of Kimmeridgian-Valanginian foraminiferal marker and "zone diagnostic" species, Canadian and New England Atlantic Shelf.

In Europe, most of the listed zonal markers and diagnostic species have their highest stratigraphic occurrence in the Kimmeridgian Stage of French and German authors (Bielecka and Pozaryski, 1954; Klingler *et al.*, 1962; Oertli, 1963; Ascoli, 1976; Jansa *et al.*, 1980). The upper limit of this stage coincides with the top of the Pseudomutabilis Zone (Arkell, 1956), or with the top of the Autissiodorensis Zone (Kilenyi, 1978). The tops of these ammonite standard zones correspond to the top of the lower Kimmeridgian in the British type-locality (Kilenyi, 1978; Lloyd, 1962). However, we follow the French-German definition of the Kimmeridgian because this allows correlation of the top of the Kimmeridgian with the base of the Tithonian and also places the top of the Jurassic at the top of the Tithonian (Jansa *et al.*, 1980). We assign the highest occurrence of *E. mosquensis* to approximately the top of the Kimmeridgian because, on the Scotian Shelf, it generally disappears about 150m (500 ft) higher than the youngest specimens of the early Kimmeridgian marker *Alveosepta jaccardi* (Schrodt).

This zone replaces the partly equivalent *Epistomina madagascariensis – E. dneprica* Zone of Ascoli (1976) and also replaces the *Lenticulina quenstedti – Epistomina mosquensis* Zone of Ascoli (1981). *Lenticulina quenstedti* has recently been identified in the lower part of the overlying *Epistomina stellicostata – E. uhligi* Assemblage Zone.

Epistomina Stellicostata – E. Uhligi Assemblage Zone. This zone was established by Ascoli (1976) and subsequently was used by Jansa *et al.* (1980). Its top is defined by the highest stratigraphic occurrence of *Epistomina stellicostata* Bielecka and Pozaryski and *Epistomina uhligi* Mjatliuk (*sensu* Ohm, 1967). Other diagnostic species having their highest stratigraphic occurrence at the top of this zone include *Epistomina madagascariensis* (Espitalié and Sigal), *Planularia beierana* (Guembel), *P. dofleini* Kasanzev, *Neobulimina varsoviensis* Bielecka and Pozaryski, Eoguttulina exserta (Berthelin), *Trocholina alpina* (Leupold), *T. elongata* (Leupold), *Trocholina* sp. 1 of Ascoli (1976), *Conorboides* aff. *marginata* Lloyd, *C. irregularis* Groiss, and possibly *Epistomina* aff. *uhligi* Mjatliuk. The base of the zone corresponds to the top of the underlying *Planularia tricarinella – Epistomina mosquensis* Assemblage Zone. Other diagnostic species which have their highest stratigraphic occurrence in the lower part of this zone include *Epistomina dneprica* Kaptarenko, *Lenticulina varians* (Bornemann), *L. quenstedti* (Guembel), *L. polonica* Wisniowski, and *Patellinella* sp. 1 (Fig. 3).

The listed zonal markers and diagnostic species have their highest stratigraphic occurrences in the Tithonian Stage (Bielecka and Pozaryski, 1954; Espitalié and Sigal, 1963; Groiss, 1967, Bielecka, 1975; Ascoli, 1976, 1981). The stratigraphic ranges of foraminiferal markers in this zone have been calibrated with coexisting calpionellid marker species in the Bonnition H-32 well (Jansa *et al.*, 1980) and in the COST G-2 well (Poag, 1982a, 1982b). The Tithonian-Berriasian boundary, which is present in the middle part of calpionellid Zone "B", is at about 7500 ft (2285m) in the Bonnition well; the highest stratigraphic occurrence of *Epistomina uhligi, Neobulimina varsoviensis* and *Trocholina alpina* have been observed between 7650 and 7700 ft (2332 and 2347m) in the same well. In the COST G-2 well, calpionellid zonation establishes the Tithonian-Berriasian boundary at about 5700ft (1737m) (Fig. 5); the highest stratigraphic occurences of *Epistomina uhligi* and *E. stellicostata* have been observed at 5856ft (1785m) in the G-2 well. Therefore, the Jurassic-Cretaceous boundary established by calcareous benthic Foraminifera nearly coincides (within 150 ft = 46m) with that established by calpionellids. *Epistomina uhligi* is the most common Upper Jurassic epistominid of the study area. It appears to have been a eurytopic species inhabiting a wide variety of substrates ranging from marginal marine to outer neritic.

Lenticulina Busnardoi-L. Saxonica Bifurcilla Assemblage Zone. The top of this zone is defined by the highest stratigraphic occurrence of *Lenticulina busnardoi* Moullade and *Lenticulina saxonica bifurcilla* Bartenstein and Brand. Other diagnostic species having highest stratigraphic occurrences at the top of this zone include *Saracenaria valanginiana* Bartenstein and Brand, *Lenticulina subtilis* Wisniowski, *Astacolus calliopsis* (Reuss), *Marginulinopsis* sp. 1, *Trocholina valdensis* (Reichel), *Paalzowella feifeli* (Paalzow), *Conorboides* aff. *hofkeri* (Bartenstein and Brand) and *Epistomina* aff. *tenuicostata* Bartenstein and Brand. The base of the zone corresponds to the top of the underlying *Epistomina stellicostata-Epistomina uhligi* Assemblage Zone. Diagnostic species which have their highest stratigraphic occurrence in the lower part of this zone include *Lenticulina collignoni* Espitalié and Sigal, *?Epistomina* sp. 1 of Jansa *et al.* (1980), *Epistomina* sp. 2 of Jansa *et al.* (1980), *Epistomina* sp. 3 and *Epistomina* sp. 4 (Fig. 3).

The listed zonal markers and diagnostic species have their highest stratigraphic occurrence at the top of the Valanginian (Bartenstein and Brand, 1951; Bartenstein and Bettenstaedt, 1962; Magné, 1965; van Hinte, 1976; Bartenstein, 1977; Moullade, 1979; and Jansa *et al.*, 1980). The *Lenticulina busnardoi-Lenticulina saxonica bifurcilla* Assemblage Zone also includes rocks of Berriasian age, because its base corresponds to the top of the *Epistomina stellicosta-E. uhligi* Zone, which has been assigned a Tithonian age.

Whether or not the top of the Berriasian Stage can be identified within this zone, is still in question. According to Magné (1965), the upper boundary of the Berriasian type-section is not identifiable on the basis of highest stratigraphic occurrences of Foraminifera, but only on the basis of Foraminifera having lowest stratigraphic occurrences at the base of the Valanginian. Lowest stratigraphic occurrences are not reliable in our rotary cutting samples, so we must rely on the upper ranges of ostracoda and calpionellids. In the COST G-2 well, the Berriasian ostracod marker *Schulerídea juddi* Neale has its highest stratigraphic occurrence at 5590 ft (1704 m), a level also dated as Late Berriasian by calpionellids (see below). The foraminifer *Epistomina* sp. 3 also has its highest stratigraphic occurrence at 5590ft (1704m) in the COST G-2 well, and its Late Berriasian disappearance has been confirmed in the Bonnition H-32 as well. In the latter well, *Epistomina* sp. 3 occurs together with *?Epistomina* sp. 1 of Jansa *et al.* (1980) as high as 7250ft (2210m), a level 83ft (25m) higher than the highest sample containing calpionellids (7333ft = 2235m), which are Early Berriasian forms (see Jansa *et al.*, 1980). Though we

PLATE I

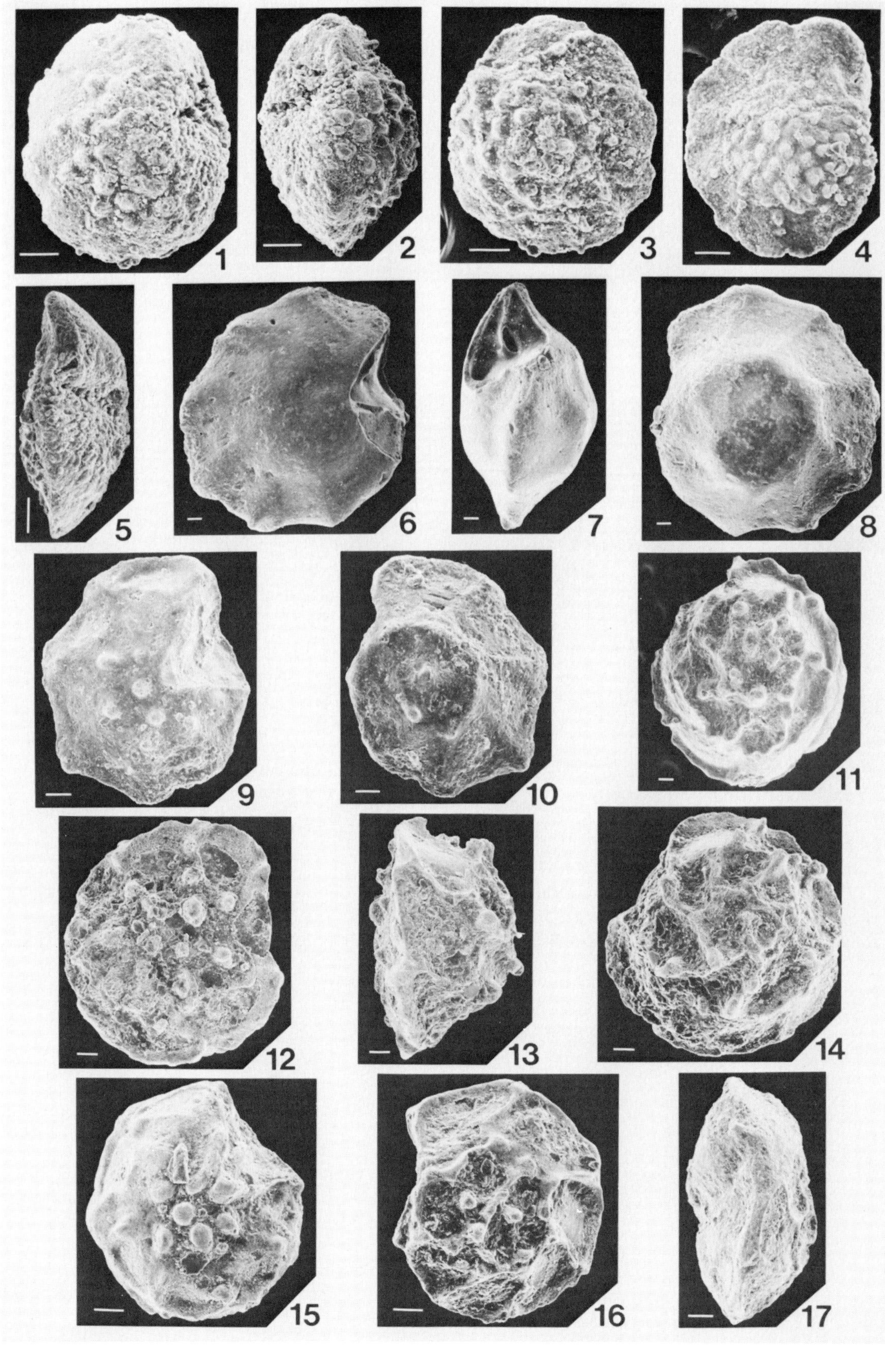

cannot be sure yet that both *Epistomina* sp. 3 and *?Epistomina* sp. 1 range as high as the top of the Berriasian in the study area, it is safe to say that both species range into the Late Berriasian. The same is true of *Epistomina* sp. 2 of Jansa *et al.* (1980) and *Epistomina* sp. 4, both of which range as high as *?Epistomina* sp. 1 and *Epistomina* sp. 3 (Fig. 3).

This zone corresponds to the *Conorboides valendisensis* – *Lenticulina busnardoi* – *L. saxonica bifurcilla* Zone of Ascoli (1976) which we update by eliminating *Conorboides valendisensis* (Bartenstein and Brand) as an Early Valanginian indicator on the Canadian Atlantic Shelf. Abundant specimens of *C. valendisensis* have now been observed as high as the upper Hauterivian in the Bonnition H-32 well, East Newfoundland Basin (H. Bartenstein, pers. commun.; Jansa *et al.*, 1980).

Among the diagnostic species listed above, *?Epistomina* sp. 1, *Epistomina* sp. 2, *Epistomina* sp. 3 and *Epistomina* sp. 4 are possibly new and are, therefore, briefly and informally described below.

Epistomina? sp. 2 (Pl. I, Figs. 1-5) is rather small (maximum diameter ranging from 270 to 390 μm). This species is only tentatively assigned to the genus *Epistomina* because all specimens available, even those very well preserved (Pl. I, Fig. 1), lack the peripheral latero-marginal openings typical of this genus. Furthermore, the obscured aperture on the umbilical side seems perpendicular (Pl. I, Fig. 2) rather than parallel to the periphery. The test is thick, strongly biconvex in most specimens (Pl. I, Fig. 2), with slightly lobate peripheral margin (Pl. I, Figs. 3, 4). A minority of specimens shows a weakly biconvex test (Pl. I, Figs. 4, 5), almost flat dorsally over the last whorl which consists of 9 to 11 chambers. Both on dorsal and umbilical sides the sutures of chambers are curved backwards, slightly raised and covered by a row of coarse pustules which are also concentrated on the umbilical area and on the central part of the dorsal side where they show a kind of nebulose pattern (Pl. I, Fig. 4). The remaining part of the test is also covered, both dorsally and umbilically, by a fine net of minute pustulations which tend to disappear around the peripheral margin. This species does not resemble any other epistominid species recognized so far, in the Late Jurassic and Early Cretaceous of the Canadian and New England Atlantic Shelf.

Occurence: Bonnition H-32, ?Acadia K-62 and COST B-3: *Epistomina stellicostata-E. uhligi* Zone (Tithonian); Bonnition H-32 and ?Acadia K-62: lower part of *Lenticulina busnardoi-L. saxonica bifurcilla* Zone (Berriasian).

Epistomina sp. 2 (Pl. I, Figs. 6-10) has medium to large, very variable size, with maximum diameter ranging from 390 to 850 μm. It is characterized by a rather flattened test and by a sharp, polygonally shaped peripheral margin. This polygonal outline is produced by a strong thickening of the peripheral ends of sutures and by an inward leaching of the peripheral margin between the latter ones. This process of leaching can vary from strong (Pl. I, Figs. 9, 10) to weak (Pl. I, Figs. 6, 8), thereby originating a more or less pronounced polygonal outline. On the umbilical side, which is smooth or very weakly pitted, most specimens appear rather flattened and only a few show a slight convexity, as the specimen represented in Plate I, Figure 7 which is much more convex than the average. Only the last whorl of 7 to 8 trapezoidal chambers is visible – the latter being separated by straight, raised, thickened sutures. The central part of the test is thickened, obscured and in most specimens (Pl. I, Fig. 9) shows a pustule right over the umbilicus and 5 to 6 other pustules concentrically disposed around it. Leached specimens have umbilical pustules corroded or even missing (Pl. I, Fig. 6), but they always show a characteristic circular depression, hardly visible in specimen of Figure 6, at the center of umbilicus. The peripheral latero-marginal openings, sometimes obscured by filling material, are often clearly visible, particularly those of the last two chambers. The dorsal surface, also smooth or very weakly pitted, is concave over the last whorl and slightly domed in the axial region. All two and a half whorls of concentrically arranged chambers can be seen in most specimens, the thickest of which exhibits strongly raised and thickened sutures. Leached specimens of *Epistomina uhligi* Mjatliuk may show a peripheral outline somewhat similar to that of *Epistomina* sp. 2. The former species, however, is much more convex umbilically (see Ascoli, 1976, Pl. 3) and never shows the umbilical pustules nor the umbilical depression in the leached specimens which are characteristic of *Epistomina* sp. 2.

Occurrence: Bonnition H-32, Venture B-13, Migrant N-20 and COST B-3: *Planularia tricarinella-Epistomina mosquensis* Zone (Kimmeridgian) and *Epistomina stellicostata-E. uhligi* Zone (Tithonian); Bonniton H-32: lower part of *Lenticulina busnardoi-L. saxonica bifurcilla* Zone (Berriasian).

Epistomina sp. 3 (Pl, I, Figs. 11-14) also has medium to

EXPLANATION OF PLATE I

(Bar scale – 50 microns)

Figures 1-5. *Epistomina?* sp. 1: 1, umbilical view, x127; 2, edge view, x127; 3, spiral view, x127; 4, spiral view, x113; 5, edge view, x113 (GSC Nos. 72240-1-2-3-4). All specimens from Mobil-Gulf Bonnition H-32 (East Newfoundland Basin): cutt. 7750-7780ft (2348-2357m), *Epistomina stellicostata* – *E. uhligi* Zone (Tithonian).

Figures 6-10. *Epistomina* sp. 2: 6, umbilical view, x48; 7, edge view, x48; 8, spiral view, x48; 9, umbilical view, x84; 10, spiral view, x81 (GSC Nos. 72245-6-7-8-9). Mobil-Gulf Bonnition H-32: specimens Figs. 6-8 from cutt. 7750-7780ft (2348-2357m), *Epistomina stellicostata* – *E. uhligi* Zone (Tithonian); spec. Figs. 9-10 from cutt. 8250-8280ft (2500-2509m), *Planularia tricarinella* – *Epistomina mosquensis* Zone (Kimmerdigian).

Figures 11-14. *Epistomina* sp. 3: 11, spiral view, x50 (GSC No. 72250); 12, umbilial view, x68; 13, edge view, x68; 14, spiral view, x68. Specimen Fig. 11from Mobil-Tetco-Pex Migrant N-20 (SW Scotian Basin): cutt. 12240-12270ft (3709-3718m), *Epistomina stellicostata* – *E. uhligi* Zone (Tithonian); specimens Figs. 12-14 from COST G-2 (Georges Bank Basin): cutt. 5710-5730ft (1730-1736m), lower part of *Lenticulina busnardoi* – *L. saonica bifurcilla* Zone (Berriasian).

Figures 15-17. *Epistomina* sp. 4: 15, umbilical view, x99; 16, spiral view, x99; 17, edge view, x83. COST G-2: cutt. 5590-5620ft (1694-1703m), lower part of *Lenticulina busnardoi* – *L. saxonica bifurcilla* Zone (Berriasian).

large, very variable size (maximum diameter ranging from 340 to 880 μm) and typically has strongly plano-convex shape and heavy ornamentation. The umbilical side is completely flat (Pl. I, Fig. 13) or very slightly convex with thickened peripheral margin and several large pustules covering not only the umbilical but often also the extra-umbilical area (Pl. I, Fig. 12) particularly along the sutures of chambers. In the last whorl 6-7 chambers appear; they are separated by straight to slightly curved, raised and thickened sutures, often obscured by more or less corroded pustulose ornamentation. The peripheral latero-marginal openings are visible with difficulty, as they are obscured mostly by filling material. The dorsal side is very strongly convex and shows two and a half whorls of concentrically arranged chambers often obscured, particularly in the central part, by a heavy ornamentation of large pustules (Pl. I, Fig. 11) and/or rib-like, overgrown chamber sutures (Pl. I, Fig. 14) which are thick, highly raised and strongly curved backwards in the last whorl. As *Epistomina?* sp. 1. *Epistomina* sp. 3 is similar to no other species of the same genus known from the Late Jurassic and Early Cretaceous of the Canadian and New England Atlantic Shelf.

Occurrence: Demascota G-32 and COST G-2: *Planularia tricarinella-Epistomina mosquensis* Zone (Kimmeridgian); Venture B-13, Migrant N-20, Demascota G-32 and COST G-2: *Epistomina stellicostata-E. uhligi* Zone (Tithonian); Bonnition H-32 and COST G-2: lower part of *Lenticulina busnardoi-L. saxonica bifurcilla* Zone (Berriasian).

Epistomina sp. 4 (Pl. I, Figs. 15-17) has average size (maximum diameter ranging from 365 to 585 μm) and appears somehow intermediate between *Epistomina* sp. 2 and *Epistomina* sp. 3. Its moderate biconvexity (Pl. I, Fig. 17) falls in fact in the middle between the flatness of *Epistomina* sp. 2 and the strong plano-convexity of *Epistomina* sp. 3. Furthermore, its pustulose ornamentation is moderate umbilically (the specimen illustrated in Pl. I, Fig. 15 has ornamentation heavier than average) and only occasionally present dorsally (Pl. I, Fig. 16). This also appears intermediate between the ornamentation of *Epistomina* sp. 2 (weak umbilically and absent dorsally) and that of *Epistomina* sp. 3 (heavy both umbilically and dorsally). The shape, arrangement, and number of chambers (7 to 8 in the last whorl) in *Epistomina* sp. 4 are close to those of *Epistomina* sp. 3 on the dorsal side. Here, however, the thickened sutures of chambers appear much heavier and rib-shaped in the latter species, which shows only 6 to 7 chambers in the last whorl. On the umbilical side, the sutures of chambers, as well as the peripheral latero-marginal openings, are often clearly visible in *Epistomina* sp. 4, the former ones appearing almost straight and perpendicular to the peripheral margin. *Epistomina* sp. 4 shows a certain affinity with *E. mosquensis* Uhlig as described and illustrated by Ohm (1967), but differs from this species in showing a more regularly concentrical trend of chambers both dorsally and umbilically. Furthermore, a pustulose ornamentation is always present in the umbilical area of *Epistomina* sp. 4 where the "umbilical collar", typical of *E. mosquensis,* is lacking.

Occurence: Bonnition H-32: *Planularia tricarinella-Epistomina mosquensis* Zone (Kimmeridgian); COST B-3: *Epistomina stellicostata-E. uhligi* Zone (Tithonian); COST G-2: lower part of *Lenticulina busnardoi-L. saxonica bifurcilla* Zone (Berriasian).

Arenaceous Benthic Foraminiferal Zones

Alveosepta Jaccardi Assemblage Zone. (Some of the "diagnostic species" in the arenaceous zones are calcareous, e.g., *Trocholina* spp.) The top of this zone is defined by the highest stratigraphic occurrence of *Alveosepta jaccardi* (Schrodt). Diagnostic species having their highest stratigraphic occurrence at the top of this zone include *Gaudryina* ex gr. *heersumensis* Lutze and *Everticyclammina* sp. 2. The base of the zone corresponds to the top of the *Trocholina conica-Haplophragmoides canui* Zone of Ascoli (1976). Diagnostic species having their highest stratigraphic occurrence in the lower part of this zone include *Gaudryina heersumensis* Lutze, *Haplophragmoides rotunda* Barnard, and *Verneuilinoides tryphera* Loeblich and Tappan (Fig. 3).

The markers and diagnostic species listed above have their highest stratigraphic occurrences in the Oxfordian to Early Kimmeridgian sections in America and Europe (Loeblich and Tappan, 1950; Maync, 1960; Lutze, 1960; Ascoli, 1976). This zone was defined on the Canadian margin by Ascoli (1976) but in that paper no marker was established for the lower part of the zone. The maximum number of specimens (acme) of *A. jaccardi* has been observed mostly in the middle to upper part of the zone approximately corresponding to the top of the Oxfordian Stage.

The arenaceous markers and diagnostic species of this and other zones occupied a variety of shelf habitats. The genus *Ammobaculites* and other small, simple arenaceous taxa range from paralic to outer neritic environments, whereas the large, complex arenaceous forms, such as *Alveosepta* and *Anchispirocyclina* apparently preferred neritic environments which have mostly coarse-grained substrates.

Ammobaculites Coprolithiformis – Anchispirocyclina Lusitanica Assemblage Zone. The top of this zone is defined by the highest stratigraphic occurrence of *Anchispirocyclina lusitanica* (Egger). Diagnostic species having highest stratigraphic occurrences at the top of this zone include *Trocholina alhina* (Leupold), *T. elongata* (Leupold), *Trocholina* sp. 1 of Ascoli (1976), *Everticyclammina* sp. 3 and *Haplophragmoides* ex gr. *canui* Cushman. The base of the zone corresponds to the top of the underlying *Alveosepta jaccardi* Zone. The zone marker *A. coprolithiformis* has its highest stratigraphic occurrence in the lower part of the zone together with *Trocholina solecensis* Bielecka and Pozaryski. In addition, *Alveosepta* aff. *jaccardi* (Schrodt), *Alveosepta* gr. *jaccardi* (Schrodt), and *Everticyclammina* sp. 1 never occur higher than the lowermost part of the zone (Fig. 3).

Most European authors (e.g., Hottinger, 1967) report that *Anchispirocyclina lusitanica* ranges no higher than the top of the Tithonian in North Africa and Europe. This seems to be true also on the Canadian Atlantic Shelf where the highest occurrence of this species corresponds to the highest occurrence of Tithonian calcareous Foraminifera, such as *Epistomina uhligi* and *E. stellicostata,* and Tithonian ostracod species, such as *Galliaecytheridea postrotunda* Oertli. A late Kimmeridgian to Tithonian age, therefore, has been assigned to the *Ammobaculites coprolithiformis – Anchispirocyclina lusitanica* Assemblage Zone. The highest stratigraphic occurrence of the latter species appears to be approximately at the top of the Tithonian, as indicated by the calpionellids in the Bonnition H-32 well (East Newfoundland Basin). There, *A. lusitanica* is present in a 10-ft-thick unit which contains calpionellids indicative of the lower part of Zone "B" (latest Tithonian age; Jansa *et al.*, 1980).

This zone is equivalent to the *"Ammobaculites coprolithiformis-Trocholina* sp. 1-*Anchispirocyclina lusitanica* Zone" of Ascoli (1976). We eliminate *Trocholina* sp. 1 from the zone name in order to avoid the confusion of having a calcareous foraminifer defining a chiefly arenaceous foraminiferal zone.

Everticyclammina Virguliana Assemblage Zone. The top of his zone is defined by the highest stratigraphic occurrence of *Everticyclammina virguliana* (Koechlin). Diagnostic species having their highest stratigraphic occurrence at the top of this zone include *Trocholina valdensis* (Reichel), *Paalzowella feifeli* (Paalzow) and *Marssonella hechti* Dieni and Massari. The base of this zone corresponds to the top of the underlying *Ammobaculites coprolithiformis-Anchispirocyclina lusitanica* Assemblage Zone. A diagnostic species having its highest stratigraphic occurrence in the lower part of the zone is *Ammobaculites alaskensis* Tappan. This zone corresponds to the *Buccicrenata italica* Zone of Ascoli (1976) and of Jansa *et al.* (1980). *Buccicrenata italica* Dieni and Massari 1966 is a junior synonymn of *Everticyclammina virguliana* Koechlin 1942 (Fig. 3).

Most of the markers and diagnostic species listed above have their highest stratigraphic occurrences in the Valanginian of Europe (Bartenstein, 1977; Dieni and Massari, 1966; and Moullade, 1979). However, *E. virguliana* ranges into the Hauterivian, Barremian and perhaps earliest Aptian (Bedoulian) in Europe (except Italy) and North Africa (Brun and Rey, 1975). *Everticyclammina virguliana* is one of the commonest species of the Kimmeridgian-Valanginian interval of the western North Atlantic basins, particularly in proximity to the Jurrassic-Cretaceous boundary. Our specimens fit well within the broad concept of *Everticyclammina virguliana* as discussed by Brun and Rey (1975). The North American specimens may constitute, however, a local variant; they appear to have a shorter stratigraphical range than their European counterparts.

Ostracod Zones

Eocytheropteron Decoratum – Cytherelloidea Weberi – Schuleridea Triebeli Assemblage Zone. The top of this zone is defined by the highest stratigraphic occurrence of *Cytherelloidea weberi* Steghaus and *Schuleridea triebeli* (Steghaus). Diagnostic species having their highest occurrence at the top of this zone include *?Hutsonia* sp. 1 Ascoli (1976), *?Hutsonia* sp. 2, Hutsonia sp. 6, *Schuleridea* sp. 5 of Jansa *et al.* (1980), *Schuleridea* sp. 6, *Cytherelloidea paraweberi* Oertli, *Schuleridea strzalkowiensis* Bielecka *et al.*, Eocytheropteron *purum* (Schmidt), *Eocytheropteron* sp. 6, *Eucytherura hetermorpha* Peterson, *Eucytherura* sp. 1, *Nodophthalmocythere?* sp. 1, and *Protocythere serpentina* (Anderson).

The base of the zone corresponds to the top of the underlying *Cytheropteron* aff. *decoratum* Zone of Ascoli (1981). Besides the zone marker *Eocytheropteron decoratum* (Schmidt), diagnostic species having their highest stratigraphic occurrences in the lower part of the zone include *Protocythere rodewaldensis* (Klingler), *Paranotacythere* sp. 1 of Ascoli (1976) and Jansa *et al.* (1980), *Procytheropteron* sp. 1 of Kilenyi (1969), and *Galliaecytheridea* sp. 2 (Fig. 4).

The listed markers and diagnostic species have their highest stratigraphic occurrences in the Kimmeridgian of Europe (Klingler, 1955; Oertli, 1959, 1963; Christensen and Kilenyi, 1970; Ascoli, 1976; Kilenyi, 1978). This zone is equivalent to the *Paranotacythere* sp. 1 – *Hutsonia?* sp. 1 Zone of Ascoli (1976). The unnamed zone markers are replaced herein by widely known European Kimmeridgian markers.

The ostracod markers and diagnostic species discussed herein are generally found in inner and outer shelf paleoenvironments associated with either coarse or fine-grained sediments. Their optimum habitat seems to have been moderate depths, not far from shore.

Hustonia gr. Collinsensis – Galliaecytheridea Postrotunda – Schuleridea sp. 1 Assemblage Zone. This zone was defined by Ascoli (1976) and referred to by Jansa *et al.* (1980). Its top is defined by the highest stratigraphic occurrence of *Galliaecytheridea postrotunda* Oertli and *Schuleridea* sp. 1 of Ascoli (1976). Diagnostic species having their highest stratigraphic occurrence at the top of this zone include *Eocytheridea eusarca* (Anderson), *Klieana* ex gr. *alata* Martin, *Eocytheropteron* aff. *brodiei* (Jones), *Schuleridea* sp. 2 Ascoli (1976), *Asciocythere* gr. *amygdaloides* (Cornuel), *Asciocythere* sp. 4, and *Asciocythere* sp. 5.

The base of the zone corresponds to the top of the underlying *Eocytheropteron decoratum – Cytherelloidea weberi-Schuleridea triebeli* Assemblage Zone. Besides the zone marker *Hutsonia* gr. *collinsensis* Swain and Brown, diagnostic species having highest stratigraphic occurrence in the lower part of this zone include *Macrodentina* aff. *lineata* Martin, *Haplocytheridea* sp. 1 Ascoli (1981) and *Cytherella index* Oertli (Fig. 4).

The markers and diagnostic species listed above have their highest stratigraphic occurrences near the top of the Tithonian (Klingler *et al.*, 1962; Kilenyi and Neale, 1978; H.J. Oertli, pers. commun.; F.M. Swain, pers. commun.). These data and the calpionellid ranges in the COST G-2 well confirm the Thithonian age of this zone. In the G-2 well, calpionellid Zone "A" (Late Tithonian) is present between 5770ft (1748m) and 6040ft (1830m) (Fig. 5). The ostracod markers and diagnostic species have their highest stratigraphic occurrences at the following depths in the G-2 well: *Hutsonia* gr. *collinsensis* at 6100ft (1848m); *Cytherella index* at 6024ft (1825m sidewall core sample); *Galliaecytheridea postrotunda* at 6010ft (1821m); *Asciocythere* gr. *amygdaloides* and *Asciocythere* sp. 5 at 5920ft (1794m); *Schuleridea* sp. 2 at 5895ft (1786m swc sample); *Schuleridea* sp. 1 at 5890ft (1785m); and *Eocytheridea eusarca* at 5856ft (1775m swc sample).

Ascioythere sp. 1-Schuleridea aff. Praethoerenensis Assemblage Zone. The top of this zone is defined in Ascoli (1976) by

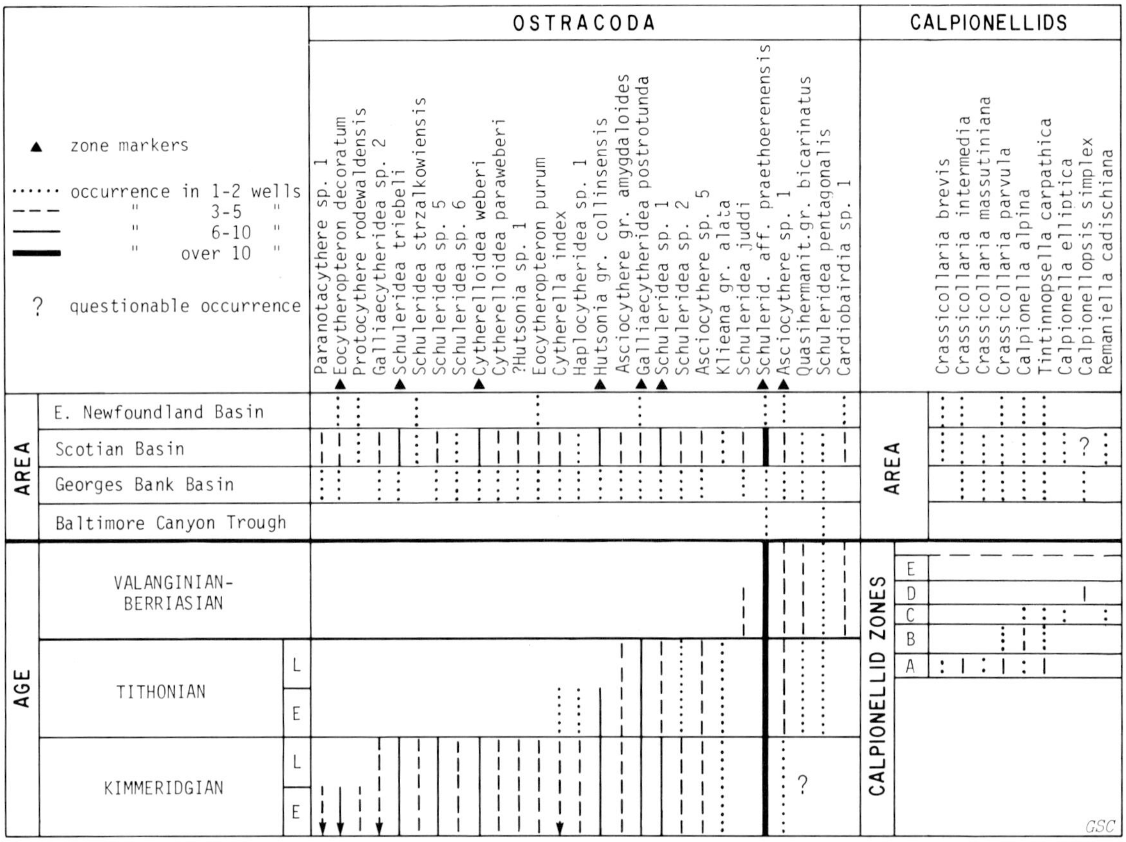

Figure 4. Stratigraphic ranges and areal distribution of Kimmeridgian-Valanginian ostracod and calpionellid marker and "zone diagnostic" species, Canadian and New England Atlantic Shelf.

the highest stratigraphic occurrence of *Asciocythere* sp. 1 and *Schuleridea* aff. *praethoerenensis* Bartenstein and Brand. Diagnostic species having their highest stratigraphic occurrences at the top of this zone include *Quasihermanites* gr. *bicarinatus* Gruendel, *Schuleridea pentagonalis* Swain, *Schuleridea praethoerenensis* Bartenstein and Brand, *Cardiobairdia* sp. 1 of Ascoli (1976) and *Asciocythere* sp. 2 of Ascoli (1976). The base of the zone corresponds to the top of the underlying *Hutsonia* gr. *collinsensis* – *Galliaecytheridea postrotunda* – *Schuleridea* sp. 1 Assemblage Zone (Fig. 4).

The markers and diagnostic species listed above have their highest stratigraphic occurrence in the Berriasian to Valanginian interval of European localities (Bartenstein and Brand, 1959; Neale, 1967, 1978; Swain, 1981). This zone, therefore, has been assigned a Berriasian to Valanginian age. *Schuleridea praethoerenensis* is a widely known European species that disappears in the Late Valanginian. It has been recognized by us only in the COST G-2 well (Georges Bank). A similar species is our "zonal marker" *Schuleridea* aff. *praethoerenensis*, which has the same highest stratigraphical occurrence as *S. praethoerenensis* Bartenstein and Brand and as Valanginian Foraminifera such as *Lenticulina busnardoi* and *L. saxonica bifurcilla*. *Schuleridea* aff. *praethoerenensis* (Ascoli, 1976) differs from *S. praethoerenensis* in being weakly instead of strongly pitted, is larger, and has a distinctive, smooth, long, transverse swelling on the left valve of female specimens; it is the most common ostracod species in the Kimmeridgian-Valanginian interval of the northwestern Atlantic margin, probably because it is eurytopic.

Schuleridea juddi and *Paranotacythere speetonensis* are good guides to the British Berriasian (Neale, 1978). Their highest stratigraphic occurrence on the western North American Atlantic margin is considered to be very close to the top of this stage, as confirmed by the rich calpionellid succession of the COST G-2 well (George Bank Basin). The calpionellids indicate a Berriasian age (possibly Zone "C") for the interval 5560-5650 ft (1685-1712m) and Late Berriasian to earliest Valanginian age for the overlying interval (5530-5560 ft = 1676- 1685m; Zone "D"; Fig. 5). The highest stratigraphic occurrences of *Schuleridea juddi* and *Schuleridea* aff. *praethoerenensis* have been found at 5590 ft (1694m); that of *Schuleridea praethoerenensis* is at 5427 ft (1645m; swc sample).

Calpionellid Zones

The first calpionellids from the Canadian Atlantic Shelf were reported by Jenkins *et al.* (1974), who pointed out the presence of a few calpionellid specimens in the Berriasian-Valanginian sediments of the Puffin B-90 well (Grand Banks). Gradstein *et al.* (1975) provided a list of tentatively identified calpionellids from the Mohican I-100 well which were later studied in more detail by Jansa *et al.* (1980). The latter authors were the first to establish a calpionellid zonation for the Canadian Atlantic Shelf, and they fully integrated it with foraminiferal and ostracod zonations. On the New England Atlantic Shelf, the presence of calpionellids was first eported from Heezen Canyon by Ryan *et al.* (1979). Later, they were noted in the COST G-2 well (Georges Bank Basin) by Poag (1982a, 1982b), who attempted no definitive zonation.

The evolutionary sequence of calpionellids on the western, North American Atlantic margin appears to be identical to that of the Mediterranean area which was described by Remane (1964) in the "Fosse Vocontienne" of southeastern France (Jansa *et al.*, 1980). This biozonation has been directly tied to the ammonite zonation in the same region (Le Hégarat and Remane, 1968; Busnardo *et al.*, 1979) and in the Subbetic region of southern Spain (Barthel *et al.*, 1966; Allemann *et al.*, 1975). Remane recognized five zones (Fig. 2): Zone "A" (middle part of Late Tithonian); to Zone "E" (Early-Middle Valanginian). The "Standard Calpionellid Zonation" of the Mediterranean area (Allemann *et al.*, 1971) is a somewhat simplified version of Remane's (1964) biozonation, in which Remane's Zones "B" and "C" are combined. The Standard Calpionellid Zonation, therefore, consists of four zones (Fig. 2). As many as eight subzones also can be recognized in the Mediterranean and in our study area (Jansa *et al.*, 1980).

Owing to the amphiatlantic distribution of the planktonic calpionellids, to their fine stratigraphic resolution and to their calibration with European ammonite zones, this group serves as the standard for calibrating stratigraphic ranges of the other microfossil groups within the Tithonian to Valanginian interval of our study area. As discussed by Remane (1980), calpionellids in Europe occur typically in micritic or marly pelagic limestones that often contain benthic fossils. Their frequent association with abundant radiolarians, coccoliths, and nannoconids clearly indicates that calpionellids are marine planktonic organisms. Heretofore, rich calpionellid assemblages have been found only in outer shelf to abyssal sediments. They have been rarely reported in coarse-grained shallow neritic sediments. However, in our study area, the rather coarse-grained calpionellid-bearing micrites, appear to have accumulated in depths shallower than normal for this group. Jansa *et al.* (1980) suggested a middle to outer shelf paleoenvironment for the calpionellid assemblages of the East Newfoundland and Scotian Basins. Poag (1982a) demonstrated that the calpionellids of the COST G-2 well accumulated in sediments deposited more than 50 km landward of the Kimmeridgian-Valanginian shelf edge.

Zone "A": Crassicollaria Assemblage Zone. The top of this zone can also be defined by the highest stratigraphic occurrences of *Crassicollaria brevis* Remane, *Crassicollaria intermedia* (Durand Delga) and *Crassicollaria massutiniana* (Colom). The base of the zone corresponds to the highest occurrence of *Chitinoidella* and to the earliest appearance of the Calpionellidae *sensu* Bonet (or forms having a hyaline microgranular wall) (Figs. 2, 4). *Chitinoidella,* which is included in the Calpionellidae by Remane (1980), does not belong to the Calpionellidae *sensu* Bonet, because it has on opaque wall.

The presence in this zone of the three listed species indicates an age of middle Late Tithonian (Remane, 1980; Jansa *et al.*, 1980) as established in Europe by means of correlation with ammonites.

Zone "B": "Calpionella alpina Acme Zone" (lower part of Calpionella Assemblage Zone). The top of this zone can be recognized by a significant upward reduction in the number of specimens of the genus *Calpionella.* It is also marked by the acme of a moderately large, spherical form of C. *alpina* Lorenz. The base of the zone corresponds to the top of the underlying Zone "A" (*Crassicollaria* Assemblage Zone). *Crassicollaria parvula* Remane disappears at the top of this zone in the Bonnition H-32 (East Newfoundland Basin) and Mohican I-100 (Scotian Basin) wells, but this species is well represented within Zone "C" in Europe (Remane, 1980). *Zone* "B" straddles the Jurassic-Cretaceous boundary, ranging in age from latest Tithonian to earliest Berriasian (Remane, 1980) (Figs. 2, 4).

The relative abundance and morphology of *Calpionella alpina* s.l. in our samples display the same stratigraphic changes as in the Mediterranean area. Its acme (not to be confused with the acme of its spherical form only) occurs in both regions at the boundary between Zone "A" and Zone "B" where there is also a transition from large, somewhat elongate forms in Zone "A" to smaller, spherical forms in Zone "B".

Zone "C": Upper part of Calpionella Assemblage Zone. The top of this zone can be determined approximately by the highest stratigraphic occurrence of *Crassicollaria parvula* Remane and of *Calpionella elliptica* Cadisch (which is restricted to this zone). The base of the zone corresponds to the top of the underlying Zone "B" and to an increase in the numbers of *Tintinnopsella carpathica* (Murgeanu and Filipescu) whose acme occurs higher within the zone. *Calpionella alpina* has its highest stratigraphic occurrence in this zone in our samples, but has been reported also from Zone "D" in the Mediterranean area. In Europe, the diagnostic species *Calpionella elliptica* and *Crassicollaria parvula* disappear in rocks of Early to Middle Berriasian age (Remane, 1980) (Figs. 2, 4).

Zone "D": Calpionellopis Assemblage Zone. The top of this zone can be defined also by the highest stratigraphic occurrence of *Calpionellopsis simplex* (Colom) which is confined to this zone. The base of this zone corresponds to the top of the underlying Zone "C". *Calpionellopsis simplex* occurs in Late Berriasian to earliest Valanginian rocks in the Mediterranean area (Figs. 2, 4).

Zone "E": Calpionellites Assemblage Zone. The top of this zone is defined by the highest stratigraphic occurrence of the family Calpionellidae *sensu* Bonet which, in Europe, is in rocks of Late Valanginian age (but not latest Valanginian). The exact stratigraphic level of this highest occurrence has not been defined precisely because its numbers diminish gradually within the *campylotoxum* Zone and it may extend into the overlying *verrucosum* Zone in several European sections. The extinction level of the Calpionellidae which Remane (1980) placed in the *verrucosum* Zone is, therefore, represented by a dashed line in

Figures 2, 4, 6, 7. The base of this zone corresponds to the top of the underlying Zone "D" (Figs. 2, 4)

Calpionellites darderi (Colom), one of the best known species confined to this zone, has not been recorded so far from our samples. Thus, this zone has not been recognized with certainty in our study area but it may be present in the Mohican I-100, the Acadia K-62, and the Puffin B-90 wells.

CALPIONELLIDS FROM THE COST G-2 WELL, GEORGES BANK

To our knowledge, the first calpionellids from the subsurface of the New England Shelf were observed by Poag (1982a, 1982b) in the COST G-2 well. Following Poag's initial observations, thin sections were made for the 72 cuttings samples from the 5170-7480ft (1567-2267m) interval of the COST G-2 well and J. Remane studied microfacies and calpionellids in detail.

The lithic components of the examined interval are variable brown-grey shale and light grey mudstone from 5170-5320ft (1567-1612m). Below that are grey, silty calcareous mudstones interbedded with a few sandstone and limestone beds (5320-5750ft = 1612-1742m). An underlying carbonate section (5750-6920ft = 1792-2097m) contains light grey and tan, clayey, micritic to chalky limestone interbedded with grey, silty calcareous shale and sparse sandstone. This is followed by a heterogeneous section of sandstone, red and grey shale, oolitic limestone and streaks of coal (6920-9580ft = 2097-2903m) (Amato and Simonis, 1980; Poag, 1982a, 1982b).

The highest cuttings sample containing *in situ* calpionellids represents the 5530-5560ft (1676-1685m) interval. It is referred to calpionellid Zone "D" (Late Berriasian-earliest Valanginian). The lowest sample containing calpionellids represents the 6010-6040ft (1821-1830m) interval and is assigned to calpionellid Zone "A" (Late Tithonian).

Since the cuttings sample at 6070-6100ft (1839-1848m) contains abundant micritic limestone without calpionellids, the very sparse calpionellid occurrences observed between 6100 and 6700 ft. (1849 and 2030m) are interpreted as cavings. On the whole, calpionellids are sparse in the 5530-6040ft (1676-1830m) interval, particularly in its upper part. However, no core material was available for study and even the cuttings samples lacked a coarse fraction. Thus we could not carry out detailed statistical analysis of calpionellid assemblages. This, in turn, limits our biostratigraphic interpretations (Fig. 5).

The presence of *Calpionellopsis* sp. at 5530-5560ft (1676-1685m) is indicative of calpionellid Zone "D" of Late Berriasian-Valanginian age (Le Hégarat and Remane, 1968). One specimen of *Calpionellopsis simplex* (Colom) at 6490-6520ft (1967-1976m) is considered to be caved, but confirms the existence of Late Berriasian calpionellid facies of Zone "D" higher in the well (Pl. II, Fig. 10).

Calpionella alpina Lorenz (5710-5920ft = 1730-1794m) ranges from Late Tithonian to latest Berriasian in age (Le Hégarat and Remane, 1968; Remane, 1980). The variable morphology of *C. alpina*, which is characteristic of calpionellid Zone "A", (middle Late Tithonian) has been recognized at 5830-5860ft (1767-1776m) (Pl, II, Figs. 1-4). *Crassicollaria parvula* Remane (5770-5920ft = 1748-1794m) ranges from Zone "A" through Zone

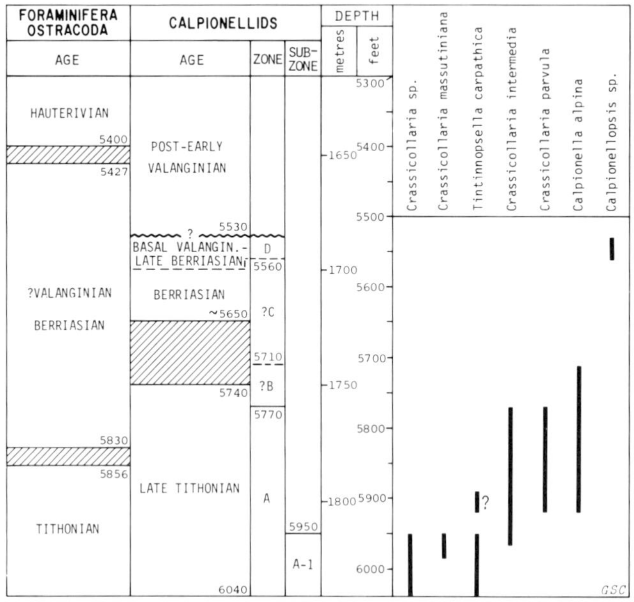

Figure 5. Range-chart of calpionellids and calpionellid biozonation of the COST G-2 well, Georges Bank Basin.

"C" (Remane, 1980) and, therefore, does not help to recognize the Jurassic-Cretaceous boundary (Pl. II, Fig. 5).

Crassicollaria intermedia (Durand Delga; 5770-5980ft = 1748-1812m) and *Crassicollaria massutiniana* (Colom; 5950-5980ft = 1803-1812m) are restricted to calpionellid Zone "A" (middle Late Tithonian) (Pl. II, Figs. 6-9).

Tintinnopsella carpathica (Murgeanu and Filipescu) (5950-6040ft = 1803-1830m) is represented by a very small variety, which is characteristic of the lower part of Zone "A" (subzone A-1). Large forms of *T. carpathica,* typical of the earliest Cretaceous elsewhere, hae not been observed in the COST G-2 well.

As far as can be determined from the examination of only cuttings samples, the succession of calpionellid species from the COST G-2 well fits the classic Mediterranean scheme, as do those from the Bonnition H-32, Moheida P-15, and Mohican I-100 wells (Jansa *et al.*, 1980). In the interval from 5530 to 6040ft (1676 to 1830m), Zones "D" and "A", as originally defined in southeastern France (Remane, 1964, 1980; Le Hégarat and Remane, 1968), are well documented. There is no evidence for Zones "B" and "C", but the lithofacies between 5560 and 5770ft (1685 and 1748m) seems not to be very favourable for calpionellids.

The Tithonian/Berriasian (Jurassic/Cretaceous) boundary lies within Zone "B". In the absence of the critical association, this boundary cannot be determined with great precision, but it is certain that the boundary lies between 5560ft (1685m) (Zone "D") and 5770ft (1748m) (Zone "A") because the cuttings sample at 5530-5560ft (1676-1685m), containing the genus *Calpionellopsis,* still belongs to Zone "D" (Late Berriasian-basal Valanginian). On the other hand, the cuttings sample at 5770-5800ft (1748-1758m), with *Crassicollaria intermedia*, is already referable to Zone "A", which does not even reach the top of the Tithonian (Remane, 1980). This position agrees well with the highest stratigraphic occurrence of the Berriasian ostracod marker *Schuleridea juddi* at 5590ft (1694m). This is somewhat higher than it has been placed previously on the basis of foraminifera (Poag, 1982a, 1982b) and calcareous nannoplankton (Valentine, 1982).

The fact that *Calpionella alpina* is missing in the lowermost calpionellid assemblage (5959-6040ft = 1803-1830m) suggests that this interval corresponds to subzone A-1, near the base of the Late Tithonian. This relationship is supported by the general aspect of *Crassicollaria* and, particularly by the presence of very small specimens of *Tintinnopsella carpathica* in the lowest assemblage.

Globochaete alpina Lombard (*incertae sedis*), which is common in the calpionellid limestones of the Mediterranean area, has been observed in the 6370-6640ft (1930-2012m) interval of the COST G-2 well. *Saccocoma,* a pelagic crinoid that normally is abundant in the Mediterranean Kimmeridgian-Early Tithonian section below the "calpionellid limestones", has not been observed in our samples. However, this crinoid was reported in samples from nearby Heezen Canyon (Ryan *et al.*, 1979), a locality that was farther seaward (at the shelf edge) than the G-2 site during accumulation of the calpionellids.

Comparison of the calpionellid sequence of the COST G-2 with that of the Bonnition H-32 well (East Newfoundland Basin), shows that calpionellid Zone "A" is well represented in the latter and its basal subzone A-1 can be identified (Jansa *et al.*, 1980). The 5770-6040ft (1748-1830m) interval of the COST G-2 well, therefore, is correlated with the 7046-7998ft (2135-2424m) interval in Bonnition H-32. However, there is no counterpart in the COST G-2 of the Zone "B" assemblages observed in Bonnition H-32.

On the other hand, post-earliest Berriasian calpionellids are missing in Bonnition H-32 and in the Mohican I-100 well (Scotian Shelf). The only calpionellids of Berriasian age reported so far from the Canadian Atlantic Shelf are present in the Moheida P-15 well (Scotian Shelf; Jansa *et al.*, 1980). This assemblage consists of large *Tintinnopsella carpathica* and of *Remaniella cadischiana* and *Calpionella elliptica* and belongs to Zone "C" (Middle Berriasian). However, these specimens are not *in situ,* as they have been reworked into Hauterivian sediments. Similarly, calpionellids of Zone "A" and Zone "B" of Mohican I-100 are reworked into Valanginian sediments (Jansa *et al.*, 1980). The calpionellids of Zone "D" in the COST G-2 well are the first Berriasian calpionellids that have been found *in situ* on the western North Atlantic shelf. Thus, the highest stratigraphic occurences of several formainiferal and ostracod species, previously referred to as Berriasian-Valanaginian in age, now can be assigned to the Berriasian on the basis of the associated calpionellids. Consequently, the top of the Berriasian Stage can now with good approximation be extrapolated using benthic Foraminifera and Ostracoda.

REGIONAL CORRELATION AND PALEOGEOGRAPHY

The four groups of microfossils we have discussed do not always occur together. A given group may tolerate or prefer environmental conditions that exclude or limit the presence of other groups. However, their biotopes overlap sufficiently to provide biozonation and correlation in a wide variety of paralic and marine lithofacies. Multiple zonation using the benthic groups (Foraminifera and Ostracoda) has provided the most consistent and areally extensive correlations, because these groups are present in all 28 wells studies (Figs. 6, 7). The planktonic calpionellids, on the other hand, were more stenotopic organisms, and have been found in only six wells. Five wells (COST G-2, Mohican I-100 Puffin B-90, Moheida P-15, Bonnition H-32) contained all four microfossil groups in sufficient quantity to provide planktonic calibration of the benthic ranges and to allow a fourfold integrated zonation. On the basis of this control, the individual elements of this tetrad can now be (and have been, with a very high degree of reliability) used confidently to extend correlations to more sparsely fossiliferous sections in the study area (see Figs. 6, 7).

The presence of the stenotopic calpionellids from the East Newfoundland Basin to the Baltimore Canyon Trough is evidence that oceanic conditions existed between Africa, Iberia and North America during the Late Jurassic and Early Cretaceous, and that warm Tethyan water masses reached as far northwest as the Newfoundland Basin.

SUMMARY AND CONCLUSIONS

Having analyzed calcareous and arenaceous benthic Foraminifera, Ostracoda, and calpionellid assemblages in 28 wells along a 2300km transect through the Baltimore Canyon Trough, Georges Bank Basin, Scotian Basin and East Newfoundland Basin, we have erected a fourfold biozonation across the Jurassic-Cretaceous boundary. The sequence of short-ranging planktonic calpionellids in five of the wells has provided a more accurate identification of the Jurassic-Cretaceous boundary than the other groups studied because the calpionellid sequence is calibrated with the ammonite zonation in Europe. The calpionellids, in turn, provide calibration of the ranges of the benthic microfossils in our study area. The benthic forms, thus, can be used to extrapolate the approximate position of the Jurassic-Cretaceous boundary in a wide variety of facies where calpionellids are not present. In fact, similarly constructed foraminiferal-ostracod biozones have been demonstrated to be indispensable in dating and correlating the entire Upper Jurassic and Lower Cretaceous interval of the study area (Ascoli, 1976; Poag, 1980, 1982a, 1982b). The calpionellids also provide calibration of palynomorph ranges, which can then be used to date and zone nonmarine facies in the study area (E.H. Davies, pers. commun.).

Our data show, moreover, that the Late Jurassic foraminiferal and ostracod species whose highest stratigraphic appearances are most useful for correlation, consistently disappear 100 to 150 ft below the top of the Tithonian Stage as defined by calpionellids. Thus, the Jurassic-Cretaceous boundary has been placed slightly too low in most previously published studies such as Jansa *et al.* (1980).

The presence of calpionellids on this segment of the western North Atlantic margin is evidence that warm, oceanic, Tethyan waters separated North America from Iberia and Africa at this time. This has been demonstrated by Jansa *et al.* (1980), who

PLATE II

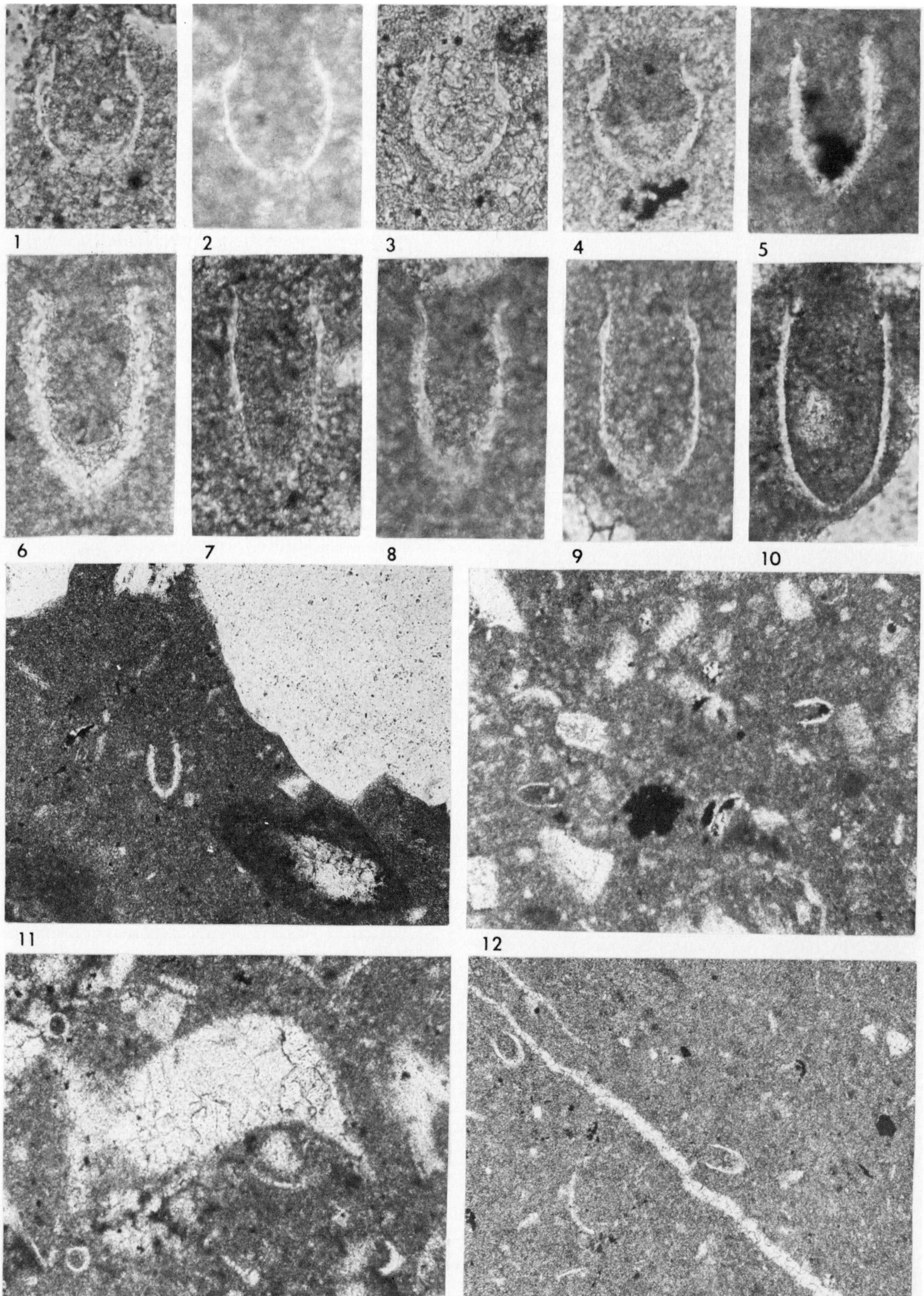

AGE	CALCAREOUS BENTHIC FORAMINIFERA ZONES	BALTIM. CANYON COST B-3	GEORGES BANK COST G-2	SW SCOTIAN BASIN MOHAWK B-93	MOHICAN I-100	NASKAPI N-30	MOHEIDA P-15	ACADIA K-62	ONEIDA O-25	DEMASCOTA G-32	CREE E-35	COHASSET D-42	ONONDAGA E-84	MIGRANT N-20	THEBAUD I-94
VALANGINIAN-BERRIASIAN	Lenticulina saxon. bifurcilla Lenticulina busnardoi		5427 5830		8335		8620 8750	?	8730	10350 11280			~11680 ~12980		
TITHONIAN	Epistomina uhligi Epistomina stellicostata	~11930 ~13400	5856 6220		8940 9240	5560	8820 8950		9445 10220	11350 11780	12400 13070	?		11690 12650	12580 ~13000
KIMMERIDGIAN	Epistomina mosquensis Planularia tricarinella	~13480 ~15400	6245 9195		9314 ~10400	~5600	9020 9050		10320 11420	12260 ~13660		10358 ~11300		~12710 13190	

(hatched) =part of section not drilled

? =part of section probably missing or non-marine

AGE	ARENACEOUS BENTHIC FORAMINIFERA ZONES	COST B-3	COST G-2	MOHAWK B-93	MOHICAN I-100	NASKAPI N-30	MOHEIDA P-15	ACADIA K-62	ONEIDA O-25	DEMASCOTA G-32	CREE E-35	COHASSET D-42	ONONDAGA E-84	MIGRANT N-20	THEBAUD I-94
VALANGINIAN-BERRIASIAN	Everticyclammina virguliana	~10700 ~11900	5427 5830	5160	8335 8940	5339 5530	8620 8750	?				10070 ~10300		10570 11630-810	11010 12520
TITHONIAN KIMMERIDGIAN L	Anchispirocyclina lusitanica Ammobaculites coprolithiformis			5250 5420				11526 ~12510	9445 10820			?		11870 ~13000	
KIMMERIDGIAN E OXFORDIAN	Alveosepta jaccardi	~14000 ≂15400 ?	7690 11290	5500 6255	9770 ~11000	5660 6595	10020 11050		10920 12120	?					

AGE	OSTRACOD ZONES	COST B-3	COST G-2	MOHAWK B-93	MOHICAN I-100	NASKAPI N-30	MOHEIDA P-15	ACADIA K-62	ONEIDA O-25	DEMASCOTA G-32	CREE E-35	COHASSET D-42	ONONDAGA E-84	MIGRANT N-20	THEBAUD I-94
VALANGINIAN-BERRIASIAN	Schulerid. aff. praethoerenensis Asciocythere sp. 1	~10700 ~11900	5427 5830	5160	8335		8620 8750	?	8730	10350 11280	10012 12330				
TITHONIAN	Schuleridea sp. 1 Galliaecytheridea postrotunda Hutsonia gr. collinsensis		5856 6220	5250	8940	5560	8820 8950		9445	11350 11780	12400 13070	?			
KIMMERIDGIAN	Schuleridea triebeli Cytherelloidea weberi Eocytheropteron decoratum		6245 9195	~6220 5500	9314 ~10000	~5600	9020 9950	11526 13681	10320 11420	11860 ~13660		10358 ~11300			

R=reworked material

U=undefined zone(s)

AGE		CALPIONELLID ZONES	COST B-3	COST G-2	MOHAWK B-93	MOHICAN I-100	NASKAPI N-30	MOHEIDA P-15	ACADIA K-62	ONEIDA O-25	DEMASCOTA G-32	CREE E-35	COHASSET D-42	ONONDAGA E-84	MIGRANT N-20	THEBAUD I-94
VALANGINIAN	L															
VALANGINIAN	E	E Calpionellites		?		?8334										
BERRIASIAN	L	D Calpionellopsis		5530 5560		?8334		8408								
BERRIASIAN	L/E	C Calpionella		? 5710		?8334		R	9075 U							
BERRIASIAN	E	B Calpionella		?		8334 8335 R		8412								
TITHONIAN	L	A Crassicollaria		5770 6040		8337 R										

Figure 6. Stratigraphic correlation of Kimmeridgian to Valanginian foraminiferal, ostracod and calpionellid assemblage zones in 14 wells from the Baltimore Canyon Trough, Georges Bank Basin and SW Scotian Basin, listed in geographic order from SW to NE. All depths are in feet.

pointed out that "the co-occurrence of Tethyan and Boreal foraminiferal species near the Jurassic-Cretaceous boundary on the Scotian Shelf and Grand Banks is interpreted to reflect paleocirculation patterns with the Tethyan species favouring warmer waters and thus occurring in shallower environments and the Boreal species flourishing in colder, deeper waters. Colder waters from North Europe and Greenland, providing favourable living conditions for Boreal foraminiferal species, probably flowed southward along the North American margin, while warmer waters rich in Tethyan (foraminiferal and calpionellid) species were outflowing from the Mediterranean into the North Atlantic. The warm water influence on the central North Atlantic is confirmed by the presence of calpionellids in five Deep Sea Drilling Project sites from eastern and western North Atlantic" and in the COST G-2 well (Georges Bank, New England Shelf), as we reported in the previous pages.

EXPLANATION OF PLATE II

Figures 1-4. *Calpionella alpina* Lorenz, x400. COST G-2: specimens Figures 1-2 from cutt. 5830-5860ft (1767-1776m), spec. Figures 3 and 4 from cutt. 5890-5920ft (1748-1757m), Zone "A" (Late Tithonian).

Figure 5. *Crassicollaria parvula* Remane, x400 (same specimen as in Fig. 12). COST G-2: cutt. 5830-5860ft (1767-1776m), Zone "A" (Late Tithonian).

Figure 6-8. *Crassicollaria intermedia* (Durand Delga), x400. COST G-2: spec. Figures 6 and 8 from cutt. 5830-5860ft (1767-1776m), spec. Figure 7 from cutt. 5950-5980ft (1803-1812m), Zone "A" (Late Tithonian). Specimen of Figure 6 is also represented in Figure 11.

Figure 9. *Crassicollaria massutiniana* (Colom), x400. COST G-2: cutt. 5950-5980ft (1803-1812m), Zone "A" (Late Tithonian).

Figure 10. *Calpionellopsis simplex* (Colom), x400. COST G-2: cutt. 6490-6520ft (1967-1976m), Zone "D" (Late Berriasian-basal Valanginian cuttings near 5530-5550ft = 1676-1682m, caved into Kimmeridgian sediments in place at 6490ft = 1967m).

Figure 11. Microfacies with calpionellids (*Crassicollaria intermedia* of Fig. 6) and quartz grain (upper right), x100. COST G-2 cutt. 5830-5860ft (1767-1776m), Zone "A" (Late Tithonian).

Figure 12. Microfacies with calpionellids (among which *Crassicollaria parvula* of Fig. 5), x 100. COST G-2: cutt. 5830-5860ft (1767-1776m), Zone "A" (Late Tithonian).

Figure 13. Microfacies with calpionellids, x100. COST G-2: cutt. 5950-5980ft (1803-1812m), Zone "A" (Late Tithonian).

Figure 14. Microfacies with calpionellids, x100. COST G-2: cutt. 6010-6040ft (1821-1830m), Zone "A" (Late Tithonian). Figures 11-14 show that typical pelagic calpionellid limestones (as that of Fig. 14) are rare in the COST G-2 well. Most microfacies show instead biomicrites of outer neritic environment (Figs. 12-13).

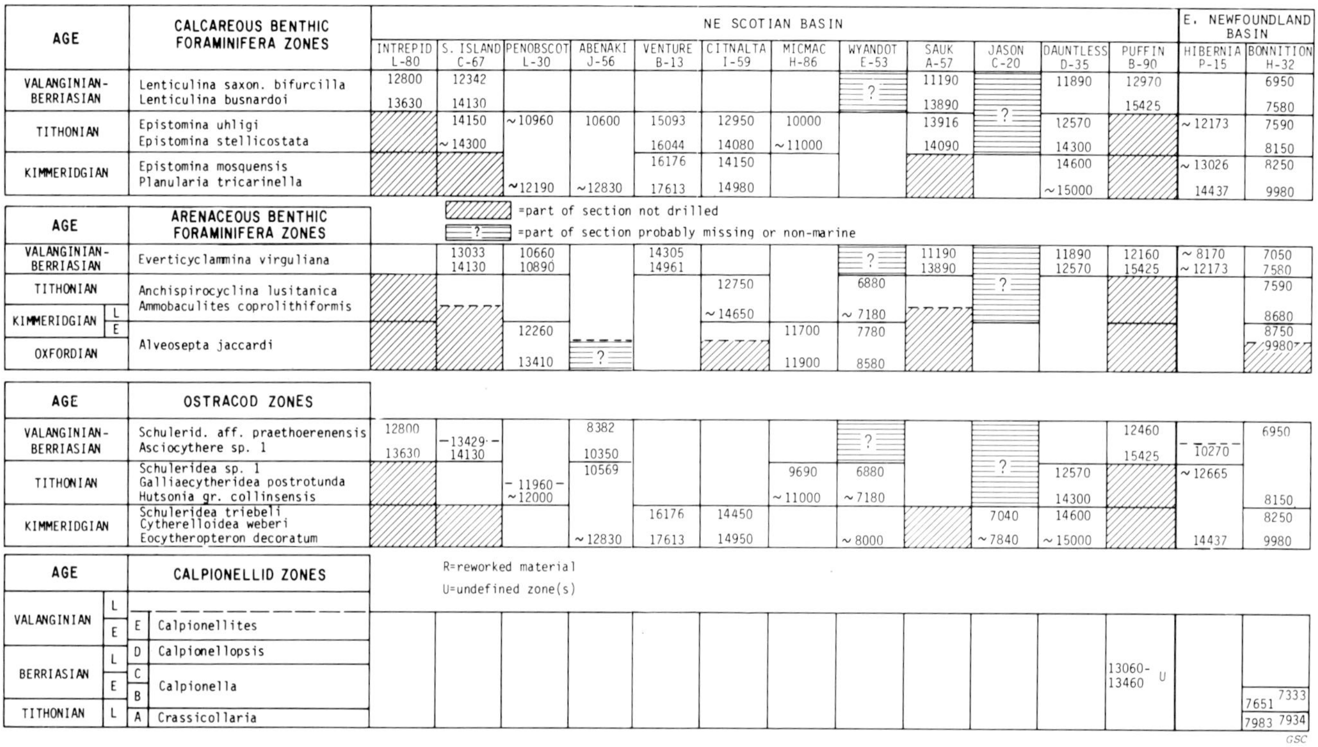

Figure 7. Stratigraphic correlation of Kimmeridgian to Valanginian foraminiferal, ostracod and calpionellid assemblage zones in 14 wells of the NE Scotian Basin and East Newfoundland Basin, listed in geographic order from SW to NE. All depths are in feet.

REFERENCES

Alleman, F., Catalano, R., Fares, F., and Remane, J., 1971, Standard Calpionellid Zonation (Upper Tithonian – Valanginian) of the Western Mediterranean Province: Proceedings of Second Planktonic Conference, Roma, 1970, v. 2, p. 1337-1340.

Allemann, F., Grun, W., and Wiedmann, J., 1975, The Berriasian of Caravaca (Prov. of Murcia) in the Subbetic Zone of Spain and its Importance for Defining This Stage and the Jurassic-Cretaceous Boundary: Colloque Limite Jurassique-Crétacé Lyon-Neuchâtel, 1973, Memoires Bureau Recherches Géologiques Minières, no. 86, p. 14-22.

Amato, R.V., and Simonis, E.K., eds., 1980, Geologic and Operational Summary, COST No. G-2 Well, Georges Bank Area, North Atlantic OCS: United States Geological Survey Open-File Report 80-269, 116 p.

Arkell, W.J., 1956, Jurassic Geology of the World: Oliver and Boyd, Edinburgh, 806 p.

Ascoli, P., 1976, Foraminiferal and Ostracod Biostratigraphy of the Mesozoic-Cenozoic, Scotian Shelf, Atlantic Canada: Maritime Sediments Special Publication 1, part B., p. 653-671.

______, 1981, Foraminiferal-Ostracod Late Jurassic Biozonation of the Scotian Shelf: Geological Survey of Canada, Open File No. 753, 32 p.

Austin, J.A., Uchupi, E., Schaughnessy, D.R., III, and Ballard, R.D., 1980, Geology of New England Passive Margin: American Association of Petroleum Geologists Bulletin, v. 64, p. 501-526.

Ballard, R.D., and Uchupi, E., 1975, Triassic Rift Structure in Gulf of Maine: American Association of Petroleum Geologists Bulletin, v. 59, p. 1041-1072.

Barss, M.S., Bujak, J.P., and Williams, G.L., 1979, Palynological Zonation and Correlation of Sixty-seven Wells, Eastern Canada: Geological Survey of Canada Paper 78-24, 118 p.

Bartenstein, H., 1977, Stratigraphic Parallelization of the Lower Cretaceous in the Northern Hemisphere. Zonation with Benthonic Index Foraminifera: Newsletters in Stratigraphy v. 6, p. 30-41.

Bartenstein, H., and Bettenstaedt, F., 1962, Marine Unterkreide (Boreal and Tethys): *in* Leitfossilien der Mikropaläontologie: Gebrüder Borntraeger, Berlin, p. 225-298.

Bartenstein, H. and Brand, E., 1951, Mikropaläontologische Untersuchungen zur Stratigraphie des Norwestdeutschen Valendis: Abhandlungen der Senckenbergischen Naturforschenden Gesellschaft, v. 485, p. 239-336.

______, 1959, Feinstratigraphisch Wichtige Ostracoden aus dem Nordwestdeutschen Valendis: Paläontologische Zeitschrift, v. 33, p. 224-240.

Barthel, K.W., Cediel, F., Geyer, O.F., and Remane, J., 1966, Der Subbetische Jura von Cehegin (Provinz Murcia, Spanien): Mitteilungen der Bayerischen Staatssammlung fur Paläontologie und Historische Geologie, v. 6, p. 167-211.

Bebout, J.W., 1980, Biostratigraphy: *in* Amato, R.V., and E.K. Simonis, eds., Geologic and Operational Summary, COST No. G-2 Well, Georges Bank Area, North Atlantic OCS: United States Geological Survey Open-File Report 80-269, p. 20-28.

Bielecka, W., 1975, Foraminifera and Brackish Ostracoda from the Portlandian of Polish Lowlands: Acta Palaeontologica Polonica, v. 2, no. 3, 295 p.

Bielecka, W., and Pozaryski, W., 1954, Micropaleontological Stratigraphy of the Upper Malm in Central Poland (with English summary): Warsaw Institute of Geology, Prace, v. 12, p. 1-206.

Brun, L., and Rey, J., 1975, Nouvelles Observations sur le Genre *Everticyclammina* Redmond, 1964 et sa Repartition Paléoécologique dans le Crétacé Inférieur du bassin de l'Estremadura (Portugal): Revista Espanola de Micropaléontologiá, Special Number 1975, p. 15-36.

Bujak, J.P., and Williams, G.L., 1977, Jurassic Palynostratigraphy of Offshore Eastern Canada: *in* Swain, F.M., ed., Stratigraphic Micropaleontology of Atlantic Basin and Borderlands: Elsevier, Amsterdam, p. 321-339.

______, 1978 Cretaceous Palynostratigraphy of Offshore Southeastern Canada: Geological Survey of Canada Bulletin 297, p. 1-19.

Busnardo, R., Thieuloy, J.-P., Moullade, M., *et al*, 1979, Hypostratotype mésogéen de l'étage Valanginien (Sud-Est de la France): Les stratotypes français, Editions du Comité Nationale de la Recherche Scientifique, v. 6, 143 p.

Christensen, O.B., and Kilenyi, T.J., 1970, Ostracod Biostratigraphy of the Kimmeridgian in Northern and Western Europe: Geological Survey of Denmark; Ser. 2, No. 95, p. 7-63.

Dieni, I., and Massari, F., 1966, I Foraminiferi del Valanginiano superiore di Orosei (Sardegna): Palaeontographia Italica, v. 66, p. 76-186.

Eliuk, L.S., 1978, The Abenaki Formation, Nova Scotia Shelf, Canada - A Depositional and Diagenetic Model for a Mesozoic Carbonate Platform: Bulletin of Canadian Petroleum Geology, v. 26, p. 424-514.

Emery, K.O., and Uchupi, E., 1965, Structure of Georges Bank: Marine Geology, v. 3, p. 349-358.

______, 1972, Western North Atlantic Ocean: Topography, Rocks, Structure, Water, Life, and Sediments: American Association of Petroleum Geologists, Memoir 17, 532 p.

Espitalié, J., and Sigal, J., 1963, Contribution a l'étude des Foraminifères (Micropaléontologie-Microstratigraphie) du Jurassique Supérieur et du Néocomien du bassin de Majunga (Madagascar): Annales Géologiques de Madagascar, v. 32, p. 1-100.

Given, M.M., 1977, Mesozoic and Early Cenozoic Geology of Offshore Nova Scotia: Bulletin of Canadian Petroleum Geology, v. 25, p. 63-91.

Gradstein, F.M., 1976, Biostratigraphy and Biogeography of Jurassic Grand Banks Foraminifera: Maritime Sediments, Special Publication 1, Pt. B, p. 557-583.

______, 1978, Jurassic Grand Banks Foraminifera: Journal of Foraminiferal Research, v. 8, no. 2, p. 97-109.

Gradstein, F.M., Williams, G.L., Jenkins, W.A.M., and Ascoli, P., 1975, Mesozoic and Cenozoic Stratigraphy of the Atlantic Continental Margin, Eastern Canada: *in* Yorath, C.J., Parker, E.R., and Glass, D.J., eds., Canada's Continental Margin and Offshore Petroleum Exploration: Canadian Society of Petroleum Geologists, Memoir 4, p. 103-133.

Groiss, J.T., 1967, Mikropaläontologische Untersuchung der Solnhofener Schichten im Gebiet um Eichstätt (Südliche Frankenalb): Erlanger Geologische Abhandlungen, v. 66, p. 75-95.

Heller, P.L., Wentworth, C.M., and Poag, C.W., 1982, Episodic post-rift subsidence of the U.S. Atlantic Continental Margin: Geological Society of America Bulletin, v. 93, p. 379-390.

Hottinger, L., 1967, Foraminifères imperforés du Mésozoique marocain: Notes Mémoires Service Géologique Maroc, v. 209, p. 1-168.

Jansa, L.F., 1981, Mesozoic Carbonate platforms and Banks of the Eastern North American Margin: *in* Cita, M.B., and Ryan, W.B.F., eds., Carbonate Platforms of the Passive-Type Continental Margins, Present and Past: Marine Geology, v. 44, p. 97-117.

Jansa, L.F., and Wade, J.A., 1975, Geology of the Continental Margin off Nova Scotia and Newfoundland. Geological Survey of Canada Paper 74-30, p. 51-105.

Jansa, L.F., and Wiedmann, J., 1982, Mesozoic-Cenozoic Development of the Eastern North American and Northwest African Continental Margins: A Comparison; *in* von Rad, U., Hinz, K., Sarnthein, M., and Seibold, E., eds., Geology of the Northwest African Continental Margin: Springer-Verlag, Berlin, p. 215-269.

Jansa, L.F., Remane, J., and Ascoli, P., 1980, Calpionellid and Foraminiferal-Ostracod Biostratigraphy of the Jurassic-Cretaceous Boundary, Offshore Eastern Canada: Rivista Italiana Paleontologia e Stratigrafia, v. 86, p. 67-126.

Jenkins, W.A.M., Ascoli, P., Gradstein, F.M., Jansa, L.F., and Williams, G.L., 1974, Stratigraphy of the Amoco IOE A-1 Puffin B-90 Well, Grand Banks of Newfoundland: Geological Survey of Canada Paper 74-61, p. 1-12.

Kilenyi, T.J., 1969, The Ostracods of the Dorset Kimmeridge Clay: Palaeontology, v. 12, no. 1, p. 112-160.

______, 1978, The Jurassic Part III Callovian-Portlandian: *in* Bate, R., and Robinson, E., eds., A Stratigraphical Index of British Ostracoda: Geological Journal Special Issue No. 8, p. 259-298.

Kilenyi, T.J., and Neale, J.W., 1978, The Purbeck/Wealden: *in* Bate, R., and Robinson, E., eds., A Stratigraphical Index of British Ostracoda: Geological Journal Special Issue No. 8, p. 299-324.

Klingler, W., 1955, Mikrofaunistische und Stratigrafisch-fazielle Untersuchungen im Kimmeridge und Portland des Weser-Aller-Gebietes: Geologisches Jahrbuch, v. 70, p. 167-246.

Klingler, W., Malz, H., and Martin, G.P.R., 1962, Malm NW-Deutschlands: *in* Leitfossilien der Mikropaläontologie: Gebrüder Borntraeger, Berlin, p. 159-190.

Klitgord, K.D., and Behrendt, J.C., 1979, Basin Structure of the U.S. Atlantic Margin: *in* Watkins, J.S., Montadert, L., and Dickerson, P.W., eds., Geological and Geophysical Investigations of Continental Margins: American Association of Petroleum Geologists, Memoir 29, p. 85-112.

Klitgord, K.D., and Grow, J.A., 1980, Jurassic Seismic Stratigraphy and Basement Structure of Western North Atlantic Magnetic Quiet Zone: American Association of Petroleum Geologists Bulletin, v. 64, p. 1658-1680.

Klitgord, K.D., Schlee, J.S., and Hinz, K., 1982, Basement Structure, Sedimentation, and Tectonic History of the Georges Bank Basin: *in* Scholle, P.A., and Wenkam, C.E., eds., Geological Studies of the COST G-1 and G-2 Wells, United States North Atlantic Outer Continental Shelf: United States Geological Survey Circular 861, p. 160-186.

Le Hégarat, G., and Remane, J., 1968, Tithonique supérieur et Berriasien de la bordure cévenole: Corrélation des Ammonities et des Calpionelles: Géobios., v. 1, p. 7-70.

Lloyd, A.J., 1962, Polymorphinid, Miliolid, and Rotaliform Foraminifera from the Type Kimmeridgian: Micropaleontology, v. 8, p. 369-383.

Loeblich, A.R., Jr., and Tappan, H., 1950, North American Jurassic Foraminifera. I. The Type Redwater Shale (Oxfordian) of South Dakota: Journal of Paleontology, v. 24, no. 1, p. 5-19.

Lutze, G.F., 1960, Zur Stratigraphie und Paläontologie des Callovien und Oxfordien in Nordwest-Deutschland: Geologisches Jahrbuch, v. 77, p. 391-532.

Magné, J., 1965, Le Stratotype du Berriasien; III, La microfaune: Mémoire Bureau Recherche Géologiques Minières, v. 34, p. 17-24.

Manspeizer, W., 1982, Early Mesozoic Basins of the Central Atlantic Passive Margins: *in* Bally, A.W., ed., Geology of Passive Continental Margins, History, Structure and Sedimentologic Record:

American Association Petroleum Geologists Education Course Note Series No. 19, art. 4, p. 1-60.

Manspeizer, W., Puffer, J.H., and Cousminer, H.L., 1978, Separation of Morocco and Eastern North America – A Triassic-Liassic Stratigraphic Record: Geological Society of America Bulletin, v. 89, p. 901-920.

Mattick, R.E., Foote, R.Q., Weaver, N.L., and Grim, M.S., 1974, Structural Framework of the U.S. Atlantic Outer Continental Shelf North of Cape Hatteras: American Association of Petroleum Geologists Bulletin, v. 58, p. 1179-1190.

Maync, W., 1960, Biocaractères et analyse morphométrique des espèces jurassiques du genre *Pseudocyclammina* (Foraminifère): II. *Pseudocyclammina jaccardi* (Schrodt): Revue Micropaléontologie, v. 3, no. 2, p. 103-118.

McIver, N.L., 1972, Cenozoic and Mesozoic Stratigraphy of the Nova Scotia Shelf: Canadian Journal of Earth Sciences, v. 9, no. 1, p. 54-70.

Moullade, M., 1979, Les Foraminifères du Valanginien hypostratotypique: *in* Busnardo, R., Thieuloy, J.P. and Moullade, M., eds., Hypostratotype Mesogéen de l'Etage Valanginien (Sud-Est de la France): Comité Français de Stratigraphie, Les Stratotypes Français, v. 6, Editions du C.N.R.S., p. 110-126.

Neale, J.W., 1967, Ostracodes from the type Berriasian (Cretaceous) of Berrias (Ardèche, France) and their significance: University of Kansas Department of Geology Special Publication, No. 2, p. 539-569.

———, 1978, The Cretaceous: *in* Bate, R., and Robinson, E., eds., A Stratigraphical Index of British Ostracoda: Geological Journal, Special Issue No. 8, p. 325-384.

Oertli, H.J., 1959, Malm-Ostrakoden aus dem schweizerischen Juragebirge: Mémoires de la Société Helvétique des Sciences Naturelles, v. 83, p. 1-44.

———, 1963, Faunes d'Ostracodes du Mésozoique de France: E.J. Brill, Leiden, 57 p.

Ohm, U., 1967, Zur Kenntnis der Gattungen *Reinholdella, Garantella* und *Epistomina* (Foraminifera): Palaeontographica, v. 127, pt. A, p. 103-188.

Poag, C.W., 1977, Foraminiferal Biostratigraphy: *in* Scholle, P.A., ed., Geological Studies of the COST No. B-2 Well, U.S. Mid-Atlantic Outer Continental Shelf Area: United States Geological Survey Circular 750, p. 35-36.

———, 1979, Stratigraphy and Depositional Environments of Baltimore Canyon Trough: American Association of Petroleum Geologists Bulletin, v. 63, p. 1452-1466.

———, 1980, Foraminiferal Biostratigraphy, Paleoenvironments, and Depositional Cycles in the Baltimore Canyon Trough: *in* Scholle, P.A., ed., Geologic Studies of the COST No. B-3 Well, United States Mid-Atlantic Continental Slope Area: United States Geological Survey Circular 833, p. 44-65.

———, 1982a, Foraminiferal and Seismic Stratigraphy, Paleoenvironments, and Depositional Cycles in the Georges Bank Basin: *in* Scholle, P.A. and Wenkam, C.R., eds., Geological Studies of the COST Nos. G-1 and G-2 Wells, United States North Atlantic Outer Continental Shelf: United States Geological Survey Circular 861, p. 43-92.

———, 1982b, Stratigraphic Reference Section for Georges Bank Basin – Depositional Model for New England Passive Margin: American Association of Petroleum Geologists, Bulletin, v. 66, p. 1021-1041.

Remane, J., 1964, Untersuchungen zur Systematik und Stratigraphie der Calpionellen in den Jura-Kreide Grenzschichten des Vocontischen Troges: Palaeontographica, v. 123, A, p. 1-57.

———, 1980, Calpionellids: *in* Haq, B.U., and Boersma, A., eds., Introduction to Marine Micropaleontology: Elsevier, New York, p. 161-170.

Ryan, W.B.F., Cita, M.B., Miller, E.L., Hanselman, D., Nesteroff, W.D., Hecker, B., and Nibbelink, M., 1979, Bedrock Geology in New England Submarine Canyons: Oceanologica Acta, v. 1, p. 233-254.

Schlee, J., Behrendt, J.C., Grow, J.A., Robb, J.M., Mattick, R.W., Taylor, P.T., and Lawson, B.J., 1976, Regional Geologic Framework off Northeastern United States: American Association of Petroleum Geologists Bulletin, v. 60, p. 926-951.

Schlee, J.S., Dillon, W.P., and Grow, J.A., 1979, Structure of the Continental Slope off the Eastern United States: Society of Economic Paleontologists and Mineralogists, Special Publication No. 27, p. 95-117.

Sheridan, R.E., 1974, Atlantic Continental Margin of North America: *in* Burke, C.A., and Drake, C.L., eds., Geology of Continental Margins: Springer-Verlag, New York, p. 391-407.

———, 1976, Sedimentary Basins of the Atlantic Margin of North America: Tectonophysics, v. 36, p. 113-132.

Steckler, M.S., and Watts, A.B., 1978, Subsidence of the Atlantic-Type Continental Margin off New York: Earth and Planetary Science Letters, v. 41, p. 1-13.

Steinkraus, W.E., 1979, Biostratigraphy: *in* Amato, R.V., and Simonis, E.K., eds., Geological and Operational Summary, COST No. B-3 Well, Baltimore Canyon Trough Area, Mid-Atlantic OCS: United States Geological Survey Open-File Report 79-1159, p. 21-31.

———, 1980, Biostratigraphy: *in* Amato, R.V., and Bebout, J.W., eds., Geologic and Operational Summary, COST No. G-1 Well, Georges Bank Area, North Atlantic OCS: United States Geological Survey Open-File Report 80-268, p. 39-52.

Swain, F.M., 1981, Mesozoic Ostracoda in Several COST Atlantic Wells: *in* Neale, J.W., and Brasier, M.A., eds., Microfossils from Recent and Fossil Shelf Seas: Ellis Horwood, Ltd., Chichester, p. 90-112.

Valentine, P.C., 1977, Nannofossil Biostratigraphy: *in* Scholle, P.A., ed., Geologic Studies on the COST No. B-2 Well, U.S. Mid-Atlantic Outer Continental Shelf Area: United States Geological Survey Circular 750, p. 35-43.

———, 1980, Calcareous Nannofossil Biostratigraphy, Paleoenvironments, and Post-Jurassic Continental Margin Development; *in* Scholle, P.A., ed., Geolgocial Studies of the COST No. B-3 Well, United States Mid-Atlantic Continental Slope Area: United States Geological Survey Circular 833, pp. 67-83.

———, 1982, Calcareous Nannofossil Biostratigraphy and Paleoenvironment of Two Deep Stratigraphic Wells in the Georges Bank Basin: *in* Scholle, P.A. and Wenkam, C.R. eds., Geological Studies of the COST No. G-1 and G-2 Wells, United States North Atlantic Outer Continental Shelf: United States Geological Survey Circular 861, p. 34-42.

van Hinte, J.E., 1976, A Cretaceous Time Scale: American Association of Petroleum Geologists Bulletin, v. 60, no. 4, p. 498-516.

Watts, A.B., 1982, Tectonic Subsidence, Flexure, and Global Changes of Sea Level: Nature, v. 297, p. 469-474.

Watts, A.B., and Steckler, M.S., 1979, Subsidence and Eustasy at the Continental Margin of Eastern North America: *in* Talwani, M., Hay, W.W., and Ryan, W.B.F., eds., Deep Drilling Results in the Atlantic Ocean: Continental Margins and Paleoenvironments: American Geophysical Union, Maurice Ewing Symposium Series 3, p. 273-310.

Jurassic-Cretaceous Biochronology and Paleogeography of North America,
edited by G.E.G. Westermann,
Geological Association of Canada Special Paper 27, 1984

MESOZOIC BIOTA AND DEPOSITIONAL SYSTEMS OF THE GULF OF MEXICO-CARIBBEAN REGION

R.W. Scott
Research Center, Amoco Production Company, 4502 East 41st Street, Tulsa, Oklahoma 74102

ABSTRACT

Rifting and volcanism in the Gulf of Mexico region began during the Triassic. The Late Triassic marine fauna in northwestern Mexico had Circum-Pacific affinities. Marine flooding of local marginal areas in eastern Mexico began in the Early Jurassic when Tethyan faunas migrated from either the Pacific or the Atlantic ocean. Middle Jurassic faunas are found in Cuba, eastern Mexico and southwestern Mexico. Widespread flooding of the Gulf deposited evaporites during the Middle Jurassic. The earliest widespread Tethyan marine biota to encircle the central Gulf of Mexico Basin is of Oxfordian age, which has Mediterranean affinities.

Five major tectonic elements controlled stratigraphic accumulations: a convergent island-arc system on the Pacific coast, a backarc Gulf of Mexico Basin, the Florida-Bahama and Yucatan platforms, the Caribbean oceanic volcanic arc system, and oceanic basins. Four depositional systems resulted: rifted basins, carbonate platforms, terrigenous clastic shelves, and trough-volcanic arc systems. The rifted basin system consists of thick continental clastics and volcanics in narrow, linear basins. The carbonate platform consists of widespread, thick limestone, dolomite and evaporites on the continental margin. The terrigenous shelf system consists of fluvial to shelf sandstone, siltstone and shale on continental margins and in interior basins. The trough-volcanic arc system consists of submarine volcaniclastic rocks, lime mudstone and chert locally intruded by ophiolite. Plate reconstructions of this region should incorporate these depositional systems.

RÉSUMÉ

Failles et volcanisme dans la région du goff de Mexique ont commencé pendant le Triassique. La faune marine du Triassique supérieur du Mexique du nord-ouest ont affinites avec la faune du circumpacifique. L'inondation marine des baies locales en Mexique orientale ont commencé du Jurassique inferieur quand les faunes Tethiques ont émigré soit de l'ocean Pacifique où de l'ocean Atlantique. Les faunes du Jurassique moyen se trouvent Cuba, au Mexique orientale et au Mexique du sud-ouest. Une grande inondation du Golf a déposé les évapores pendant le Jurassique moyen. La faune premiere marine Tethique qui encercle le bassin du Golf du Mexique est de l'âge Oxfordien, qui a les affinites avec la faune du Méditerrané.

Cinq éléments tectonique majeur ont dirigé les accumulations stratigraphiques: le systeme île-arc convergente sur la côté Pacifique, le bassin arc-derriere Golf du Mexique, les plate-formes Floride-Bahama-Yucatan, le systeme arc-volcanique oceanique Caraïbe, et les basins oceaniques. Quatre systeme de déposition en resulterent. Les bassins faillé, les plate-formes carbonates, plateau clastique, et les systemes fossés-arc-volcanique. Le systeme du bassins faillé consiste du clastiques continentales et volcaniques epaisses en bassins etroits et linéaires. Le plate-forme carbonate consiste de la calcaire, de la dolomite et des les evaporites entendus et epaisses sur le plate-forme continentale. Le systeme du plateau clastique consiste du gres, et schiste fluvial aux plateau de la plate-forme continentale et les bassins interieurs. Le systeme fossé-arc-volcanique consiste des roques des volcaniclastiques sous marines, le chaux de boue et le silex noir, qui est introduite localement par l'ophiolite. Le Reconstruction du plates du cette région te incorporérait ces systeme de déposition.

INTRODUCTION

The Mesozoic history of Central America, the Gulf of Mexico and the Caribbean region is fascinating in its complexity and yet enigmatic in its details. Authoritative encyclopedic syntheses, summaries and reviews have been published by Schuchert (1935), Mülleried (1942) and Woodring (1954). Woodring's caveat must be posted again at a time when the exciting theories of plate tectonics are producing a renaissance in paleogeographic studies of this region. "Paleogeography starts as a concoction of essential ingredients, generally too meager, and

IUGS UNESCO
Project 171:
Circum-Pacific Jurassic

TABLE I

TIMING OF THE "OPENING" OF THE GULF OF MEXICO BASIN. SELECTED AUTHORS AND A SYNTHESIS

Time	Authors	Hallam 1971	Barr 1974	Viniegra 1971	Buffler 1980	Schmidt-Effing 1980	Salvador 1981	Hillebrandt 1981	Synthesis
EARLY CRETACEOUS	Albian Aptian Barremian Hauterivian Valanginian Berriasian								
LATE JURASSIC	Tithonian Kimmeridgian Oxfordian	Oxfordian: Gulf Mex. Open to Atlantic	Gulf Mex. Open Seaway Between Caribbean and So. Am.	Gulf Mex. Shield Subsided			Kimmerid.: Atlantic Connection		Gulf Basin Open E and W
MIDDLE JURASSIC	Callovian Bathonian Bajocian				Late Rift Open To Sea: Evaporites	Central Gulf Open to Sea: Evaporites	Pacific Connection: Evaporites		Gulf Rift Basin Restric. Sea Connec.
EARLY JURASSIC	Aalenian Toarcian Pliensbachian Sinemurian Hettangian	North Atlantic Seaway (Toarcian)				Sinemurian: Huayacocot. Basin		Pliensbach: Tethyan Connections to W.C. Am. and W.S. Am.	Local Rift Basins Open to Sea and Cosmop. Biota
LATE TRIASSIC									Intra-Continent Rifting

winds up as a heady essence distilled through the imagination of the perpetrator. It therefore is an art ... too frequently abused. ..." (Woodring, 1954, p. 719).

Much progress has been made since 1954, particularly as a result of geologic mapping, geophysical studies, drilling in the oceanic basins, radiometric dating, paleontologic and sedimentologic studies. We know now that oceanic crust underlies the deep Gulf of Mexico, the Yucatan Basin and the Colombian Basin (Martin and Case, 1975; Arden, 1975; Buffler *et al.*, 1981), thereby virtually disproving the hypothesis that the Gulf of Mexico crust was once continental crust that has been converted to oceanic crust and has subsided. In its place is the prediction by plate tectonics that Yucatan once occupied the Gulf (summary by Pindell and Dewey, 1982). Alternatively, the Gulf may have been an ocean basin throughout Phanerozoic time as Schuchert (1935) thought, but recent DSDP (Deep Sea Drilling Project) holes have not penetrated Paleozoic sedimentary rocks.

The paleontologic data from this region provide important controls on timing of the initial opening of the Atlantic Sea and the origin of the Gulf and Caribbean basins. The biogeographic history was directly dependent upon the establishment of migration pathways between the European Tethys and the Gulf of Mexico. Hence, the stratigraphic sequence is a direct test of tectonic hypotheses for the origin of these basins.

A comprehensive and thorough review is not practical in this space. Furthermore, such a task should be accomplished by a team of the many active experts in the region. R.W. Imlay gathered a large part of the basic paleontologic and stratigraphic data and his stratigraphic syntheses (Imlay, 1943, 1944a, 1944b, 1952, 1980) have suffered few major changes.

Much of the new paleontologic data are of local assemblages or of a succession of a single fossil group in a few sections. Few syntheses have been attempted because so many unexplored stratigraphic exposures yet remain. Consequently, the precision of correlation among the disjunct outcrops is still rather low. However, the rampant speculation on the tectonic history might be tamed by a review of the most important stratigraphic events represented by biogeographic, evolutionary, depositional, and environmental data.

MESOZOIC OPENING OF THE GULF OF MEXICO

The margins of the Gulf of Mexico region were open to marine waters as early as the Early Jurassic, which is indicated by Sinemurian ammonites in the Huayacocotla Basin of eastern Mexico (Erben, 1957b; Schmidt-Effing, 1980; von Hillebrandt, 1981). However, the earliest widespread Tethyan marine biota to encircle the central Gulf of Mexico Basin immigrated during the Oxfordian, earliest Late Jurassic. These organisms are found in the widespread Smackover, Zuloaga, Chinemeca and Jagua formations. Eastward connections with the Atlantic dur-

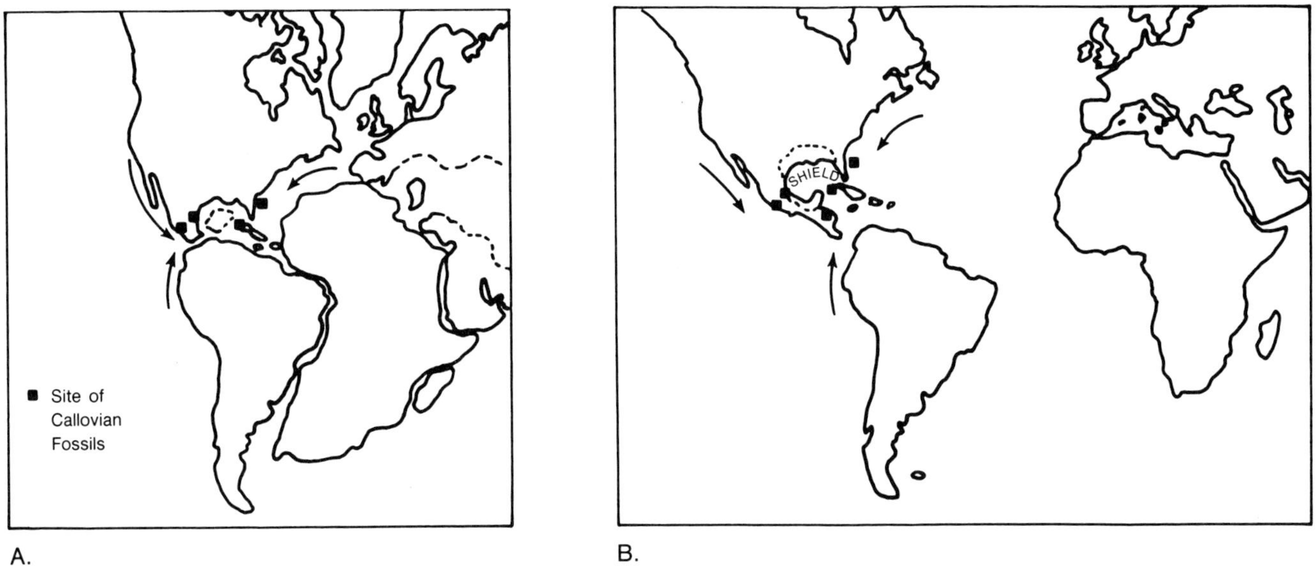

Figure 1. Middle Jurassic (Callovian) migration paths into the Gulf of Mexico. A. Plate reconstruction (Smith and Briden, 1977); B. Fixed continents (Meyerhoff and Meyerhoff, 1974; Viniegra, 1971). Mercator projection.

ing the Oxfordian are indicated by ammonite occurrences in Cuba (Meyerhoff, 1964; Khudoley and Meyerhoff, 1971).

Opinions and hypotheses regarding the timing of the first opening of the Gulf of Mexico are diverse (Table I), and only a few key ideas will be reviewed. The "opening" of the Gulf must be carefully defined because at least four meanings exist: 1) first stage of rifting and deposition, 2) first connection with sea water to form evaporites in the central Gulf, 3) first connection with Circum-Pacific biotas, and 4) first biota from the Atlantic and the Mediterranean. This divergent useage is one reason for the diversity in opinions on the timing. Another reason is the uncertainty and imprecision in dating the Louann, Punta Alegre and Salina salt formations.

Early Jurassic rifted basins were of local extent and probably were open southwards to the immigration of Tethyan biota. Sinemurian ammonites in the Huayacocotla Basin (Erben, 1957b; Schmidt-Effing, 1980) are part of a cosmopolitan fauna (Imlay, 1965), which could have migrated from either the Pacific or the Atlantic, through a passage between North and South America (Fig. 1), not through the eastern Gulf. Mediterranean late Early Jurassic (Pliensbachian) ammonites, bivalves and sponges on the western margins of North and South America also could have migrated through an eastward Tethyan connection between these two continents (Hillebrandt, 1981) or could be part of a warm-water Pacific fauna. The meager stratigraphic evidence on the northern margin of South America does not clarify this problem. However, an eastward opening of the Gulf is not likely because of the absence of marine Lower Jurassic rocks in other parts of the Gulf of Mexico and just east of the Florida-Bahama Platform (Sheridan *et al.*, 1981). Furthermore, east of the Blake Plateau marine rocks no older than Callovian have been drilled at the western edge of the Atlantic (Sheridan *et al.*, 1982). This is consistent with an open southern North Atlantic by Callovian time as suggested by Smith and Briden (1977). The northernmost Atlantic basin was first flooded with marine waters from the Tethys during the Early Jurassic Pliensbachian and Toarcian epochs (Hallam, 1971; Sheridan *et al.*, 1982), and platform carbonates first accumulated along the northeastern margin of North America during the Pliensbachian-Bathonian (Jansa, 1981). The Mediterranean biota could then have moved southward as well and between North and South America. In fact, the fixist model of continental positions would predict an Early Jurassic connection between the north Atlantic and the Pacific through an east-west straits between North and South America (Khudoley, 1974) which would have fostered the development of cosmopolitan ammonites (Imlay, 1965), belemnoids (Stevens, 1963), and corals (Beauvais, 1979). On the other hand, a mobilist model can readily accommodate the cosmopolitan biogeographic data equally well, if the Atlantic were connected with the Pacific by means of a strait between North and South America as early as Pliensbachian time (Westermann, 1981; von Hillebrandt, 1981; Schmidt-Effing, 1980).

By Middle Jurassic the Gulf of Mexico was connected with either the Atlantic or Pacific Oceans or both, which is indicated by the pre-Oxfordian Gulf evaporite complex. These evaporites including salt were deposited in rifted basins along the northern margin of the Gulf in Cuba and the U.S., and along the southern margin in Chiapas and Yucatan. Geophysical evidence (Martin and Case, 1975) does not confirm the lateral continuity of the salt among these areas; consequently, Salvador (1979) and Humphris (1979) suggested that the Yucatan platform occupied the central Gulf to form a narrow rifted basin between it and the southern United States. Viniegra (1971, 1981) and others suggested instead, that sialic crust subsided and was altered into transitional oceanic crust.

MAJOR TECTONIC ELEMENTS

Five major types of tectonic features influenced Mesozoic sedimentation in the Gulf of Mexico and Caribbean region (Fig. 2). A convergent island-arc system formed along the Pacific margin of North America beginning in the Triassic.

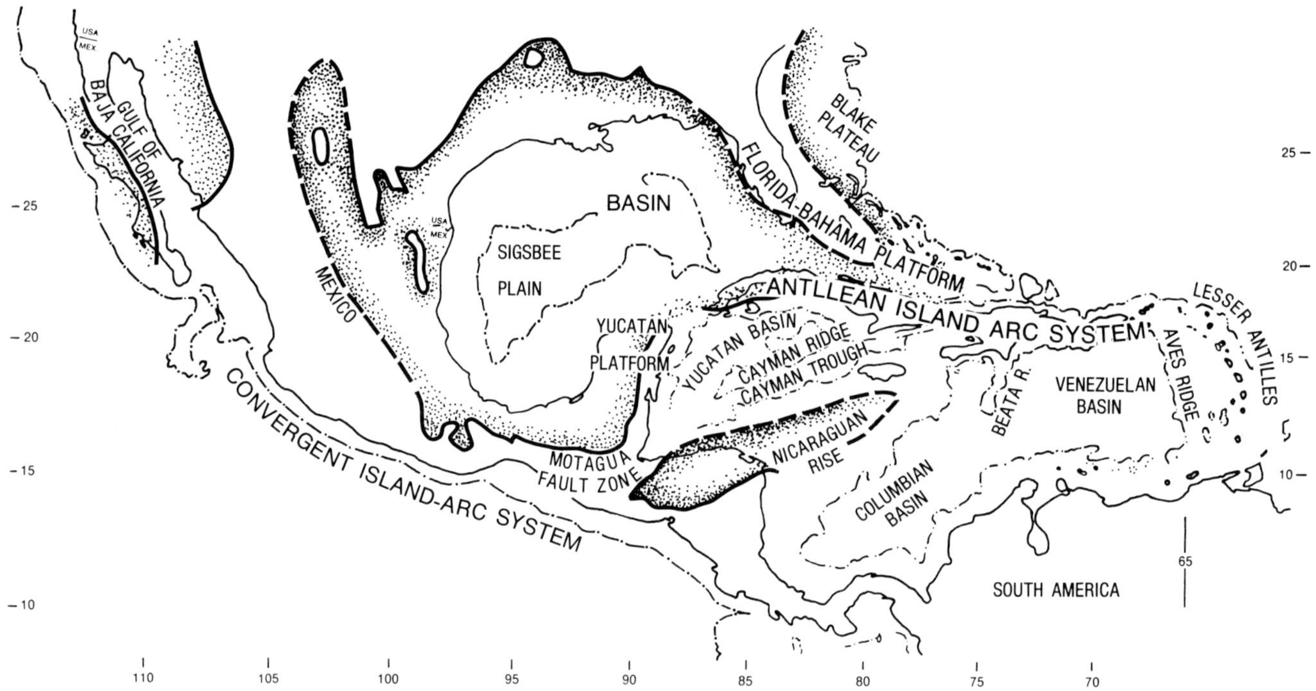

Figure 2. Major features, tectonic elements and distribution of Jurassic sedimentary rocks (from Cook and Bally, 1975) in the Central American-Caribbean region. Dotted-dashed line is the 3000m isobath.

During the Jurassic trench sedimentation extended southwards along the Mexican coast. By Late Cretaceous time volcanic rocks and intrusives accumulated as far south as Panama, the Nicoya complex. An alternate hypothesis is that this complex originated as an eastern Pacific arc during the Late Cretaceous and was accreted to the Chortis block by Paleogene time (Pindell and Dewey, 1982; Campa and Coney, 1983). During the Late Cretaceous granitic plutons were emplaced within this arc system in Mexico (Mullan, 1978; Gastil *et al.*, 1978).

The Gulf of Mexico Basin lay behind this trough-island arc system. The shelves had gentle slopes characteristic of trailing plate margins. Depositional systems of a terrigenous shelf or a carbonate ramp developed on these margins. Partly enclosing this basin were two positive platforms. The Florida-Bahama platform on the northeast overlies continental crust in part (Meyerhoff and Hatten, 1974). At its southeastern termination, the configuration of blocks separated by deep canyons suggests partial breakup of the platform by block faulting. However, nearly continuous carbonate-evaporite deposition since at least the Late Jurassic has masked the original structure. On the southeast side of the Gulf is the Yucatan Platform that also overlies Paleozoic continental basement (Buffler *et al.*, 1981). This platform is bordered on the east by a fault system at least as old as Miocene, and probably of Mesozoic age (Uchupi, 1975). The southern margin is the complex Montagua-Polochic fault zone. The structural relation between Yucatan and the Honduran-Nicaraguan Rise is more complex than the fault system implies, and is not fully known. The Honduran-Nicaraguan Rise is mostly underlain by continental basement on the west and oceanic crust on the east (Arden, 1975); it acted as a platform during the Mesozoic.

The major Antillean island arc system forms the northern margin of the Caribbean area. It extends from western Cuba eastward to the Virgin Islands and then turns southward as the Lesser Antillean chain. That part along the Greater Antilles has been the site of Mesozoic volcanism and trench sedimentation. It probably was a Jurassic-Cretaceous subduction zone, which became younger eastwards.

Oceanic basins occupy the Yucatan Basin, the Columbian Basin and the Venezuelan Basin. These are underlain by oceanic crust mainly of Cretaceous age, that is covered by Late Cretaceous and Cenozoic pelagic sediments (Uchupi, 1975).

BIOGEOGRAPHY OF GULF OF MEXICO – CARIBBEAN

Distribution and migration pathways of various organisms indicate that a Circum-Pacific Tethyan biota was established along the northwestern side of Mexico as early as Late Triassic (Fig. 3). By the late Middle Jurassic marine organisms migrated to the eastern parts of the Bahama Platform and to Cuba, and by the beginning of the Late Jurassic a marine biota was widespread.

Plants and palynomorphs provide background data for understanding the biogeographic relations between the Mexican Gulf-Caribbean area and the rest of the world. Numerous megaplant fossils have been found in Triassic and Early-Middle Jurassic rocks. Late Triassic plants of North American affinities are known from the La Boca Formation in eastern Mexico (R.W. Brown *in* Mixon *et al.*, 1959, p. 769) and in northwestern Sonora (de Cserna, 1961) (Fig. 3). Late Triassic or Early Jurassic plants from the El Plan Formation in Honduras (Mills *et al.*, 1967) represent the same assemblage. Plants that are widespread in eastern and western North American Triassic basins occur in the Eagle Mills red beds in the northern Gulf Coast (Scott *et al.*,

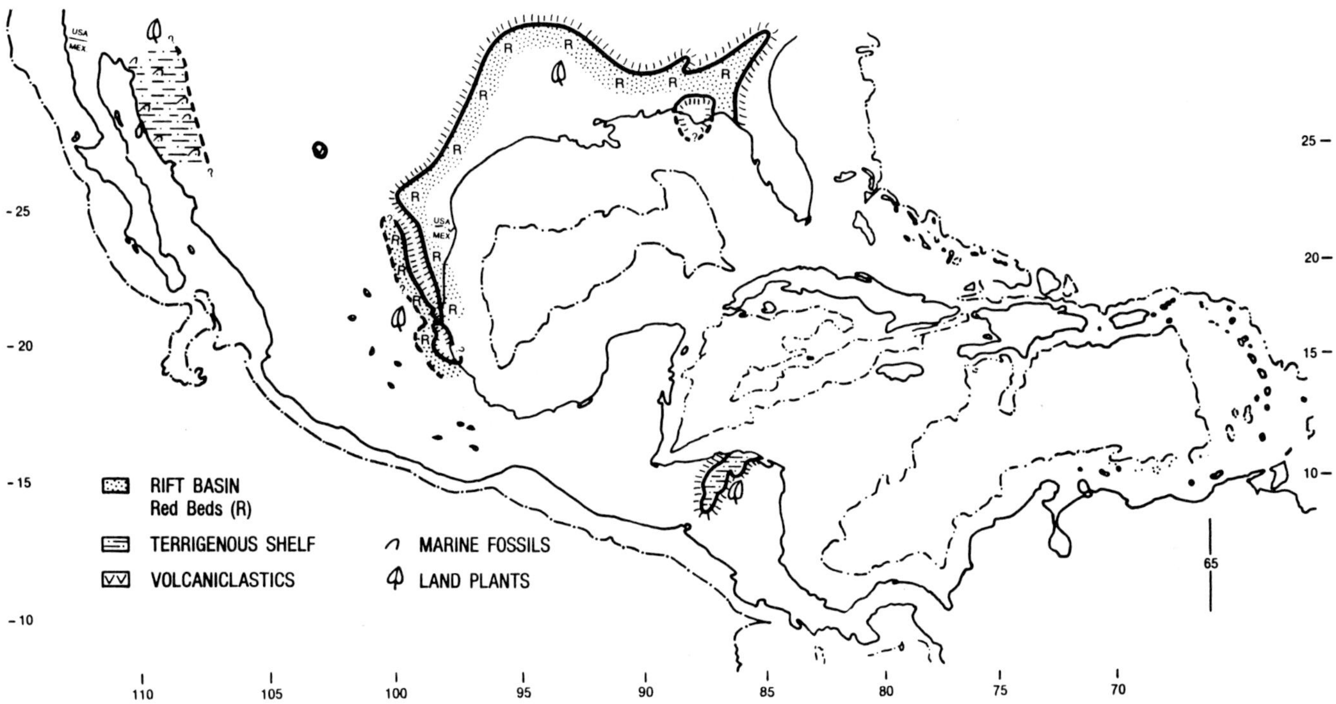

Figure 3. Distribution of Triassic sedimentary and volcaniclastic rocks (in part from Mixon *et al.*, 1959; de Cserna, 1961; Cook and Bally, 1975).

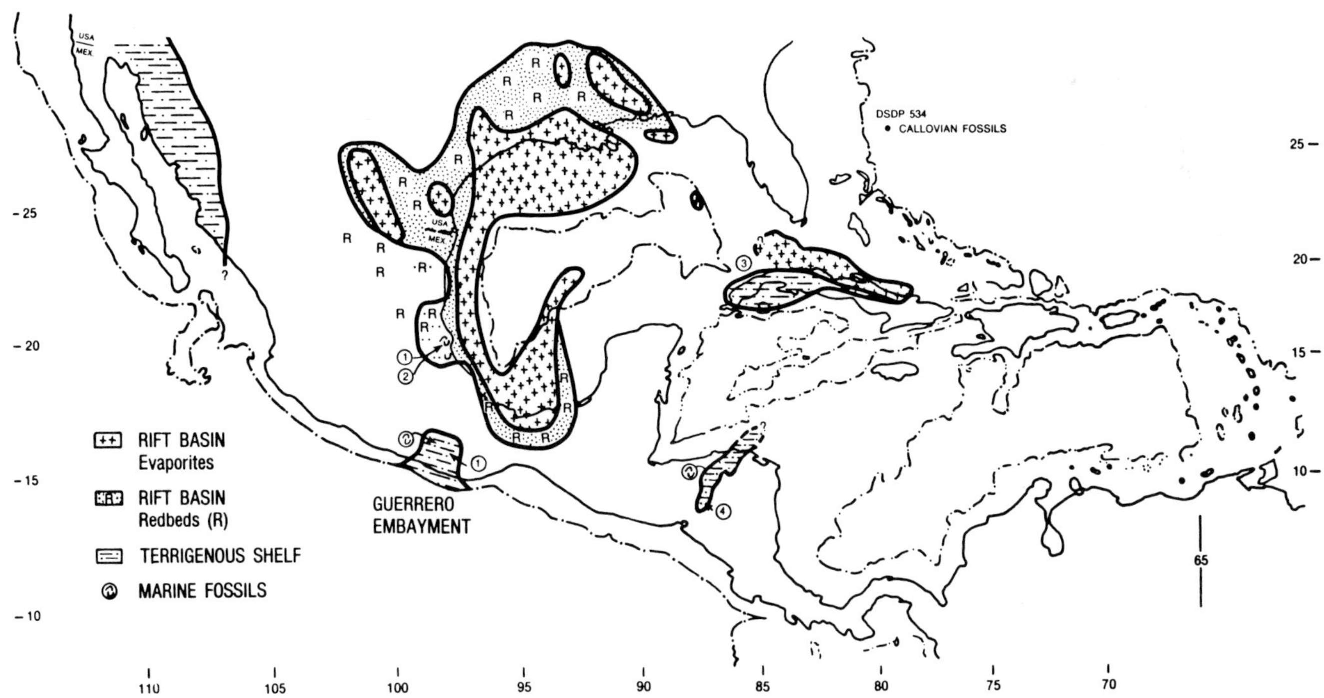

Figure 4. Distribution of Middle Jurassic sedimentary rocks. 1) Erben, 1957a; Humphries, 1979; 2) Imlay, 1952; 3) Meyerhoff, 1964; Meyerhoff and Hatten, 1968; 4) Mills *et al.*, 1967; Mixon *et al.*, 1959; Murray, 1961.

1961). Early Jurassic plants also occur in eastern and western Mexico, which have affinities with plants in England and Germany (Erben, 1957a). Younger Jurassic subtropical and temperate plants are widely distributed in the Middle Jurassic paralic facies in the Guerrero Basin (Person and Delevoryas, 1982) (Fig. 4) and in the Upper Jurassic Todos Santos Formation in Guatemala (Burkart *et al.*, 1973). Palynomorphs from the Louann Salt (Jux, 1961) are similar to Middle Jurassic North African assemblages (D.N. Beju, pers. commun., 1982) and some Upper Jurassic genera from the caprock on Challenger Knoll have nearly worldwide distribution (Kirkland and Gerhard, 1971).

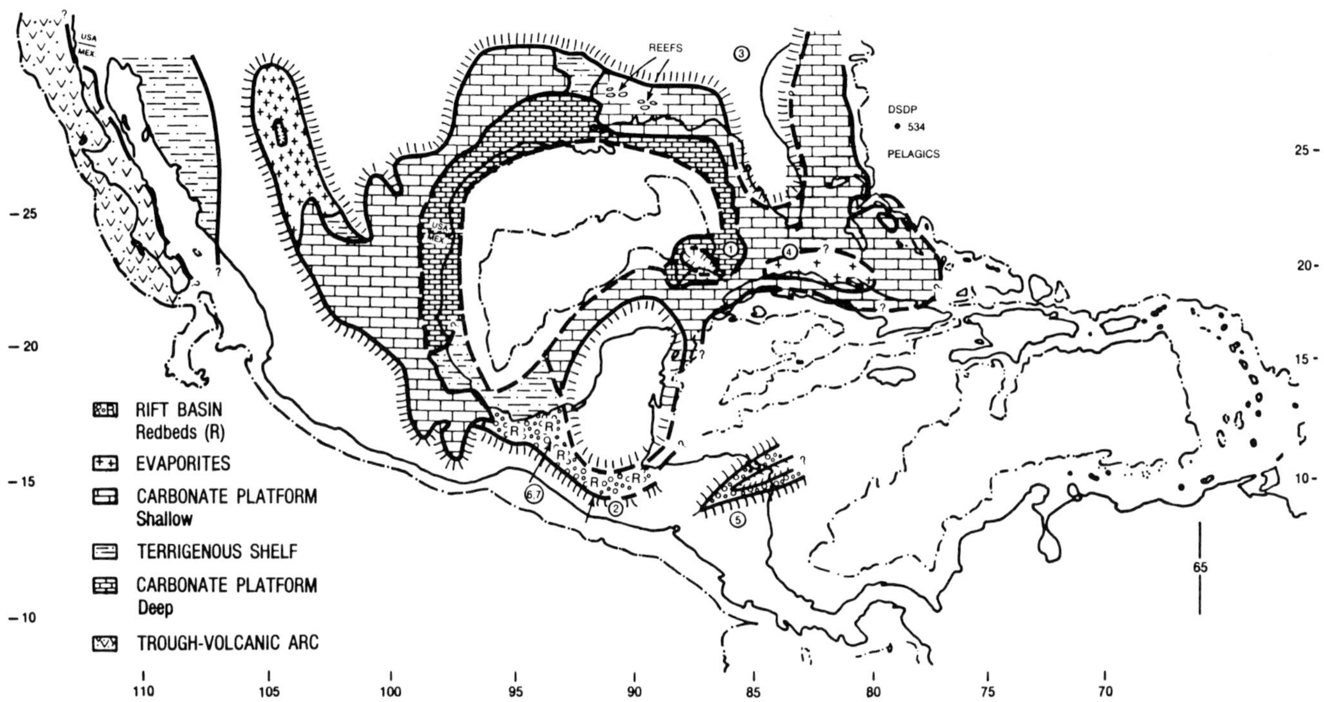

Figure 5. Distribution of Upper Jurrassic (Oxfordian) sedimentary rocks. 1) Buffler and Phair, 1982; 2) Burkhart *et al.*, 1973; 3) Baria *et al.*, 1982; 1982; 4) Meyerhoff and Hatten, 1974; 5) Mills and Hugh, 1974; 6) Richards 1963; 7) Viniegra, 1971.

Distribution of marine megafossils places tighter constraints on biogeographic and tectonic hypotheses regarding the origin of the Gulf of Mexico. Late Triassic marine bivalves and ammonoids in northwestern Sonora indicate connections with the Circum-Pacific eastern Tethys (Fig. 3). A majority of the taxa are similar to those found in the Tethyan Far East, Asia, Mediterranean and Europe, although some of the taxa are endemic to the west coast of North America (de Cserna, 1961).

Lower Jurassic ammonites in Mexico indicate that the eastern Huayacocotla Basin had biogeographic connections with the Tethys during the Sinemurian and early Pliensbachian ages (Erben, 1957a; Imlay, 1980; Schmidt-Effing, 1980). A graben-like basin is suggested by the more than 1750 m of coarse- to fine-grained clastics in the state of Hidalgo that indicate two transgressive-regressive cycles (Schmidt-Effing, 1980, Fig. 3; von Hillebrandt, 1981). However, the migration route of these ammonites is not clear because during the Early Jurassic cosmopolitan faunas were developed (Imlay, 1965; Khudoley, 1974). Direct connections with the Pacific through the Guerrero Basin in southwestern Mexico would seem unlikely in view of the eastward paralic facies distribution pattern there (Erben, 1957a) and the absence of intermediate outcrops or subcrops (Schmidt-Effing, 1980). However, a westward connection is supported by Salvador (1981) and Imlay (1980), which is the only possible direction if South America were joined with North America. The alternative is a strait between North and South America (Taylor *et al.*, this volume). The Huayacocotla Basin would open southward into this east-west passage between North and South America, which requires either a loose fit or early separation of South America.

Late Middle Jurassic fossils are known only from four places within the Gulf-Caribbean region (Fig. 4) and in the bottom of the DSDP site 534 in the Blake-Bahama Basin. At the latter site Callovian pelagic radiolarian claystone and peloid limestone overlie basalt (Sheridan *et al.*, 1982). In western Cuba a 1500-3000 m thick sequence of terrigenous clastic sediments accumulated during Middle Jurassic time. This unit, the San Cayetano Formation *s.s.* of Sieglie, contains Bajocian-Callovian bivalves according to Imlay (*in* Meyerhoff, 1964) and Oxfordian ammonites near the top (Imlay, 1980, p. 39). The dark colored shale, sandstone, graywacke, and phyllite are similar to the El Plan Formation in Honduras, which consists of dark gray shale, siltstone and sandstone (Finch, 1981) and also yields Jurassic bivalves (Mills *et al.*, 1967). Trigonids and palynomorphs from both localities should be compared, in order to narrow the age of the Honduran section, if possible. Source of the San Cayetano clastics, relative to the present position of Cuba, was either to the west and south (Meyerhoff and Hatten, 1974) or to the west and north (Pardo, 1975). Plate reconstructions of western Cuba must account for this source area. Overlying the San Cayetano in Cuba is the Pan (=Azucar) Formation, which is a shallow shelf oolite limestone with mollusks, miliolids and *Conicospirillina* sp. aff. *C. basiliensis*. This assemblage indicates a Callovian-Oxfordian age (Meyerhoff, 1964), although the named benthic foraminifer ranges from Pliensbachian to Kimmeridgian. In eastern Mexico a Callovian assemblage of Tethyan ammonites was recorded from the shallow water Tepexic Limestone (Erben, 1957b; Imlay, 1980; but see Westermann in Taylor *et al.*, this volume). This unit overlies the La Joya Formation, which may be a facies of the Louann-Salina evaporites. The report of a late Bathonian ammonite from a well in the Huayacocotla Basin (Cantu Chapa, 1969) requires documentation in order to understand its biogeographic significance. The fourth locality, the Guerrero Basin in southwestern Mexico, contains an incomplete succession of Middle Jurassic Andean and Tethyan ammonites (Erben, 1957a; Imlay, 1980; Westermann, 1981; and Westermann *et al.*, this volume) as well as bivalves and brachiopods (Ager, 1967), which indicate Tethyan connections. The absence of widespread Middle Jurassic marine fossils in the Gulf of Mexico supports the hypothesis that it was restricted by closure or inhospitable conditions, such as hypersalinity, or both. The Louann Salt probably accumulated during some part of the Middle Jurassic. By the beginning of the Late Jurassic the Gulf was flooded by normal marine waters, and a Tethyan Oxfordian biota became widespread. Direction of flooding could have been either from the Pacific on the west (Salvador, 1979, 1981) or from the Atlantic on the east, or perhaps from both directions. However, no biogeographic differences between eastern and western Tethyan organisms are apparent at this time.

Upper Jurassic marine sequences are well developed and widespread throughout the Gulf of Mexico and its marginal areas (Fig. 5). Oxfordian marine faunas occur around the margins of the Gulf. In Cuba the Jagua Formation of the San Cayetano Group contains well known Upper Oxfordian ammonites (Meyerhoff, 1964; Imlay, 1942, 1980; Myczyinski and Brochwicz-Lewinski, 1981), many of which have European-Mediterranean affinities. Three genera, however, are endemic to Cuba (Khudoley and Meyerhoff, 1971, p. 108), which must be accounted for by plate reconstructions. In the US Gulf Coast the Smackover Formation has yielded numerous Oxfordian ammonites (Imlay, 1941, 1980), bivalves (Imlay, 1940), and corals (Wells, 1942). Most of these genera are widespread in the Mediterranean to Japan area, and some of the species are similar to western Tethyan species, although a few seem to be endemic (see Imlay, 1943, list on p. 1449). Similar ammonite faunas occur in eastern Mexico in the upper part of the Zuloaga and La Gloria formations (Imlay, 1943, 1980). Tithonian ammonites are well developed in eastern and northeastern Mexico (Cantu Chapa, 1969, 1971). Many of the basic paleontologic studies of Upper Jurassic fossils were done by Imlay (1940, 1941, 1942, 1945), Verma and Westermann (1973), Wells (1942), Burckhardt (1930), and Cragin (1905).

TECTONIC RELATIONS OF DEPOSITIONAL SYSTEMS

The depositional system is a unifying concept for process-related assemblages of lithofacies (Fisher *et al.*, 1969). These sets of facies are three-dimensional packages of lithosomes that represent a physiographic environmental complex. Contacts between subjacent and superjacent depositional systems are

TABLE II

SUMMARY OF DEPOSITIONAL SYSTEMS

Features	Rifted Basin	Terrigenous Clastic Shelf	Shallow Carbonate Shelf	Deep Carbonate Basin	Trough Volcanic-Arc
Characteristic Lithology	Conglomerate Sandstone Siltstone Shale Intrusive Volcanic Anhydrite Salt	Sandstone Siltstone Shale Bituminous Rx	Limestone Dolomite Anhydrite Shale Reef Facies	Limestone Chalk Marl Shale	Volcaniclastics Limestone Chert Shale Ophiolites Mafic Extrusives
Rock Color	Red, Green, Light Grey	Dark Grey Black Brown	Tan Light Grey Dark Grey	Dark Grey Black Light Grey	Dark Grey Black Dark Green
Fossils	Land Plants Vertebrates	Land Plants Ammonites Bivalves Gastropods	Foraminifers Marine Algae Mollusks Coelenterates	Nannoplankton Foraminifers Radiolaria Tintinnids	Radiolaria
Tectonic Significance	Tensional Block Faulting Rifting	Rift Basin with Marine connection, Early Uplift of Source Backarc Basin	Subsidence = Sediment Rate or Sea Level Rise, Progradation, Bahaman-Yucatan Platform	Subsidence > Sediment Rate or Rapid Sea Level Rise, Oceanic Basin	Rapid Subsidence Submarine Volcano Subduction, Rifting to Oceanic Crust, Convergent Island and Oceanic Volcanic Arc
Examples Formations	Louann Salt Werner Eagle Mills La Boca	San Cayetano La Casita Rosario Huayacocotla Cotton Valley Mendez	Smackover Zuloaga Coban Sligo Stuart City	Austin Chalk San Felipe Tamaulipas	Eugenia El Zapote

either regional unconformities or virtually synchronous conformities (Mitchum *et al.*, 1974).

Four major depositional systems are recognized in the Gulf and Caribbean region (Table II). The rifted basin, filled with clastics in a faulted depression is a response to tensional tectonics. The carbonate platform-shelf reflects shallow- or deep-water, tectonically stable conditions. The terrigenous shelf indicates a subsiding depocenter near a rejuvenated source area; rejuvenation may have resulted from uplift, sealevel drop, increased runoff, or a combination of factors. The trough and volcanic arc-trench systems developed either at sites of crustal subduction or deep-seated rifting that affected oceanic crust.

The present tectonic framework of these depositional systems was not necessarily their setting at the time of accumulation. Mexico consists of two tectonic domains: the eastern Mesogean Sierra Madre Oriental and the western Cordilleran domains (Cordoba *et al.*, 1980). The Cordilleran complex comprises volcaniclastic sequences with local ophiolite and melange that originated in the Triassic and continued through the Cretaceous. The Mesogean complex consists of Jurassic and Cretaceous carbonates and fine-grained clastics with a Caribbean biota. Each domain had separate structural histories until the Paleocene Laramide event when both domains were subjected to the same tectonic processes (Cordoba *et al.*, 1980). Some ten tectono-stratigraphic terranes having sharp boundaries between areas of unique basement and sedimentary cover characterize the Mesozoic portion of western Mexico (Campa and Coney, 1983). These terranes may have been accreted to North America by the early Tertiary. Some terranes may have moved laterally great distances. Translational displacement of these terranes eliminates the overlap of Mexico and South America in the tight plate fit (Pindell and Dewey, 1982; Campa and Coney, 1983). However, this hypothesis needs to be tested by careful field work. Eastern Mexico consists of four tectono-stratigraphic terranes having Precambrian basement overlain by Paleozoic sediments of various ages (Campa and Coney, 1983). These terranes may have been accreted to North America during their pre-Mesozoic history. Nonetheless, the four depositional systems explain sedimentation histories within these tectono-stratigraphic terranes, and plate reconstructions must place these systems into rational positions relative to provenance and tectonic processes.

Rifted Basin

The rifted basin depositional system is characterized by thick continental clastics in narrow, linear, fault-controlled basins (Fig. 6). The clastic rocks generally are coarse to fine, poorly sorted, immature, and reddish to greenish in color. Associated volcanic rocks are intrusive mafic rocks and extrusive lavas and volcaniclastic rocks. This depositional complex accumulated during the Triassic and Early Jurassic in small block-faulted basins, and during Middle Jurassic it accumulated in large-scale rifted basins having restricted connections with the open ocean.

Block faulting and volcanism during Triassic and Early Jurassic were widespread in the interior of Mexico (Fig. 3). Numerous isolated outcrops consist of sequences of volcanic flows, pyroclastics and volcaniclastics. In northern Zacatecas these facies comprise the Nazas Formation, which is overlain unconformably by the Oxfordian Zuloaga Formation. The Nazas is tentatively correlated with Late Triassic units in Mexico and the United States, although some volcanic rocks are older and dated as 230 ± 20Ma by radiometry (Blickwede, 1981).

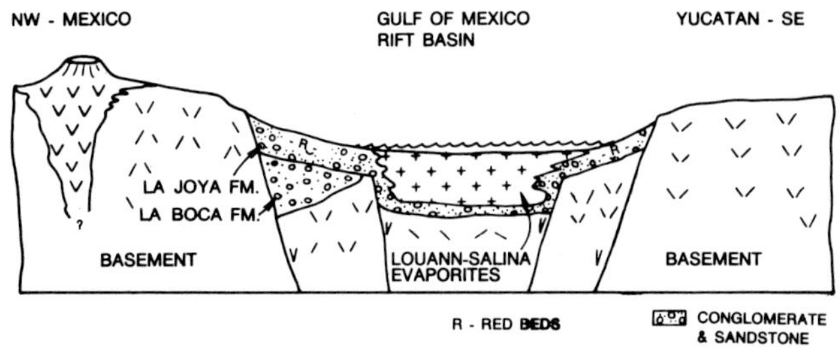

Figure 6. Model of a rift basin depositional system for the Middle Jurassic of the Gulf of Mexico. Adapted from Humphris (1979), Contreras and Castillon (1968), Erben (1957a), Weidie and Murray (1967).

In eastern Mexico (Fig. 3), more than 1000 meters of red and green clastics called the La Boca Formation, were deposited in the Huayacocotla Basin (Erben, 1957a) and contain Late Triassic plants (Mixon *et al.*, 1959). The La Boca overlies with angular unconformity Lower Permian and older rocks and is intruded by basic igneous rocks. It is overlain with an angular unconformity by the Middle Jurassic La Joya Formation.

During Middle Jurassic time, a large-scale rift basin developed between the northern Gulf of Mexico and the Yucatan Platform (Figs. 4, 6) (Humphris, 1979; Salvador, 1979, 1981). Thick sequences of anhydrite and salt grade laterally into or overly red beds. The Werner Formation, for example, in the northern Gulf Coast, is a lateral equivalent of the Louann Salt (Murray, 1961; Scott *et al.*, 1961). This clastic-evaporite depositional system on both sides of the Gulf of Mexico is evidence of the original pre-drift position of the Yucatan Platform and the northern Gulf according to Humphris (1979). Smaller rifted basins extended laterally to the major basin and are indicated by the presence of salt in northern Cuba (Punta Alegre Formation; Meyerhoff and Hatten, 1968) and northeastern Mexico (Minas Viejas Formation; Weidie and Wollenben, 1969).

The age of the salt is not tightly controlled by any chronostratigraphic method. The unit underlies well-dated Oxfordian carbonates and overlies Upper Triassic red beds. Poorly preserved palynomorphs from the Louann Salt suggest a Late Triassic-Early Jurassic age (Jux, 1961) and better preserved palynomorphs from the cap rock on the Challenger Knoll salt dome in the Sigsbee Plain indicate a Middle-Upper Jurassic age (Kirkland and Gerhard, 1971). Palynomorphs from the Punta Alegre Formation suggest a Triassic-Jurassic age (Meyerhoff and Hatten, 1968). In the southern part of Mexico, the Salina Formation in the Isthmian Basin consists mainly of salt and underlies the Chinameca Limestone, which has yielded Kimmeridgian to Early Cretaceous ammonites (Contreras and Castillon, 1968). Palynomorphs from the Salina Formation suggest a Middle-Late Jurassic age (Cousminer, 1978). The Salina apparently correlates with red clastics of part of the Todos Santos Formation (Contreras and Castillon, 1968). Although such a correlation has not been proven, red beds are likely to be found on the margin of a rifted salt basin (Fig. 6). A probable candidate for marginal redbeds to the Louann-Minas Viejas salt is the Cahuasas Formation in eastern Mexico (Salvador, pers. commun., 1982). The Cahuasas is Middle Jurassic because it underlies limestone with Callovian ammonities and overlies the Rosario Formation which elsewhere contains Middle Jurassic plants (Erben, 1957a; Imlay, 1980). The Cahuasas was deposited in the Huayacocotla Basin (Carillo-Bravo, 1965) and consists of more than 40 to 1200 m of varicolored clastics that disconformibly underlie the late Middle to Upper Jurassic Santiago Formation. The Cahuasas represents a Mid-Jurassic depositional event that followed the Triassic-Early Jurassic block faulting. In summary, the greatest volume of salt in the Gulf of Mexico and marginal areas is most likely Middle Jurassic in age and represents a large-scale rifting event which ultimately led to the formation of the present Gulf of Mexico. A slightly younger late Callovian-middle Oxfordian age is preferred by Imlay (1980, p. 47) in order to bury the salt quickly protecting it from erosion.

Younger evaporites were deposited upon the stable carbonate platform in both Yucatan and Florida-Bahamas. These salts can be dated as Lower Cretaceous and in Guatemala as Aptian because of associated fossiliferous limestones. Because these evaporites in northern Guatemala are younger than the Salina Formation in the Isthmian Basin, they did not form the Chiapas Salt Basin (Viniegra, 1971, 1981) which extends southeast of the Great Campeche Salt Basin. These will be discussed further in the section on stable carbonate platforms.

Carbonate Platform

The carbonate platform depositional system consists of widespread, thick complexes of limestone, dolomite, evaporites, and subordinate amounts of marl, shale and sandstone (Fig. 7). The carbonate system represents a platform on the continental margin or on volcanic atolls and pinnacles, which has a broad horizontal surface and a margin inclined seaward (Read, 1982).

A carbonate platform first developed in the Gulf of Mexico during the Oxfordian when the Smackover and Zuloaga limestones were deposited (Fig. 5). These units accumulated upon a platform-ramp that sloped gently seaward. Prevalent facies are bioclast and oolite grainstone of high energy shoals, peloidal to laminated algal limestone-dolomite-evaporites of tidal environments, deeper shelf pelagic-fossil mudstones (Bishop, 1972; Oivanki, 1974; Budd and Loucks, 1981), and local buildups of corals, algae, and sponges (Beauvais and Stump, 1976; Baria *et al.*, 1982). In Cuba the Jagua Formation indicates a local Oxfordian carbonate platform (Meyerhoff, 1964). This platform-ramp system nearly encircled the Gulf of Mexico except for possible breaks at the Isthmus of Tehuantepec. These carbonates generally have a basal conglomerate that uncomformably overlies clastics and evaporites of the Middle Jurassic rifted basin.

The second time of platform development was during the Kimmeridgian-Valanginian when the Chinameca Limestone was deposited in southern Mexico and the Knowles Limestones was deposited in eastern Texas and northwestern Louisiana (Finneran *et al.*, 1984). This platform was discontinuous in time and space around the Gulf. The Chinameca Limestone in the

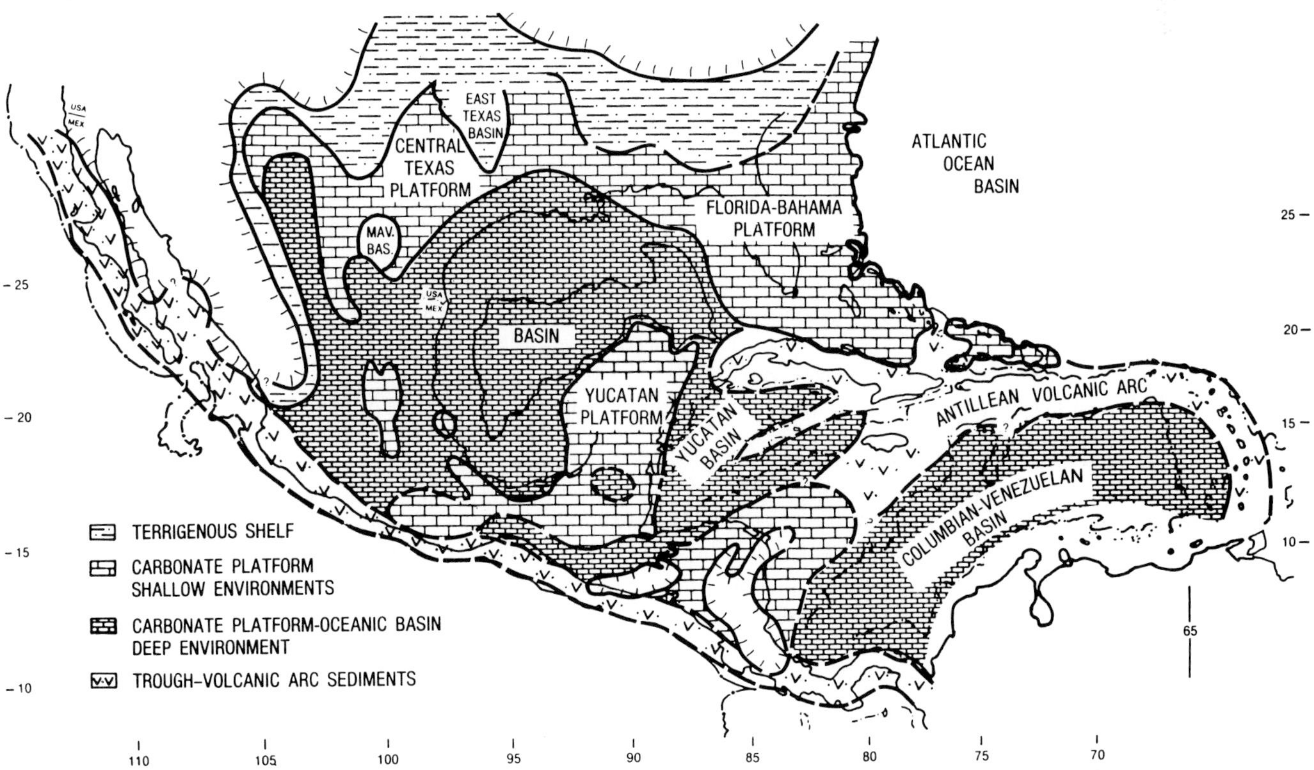

Figure 7. Distribution of Lower Cretaceous Aptian-Cenomanian sedimentary rocks (Bay, 1977; Bryant *et al.*, 1969; Uchupi *et al.*, 1971; Viniegra, 1981).

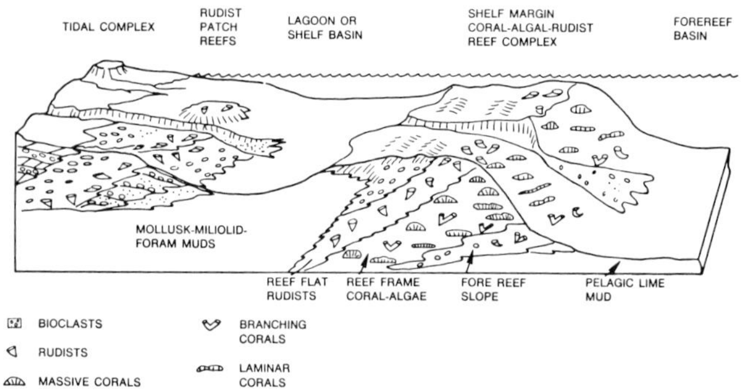

Figure 8. Model of a carbonate shelf-shelf margin depositional system for the Aptian-Cenomanian of the Gulf of Mexico.

Isthmus of Tehuantepec (Contreras and Castillon, 1968) and its updip, nearshore counterpart, the San Ricardo Formation (Richards, 1963) is Kimmeridgian in age, or younger. The Knowles is Berriasian-Valanginian and indicates an areally restricted, short-lived ramp with a spectrum of facies from deeper shelf mudstone, reefal coral-algal-sponge boundstone, and high energy bioclast and oolitic grainstone.

During the Barremian to Early Aptian a third carbonate platform-ramp system developed around the entire Gulf of Mexico, which is represented by the Cupido and Sligo Formations in the northern and western parts of the Gulf (Fig. 7). This platform was much wider than the previous ones and produced over 600 m of limestone and dolomite at its margins. A diverse set of paleoenvironments developed upon the interior shelf with the resulting variety of tidal, lagoon, and shoal facies (Wooten and Dunaway, 1977; Bebout *et al.*, 1977). At the shelf margin ramp reefs and shoals developed around the embayments, such as East Texas (Bay, 1977), and in Coahuila, Mexico, the ramp developed into a rimmed margin (Conklin and Moore, 1977). Both coral-algal and caprinid paleocommunities formed buildups at this margin. The Sligo consists of caprinid bioherms and intercalated bioclast sands and gravels, deposited on a shallow ramp margin (R. Hinote, pers. commun., 1982). Coral reefs developed at the margin in Mexico (Conklin and Moore, 1977; Wilson and Pialli, 1977).

Carbonate platform development reached its maximum extent during the Albian (Fig. 7). A continuous carbonate shelf margin extended from the eastern side of the Blake-Florida-Bahama platform across the northern Gulf to northeastern Mexico (Meyerhoff and Hatten, 1968; Coogan, 1977). South of the Coahuila-Tamaulipas platforms carbonates developed upon smaller discrete structures such as Golden Lane, Valles and El Doctor platforms (Coogan *et al.*, 1972; Enos, 1974). From the Isthmus of Tehuantepec a continuous shelf margin encircled the Yucatan Platform and was developed at least locally on the Honduran-Nicaraguan platform. In the Caribbean local carbonate platforms developed as atolls or pinnacles upon the volcanic islands of Jamaica, and Hispanola (Kauffman and Sohl, 1974; Coates, 1977; Bowin, 1975; Burke *et al.*, 1968).

A wide range of Albian depositional environments resulted in diverse facies (Fig. 8). The shelf margin reefal facies consists of coral-algal-rudist paleocommunities, and the buildups encircling the intrashelf basins and fringing, high salinity shelf environments consist of rudist-dominated paleocommunities (Scott, 1979). The interior shelf basins, such as the South Florida Basin, East Texas Basin, Maverick Basin, and Chiapas Embayment were the sites of organic-rich carbonates alternating with evaporites. On the intervening broad shallow shelves tidal sediments, shoal banks, and subtidal low energy carbonates accumulated.

During the Late Cretaceous, drowned carbonate shelves developed. Across the northern part of the Gulf chalk and marl accumulated during the Coniacian-Santonian and locally persisted into the Maastrichtian. In Mexico pelagic, cherty limestones were deposited during the Albian to the Campanian. In northwest Mexico and West Texas Cenomanian-Coniacian, thin-bedded limestone and marl buried the older shallow platform. However, shallow platform sediments continued to accumulate on Yucatan and Florida until the end of the Cretaceous. Locally, Maastrichtian rudist buildups developed there and around the islands of Jamaica, Hispanola, Puerto Rico, and the Virgin Islands (Kauffman and Sohl, 1974).

Terrigenous Shelf

The terrigenous shelf depositional system consists of widespread complexes of sandstone, siltstone and shale (Fig. 9). This complex was the product of fluvial, paralic, strandline, deltaic, and shelf environments. Locally oxygen circulation into interior shelf basins from the open sea was reduced and these basins, such as the East Texas Basin, became the site of organic-rich muds. This shelf system resulted from flooding of rifted basins by the sea, or by pre-orogenic uplift of the sediment source for backarc basins and arc shelves.

Terrigenous shelves developed locally during the Early and Middle Jurassic in southwestern Mexico (Guerrero Basin; Fig. 4), where rifted basins became connected with the sea. The first widespread terrigenous shelf in the Gulf developed during the early Kimmeridgian, when the Bossier Formation and its lateral equivalents formed.

The most widespread terrigenous shelf in the Gulf is Tithonian to at least Hauterivian in age and produced the Cotton Valley Group (Collins, 1980) and the overlying Hosston-Travis Peak units. Downdip marine facies comprise the Taraises and La Casita formations in Mexico, which yield ammonites, bivalves, foraminifers, tintinnids, and nannoplankton. This Jurassic-Cretaceous clastic wedge contains non-marine fluviatile sediments that grade below and above into marginal marine facies.

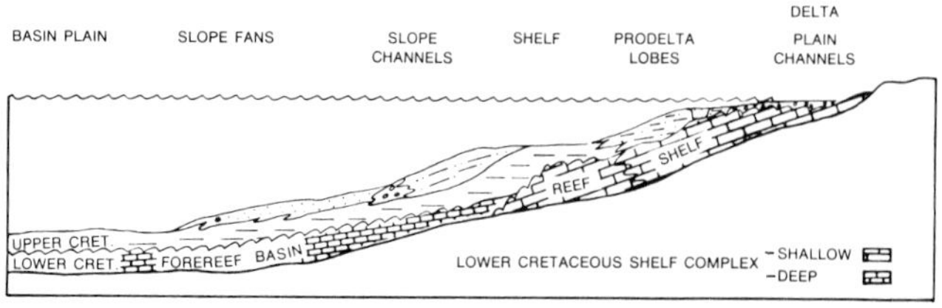

Figure 9. Model of a terrigenous shelf depositional system for the Mesozoic of the Gulf of Mexico. Based upon a model for the Upper Cretaceous Tuscaloosa Formation unconformably overlying Lower Cretaceous carbonate platform. Adapted from Siemers (1981).

Periodically during the Cretaceous dark gray muds were deposited upon the shelf. These sediments were widespread and immediately followed carbonate platform development. The oldest is the Aptian Pearsall-La Pena formations in the northern and western Gulf Coast. During the Late Albian numerous cycles of dark gray shale and shelf carbonates were deposited locally in the East Texas Basin, while downdip, deeper water carbonates accumulated. During the Cenomanian another widespread clastic unit is coarser grained updip, represented by the Woodbine and Tuscaloosa formations, and downdip is finer grained, represented by the Eagleford Formation. Late Cretaceous clastic wedges are represented by the Mendez Shale and Difunta Group in Mexico and the Escondido Formations in southwest Texas (Weidie and Murray, 1967; Murray, 1961).

This clastic system indicates tectonically active intervals in the Gulf basin. Source areas were renewed or sea level dropped somewhat to a lower level at which extensive clastic wedges prograded seawards.

Trough-Volcanic Arc Depositional Systems

Thick accumulations of submarine volcanic and volcaniclastic rocks, lime mudstone and radiolarian chert, locally intruded by ophiolite and serpentinite characterize this depositional system (Fig. 10). This complex is best developed in three major belts, one in the Caribbean, a second in southern Guatemala, and the third along the western margin of Mexico and Central America.

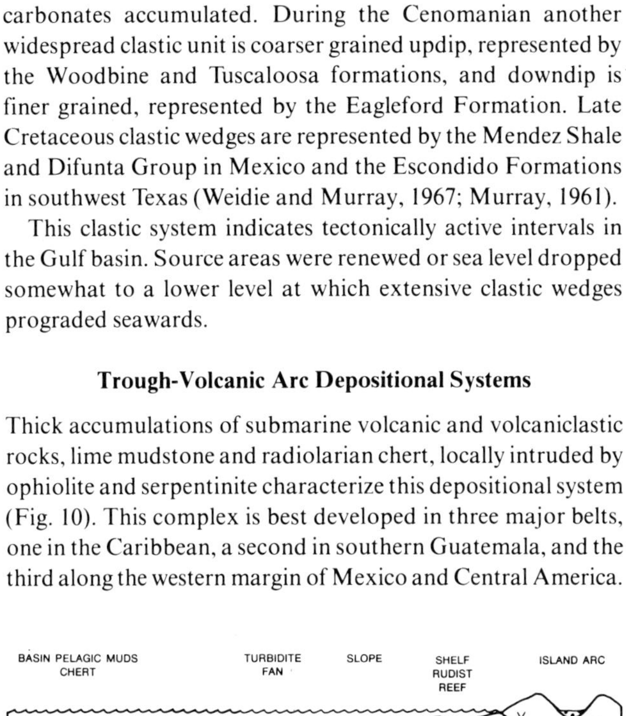

Figure 10. Model of a trough-volcanic arc depositional system for the Mesozoic of the Gulf of Mexico. Adapted from Rangin (1978); and Gastil *et al.*, (1978).

A trough-volcanic arc depositional system extends along the western coast of Mexico continuing northward into California and southward to Panama (Fig. 2). This complex represents volcanic arc and backarc basin settings that become younger southward. These sedimentary and metasedimentary rocks in Mexico together with plutonic igneous rocks comprise part of the Nevadan Orogen (Mullan, 1978). Deposition in the backarc basin began during Triassic time in Baja California, and by the Middle Jurassic deep-water flysch, turbidite and submarine slide deposits accumulated in the trough with ophiolites (Gastil *et al.*, 1978). Pluton emplacement began near the end of the Early Cretaceous along the trend in Baja California, and in Costa Rica and Panama it began in the Late Cretaceous (Henningsen, 1968; Weyl, 1980, p. 121). Subduction along the western margin of Mexico persisted from Middle Jurassic to Cenomanian-Turonian time (Rangin, 1978). Along the western margin of Costa Rica-Panama it began in the Late Cretaceous, when the basic-ultrabasic volcanic Nicoya Complex of a Pacific submarine ridge was obducted to the Caribbean plate (Schmidt-Effing, 1979). Biostratigraphy of this belt is piecemeal and based on mollusks, foraminifers and radiolaria (Pessagno, 1960, 1963; other references in Weyl, 1980, p. 154). An alternate origin of the Nicoya Complex in Costa Rica and Panama is as a separate volcanic arc in the east Pacific that migrated into position between Central and South America during the Tertiary (Pindell and Dewey, 1982).

The second trough system transects southern Guatemala and Honduras parallel with the complex Polochic-Motagua fault system in Guatemala. This fault system may represent the suture-subduction zone between the Yucatan Platform and the Chortis block of the Honduran-Nicaragua platform (Perfit and Heezen, 1978; B. Burkart, pers. commun., 1982), or it may be a deep rift zone. In Guatemala it contains moderately deep-water, thin-bedded pelagic limestone, shale, green and black radiolarian chert of the Aptian-Albian Cantarranas Formation (Mills *et al.*, 1967; Burkart *et al.*, 1973). Apparently, these rocks were deposited in a trough bounded by the Jocotan Fault on the north, where fanglomerate is interbedded with the marine limestone (Burkart *et al.*, 1973). The trough system is now bordered by shallow-water, coral-rudist, Albian limestone of the Coban Formation on the north and the Atima Formation on the south (Mills *et al.*, 1967). In Honduras the Cantarranas consists of shallow water lagoonal facies (Finch, 1981). The Late Jurassic and Cretaceous sedimentary histories of Guatemala and Honduras-Nicaragua were dominated by similar carbonate platforms. During the Late Cretaceous, however, serpentinite was intruded into these rocks, presumably along fault zones, and then exposed and eroded, contributing clasts to the overlying Campian Sepur Formation (Bonis, 1968). This short-lived Early Cretaceous trough coincides with a zone of younger faults continuous with the Bartlett-Cayman Trough (Donnelly *et al.*, 1968) and was the subduction-suture zone between the two platforms (Perfit and Heezen, 1978). Evidence of large-scale strike-slip or left-lateral movement during the Mesozoic is not compelling (B. Burkart, pers. commun., 1982), but since the middle Tertiary 130 km of lateral displacement has occurred on the Polochic Fault (Burkart, 1982). The Montagua Fault apparently continues into the Cayman Trench (Burkart, 1982). This depositional system may represent either the trough-subduction zone along the south margin of the Yucatan Platform or a deep rift that intersected oceanic crust, which was the source of the serpentinites.

The third trough-volcanic arc system extended for over 3200 km from Cuba to the Virgin Islands (Fig. 2). This system represents a Mesozoic Benioff zone (Khudoley and Meyerhoff, 1971, p. 108) or a subduction zone (Pindell and Dewey, 1982). It first developed in Cuba during the Tithonian as indicated by the Artemisa Formation, a dark colored, thin-bedded cherty lime mudstone (Khudoley and Meyerhoff, 1971, p. 65, p. 71). The trough depositional regime continued into the Early Cretaceous. Deep-water limestones contain diverse pelagic tintinnids and ammonites and overly the Late Jurassic section in Cuba (Khudoley and Meyerhoff, 1971, p. 68, p. 112; Pop, 1976). Most of the taxa are widely distributed throughout the Tethys. At the northwestern end of Cuba during the Late

Jurassic, a carbonate shelf was developed over a basement of presumed Paleozoic metamorphic rocks (Meyerhoff and Hatten, 1968; Pardo, 1975). To the east the basement rocks of Puerto Rico and the Virgin Islands appear to have formed in a Late Jurasic oceanic volcanic arc. This basement in the Virgin Islands consists of quartz keratophyre submarine flows, which are mapped as the Water Island Formation, and are older than Albian (Donnelly *et al.*, 1971). This basement is also known in the Lesser Antilles on the Island of Désirade. Here submarine pillow lavas, quartz-keratophyre flows and chert are intruded by Late Jurassic quartz diorite dated at 142 ± 10 ma (Fink, 1972). The full implications of this complex are not realized, however, it indicates that mafic and intermediate igneous activity prevailed near the eastern end of the Antilles during Late Jurassic time, although Tomblin (1975) suggested that this complex may have been displaced from eastward of the Virgin Islands.

Outside of Cuba the oldest dated rocks in the Greater Antillian arc are no older than Aptian-Albian; these consist of submarine basalt, andesite, quartz keratophyre, and volcanic conglomerate. The age is based on scattered occurrences of *Orbitolina, Coskinolina* and rudists in the Dominican Republic (Bowin, 1966), on rudists in Jamaica (Burke *et al.*, 1968; Arden, 1975), ammonites, rudists and radiolaria in the Virgin Islands (Donnelly, 1966; Donnelly *et al.*, 1971) and Puerto Rico (Briggs, 1968).

The composition of the igneous rocks of the Greater Antilles changed during the Late Cretaceous from mafic to felsic (Nelson, 1968; Weyl, 1968). It may be significant that emplacement of sialic intrusives began earlier during the Late Cretaceous in the Cordillera of Mexico. Pyroclastic tuffs and flows became prominent in the Caribbean also during the Late Cretaceous (Nelson, 1968) indicating emergence of volcanoes above sea level. Consequently, Campanian and Maastrichtian rudist reefs fringing volcanic islands became widely developed in this area (Kauffman and Sohl, 1974). Some of these rudists were transported downslope as part of turbidites; for example, large specimens of *Barretia* are concentrated at the base of turbidites in the Judith Fancy Formation on St. Croix.

CONCLUSIONS

The marginal parts of the Gulf of Mexico region were open to Early Jurassic faunas that had cosmopolitan distribution. Ammonites could have migrated into the Huayacocotla Basin from the Pacific or the Atlantic through a passage between North and South America, not the eastern Gulf. During the Middle Jurassic large-scale rift basins became the site of evaporite deposition. By Oxfordian time the Gulf of Mexico was open to normal marine waters and a widespread carbonate shelf-ramp developed in many places around the Gulf.

Mesozoic sedimentation was characterized by four major depositional systems. The rifted basin system consists of continental clastics of Triassic to early Late Jurassic age. The carbonate platform-shelf began during the early Late Jurassic, became widespread during the Early Cretaceous and persisted mainly as deep shelves to the end of the Late Cretaceous, when its areal extent was greatly reduced. Terrigenous shelf system was dominated by marine clastics and became prevalent during the Late Jurassic. The trough-volcanic arc system consists of deep water pelagic sediments and submarine volcaniclastic rocks in subduction zones or deep-seated rifts.

Several facts must be accounted for in any plate reconstruction of the Gulf-Caribbean region.

1) Warm, dry and humid climatic cycles prevailed during the Triassic and Jurassic.

2) Pre-Oxfordian depositional systems in Mexico accumulated in local graben-like basins and generally reflect single transgressive-regressive cycles.

3) From Oxfordian time on broad shelves were widespread throughout the Gulf.

4) Mediterranean Oxfordian and younger reef communities were present in the Gulf.

5) The western end of Cuba was adjacent to a continental source of terrigenous clastics during Middle and Late Jurassic time.

6) Oceanic crust now accreted to the Antilles is Tithonian to Albian.

7) During the Late Jurassic a large continental source is required for the Todos Santos Formation in Guatemala and equivalent facies in Honduras.

8) Cretaceous reefing is older in the Antillean arc than in the Costa Rica-Panama arc.

9) Oceanic crust in the Colombian-Venezuelan Basin is at least as old as Late Cretaceous, probably Coniacian or older.

ACKNOWLEDGEMENTS

This summary is based on the work of many geologists who have devoted much time to field work in the region and have shared their observations and opinions. Amoco Production Company provided help to find the many references, to draft the illustrations, to type the manuscript, and to release this study.

REFERENCES

Ager, D.V., 1967, Some Mesozoic Brachiopods in the Tethys Region: *in* Adams, C.G. and Ager, D.V. eds., Aspects of Tethyan Biogeography: Systematics Association Publication 7, p. 135-151.

Arden, D.D., Jr., 1975, Geology of Jamaica and the Nicaragua Rise: *in* Nairn, A.E.M. and Stehli, F.G., eds., The Ocean Basins and Margins, v. 3, The Gulf of Mexico and The Caribbean; Plenum Press, New York, p. 617-661.

Baria, L.R., Stoudt, D.L., Harris, P.M., Crevello, P.D., 1982, Upper Jurassic Reefs of Smackover Formation, United States Gulf Coast: American Association of Petroleum Geologists Bulletin, v. 66, p. 1449-1482.

Barr, K.W., 1974, The Caribbean and Plate Tectonics: Some Aspects of the Problem: Naturforschung Gesellschaft Basel, Verhandlung, v. 84, p. 45-67.

Bay, T.A., Jr., 1977, Lower Cretaceous Stratigraphic Models from Texas and Mexico: Bureau of Economic Geology, University of Texas at Austin, Report of Investigation 89, p. 12-30.

Beauvais, L., 1979, Paleobiogeography of the Middle Jurassic Corals: *in* Gray, J. and Boucot, A.J., eds., Historical Biogeography, Plate Tectonics, and the Changing Environment: Oregon State University Press, Corvallis, p. 289-304.

Beauvais, L., and Stump, T.E., 1976, Corals, Molluscs, and Paleography of Lower Jurassic Strata of the Cerro Pozo Serna, Sonora, Mexico: Palaeogeography, Palaeoclimatology, Palaeoecology, v. 19, p. 275-301.

Bebout, D.G., Schatzinger, R.A., and Loucks, R.G., 1977, Porosity Distribution in the Stuart City Trend, Lower Cretaceous, South Texas: Bureau of Economic Geology, University of Texas at Austin, Report of Investigation 89, p. 234-256.

Bishop, B.A., 1972, Petrography and Origin of Cretaceous Limestones, Sierra de Picachos Vicinity, Nuevo Leon, Mexico: Journal of Sedimentary Petrology, v. 42, p. 270-286.

Blickwede, J.F., 1981, Petrology and Stratigraphy of Triassic(?) "Nazas Formation": Sierra de San Julian, Zacatecas, Mexico (Abst.): American Association of Petroleum Geology Bulletin, v. 65, p. 1012.

Bonis, S.B., 1968, Age of Guatemalan Serpentine (Abst.): Abstracts with Programs, Geological Society of America Special Paper 115, p. 18.

Bowin, C.O., 1966, Geology of Central Dominican Republic (A Case History of Part of an Island Arc): Geological Society of America Memoir 98, p. 11-84.

______, 1975, The Geology of Hispaniola: *in* Nairn, A.E.M. and Stehli, F.G., eds., The Ocean Basins and Margins, v. 3, The Gulf of Mexico and The Caribbean: Plenum Press, New York, p. 501-552.

Briggs, R.P., 1968, The Geologic History of Puerto Rico: *in* Saunders, J.B., ed., Transactions of the Fourth Caribbean Geologic Congress; Caribbean Printers, Trinidad, p. 234.

Bryant, W.R., Meyerhoff, A.A., Brown, N.K., Jr., Furrer, M.A., Pyle, T.E., and Antoine, J.W., 1969, Escarpments, Reef Trends and Diapiric Structures, Eastern Gulf of Mexico: American Association of Petroleum Geologists Bulletin, v. 53, p. 2506-2542.

Budd, D.A. and Loucks, R.G., 1981, Smackover and Lower Buckner Formations, South Texas: Depositional Systems on a Jurassic Carbonate Ramp: Bureau of Economic Geology, University of Texas at Austin, Report of Investigation 112, 38 p.

Buffler, R.T., 1980, Geologic History of Deep Gulf of Mexico Basin (Abst.): American Association of Petroleum Geologists Bulletin, v. 64, p. 1277-1278.

Buffler, R.T., Watkins, J.S., Shaub, F.J., and Worzel, J.L., 1981, Structure and Early Geologic History of the Deep Central Gulf of Mexico Basin: *in* Pilger, R.H., Jr., ed., The Origin of the Gulf of Mexico and the Early Opening of the Central North Atlantic Ocean: Louisiana State University, School Geoscience Baton Rogue, La., p. 3-16.

Buffler, R.T. and Phair, R.L., 1982, Pre-Cretaceous geologic history of deep southeastern Gulf of Mexico (Abst.): American Association of Petroleum Geologists Bulletin, v. 66, p. 554.

Burckhardt, C.E., 1930, Etude synthétique sur le Mesozoïque Mexicain: Societé Paleóntologie Suisse Mémoir, v. 49, p. 1-123; v. 50, p. 125-280.

Burkart, B., 1983, Neogene North American-Caribbean Plate Boundary across Northern Central America: Offset Along the Polochic Fault: Tectonophysics, v. 99, p. 251-270.

Burkart, B., Clemons, R.E., and Crane, D.C., 1973, Mesozoic and Cenozoic Stratigraphy of Southeastern Guatemala: American Association of Petroleum Geologists Bulletin, v. 57, p. 63-73.

Burke, Kevin, Coates, A.G., and Robinson, E., 1968, Geology of the Benbow Inlier and Surrounding Areas, Jamaica: *in* Saunders, J.B., ed., Transactions of the Fourth Caribbean Geological Congress: Caribbean Printers, Trinidad, p. 299-308.

Campa, M.F., Campas, M., Flores, R., and Oviedo, R., 1974, La Secuencia Mesozoica Volcanico-Sedimentaria Metamorfizada de Ixtapan de la Sal, Mex.-Teloloapan, Gro.: Boletin Societe Geológica Mexicana, v. 35, p. 7-28.

Campa, M.F., and Coney, P.J., 1983, Tectono-Stratigraphic Terranes and Mineral Resource Distributions in Mexico: Canadian Journal of Earth Sciences, v. 20, p. 1040-1051.

Cantu Chapa, A., 1969, Estratigrafia del Jurásico Medio-Superior del Subsuelo de Poza Rica, Ver. (Area de Soledad-Miquetla): Instituto Mexicano Petróleo Revista, v. 1, p. 3-9.

______, 1971, La Serie Huasteca (Juràsico Medio-Superior) del Centro Este de México: Instituto Mexicano Petróleo Revista, v. 3, p. 17-40.

Carillo-Bravo, Jose, 1965, Estudeo Geológico de Una Parte del Anticlinorio de Huayacocotla: Asociación Mexicana Geólogos Petroleros Boletin, v. 17, p. 73-96.

Coates, A.G., 1977, Jamaican Cretaceous Coral Assemblages and Their Relationships to Rudist Frameworks: Second Symposium International sur les Coraux et Récifs Coralliens Fossiles: Bureau de Reserches Géologiques et Miniéres, no, 89, p. 336-342.

Collins, S.E., 1980, Jurassic Cotton Valley and Smackover Reservoir Trends, East Texas, North Louisiana, and South Arkansas: American Association Petroleum Geologists Bulletin, v. 64, p. 1004-1013.

Conklin, J. and Moore, C., 1977, Paleoenvironmental Analysis of the Lower Cretaceous Cupido Formation, Northeast Mexico: Bureau of Economic Geology, University of Texas at Austin, Report of Investigation 89, p. 302-323.

Contreras, V.H. and Castillon, M.B., 1968, Morphology and Origin of Salt Domes of Isthmus of Tehuantepec: American Association of Petroleum Geologists Memoir 8, p. 244-260.

Coogan, A.H., 1977, Early and Middle Cretaceous Hippuritacea (Rudists) of the Gulf Coast: Bureau of Economic Geology, University of Texas at Austin, Report of Investigation 89, p. 32-70.

Coogan, A.H., Bebout, D.G., and Maggio, C., 1972, Depositional Environments and Geologic History of Golden Lane and Poza Rica Trend, Mexico, An Alternate View: American Association of Petroleum Geologists Bulletin, v. 56, p. 1419-1447.

Cook, T.D. and Bally, A.W., ed., 1975, Stratigraphic Atlas of North and Central America: Princeton University Press, Princeton, N.J., 272 p.

Cordoba, D.A., Tardy, M., Carfantan, J.-C., Campa, M.F., and Rangin, C., 1980, Le Mexique Mesogeen et le Passage au Systeme Cordillerain de Type Californie: Mémoir 26th Congress Géologie International, Colloque C5, p. 18-29.

Cousminer, H.L., 1978, Palynoflorules of Jurassic Circum-Gulf Salt: Age, Environment and Source (Abst.): Palynology (USA), v. 2, p. 216.

Cragin, F.W., 1905,Paleontology of the Malone Jurassic Formation of Texas: United States Geological Survey Bulletin 266, 172 p.

de Cserna, G.A., 1961, Paleontologia del Triassico Superior de Sonora Pt. III: Paleontologia Mexicana, v. 11, p. 1-38.

Donnelly, T.W., 1966, Geology of St. Thomas and St. John, U.S. Virgin Islands: Geology Society of America Memoir 98, p. 85-176.

Donnelly, T.W., Crane, D., and Burkart, B., 1968, Geologic History of the Landward Extension of the Bartlett Trough-Some Preliminary Notes: *in* Saunders, J.B., ed., Transactions of the Fourth Caribbean Geological Congress: Caribbean Printers, Trinidad, p. 225-228.

Donnelly, T.W., Rogers, J.J.W., Pushkar, P., and Armstrong, R.L., 1971, Chemical Evolution of the Igneous Rocks of the Eastern West Indies: An Investigation of Thorium, Uranium, and Potassium Distributions, and Lead and Strontium Isotopic Ratios: Geological Society of America Memoir 130, p. 181-224.

Enos, P., 1974, Reefs, Platforms, and Basins of Middle Cretaceous in Northeast Mexico: American Association of Petroleum Geologists Bulletin, v. 58, p. 800-809.

Erben, H.K., 1957a, A Paleogeographic Reconstruction for the Lower and Middle Jurassic and for the Callovian of Mexico: Congress Geologie International XX Sesiõn, Section II, p. 35-43.

———, 1957b, New Biostratigraphic Correlations in the Jurassic of Eastern and South-Central Mexico: Congress Geologie International XX Sesiõn, Section II, p. 44-52.

Erdlac, R.J., Jr. and Anderson, T.H., 1982, The Chixoy-Polochic Fault and Its Associated Fractures in Western Guatemala: Geological Society of America Bulletin, v. 93, p. 57-67.

Finch, R.C., 1981, Mesozoic Stratigraphy of Central Honduras: American Association of Petroleum Geologists Bulletin, v. 65, p. 1320-1333.

Fink, L.K., Jr., 1972, Bathymetric and Geologic Studies of the Guadeloupe Region, Lesser Antilles Island Arc: Marine Geology, v. 12, p. 267-288.

Finneran, J.M., Scott, R.W., Taylor, G.A., and Anderson, G.H., 1984, Lowermost Cretaceous Ramp Reefs: Knowles Limestone, SW Flank of the East Texas Basin: Gulf Coast Section Society of Economic Paleontologists and Mineralogists Foundation, Third Annual Research Conference proceedings, p. 125-133.

Fisher, W.L., Brown, L.F., Scott, A.J., and McGowen, J.H., 1969, Delta Systems in the Exploration for Oil and Gas: Bureau of Economic Geology, University of Texas at Austin, Research Colloquium, 78 p.

Gastill, G., Morgan, G.J., Krummenacher, D., 1978, Mesozoic History of Peninsular California and Related Areas East of the Gulf of California: *in* Howell, D.G. and McDougall, K.A., eds., Pacific Coast Paleogeography Symposium 2 – Mesozoic Paleogeography of the Western United States: Society of Economic Paleontologists and Mineralogists, Pacific Section, p. 107-115.

Hallam, A., 1971, Mesozoic Geology and the Opening of the North Atlantic: Journal of Geology, v. 79, p. 129-157.

Henningsen, D.W., 1968, Stratigraphy and Paleogeography of Upper Cretaceous and Tertiary Sediments in Southern Costa Rica: *in* Saunders, J.B., ed., Transactions of the Fourth Caribbean Geological Congress: Caribbean Printers, Trinidad, p. 353-356.

Hillebrant, A., von, 1981, Kontinentalverschiebung und die Paläozoogeographischen Beziehungen des Süd-Amerikanischen Lias: Geologische Rundschau, v. 70, p. 570-582.

Humphris, C.C., Jr., 1979, Salt Movement on Continental Slope, Northern Gulf of Mexico: American Association of Petroleum Geologists Bulletin, v. 63, p. 782-798.

Imlay, R.W., 1940, Jurassic Pelecypods from Mexico: Journal of Paleontology, v. 14, p. 393-411.

———, 1941, Jurassic Fossils Gulf Coast Subsurface: Journal of Paleontology, v. 15, p. 256-277.

———, 1942,Late Jurassic Fossils from Cuba and Their Economic Significance: Geological Society of America Bulletin, v. 53, p. 1417-1478.

———, 1943, Jurassic Formations of Gulf Region: American Association of Petroleum Geologists Bulletin, v. 27, p. 1407-1533.

———, 1944a, Correlation of the Cretaceous Formations of the Greater Antilles, Central America and Mexico: Geological Society of America Bulletin, v. 55, p. 1005-1046.

———, 1944b, Cretaceous Formations of Central America and Mexico: American Association of Petroleum Geologists Bulletin, v. 28, p. 1077-1195.

———, 1945, Jurassic Fossils from the South States: Journal of Paleontology, v. 19, p. 253-276.

———, 1952, Jurassic Correlation Charts: Geological Society of America Bulletin, v. 63, p. 953-992.

———, 1965, Jurassic Marine Faunal Differentiation in North America: Journal of Paleontology, v. 39, p. 1023-1038.

———, 1980, Jurassic Paleobiogeography of the Conterminous United States in its Continental Setting: United States Geological Survey Professional Paper 1062, 134 p.

Jansa, L.F., 1981, Mesozoic Carbonate Platforms and Banks of the Eastern North American Margin: Marine Geology, v. 44, p. 97-117.

Jux, Ulrich, 1961, The Palynologic Age of Diapiric and Bedded Salt in the Gulf Coastal Province: Louisiana Geological Survey Bulletin 38, p. 1-46.

Kauffman, E.G. and Sohl, N.F., 1974, Structures and Evolution of Caribbean Cretaceous Rudist Frameworks: Naturforschung Gesellschaft Basel, Verhandlung, v. 84, p. 399-467.

Khudoley, K.M., 1974, Circum-Pacific Mesozoic Ammonoid Distribution: Relation to Hypotheses of Continental Drift, Polar Wandering, and Earth Expansion: *in* Kahle, C.F., ed., Plate Tectonics – Assessments and Reassessments: American Association of Petroleum Geologists Memoir 23, p. 295-330.

Khudoley, K.M. and Meyerhoff, A.A., 1971, Paleogeography and Geological History of Greater Antilles: Geological Society of America Memoir 129, 199 p.

Kirkland, D.W. and Gerhard, J.E., 1971, Jurassic Salt, Central Gulf of Mexico, and its Temporal Relation to Circum-Gulf Evaporites: American Association of Petroleum Geologists Bulletin, v. 55, p. 660-686.

Martin, R.G. and Case, J.E., 1975, Geophysical Studies in the Gulf of Mexico: *in* Nairn, A.E.M., and Stehli, F.G., eds., The Ocean Basins and Margins, v. 3, The Gulf of Mexico and the Caribbean: Plenum Press, New York, p. 65-106.

Meyerhoff, A.A., 1964, Review of Bermudez, P.J., 1961, Las Formaciones Geologicas de Cuba: International Geological Review, v. 6, p. 149-156.

Meyerhoff, A.A. and Hatten, C.W., 1968, Diapiric Structures in Central Cuba: American Association of Petroleum Geologists Memoir 8, p. 315-357.

———, 1974. Bahamas Salient of North America: Tectonic Framework, Stratigraphy, and Petroleum Potential: American Association of Petroleum Geologists Bulletin, v. 58, p. 1201-1239.

Meyerhoff, A.A. and Meyerhoff, H.A., 1974, Tests of Plate Tectonics: *in* Kahle, C.F., ed., Plate Tectonics – Assessments and Reassessments: American Association of Petroleum Geologists Memoir 23, p. 43-145.

Mills, R.A. and Hugh, K.E., 1974, Reconnaissance Geologic Map of Mosquitia Region, Honduras and Nicaragua Caribbean Coast: American Association of Petroleum Geologists Bulletin, v. 58, p. 189-207.

Mills, R.A., *et al.*, 1967, Mesozoic Stratigraphy of Honduras: American Association of Petroleum Geologists Bulletin, v. 51, p. 1711-1786.

Mitchum, R.M., Jr., Vail, P.R., and Sangree, J.B., 1974, Regional Stratigraphic Framework from Seismic Sequences (Abst.): Geological Society of America, Abstracts with Programs, v. 6, no. 7, p. 873.

Mixon, R.B., Murray, G.E., Diaz, G., 1959, Age and Correlation of Huizachal Group (Mesozoic), State of Tamaulipas, Mexico: American Association of Petroleum Geologists Bulletin, v. 43, p. 757-771.

Mullan, H.S., 1978, Evolution of Part of the Nevadan Orogen in Northwestern Mexico: Geological Society of America Bulletin, v. 89, p. 1175-1188.

Müllerried, F.K.G., 1942, The Mesozoic of Mexico and NW Central America: Proceedings of the Eighth American Scientists Congress, v. 4, p. 125-147.

Murray, G.E., 1961, Geology of the Atlantic and Gulf Coastal Province of North America: Harper and Bros., New York, 692 p.

Myczynski, R. and Brochwicz-Lewinzki, W., 1981, Cuban Oxfordian

Aspidoceratids: Their Relation to the European Ones and Their Stratigraphic Values: Bulletin l'Academie Polonaise des Sciences, Series Sciences Terre, v. 28, p. 325-330.

Nelson, A.E., 1968, Trends of Cretaceous Volcanism in North-Central Puerto Rico: *in* Saunders, J.B., ed., Transactions of the Fourth Caribbean Geological Congress: Caribbean Printers, Trinidad, p. 235-242.

Oivanki, S.M., 1974, Paleodepositional Environments in the Upper Jurassic Zuloaga Formation (Smackover), Northeastern Mexico: Gulf Coast Association Geological Society Transactions, v. 24, p. 258-278.

Pardo, Georges, 1975, Geology of Cuba: *in* Nairn, A.E.M. and Stehli, F.G., ed., The Ocean Basins and Margins, v. 3: The Gulf of Mexico and the Caribbean: Plenum Press, New York, p. 553-616.

Perfit, M.R. and Heezen, B.C., 1978, The Geology and Evolution of the Cayman Trench: Geological Society of America Bulletin 89, p. 1155-1174.

Person, C.P. and Delevoryas, T., 1982, The Middle Jurassic Flora of Oaxaca, Mexico: Palaeontographica, Abteilung B, v. 180, p. 82-119.

Pessagno, E.A., 1960, Stratigraphy and Micropaleontology of the Cretaceous and Lower Tertiary of Puerto Rico: Micropaleontology, v. 6, p. 87-110.

______, 1963, Upper Cretaceous Radiolaria from Puerto Rico: Micropaleontology, v. 9, p. 197-214.

Pindell, J., and Dewey, J.F., 1982, Permo-Triassic Reconstruction of Western Pangea and the Evolution of the Gulf of Mexico / Caribbean Region: Tectonics, v. 1, p. 179-211.

Pop, G., 1976, Tithonian-Valanginian Calpionellid Zones from Cuba: Institutl de Geologie Geofizica, Dari de seama ale sedintelor, v. 62, 3: Paleontologische, p. 237-266.

Rangin, Claude, 1978, Speculative Model of Mesozoic Geodynamics, Central Baja California to Northeastern Sonora (Mexico): *in* Howell, D.G. and Kristin, A.M., eds., Pacific Coast Paleogeography Symposium 2-Mesozoic Paleogeography of the Western United States: Pacific Section, Society of Economic Paleontologists and Mineralogists, p. 85-106.

Read, J.F., 1982, Carbonate Platforms of Passive (Extensional) Continental Margins: Types, Characteristics and Evolution: Tectonophysics, v. 81, p. 195-212.

Richards, H.G., 1963, Stratigraphy of Earliest Mesozoic Sediments in Southeastern Mexico and Western Guatemala: American Association of Petroleum Geologists Bulletin, v. 47, p. 1861-1870.

Salvador, A., 1979, Late Triassic-Jurassic Paleogeography and Origin of Gulf of Mexico (Abst.): American Association of Petroleum Geologists Bulletin, v. 63, p. 520.

______, 1981, Late Triassic-Jurassic Paleogeography and the Origin of the Gulf of Mexico (Abst.): American Association of Petroleum Geologists Bulletin, v. 65, p. 1358.

Schmidt-Effing, R., 1979, Alter und Genese des Nicoya-Komplexes, Einer Ozeanischen Paläokruste (Oberjura bis Eozän) in Südlichen Zentralamerika: Geologische Rundschau, v. 68, p. 457-494.

______, 1980, The Huayacocotla Aulacogen in Mexico (Lower Jurassic) and the Origin of the Gulf of Mexico: *in* Pilger, R.H., Jr., ed., The Origin of the Gulf of Mexico and the Early Opening of the Central North Atlantic Ocean: Louisiana State University, School Geoscience, Baton Rouge, La., p. 79-86.

Schuchert, C., 1935, Historical Geology of the Antillean-Caribbean Region: John Wiley and Sons, Inc., New York, 811 p..

Scott, K.R., Hayes, W.E. and Fietz, R.P., 1961, Geology of the Eagle Mills Formation: Gulf Coast Association Geological Society Transactions, v. 11, p. 1-14.

Scott, R.W., 1979, Depositional Model of Early Cretaceous Coral-Algal-Rudist Reefs, Arizona: American Association of Petroleum Geologists Bulletin, v. 63, p. 1108-1127.

Sheridan, R.E., Crosby, J.T., Bryan, G.M., and Stoffa, P.L., 1981, Stratigraphy and Structure of Southern Blake Plateau, Northern Florida Straits and Northern Bahama Platform from Multichannel Seismic Reflection Data: American Association of Petroleum Geologists Bulletin, v. 65, p. 2571-2593.

Sheridan, R.E. *et al.*, 1982, Early History of the Atlantic Ocean and Gas Hydrates on the Blake Outer Ridge: Results of the Deep Sea Drilling Project Leg 76: Geological Society of America Bulletin, v. 93, p. 876-885.

Siemers, C.T., 1981, Recognition of "Shallow-Water" Updip Shelf and "Deep-Water" Downdip Slope Deposits of the Subsurface Woodbine-Eagleford Interval (Upper Cretaceous) in East Texas: Gulf Coast Section, Society of Economic Paleontologists and Mineralogists, 2nd Annual Research Conference, p. 57-61.

Smith, A.G. and Briden, J.C., 1977, Mesozoic and Cenozoic Paleocontinental Maps: Cambridge University Press, Cambridge, U.K., p. 1-63.

Stevens, G.R., 1963, Faunal Realms in Jurassic and Cretaceous Belemnites: Geological Magazine, v. 100, p. 481-497.

Tomblin, J.F., 1975, The Lesser Antilles and Aves Ridge: *in* Nairn, A.E.M. and Stehli, F.G., eds., The Ocean Basins and Margins, v. 3, The Gulf of Mexico and the Caribbean: Plenum Press, New York, p. 467-500.

Uchupi, E., 1975, Psysiography of the Gulf of Mexico and Caribbean Sea: *in* Nairn, A.E.M. and Stehli, F.G., eds., The Ocean Basins and Margins, v. 3, The Gulf of Mexico and the Caribbean: Plenum Press, New York, p. 1-64.

Uchupi, E., Milliman, J.D., Luyendyk, B.P., Bowin, C.O., and Emery, K.O., 1971, Structure and Origin of Southeastern Bahamas: American Association of Petroleum Geologists Bulletin, v. 55, p. 687-704.

Verma, H.M. and Westermann, G.E.G., 1973, The Tithonian (Jurassic) Ammonite Fauna and Stratigraphy of Sierra Catorce, San Luis Potosi, Mexico: Bulletins of American Paleontology, v. 63, p. 107-278.

Viniegra, F., 1971, Age and Evolution of Salt Basins of Southeastern Mexico: American Association of Petroleum Geologists Bulletin, v. 55, p. 478-494.

______, 1981, Great Carbonate Bank of Yucatan, Southern Mexico: Journal of Petroleum Geology, v. 3, p. 247-278.

Weidie, A.E. and Murray, G.E., 1967, Geology of Parras Basin and Adjacent Areas of Northeastern Mexico: American Association of Petroleum Geologists Bulletin, v. 51, p. 678-695.

Weidie, A.E. and Wollenben, J.A., 1969, Upper Jurassic Stratigraphic Relations Near Monterrey, Nuevo Leon, Mexico: American Association of Petroleum Geologists Bulletin, v. 53, p. 2418-2420.

Wells, J.W., 1942, Jurassic Corals From the Smackover Limestone, Arkansas: Journal of Paleontology, v. 16, p. 126-129.

Westermann, G.E.G., 1981, Ammonite Biochronology and Biogeography of the Circum-Pacific Middle Jurassic: Systematics Association Special Volume 18: Academic Press, London, p. 459-498.

Weyl, Richard, 1968, Volcanoes and Volcanic Rocks in Central America and the West Indies: *in* Saunders, J.B., ed., Transactions of the Fourth Caribbean Geological Congress: Caribbean Printers, Trinidad, p. 357-360.

______, 1980, Geology of Central America: Gebrüder Borntraeger, Berlin, 371 p.

Wilson, J.L. and Pialli, G. 1977, A Lower Cretaceous Shelf Margin in Northern Mexico: Bureau of Economic Geology, University of Texas at Austin, Report of Investigation 89, p. 286-294.

Woodring, W.P., 1954, Caribbean Land and Sea Through the Ages:

Geological Society of America Bulletin, v. 65, p. 719-732.
Wooten, J.W. and Dunaway, W.E., 1977, Lower Cretaceous Carbonates of Central South Texas: A Shelf-Margin Study: Bureau of Economic Geology, University of Texas at Austin, Report of Investigation 89, p. 71-78.

Jurassic-Cretaceous Biochronology and Paleogeography of North America,
edited by G.E.G. Westermann,
Geological Association of Canada Special Paper 27, 1984

MESOZOIC TECTOSTRATIGRAPHIC DOMAINS IN EAST-CENTRAL MEXICO

Jose F. Longoria
Geosciences Programs, The University of Texas at Dallas, Box 688, Richardson, Texas 75080

ABSTRACT

Six tectostratigraphic domains are established for the Mesozoic of east-central Mexico: 1) Late Triassic-earliest Jurassic rifting; 2) Early Jurassic (Sinemurian-Pliensbachian) intracratonic basin; 3) Mid-Jurassic (Toarcian-Bathonian) continental deposition; 4) (Callovian) Late Jurassic-Early Cretaceous (Valanginian-Barremian) geosyncline; 5) Mid-Cretaceous (Aptian-Cenomanian) carbonate platform; and 6) Late Cretaceous (Cenomanian)-Early Eocene flysch. Eleven major sea level changes (Paleobathymetric events P-0 to P-10) are postulated. The contrast between deepening and shallowing of the Mesozoic sequence is interpreted geotectonically. Six tectonic events (T-I to T-VI) are established for this region and tied into the platetectonic history of the Gulf of Mexico. The geosynclinal domain is clearly differentiated into a miogeocline to the east (Tamazunchale area), and an eugeocline in the west (Teloloapa-La Trancas sequence). Tectonic event IV affecting this domain corresponds to the overriding of the eugeocline on the miogeocline, thus, developing the paleogeographic conditions for the flourishment of the calcareous (rudist) platforms (El Doctor-Valles San Luis Potosi) in Aptian-Albian. Tectonic event IV3 is responsible for the extinction of these calcareous platforms. Major plate re-ordering in early Tertiary resulted in the juxtaposition of Mesozoic tectostratigraphic terranes in east-central Mexico.

RÉSUMÉ

Six domaines tectonostratigraphiques sont établies pour le Mésozoique de la région est-centrale du Mexique: 1) Rifting du Trias supérieur-Jurassique inférieur; 2) La bassin infracratonique du Jurassique inférieur (Sinemurien-Pliensbachien); 3) Des dépôts continentaux du Jurassique moyen (Toarcien-Bathonien); 4) Géosynclinal du Jurassique superieur (Callovien)-Crétacé inférieur (Valanginien-Barremien); 5) Plateforme calcaire du Crétacé moyen (Aptien-Cenomanien); et 6) du Flysch Crétacé supérieur (Cenomanien)-Eocene inférieur. On postule onze changements majeurs du niveau de la mer Mésozoïque (Évenements paléobathymétriques P-0 au P-10). Le contraste entre la bathymétrie profonde et superficielle de la série Mésozoïque est interprètée du point de vue géotectonique. Six phases tectoniques (T-I au T-VI) sont établies pour cette région et corrélées à l'histoire de la tectonique des plaques du Golfe du Mexique. Le domaine géosynclinal est clairement divisé en un miogéoclyne à l'est et un eugéoclyne à l'ouest (La série Teloloapa-Las Trancas). La phase tectonique IV affectant ce domaine correspond à un chevauchemet du eugeoclyne sur le miogeoclyne, permettant ainsi le développement de certaines conditions paléogéographiques favorables à la croissance des plateformes calcaires (rudistes) en Aptian-Albien. La phase tectonique IV3 a causé la disparition de ces plateformes calcaires. Les réarrangements majeurs des plaques pendant le Tertiaire inférieur ont mené à la juxtaposition des terrains tectono-stratigrafiques du Mésozoïque dans la région est-centrale du Mexique.

INTRODUCTION

The combined use of microfacies analysis and physical stratigraphy to postulate relative changes in paleobathymetry of Mesozoic sequences in northern Mexico has proved to be a reliable technique in regional studies (Longoria, 1975; Longoria and Davila, 1979; Gamper, 1977).

In previous reports, the change in paleobathymetry was related for the most part to geotectonics (Longoria and Davila, 1979; Gamper, 1977). However, in these earlier reports, very little connection between regional and global tectonics was postulated. The study of the Mesozoic succession of the Huayacocotla Segment of the Sierra Madre Oriental demonstrates that the majority of the postulated Mesozoic paleobathymetric changes can be linked to global tectonics, that is, they are related to the plate-tectonic history of the Gulf of Mexico and Caribbean region. Thus, the postulated paleobathymetric

Project 171:
Circum-Pacific Jurassic

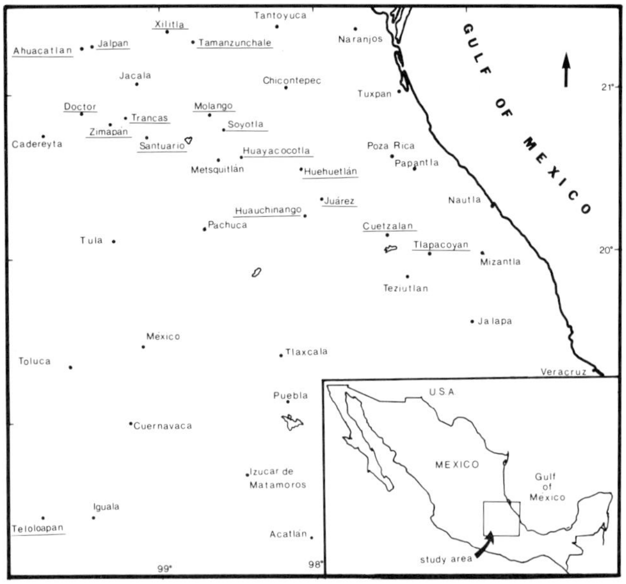

Figure 1 Index map showing geographic location of the towns mentioned in the text. Underlined towns indicate areas of field work.

history for the Mesozoic sequence of the Huayacocotla Segment of the Sierra Madre Oriental is considered to be of fundamental significance in unravelling the geotectonic history and paleogeography of this region.

The area under consideration is part of the Sierra Madre Oriental geologic province of eastern Mexico and lies between meridians 98° 00' and 100° 00' and parallels 20° 00' and 22° 00'. It includes parts of the states of Queretaro, San Luis Potosi, Hidalgo, and Puebla of east-central Mexico (Fig. 1). The Sierra Madre Oriental is the most prominent physiographic feature in eastern Mexico. It is formed mainly of Mesozoic sedimentary rocks which have been related classically to the Mexican Geosyncline (Cserna, 1967, 1971, 1976). The broad lithostratigraphic setting of this region is known from the pioneering work of Burckhardt (1930), Erben (1956a,b,c,d), Bonet (1956), Bondelos (1956), Segerstrom (1961), and Carrillo Bravo (1965).

GEOLOGICAL SETTING

The Sierra Madre Oriental of eastern Mexico covers a vast area, forming an almost continuous structural belt from northern Veracruz to southern Nuevo Leon (Fig. 1). Baisically, it has a northwest-southwest trend. The general tendency is to consider the Sierra Madre Oriental as a single uniform geological province. From the stratigraphic and structural viewpoint, however, this extensive belt can easily be separated into two segments – a northern segment, herein termed the "Victoria Segment", and a southern segment, herein named the "Huayacocotla Segment". Both segments contrast in their lithostratigraphy and structural style of deformation yet their boundaries are difficult to define. It is likely that these segments of the Sierra Madre Oriental reflect the Permo-Triassic segmentation of the Mexican cratonic area as postulated by Cserna (1976). The role of these left-lateral (megashears in) Mesozoic paleogeography is completely unknown. Recent paleomagnetic studies in northern Mexico by Gose *et al.* (1982) demonstrated that a significant part of northern Mexico rotated 130 degrees counterclockwise during early Mesozoic time. Their results yielded new insight into the paleogeographic and tectonic reconstruction of east-central Mexico which may have direct implications on the difference in lithostratigraphy on both segments of the Sierra Madre Oriental.

The geological evolution of the Sierra Madre Oriental has been interpreted from the structural and tectonic studies of the northern segment resulting in a generalization of the tectonic style of deformation. Early geologic mapping by Cserna (1956, 1967, 1976) related the structure of the Sierra Madre Oriental to the Mexican Geotectonic Cycle ending with the Hidalgoan Orogeny. According to Cserna (1971) this tectonic event is characterized by structures varying from broad open folds to isoclinal folds and overthrusts. He also stated that thrusting was invariably related to the rupture of the folds and that the structural axes of the folds are characteristically NNW-SSE oriented.

Through tectonic studies, Tardy (1980) postulated that the Mesozoic sequence of the Sierra Madre Oriental was involved in two tectonic phases resulting in two different styles of fold orientation. According to Tardy (1980) the NNW-SSE oriented folds resulted from a Miocene tectonic event whereas the E-W oriented folds are related to the main Laramide Orogeny. A controversy has thus developed pertaining to the origin of these two styles of deformation observed in the Mesozoic sequence of the Sierra Madre Oriental.

Recent structural studies by Suter and Carrillo (1981) in the Huayacocotla Segment demonstrate that the Mesozoic rocks are characterized by their isoclinal folding with sub-parallel structural axes. At the same time, they concluded that the thrusts in this area are not related to the rupturing of eastward overturned folds. In spite of these marked differences in tectonic styles, Suter and Carrillo (1981) related the structure of east-central Mexico to the Hidalgoan (Laramide) Orogeny.

Preliminary field observations on the style of deformation of the Huayacocotla Segment (Longoria, 1982) show that both eastern and western terranes of this segment vary in structural styles. The western terrane is characterized by low angle thrusts involving the Las Trancas and Santuario formations. In contrast, the eastern terrane is characterized by the tightly folded sequences of the Santiago, Taman, Pimienta and Chapulhuacan formations having only minor low-angle faults of short displacement whereby the Pimienta over-rides the Chapulhuacan Formation. The uncomfortable relationships between the Las Trancas and Santuario formations and the overlying El Doctor, as well as the uncomfortable relationship between the Chapulhuacan and Xilitla formations, suggest that pre-Laramide tectonism occurred in the area.

STRATIGRAPHIC FRAMEWORK

Eastern Terrane

Pre-Callovian Rocks. The pre-Callovian sequence includes

	MESOZOIC CHRONOSTRATIGRAPHY		SYNOPSIS OF LITHOSTRATIGRAPHY	
	Stage	Age (Ma)	Western terrane	Eastern Terrane
CRETACEOUS	Maestrichtian	70	**Mexcala Formation** (Fries) : Rhythmic alternation of marls, ,limestone, sandy limestone, and sandstone	**Mendez Shale** (Stephenson,1921) : Massive bedded yellowish-brown mudstone, and limestone; 1090 m
	Campanian	76		
	Santonian	82	**Soyatal Formation** (Segerstrom, 1961): Thinbedded, dark-gray limestone, and shaly limestone.	**Xilitla Group** (Bondelos,1956): Rhythmic alternation of shale, limestone, tuffs, and sandy limestone; erratic blocks; 880 m.
	Conician	88		
	Turonian	94		
	Cenomanian	100	**El Doctor Limestone** (Seger 1961): thin beddedshales, and chert. Limestone, dolomitic limestone,	
	Albian	106		**Ahuacatlan Formation** (Bondelos 1956): Dark-gray, medium-bedded limestone, black chert, and marls; 350 m.
	Aptian	112	**Santuario Formation** (Segerstrom, 1961): Dark-gray limestone, phyllitic shale with calcareous concretionns, and graywacke.	
	Barremian	118		**Chapulhuacan Formation** (Bondelos,1956): Thick to medium-bedded, light brown limestone (mudstone) and chert nodules; 400 m.
	Hauterivian	124		
	Valanginian	130		
	Berriasian	136		**Pimienta Formation** (Heim,1940): Black,medium to thin-bedded limestone, chert, and tuffs; 450 m.
JURASSIC	Purbeckian / Tithonian	140	**Las Trancas Formation** (Segerstrom, 1961): Shales, calcareous shale, phyllite; thin-bedded graywacke, chert	**Taman Group** (Heim,1940): Blackshale, limestone, and nodular limestone; 600 m. calcareous shale, medium-bedded
	Portlandian	146		
	Kimmeridgian	151		
	Oxfordian	157		**Tepexic Calcarenite** (Erben,1956): Mudstone, packstone, conglomeratic limestomestone; 45 m.
	Callovian	162		
	Bathonian	167		**Cahuasas Formation** (Carrillo-Bravo,1965): 'Continental red beds', shale,sandstone, and conglomerate; 800 m.
	Bajocian / Aalenian	172		
	Toarcian	178		
	Pliensbachian	183		**Huayacocotla Group** (Imlay,1948): Brown to yellowishshale; sandy shale, massive siltstone, sandstone; 1500 m.
	Sinemurian	188		
	Hettangian	190-195		
TRIASSIC	Rhaetian			**Huizachal Formation** Imlay,1948): Red shale, sandy shale, sandstone,and conglomerate; 2000 m.
	Norian	(205)		
	Carnian			
	Ladinian	(215)		
	Anisian			
	Olenekian / Scythian	225		
	Induian			

Figure 2 Mesozoic stratigraphy of the Huayacocotla Segment of the Sierra Madre Oriental (east-central Mexico).

the Huizachal, Huayacocotla and Cahuasas. The Triassic rocks (Fig. 2) have been regarded as continental "red beds" and they are assigned to the Huizachal Group, originally described from the northern segment of the Sierra Madre Oriental (Imlay, 1938; Carrillo Bravo, 1965). At Peregrina Canyon (NW Ciudad Victoria, Tamaulipas state), the Huizachal overlies the Permian flysch deposits (Guacamaya Formation) with a marked angular unconformity. In the Pahuatlan area (Puebla, state), the Huizachal is unconformably overlain by the Early Jurassic Huayacocotla Group, which in turn is overlain by the Cahuasas Formation. The Huayacocotla-Cahuasas contact is controversial and needs to be documented.

The Huizachal contains plant remains which were assigned by Mixon (1963) to the Rhaetian stage. The Huayacocotla contains abundant ammonite asemblages indicative of the Sinemurian and the Cahuasas is regarded as Middle Jurassic.

Late Bathonian-Aptian Rocks. Jurassic (Late Bathonian-Callovian) and Lower Cretaceous (Aptian) rocks from the eastern terrane are referred to the following lithostratigraphic units in ascending order: Tepexic Calcarenite, "Santiago Formation", Taman Formation, San Andres Formation, Pimienta Formation and Chapulhuacan Formation (Fig. 3).

Tepexic Calcarenite. The Tepexic is underlain by the Cahuasas red-beds. This contact marks the initiation of an Upper Jurassic marine transgression. The lower Tepexic is formed of bivalve-gastropod packstone and grainstone. Near the town of Tepexic (Puebla state), the upper part of the Tepexic Calcarenite

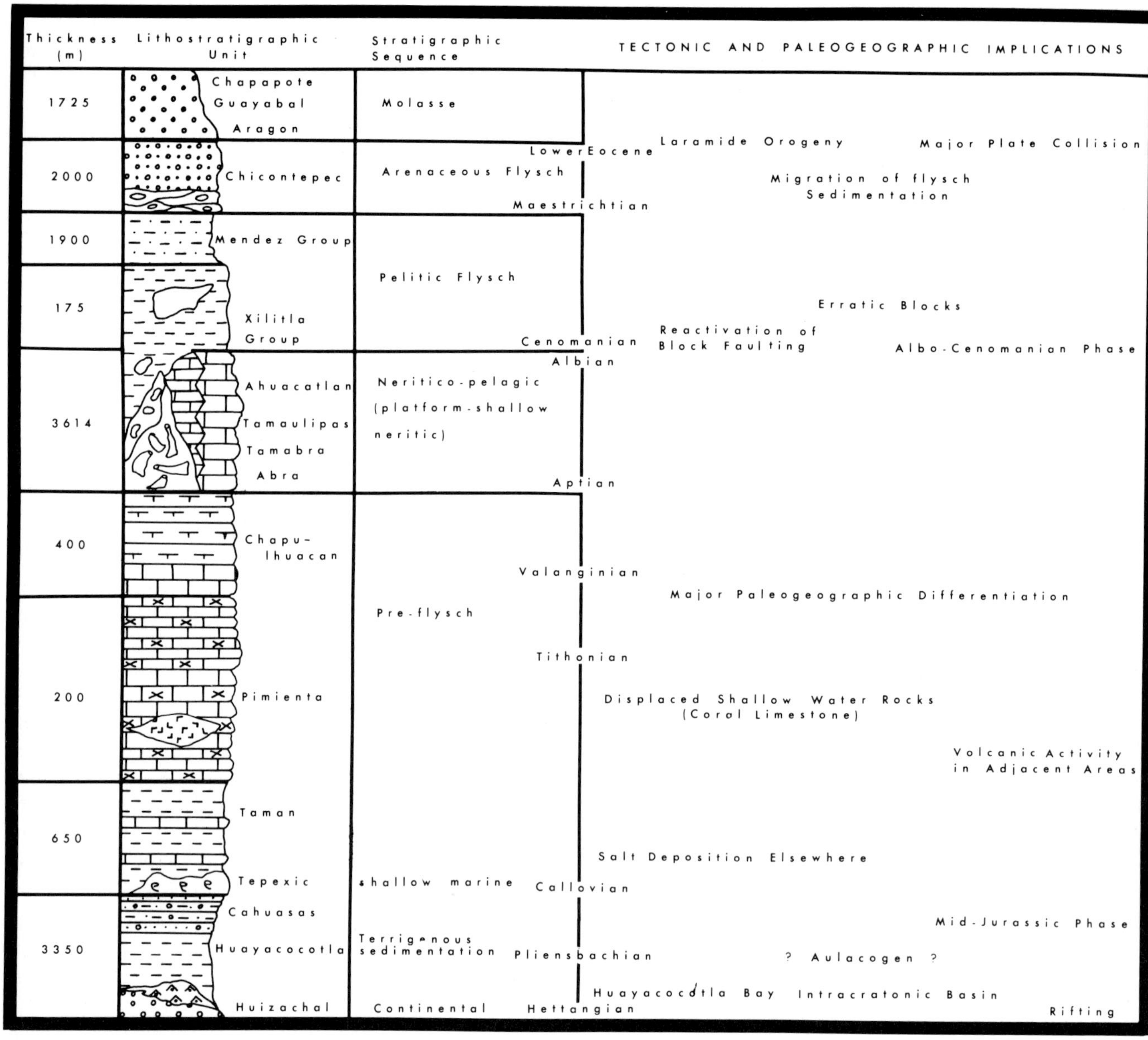

Figure 3 Tectonic and paleogeographic implications of the Mesozoic sequence, eastern tectonostratigraphic terrane.

changes abruptly into a dark mudstone containing ammonites assigned to *Reineckia (Neuqueniceras) neogaea* Burckhardt indicative of the Callovian (Erben, 1956a, p. 22) or latest Bathonian according to Westermann (Westermann *et al.*, this volume). This abrupt change in faunal and lithological character in the Tepexic Calcarenite is interpreted as a rapid change in sea level.

The shaly facies containing calcareous nodules was referred to as the "Santiago Formation" by Cantu Chapa (1971). As stated by Lopez Ramos (1979), however, this term has previously been applied in eastern Mexico to a Paleozoic unit. A detailed revision of the Jurassic lithostratigraphy of this region is needed in order to establish lithologic characterization of these units. For the moment, the black shaly facies which lacks calcareous nodules is included in the Tepexic. The unit consisting of an alternation of black shale ("Santiago" sensu Cantu Chapa, 1971.) with calcareous nodules and argillaceous black limestones is placed in the Taman Group. The maximum observed thickness of the Tepexic is about 50 m. However, near the town of Cahuasas the Tepexic is only about 0.70 m thick. In the vicinity of Tlacolula (ESE Tlapoacoyan), Veracruz, the Cahuasas Formation is overlain by the Lower Taman. This contact is sharp and is marked by the presence of a conglomeratic sandy dark limestone. In the Huehuetla area (Hidalgo state, Fig. 1), the Tepexic-Taman contact is transitional and is marked by a predominance of black shales with abundant calcareous nodules intercalated with argillaceous black limestone at the base of the Taman (often referred as to the "Santiago

Formation", see below). At this locality ammonites of Middle Callovian to Upper Oxfordian were identified by Cantu Chapa (1971).

Taman Group. The Taman Group is proposed herein to include all the strata between the Tepexic (below) and the Pimienta (above), including the "Santiago" Formation as defined by Cantu Chapa (1971). This group is subdivided into three parts. From bottom to top; they are 1) black shale with calcareous nodules and argillaceous black limestone, 2) medium to thick-bedded black limestone and shale, and 3) thin to medium-bedded black clayey limestone, shale, and tuffaceous layers.

The lower part of the Taman Group corresponds to the Santiago Formation as described by Cantu Chapa (1971). However since the term Santiago has previously been used in Paleozoic lithostratigraphy of eastern Mexico and in order to avoid further nomenclatural conflicts for the moment, the Taman Group is used.

The lower-middle Taman contact is transitional. It is placed at the first occurrence of homogeneously bedded limestone and shale. The middle part of the middle Taman often contains thick calcarenitic horizons made of shell fragments derived from shallow waters (referred as to the San Andres Member), while the upper part is remarkably similar to the lower part. The presence of displaced shallow-water fauna and rocks in the middle part of the Taman is interpreted here as an indication of the progressive advance of the Upper Jurassic transgressive cycle in east-central Mexico. The unconformable contact of the lower and middle Taman with the continental Cahuasas red beds at different localities (i.e., Tamazunchale area, Fig. 1) is an indication of an irregular topography. At the same time, this uncomfortable contact is an indication of the progressive incursion of marine waters on the continent. It is quite clear that the major Triassic taphrogenic (block-faulting) event led to the formation of rift topography. However, there is no evidence to suggest that the paleorelief resulting from Triassic block faulting was still present during the Middle Jurassic. It is possible, in fact, that a second episode of block-faulting occurred during mid-Jurassic time reactivating the rift topography. Both possibilities need to be examined with extensive field work and lithofacies analyses. In any case, field relationships show the diachronous contact between the Cahuasas Formation and the Upper Jurassic units (e.g., at the Cahuasas type locality and in the Huehuetla area, Hidalgo state; Fig. 1). This is perhaps the best evidence documenting an east to west transgression of the Callovian sea into east-central Mexico.

The Taman contains abundant ammonites which have been extensively studied by Burckhardt (1930), Erben (1956a, b, c, d) and Cantu Chapa (1971). According to its ammonite content the Taman ranges from early Kimmeridgian to early Tithonian.

San Andres Formation. The calcarenite horizons in the middle Taman have been referred to as the San Andres Formation by Cantu Chapa (1971). This unit consists of dark grey limestones (grainstone, packstone) with shell fragments (mainly nerineids). Burckhardt (1930) first reported the existence of this unit in the Huayacocotla area (Veracruz state) and referred to it as the "calcaire a Nerinees". The San Andres was regarded as a member of the Taman Formation by Cantu Chapa (1971). It overlies the Cahuasas Formation in the Rio Apulco area (Cuetzalan, Puebla state, Fig. 1)). Lithologically the San Andres is remarkably similar to the Tepexic Calcarenite from which it is sometimes difficult to distinguish. Field relationships with the overlying and underlying units seem to be the only means of differentiation these units. The San Andres was regarded as Kimmeridgian by Cantu Chapa (1971) mainly on the basis of stratigraphic relationships with the overlying Taman Formation. In the Tlacolula area (Veracruz state, Fig. 1)) the San Andres is well exposed and consists of about 80 m of calcarenites and fine-grained limestone. At this locality the San Andres is both underlain and overlain by the Taman.

Pimienta Formation. The Pimienta Formation overlies the Taman. This contact is conformable and transitional and is placed at the base of the first stratigraphic occurrence of thin-bedded limestone and chert. The lower Pimienta Formation is characterized by an alternation of thin-bedded limestone, shale and yellow-green tuffs. Excellent exposures of this unit can be studied in its type region (W of Tamazunchale, San Luis Potosi state, Fig. 1).

In the Tamazunchale area (San Luis Potosi state) the middle part of the Pimienta consists of thick- to medium-bedded calcarenites which contain reworked shallow water fossils (mainly corals), and the contact of the Pimienta with the overlying Chapulhuacan Formation is sharp but conformable. This boundary is marked by the first occurrence of brown, thick-bedded limestone and chert of the Chapulhuacan. The Pimienta-Chapulhuacan contact can be studied at several localities within the Tamazunchale-Pisaflores area (W of Tamazunchale). According to Cantu Chapa (1971) the Pimienta Formation contains Lower to Upper Tithonian ammonites. Calpionellids and radiolarians are also abundant in the Pimienta Formation.

Chapulhuacan Formation. The Chapulhuacan Formation consists of a homogeneous sequence of brown colored, thick-bedded limestone with chert nodules and layers. It overlies the Pimienta Formation and is overlain by the Ahuacatlan Formation. The upper contact is marked by a slight angular unconformity which can be observed at several localities in the Xilitla area (Fig. 1). Microfossils (calpionellids, nannoconids and planktonic foraminifera) occurring in the Chapulhuacan indicate that this unit includes strata ranging in age from Berriasian to Aptian.

Western Terrane

Jurassic-Lower Cretaceous strata of the Hidalgo state are included in the Las Trancas, Santuario and El Doctor formations (Fig. 2).

Las Trancas Formation. The Las Trancas consists of a thick, rhythmic alternation of thin-bedded limestone, shale, greywakes, and tuffs. It is well exposed along the highway from Zimapan to Jacala (Fig. 1). The lower contact is not exposed and has never been observed. Segerstrom (1961) reported the Tithonian ammonite *Paradontoceras*. Locally the shales and

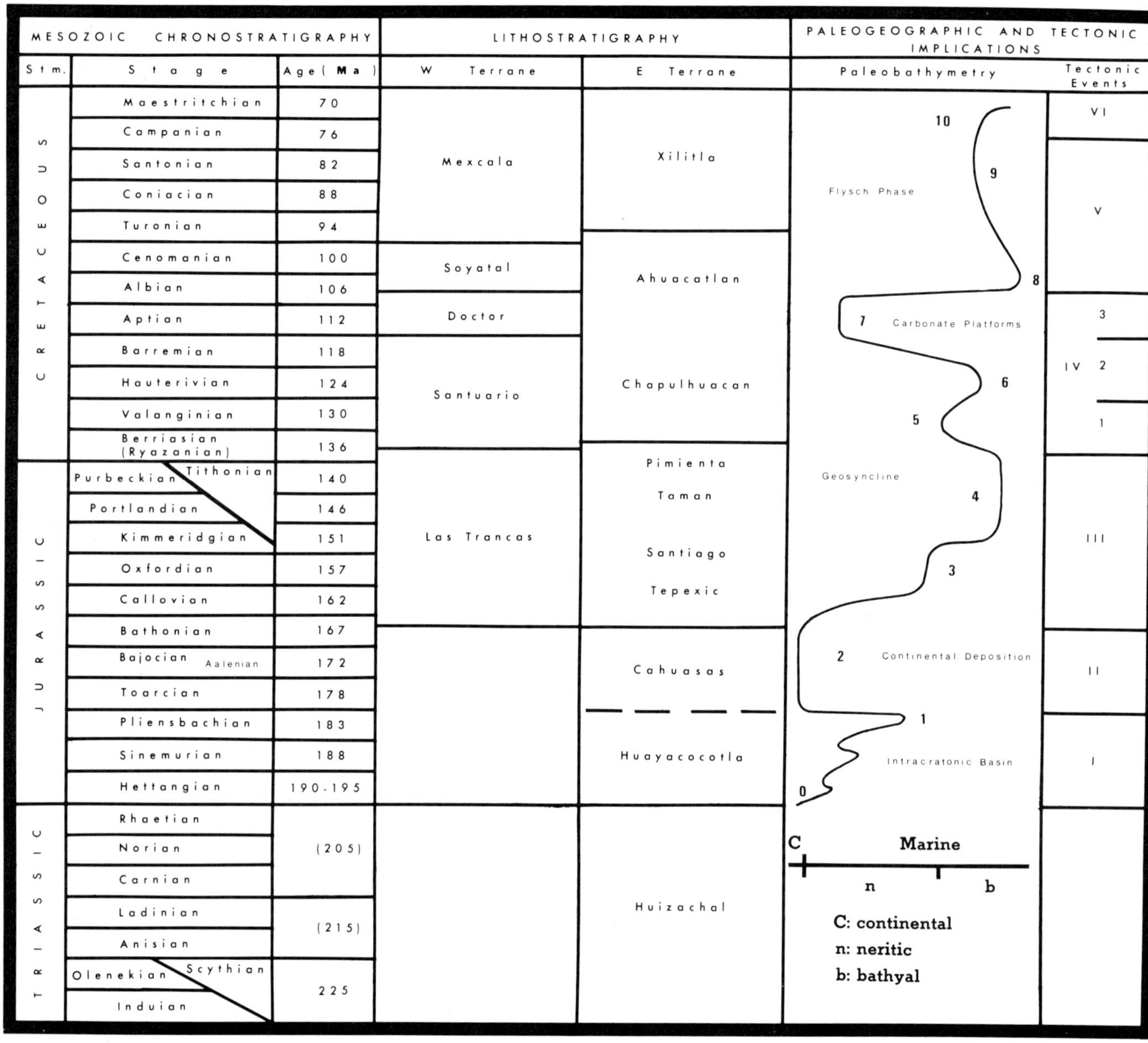

Figure 4 Sea level fluctuations and tectonic interpretations of both western and eastern tectonostratigraphic terranes of the Huayacocotla Segment, Sierra Madre Oriental (east-central Mexico).

limestone contain abundant radiolarians. The Las Trancas Formation is overlain by the Santuario Formation and locally by the El Doctor Formation. The Las Trancas-Santuario contact is sharp but conformable, corresponding with the predominance of limestone at the base of the Santuario. In the Pisaflores and Jalpan areas (Fig. 1), the El Doctor Formation overlies the Las Trancas Formation without evidence of a major angular unconformity.

Santuario Formation. The Santuario Formation is characterized by a rhythmic alternation of limestone, shale, chert and a minor amount of greywacke and it underlies the El Doctor Formation. This contact is transitional and is placed at the base of the first stratigraphic occurrence of the medium-bedded limestone characteristic of the El Doctor Formation. The Santuario Formation contains abundant Valanginian-Hauterivian microfossils. Segerstrom (1961) reported ammonites (*Olcostephanus* and *Acanthodiscus*) indicative of the same time interval.

El Doctor Formation. El Doctor is predominantly a carbonate unit whose lateral relationships are poorly understood. At least four different facies can be recognized: 1) biostromal, 2) conglomeratic, 3) thick-bedded limestone and chert, and 4) thin-bedded limestone and chert. It is postulated that, based mainly on the presence of abundant pelagic faunas, facies 3 and 4 were deposited in deeper waters. On the other hand, facies 1 and 2 are typical shallow platform facies and contain abundant rudist and miliolids. Deeper water facies 3 and 4 of the El Doctor conformably overlie the Santuario Formation. However, the shallow water facies 1 and 2 overlie the Las Trancas Formation. This contact does not represent a major angular unconformity. Fossils present in the El Doctor limestone (mainly planktonic foraminifera of facies 3 and 4) indicate that the base of this unit is Aptian.

PALEOGEOGRAPHIC INTERPRETATION AND TECTONICS

Paleobathymetric Changes

The first Paleobathymetric Level (P-0, Fig. 4) is related to the Sinemurian marine transgression in the area. The Huizachal-Huayacocotla contact in the Honey-Paxhuatlan area (northern Puebla state) is marked by a basal conglomerate, mainly composed of sedimentary and igneous fragments derived from Paleozoic and Triassic rocks. Thus, the initial sedimentary marine cycle is expressed by an uncomformity. The Lias in the area is characterized by a gradual marine transgression which terminates with the deposition of the Cahuasas continental red-beds, corresponding to a major fall in sea level (P-2). This drop in sea level is linked to the late rifting phase which occurred in the region during the Middle Jurassic. By late Bathonian-early Callovian time, however, marine deposition resumes (Paleobathymetric Level P-3; the Tepexic Calcarenite). The marine sedimentary cycle beginning in the late Bathonian-early Callovian can be characterized as a continuous transgression (P-4, Figs. 3 and 4).

The Late Jurassic marine transgression in the region reached its maximum advance during the Tithonian (Pimienta Formation), while in the Early Cretaceous (Valanginian) a significant fall in sea level took place (P-5). In the Hauterivian-Barremian (Chapulhuacan Formation) a rise in sea level is evident (P-6), followed by a fall in sea level during Aptian-early Albian (El Doctor Formation) (Paleobathymetric Level P-7). The abrupt change in sea level (El Doctor-Ahuacatlan) from late Albian to Cenomanian corresponding to a major transgressive episode is defined by Paleobathymetric Level P-8. Only minor data on the subsequent sea level changes could be obtained from the Huayacocotla Region.

Tectonic Events

Five major and one minor tectonic episodes can be defined from the sequences exposed in the Huayacocotla Segment of the Sierra Madre Oriental (Fig. 4). It is difficult, however, to correlate these six episodes of tectonic activity to the geologic evolution of North America due to the lack of reliable information on the stratigraphic and structural evolution of this part of the Sierra Madre Oriental.

Tectonic Event O: Early Rifting Phase. From the stratigraphic analysis it is evident that the rocks of Huayacocotla segment underwent a major taphrogenic event during Permo-Triassic times. This episode of block faulting in the area resulted in an abrupt irregular topography, allowing the deposition of thick sequences of continental and fluvial deposits (Huizachal Formation) corresponding to an early rifting phase.

Tectonic Event I: Advanced Rifting Phase. The incursion of marine water in the area took place in the Sinemurian corresponding to a late phase of rifting (Tectonic Event I). The Huayacocotla Group was deposited during this late rifting phase. Due to the serious geologic constraints in the tectonic reconstruction of the Gulf of Mexico area (Dickinson and Coney, 1980) it is, at this time, difficult to assess the existence of an aulacogen in the Huayacocotla region as postulated by Schmidt-Effing (1980).

Tectonic Event II: Shearing Phase. My own structural and stratigraphic field observations indicate that the Huayacocotla Group is affected by an earlier tectonic event which remains difficult to relate to other tectonic deformations in the area. Furthermore, ammonite faunas suggest the Pacific connection of this region during Sinemurian and Pliensbachian times and a lack of marine early Jurassic in Cuba and the eastern coast of North America (Imlay, 1980, p. 7-8). Thus, it is suggested, herein, that this part of the sequence is more related to the northwestern Lower Jurassic terrane than to either the Gulf or Caribbean domains. The Huayacocotla Group would, therefore, be unrelated to the early rifting in the Gulf area. Further work involving comparative studies of both the Huayacocotla Group and the Antimonio section (Caracahui Formation) in northern Sonora (Sierra de Santa Rosa) may solve this problem. This horizontal displacement via a left-lateral transform fault would thus correspond to Tectonic Event II and to the Callovian opening of the Gulf of Mexico (Fig. 5). The translation along this fault will account for the displacement of Paleozoic rocks of the Ouachita-Cordilleran belts from central Sonora (Hermosillo area) into western Tamaulipas (Cuidad Victoria area).

Tectonic Event III: Drifting Phase. The Upper Jurassic marine cycle of the Huayacocotla Segment was initiated by the deposition of the Tepexic Calcarenite. This cycle of deposition represents a continuous marine transgression resulting in a complete Upper Jurassic-Lower Eocene stratigraphic succession. The contrast in sedimentary environments between the Tepexic Calcarenite (Upper Bathonian – Lower Callovian) and the Cahuasas marks the beginning of Tectonic Event III. This event is related to active sea-floor spreading in the central Gulf of Mexico during the Late Jurassic corresponding to a continuous transgressive event (P-4). This event terminated in the lower part of the Chapulhuacan Formation (Valanginian-Barremian) by a falling of sea level. In this area this change in sea level is expressed in the reworking of shallow marine debris from the platforms, which coincides with Tectonic Event IV (Fig. 4).

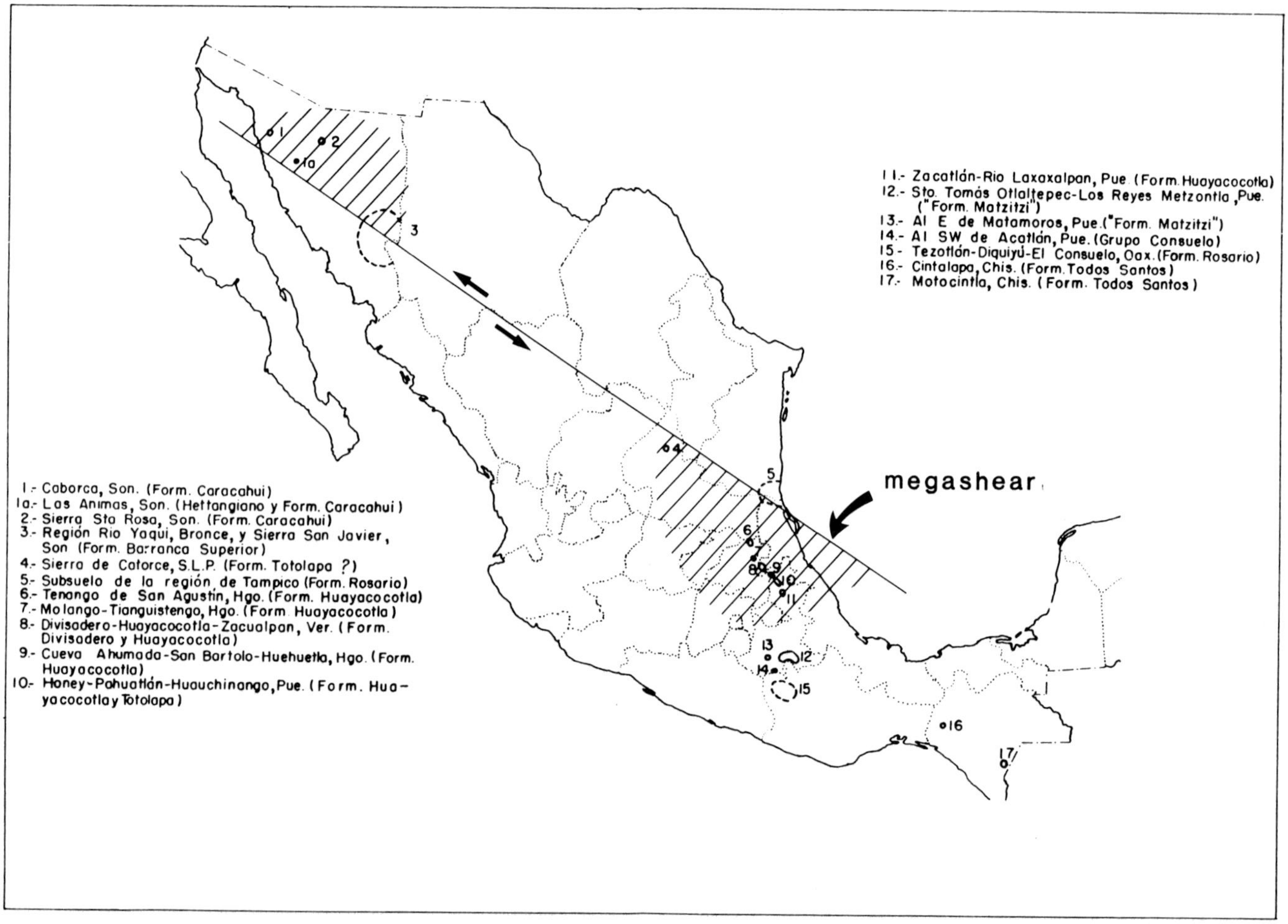

Figure 5 Distribution of early Jurassic marine deposits in Mexico (after Erben, 1956b), and postulated extension of the left-lateral transcurrent fault in Middle Jurassic times. Movement along the megashear (Tectonic Event II) initiates the Callovian marine cycle in east-central Mexico.

Tectonic Event IV: Reconfiguration Phase. Figure 6 illustrates schematically the paleogeographic setting postulated in the region during Late Jurassic (Kimmeridgian)-Valanginian time. The spatial relationships between both the eastern and western sequences are shown. It is assumed that an east-dipping subduction zone existed during Kimmeridgian-Tithonian and active sea floor spreading in the Gulf of Mexico was taking place (Tectonic Event III). However, by Valanginian-Hauterivian time the spreading in the Gulf ceased whereas the subduction in the west continued with Tectonic Event IV, that is, the over-riding of the eugeosynclinal (western) sequence on the eastern (miogeosynclinal) sequence produced the tight folding of the miogeosynclinal sequence and the thrusting of the euogeosynclinal (back-arc) sequence. Thus, plate collision on the west (paleopacific) led to partial closing of the back-arc basin (eugeosynclinal tectonostratigraphic domain) with a eugeosynclinal sheet thrust of more than 100 km eastward onto the shelf, over-riding the miogeosynclinal tectostratigraphic domain.

The unconformable relationship between the Santuario-Las Trancas formations and the overlying El Doctor Formation in the western terrane on one hand, and the non-deposition of the Ahuacatlan Formation in the Tamazunchale area on the other, offer the best evidence for this tectonic event during Early Cretaceous time. This paleogeographic scenario will help also to understand the development in the central part of the Huayacocotla Segment of Aptian-Albian platform (reef) facies (Laguna Colorada area) which in some localities unconformably overlaps the deeper water facies of the Santuario and Chapulhuacan formations. Burckhardt (1930) was the first to record this unconformable relationship between the Upper Jurassic-Valanginian strata and Albian-Aptian reef facies in western Mexico (Guerrero state). At the same time, the paleogeographic model proposed herein (Fig. 6) should explain facies differences within the El Doctor Formation. As noted earlier by Wilson *et al.* (1955), the El Doctor Formation displays at least four different facies varying from shallow biostromes to deeper pelagic limestones.

A major rise in sea level (P-8) marks the beginning of another tectonic episode in the area which corresponds to the deposition of the Soyatal and Ahucatlan limestones. This deepening is clearly seen in the entire Sierra Madre. In the northern part (Monterrey area) it corresponds to the Cenomanian (Cuesta del Cura Formation) transgression. Clearly this rise in sea level

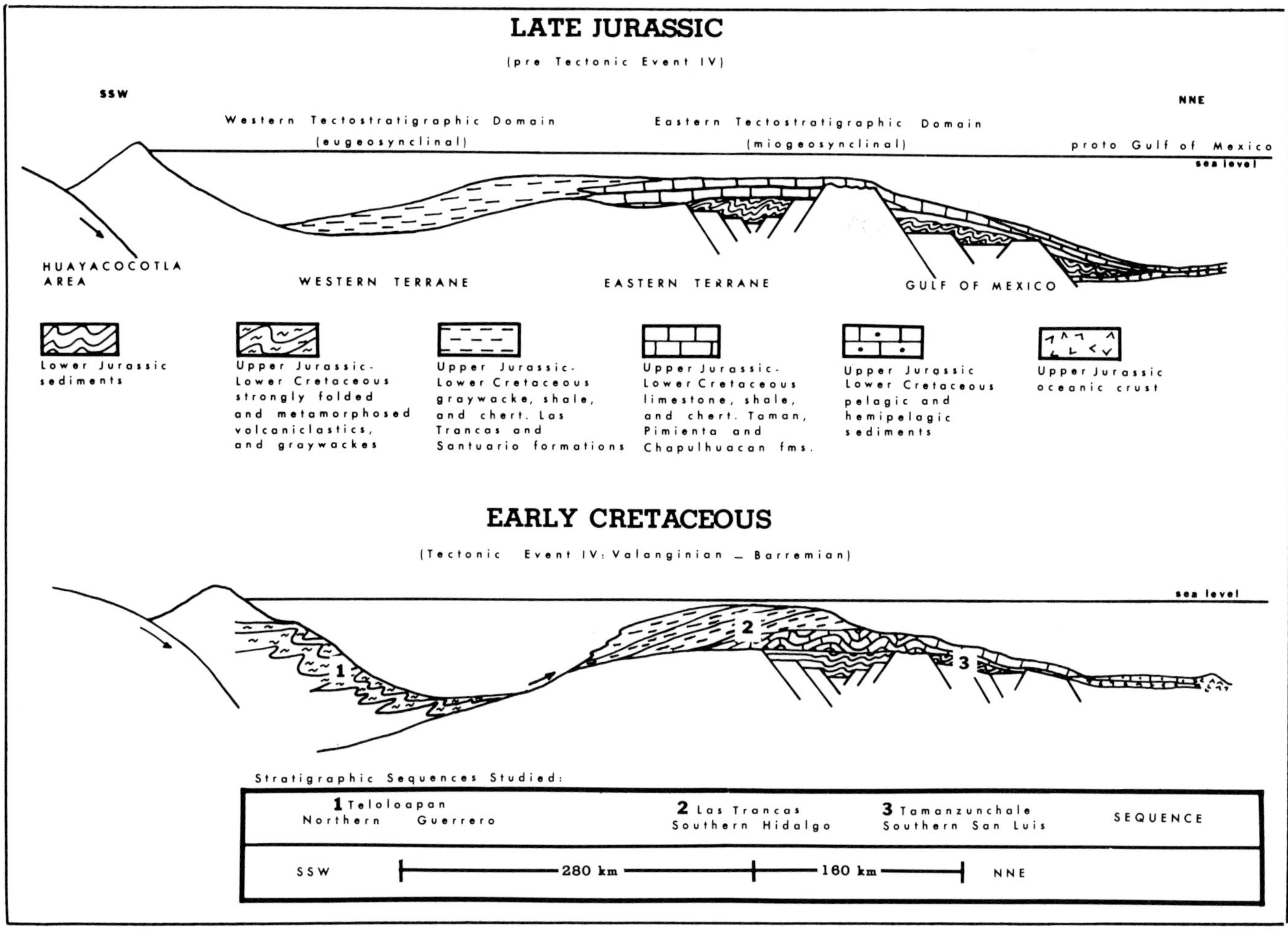

Figure 6 Late Jurassic-Early Cretaceous paleogeographic development in east-central Mexico.

represents a major transgressive episode in Mexico since the paleogeographic reconstruction of northern Mexico by Imlay (1943), Humphrey (1956) and Böse (1923) demonstrated that the paleopeninsulas of Tamaulipas and Coahuila were almost entirely covered by the marine transgression. This event corresponds to active sea-floor spreading in the Gulf of Mexico.

Tectonic Event IV is defined by two major contrasts in bathymetry – the Valanginian fall and the Cenomanian rise (Fig. 4). This event took place in three phases. Phase IV-1 corresponds in time with the cessation of sea-floor spreading in the Gulf of Mexico. Phase IV-2 is difficult to relate to the history of the Gulf since there is no information on the geological history of the Gulf at this time. Phase IV-3 corresponds to the development of shallow carbonate platforms in east-central Mexico (Enos, 1974).

Tectonic Event V: Reactivation of Block-Faulting. As it was stated before, there is little reliable information concerning the tectonic history of the late Cretaceous. It is likely, however, that by Albo-Cenomanian time a reactivation of the block-faulting occurred in the region, corresponding to Tectonic Event V. This may be reflected by displaced shallow-water fossils (e.g., rudists) occurring in the San Felipe and Mendez (Figs. 3, 4).

TOWARD A PALEOGEOGRAPHIC MODEL

The paleogeographic reconstruction of east-central Mexico is of fundamental significance in understanding the evolution of cratonic North America. However, paleogeographic analysis dealing with field and laboratory studies are still rare in this vast region of Mexico. My preliminary interpretation of the Mesozoic paleogeography of east-central Mexico is based on a combination of field stratigraphic studies and laboratory analyses of microfacies. Pre-Mesozoic and post-early Cenozoic paleogeographies are analyzed only as a background for the Late Jurassic to Early Cretaceous evolution of this region.

Tectostratigraphic Domains

From the foregoing stratigraphic synthesis, it is evident that the Mesozoic sequences of the Huayacocotla Segment were deposited in six tectostratigraphic domains: 1) late Triassic-earliest Jurassic (Hettangian) rifting, 2) early Jurassic (Sinemurian-Pliensbachian) intracratonic basin, 3) mid-Jurassic (Toarcian-Bathonian) continental deposition, 4) (Callovian) Late Jurassic-Early Cretaceous (Valanginian-Barremian) geosyncline, 5) mid-Cretaceous (Aptian-Cenomanian) shallow platform, and 6) late Cretaceous (Cenomanian)-early Eocene flysch.

These tectostratigraphic domains are related to the six tectonic events described previously (Fig. 3). The first three domains have not been documented directly from the western terrane of the Huayacocotla Segment. It is likely that the lack of stratigraphic studies in this area has impeded the recognition of these features, or else the western terrane corresponds to a suspect terrane. Paleogeographic analysis of selected areas is needed before arriving at definite conclusions on the paleogeographic affinities of the western terrane during the time corresponding to these domains.

1) Late Triassic to Hettangian (205 to 190 Ma; Harland et al. scale) Rifting. A thick sequence of continental red-beds now comprising the Huizachal accumulated in the rift valleys. In the Huayacocotla Segment (Honey, Puebla) its upper contact is marked by a basal conglomerate in the Huayacocotla Group. According to Belcher (1979) the Huizachal represents the infilling of grabens by alluvial deposits. The incursion of marine waters in this region is marked by the presence of Sinemurian ammonites (Erben, 1956a,b). Thus, the early rifting phase in this area occurred in pre-Huizachal time (i.e., pre-Carnian time–215 Ma). Although this may be the case, the sedimentological and stratigraphic evolution of the Huizachal Formation needs to be better documented to draw definitive conclusions bearing on the iniciation of rifting in the region.

2) Sinemurian-Pliensbachian (188 to 177 Ma) Intracratonic Basin. Marine sedimentation prevailed in an intracratonic tectostratigraphic domain during most of the Early Jurassic. The lack of Toarcian (178-171 Ma) marine faunas in east central Mexico was noted by Erben (1956b) and Imlay (1980). This can be interpreted as a result of tectonic event occurring at this time. By this time the Atlantic rift had reached the Gulf of Mexico region but it is difficult to determine the role of this rifting episode in the paleogeographic setting of east-central Mexico. The similarities in ammonite faunas and gross lithostratigraphy between the Sinemurian rocks of northwestern Mexico (Sonora) and east-central Mexico (Huayacocotla Segment) may, nonetheless, bear important information concerning the horizontal southeast displacements of large terranes from northwest Mexico to east-central Mexico (Fig. 5).

3) Toarcian-Bathonian (178 to 166 Ma) Continental Deposition. From Toarcian to Bathonian times a tectonic episode took place resulting in the opening of the Gulf of Mexico by transform faults in northern Mexico such as the Caltam megashear (Tardy, 1980) and the Mojave-Sonora megashear of Anderson (1982). However, the role of these megashears in the geodynamics of the Mesozoic history of Mexico remain on purely speculative grounds. The uncomformable relationship of the Huayacocotla-Cahuasas contact and the continental nature of the sediments of the latter unit support the presence of a tectonic episode in the region during mid-Jurassic time (Fig. 5).

4) Bathonian to Valanginian-Barremian (167 to 118 Ma) Geosyncline. A major paleogeographic differentiation occurred at this time in east-central Mexico and an E-W progressive transgression is evident throughout the region. This marine cycle marked the incursion of marine waters from the Tethys sea into continental Mexico. The first evidence of such a transgression is related to the deposition of Tepexic Calcarenite strata uncomformably overlying Cahuasas continental deposits. The gradual advance of the marine waters on the continental mass of east-central Mexico can be traced from Tamazunchale to Las Trancas (Hidalgo state) whereby the stratigraphic relationships indicate the time-transgressive nature of the Upper Jurassic sequence. During this time the Huayacocotla segment of the Sierra Madre Oriental reached its maximum paleogeographic differentiation. As the marine waters from the Tethys advanced into central Mexico (Tardy, 1980) the western side underwent the development of an island-arc system with a well-established back-arc basin to the east (Fig. 6). This paleogeography prevailed for about 40 Ma until a major plate reordering occured during Valanginian to Barremian times, resulting in the eastward thrusting of the back-arc sequence (eugeoclinal – Las Trancas and Santuario formations) and of the isoclinally folded western terrane (miogeoclinal – Taman, Pimienta, and Chapulhuacan formations). This phenomenon should account for the different styles of deformation in both western and eastern terranes of the Huayacocotla Segment of the Sierra Madre Oriental (Fig. 6).

From the structural and stratigraphic relationships, however, it is clear that marine conditions resumed immediately after the forementioned tectonic event. In the western terrane of the Huayacocotla Segment (Las Trancas area, Hidalgo state and Teloloapan area, Guerrero state) the reef-facies limestone (El Doctor Formation) rests with angular unconformity on the Las Trancas and Santuario formations in the Nacori area of Queretaro state. However, in the Pisaflores area the Santuario Formation grades upwards into the deep-water facies of the overlaying El Doctor Formation. This paleogeographic scenario served as foundation for the later development of the Aptian-Albian carbonate platform (El Doctor-San Luis Potosi-Valles).

The Upper Jurassic-Lower Cretaceous stratigraphic sequence studied from both the eastern and western terranes of the Huayacocotla Segment contains characteristic features which can be used to postulate the existence of a geosyncline. According to the distribution of these rocks, I can postulate that during late Bathonian-Callovian to Barremian time east-central Mexico was differentiated into two tectostratigraphic domains: 1) a miogeoclinal domain to the east, and 2) an eugeoclinal domain to the west (Fig. 6). The westernmost extension of the eugeoclinal domain may correspond to the island-arc setting of Guerrero and Michoacan states. Thus, an east-dipping subduction zone is envisioned in the westernmost portion of this segment corresponding to the metamorphic terrane of Guerrero (Teloloapan-Taxco area). This paleogeographic setting prevailed from the Kimmeridgian to the Valanginian (Fig. 6). In Valanginian to Barremian, however, a major plate collision took place leading to the partial closing of the back-arc basin resulting in the eastward thrusting of the eugeoclinal domain over the miogeoclinal domain (Fig. 6). Thus, it is clear that the unconformable relationship between the Jurassic and Aptian rocks, as seen in the Teloloapan (Guerrero state), Las Trancas (Hidalgo state), and Tamazunchale (San Luis Potosi state) areas (Fig. 6), is strong evidence for this Valanginian-Barremian tectonic episode.

5) Aptian to Cenomanian (117 to 110 Ma) Carbonate Platform. In Aptian-Albian time large carbonate platforms flourished on submarine topographic highs produced by the Valanginian-Barremian event (Tectonic Event IV). This stable condition lasted for about 15 Ma with only minor fluctuations in sea level. By Cenomanian time (100 Ma), however, transgressive episode P-8 took place. This episode, however, was only brief. By Santonian time (about 82 Ma) the stratigraphic sequence of the Huayacocotla Segment underwent a reactivation of block-faulting as can be demonstrated by the presence of large erractic blocks in the Xilitla Group of the Pisaflores region (W of Tamazunchale, Figs. 1,3).

6) Cenomanian to early Eocene (100 to 55 Ma) Flysch. This is the time of flysch generation in east-central Mexico with more than 4000 m of flysch sediments. Two sequences are recognized in the eastern terrane (Fig. 3): 1) the pelitic flysch (Xilitla and Mendez), and 2) the arenaceous flysch (Chicontepec Formation). The flysch sequence of the Huayacocotla Segment is older to the west (Soyatal Formation) than the pelitic flysch of the eastern terrane (Xilitla Group). Both the Soyatal and the Xilitla represent a pelitic flysch, however, the Xilitla developed into an arenaceous flysch (Chicontepec Group) in the late Paleocene-early Eocene (Fig. 2).

CONCLUSIONS

The superposition of tectonic events and changing paleogeographic scenarios through time make it difficult to correlate all of the observed events. At the same time, it is obvious that paleogeographic studies combining both field and microfacies analyses are needed before one can hope to understand the geotectonic processes that occurred in east-central Mexico. Paleogeographic studies are vital in unravelling the complex geologic history of this region and, at the same time, in establishing the role of this region in the evolution of cratonic North America. Finally, paleogeographic studies are fundamental in determining the paleogeographic affinities of large tectonostratigraphic terranes in east-central Mexico that may well be suspect terranes. However, the existence of suspect terranes in east-central Mexico should be documented via: 1) physical stratigraphy, 2) microfacies analysis, 3) paleomagnetic studies, or 4) paleontological studies. The use of more than one of these parameters is necessary to postulate the existence of suspect terranes in east-central Mexico. Then, hopefully, the observed tectonic phenomena can be related to orogenies and other tectonic activity as described from other regions of North America, particularly along its western margin.

ACKNOWLEDGEMENTS

The present report was founded in part by the National Science Foundation (Grant EAR-8121550). Discussions with Emile A. Pessagno, Jr. (The University of Texas at Dallas) and Jaime Rueda Gaxiola (Instituto Mexicano del Petroleo) have been encouraging and very stimulating. I also thank Emile A. Pessagno, Jr. and Prof. G.E.G. Westermann (McMaster University) for critically reviewing the manuscript. The present paper is the contribution no.. 435 of the Geosciences Programs, The University of Texas at Dallas. Special thanks are due to Mr. Manuel Valenzuela for his assistance in the field and for drafting the figures.

REFERENCES

Anderson, T., 1982, Geologic constraints on early Mesozoic reconstructions of the Gulf of Mexico: Geological Society of America Abstracts with Programs, v. 14, p. 433.

Belcher, R.C., 1979, Depositional environments, paleomagnetism, and tectonic significance of the Huizachal Red Beds (Lower Mesozoic) northern Mexico: Ph.D. Dissertation, Austin, University of Texas, 286 p.

Bondelos, A.J., 1956, Ruta Tamazunchale, S.L.P, Zimapan, Hgo. Tectónica de la Sierra Madre Oriental, Vulcanismo en el Valle de México: in Estratigrafia del Cenozoico y del Mesozoico a lo largo de la carretera entre Reynosa, Tamps y Mexico, D.F.: XX Congreso Geológico Internacional, Excursiones A-14 y C-6. Libro Guia, p. 119-152.

Bonet, F., 1956, Ruta: Taninul, San Luis Potosi-Tamazunchale, San Luis Potosi: XX Congreso Geológico Internacional, Excursión A-14, p. 93-117.

Böse, E., 1923, Vestiges of an ancient continent in Northeast Mexico: American Journal of Science, v. 6, p. 127-136.

Burckhardt, C., 1930, Etudé synthétique sur le Mesozoique Mexican: Sociéte Paléontologie Suisse, Mémoire 49-50, 280 p.

Burckhardt, C., 1930, Etude synthétique sur le Mesozoique Mexican: Societé Paléontologie Suisse, Mémoire 49-50, 280 p.

Cantú Chapa, A., 1971, La Serie Huasteca (Jurásico Medio-Superior) del Centro Este de Mexico: Instituto Mexicano del Petróleo Revista III, No. 2, p. 17-40.

Carrillo Bravo T., 1965, Estudio Geológico de una Parte del Anticlinorio Huayacocotla: Asociación Mexicana de Geologos Petroleros Boletin XVII, Nos. 5-6, p. 73-96.

Cserna, Z., De., 1956, Tectónica de la Sierra Madre Oriental de México, entre Torréón y Monterrey: XX Congreso Geológico Internacional (México), Monographia, 87 p.

______, 1967, Tectonic Framework of Southern Mexico and its Bearing on the Problems of Continental Drift: Sociedad Geológica Mexicana Boletin 30, p. 159-168.

______, 1971, Development and Structure of the Sierra Madre Oriental of Mexico: Geological Society of America Abstracts with Programs, Vol. 3, p. 377-378.

______, 1976, Mexico-Geotectonics and Mineral Deposits: New Mexico Geological Society, Special Publication No. 6, p. 18-25.

Dickinson, W.R., 1981, Plate Tectonic Evolution of the Southern Cordillera: in Dickinson, W.R., and Payne, W.D., eds., Relations of Tectonics to Ore Deposits in the Southern Cordillera: Arizona Geological Society Digest XIV, p. 113-135.

Dickinson, W.R. and Coney, P.J., 1980, Plate Tectonic Constraints on the Origin of the Gulf of Mexico: in Pilger, R.H., ed., The Origin of the Gulf of Mexico and the Early Opening of the Central North Atlantic Ocean: Louisiana Geological Survey, p. 27-36.

Enos, P., 1974, Reefs, Platforms and Basins of Middle Cretaceous in Northeast Mexico: American Association of Petroleoum Geologists, Bulletin 58, p. 800-809.

Erben, H.K., 1956a, El Jurásico Inferior de México y Sus Amonitas: XX Congreso Geológico Internacional México, Serie Especial, 393 p.

______, 1956b, El Jurásico Medio y El Caloviano de México: XX Congreso Geológico Internacional México, Serie Especial, 140 p.

———, 1956c, Estratigrafia y Paleontologia del Juràsico Inferior y Medio Marino de la Region Central de la Sierra Madre Oriental: XX Congreso Geológico Internacional México, Excursión C-8, 53 p.

———, 1956d, Paleogeographic reconstruction for the lower and middle Jurassic and from the Callovian of Mexico: in El Mesozoico de Hemisferio Occiental: XX Congreso Geológico Internacional México, p. 35-40.

Gamper, M.A., 1977, Estratigrafia y microfacies Cretàcicas del Anticlinorio Huizachal-Peregrina (Sierra Madre Oriental): Sociedad Geológica Mexicana, Boletin XXXVIII, No. 2, p. 1-17.

Gose, W., Belcher, R.C., and Scott, G.R., 1982, Paleomagnetic Results from Northeastern Mexico: Evidence for Large Mesozoic Rotations: Geology, vol. 10, p. 50-54.

Humphrey, W.E., 1956, Tectonic Framework of Northeast Mexico: Transactions of the Gulf Coast Association of Geological Societies, 6, p. 25-35.

Imlay, R.W., 1938, Studies of the Mexican Geosyncline. Geological Society of America Bulletin 49, p. 1651-1694.

———, 1943, Jurassic Formations of the Gulf Region: American Association of Petroleum Geologists Bulletin 27, p. 1407-1533.

———, 1980, Jurassic Paleobiogeography of the Conterminous United States in its continental Setting: United States Geological Survey Professional Paper 1062, 134 p.

Longoria, J.F., 1975, Estratigrafia de la Serie Comancheana del Noreste de México: Sociedad Geologica Mexicana, XXXVI, p. 31-59.

———, 1982, Anatomy of the Huayacocotla Segment of the Sierra Madre Oriental: Its Stratigraphic, Paleogeographic, and Tectonic Implications on the Origin of the Gulf of Mexico: Geological Society of America Abstracts with Program, v. 14, p. 549.

Longoria, J.F., and Davila, V.M., 1979, Estratigrafia y Microfacies del Cerro de la Silla, SE de Monterrey: Departamento de Geologia, Universidad de Sonora, Boletin 2, No. 1, p. 65-95.

Lopez-Ramos, E., 1979, Geologia de México: Edición Escolar, 446 p.

Mixon, R.B., 1963, Geology of the Huizachal Red Beds, Sierra Madre Oriental, Mexico: M.A. Thesis, University of Texas, Austin, 105 p.

Segerstrom, K., 1961, Geologia del Suroeste del Estado de Hidalgo y del Noreste del Estado de México: Asociación Mexicana de Geólogos Petroleros, Boletin XIII, Nos. 3-4, p. 147-168.

Schmidt-Effing, R., 1980, The Huayacocotla Aulacogen in Mexico (Lower Jurassic) and the Origin of the Gulf of Mexico: in: The Early Opening of the Central North Atlantic: Louisiana Geological Survey, p. 75-86.

Suter, M. and Carillo, M., 1981, Structural Traverse Across the Fold and Thrust Belt of the Sierra Madre Oriental of Mexico, Between Tamazunchale and Toliman: Geological Society of America Abstracts with Programs, Vol. 13, p. 109.

Tardy, M., 1980, Contribution a L'etude Géologique de la Sierra Madre Oriental de Mexique: Department de Geotectonique de l'Université Pierre et Marie Curie, Travaux, 459 p.

Wilson, B.W., Hernandez, P., and Meave-Torrescano, E., 1955, Un Arrecife Cretàcico en la Parte Oriental del Estado de Querétaro: Sociedad Geológica Mexicana, Boletin 18, p. 1-10.

Jurassic-Cretaceous Biochronology and Paleogeography of North America,
edited by G.E.G. Westermann,
Geological Association of Canada Special Paper 27, 1984

LATE JURASSIC-CRETACEOUS MOLLUSCAN PALEOGEOGRAPHY OF THE SOUTHERN HALF OF MEXICO

Gloria Alencaster
Instituto de Geologia, Universidad Nacional Autónoma, México 20, D.F., México

ABSTRACT

The paleogeography of the southern half of Mexico, reconstructed on the basis of the distribution of molluscan faunas, indicates that near the beginning of the Late Jurassic a gradual and progressive marine transgression advanced from the east toward the west covering high-relief topography. Islands, platforms, and basins were inhabited by pelagic and benthic, neritic and littoral molluscs. The deeper facies with ammonites were abundant from Late Jurassic to Early Cretaceous. Since the Late Jurassic, the Mexican Sea was connected with oceans to the east and west. Molluscs genera and even species have affinities with the Mediterranean Province of the Tethys Realm, and also with the northern and southern part of the Circumpacific Geosyncline. During Neocomian and Aptian times, the benthic communities of nearshore facies are very widespread. During Albian and Cenomanian times, the sea reached its maximum extent. Shallow platforms were very extensive where reefs, banks and patches of diverse communities of rudists and other molluscs proliferated. During the Turonian the slow regression of the sea toward the east began. The biocenosis of hippuritids and inoceramids thrived in a shallow sea. In Maastrichtian time, only a shallow platform extended around the southern Mexican Gulf Coast, with a rich endemic fauna in Chiapas and the Caribbean region.

Subsurface studies of the Yucatan Peninsula show that it was a low-relief platform covered by a shallow sea since at least the Albian to the end of the Tertiary, when it was again above sea level. Cretaceous limestone lenses with rudists are intercalated with evaporites.

RÉSUMÉ

La paléogeographie de la moitié sud du Méxique, établie sur la distribution des localités marines portant des mollusques, révele que, aux cours du commencement du Jurassique Supérieur, une mer transgressive a venu a envahir le continent, de maniere graduelle et progressive des l'est jusqu'a l'ouest. Dans le bassin, les îles et les plateformes resultantes ont veçu mollusques péllagiques et benthoniques. Le facies de profondeur modérée avec ammonites a dominé pendant le Jurassique Supérieur jusqu'a le Cretacique Inférieur. La communication entre la mer mexicaine et les mers de l'ouest et de l'est depuis le Jurassique Supérieur, est indéniable, a cause des affinités que presentent les mollusques, en genres et méme en especes, tant entre la Province Mediterraniénne du Tethys, comme des parties Nord et Sud du Géosynclinal Circumpacifique. Au Néocomien et Aptien, on remarque par leur abondance des faunes littorales près des cotes. A l'Albien et au Cénomanien, la mer atteignait son amplitude maxime. Les plateformes étaient tres étendues et permettait le développement des récifs et des bancs du rudistes et d'autres mollusques. Au Turonien la mer commence a se retirer lentement vers l'est. Les biocenosis des hippuritides et des inocéramides fleurirent dans eaux peu profondes. Au Maastrichtien la mer se limite a une frange bordant la Cote du Golfe, tres large au Chiapas, ou se développe une faune endémique de rudistes que s'étend jusqu'au les Caraibbes.

Les études du sous-sol de la Presqu'île de Yucatán, indiquent qui une plateforme s'est formée depuis l'Albien, quand sont deposées couches marines avec rudistes intercalés d'evaporites. Cette terre a emergait a nouveau a la fin du Tertiare.

INTRODUCTION

The purpose of this paper is to present the Late Jurassic to Cretaceous paleogeography of southern Mexico, reconstructed on the basis of the distribution of molluscan faunas. The time interval selected corresponds to the principal period of marine sedimentation in Mexico, starting with the progressive flooding of the country and ending with the gradual withdrawal of the seas.

Molluscs are the most abundant marine megafossils in the Mesozoic rocks in Mexico. Among these, the ammonites are the most numerous and reasonably well preserved both in the Jurassic and the Cretaceous rocks. The rudistids are numerous and well distributed only in the Cretaceous. In addition, other pelecypods and the gasteropods are widespread and offer valuable information concerning paleoenvironmental conditions. It is worthy to note, however, that in many localities the molluscs are indeed poorly preserved, scarce or fragmented. These are generally associated with other fossils of known age, such as tintinnids, nannoconids, radiolarids, echinoids and foraminifers. Localities with this kind of fossil material were also studied in order to be able to complete the record, specially among the Gulf Coast of Mexico where information has been obtained by drilling for oil.

Although, in the past, excellent paleogeographic reconstructions for the Mesozoic of Mexico have been presented by Burckhardt (1930), Imlay (1943, 1944, 1953, 1980) and Erben (1956b, 1957a) the new geological information gathered through drilling in east and south Mexico by Petroleos Mexicanos (PEMEX), in my opinion, deserves certain reconsiderations of southern Mexico. Consequently, I have concentrated my research on this part of the country during the past few years.

The localities shown on the maps in this paper are not all that have been cited in the geological literature, but are the most important ones from paleontological, stratigraphical and paleoenvironmental stand points. The numerous citations of the papers of Burckhardt and Imlay show clearly that these contain the most information to the subject of this paper.

When two or more stages are shown together on a map, it is because these occur together at most of the localities and thus their separate presentation would lengthen the paper but would not change the interpretation substantially.

DATA BANK OF FOSSIL LOCALITIES

The complete set of locality references for Figures 1 to 7 is available from the Departamento de Paleontologia, Instituto de Geologia, Universidad Nacional de México, Ciudad Universitaria, 04510 México, D.F. México.

PRE-OXFORDIAN

During the Early and Middle Jurassic most of Mexico was a landmass and in its southern half there were only two regions that received marine sediments. One of these was the central part of the present Gulf Coastal Plain where Lower Jurassic marine deposition occured (the Huayacocotla Embayment) (Burckhardt, 1930, p. 12-20; Erben, 1956a) and the other (the Guerrero-Oaxaca Embayment) was in south-central Mexico where the marine Middle Jurassic interfingers with continental coal and plant bearing strata (Burckhardt, 1927) (Fig. 1). It is worthy to note that in the Huayacocotla Embayment, Lower Jurassic marine beds are overlain unconformably by Middle Jurassic plant-bearing continental deposits which are in turn covered by marine Callovian sediments, whereas in the Oaxaca-Guerrero Embayment the Middle Jurassic marine deposition continued without interruption into the Late Jurassic. These embayments, although apparently separated in time and space, could have been isolated either from each other or

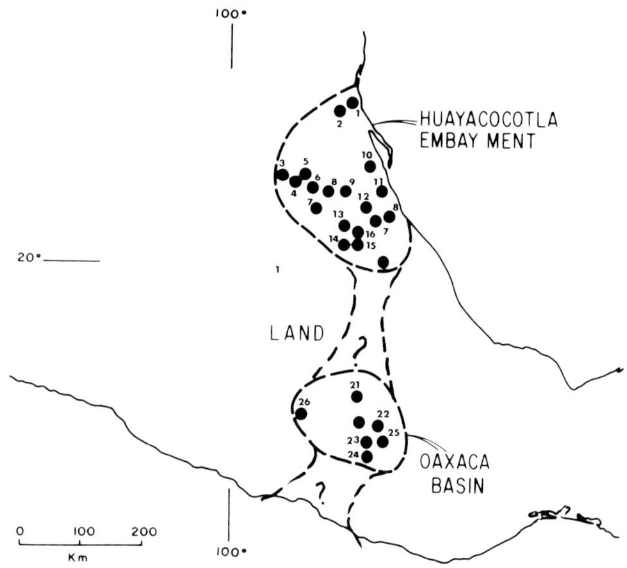

Figure 1. Pre-Oxfordian molluscan localities and paleogeography of Mexico.

connected. However, they probably had connections with the Pacific for there are sufficient arguments against the existence of the Gulf of Mexico at that time (Imlay, 1980; Viniegra, 1971, 1981).

Along the Gulf Coastal Plain, ammonites and oysters are abundant, with the more typical Callovian forms *Reineckeia, Neuqueniceras neogaeum* (Burckhardt) and *Liogryphaea nebrascensis* (Meek and Hayden) (Cantu Chapa, 1969; Westermann in Taylor *et al.*, this volume). This last species is present in the west of the United States of America and is similar to species of Argentina, while *Neuqueniceras* only exists in South and Middle America (Cantu Chapa, 1971, p. 19; Westermann *et al.*, this volume).

In northwestern Oaxaca (Fig. 1, locs. 21-24) Westermann (1983) has distinguished two ammonite faunas, named the *Parastrenoceras* and the *Zigzagiceras* assemblages, of late-Bajocian and supposed early Bathonian ages respectively which are more closely affiliated to the West-Tethyan than to the circum-Pacific fauna and are good evidence for a "Central Atlantic" marine sporadic connection between the Pacific and Tethys. The alleged *Zigzagiceras* subgenus *Duashnoceras* Westermann, however, has presently been recognized as a Late Bajocian stephanoceratid, so that Bathonian remains unknown (Westermann in Taylor *et al.*, this volume). In contrast, in the near-by Cualac area (Fig. 1, loc. 26) of northeastern Guerrero, the late Bathonian-early Callovian *Neuqueniceras* assemblage, shows a closer conspecific affinity with the Andean ammonite fauna of Argentina and Chile than with the Tethyan ammonites (Westermann *et al.*, this volume).

OXFORDIAN

The distribution of marine Oxfordian outcrops in southern Mexico (Fig. 2) shows the location and extent of the epicontinental seas at that time. There is as yet no faunal evidence to suppose that these marine areas were connected.

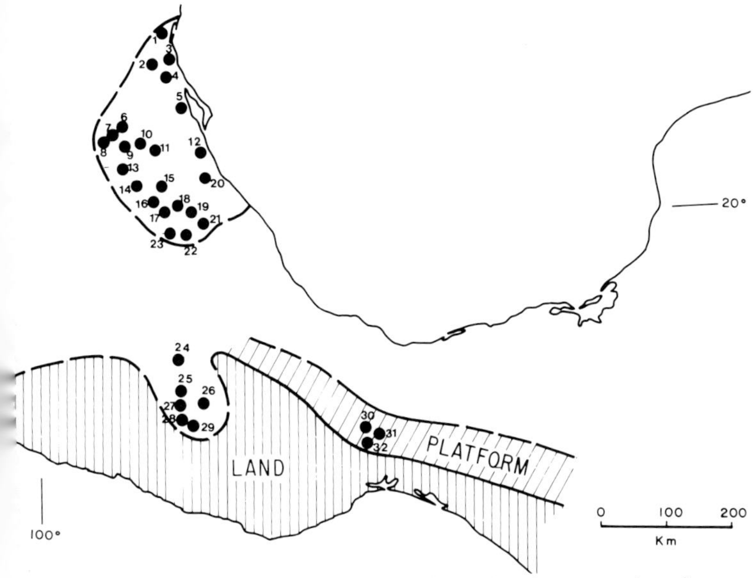

Figure 2. Oxfordian molluscan localities and paleogeography of Mexico.

In the lower and middle Oxfordian rocks, as in the upper Callovian, the fossils are absent or very scarce. *Fehlmannites* from northeastern Puebla (Cantu Chapa, 1971, p. 23) and *Creniceras renggeri* (Oppel) from Huayacocotla, Veracruz. (Erben, 1957a, p. 47) are the only two records of an early Oxfordian fauna. Imlay (1980, p. 50) and Viniegra (1971) assume that the major evaporitic sequences in the Gulf region were deposited during this time. This would explain the scarcity of life in a hypersaline environment. In contrast with the lower and middle Oxfordian, the upper Oxfordian rocks contain a very abundant fauna. *Discosphinctes, Dichotomosphinctes, Campylites* and *Euaspidoceras* are some of the common genera in these strata, which are found in the subsurface of the Tampico Embayment (Fig. 2, locs. 1-5), the Poza Rica area (Fig. 2, locs. 12-20) and in the subsurface and surface of the Huasteca region (Fig. 2, locs. 6-14) and the adjacent area of northern Puebla (Fig. 2, locs. 15-23) (Cantu Chapa, 1969, 1971, 1979).

The middle and late Oxfordian ammonite faunas of Cuba (Wierzbovski, 1976) are closely related to the Oxfordian faunas of Mexico. Some taxa which were considered as endemic of Cuba are now known from Mexico and from South America (*Cubaochetoceras mexicanum* (Burckhardt), *Discoscaphites caribeanus* (Jaworski), *Vinalesphinctes*, etc.). After Wierzbovski (1976, p. 163) the whole area of Middle and South America should be treated as biogeographically uniform.

In California and Oregon, as in the southern United States, the late Oxfordian is represented by species of *Dichotomosphinctes* and *Discosphinctes* that are similar to species in Mexico and Cuba (Imlay, 1980, p. 30). The same genera are present also in the contemporaneous strata of the Mediterranean region (Cantu Chapa, 1979, p. 19).

In southern Puebla and northern Oaxaca (Fig. 2, locs. 24-29) Oxfordian faunas which include bivalves, gastropods, brachiopods and spines of cidarid echinoids (Alencaster and Buitron, 1965) are closely related to species of the West Tethys.

The upper Oxfordian rocks of the Huasteca region and adjacent areas pertain to a marine sequence with middle Callovian beds at the base and upper Tithonian rocks at the top. The sequence was deposited during gradual and successive advances of the sea from the Gulf region toward the west. The land that was progressively being covered had rather abrupt relief in places, and was formed by granitic, metamorphic and sedimentary rocks of different composition and ages. Thus the Callovian strata have an unconformable relation with the underlying units and their basal terrigenous interval is of different ages from place to place due to the geometry of the transgression. The chronostratigraphic package that was deposited during this transgression forms the Huasteca Series (Cantu Chapa, 1969, 1971).

In the Huayacocotla area, the upper Oxfordian deposits are transgressive over the continental Liassic rocks; here, over a basal conglomerate and red shales, is a black compact limestone with *Nerinea* and above this is a thin bed with brackish water molluscs and coal denoting the proximity of a landmass (Burckhardt, 1930, p. 91).

The Oxfordian beds of southern Puebla and Oaxaca as well as of the Isthmus of Tehuantepec (Fig. 2, locs. 30-32) probably were deposited in a sea that transgressed northward or northeastward from the Pacific Ocean (Imlay, 1980, p. 51).

KIMMERIDGIAN

The area that became covered by marine Kimmeridgian sediments is considerably greater than that covered by marine Oxfordian sediments. The sea extended farther to the west, south and southeast (Fig. 3). In the northeastern part of the map the most widespread lithology is that of the Taman Formation (Heim, 1926) which consists of well bedded black, fine-grained limestone that alternates with black shale. Ammonites are abundant in this formation. Cantu Chapa (1969, p. 7; 1971, p. 25-28) recognized three lower Kimmeridgian biozones in this unit, which from the base upward includes: 1) the *Ataxioceras* Zone characterized by various species of this genus, as well as *Taramelliceras, Rasenia, Parataxioceras* and other genera, 2) the *Idoceras* Zone, widely recognized in east-central Mexico, contains many species of this genus, and rare species of *Aspidoceras;* and 3) the *Glochiceras* gr. *fialar* (Oppel) Zone, is very widespread throughout Mexico, and is associated mainly with the genera *Ochetoceras, Taramelliceras* (*Metahaploceras*), *Aspidoceras* and *Subdichotomoceras.*

The species of the *Ataxioceras* Zone are closely related to species of lower Kimmeridgian strata of western Europe, mainly from the south of Germany (Cantu Chapa, 1971, p. 37). As to *Idoceras,* Mexico is considered the dispersal centre of the genus, due to the great abundance and diversity that are present in the country. Outside Mexico it is found in Lower Kimmeridgian rocks of the southeastern United States, together with *Ataxioceras, Metahaploceras* and *Glochiceras* (Imlay, 1980, p. 31). It is also common in California, Argentina, Chile and Bolivia (Cantu Chapa, 1979, p. 20) as well as in the early Kimmeridgian of Italy and Germany (Cantu Chapa, 1971, p. 39).

In the Gulf Coastal Plain between the Tampico region and the areas of Tuxpan and Poza Rica (Fig. 3, locs. 1-16, 28-30) subsurface studies revealed the presence of a variety of facies of

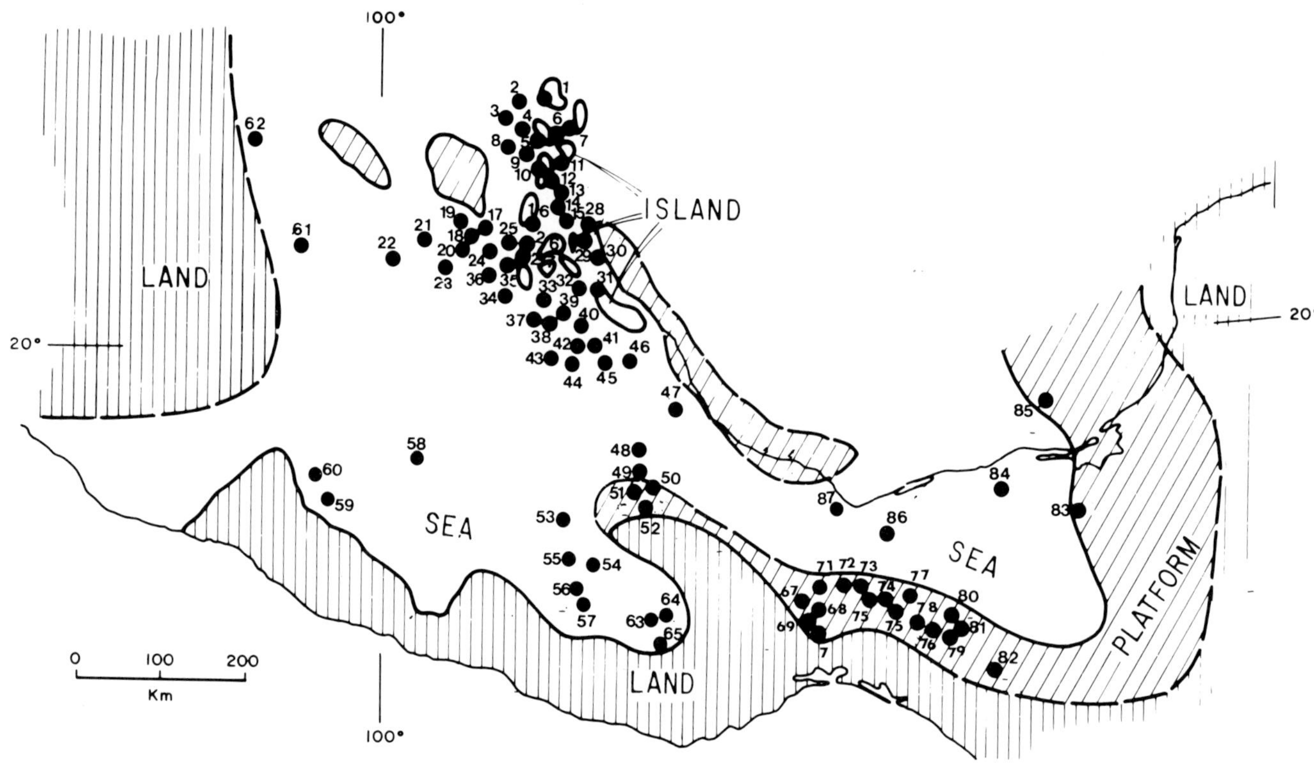

Figure 3. Kimmeridgian-Tithonian molluscan localities and paleogeography of Mexico.

the Kimmeridgian Stage included in several formations. Microfacies studies coupled with electric log correlations contributed a great deal of information about the shape and size of the islands that formed the Tampico-Tuxpan archipelago during this age. Several greater islands were located near Tampico (Stabler, 1972; Sânchez-Rosas, 1972) (Fig. 3). These studies show that some of the higher land areas were not covered yet by the sea during this time, which influenced the development of 11 facies in this region (Aguilera, 1972).

The localities in eastern Oaxaca and Chiapas (Fig. 3 locs. 66-82) contain a fauna of bivalves and gastropods, whose species show a clear relation with forms of the West Tethys (Alencaster, 1977, 1978).

TITHONIAN

During the Tithonian the sea extended farther to the west and to the south. Localities shown in the west and southwest of the map (Fig. 3, locs. 21-23, 58-62) consist of Tithonian deposits only.

The ammonites of south Zacatecas (Fig. 3, loc. 62) are scarce and badly preserved; nevertheless, early Tithonian beds are present with *Virgatosphinctes mexicanus* Burckhardt, *Suplanites* and *Pseudolissoceras* (Imlay, 1980, p. 31). In central Guanajuato (Fig. 3, loc. 61) *Himalayites* (*Micracanthoceras*) is present (López-Ramos, 1979a, p. 428; 1979b, p. 352).

In the southwest the Tithonian deposits consist of marine shales, sandstones and conglomerates (Fig. 3, locs. 59, 60) of metasediments and metavolcanics (Fig. 3, loc. 58), with poorly preserved *Gervillia* and probably *Gryphaea mexicana* Felix (Fig. 3, loc. 60), and very poorly preserved himalayitids *Micrancanthoceras, Wichmanniceras* and other forms similar to *Durangites* and *Corongoceras (*Campa *et al.*, 1974).

At the localities in the east (Fig. 3, locs. 21-23), the sequence of Tithonian-Neocomian age includes *Parodontoceras* and *Protancyloceras,* and consists of calcareous shales, greywackes and thin beds of chert, and represents a deep water facies.

In the Huasteca region, Cantu Chapa (1969, 1971) selected several species as zone markers. At the base of the Tithonian, he defined the *Virgatosphinctes mexicanus* (Burckhardt) Zone with *Aulacomyella neogaea* Burckhardt (Cantu Chapa, 1971, p. 27). Other species of *Aulacomyella* occur elsewhere in Mexico in beds with *Idoceras* and also in beds with *Glochiceras* gr. *fialar* (Oppel), but the genus ranges higher into beds containing the lowest occurrence of *Virgatosphinctes* (Imlay, 1980, p. 31). The zone of *V. mexicanus* which contains the early Tithonian *Taramelliceras* and *Subdichotomiceras,* is correlated with the upper part of the Taman Formation (Heim, 1926, 1940).

The *Kossmatia victoris* and *Pseudolissoceras zitteli* Zone, of middle Tithonian age, is correlated with the Pimienta Formation of Heim (1926). The *Suarites bituberculatum* Zone, which is of late Tithonian age, has a wide distribution and contains *Mazapilites, Kossmatia, Haploceras, Aulacosphinctoides.* The upper Tithonian is marked by the *Parodontoceras* aff. *callistoides* Zone.

In this region, the Upper Jurassic rocks are represented by a deep facies (Taman Formation), by a shallow water and lagoonal facies (San Andres Member) and by an open sea neritic facies (Pimienta Formation).

In general, there is a strong resemblance between the Titho-

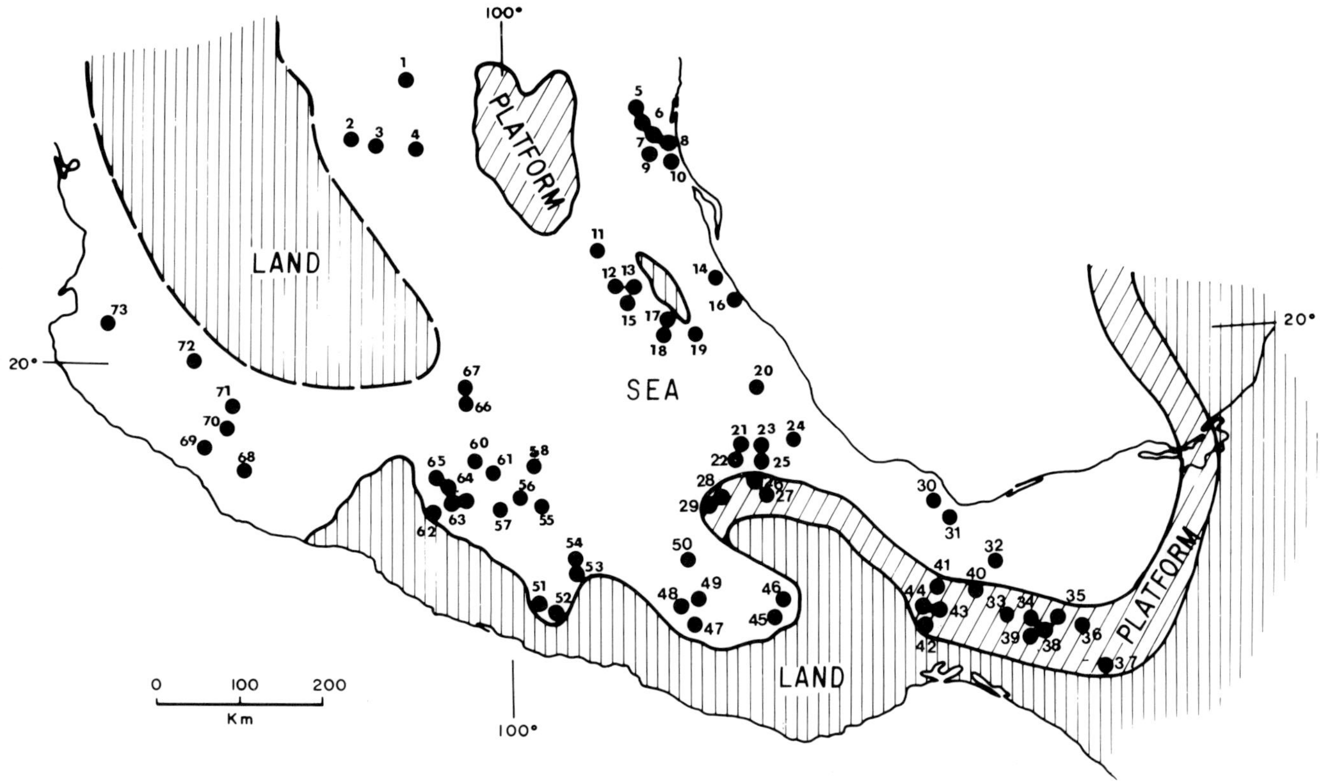

Figure 4. Neocomian-Aptian molluscan localities and paleogeography of Mexico.

nian ammonite faunas of Mexico and Cuba. The similarity is still more remarkable with the faunas located in the Gulf region and in the Sierra Madre Oriental, where the Pimienta Formation can be compared with the Artemisa Formation of Cuba. Both formations have the same lithological and faunistic sequences and in their tops are tintinnids in the Zone of *Parodontoceras* aff. *callistoides* (Cantu Chapa, 1976, p. 68).

In the southern parts of the Gulf Coastal Plain, the Upper Jurassic rocks have been referred to the Chinameca Limestone (Burckhardt, 1930, p. 97) and Mogoñé Limestone (Böse, 1905). In this southernly region, during the Late Jurassic there were several massifs formed by granitic, metamorphic and volcanic rocks which remained above sea level. In the west, apparently, a considerable region was not covered by the sea during this time, as deduced by the absence of marine sedimentary rocks or fossils (Fig. 3).

In southeastern Mexico, a thick sequence of Kimmeridgian-Tithonian rocks covers a vast region that includes the western half of Chiapas and parts of Oaxaca and Veracruz. These strata contain abundant faunas which include among many other species *Inoperna plicata, Pinna suprajurensis, Placunopsis suprajurensis, Myophorella clavellata, M. hudlestoni, "Cyprina" brongniarti,* and many species of *Pholadomya, Pleuromya, Gresslya, Anisocardia,* as well as *Harpagodes oceani* and other gastropods. These are benthic forms of neritic to litoral environments. The species of this fauna are identical to those of the Tethyan Mediterranean Province, indicating the presence of the same biozones in this region in similar stratigraphic position (Alencaster, 1977, 1978; Alencaster and Quezada, in prep.) and in a direct and close relation to that province. The shallow platform over which the sediments were deposited that contains this fauna deepens gradually to the north (Sânchez-Montes de Oca, 1969), as detected by the lithology and by the ammonite fauna of Cerro Pelón and Chinameca (Fig. 3, locs. 86, 87) as well as those close to the City of Oaxaca (Fig. 3, loc. 64). The presence of abundant fossil fish remains at these localities as well as the Tampico region within this stratigraphic interval is especially noteworthy.

The Mexican sea at this time had a direct and effective communication with the Tethys, as shown by the molluscs present in these localities. The fauna, to a lesser extent, contain some Andean and Californian elements also, implying certain communication with the Circum-Pacific Geosyncline toward the north and south (Cantu Chapa, 1969, 1971).

NEOCOMIAN-APTIAN

The gradual flooding of Mexico continued during the Early Cretaceous and an intracratonic geosyncline developed (Fig. 4). The ammonites are very abundant and widely distributed, as shown by their presence in almost all localities. The areas devoid of fossils were probably positive areas, like the western landmass of Mexico. In the south, the massifs of Guerrero, Oaxaca and Chiapas remained as positive areas and formed a barrier preventing the extending of the sea in that region. The Late Jurassic islands of the northeast became flooded by this Early Cretaceous transgression; however, like the one that existed in the Huayacocotla-Huachinango region (Imlay, 1944, Fig. 2), some still remained above sea level.

In the Neocomian-Aptian sequence a variety of depositional

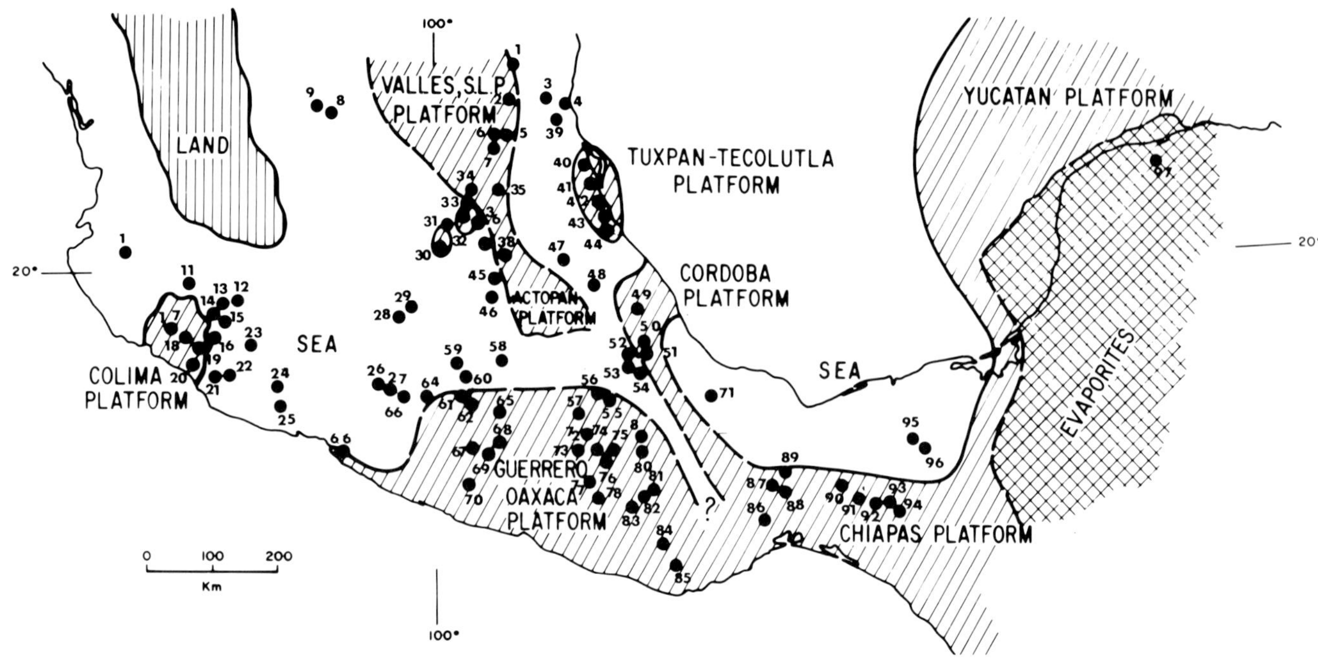

Figure 5. Albian-Cenomanian molluscan localities and paleogeography of Mexico.

environments can be recognized, as evidenced by the lithology and fauna. The localities shown within the geosyncline correspond (Fig. 4) to basin facies of thin-bedded limestones, commonly with chert interbedded with shales, and commonly containing ammonites. Thus, at Mazatepec (Cantu Chapa, 1967) (Fig. 4, loc. 19) the Berriasian is represented by limestones of the Lower Tamaulipas Formation, containing several species of *Subthurmannia, Groebericeras, Berriasella, Taraisites, Spiticeras* and tintinnids; it is a deep-water community related to Berriasian faunas from Argentina, France and India (Cantu Chapa, 1967). Similar associations are present in the north of Mexico (San Pedro del Gallo, Dgo.) as well as in the Oaxaca localities.

Deep-water localities (Fig. 4, locs. 1-4) in the north-central area, in the Tampico region (Fig. 4, locs. 5-10), in central Veracruz (Fig. 4, locs. 20-21) and in central Oaxaca (Fig. 4, locs. 45-46) contain *Hoplites, Astieria, Acanthodiscus, Crioceratites, Parahoplites, Olcostephanus, Neocomites,* etc.

The shallow-water facies has a wide distribution in the Huasteca region (Fig. 4, locs. 11-15), consisting of calcareous and carbonaceous shales with ammonites, and coal beds and ostreids near Huayacocotla (Imlay, 1944, p. 1079).

The rich Neocomian-Aptian faunas of southern Puebla deserve a special comment (Fig. 4, locs. 28-29). The Hauterivian and Barremian deposits consist of nearshore clastics, carbonaceous shale and limestone, with corals, monopleurid rudists, gastropods and bivalves. Similar lithology and fauna are found in localities 61-65 (Fig. 4), with *Nerinea titania* Felix in all outcrops. At some of these localities (Huetamo, Fig. 4, loc. 65 and Coyuca de Catalan, Fig. 4, loc. 62), a pelagic Barremian fauna is suggested by the presence of *Pulchellia, Holcodiscus, Barremites,* etc. (Burckhardt and Müllerried, 1936). The late Aptian rocks at San Juan Raya (Fig. 4, loc. 29) consist of near-shore facies, shaly marl, thin-bedded limestone, some conglomerate, and locally, beds of salt and gypsum. Bivalves and gastropods (Müllerried, 1933, 1934; Alencaster, 1956), as well as echinoids (Buitrón, 1970), ostreid banks and coral reefs are extraordinarily abundant. Remarkable is the strong affinity of the fauna with Aptian faunas of the Mediterranean Province of the Tethys – many species are common to both regions. There is also some relation with species of Lebanon and Japan.

In the southwest, a large region comprising the States of Colima and the southern part of the State of Jalisco and Michoacán (Fig. 4, locs. 68-73), the late Aptian benthic communities formed primarily by gastropods of the genera *Cassiope, Nerinea, Uchauxia* and *Craginia,* and monopleurid and requienid rudists, suggest that the region was covered by shallow seas.

Subsurface information from the Yucatán Peninsula indicates the absence of Neocomian-Aptian marine deposits, implying that at that time this region was probably a flat land area or a very shallow platform over which evaporites precipitated (López-Ramos, 1973, 1979b).

ALBIAN-CENOMANIAN

During the Albian-Cenomanian the sea reached its maximum extent in Mexico (Fig. 5). The massifs of Chiapas, Oaxaca and Guerrerro became flooded by the sea that advanced gradually toward the south-southwest, covering older terrains that were not inundated earlier. The epicontinental sea was relatively shallow and the organisms lived mostly in a neritic to littoral environment. The western landmass became more reduced and the sea invaded its western and southern margins (Fig. 5).

The rudists attained their maximum development during this time. They were the dominant invertebrates and with them abundant shallow-water benthic faunas flourished. The ideal conditions for this type of biota were the large platforms that had their widest development at this time. Thus, the southern half of the Valles-San Luis Platform (Carrillo, 1971) and its narrow southern prolongation called the Actopan Platform (Carrasco, 1971) are in the region discussed in this paper. To the west of this, two small platforms, Toliman and El Doctor (Segerstrom, 1961b), were sites of reef-facies limestones. Another great elliptical platform extended between Tuxpan and Tecolutla (Pedrazzini, 1978) upon which the oil fields of the Golden Lane are located, and farther south was the Cordoba Platform (González-Alvarado, 1976). This last one possibly continued to Chiapas, bordering the more or less deeper Veracruz Basin and the Gulf of Chiapas (Sánchez-Montes de Oca, 1969) (Fig. 5).

In south-central Mexico, the Morelos-Guerrero Platform extended roughly east of the meridian of Taxco, whereas toward the west a basin separated this platform from that of the Huetamo area (de Cserna *et al.*, 1978).

For the early Albian, *Coalcomana ramosa* (Boehm) is a very valuable guide fossil, for it has a very wide distribution. It is present in different faunal assemblages with monopleurids (Scott, 1981), caprotinids (Alencaster, 1984) being the most frequent faunal community made up of *Caprinuloidea and Toucasia,* as well as *Nerinea, Actaeonella* and *Orbitolina texana.* This assemblage is present at all shallow-water localities shown on the map as well as in Sinaloa (Bonneau, 1971), Baja California (Allison, 1955), Texas (Coogan, 1977), and in the Caribbean region (MacGillavry, 1937). For a long time this rudist facies was considered as Cenomanian (Boehm, 1898; Douvillé, 1900; Burckhardt, 1930), and even Cenomanian-Turonian (Palmer, 1928). Müllerried (1939) placed it into the middle Albian, whereas Imlay (1944, p. 1098) expressed his doubts as to its Cenomanian consideration because of the presence of *Orbitolina texana* in the same beds. The accepted age, presently, of the beds carrying this fauna is early Albian; it is based on the correlation of the *Coalcomana* beds with other strata of an age well established by microfossils, ostreids and ammonites (Bonet, 1956b, 1969; Coogan, 1973).

Toucasia is a very abundant requienid rudist, which forms part of various Albian assemblages or forms banks of one single species (Alencaster, 1973).

The basin facies deposits of Albian-Cenomanian age are generally present in areas between platforms and around them (Fig. 5). They contain pelagic faunas, principally ammonites together with planktonic microfossils, for instance, in the subsurface of the Tampico region (Fig. 5, locs. 3, 4, 39), in the Zimapan Basin (Carrasco, 1977) (Fig. 5, locs. 36, 37), and in the Apizaco-Zongolica Basin (Fig. 5, loc. 54) (Ontiveros, 1973). In those areas the presence of ammonites suggests a depositional environment transitional from neritic to moderately deep for the Upper Tamaulipas Formation (Muir, 1936; Ontiveros, 1973; Carrasco, 1977). Generally, the Albian beds are transitional with the Cenomanian. However, at a few localities the lithologies are indicative of a late Albian-Cenomanian deepening of the sea. At Orizaba (Fig. 5, loc. 53) and Coquimatlán (Fig. 5, loc. 18), the lower Albian rudist limestones are overlain by marly limestones containing banded chert and carrying *Mantelliceras, Mortoniceras* and *Uhligella.* In central Oaxaca (Fig. 5, locs. 81-84), the basin facies deposits of the upper Albian-Ceno-

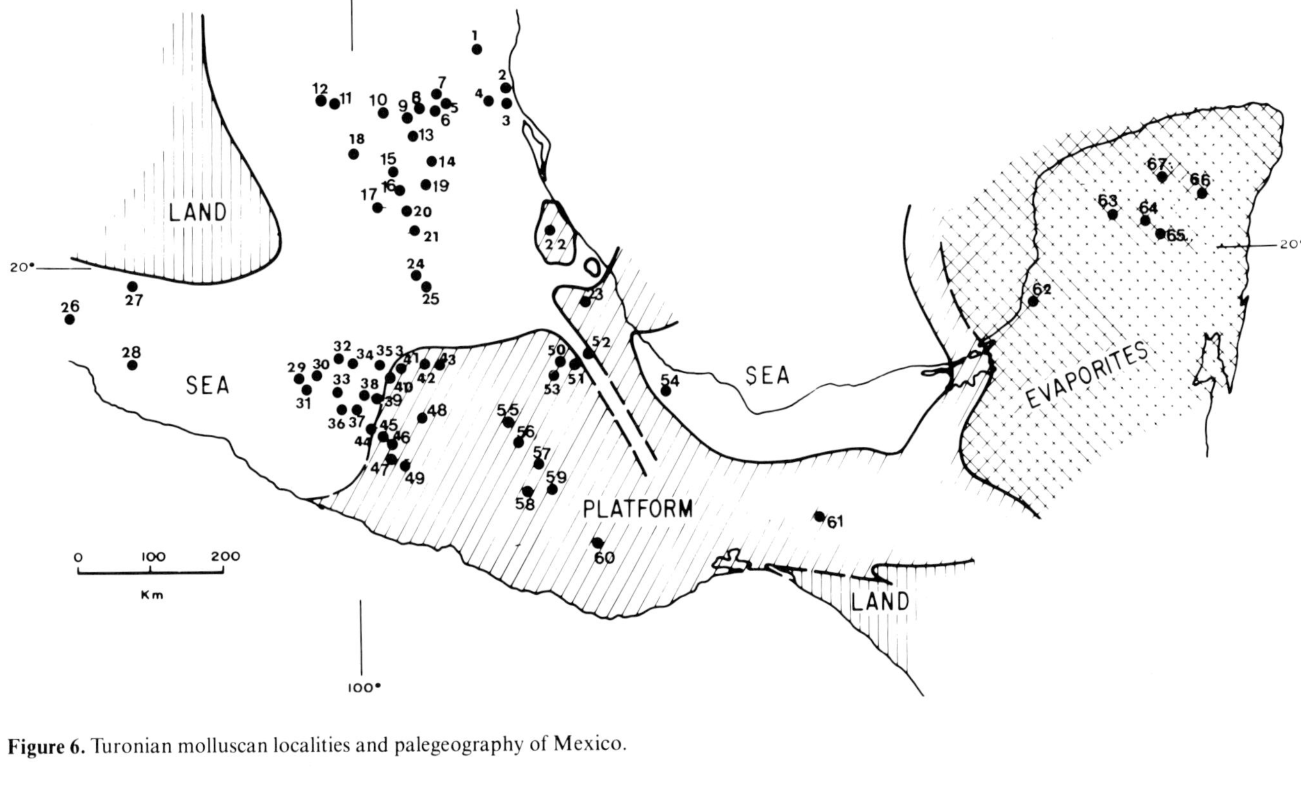

Figure 6. Turonian molluscan localities and palegeography of Mexico.

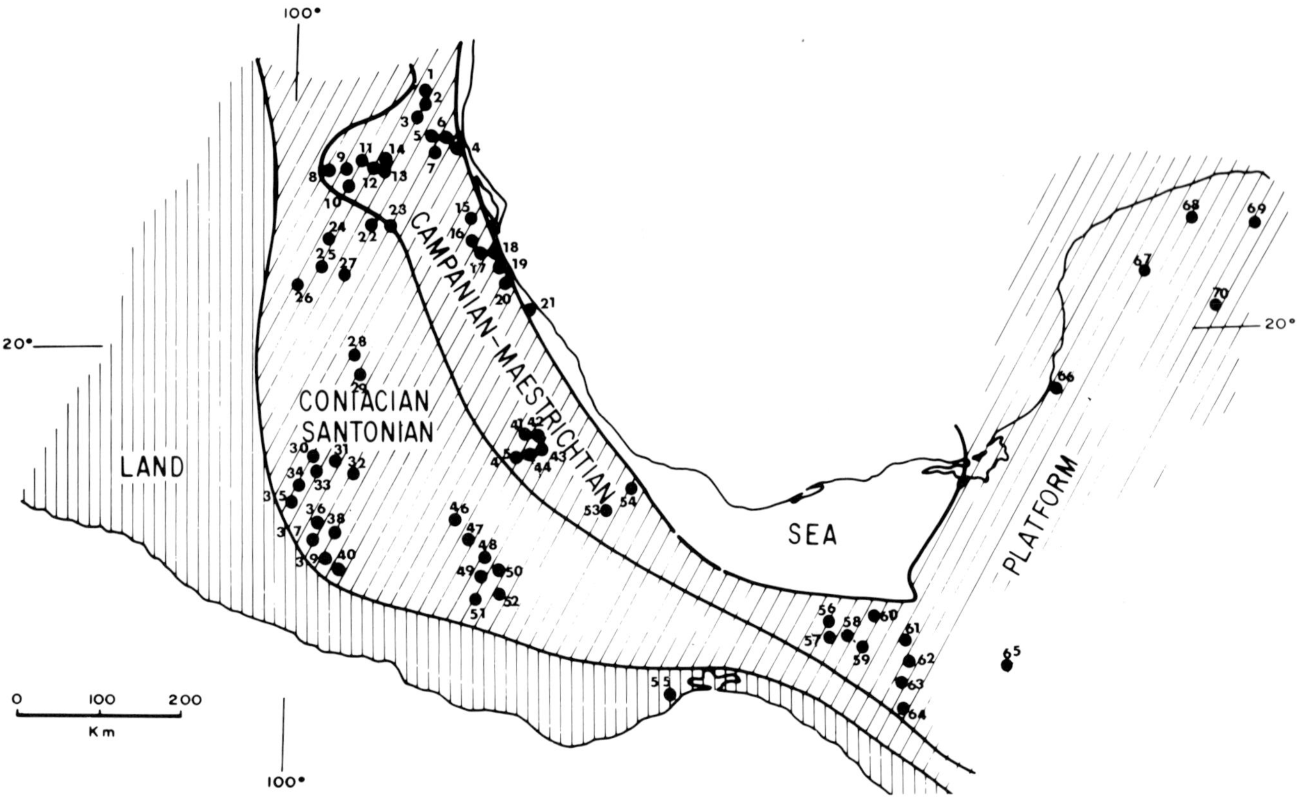

Figure 7. Senonian-Maastrichtian molluscan localities and paleogeography of Mexico.

manian contains *Hamites, Acanthoceras* and *Mantelliceras,* whereas in nearby localities (Fig. 5, locs. 72-80) of the lower and middle Albian, the faunal assemblages indicate littoral conditions of deposition of a shallow platform, similar to those that prevailed around the Chiapas Platform. In southwestern Mexico, the deposition upon the Colima Platform and nearby areas (Fig. 5, locs. 10-25) took place in shallow waters, as deduced by the presence of rudist banks with *Coalcomana.* In this latter region the carbonate rocks are interbedded with volcanics, indicating clearly eugeosynclinal conditions (Pantoja *et al.*, 1978).

TURONIAN

The distribution of the Turonian localities (Fig. 6) shows that the sea became somewhat reduced in extent in the west, as compared with the Cenomanian. This reduction becomes even more noticeable if it is kept in mind that localities 26-28 (Fig. 6) in the southwest are doubtfully Turonian. To date no certain Turonian fauna has been found in that region. The faunas reported (Palmer, 1928; Müllerried, 1950; López-Ramos, 1979b, p. 361) pertain to the Cenomanian.

The Turonian faunas show two wide-spread associations which are characterized by uniform and little varied composition. The inoceramid assemblage includes the characteristic *Inoceramus labiatus* Schlotheim, *I. hercynicus* (Petraschek), as well as other species of this genus, other bivalves and gastropods. This assemblage is found at eastern localities (Fig. 6, locs. 1-10, 13, 14, 20, 21) and in south-central Mexico (Fig. 6, locs. 55-59). The other assemblage is of hippuritids, whose principal form is *Hippurites resectus mexicanus* Bârcena (Müllerried, 1930, 1949). There are also other species of this genus in this assemblage, as well as various genera of radiolitids, such as *Durania* and *Sauvagesia.* This assemblage is found in the most eastern localities (Fig. 6, locs. 11, 12, 15-17, 24, 25, 29-44). It should be noted, however, that although these two facies generally occur separately, they do occur together in the Tampico region and north of Oaxaca.

In the central part of the State of Guerrero (Fig. 6, locs. 45-49), gastropods and pelecypods are very abundant, and *Actaeonella* forms banks at the top of the Turonian which are concordantly overlain by beds containing ammonites of early Coniacian age (Burckhardt, 1930, p. 237).

Solely because of the characteristics of the faunas, one can deduce that the seas during the Turonian were shallow to moderately deep, and that the animals inhabited the neritic and littoral zones. The hippuritids, just like all rudists, required a warm climate, well-oxygenated waters of high enery and shallow depths – conditions which are present in the littoral zone of a shallow platform (Fig. 6).

As to the inoceramids, Kauffman (1965, p. 213, 214) showed that these occur in all major different lithologies of the Western Interior Basin of the United States, but that the shallow shelf areas are the principal inoceramid habitats throughout the world. In the case of the Turonian sea of Mexico, the proximity of the hippuritid and inoceramid localities and also their association in some places show that they lived in a similar habitat.

SENONIAN-MAASTRICHTIAN

The Coniacian, Santonian, Campanian and Maastrichtian localities were grouped together and are shown on one single map (Fig. 7) in spite of the fact that these do not form complete sequences. This was done to show in a contrasting manner the extent of the Coniacian-Santonian seas with respect to the Campanian-Maastrichtian seas. As can be observed on the map (Fig. 7), the localities of these stages are toward the east and the region on the west is devoid of outcrops of marine rocks of these ages. In other words, the eastward migration of the coast that started toward the end of the Turonian continued advancing.

The Senonian-Maastrichtian faunas have a highly varied composition of rudists, ammonites, gastropods, and bivalves. However, it should be kept in mind that Upper Cretaceous outcrops at many localities are devoid of fossils and thus their more precise age assignment is very difficult at times. In general, the Upper Cretaceous fossils of this interval, where they are present, are abundant and well preserved.

The important Coniacian-Santonian faunas are those that occur in the westernmost parts of the paleogeographic map (Fig. 7). They consist of rudists of the same genera as in the Turonian, but different species. There are also *Inoceramus* like *I. undulatoplicatus* and some very large species of this genus. Ammonites are present at many localities, but generally they are not abundant. *Texanites* and *Nowakites* are indicative of the Coniacian-Campanian in the Zimapan region (Fig. 7, loc. 27; Segerstrom, 1956) and in northwestern Oaxaca (Fig. 7, loc. 48; Ferrusquia, 1976, p. 27).

The most important Coniacian fauna is in the central part of the State of Guerrero at Zumpango del Rio and Tixtla (Fig. 7, locs. 39, 40). It contains ammonites, gastropods, bivalves, echinoids, rare inoceramids, and *Didymotis* (Burckhardt, 1921; 1930, p. 237, 238). The ammonites are the most numerous and are represented by various species of *Barroisiceras, Peroniceras* and *Otoscaphites;* the latter was thought to be *Scaphites* of late Turonian age, until Imlay (*in* Fries, 1960, p. 84) showed that it was a different genus belonging to the lower Coniacian, just like the rest of the fauna.

In Central Guerrero, at the end of the Santonian the marine rocks became covered by continental deposits, indicating the withdrawal of the sea eastward.

As to the environments, the faunas are composed of benthic and pelagic molluscs, that lived in the littoral and neritic zones.

During the Campanian-Maastrichtian the withdrawal of the sea continued, and the sea occupied mainly the areas adjacent to the Gulf of Mexico (Fig. 7). It was probably somewhat deeper in its eastern parts. The faunas are isolated, but very abundant where they are preserved. In the region of Cárdenas (Fig. 7, loc. 8) the abundant fauna of the Maastrichtian contains gastropods of which *Actaeonella* is the most numerous, bivalves with *Exogyra costata* Say and *Arctostrea aguilerae* (Böse), and rudists with species in common with the fauna of Chiapas (Myers, 1968). Most of the fauna consists of benthic forms which lived in the infralittoral zone. The shallow depth of the sea is suggested by the presence of corals together with rudists and by the presence of ostreid banks and mud cracks in some of the beds (Myers, 1968, p. 26).

Another important fossiliferous region of the Campanian-Maastrichtian stages is located in the west-central part of the State of Chiapas. The fauna of these ages are very abundant and are found at many localities (Fig. 7, locs. 56-65). Here the rudists predominate, however, there are gastropods (*Actaeonella*) and bivalves including some inoceramids. Ammonites are very scarce at some localities, but the foraminiferal fauna is rich. Among the rudists, the most frequent are species of *Durania, Titanosarcolites, Barretia monilifera, Praebarrettia, Plagyoptichus, Chiapasella* and many other forms. Many of the genera attained gigantic sizes and the major part of the fauna is endemic (Alencaster, 1971). The fauna constitutes the Caribbean-Subprovince which is an extension of the Mediterranean Province of the Tethys Realm (Kauffman, 1973). It comprises the south of Mexico, the Antilles (MacGillavry, 1937; Chubb, 1971), and the northern parts of Central America. Toward the north it extends to the region of Cardenas; however, to date no intermediate localities have been found between these two regions.

In the Chiapas region, apparently the faunas were very scarce from the Turonian until the Campanian, since only a few specimens have been found so far indicating certain doubts as to these ages. It appears that in this region there is an unconformity between the "middle" and Upper Cretaceous and the corresponding discordant relations have been observed only at a few localities (Sânchez-Montes de Oca, 1969, p. 13). What is really surprising is that a scarce and impoverished as well as little varied fauna at a few localities was succeeded suddenly by an abundant, varied and very specialized endemic fauna that developed and became distributed in a relatively large region.

As to the environmental conditions, it was a sea of high energy with warm and well-oxygenated waters which permitted the development of gigantic rudists on banks and reefs, in shallow waters, over the internal and marginal portions of a platform. To the north of this region, the oil wells of the southern parts of Tabasco and Veracruz yielded information regarding the northerly progradation of this platform which bordered a sea of moderate depth (Sânchez-Montes de Oca, 1969; López-Ramos, 1979b, p. 367). The planktonic and benthic foraminiferal faunas are very abundant in this region as well as in the entire belt that surrounds this platform.

Subsurface studies in the Yucatân Peninsula show that this region was a lowland until the Albian, when it started to be covered by marine deposits consisting of a few beds and lenses of limestone with rudists and foraminifera, interbedded with thick intervals of evaporites which range from the Albian to the Turonian. The shallow platform conditions of this region lasted until the end of the Tertiary, when it began to rise to form a positive lowland.

In summation, throughout the Cretaceous, the molluscan faunas of the southern half of Mexico belong to a tropical biotic province characterized by "assemblages rich in nerineid and actaeonellid gastropods, associated with rudist reef facies" (Sohl, 1971, p. 1631). The predominant affinity of the faunas during most of the period with the Mediterranean Tethyan region is evident.

Paleogeographic interpretation are contingent on stratigraphy and paleontology which in turn depend on new geological information from the field. As geologic mapping progresses in the country, and as new fossil material is studied the interpretation will be updated, modified or corrected. Therefore, the present approach to this problem should be considered with an open mind, and as a preliminary step.

ACKNOWLEDGEMENTS

I am very grateful to Dr. Gerd Westermann who kindly invited me to participate in this symposium. I express a sincere admiration to the late Carl Burckhardt, who gave us an immense heritage through his basic work on the Mexican Mesozoic. To Dr. Ralph W. Imlay I want to extend my deep gratitude, also in the name of the paleontologists and geologists of Mexico, for carrying out superb research in Mexico, and for collaborated with Mexican professionals and scientists and for helping to solve their problems. The cordial collaboration of I. Ernesto López-Ramos, who through his compilations placed at our disposal newly published and even unpublished stratigraphic information, is duly acknowledged.

With thanks to the suggestions of the reviewers, Drs. Ralph W. Imlay, Gerd Westermann and Robert W. Scott, I improved the manuscript in many ways.

REFERENCES

Aguilera, E.H., 1972, Ambientes de Depósito de las Formaciones del Jurásico Superior de Tampico, Tamps.: Boletin de la Asociación Mexicana de Geólogos Petroleos, v. 24, p. 129-146.

Alencaster, Gloria, 1956, Pelecipodos y Gasterópodos del Cretácico Inferior de la Región de San Juan Raya-Zapotitlán, Estado de Puebla: Universidad Nacional Autónoma de México, Instituto de Geologia, Paleontologia Mexicana 2, 47 p.

——, 1971, Rudistas del Cretácico Superior de Chiapas, México: Universidad Nacional Autónoma de México, Instituto de Geologia, Paleontologia Mexicana 34, 91 p.

——, 1973, Una nueva especie de *Toucasia* en el Albiano de Oaxaca y Puebla: Universidad Nacional Autónoma de México, Instituto de Geologia, Paleontologia Mexicana 37, 18 p.

——, 1977, Moluscos y Braquiópodos del Jurásico Superior de Chiapas: Universidad Nacional Autónoma de México, Instituto de Geologia, Revista, v. 1 p. 151-166.

——, 1978, Presencia del Tethys en el Sureste de México durante del Jurásico Superior: Actas, II Congreso Argentino de Paleontologia y Bioestratigrafia y I Congreso Latinoamericano de Paleontologia, Buenos Aires, v. 5, p. 9-14.

——, 1984, Rudista Caprotinido (Bivalvia Hippuritacea) del Cretácico Inferior de Pihuamo, Jalisco: Universidad Nacional Autónoma de México, Instituto de Geologia, in press.

Alencaster, Gloria and Buitrón, B.E., 1965, Fauna del Jurásico Superior de la Región de Petlalcingo, Estado de Puebla: Universidad Nacional Autónoma de México, Instituto de Geologia, Paleontologia Mexicana 21, p. 1-53.

Alencaster, Gloria and Quezada-Muñetón, J.M. in preparation. Bioestratigrafia del Mesozoico de la Parte Occidental del Estado de Chiapas y del la Parte Oriental de los Estados de Veracruz y Oaxaca: Universidad Nacional Autónoma de México, Instituto de Geologia, Paleontologia Mexicana.

Allison, Edwin, C., 1955, Middle Cretaceous Gastropoda from Punta China, Baja California, México: Journal of Paleontology, v. 29, p. 400-432.

Boehm, G., 1898, Ueber Caprinidenkalke aus Mexico: Zeitschrift der Deutschen Geologischen Gesellschaft, v. 50, p. 323-332.

Bonet, Federico, 1956b, Zonificación Microfaunistica de las Calizas Cretácicas del Este de México: Congreso Geológico Internacional, Sesión 20, México, D.F., Sistema Cretácico T. III, 102 p.

——, 1969, Microfacies de las Calizas Cretácicas de la Región de Córdoba y Orizaba: Seminario sobre Exploración Petrolera, Instituto Mexicano del Petróleo, Mesa Redonda 4, 24 p.

Bonneau, M., 1971, Una nueva área fosilifera en el Estado de Sinaloa: Boletin de la Sociedad Geológica Mexicana, v. 32, p. 159-167.

Böse, Emil, 1905, Algunos datos sobre la Geologia del Istmo de Tehuantepec: Boletin del Instituto Geologico de México, v. 20, p. 75-85.

Buitrón, B.E., 1970, Equinoides del Cretácico Inferior de la Región de San Juan Raya-Zapotitlán, Estado de Puebla: Universidad Nacional Autónoma de México, Instituto de Geologia, Paleontologia Mexicana 30, 46 p.

Buitrón, B.E., 1983, Gasterópodos del Cretácico Inferior del Cerro de Tuxpan, Jalisco: Universidad Nacional Autónoma de México, Instituto de Geologia, Revista, in press.

Burckhardt, C., 1921, Faunas Jurásicas de Symon (Zacatecas) y Faunas Cretácicas de de jZumpango del Rio (Guerrero): Boletin del Instituto Geológico de México, v. 13, 135 p.

——, 1927, Cefalópodos del Jurásico Medio de Oaxaca y Guerrero: Boletin del Instituto Geológico de México, v. 47, 108 p.

——, 1930, Etude Synthétique Sur le Mésozoique Mexicain: Mémoire Societé Paléontologique Suisse, v. 49-50, 280 p.

Burckhardt, Carl and Müllerried, F.K.G., 1936, Neue Funde in Jura und Kreide Ost- und Süd- Mexicos: Eclogae Geologicae Helvetiae, v. 29, p. 309-324.

Campa, M.F.; Campos, Mario; Flores, Roberto and Oviedo, Ramón, 1974, La Secuencia Mesozoica Volcánico-Sedimentaria Metamorfizada de Ixtapan de la Sal, México-Teloloapan, Guerrero: Boletin de la Sociedad Geológica Mexicana, v. 35, p. 7-28.

Cantu Chapa, A., 1967, El Limite Jurásico-Cretácico en Mazatepec, Puebla (México): Instituto Mexicano del Petróleo, Sección Geologia, p. 3-24, pl. 1-7; also *in* 1968, Ingenieria del Petróleo, v. 8, p. 5-26.

——, 1969, Estratigrafia del Jurásico Medio-Superior del Subsuelo de Poza Rica, Ver. (Area de Soledad-Miquetla): Revista del Instituto Mexicano del Petróleo, v. 1, p. 3-9.

——, 1976, El Contacto Jurásico-Cretácico, la Estratigrafia del Neocomiano, el Hiato Hauteriviano Superior-Eoceno Inferior y las Amonitas del Pozo Bejuco 6 (Centro-Este de México): Boletin de la Sociedad Geológica Maxicana, v. 37, p. 60-83.

Cantú Chapa, A., 1976, Bioestratigrafia de la Serie Huasteca (Jurásico Medio) y Superior en el subsuelo de Poza Rica, Veracruz: Revista del Instituto Mexicana, v. 37, p. 60-83.

——, 1971, La Serie Huasteca (Jurásico Medio-Superior) del Centro Este de México: Revista del Instituto Mexicano del Petróleo, v. 3, p. 17-39.

Carrasco, V., Baldomero, 1971, Litofacies de la Formación El Abra en la Plataforma de Actopan, Hgo.: Revista del Instituto Mexicano del Petróleo, v. 3, p. 5-26.

——, 1977, Albian Sedimentation of Submarine Autochthonous and Allochthonous Carbonates, East Edge of the Valles-San Luis Potosi Platform, México: *in* Cook, H.E. and Enos, Paul, eds., Deep-Water Carbonate Environments: Society of Economic Paleontologist and Mineralogists, Special Publication No. 25, p. 263-272.

Carrillo, B.J., 1971, La Plataforma Valles-San Luis Potosi: Boletin de la Asociación Mexicana de Geólogos Petroleros, v. 23, 102 p.

Coogan, A.H., 1973, Nuevos Rudistas del Albiano y Cenomaniano de México y del Sur de Texas: Instituto Mexicano del Petróleo, v. 5, p. 51-83.

______, 1977, Early and Middle Cretaceous Hippuritacea (Rudists) of the Gulf Coast: *in* Bebout, D.G. and Loucks, R.G., eds., Cretaceous Carbonates of Texas and Mexico, Applications to Subsurface Exploration: Report Investigations No. 89, Bureau of Economic Geology, University of Texas, Austin, p. 32-70.

Cserna, Zoltan de, 1978, Notas sobre la Geologia comprendida entre Iguala, Ciudad Altamirano y Temascaltepec, Estado de Guerrero, México: Sociedad Geológica Mexicana, Libro-Guia de la Excursión a Tierra Caliente, Estados de Guerrero y México, p. 1-25.

Chubb, L.J., 1971, Rudists of Jamaica: Paleontographica Americana, v. 7, p. 161-257.

Douvillé, H., 1900, Sur Quelques Rudistes Américains: Bulletin Societé Geologique de France, Ser. 3, v. 28, p. 205-221.

Erben, H.K., 1956a, El Jurásico Inferior de México y sus Amonitas: Universidad Nacional Autónoma de México, Instituto de Geologia, Congreso Geológico Internacional, Sesión 20, 393 p.

______, 1956b, El Jurásico Medio y el Calloviano de México: Universidad Nacional Autónoma de México, Instituto de Geologia, Congreso Geológico Internacional, Sesión 20, 140 p.

______, 1957, Paleogeographic Reconstruction for the Lower and Middle Jurassic and for the Callovian of México: *in* El Mesozoico del Hemisferio Occidental y sus Correlaciones Mundiales: Congreso Geologógico Internacional, Sesión 20, México, D.F., Sección 2, p. 35-41.

Ferrusquia-V., Ismael, 1976, Estudio Geológico-Paleontológico de la Región Mixteca: Universidad Nacional Autónoma de México, Instituto de Geologia, Boletin 97, 106 p.

Fries, Carl, 1960, Geologia del Estado de Morelos y Partes Adyacentes de México y Guerrero, Región Central Meridional de México: Universidad Nacional Autónoma de México, Instituto de Geologia, Boletin 60, 236 p.

González-Alvarado, J., 1976, Resultados Obtenidos en la Exploración de la Plataforma Córdoba y Principales Campos Productores: Boletin de la Sociedad Geológica Mexicana, v. 37, p. 53-59.

Heim, A., 1926, Notes on the Jurassic of Tamazunchale (Sierra Madre Oriental, México): Eclogae Geologicae Helvetiae, v. 20, p. 84-89.

______, 1940, The Front Ranges of Sierra Madre Oriental, from Ciudad Victoria to Tamazunchale: Eclogae Geologicae Helvetiae, v. 33, p. 313-362.

Imlay, R.W., 1943, Jurassic Formations of the Gulf Region: Bulletin of the American Association of Petroleum Geologists, v. 27, p. 1407-1533.

______, 1944, Cretaceous Formations of Central America and México: Bulletin of the American Association of Petroleum Geologists, v. 28, p. 1077-1195.

______, 1953, Las Formaciones Jurásicas de México: Boletin de la Sociedad Geológica Mexicana, v. 16, 65 p.

______, 1980, Jurassic Paleobiogeography of the Conterminous United States in its Continental Setting: United States Geological Survey, Professional Paper 1062, 134 p.

Kauffman, E.G., 1965, The Upper Cretaceous *Inoceramus* of Puerto Rico: Fourth Caribbean Geological Conference, Trinidad, p. 203-218.

______, 1973, Cretaceous Bivalvia: Atlas of Paleobiogeography: Elsevier, Amsterdam, p. 353-383.

López-Ramos, Ernesto, 1973, Estudio Geológico de la Peninsula de Yucatán, México: Boletin de la Asocación Mexicana de Geólogos Petroleros, v. 25, p. 24-76.

______, 1979a, Geologia de México, Tomo II: Second Edition, México, D.F., 454 p.

______, 1979b, Geologia de México, Tomo III: Second Edition, México, D.F., 446 p.

MacGillavry, H.J., 1937, Geology of the Province of Camagüey, Cuba, with Revisional Studies in Rudist Paleontology: Ph.D. Thesis, University of Utrecht, 168 p.

Muir, J.M., 1936, Geology of the Tampico Region, México: American Association of Petroleum Geologists: Tulsa, Oklahoma, 280 p.

Müllerried, F.K.G., 1930, El *Hippurites calamitiformis* Bárcena: Universidad Nacional Autónoma de México, Anales del Instituto de Biologia, v. 1, p. 169-174.

______, 1933-1934, Estudios Paleontológicos y Estratigráficos en la Región de Techuacán, Puebla: Universidad Nacional Autónoma de México, Anales del Instituto de Biologia, v. 4, p. 33-46, 79-93, 309-330; v. 5, p. 55-79.

______, 1939, Apuntes Paleontológicos y Estratigráficos sobre el Valle de Mezquital, Estado de Hidalgo: Anales de la Escuela Nacional de Ciencias Biológicas, México, v. 1, p. 225-252.

______, 1949, Estratos del Cretácico Medio y Superior en la Región de Apaxco, Estado de México, y Terrenos Litmitrofes de Hidalgo: Memorias de la Comisión Impulsora para la Coordinación de la Investigación Cientifica, p. 31-34.

______, 1950, La Facies Arrecifal de Paquiodontos del Cretácico Superior, en el Sur de México, de Chiapas a Colima: Memorias de la Comisión Impulsora para la Coordinación de la Investigación Cientifica, p. 13-16.

Myers-L., Ralph, 1968, Biostratigraphy of the Cardenas Formation (Upper Cretaceous), San Luis Potosi, México: Universidad Nacional Autónoma de México, Instituto de Geologia, Paleontologia Mexicana 24, p. 1-89.

Ontiveros, T.G., 1973, Estudio Estratigráfico de la Porción Nororiental de la Cuenca Morelos-Guerrero: Boletin de la Asociación Mexicana de Geólogos Petroleros, v. 25, p. 190-234.

Palmer-H., Robert, 1928, The Rudistids of Southern Mexico: California Academy of Sciences, San Francisco, Occasional Paper N° 14, 137 p.

Pantoja-Alor, Jerjes; Estrada-B; Samuel, and Alencaster, Gloria, 1978, Estratigrafia del Cretácico Inferior de la Mina El Encino, Mpio. de Pihuamo, Jalisco: Boletin de la Sociedad Geológica Mexicana, IV Convención Geológica Nacional, Resúmenes, Tomo 39, p. 28.

Pedrazzini, Carmen, 1978, Facies Ambientales, Desarrollo de la Plataforma Cretácica de Tuxpan-Tecolutla (Faja de Oro): Boletin de la Sociedad Geológica Mexicana, v. 39, p. 118-125.

Sánchez Montes de Oca, Rafael, 1969, Estratigrafia y Paleogeografia del Mesozoico de Chiapas: Seminario sobre Exploración Petrolera, Mesa Redonda No. 5, Problemas de Exploración de la Zona Sur, Instituto Mexicano del Petróleo, p. 23-29.

Sánchez-Rosas, G.J., 1972, Estudio Estratigráfico del Area Los Cues-Salinas, SE de Tamaulipas y N de Veracruz: Boletin de la Asociación Mexicana de Geólogos Petroleros, v. 24, p. 65-81.

Scott, R.W., 1981, Biotic relations in Early Cretaceous coralalgal-rudist reefs, Arizona: Journal of Paleontology, v. 55, p. 463-478.

Segerstrom, Kenneth, 1956, Estratigrafia y Tectónica del Cenozoico entre México, D.F. y Zimapán, Hidalgo: Congreso Geológico Internacional, Sesión 20, México, D.F., Excursiones A-14 y C-6, p. 311-323.

Stabler, C.L., 1972, Secuencia de una Trampa Estratigráfica en el Jurásico Superior, Campo Tamaulipas-Constituciones: Boletin de la Asociación Mexicana de Geólogos Petroleros, v. 24, p. 165-178.

Viniegra-O., Francisco, 1971, Age and Evolution of Salt Basins of South-Eastern México: American Association of Petroleum Geologists Bulletin, v. 55, p. 478-494.

______, 1981, Carbonate Bank of Yucatán, Southern México: Journal of Petroleum Geology, v. 3, No. 3, p. 247-278.

Westermann, G.E.G., 1981, Ammonite Biochronology and Biogeography of the Circum-Pacific Middle Jurassic: in House, M.R. and Senior, J.R., The Ammonoidea: Systematics Association Special Volume No. 18, p. 459-498.

———, 1983, The Upper Bajocian and Lower Bathonian (Lower Jurassic) ammonite fauna of Oaxaca, México, and West Tethyan affinities: Universidad Nacional Autónoma de México, Instituto de Geologia, Paleontologia Mexicana 46, for 1981, 42 p.

Wierzbowski, A., 1976, Oxfordian ammonites of the Pinar del Rio Province (Western Cuba), their revision and stratigraphical significance: Acta Geológica Polónica, v. 26, n. 2, p. 137-260.

Jurassic-Cretaceous Biochronology and Paleogeography of North America,
edited by G.E.G. Westermann,
Geological Association of Canada Special Paper 27, 1984

LATE JURASSIC BIVALVES AND GASTROPODS FROM NORTHERN ZACATECAS, MEXICO, AND THEIR BIOGEOGRAPHIC SIGNIFICANCE

B.E. Buitrón
Instituto de Geologia, Universidad Nacional Autónoma de México, 04510 México

ABSTRACT

The Upper Jurassic of Sierrecilla de San Antonio in the Mazapil region of Zacatecas includes the Zuloaga Formation (Oxfordian-Kimmeridgian) and the La Caja Formation (Kimmeridgian-Tithonian). The Zuloaga consists of 100 m of limestone with Pholadomya *cf.* P. paucicosta *Roemer,* Nerinea goodellii *Cragin,* N. circumvoluta *Cragin and* Aptyxiella stantoni *(Cragin). The La Caja Formation includes 55 m of phosphoric limestone, marl, sandstone and shale with* Grammatodon delicatulus *Imlay,* Buchia (Anaucella) concentrica *(Sowerby),* B. tenuistriata *(Lahusen),* Codakia (Epilucina) potosina metrica *(Cragin),* Arctica? coteroi *(Castillo and Aguilera),* Eocallista? cuneovatus *(Cragin),* Nanogyra striata *(Smith), ammonoids* (Idoceras, Haploceras) *and belemnoids* (Cylindroteuthis?). *These faunas indicate paleobiogeographic relationships with the Mexican states of Coahuila, Durango, San Luis Potosi and Puebla, the western United States (Alaska, Washington, Oregon, California, Texas, Arkansas and Louisiana), Canada, and Greenland. The bivalves and gastropods contain a mixture of Tethyan and Boreal genera.*

RÉSUMÉ

La Sierrecilla de San Antonio, située dans la Région de Mazapil au nord de l'Etat de Zacatecas, présentent des gisements fossilifères d'âge Jurassique Supérieur et les quelles correspondent à deux unités litostratigraphiques: La Formation Zuloaga (Oxfordien-Kimmeridgien) et la Formation La Caja (Kimmeridgien-Tithonien). La Formation Zuloaga comprend des calcaires avec des spécimens de Pholadomya *cf.* P. paucicosta *Roemer,* Nerinea goodellii *Cragin,* N. circumvoluta *Cragin et* Aptyxiella stantoni *(Cragin). La Formation La Caja se caractérisse dans cette zone par des argilles et des calcaires phosphoritiques avec des ammonites* (Idoceras, Haploceras) *des belemnites* (Cylindroteuthis?) *et abondants bivalves appartenant aux spècies* Grammatodon delicatulus *Imlay,* Buchia (Anaucella) concentrica *(Sowerby),* B. tenuistriata *(Lahusen),* Codakia (Epilucina) potosina metrica *(Cragin),* Arctica? coteroi *(Castillo et Aguilera),* Eocallista? cuneovatus *(Cragin) et* Nanogyra striata *(Smith). L'examen de la répartition paléogéographique de cette faune a permis de constater l'existence d'spèces communes dans localités differentes du Mexique: Coahuila, Durango, San Luis Potosi et Puebla, des Etat Unis (Alaska, Washington, Oregon, California, Texas, Arkansas, et Louisiana), du Canada et du Groeland. La faune du Jurassique Supérieur, dans le nord de Zacatecas, combinant des genres caractéristiques, du domaine téthysien et boreal.*

INTRODUCTION

In the Sierrecilla de San Antonio, at a site located about 15 km south of the Camacho Railroad Station and 8 km west of the Opal Railroad Station in the Mazapil region of the State of Zacatecas, northern Mexico (Fig. 1), bivalves, gastropods, ammonoids, and belemnoids have been discovered in fossiliferous Upper Jurassic sediments.

Geological and paleontological papers on the Upper Jurassic of the Mazapil region include those by Burckhardt (1930), Imlay (1938, 1939, 1940), Ledezma (1981) and Ledezma and Buitrón (in press). This paper describes and illustrates a fauna collected in Sierrecilla de San Antonio and compares it with faunas of Greenland, Canada, Alaska, the western and southern regions of the United States and several localities of Mexico, as well as various other regions of the world.

STRATIGRAPHY

The Sierrecilla de San Antonio is composed of a thick sequence of predominantly Upper Jurassic and Cretaceous marine limestone. Two Jurassic lithostratigraphic units are present in this region – the Zuloaga Formation (Oxfordian-

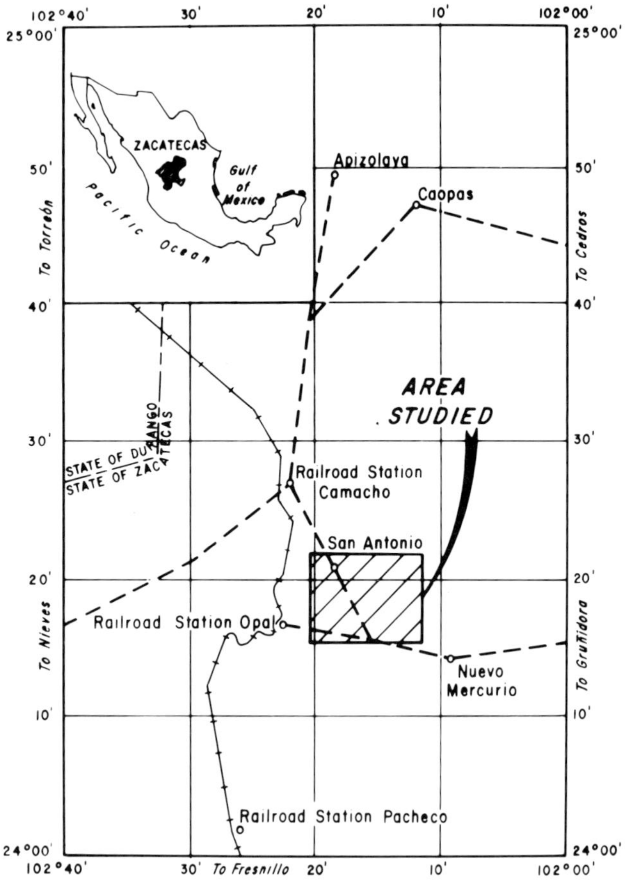

Figure 1. Index Map of the Sierrecilla de San Antonio area, Northern Zac, Mexico (from G.O. Ledezma, 1976).

Kimmeridgian) and the La Caja Formation (Kimmeridgian-Tithonian).

The Zuloaga Formation, named by Imlay (1938, p. 1657), includes 50 to 550 m of thin-bedded grey limestones. The type section can be found at Sierra Sombreretillo in the north of Melchor Ocampo, Zacatecas. In the Sierrecilla de San Antonio this formation is composed mainly of limestone with approximately 100 m of continuously exposed section. The lower part consists of thick-bedded (0.40 m to 0.60 m) light grey, fine-grained limestone beds. The upper part includes a layer of thin bedded (0.50 m to 1.5 m) dark limestone beds with pure microcrystalline calcite with many straight veins of calcite. The top of Zuloaga Formation has yielded some bivalves *(Pholadomya* cf. *P. paucicosta* Roemer) and gastropods *(Nerinea goodellii* Cragin, *N. circumvoluta* Cragin and *Aptyxiella stantoni* (Cragin)) with modest preservation but rarely complete. These species are remarkably similar to Oxfordian-Kimmeridgian species of Texas (Malone Formation) and Arkansas (Smackover Limestone).

The La Caja Formation was named by Imlay (1938, p. 1659) and it consists of 43 to 91 m of thick-bedded limestones and marls. The type section is in The Sierra de La Caja, near Mazapil, Zacatecas (Burckhardt, 1930; Verma and Westermann, 1973). In the Sierrecilla de San Antonio, this formation is composed of 55 m of interbedded phosphoric limestone and marl with some beds of sandstone and shale. A thick section at the southeast side of this hill is composed of 4 m interbedded thin (0.10 m) beds of light grey fine-grained sandstone and dark grey limestone (0.40 m to 0.60 m) with thin shale beds. At approximately 3 m from the base *Idoceras* spp. are associated with *Buchia (Anaucella) concentrica, Grammatodon delicatulus, Arctica? coteroi, Eocallista? cuneovatus* and *Nanogyra striata.* Overlaying this fossiliferous bed are 15 m of interbedded thin (0.15 m) sandstone and shales. The upper La Caja Formation includes approximately 34 m of interbedded thin (0.20 m to 0.40 m) massive hard grey limestone and yellow fined-grained shale and marl with *Haploceras, Buchia tenuistriata, Codakia (Epilucina) potosina metrica* and *Cylindroteuthis?*. The presence of *Idoceras* is particularly useful for an age determination as the genus is known elsewhere from the early to ?mid-Kimmeridgian of Mexico (Verma and Westermann, 1973; Imlay, 1980, etc.).

In the Sierrecilla de San Antonio, the ammonite and bivalve evidence for a later Kimmeridgian age consists of *Haploceras* and *Buchia tenuistriata* (Fig. 2).

BIOGEOGRAPHY

The Zuloaga (Oxfordian-Kimmeridgian) and the La Caja (Kimmeridgian-Tithonian) formations in the Sierrecilla de San Antonio, northern Mexico, are distinguished by two vastly differing faunal assemblages of the Upper Jurassic. The invertebrates from the Zuloaga Formation include mainly gastropods which belong to the genus *Nerinea* Defrance, 1825 and one species of the bivalve *Pholadomya* Sowerby, 1923. The presence of the genus *Nerinea* in this region suggests a shallow water tropical environment (Delpey, 1940; Sohl, 1969, Hölder, 1979) which indicates affinities with fauna of the Tethyan Realm. In the Upper Jurassic, the corals (Beauvais, 1974), some ammonoids (Imlay, 1965; Imlay and Jones, 1970; Imlay, 1980; Westermann and Riccardi, 1982), belemnoids (Hölder, 1979, etc.), and some other groups show a geographical distribution which indicates a westward extension of the Tethys Sea.

The three species of *Nerinea* that are reported in this paper, are restricted in their distribution (aside from those found in Zacatecas) to the Malone Formation (Kimmeridgian-Tithonian) of northern and western Texas, U.S.A. (Cragin, 1905). Two species, *Nerinea goodelli* and *Aptyxiella stantoni*, are present in the Smackover Formation (Upper Jurassic) of southern Arkansas (Imlay, 1945). *Pholadomya paucicosta* is known from the La Casita Formation (Kimmeridgian-Tithonian) of Villa Juárez, Durango and Sierra Jimulco, Coahuila, northern Mexico (Imlay, 1940) and from the Malone Formation of Texas (Cragin, 1905) (Table I, Fig. 3).

The fauna of latest Jurassic (Kimmeridgian-Tithonian) age in northern Zacatecas is dominated by two species of the genus *Buchia* Rouillier 1945, and one species of each of the following genera: *Grammatodon* Meek and Hayden, 1981; *Codakia* Scopoli, 1777; *Arctica?* Schumacher, 1817; *Eocallista?* Douville, 1921, and *Nanogyra* Beurlen, 1858. The genus *Buchia* is characteristic of the Boreal Realm. Jeletzky (1965) concluded that the marine invertebrates (principally the assemblages of ammonoids, belemnoids and buchias) of the Upper Jurassic-Lower Cretaceous found in the Western Cordilleran belt of North America from Alaska to California represent a remarkable

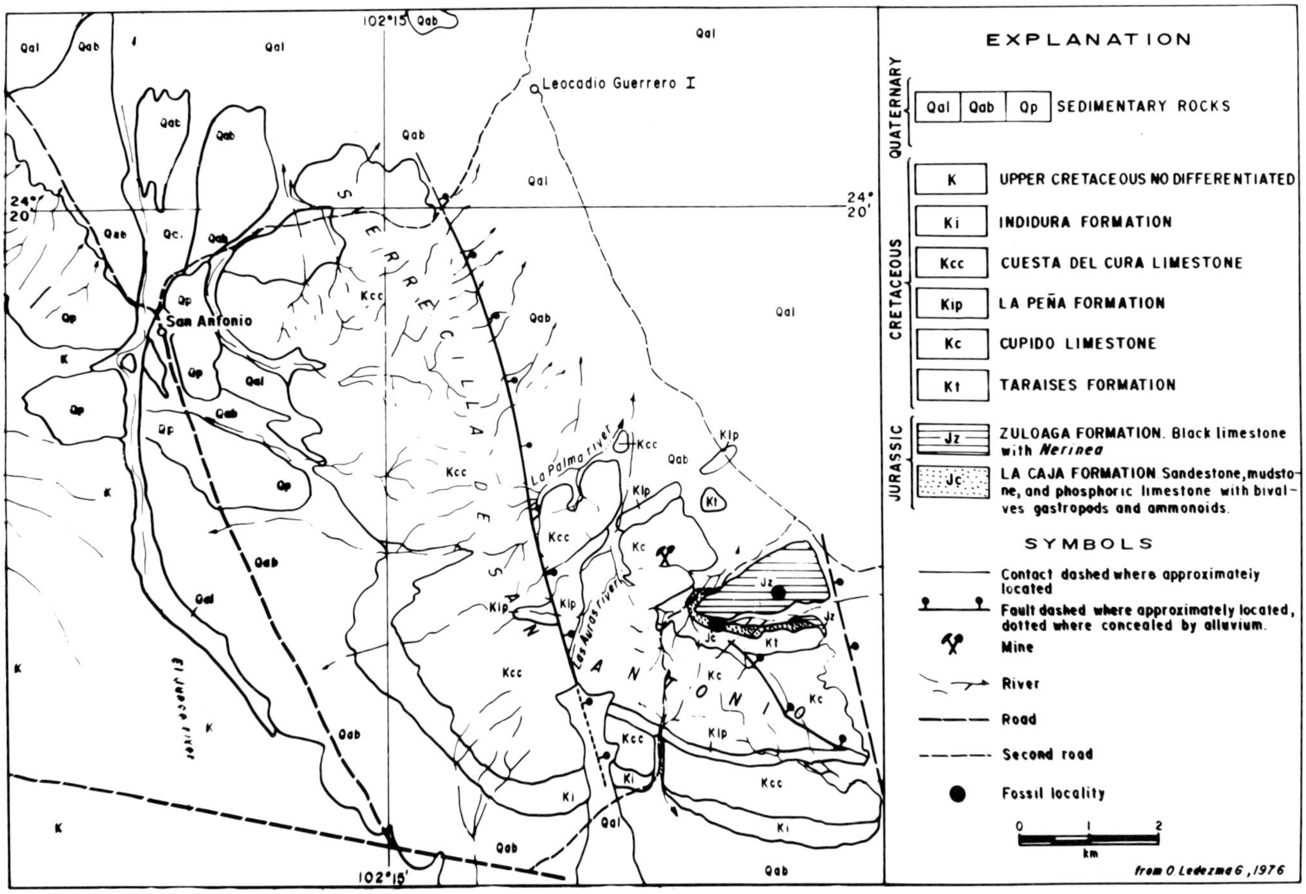

Figure 2. Areal distribution of Upper Jurassic rocks in the Sierrecilla de San Antonio, Northern Zacatecas, Mexico.

TABLE I
FOSSILS OCCURRENCES

STRATIGRAPHIC UNIT	SPECIES	MEXICO: ZACATECAS	COAHUILA	DURANGO	SAN LUIS POTOSI	PUEBLA	UNITED STATES: ALASKA	WASHINGTON	OREGON	CALIFORNIA	TEXAS	ARKANSAS	LOUISIANA	CANADA	GREENLAND
ZULOAGA FORMATION	*Pholadomya cf. P. paucicosta* Roemer	●	●	●							●				
	Nerinea goodellii Cragin	●									●	●			
	Nerinea circumvoluta Cragin	●									●				
	Aptyxiella stantoni (Cragin)	●									●	●			
LA CAJA FORMATION	*Eocallista ? cuneovatus* (Cragin)	●	●			●					●				
	Grammatodon delicatulus Imlay	●	●	●											
	Buchia concentrica (Sowerby)	●			●		●	●	●	●				●	●
	Buchia tenuistriata (Lahusen)	●													
	Codakia (Epilucina) potosina metrica (Cragin)	●				●					●				
	Arctica ? coteroi (Castillo and Aguilera)	●	●		●	●									
	Nanogyra striata (Smith)	●	●										●		

Figure 3. Distribution of Oxfordian fossils mentioned in the text.

association of Tethyan elements together with Boreal elements. The invertebrate fauna of the Upper Jurassic of northern Mexico, described in this report, is also of a similarly Tethyan-Boreal character. *Buchia* is an important indicator for marine connection between Boreal, Pacific and Tethys areas (Jeletzky, 1964; Jeletzky and Tipper, 1968; Jones *et al.*, 1969; Jeletzky, 1973; Hölder, 1979).

The distribution and ages of the species are as follows. *Grammatodon delicatulus* Imlay is known from Kimmeridgian-Tithonian of Coahuila and Durango (Imlay, 1940). *B. concentrica* (Sowerby) is represented in the Kimmeridgian of San Luis Potosi, Mexico. This species is also known from the upper Oxfordian? to lower Kimmeridgian in the states of California and Oregon and from upper Oxfordian to middle Kimmeridgian of Alaska and northern Eurasia (Imlay, 1955, 1959), and it has been found in the upper Oxfordian to (?) lower Kimmeridgian in southwestern British Columbia (Jeletzky and Tipper, 1968). *B. tenuistriata* (Lahusen) is known from the upper Kimmeridgian of northern Eurasia and Alaska (Zakharov, 1981). *Codakia (Epilucina) potosina metrica* (Cragin) was found in the Upper Jurassic of North America, in San Luis Potosi (Castillo and Aguilera, 1895), in Puebla (Alencaster and Buitron, 1965), and in Texas (Cragin, 1905). *Arctica? coteroi* (Castillo and Aguilera, 1895) has largely the same geographical range as the aforementioned species but is also found in Coahuila, Mexico (Imlay, 1940). *Eocallista* cf. *E. cuneovatus* (Cragin) has been described from the Upper Jurassic of Coahuila (Imlay, 1940), Puebla (Alencaster and Buitrón, 1965), and Texas (Cragin, 1905). Loriol (1872) described and Arkell (1933) discussed *Nanogyra striata* (Smith) ["*Ostrea virgula* Defrance" (Stenzel, 1971, p. 1121)] from many Upper Jurassic localities in Europe. Burckhardt (1930) reported that this species came from Coahuila, Mexico and Imlay (1945) reported it from Louisiana, USA (Table I, Fig. 4).

Conclusions

The knowledge of the Zacatecas fauna permits paleontologists to establish paleobiogeographic relationships with other Late Jurassic faunas in Mexico (Coahuila, Durango, San Luis Potosi and Puebla). Elsewhere, it is related to the Late Jurassic faunas of Alaska, Washington, Oregon, California, Louisiana, Texas and Arkansas, Canada and Greenland. This assemblage of bivalves and gastropods is characterized by a mixture of genera with Tethyan and Boreal affinities.

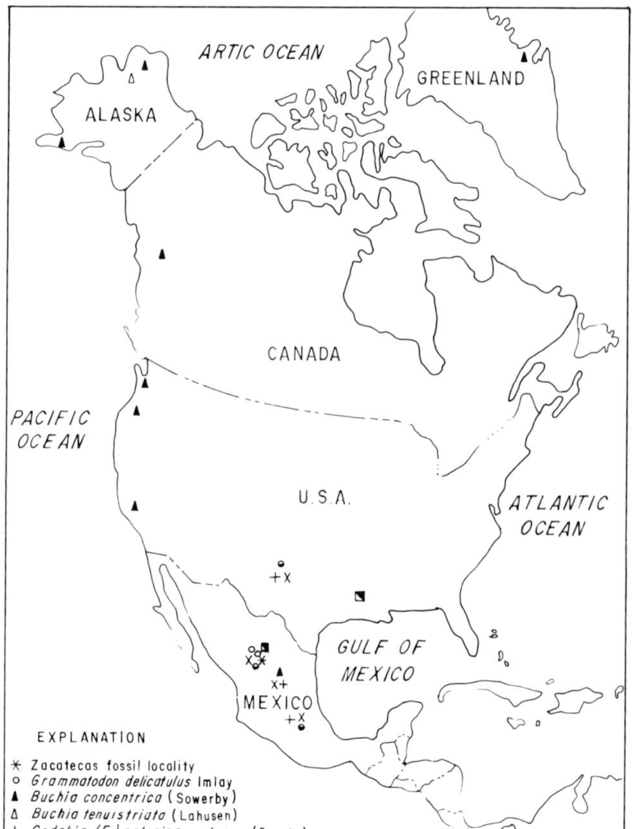

Figure 4. Distribution of Kimmeridgian to Tithonian fossils mentioned in the text.

SYSTEMATIC PALEONTOLOGY

The specimens studied herein are deposited in the Paleontological Museum of the Instituto de Geologia (IGM), Universidad Nacional Autónoma de México.

Class Bivalvia – Family Paralellodontidae Dall, 1898

Grammatodon delicatulus Imlay (Pl. I, Figs. 1-3)

Grammatodon delicatulus Imlay, 1940, p. 400, pl. 53, Figs. 22-25

Material and Stratigraphic Distribution. The three figured specimens (IGM-3327, 3328, 3329) come from the La Caja Formation.

Discussion. G. delicatulus is caracterized by its shell which is subquadrate and strongly convex with prominent beaks and fine radial ribs. According to Imlay (1940, p. 400) this species occurs also in the La Casita Formation (Kimmeridgian) of the Sierra de Parras, Coahuila and in the La Caja Formation (Kimmeridgian-Tithonian) of Cañón del Toboso, Durango.

Family Buchiidae Cox, 1953

Buchia (Anaucella) concentrica (Sowerby) (Pl. I, Figs. 4-9)

Plagiostoma concentrica Sowerby, 1827, p. 113, Pl. 559, Fig. 1.
Aucella concentrica (Sowerby) Lahusen, 1888, p. 32, 33; Pavlow, 1907, p. 14-16; Sokolov, 1908, p. 8-10, 27, 28; Jeletzky, 1950, p. 27-29; Waterston, 1951, p. 40, 41, Pl. 1, Figs. 2a-c; Imlay, 1955, p. 83, Pl. 9, Figs. 11-16; Imlay, 1959, p. 157, Pl. 16, Figs. 1-10.
Aucella bronni (Rouiller) Sokolov and Bodylevsky, 1931, p. 34, 35.
Buchia concentrica Zakharov, 1981, p. 64, Pl. 3, Figs. 1-12.

Material and Stratigraphic Distribution. The six figured specimens (IGM-3330, 3331, 3332, 3333, 3334, 3342) were found in the La Caja Formation and *B. concentrica* was found in association with *Idoceras* (early to ?middle Kimmeridgian).

Discussion. Buchia concentrica is common in the Sierrecilla de San Antonio. It is characterized by a medium to large, nearly equivalve shell which is elongated obliquely. The left valve is convex on the umbonal area and flattened posteriorly with a small low inwardly-curved beak. The shells have many sharp concentric ribs that are crossed by very fine dense radial striae.

B. concentrica, according to Imlay (1955, p. 83; 1959, p. 157, Table 1), is present in the late Oxfordian to early Kimmeridgian of Oregon and California and is common from the late Oxfordian to middle Kimmeridgian of Alaska and Northern Eurasia and from the late Oxfordian to (?) lower Kimmeridgian in the Southwestern British Columbia (Jeletzky and Tipper, 1968, p. 100, Table 1).

In addition, Castillo and Aguilera (1895, Pl. 3, Figs. 4, 9, 10 and 14) placed four specimens from Catorce, San Luis Potosi, México as *Aucella fischeriana* but Jeletzky placed them in *B. concentrica* when he in 1968 studied this material at the Instituto de Geologia of the Universidad Nacional Autónoma de México.

Buchia tenuistriata (Lahusen) (Pl. I, Fig. 10-14)

Aucella pallasi var. *tenuistriata* Lahusen, 1888, p. 9, Pl. 4, Fig. 25-27; Pompeckj, 1901, p. 25, Pl. 4, Figs. 5-6; Schmidt, 1905, p. 156, Pl. 5, Figs. 1-6.
Aucella mosquensis Keyserling, 1846, Pl. 16, Fig. 8; Pl. 7, Fig. 20 (*non* Buch, 1844).
Aucella tenuistriata Ravn, 1911, p. 458, Pl. 32, Fig. 7.
Aucella lindstroemi Sokolov, 1928 (*non* Sokolov, 1908), p. 34, Pl. 4, Figs. 3-10.
Aucella bononiensis Bodylevskii, 1928, p. 706, Figs. A, B (*non* Pavlow, 1907).
(?) *Buchia vuquaamensis* Paraketsov, 1975, p. 19, Figs. 1-9.
Buchia tenuistriata Zakharov, 1981, p. 74, Pl. 5, Figs. 5-8; Pl. 6, Figs. 1-11; Pl. 7, Fig. 1-20; Pl. 8, Figs. 1-18.

Material and Stratigraphic Distribution. The five figured specimens (IGM-3335, 3336, 3337, 3346 and 3347) come from the La Caja Formation (late Kimmeridgian). *Buchia tenuistriata* was found in association with *Haploceras* (late Kimmeridgian) and *Cylindroteuthis?*

Discussion. This species resembles *B. tenuistriata* (Zakharov, 1981, p. 74, Pl. 5, Figs. 5-8; Pl. 6, Figs. 1-11; Pl. 7, Fig. 1-20; Pl. 8, Figs. 1-18) from the late Kimmeridgian of northern Eurasia and Alaska and is distinguished by a medium to large inequivalve shell that is elongated posteriorly. The left valve is strongly convex on the umbonal area with a strong incurved beak. The right valve is gently convex with a well developed byssus ear and notch and the beak low. The surface of the valves has concentric ribs of irregular strength and irregularly spaced conspicuous undulations and constrictions. Striae are fairly fine, scattered and are not evident on some specimens.

Family Lucinidae Fleming, 1828

Codakia (Epilucina) potosina metrica (Cragin) (Pl. I, Figs. 15-18)

Lucina metrica Cragin, 1898, p. 817
Lucina potosina? Cragin, 1898, p. 817, 819
Lucina potosina var. *metrica* Cragin, 1905, p. 73, Pl. 13, Figs. 6-10; Alencaster and Buitrón, 1965, p. 29, Pl. 7, Figs. 19-20.

Material and Stratigraphic Distribution. The two figured specimens (IGM-3348, 3349) come from the La Caja Formation.

Discussion. This subspecies has a compressed shell with a nearly equilateral contour, and subcentral and very small beaks. The surface has concentric and conspicuous ribs that are separated by broad spaces with fine concentric growth lines.

Cragin (1905, p. 73, Pl. 13) first described this variety from the Malone Formation (Kimmeridgian-Tithonian) of Texas. Alencaster and Buitrón (1965, p. 29, Pl. 9, 10) reported it as coming from the Mapache Formation (Kimmeridgian-Tithonian) of Puebla.

Family Arcticidae Newton, 1891

Arctica? coteroi (Castillo and Aguilera) (Pl. II, Figs. 1-2)

Cyprina coteroi Castillo and Aguilera, 1895, p. 8, Pl. 5, Figs. 4-10.
Arctica coteroi Imlay, 1940, p. 409, Pl. 55, Figs. 9, 10; Alencaster and Buitrón, 1965, p. 27, Pl. 7, Figs. 4-6.

Material and Stratigraphic Distribution. The two figured specimens (IGM-3350, 3351) come from the La Caja Formation.

Discussion. This species is distinguished by its cordiform or subovated and inflated shell. The beaks are even and terminal. The surface has conspicuous concentric growth lines that are separated by deep and broad spaces which are more prominent in the central part of the shell.

Castillo and Aguilera (1895, p. 8) described *A. coteroi* from the Sierra de Catorce, San Luis Potosi, and Alencaster and Buitrón (1965, p. 27, Pl. 7, Figs. 4-6) described it from the Kimmeridgian of Puebla. Imlay (1940, p. 409) reported this species from the Kimmeridgian-Tithonian of the Sierra de la Ventura, Coahuila, México.

PLATE I

Family Corbiculidae Gray, 1847

Eocallista? cuneovatus (Cragin) (Pl. II, Figs. 3-4)

Tapes? cuneovatus Cragin, 1905, p. 79, Pl. 13, Fig. 12.
Tapes cuneovatus Cragin, Alencaster and Buitrón, 1965, p. 32, Pl. 6, Figs. 10, 11.
Eocallista? cf. *E. cuneovatus* (Cragin) Imlay, 1940, p. 409, Pl. 55, Figs. 2-6.

Material and Stratigraphic Distribution. The two figured specimens (IGM-3355, 3356) come from the La Caja Formation.

Discussion. This species is represented by the external and internal casts of the specimens. The figured specimens illustrate the form and ornamentation. The little beaks are placed in the anterior third and have many finely concentric growth lines and undulations.

Tapes? cuneovatus were described by Cragin (1905, Pl. 79) from the Malone Formation (Kimmeridgian-Tithonian) of Texas and by Alencaster and Buitrón (1965, p. 32) from the Mapache Formation Kimmeridgian-Tithonian) of Puebla. Imlay (1940, p. 409) reported *E.*? cf. *cuneovatus* (Cragin) from La Casita Formation (Kimmeridgian-Tithonian) of the Sierra de Parras, Coahuila, México. He stated "The generic position of this species is uncertain due to lack of knowledge of its internal structures. Except for the presence of a small lunule, it greatly resembles *Corbicella*".

Family Pholadomyidae Gray, 1847

Pholadomya cf. *P. paucicosta* Roemer (Pl. II, Figs. 5-6)

Material and Stratigraphic Distribution. The two figured specimens (IGM-3358, 3359) come from the Zuloaga Formation.

Discussion. This species is represented by two poorly preserves specimens. It is characterized by cuneate form, elevated umbones with salient beaks, and with unequal concentric ribs and furrows. The Zacatecas species exhibits undeveloped radial folds from the beaks to the ventral margin.

The species was described by Cragin (1905, p. 81, Pl. 16, Figs. 5, 6) from the Malone Formation (Kimmeridgian-Tithonian) of Texas, and by Imlay (1940, p. 409, Pl. 54, Fig. 8) from the La Casita Formation (Kimmeridgian-Tithonian) of Durango and Sierra Jimulco, Coahuila.

Family Gryphaeidae Vyalov, 1936

Nanogyra striata (Smith) (Pl. II, Figs. 7-10)

Exogyra virgula (Defrance) Loriol, 1872, p. 399, Pl. 23, Figs. 8-14; Jourdy, 1924, Pl. 68, Pl. 5, Figs. 1, 8; Pl. 6, Fig. 4; Pl. 8, Figs. 9, 10; Pl. 9. Figs. 17, 18; Burckhardt 1930, p. 84, 143.
Exogyra cf. *E. virgula* (Defrance) Imlay, 1945, p. 262, Pl. 39, Figs. 16, 17, 19.
Nanogyra striata Smith, Stenzel, 1971, p. N1122, Fig. J95, 1a-m; Arkell, 1933, p. 440.

Material and Stratigraphic Distribution. The two figured specimens (IGM-3360, 3361) come from the La Caja Formation.

Discussion. The specimens of *Nanogyra striata* (Smith) resembling *Ostrea virgula* Defrance (Stenzel, 1971, N1121, N1122) are not well preserved, however the external form of the shell, with fine radial ribs and prominent concentric constrictions, clearly place them within this species.

This species has been described by Loriol (1872, p. 399) and by Arkell (1933, p. 385, 440, 449) from many localities in Europe. Burckhardt (1930, p. 84, 143) reported similar forms from Coahuila, Mexico, as did Imlay (1945, p. 263) from Louisiana, U.S.A.

Class Gastropoda – Family Nerineidae Zittel, 1873

Nerinea goodellii Cragin (Pl. II, Figs. 11, 12)

Nerinea goodellii Cragin, 1905, p. 96, Pl. 21, Figs. 1-3.
Nerinea aff. *N. goodellii* Imlay, 1941, p. 275, Pl. 39, Fig. 10.

Material and Stratigraphic Distribution. The two figured specimens (IGM-3362, 3363) come from the Zuloaga Formation.

Discussion. This species is represented by two fragments of the shell which have folds that are well defined. The shell fragments have thin walls, a slender columella, and three folds (a long acute labial fold, a slend columella fold and a small thin parietal fold).

Cragin (1905, p. 96) and Imlay (1941, p. 275) described this species from the Malone Formation (Kimmeridgian-Tithonian) of Texas and the Smackover limestone (Upper Jurassic) of Columbian Country, Arkansas, U.S.A., respectively.

Nerinea circumvoluta Cragin (Pl. II, Fig. 13)

Nerinea circumvoluta Cragin, 1905, p. 97, Pl. 21, Figs. 4, 5.

EXPLANATION OF PLATE I

Grammatodon delicatulus Imlay

Figure 1. Lateral view of right valve of hypotype IGM-3327; La Caja Formation: X 2.
Figure 2. Lateral view cast of left valve of hypotype IGM-3328; La Caja Formation: X 1.5.
Figure 3. Lateral view of left valve of hypotype IGM-3329; La Caja Formation: X 1.

Buchia (Anaucella) concentrica (Sowerby)

Figure 4. Lateral view of left valve of hypotype IGM-3330; La Caja Formation: X 2.
Figure 5. Lateral view of left valve of hypotype IGM-3331; La Caja Formation: X 2.
Figure 6. Lateral view of left valve of hypotype IGM-3332; La Caja Formation: X 1.5.
Figure 7. Lateral view of left valve hypotype IGM-3333; La Caja Formation: X 1.
Figure 8. Lateral view of left valve of hypotype IGM-3334; La Caja Formation: X 1.
Figure 9. Lateral view of left valve of hypotype IGM-3342; La Caja Formation: X 1.

Buchia tenuistriata (Lahusen)

Figure 10. Lateral view of left valve of hypotype IGM-3335; La Caja Formation: X 1.
Figure 11. Lateral view of left valve of hypotype IGM-3336; La Caja Formation: X 1.
Figure 12. Lateral view of left valve of hypotype IGM-3337; La Caja Formation: X 1.
Figure 13. Lateral view of right valve of hypotype IGM-3346; La Caja Formation: X 1.
Figure 14. Lateral view of right valve of hypotype IGM-3347; La Caja Formation: X 2.5.

Codakia (Epilucina) potosina metrica Cragin

Figures 15-16. Ventral view of left valve and lateral view of hypotype IGM-3348; La Caja Formation: X 1.
Figures 17-18. Lateral view of right valve and ventral view of hypotype IGM-3349; La Caja Formation: X 1.

PLATE II

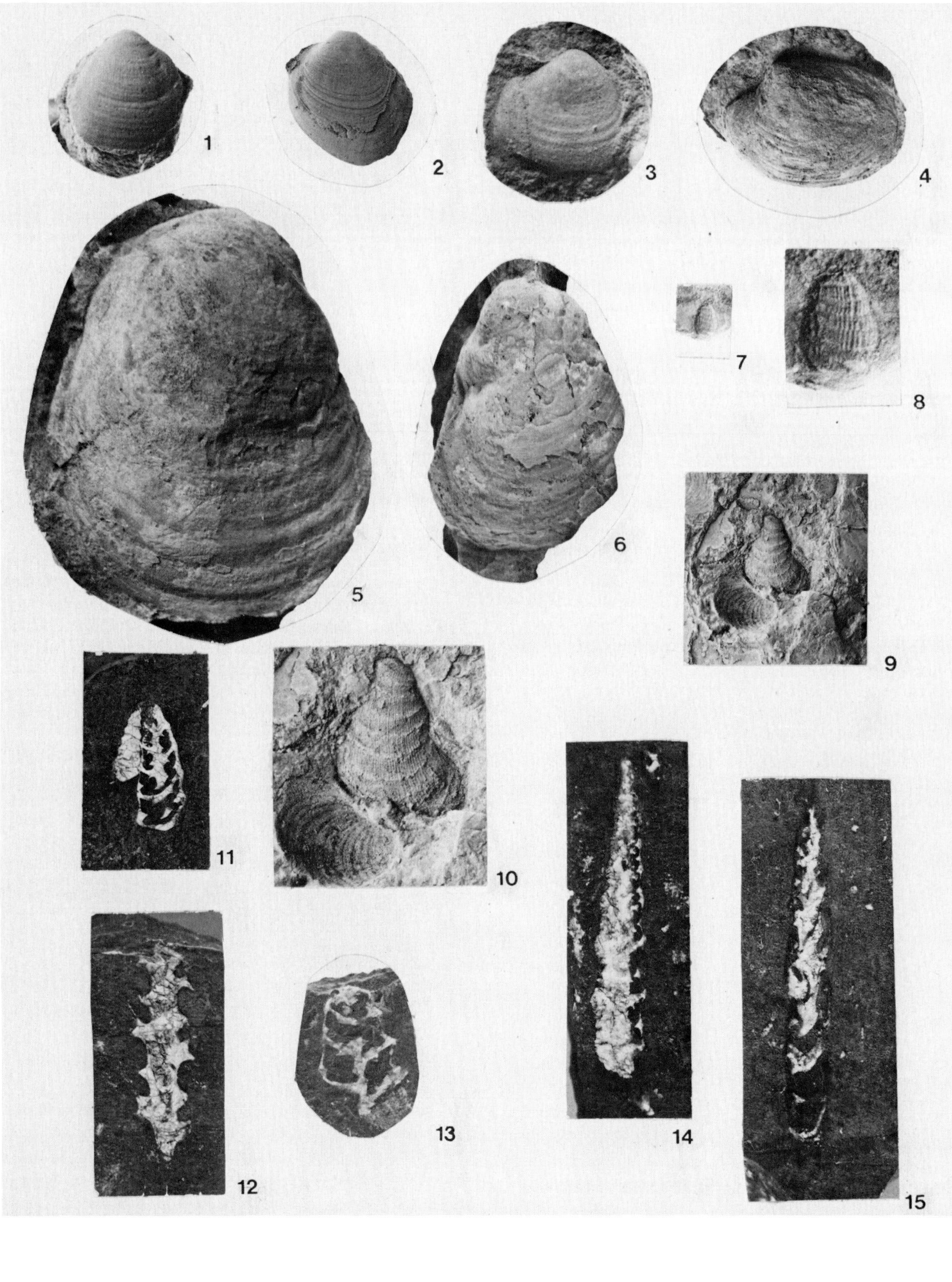

Material and Stratigraphic Distribution. The one figured specimen (IGM-3364) comes from the Zuloaga Formation.

Discussion. This species is characterized by the cylindrical shell, thin walls and three folds (a low and acute labial fold, a very short, slender, and acute columellar fold and an acute slender parietal fold).

Nerinea circumvoluta Cragin (Cragin, 1905, p. 97) from the Malone Formation (Kimmeridgian-Tithonian) of Texas is comparable with *N. goodellii* Cragin (Cragin, 1905, p. 96). Both have similar aspects of internal morphology, but, Cragin's *N. circumvoluta* is distinguished by all the delicate folds which are in a different position than the Mexican specimen.

Aptyxiella stantoni (Cragin) (Pl. II, Figs. 14, 15)

Nerinella stantoni Cragin, 1905, p. 98, Pl. 21, Figs. 6-9.
Aptyxiella (Nerinoides)? aff. *A. stantoni* Imlay, 1945, p. 270, Pl. 41, Figs. 2, 3.

Material and Stratigraphic Distribution. The two specimens (IGM-3365, 3366) come from the Zuloaga Formation.

Discussion. Only one slab with two not very well preserved specimens was collected at Sierrecilla de San Antonio. These specimens exhibits internal ornamentation similar to the Cragin species (1905, p. 98) from the Malone Formation (Kimmeridgian-Tithonian) of Texas: it consists of one short and obtuse labial fold. Imlay (1945, p. 270) reported that this species came from the Smackover Limestone from Lafayette County, Arkansas.

ACKNOWLEDGEMENTS

The author is greatly indebted to Dr. G.E.G. Westermann and Dr. R.W. Imlay for their critical reviews of the manuscript. Special thanks is extended to Dr. J.A.Jeletzky of the Geological Survey of Canada for his many suggestions, his discussion of some of the taxonomic problems, and for supplying comparative material of *Buchia* species for this study. I also thank Nancy Poss for her help and encouragement in preparing this manuscript.

REFERENCES

Alencáster, Gloria and Buitrón, B.E., 1965, Fauna del Jurásico Superior de la Región de Petlalcingo, Estado de Puebla: Universidad Nacional Autónoma de México, Instituto de Geologia, Paleontologia Mexicana 21, p. 1-53.

Arkell, W.J., 1933, The Jurassic System in Great Britain: Oxford University Press, 681 p.

Beauvais, Louise, 1974, Quelques Exemples Pris dans le Malm Montrant le Role des Madreporaires dans la Théorie de la Mobilité Continentale: Bulletin de la Société Geologique de France, t. 16, no. 4, p. 465-469.

Bodylevsky, V.J., 1928, *Aucella* from the Basin of the Siruriukt River on the Cherski Mountains: Memoir de Commission Geologique, v. 46, No. 6, p. 705-710 (in Russian).

Burckhardt, Carl, 1930, Étude Synthétique sur le Mésozoique Mexicain: Société Paleontologique de Suisse, Mémoire 49-50, 280 p.

Castillo, Antonio and Aguilera, J.G., 1895, Fauna Fósil de la Sierra de Catorce, San Luis Potosi: Instituto Geológico de México, Boletin 1, 55 p.

Cragin, F.W., 1898, Discovery of Marine Jurassic Rocks in Southwestern Texas: Journal of Geology, v. 5, 813-820.

______, 1905, Paleontology of the Malone Jurassic Formation of Texas: United States Geological Survey Bulletin 266, 172 p.

Delpey, Genèviève, 1940, Les Gastéropods Mésozoiques de la Région Libanaise et les Pays Voisins: Notes et Mémoires, Société d'Études Géologiques Haut Commissariat de la République Francaise en Syrie et au Liban, t. 3, p. 5-324.

Hallam, Anthony, 1969, Faunal Realms and Facies in the Jurassic: Paleontology, v. 12, p. 1-18.

Hölder, H., 1979, Jurassic: *in* Moore, R.C., ed., Treatise on Invertebrate Paleontology, Pt. A: Fossilization (Taphonomy) Biogeography and Biostratigraphy: Geological Society of America and University of Kansas Press, p. A390-A417.

Imlay, R.W., 1938, Studies of the Mexican Geosyncline: Geological Society of America Bulletin 49, p. 1651-1694.

______, 1939, Jurassic Ammonites from Mexico: Geological Society of America Bulletin 50, p. 1-78.

EXPLANATION OF PLATE II

Arctica? coteroi (Castillo and Aguilera)

Figure 1. Lateral view of left valve of hypotype IGM-3350; La Caja Formation: X 1.

Figure 2. Lateral view of left valve of hypotype IGM-3351; La Caja Formation: X 1.

Eocallista? cuneovatus (Cragin)

Figure 3. Lateral view of right valve of hypotype IGM-3355; La Caja Formation: X 1.

Figure 4. Lateral view of right valve of hypotype IGM-3356; La Caja Formation: X 1.

Pholadomya cf. *P. paucicosta* Roemer

Figure 5. Lateral view of left valve of hypotype IGM-3358; Zuloaga Formation: X 1.

Figure 6. Lateral view of right valve of hypotype IGM-3359; Zuloaga Formation: X 1.

Nanogyra striata (Smith)

Figures 7-8. Lateral view of left valve of hypotype IGM-3360; La Caja Formation: X 1 and X 4.

Figures 9-10. Lateral view of left valve of hypotype IGM-3361; La Caja Formation: X 1 and X 2.

Nerinea goodellii Cragin

Figure 11. Internal view with folds of hypotype IGM-3362; Zuloaga Formation: X 1.

Figure 12. Internal view of fragmentary hypotype IGM-3363; Zuloaga Formation: X 1.

Nerinea circumvoluta Cragin

Figure 13. Internal view with folds of hypotype IGM-3364; Zuloaga Formation: X 1.

Aptyxiella stantoni (Cragin)

Figure 14. Internal view with folds of hypotype IGM-3365; Zuloaga Formation: X 1.

Figure 15. Internal view of fragmentary hypotype IGM-3366; Zuloaga Formation: X 1.

———, 1940, Upper Jurassic Pelecypods from Mexico: Journal of Paleontology, v. 14, p. 393-411.

———, 1941, Jurassic Fossils from Arkansas, Louisiana, and Texas: Journal of Paleontology, v. 15, p. 256-277.

———, 1945, Jurassic Fossils from the Southern States, No. 2: Journal of Paleontology, v. 19, p. 253-276.

———, 1955, Characteristic Jurassic Mollusks from Northern Alaska: United States Geological Survey Professional Paper 274-D, p. 69-93.

———, 1959, Succession and Speciation of the Pelecypod *Aucella*: United States Geological Survey Professional Paper 314-G, p. 155-169.

———, 1965, Jurassic Marine Faunal Differentiation in North America: Journal of Paleontology, v. 39, p. 1023-1038.

———, 1980, Jurassic Paleobiogeography of the Conterminous United States in its Continental Setting: United States Geological Survey Professional Paper 1062, p. 1-134.

Imlay, R.W. and Jones, D.L., 1970, Ammonites from the *Buchia* Zones in Northwestern California and Southwestern Oregon: United States Geological Survey Professional Paper 647-B, p. B1-B59.

Jeletzky, J.A., 1950, Stratigraphy of the West Coast of Vancouver Island Between Kyuquot Sound and Esperanza Indet, British Columbia: Geological Survey of Canada Paper 50-57, 52 p.

———, 1964, Lower Cretaceous Marine Index Fossils of the sedimentary Basins of Western and Arctic Canada: Geological Survey of Canada Paper 64-11, 100 p.

———, 1965, Late Jurassic and Early Lower Cretaceous Fossil Zones of the Canadian Western Cordillera, British Columbia: Geological Survey of Canada Bulletin 103, 70 p.

———, 1973, Biochronology of the Marine Boreal Latest Jurassic, Berriasian and Valanginian in Canada: *in* Casey, R. and P.F. Rawson, ed., The Boreal Lower Cretaceous: Geological Journal, Special Issue No. 5.

Jeletzky, J.A., and Tipper, H.W., 1968, Upper Jurassic and Cretaceous of Taseko Lakes Map-Area and Their Bearing on the Geological History of Southwestern British Columbia: Geological Survey of Canada Paper 67-54, 217 p.

Jones, D.L., and Bailey, E.H., and Imlay, R.W., 1969, Structural and Stratigraphic Significance of the *Buchia* Zones in the Colyear Springs-Paskenta Area, California: United States Geological Survey Professional Paper 647-A, p. A1-A24.

Jourdy, le Général E., 1924, Histoire Naturelle des Exogyres: Annales de Paléntologie, v. 13, p. 1-104.

Keyserling, A. von, 1846, Wissenschaftliche Beobachtungen auf einer Reise in des Petschora im Jahre 1843, 2, Reste der Jura-Periode: Petersburg, C. Kray, p. 293-336.

Lahusen, J., 1888, Über die russischen Aucellen: Commision Géologieque St. Pétersbourg, Mémoir, v. 8, no. 1, p. 1046.

Ledezma, G.O., 1981, Hoja Camacho 13R-1 (12), con Resumen de la Geologia de la Hoja Camacho, Estados de Zacatecas y Durango: Universidad Nacional Autónoma de México, Instituto de Geologia, Carta Geológica de México, Serie de 1:100,000, mapa con texto.

Ledezma, G.O. and Buitrón, B.E., in press, Geologia y Estratigrafia de la Sierrecilla de San Antonio, Municipio de Mazapil, Zacatecas, México: Acapulco (México), III Congreso Latinoamericano de Geologia, 1976.

Loriol, P. de, 1872, Monographic Paléontologique et Géologique des Étages Superieurs de la Formation Jurassique du Département de la Haute Marne: Société Linnéenne de Normandie, Mémoir 16, p. 1-484.

Orbigny, Alcide d', 1845, Géologie de la Russie d'Europe, Terrains Secondaires, Systema Jurassique Mollusques: v. 2, p. 419-488.

Paraketsov, K.V., 1975, New species of *Buchia* from the Kimmeridgian Beds in NO fo U.R.S.S.: Materials on the Geology and Minerals, NO., U.R.S.S., Magadan, Ed. 22, p. 19-20.

Pavlow, A.P., 1907, Enchainement des Aucelles et Aucellines du Crétacé Russe: Société Imperial Naturalistes Moscou, Nouveaux Mémoir, t. 17, 93 p.

Pompeckj, J.P., 1901, Ueber Aucellen und Aucellen-ähnliche Formen: Neues Jahrbuch, Beilage-Band 14, p. 319-368.

Ravn, J.P.J., 1911, On Jurassic and Cretaceous fossils from Northeast Greenland: Meddelelsen om Grönland, bind 45, p. 433-500.

Schmidt, M., 1905, Über Oberen Jura in Pommern: Beiträge zur Stratigraphie und Paleontologie, Abhandlungen der Geologischen Landesanstalt, Heft 41, 222 p.

Sohl, N.F., 1969, North American Cretaceous Biotic Provinces Delineated by Gastropods: Chicago, North American Paleontological Convention, Proceedings 2, pt. L, p. 1610-1638.

Sokolov, D., 1908, Aucellen von Timan und von Spitzenbergen: Commission Geólogique St. Pétersbourg Mémoir, Nouv. sér., livr. 36, p. 1-29.

———, 1928, Mesozoiskie okamenelosti iz Bol'schezemel' skoi Tundry i Kashpura (Masozoic fossils from Bol'schzemel' skaya Tundra and Kaspur): Travaux, Musée Géologie Academie Sciences U.R.S.S., t. 3, p. 15-62 (in Russian).

Sokolov, D., and Bodylevsky, W., 1931, Jura-und-Kreidefaunen von Spitzenbergen: Skrifter om Svalbard og Ishavet, Nr. 35, 151 p.

Sowerby, James, and Sowerby, J. de C., 1812-46, Mineral Conchology, 7V, Pls. 1-337 (1812-22) by J. Sowerby, Pls. 338-648 (1822-46) by J. de C. Sowerby.

Stenzel, H.B., 1971, Oysters: *in* Moore, R.C., ed., Treatise on Invertebrate Paleontology, pt. N. Bivalvia: Geological Society of America and University of Kansas Press, p. A390-A417.

Stevens, G.R., 1963, Faunal Realms in Jurassic and Cretaceous Belemnites: Geological Magazine, v. 100, p. 481-497.

———, 1965, The Jurassic and Cretaceous Belemnites of New Zealand and a review of the Jurassic and Cretaceous Belemnites of the Indo-Pacific region: New Zealand Geological Survey, Paleontological Bulletin 36, 283 p.

Verma Harish, M., and Westermann, G.E.G., 1973, The Tithonian (Jurassic) Ammonite Fauna and Stratigraphy of Sierra Catorce, San Luis Potosi, México: Bulletins of American Paleontology, v. 63, no. 277, 314 p.

Waterston, C.D., 1951, The Stratigraphy and Paleontology of the Jurassic rocks of Eathie (Cromarty): Royal Society Edinburgh Transactions, v. 62, p. 33-52.

Westermann, G.E.G. and Riccardi, A.C., 1982, Ammonoids fauna the Early Middle Jurassic of Mendoza Province, Argentina: Journal of Paleontology, v. 56, pt. 1 of 2, p. 11-41.

Zakharov, B.A., 1981, Buchiids and the Biostratigraphy of Boreal Upper Jurassic and Neocomian: Academy of Sciences of U.S.S.R., Siberian Division, Institut of Geology and Geophysics, Trudy No. 458, 240 p.

Jurassic-Cretaceous Biochronology and Paleogeography of North America,
edited by G.E.G. Westermann,
Geological Association of Canada Special Paper 27, 1984

THE ANDEAN MID-JURASSIC *NEUQUENICERAS* AMMONITE ASSEMBLAGE OF CUALAC, MEXICO

G.E.G. Westermann
Department of Geology, McMaster University, Hamilton, Ontario L8S 4M1
Rodolfo Corona
Uramex, Mexico, D.F., México 03720
Raúl Carrasco
Department of Geology, McMaster University, Hamilton, Ontario L8S 4M1

ABSTRACT

Re-examination of the classic ammonite localities near Cualac in the state of Guerrero has revealed that: 1) the majority of species described by Burckhardt (1927) under Macrocephalites [Eurycephalites *s.l.*], Reineckeia, Peltoceras [*both ?* Neuqueniceras *s.l.*] *and* Perisphinctes [Choffatia *s.l.*] *occur within a relatively small stratigraphic interval of the ca. 550 m thick Yucuñuti Formation. 2) They overlie beds with scarce* Epistrenoceras *dated as middle Late Bathonian and, 3) range upward into a level with a latest Bathonian – earliest Callovian fauna yielding* Clydoniceras inflatum *n.sp., a genus hitherto known only from the Tethyan Bathonian,* Reineckeia (Rehmannia) *ex. gr.* rehmanni, *and* Eurycephalites (Lilloettia) *closely allied to* E. tipperi *from British Columbia and to the Andean* E. steinmanni *which at Caracoles occurs together with the Bathonian* Prohecticoceras *subjacent to* Neuqueniceras (Frickites). *4) The highest beds yield what are probably* Reineckeia *s.s.*, Hecticoceras *(*H.*) ex gr.* hecticum *indicating latest Early Callovian, and late* Eurycephalites. *We suggest that the* Neuqueniceras *assemblage is of latest Bathonian to earliest Callovian age, that the eastern Pacific* E. (Lilloettia) *species may be long-ranging, and that* Reineckeia *s.l. may have appeared earlier here than in Tethys and derived from* Neuqueniceras. *This assemblage occurs again with largely identical species in the Chile-Argentine Andes.*

RÉSUMÉ

Un second examen des localités classiques d'ammonites près de Cualac dans l'état de Guerrero a révélé que: 1: la majorité des espèces décrites par Buckhardt (1927) comme Macrocephalites (Eurycephalites *s.l.*), Reineckeia, Peltoceras *(à la fois (?)* Neuqueniceras *s.l.) et* Perisphinctes (Choffatia *s.l.*) *sont présentes dans l'intervalle statigraphique relativement petit de la formation de Yucunuti, d'une épaisseur d'environ 550 m; 2) Elles reposent sur des couches litées contenant peu de* Epistrenoceras *datant de la fin du Bathonien moyen et 3) Elles évoluent vers le sommet de la formation en un niveau contenant une faune de la fin du Bathonien – début Callovien, produisant:* Clydoniceras inflatum *n. sp., un genre qui n'est connu, jusqu'à maintenant, que dans le Bathonien Tethyen;* Reineckeia (Rehmannia) *ex. gr.* rehmanni; *et* Eurycephalites (Lilloettia) *apparentées à* E. tipperi, *de la Colombie-Britanique, de même qu'à* E. steinmanni *Andéenne laquelle, à Caracoles, est associée à* Prohecticoceras *sousjacente à* Nuequeniceras (Frickites); *4) Les couches litées supérieures produisent probablement:* Reineckeia *s.s.*, Hecticoceras *(H.) ex. gr.* hecticum *indiquant la fin du Callovien inférieur, et* Eurycephalites. *Nous suggérons que l'assemblage* Neuqueniceras *est principalement, sinon entièrement, de la fin du Bathonien; que les espèces du Pacifique oriental* E. (Lilloettia) *semblent être à grande portée; et que* Reineckeia *s.l. peuvent être apparues, ici, avant Tethys et qu'elles proviennent de* Neuqueniceras. *Cet assemblage se répète, avec des espèces largement identiques, dans les Andes Argentino-Chilienne.*

INTRODUCTION

The present study concerns the biostratigraphic revision of the classic Jurassic ammonoid outcrops near Cualac in the northeastern part of the state of Guerrero (Fig. 1) made known by Burckhardt (1927) and Erben (1956). From an approximately 550 m thick shaley series (Yucuñuti Formation of Erben, 1956), a number of new species described and assigned to the genera *Macrocephalites, Reineckeia* and *Peltoceras* by Burckhardt (1927) are now considered to belong to genera and sub-

IUGS UNESCO
Project 171:
Circum-Pacific Jurassic

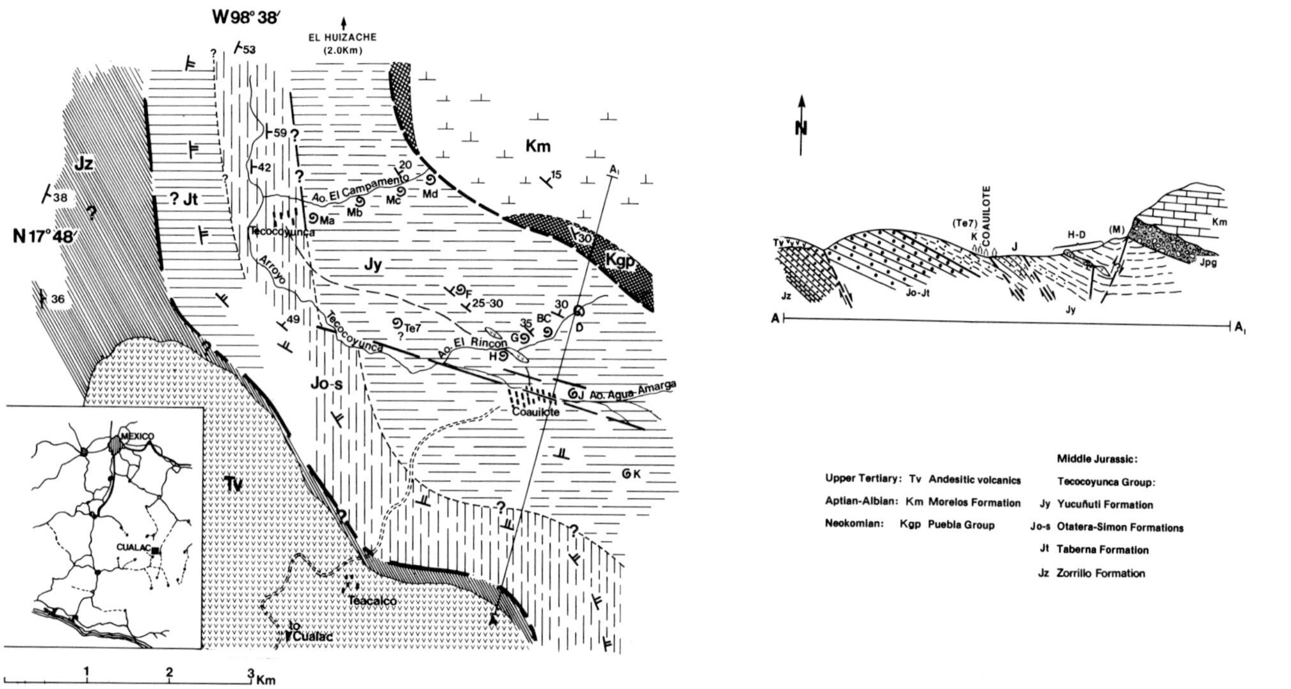

Figure 1. Geological sketch map and section of the Coauilote – Tecocoyunca area, NE Guerrero (modified from Guzman, 1950 and Erben, 1956). Our ammonoid localities marked by spirals and letters, Erben's locality by Te 7.

genera endemic to the eastern Pacific (Thierry, 1978; Cariou, 1980; Westermann, 1981; Taylor *et al.*, this volume). A current revision of the Andean late Middle Jurassic ammonite faunas (Riccardi and Westermann, in progress), shows that the Late Bathonian-earliest Callovian faunas known mainly from west-central Argentina and northern Chile appear to be largely conspecific with the Cualac faunas. The age of these Andean faunas in terms of the (European) standard stages and zones, however, is difficult to ascertain since this time was marked by maximal biogeographic differentiation of the eastern Pacific region, the East-Pacific Realm of Westermann (1981). The classical sections of Argentina (Gottsche, 1878; Tornquist, 1898; Stehn, 1923) and Chile (Steinmann, 1881; but see Hillebrandt, 1970) have so far yielded no reliable (Tethyan) guide fossils useful for inter-realm correlation.

The Yucuñuti Formation at Coauilote (Cuahulote, etc.) near Cualac has yielded the Tethyan guide ammonite *Epistrenoceras* [*"Cosmoceras paracontrarium"* Burckhardt] elsewhere characteristic of Late Bathonian. Its stratigraphic position relative to the (East) Pacific "macrocephalitids" [recte Eurycephalitinae] and the Andean early reineckeiids [*Neuqueniceras*], however, has remained uncertain. Even Erben (1956), who revised the Jurassic of Guerrero and Oaxaca giving the most up-to-date Jurassic litho- and biostratigraphy of the region, made little attempt to determine the vertical ranges of the ammonites within the Yucuñuti Formation which is some 550 m thick. Furthermore, his ammonite collections were neither individually labelled nor catalogued and have never been thoroughly examined taxonomically. Only parts of his large collections housed in the Instituto de Geologia carry satisfactory stratigraphic information but the specimens are not catalogued individually. A re-examination of the classic sections as well as collecting new samples were, therefore, highly desirable.

This paper reports the results of brief field trips by the authors in 1981-82. Fuller stratigraphic and taxonomic revisions are in progress.

The area of the sections can be reached by car in dry weather only, either from Chilpacingo on Highway No. 95 (Mexico-Acapulco) via Chilapa and continuing by dirt road over mountainous terrain past Olinalá and Chaucingo, or on similar roads from Matamoros and Huamuxtitlán. The outcrop area at Coauilote-Tecocoyunca is 7 km by dirt road from Cualac. The dirt roads are in shale, ungraded, and partly with gradients steep enough to make their use hazardous or impossible in wet weather (Fig. 1). Cualac can also be reached by a small plane of the regional aviation service from Tlapa, Huamuxtitlán, Chilpancingo, Izúcar de Matamoros and Cuautla.

GEOLOGY

The rocks and geologic structure of the area of Cualac were studied by Jenny (1933), Salas (1949), Guzman (1950) and Benavides (1978), while Burckhardt (1927, 1930), Erben (1956), Alencaster (1963) and Ochoterena (1966) examined the Jurassic molluscs.

The oldest rocks are schists of the Acatlan Complex which outcrop south and west of Cualac and are dated as early Paleozoic (Ortega, 1978). Corona has recently found Permian ammonoids (dated by Westermann and confirmed by Boardman, unpublished) in the superjacent sequence of limestones and detrital sediments which were formerly placed in the Jurassic. These are followed in sequence by the Cualac Conglomerate and the Tecocoyunca Group of the Middle (-Upper) Jurassic, Neocomian limestones, sandstones and conglomerates of the Puebla Group, and unconformably by the Middle Cretaceous limestones of the Morelos Formation. On top lie early to middle Tertiary conglomerates of the Balsas Group and andesitic rocks. Quarternary alluvia fill streams and river valleys.

The object of this paper is the Tecocoyunca Group – a clastic sequence approximately 750 m thick, of marine and continental origin. It is divided into five formations (Erben, 1956), all with conformable and transitional boundaries. From top to bottom these formations are as follows.

Yucuñuti Formation. Ca. 550 m of calcarous shales with calcareous concretions and some interbedded sandstones, particularly near the base; rich invertebrate fauna at several levels. Late Bathonian-Early (?Middle) Callovian.

Otatera Formation. 50 to 70 m of sandstones with cross-stratification and shales with ferruginous concretions; some beds with oysters.

Simon Formation. 80 to 100 m of sandstones, conglomerates and beds of coal; plant remains.

Taberna Formation. 50 to 60 m of shales and siltstones with ferruginous concretions; moderately rich marine invertebrate fauna. Late Bajocian.

Zorrillo Formation. 20 to 80 m of ferruginous sandstone and shales with some coal and plant remains.

The region lies in the Sierra Madre del Sur (Morelos-Guerrero Basin). The Middle Jurassic between Coauilote and Tecocoyunca, east of Cualac (Fig. 1) belongs to the western limb of a large synclinorium. Photogeology and field work (Corona and Carrasco) indicate the presence of major faults, some of which seem to be strike-slip faults. One major fault or fault-zone runs east-south-east through the village of Coauilote. East of the village, the southern fault block dips locally 60 to 70° NE, while bedding is almost horizontal on the northern side of the fault. Other major faults are believed to be present west of Tecocoyunca, curving around the Tertiary volcanics to the south, and at the base of the Cretaceous escarpment. Small lenticular bodies of andesite form sills and probably dykes (and ? flows) in the basal sandstones of the Tecocoyunca Group and in the Yucuñuti Formation (Campa, pers. commun.).

The rate of sedimentation for the Yucuñuti Formation was high, i.e., ca. 550 m during 3 to 4 ammonoid standard zones or 20 cm. Ka^{-1}. The basin may have subsided even more rapidly. The basal nonmarine-marine transition was followed by sandstones and shales with rich benthic brachiopod-bivalve fauna. These were superposed by the thick shales yielding almost only the pelagic ammonites described herein as well the abundant phylloceratid *Ptychophylloceras* and *Rhaxella*-like sponge spicules (Beds B-C of El Rincon section) which E. Pessagno (pers. commun.) believed to be typically abyssal. Shallowing was indicated at the top of the formation.

The principal measured section (Fig. 2) is on the ridge and along the creek bed of Arroyo El Rincon just north of Coauilote and probably equals Erben's section (1956, lam. 6, Fig. 2). The reference level is the andesite sill or flow 3 to 4 m thick at the disused road from Coauilote to Tecocoyunca (shown by Erben as sandstone with solid rectangle indicating house). The 250 m of shales with siltstones, concretions, and some lenticular sandstone beds above the reference level bear abundant ammonites between about 30 to 65 m and scarcer ammonite fauna still higher. A similar rock sequence, very approximately 200 m thick, is exposed in the lower El Rincon and upper Tecocoyunca creeks below the reference level, but it in part is repeated and possibly separated by a minor fault. Most of this lower section yielded no ammonites during our visit, but an abundant brachiopod and bivalve fauna occurred in the lower part (Alencaster, 1963). At approximately 100 to 150 m below reference is Erben's locality Te 7 from which he recorded *Epistrenoceras* – material that cannot be found now.

East of the village of Coauilote, along the banks and near the head of Agua Amarga creek (localities J,K), we obtained several *Epistrenoceras* from steeply dipping, yellowish siltstone and marl beds probably indicating a fault zone (Fig. 1). The rich shallow-infaunal bivalve fauna included *Vaugonia kobayashii* Alencaster, *Anisocardia coxi* Alencaster and *Unicardium* cf. *varicosum* (Sowerby) (Alencaster, 1963). The specimens we obtained from the local villagers were said also to come from this small area.

Another important section, comprising a 350 to 400 m thick sequence which matches most of the measured section above road at El Rincon, is east of the village of Tecocoyunca along El Campamento Creek. The highest 50 to 100 m at El Campamento containing shell and sandstone beds, together with the upper limit of the formation (? faulted) are not exposed in El Rincon Creek. They comprise the youngest fossil-bearing Jurassic (Callovian) horizons known from the Coauilote-Tecocoyunca area. The top of the El Campamento section therefore, has, been projected above the El Rincon section (Fig. 2).

AMMONITE SYSTEMATICS (WESTERMANN)

Family Phylloceratidae

Ptychophylloceras plasticum (Burckhardt), a typically Tethyan genus, is quite abundant in the middle, richly ammonitiferous part of the section.

Family Oppeliidae

Oxycerites and Paroecotraustes. Ox. (Paroxycerites) n.sp. ex gr. *subdiscus (d'Orbigny)* ♀ (Pl. I, Figs. 1a, b) and *Paroecotraustes* (?) cf. *davaicensis* (Lissajous) ♂, a possible dimorphic pair, have not been described from Mexico previously. The large macroconch with a tricarinate venter and rounded body chamber is known from the Upper Bathonian-Lower Callovian of western Europe (see e.g., Cariou, 1980, p. 16) and the Argentine-Chilean Andes. In fact, the rare Cualac form is indistinguishable from a new species currently being described from "beds 3 to 5" of the Cemetery section at Caracoles, Antofagasta province, Chile ("*Trimarginites*" *part.* in Westermann, 1967) and from the upper part of the Chacay Melehue section in Neuquén province, Argentina, where it is also associated with a similar smooth microconch (Westermann and Riccardi, unpublished). *O. subdiscus* and *O. oxynotus* (Leanza)

PLATE I

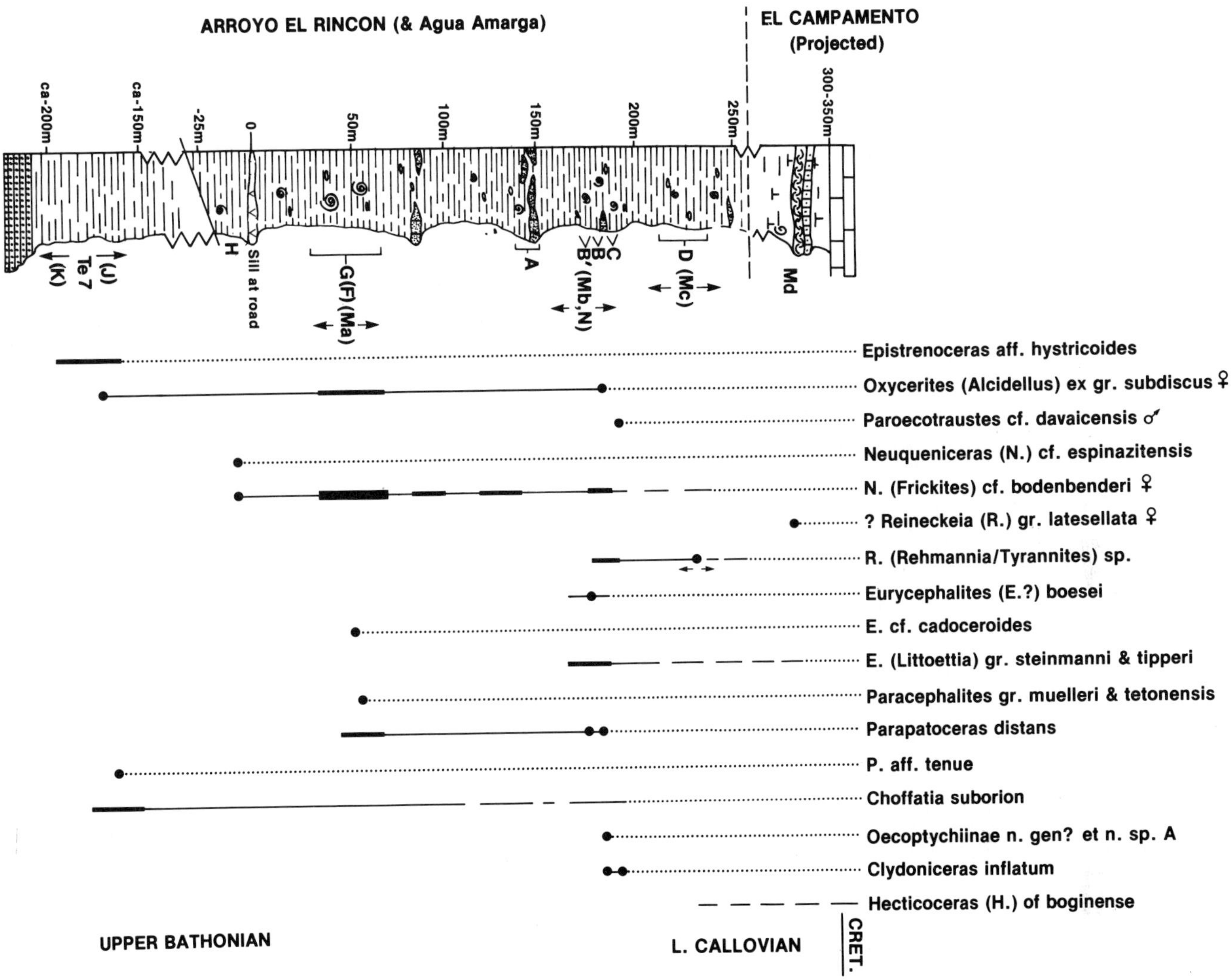

Figure 2. Columnar section of the Yucuñuti Formation at Coauilote with ammonoid occurrances, based on our new findings at the localities indicated in Figure 1. Dots indicate single occurrances.

from the ?Middle Callovian of Chacay Melehue differ in the somewhat wider umbilicus.

Clydoniceras inflatum Westermann n.sp., ♂ (Pl. I, Fig. 2a-c; Pl. III, Figs. 6, 7). This species is known in several specimens from levels B and C of the El Rincon section of which the figured specimen is the holotype, 2 specimens from bed B of the El Campamento section, and in several loose specimens. The diagnostic character is the involute, inflated, and bisulcate shell with prominent ribbing. The species is, in fact, homeomorphic with the Lower Bajocian sonniniid *Poecilomorphus* except for the septal suture which is even simplier (i.e., with two-pronged lateral lobe and multiple, small umbilical elements typical of

EXPLANATION OF PLATE I

Figures 1 a,b. *Oxycerites* (*Paroxycerites*) ex gr. *subdiscus* (d'Orbigny); lateral and ventral views; McMaster J 2109a, middle part of Yucuñuti Formation, locality F, N.W. of Coauilote. (X 0.9).

Figures 2a-c. *Clydoniceras inflatum* Westermann n.sp., holotype, lateral and ventral views of incomplete specimen; McMaster J 2110a, middle part of Yucuñuti Formation, level B of El Rincon section. (X 1.5). (See also Pl. III)

Figures 3a-c. Oecoptychiinae n.sp. A, lateral and ventral views; McMaster J 2110b, middle part of Yucuñuti Formation, level B of El Rincon section. (X 1.5).

Figures 4 a,b. *Neuqueniceras* (*Neuqueniceras*) sp. ♂, lateral and ventral views of almost complete microconch; McMaster J 1209b, middle Yucuñuti Formation, locality F, N.W. of Coauilote. (X 0.9).

Figures 5 a,b. *Neuqueniceras* (*Neuqueniceras*?) cf. "*neogaeum*" (Burckhardt) ♀, lateral and ventral views of macroconch with 1/2 whorl incomplete body chamber; McMaster J 2119a, loose from Yucuñuti Formation, Coauilote (purchase Corona). (X 1).

Figures 6 a,b. *Reineckeia* (*Rehmannia*/?*Tyrannites*) sp. ♀, lateral and ventral views of entirely septate macroconch; McMaster J2114a, upper part of Yucuñuti Formation, level D of El Rincon section. (X 1).

PLATE II

clydoniceratids). Some specimens are septate up to 52 mm in diameter (Pl. III, Figs. 7a,b) while body chamber fragments indicate only 40 to 55 mm maximum size. They show no reduction in the raised ventral shoulders defining the ventral furrows, nor in the secondary ribbing. These body chambers appear therefore to be the microconchs of *Cydoniceras,* i.e., *"Delecticeras"* (cf. Westermann, 1956). This species is much more inflated, particularly ventrally, than any known species of *Clydoniceras,* macroconch or microconch, and has more prominent and more flexed ribbing. The known range of *Clydoniceras* is from late Middle to latest Bathonian of western Europe, north Africa, Madagascar and Baluchistan. Bisulcate forms, however, are known only from the "Aspidoides" Zone (*recte* Orbis Zone) and Discus Zone of Europe (Callomon, pers. commun.; cf. Dietl, 1982) and from Madagascar (Collignon, 1958).

Hecticoceras. The single specimen of *H.* (*Hecticoceras*) cf. *boginense Petitclerc of the H. hecticum* group (Pl. III, Figs. 9 a-c) from the upper part of the El Campamento section, appears to be the first find of Hecticoceratinae in North America (specimen identified by S. Elmi and E. Cariou, pers. commun.; cf. Elmi, 1967, Pl. II). The only other known eastern Pacific occurrence of this important Tethyan subfamily is in the Neuquén Basin, Argentina (Stehn, 1923; Westermann, 1981). The specimen was found loose at level Mc and according to surface gradient, strike and dip, must have been derived from above. The matrix of yellow mudstone resembled that of level M near the top of the section. The somewhat distorted body chamber has a compresses subrectangular whorl section with blunt ventral fastigation and tuberculate dichotomous ribbing. This form is restricted to the uppermost Lower Callovian, *Proximum* Horizon and Patina Subzone of the Gracilis Zone (Cariou, 1980) and abounds in the western Subtethyan province (Elmi and Cariou, pers. commun.).

Family Sphaeroceratidae

Eurycephalites, Lilloettia, and Xenocephalites. The subfamily Eurycephalitinae Thierry and its Andean taxa are currently being revised by Riccardi and Westermann. We follow Callomon (in Donovan *et al.,* 1981) in classifying the macrocephalitins as a subfamily of the Sphaeroceratidae. *Lilloettia* is placed as a subgenus in *Eurycephalites.*

Eurycephalites (*E.*?) *boesei* (Burckhardt) ♀ was probably based on the single, figured body chamber from the Cualac area which is therefore the holotype. (A second specimen in the old collection of the Instituto de Geologia is unlabelled and was apparently not available to Burckhardt.) This macroconch is intermediate in size, whorl shape, and ornamentation between the typically small nominate subgenus with subquadrate and rather ornate whorls and the typically large subgenus *Lilloettia* with ovate to subtriangular and more finely ornate whorls. We have single specimens from ? level B of the El Rincon section and from the approximately equivalent level C of the El Huizache section 4 km north of Tecocoyunca (Pl. III, Fig. 3). At both localities, the species is associated with a more compressed and finer ribbed form with even more reduced primaries, closely resembling *E.* (*Lilloettia*) *tipperi* Frebold (Pl. III, Figs. 1a,b) [= *"Macrocephalites* aff. *macrocephalus"* of Burckhardt 1927, Pl. 16, Figs. 1-3]. Callomon (this volume) suggests that *E. boesei* may be conspecific with the North-Cordilleran *E. buckmani* (Crickmay), and furthermore, *E. lilloetensis* Crickmay (including *L. mertenyarwoodi* Crickmay) and *E. buckmani* Crickmay may be mere variants of the same species. Crickmay's and Frebold's North-Cordilleran species were originally described from isolated outcrops in southern British Columbia; only the *E. lilloetensis-buckmani* association can now be dated as clearly early Callovian from occurrances above the *Cadoceras* range zone in South Alaska (Imlay, 1953; Callomon, this volume). According to the plastercasts available to me, *E. tipperi* has consistently smooth inner flanks on the internal mould, at least as far back as 25 mm diameter while the inner whorls at a similar diameter to *E. lilloetensis* on the holotype and on the dated specimens illustrated by Imlay (1953b, Pl. 30) from the Chinitna Formation of southern Alaska, have well developed primaries at the same growth stage. Similarly, the specimens of *E.*(*L.*) *buckmani* from the Chinitna, illustrated by Imlay (Pl. 27) are partly from the same beds as *E. lilloetensis,* and are much more coarsely ribbed on the flank than the Mexican topotypes of *E. boesei* found together with *E.* gr. *steinmanni* and *tipperi.* Thus I conclude that the two Cualac forms mentioned, on the one hand, and the North-Cordilleran

---

EXPLANATION OF PLATE II

Figures 1a-c. *Neuquenicera*s (*Frickites*) cf. *bodenbenderi* (Tornquist), lateral and ventral views of entirely septate macroconch; McMaster J 2160a, middle part of Yucuñuti Formation, locality F, N.W. of Tecocoyunca. (X 0.85).

Figure 2. *Parapatoceras* cf. *distans* (d'Orbigny), in umbilicus of *Neuquenicera*s sp., lateral view of incomplete specimen; McMaster J 2110a, from middle part of Yucuñuti Formation, level B of El Rincon section. (X 1).

Figures 3-7. *Epistrenoceras* aff. *hystricoides* Rollier, XI. 3a,b) the holotype of *Cosmoceras paracontrarium* Burkhardt, (*nomen dubium*), lateral views of somewhat crushed and incomplete body chamber; Inst. Geol. Mex., lower part of Yuruñuti Formation, area of Cualac (?Coauilote-Tecocoyunca).
4) lateral view of almost complete, coarsely ribbed specimen; McMaster J 2108b, loose from (lower part of) Yucuñuti Formation, Coauilote.
5-6c) lateral and ventral views of entirely septate specimens, with spinose intermediate stages; McMaster J 2108c-d, loose from (lower part of) Yucuñuti Formation, Cuaolote. (X 1).
7) lateral view of small, almost complete specimen with 1/2 whorl body chamber, ? microconch; McMaster J 2108e, loose from (lower part of) Yucuñuti Formation, Coauilote. (X 1).

Figures 8 a,b. *Bullatimorphites* (*Kheraiceras*) *bullatus* (d'Orbigny) ♀, lateral and ventral views of almost complete macroconch; McMaster J 2108a, loose from Yucuñuti Formation, Coauilote. (X 1).

PLATE III

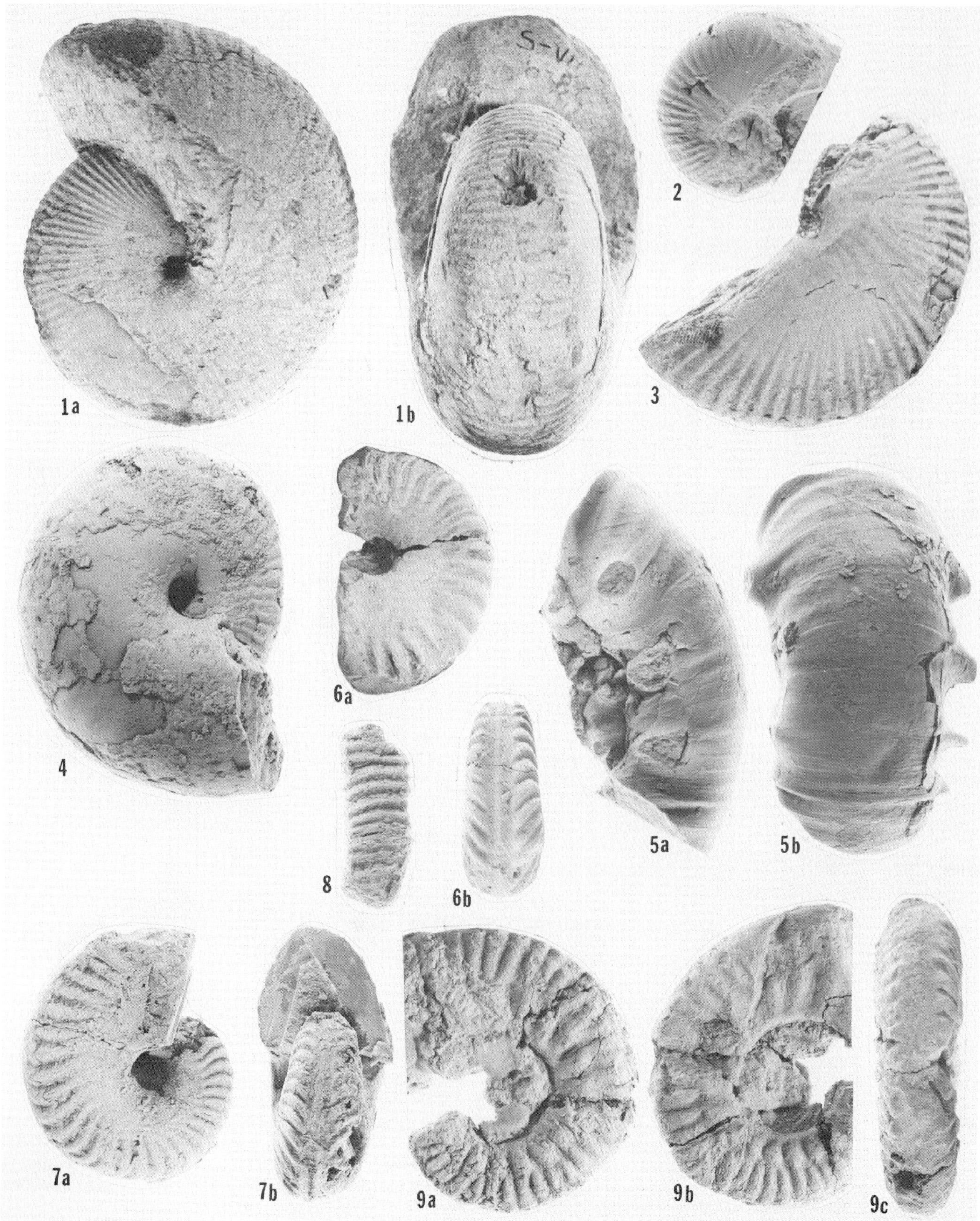

forms, on the other, may respectively be variants of single species as suggested by Callomon, but that the Mexican species is not conspecific with the North-Cordilleran species. A single minute *E.* (*Lilloettia*?) sp. indet. juv.(?) (Pl. III., Fig. 3), however, has been found as high as the top of the Yucuñuti Formation (level Md of El Campamento section).

In the Andean Province, *E.* (*L.*) *steinmanni* Spath closely resembles our compressed form. It occurs in the lower part of the Cemetery section at Caracoles, Chile (beds 2-3) below *Neuqueniceras* (*Frickites*) and together with *E.* aff. *boesei* (Westermann, 1967), and in the upper part of the thick shaley section of Chacay Melehue (McM. J 1204), Neuquén (bed 4) mainly above *Neuqueniceras* s.s. and together with *E.* (*Imlayoceras*?) cf. *cadoceroides* (Burckhardt) (Riccardi and Westermann, unpublished). Significantly, the same bed 2 at Caracoles also contains rare *Prohecticoceras* cf. *ochraceum* Elmi, ♀ and ♂ (unpublished), a species known previously only from the Lower to lower Upper Bathonian of western Mediterranean and Poland (Elmi, 1967, 1971).

Eurycephalites (*Imlayoceras* ?) *cadoceroides* (Burckhardt), based on the monotypic holotype supposedly from Mixtepec in northwest Oaxaca, also occurs at Coauilote in the middle part of the El Rincon section (ca. our level G) according to our inspection of Erben's collection Te 6. *Imlayoceras* Frebold is, however, based on a single isolate fauna from Alberta and could merely be an extremely inflated *Lilloettia* or *Iniskinites* (Callomon, this volume; Riccardi and Westermann, unpublished).

The rare microconchiate *Xenocephalites nikitini* (Burckhardt, 1927; Pl. XVI, Figs. 8, 9 here designated the neotype, holotype Figs. 4-7 lost) resembles *X. vicarius, X. hebetus,* and *X. hartsocki* Imlay spp. from South Alaska where they belong to the late Bathonian and Early Callovian *Iniskinites* and *Lilloettia* faunas (B5-B8 of Callomon, this volume). The microconchs, however, show much smaller interspecific variation than the macroconchs and cannot be dated as precisely as the macroconch (Callomon, this volume and pers. commun.). Another closely affiliated species is *X. gottschei* (Tornquist) from the *Neuqueniceras* (*Frickites*) assemblage of the ? basal Callovian of the southern Andes. *X.* cf. *phillipsi* Imlay, originally described from Montana (fauna A8 of Callomon, this volume) and *X. "herrero-duclouxi"* Stipanici (1966), a relative of *X. neuquensis* Stehn from Neuquén, occur at unknown levels also at Coauilote. Unfortunately, all Eurycephalitinae are relatively rare in Mexico so that their ranges are difficult to establish.

Of particular interest is also the probable occurrence in the *Neuqueniceras* assemblage north of Coauilote of *Paracephalites* gr. *muelleri* and *tetonesis* (Imlay) ♀ (Pl. III, Fig. 4), another East-Pacific Eurcephalitinae genus hitherto known only from the Upper Bathonian of the Western Interior United States (Imlay 1955a; faunas A5, 6 of Callomon, this volume). It is even possible that the *ex situ* ?*Xenocephalites* cf. *"herrero-declouxi"* ♂ mentioned above is the matching microconch, i.e., *X.*? *shoshonense* (Imlay, 1953a; Callomon, this volume). Closely similar Eurycephalitinae species also occur in southern Peru (Westermann *et al.,* 1980).

Family Perisphinctidae

Choffatia. The genus is understood conservatively to include *Subgrossouvria, Parachoffatia* and *Homoeoplanulites. C. suborion* (Burckhardt, 1927; lectotype here designated Pl. XXXI, Figs. 2-4, 104 mm diameter; ? synon. *"Perisphinctes" cualacensis* Burckhardt, 1927) occurs at an unknown level at Cuauilote. An identical form occurs in the Neuquén Basin at Chacay Melehue, and perhaps at Caracoles (? *"P." gottschei* Steinmann); similar species also abound around the Bathonian/Callovian boundary of Europe. Forms reminiscent of the Andean *C.*? *gleimi* (Steinmann), *C. jupiter* (Steinmann), and *C.*? *pseudoeuryptycha* (Tornquist) also occur at Cuauilote, but cannot yet be clearly identified.

Family Perisphinctidae or Reineckeiidae

?*Neuqueniceras.* Of particular interest are several small forms of Burckhardt (1927), i.e., *"Perisphinctes" weitzi* and *"Peltoce-*

EXPLANATION OF PLATE III

Figure 1. *Eurycephalites* (*Lilloettia*) ex gr. *steinmanni* Spath and *tipperi* Frebold ♀, lateral and ventral views of mostly septate macroconch; McMaster J 2121, upper third of Yucuñuti Formation, level C of El Huizache section, N. of Tecocoyunca. (X 1.1).

Figure 2. *Eurycephalites* (*Lilloettia*?) sp. indet. juv.(?), body chamber and mostly destroyed phragmocone; McMaster J 2122, top of Yucuñuti Formation, level Md of El Campamento section. (X 1.5).

Figure 3. *Eurycephalites* (*E.*?) *boesei* (Burckhardt) ♀, body chamber body chamber and incomplete phragmocone; McMaster J 23123, upper part of Yucuñuti Formation, level C of El Huizache section. (X 1).

Figure 4. ?*Paracephalites* ex gr. *muelleri* and *tatonensis* Imlay ♀, lateral view of almost complete specimen; McMaster J 2127, middle part of Yucuñuti Formation, near Tecocyunca. (X 1).

Figures 5 a,b. *Reineckeia* (*Reineckeia* ?) sp. indet. ["No. 2" of Burckhardt, 1927], lateral and ventral views of fragment from last 3 whorls, including body chamber; McMaster J 2126, top of Yucuñuti Formation, level Md of El Campamento section. (X 1).

Figures 6-7. *Clydoniceras inflatum* Westermann n.sp., Yucuñuti Formation, *ex situ* near Coauilote. (Approximately X 1).
6 a,b) lateral and ventral views of small body chamber, microconch; McMaster J 2124,
7 a,b) lateral and ventral views of large phragmocone, septate; McMaster J 2125. (See Pl. I)

Figure 8. Fragment of ?*Parapatoceras* aff. *tenue* (Baugier and Sauzé) (?or a re-worked Late Bajocian *Spiroceras annulatum* (Desh.)), lateral view of fragment; McMaster J 2129, from the lower part of the Yucuñuti Formation, *Epistrenoceras* level, basal part of Agua Amarga section, Coauilote. (X 1).

Figures 9 a-c. *Hecticoceras* (*Hecticoceras*) cf. *boginense* Petitclerc, both lateral views and ventral view of body chamber with end of phragmocone; McMaster J 2128, at or near top of Yucuñuti Formation *ex situ,* levels c-d of El Campamento section. (X 1).

ras" constrictum, which appear to be intermediate between Perisphinctidae and Reineckeiidae and resemble the inner whorls of *Neuqueniceras* (*Frickites*) cf. *bodenbenderi* (Tornquist) ♀ as described below. *"P." weitzi* also resembles the *"Simoceras* sp." from Espinacito Pass, Argentina, described by Gottsche (1878, Pl. III, Figs. 5a-b), while *"P." constrictum* is close to *"Perisphinctes" boehmi* (Steinmann) from Caracoles, Chile.

Family Reineckeiidae

Neuqueniceras. Current investigations at the Argentine type localities of *Neuqueniceras* spp. (Chacay Melehue, Neuquén province) and of the subgenus *Frickites* (Espinacito Pass, San Juan province) by Riccardi and Westermann, indicate that the two taxa are intimately related through transitional forms. Their distinction at the generic level was recently upheld, however, by Cariou (1980) who also erected the new subfamily Neuqueniceratinae. The group including the type species *N. steinmanni* (Stehn), all possibly belonging to a single variable species, is characterized by the small to medium-sized macroconchs bearing blade-like primaries on the entire phragmocone and a more or less distinct ventral costal interruption, while the body chamber has prominent mid-lateral to ventro-lateral bullae which usually bear spines. The group including the type species of *N.* (*Frickites*) *bodenbenderi* (Tornquist) has large macroconchs with a similar "perisphinctid" juvenile stage followed by an intermediate growth stage bearing moderately short primaries with mid-lateral turbercles and an outer stage of approximately two whorls developing very prominent and conical mid-lateral spines; the mid-ventral interruption is only faint and irregularly developed (obsolescent) intraspecifically and morphogenetically. The inner whorls of the macroconch and the poorly known microconch of *N.* (*Frickites*) cf. *bodenbenderi* from Cualac thus tend to resemble Burckhardt's *"Perisphinctes" weitzi,* with perisphinctid ribbing except for the tiny mid-lateral tubercles. Because of the difficulty of placing many specimens (and "species") into either taxon, *Neuqueniceras* and *Frickites* are distinguished here at the subgeneric level only. By what appears to be phyletic acceleration of development, i.e., palingenetic reduction of the juvenile perisphinctid stage and its progressive replacement by the reineckeiid stage, *N.* (*Frickites*) seems to have developed into the early subgenera *Rehmannia* and *Tyrannites* (Cariou, 1980) of *Reineckeia,* which we believe are present in the upper third of the Coauilote section (Fig. 2).

Neuqueniceras s.s. is probably represented in Mexico by *"Peltoceras neogaeum"* Burckhardt (1927, Pl. II, Figs. 4,5) which, unfortunately, is based on a single very fragmentary and partly crushed psecimen, i.e., the holotype, from the Cualac area. Even the plastotype does not permit safe identification of topotypes so that the specific name should be considered a *nomen dubium.* Similarly, the seemingly very close *"Peltoceras monacanthum"* Burckhardt is based on fragments of a single specimen from an unknown stratigraphic level at the old El Consuelo coal mine in NW Oaxaca province, and the important septate fragments have been lost. These forms are known from Coauilote only in rare, unlocated specimens. *"P. neogaeum"* has also been recorded from localities in Northwest Oaxaca, e.g., Yucuñuti, as well as from the Huayacocotla Mountains in east-central Mexico (Erben, 1956; Cantu Chapa, 1979; cf. Taylor *et al.,* this volume).

The subgenus *Frickites* is represented by the most abundant species in the highly fossiliferous middle part of the Yucuñuti Formation at Coauilote-Tecocoyunca, but apparently was not described in Burckhardt's monograph on Cualac. *N.*(*F.*) cf. *bodenbenderi* (Tornquist) (Pl. II, Figs. 1a-c) is very close to the Andean species from Espinacito Pass, San Juan province, and partly indistinguishable from Caracoles material above the *Lilloettia* beds (Cemetery section, beds 5, 8). The only possible difference may be in the even less developed mid-ventral interruption which is generally considered as a highly variable feature of low taxonomic value. The fossiliferous levels at Coauilote (El Rincon G/F, El Campamento Ma) also contain a large strongly depressed form resembling *Reineckeia aberrans* Burckhardt. This species is also poorly founded on an originally incomplete holotype only – part of which is lost. The remaining fragments resemble the highly ornate Andean *N.*(*F.*) *antipodum* (Gottsche) from Espinacito Pass and the closely similar *N.*(*F.*) *euactis* (Steinmann) from Caracoles and also extremely depressed *Choffatia.* The affinity of *"Reineckeia" plicata* Burckhardt is also poorly known because the holotype, the only specimen originally placed in that taxon, is lost.

According to the recent monograph by Cariou (1980), the Tethyan Reineckeiinae evolved from *Choffatia* (*Homoeoplanulites*) in the upper Macrocephalus Zone leading to *Rehmannia* and *Reineckeia* (*Tyrannites*). Both are said to have "perisphinctoid" inner whorls with minute tubercles only, followed by the outer coronate reineckeiid stages. *R.* (*Tyrannites*) is said to be distinguished from *Rehmannia* by the appearance of a reineckeiid stage already on the mature phragmocone, while it is restricted to the body chamber in the latter. Dr. Carious has kindly sent us a plastotype of *R.* (*Tyrannites*) *franconica.* It is almost indistinguishable from specimens from beds B to D in the upper part of the El Rincon section (Pl. I, Figs. 6a,b). Other Coauilote forms are very close to *Rehmannia rehmanni* (Oppel) and *R.* (*R.*) *grossouvrei* (Petitclerc), apparently the earliest known Tethyan reineckeiids. We thus consider the taxa *Tyrannites* and *Rehmannia* as closely interrelated and prefer to place both as subgenera, or as one subgenus, in the genus *Reineckeia.* All species are said to be restricted to the Lower Callovian (Cariou, 1980).

Reineckeia s.s. has now been tentatively identified from the uppermost Yucuñuti Formation in the El Campamanto section, i.e., ?*R.*(*R.*) *ex. gr. latesellata* Burckhardt and *"Reineckeia* sp. indet. 2" of Burckhardt (1927, Pl. 21). Both forms were originally described from the long abandoned El Consuelo coal mine, northwest Oaxaca. The majority of the 20-odd *Reineckeia* "species" described by Burckhardt from that locality, including *S.* [*Erymnoceras auct.*] *mixtecorum,* all with coronate inner and outer whorls, could belong to but a single species; *R. latesellata* represented the best material. Our specimen is mostly septate and is coronate at least as far back as the earliest preserved whorl at *ca.* 15 mm in diameter. The large fragment of "*R.*? sp. indet. 2" of Burckhardt (Pl. III, Figs. 5a,b) bears distant prominant spines on all of the three last whorls preserved. *Reineckeia* s.s. is said to range from the topmost Lower to the lower Upper Callovian (Cariou, 1980).

Family Tulitidae

Bullatimorphites. The circum-global Tethyan subgenus *Bullatimorphites* (*Kheraiceras*) is represented by the similarly widespread species *B.* (*K.*) *bullatus* (d'Orbigny) s.l. Our single macroconch (Pl. II, Fig. 8a,b) probably came from the basal or middle Yucuñuti Formation. It was found loose and said to have come from locality J. The monotypic holotype, now lost, of *B.* (*K.*) *v-costatus* (Burckhardt) from the Cualac area differs only in the curved secondaries on the intermediate whorls forming blunt obtuse ventral chevrons. Erben (1956, p. 121) recorded this species from the Otatera or basal Yucuñuti Formation of Tezoquio, but the specimen is lost. Although one cannot decide from such single specimens whether this difference is intra-or inter-specific, the same two forms occur again at Caracoles, Chile (beds 3-5 of the Cemetery section).

Our specimen of *B. bullatus* ♀ is fully grown and almost complete. The ribbing on the body chamber becomes obsolete at about 70 mm diameter and the complete diameter is estimated at only approximately 85 mm. The number of secondaries on the penultimate half-whorl is 25. This agrees more closely with the mainly Lower Callovian *B. bullatus* s.s. than with the mainly Upper Bathonian *B. bullatus hannoveranus* (Roemer), which differs in larger size and coarser ornament (Gabilly, 1964; Westermann, 1958). An inspection of Gabilly's large collections from western France, however, showed that these chrono-subspecies (or transients *sensu gallico*) have appreciable morphologic overlap. Furthermore, both forms are said to have similar ranges in southern Germany (Hahn, 1971) so that their stratigraphic usefulness is limited.

?Family Tulitidae

Oecoptychiinae. A single small specimen from bed B of the El Rincon section (Pl. I, Fig. 3a-c) is identified only with Oecoptychiinae n.sp. A. It has the characteristic geniculate body chamber of *Oecoptychius,* a micromorph genus known only from the (? entire) Callovian (and ? Oxfordian) of western Europe. Our specimen differs, however, in the absence of the mid-ventral interruption and backward curvature of the costae that are consistently present in the European genus. Our specimen resembles the *Bullatimorphites* microconchs *Bomburites* and *Treptoceras* in all features but the extremely small size and the strong geniculation and appears thus to be intermediate between *Bullatimorphites* ♂ and *Oecoptychius* in morphology and age. This supports Schindewolf's (1965) opinion which tentatively placed th Oecoptychiinae Westermann in the Tulitidae, and also is consistent with the original analysis of the sutural ontogeny (Westermann, 1956). On the other hand the recent classification by Callomon (in Donovan *et al.,* 1981) includes *Oecoptychius* in the Strigoceratidae of the Haplocerataceae. Significantly, the appearance of *Oecoptychius* in Europe is cryptogenic and the closest possible known ancestor in Europe seems to be the middle Bathonian tulitid *Sphaeroptychius* (*cf.* Enay, 1959).

?Family Spiroceratidae

Parapatoceras. The species *Parapatoceras distans* (d'Orbigny) from Coualote has been described previously under the name of *Infrapatoceras biserratum* Ochoterena, gen. et sp. nov. (1966) based on Erben's collection from locality Te 3. *Infrapatoceras* was said to be distinguished from *Parapatoceras,* a well-known late Bathonian to early Callovian genus found in western Europe, Oregon and Chile, by the somewhat simpler septal suture; L and U are without frilling, I rounded – features which today are not considered to be of taxonomic significance at the genus-level (Dietl, 1978; Donovan *et al.,* 1981). In fact the Mexican species (Pl. II, Fig. 2) may not be distinct from the European Late Bathonian to Early Callovian *P. distans* which, according to Hillebrandt (1973, p. 174) and our field work, also occurs in the Sierra Domeyko of N. Chile.

From the *Epistrenoceras* beds of the Agua Amarga section comes a single fragment (Pl. III. Fig. 8) which resembles *Parapatoceras tenue* (Baugier and Sauzé) as recently re-described by Dietl (1978, p. 44, Pl. 7, Fig. 10). Our specimen differs in the somewhat more prominent and less strongly oblique ribs, although that perhaps could be due to the slight distortion. Significantly, *P. tenue* is known mainly from the Upper Bathonian of northern Chile where it is associated with *P. distans,* but very rare in the Upper Bathonian of France and Switzerland (Dietl, 1978). There is, however, also the possibility that our fragment is a re-worked Late Bajocian *Spiroceras annulatum* (Desh.), i.e., from the Taberna Formation also present in the area; there is close resemblance in the prominence and weak obliquity of the ribs (*cf.* Dietl, 1978, Pl. 6, 7).

Epistrenoceras. The holotype of *Epistrenoceras paracontrarium* (Burckhardt, 1927), the only specimen originally described from the "basal beds of the upper Dogger" of Cuauilote at Cualac, is too incomplete for interpretation of the species (Pl. II, Figs. 3a,b). The present sample includes several almost complete specimens (Pl. II, Fig. 4). The author said that the holotype was different from the European *E. contrarium* (d'Orbigny, 1842-1849, Pl. 145, Figs. 1, 2 only) from the Upper Bathonian of Niort, northern France, because the whorl was more inflated with maximum width near mid-flank and because of the lower position of the inflexion point of the ribs, i.e., at mid-flank instead of the upper third. According to d'Orbigny's illustration (which may not be reliable), *E. contrarium* has large evolute septate whorls and isocostate, moderately coarse ribbing with ventrolateral bullae. Our well preserved topotypes of E. *paracontrarium* show, however, that the supposed differences of whorl section and ornament are due partly to the lateral crushing of the holotype of *E. paracontrarium* and partly to the slightly variocostate modification of the body chamber; the geniculate inflexion is at about 3/5 whorl height as in the European form, while the section varies from subcircular to ovate with maximal whorl width usually at about 2/5 height. Our septate specimens, 30 to 35 mm diameter, are all much more involute than d'Orbigny's illustration and most have typically variocostate secondaries. In some, the penultimate septate half-whorl of about 20 mm diameter has distant ventrolateral spines which discontinue abruptly on the ultimate half-whorl and change into dense secondaries ending in blunt nodes (Pl. II, Figs. 6a-c). On other specimens, the spinose stage continues to the body chamber. On one incomplete body chamber of 35 mm in diameter, the primaries shorten with the inflexion moving to mid-flank, thus resembling the holotype.

The narrow umbilicus and the presence of prominent ventrolateral spines at some stage in the majority of our Cualac specimens indicate close affinity to the Upper Bathonian *E. hystricoides* Rollier (for d'Orbigny, 1842-1849, Pl. 145, Figs. 3, 4) from western Europe except that the ribbing is denser than in d'Orbigny's illustration.

The holotype of *Cosmoceras paracontrarium* may thus belong to either species. Most topotypes sway the balance toward *E. hystricoides* but this must remain uncertain because its type lacks the entire phragmocone.

One specimen (Pl. II, Fig. 8) appears to be fully grown at 30 mm in diameter. It has a much more compressed body chamber than the larger specimens and has ventrolateral bullae of medium strength persisting to the end. This is either another species or possibly the (matching) microconch of the form described above.

Of the 20 specimens of *Epistrenoceras* available to us, we collected 7 from beds with yellow siltstone and marl, together with the bivalves *Vaugonia, Anisocardia* and *Unicardium* and the ammonite *Choffatia,* in the lowest part of the Agua Amarga section, Coauilote (*ca.* locality J), and one at locality K.

E. paracontrarium was reported together with *Prohecticoceras* from Sierre Domeyko in northern Chile (Hillebrandt, 1973), but the specific identification needs confirmation.

AGE AND DISTRIBUTION OF THE NEUQUENICERAS ASSEMBLAGE

This moderately diverse assemblage occurs in great abundance through 30 to 35 m (interval G/F) of shales in the middle of the Yucuñuti Formation at Caouilote (Fig. 2). It includes the dominant *Neuqueniceras* (*Frickites*) cf. *bodenbenderi* (Tornquist), the somewhat less common *N.*(*F.*)? aff. *aberrans* (Burckhardt) and *Ptychophylloceras plasticum* (Burckhardt), and the accessories *Eurycephalites* (s.s. + ?*Lilloetia*) spp., *Choffatia* sp., *Oxycerites* (*Paroxycerites*) gr. *subdiscus* (d'Orbigny) and *Parapatoceras distans* (Baugier and Sauzé). This interval was presumably also the principal source of Burckhardt's type specimens. The entire range of genus *Neuqueniceras* is, however, through at least 200 m of the Caouilote section (levels H to B), while the ranges of most of the other taxa cannot yet be established because of relative scarcity outside the interval G/F. There has been, to our knowledge, no evidence for positioning the *Neuqueniceras* assemblages above an assemblage of *Eurycephalites boesei* (Burckhardt) and *Xenocephalites* as recently tabulated by Imlay (1980, Fig. 14), nor for his dating of this (and his lower) assemblage as (late) Early Callovian following previous authors (Burckhardt, 1927, 1930, Erben, 1956). On the other hand, we now confirm the stratigraphic position of the *Neuqueniceras* assemblage above the mid-Late Bathonian *Epistrenoceras.*

The lower ca. 200 m of the Yucuñuti Formation (below bed H near the reference level in Figs. 1, 2) as well as the probably coeval beds outcropping east of Coauilote have yielded to us neither *Neuqueniceras* nor *Eurycephalites.* Near the base we found, however, a single *Epistrenoceras* aff. *hystricoides* (Rollier) (locs. J, K) together with *Choffatia suborion* and an abundant bivalve and locally, brachiopod fauna. This is at approximately the same level as Erben's locality Te 7 near Tecocoyunca from which he recorded *Epistrenoceras paracontrarium.* In the better documented sections in Mediterranean and Submediterranean Europe, the range of *Epistrenoceras* is apparently restricted to the middle Upper Bathonian, i.e., the Retrocostatum Zone *auct., recte* Aspidoides Standard Zone (Torrens, 1980), most recently renamed the Orbis Zone (Dietl, 1982). The relative scarcity in Mexico of ammonites in these lower beds yielding, in part, abundant bivalves and brachiopods, is presumably due to ecologic conditions unfavorable to ammonites, so that the first appearance of the *Neuqueniceras-Eurycephalites* fauna may be diachronous. This sequence, the first reineckeiids following well above *Epistrenoceras,* is nevertheless the same as in northern Chile and (?) southern Peru (Hillebrandt, 1970 and pers. commun.; Westermann *et al.*, 1980).

The age of the upper range limit of the *Neuqueniceras* assemblage is now suggested by the first circum-Pacific find of *Clydoniceras,* i.e., *C. inflatum* n.sp., at levels B and C of the El Rincon section and the approximately equivalent level Mb of the El Campamento section (Fig. 2). In the better documented European sections, bisulcate *Clydoniceras* species are limited to the middle and upper Upper Bathonian, i.e., Orbis and Discus Zones, and the family became extinct. However, this does not preclude survival elsewhere.

Level B of El Rincon has also yielded representatives of what seemed to be the earliest tue *Reineckeia,* i.e., subgenus *Rehmannia* (and/or *Tyrannites*), which in the Tethys appears only in the upper Macrocephalus Zone, Rehmanni Subzone, of the Lower Callovian (Cariou, 1980). At this level *Eurycephalites* (*Lilloettia*) ex gr. *tipperi* Frebold and *steinmanni* Spath as well as *E.*(*E.*?) *boesei* Burckhardt also occur. The very closely affiliated or conspecific *E.* (*L.*) *steinmanni* Spath of the Andean Province occurs at Caracoles (below *Neuqueniceras*!) together with *Prohectiococeras* cf. *ochracensis* Elmi, a West-Tethyan species known only from the Bathonian up to the Hodsoni (or ?Orbis) Zone. The Late Bathonian, rather than Early Callovian, age of the *Neuqueniceras* assemblage is also supported by the first Mexican find of ?*Paracephalites* ex. gr. *muelleri* and *tetonensis* Imlay hitherto known only from the Upper Bathonian of the Western Interior, United States (Callomon, this volume). The tentative age of the *Neuqueniceras* assemblage is latest Bathonian to (?) earliest Callovian.

The *Clydoniceras inflatus* assemblage (El Rincon levels B, C) lies approximately at the Bathonian/Callovian boundary so that the first appearance of *Reineckeia* (*Rehmannia*/*Tyrannites*) sp. appears to be here slightly earlier than in the Mediterranean Province. This and the apparent morphological intergradation into *Neuqueniceras* (*Frickites*) ex. gr. *bodenbenderi* suggest an East-Pacific, rather than Mediterranean, origin of the Reineckeiidae.

Latest Lower Callovian age is evident for the uppermost Yucuñuti Formation (El Campamento section) by the first find of the *Hecticoceras* (*Hecticoceras*) cf. *boginense* Petitclerc, hitherto known from the West-Tethyan uppermost Lower Callovian (Elmi, pers. commun.). ? *Reineckeia* (*Reineckeia*) ex gr.

latesellata Burckhardt even suggests Middle to Upper Callovian. Rare probable *Eurycephalites* (*Lilloettia*?) sp. also ranges up into the topmost Yucuñuti Formation (level Md).

PALEOBIOGEOGRAPHY

The exceptionally close taxonomic affinity of the *Neuqueniceras* assemblage of Cualac to the Andean fauna of Chile and Argentina has been discussed elsewhere by me in this volume (Taylor *et al*). At least half and possibly two-thirds of the ammonite species appear to be in common with the rich faunas of Caracoles in Antofagasta province, Chacay Melehue in Neuquén province, Passo del Espinacito in San Juan province, the Sierra Domeyko in Antofasgasta province (collected by von Hillebrandt), and probably southern Peru. This Andean assemblage overlies beds with *Epistrenoceras* as in Mexico, while *Hecticoceras* and true *Reineckeia* occur at a higher level. Such a close faunistic affinity is compatible with the recent plate-tectonic hypothesis, that the Mixteca terrane (northeast Guerrero and northwest Oataca states) carrying the *Neuqueniceras* assemblage is allochthonous with respect to the North American craton (Coney and Campa, 1983).

Parts of the *Neuqueniceras* assemblage, e.g., *Neuqueniceras neogaeum* (Burckhardt), were also reported from the Tepexic Formation of the Huayacocotla Mountains in the Sierra Madra Oriental, east-central Mexico (Erben, 1956; Cantu Chapa, 1979 and earlier). The old material, however, is either lost or unavailable for confirmation and recently one of us (Westermann) was unable to duplicate the findings (see Taylor *et al.*, this volume).

REFERENCES

Alencaster de Cserna, G., 1963, Pelecipodos del Jurásio Medio del Noroeste de Oaxaca y Noreste de Guerrero: Paleontologia Méxicana, No. 15, 52 p.

Benavides, M.M.E., 1978, Estudio Geológico del Municipio de Cualac, Estado de Guerrero: Tesis prof. Instituto Politecnico Nacional, Escuela Superior de Ingenieria y Arquitectura, Inédita.

Burckhardt, C., 1927, Cephalópodos del Jurásico medio de Oaxaca y Guerrero: Instituto de Geologia de México, Boletin 47, 108 p.

______, 1930, Etude Synthétique sur l'Mesozoique Mexicain: Societé Paléontologie Suisse, Memoire 49-50, 280 P.

Campa, M.F., and Coney, P.J., 1983, Tectono-Stratigraphic Terranes and Mineral Resource Distributions in Mexico: Canadian Journal of Earth Sciences, vol. 20, p. 1040-1051.

Cantu Chapa, A., 1979, Bioestratigrafia de la Serie Huasteca (Jurasico Medio y Superior) en el Sudsuelo de Poza Rica, Veracruz: Instituto Mexicana de Petrol, Revista XI, No. 2, p. 14-24.

Cariou, E., 1980, L'Etage Callovien dans le Centre-Ouest de la France: These a l'Université de Poitiers, Fascilles 1-3, 790 p.

Collignon, M., 1958, Atlas des fossils caraceristiques de Madagascar: Fascicule II (Bathonien Callovien): Service Geologique Tananarive, Pls. IV-XXXIII with legends.

Dietl, G. 1978, Die heteromorphen Ammoniten des Dogger: Stuttgarter Beiträge zur Naturkunde, Serie B (Geologie und Paläontologie), Nr. 33, 97 p.

______, 1982, The Real Stratigraphic Position of *Ammonites Aspidoides* (Ammonoidea, Middle Jurassic) at the Locus Typicus: Stuttgarter Beiträge zur Naturkunde, Serie B (Geologie und Paläontologie), Nr. 87, 21 p.

Donovan, D.T., Howarth, D.T., and J. Callomon, 1981, Classification of the Jurassic Ammonites: in M.R. House and J.R. Senior, eds., The Ammonoidea: Systematic Association Special Volume 18, p. 101-155.

Elmi, S., 1967, Le Lias et le Jurassique Moyen de l'Ardèche. Laboratoirs de Géologie de la Faculté de Sciences de Lyon, Documents 19, 845 p.

______, 1971, Les Faunas a *Prohecticoceras* (Oppeliidae, Ammonitina) du Bathonien Inférieur et Moyen des Confins Algero-Marocains: Geobios, v. 4, p. 243-264.

Enay, R., 1959, Note sur Quelques Tulitides (Ammonitina) du Bathonien: Societé Géologique de France, 7e Séries, Bulletin 1, p. 252-259.

Erben, H.K., 1956, El Jurásico Medio y el Calloviano de Mexico: XX Congreso Geológico International (Mexico 1956), 140 p.

Gabilly, M.J., 1964, Le Jurassiques Inferieur et Moyen sur le Littoral Vendéen: Institut de Géologie et d'Anthropoligie Préhistoire de la Faculté de Poitiers, Travaux V, p. 63-107.

Gottsche, C., 1978, Über Jurassiche Versteinerungen aus der Argentinischen Cordillere: Paläontographica Supplement 3, Lieferung 2, Abteilung, 3, 50 p.

Guzman, E.J., 1950, Geologia al Noreste de Guerrero: Asociation Mexicana de Geologia y Petrologia, Boletin 2, p. 95-156.

Hahn, W., 1971, Die Tulitidae S. Buckman, Sphaeroceratidae S. Buckman und Clydoniceratidae S. Buckman (Ammonoidea) des Bathoniums (Brauer Jura epsilon) im Südwest-Deutschen Jura: Jahrbuch des Geologischen Landesamtes von Baden-Württemberg, vol. 13, p. 55-122.

Hillebrandt, A. von, 1970, Zur Biostratigraphie und Ammoniten-Fauna des südamerikanischen Jura (insbes. Chile): Neues Jahrbuch für Geologie und Paläontologie, Abhandlungen, vol. 136, p. 166-211.

______, 1973, Neue Ergenbnisse über den Jura in Chile und Argentinien: Münstersche Forschungen in Geologie und Paläontologie, Heft 31/32, p. 167-199.

Jenny, H., 1933, Geological Reconnaissance Survey of the Northeastern Part of the State of Guerrero: Archivo Petroleos Mexicanos, Reporte Geologico no. 418.

Imlay, R.W., 1953a, Callovian (Jurassic) Ammonites from the United States and Alaska: Part 2, Alaska and Cook Inlet Region: Geological Survey Professional Paper 249-B, p. 37-108.

______, 1953b, Callovian (Jurassic) Ammonites from the United States and Alaska: Part 2, Alaska Peninsula and Cook Inlet Region: Geological Survey Professional Paper 249-B, p. 37-108.

______, 1980, Jurassic Paleobiogeography of the Conterminous United States in its Continental Setting: Geological Survey Professional Paper 1062, 134 p.

Ochoterena, H., 1966, Amonitas del Jurasico Medio de Mexico. II. *Infrapatoceras* gen. nov.: Paleontologia Mexicana 23, 18 p.

Orbigny, A. d', 1842-51, Paléontologie Francaise; Terrains Jurassiques, I. Céphalopodes: Masson, Paris, 642 p., 234 pl.

Ortega, G.F., 1978, Estratigrafia del Complejo Acatlán en la Mixteca Baja, Estados de Puebla y Oaxaca: Univ. Nal. Autón. de México, Instituo de Geologia, Revista 2.

Salas, G.P., 1949, Bosquejo Geologico de la Cuenca Sedimentaria de Oaxaca: Asociacion Mexicana de Geologia y Petrologia, Boletin 1, p. 79-156.

Schindewolf, O.H., 1965, Studien zur Stammesgeschichte der Ammoniten, Lieferung IV: Akademie der Wissenschaften und der Litteratur, Abhandlungen der Mathematisch-Naturwissenschaftlichen Klasse, Jahrgang 1965, Nr. 3, p. 140 (409)-238 (508).

Stehn, E., 1923, Beiträge zur Geologie und Paläontologie von Südamerika. XXV, Beiträge zur Kenntnis des Bathonien und Callovien in Südamerika: Neues Jahrbuch für Mineralogie, Geologie und Paläontologie, Beilage-Band XLIX, p. 52-158.

Steinmann, G., 1881, Zur Kenntnis der Jura- und Kreideformation von Caracoles (Bolivia): Neues Jahrbuch für Geologie, Mineralogie und Paläontologie, Beilage-Band 1, p. 239-301.

Stipanicic, P.N., 1966, El Jurásico en Vega de la Veranada (Neuquén), el Oxfordense y el Diastrofismo Divesiano (Agassiz-Yaila) en Argentina. Asociácion Geológica Argentina, Revista XX, p. 403-478.

Thierry, J., 1978, Le Genre Macrocephalites ou Callovien Inferieur (Ammonites, Jurassic Moyen): Mèmoire Géologique de Université de Dijon 4, 490 p.

Tornquist, A., 1898, Der Dogger am Espinazito-Pass: Palaeontologische Abhandlungen, Neue Folge, Band VIII, 2, p. 3(135)-69(201).

Torrens, H.s., 1980, Bathonian correlation chart, In Cope, J.C.W., A correlation of Jurassic rocks in the British Isles, Part two: Middle and Upper Jurassic: Geological Society of London Special Report No. 15, p. 21-45.

Westermann, G.E.G., 1956, Phylogenie der Staphanocerataceae und Perisphinctacea des Dogger: Neues Jahrbuch für Geologie und Paläontologie, Abhandlungen 103, p. 233-279.

______, 1958, Ammoniten-Fauna und Stratigraphie des Bathonien NW-Deutschlands: Beihefte zum Geologischen Jahrbuch 32, 152 p.

______, 1967, Sucesion de Ammonites de Jurasico Medio en Antofagasta, Atacama, Mendoza y Neuquén: Asociacion Geologica Argentina, Revista XIII, p. 65-73.

______, 1981, Ammonite Biochronology and Biogeography of the Circum-Pacific Middle Jurassic. In M.R. House and J.R. Senior, editors, The Ammonoidea: Systematic Association, Special Volume 18, p. 459-498.

Westermann, G.E.G., A.C. Riccardi, O. Palacios and C. Rangel, 1980, Jurásico Medio in el Perú: Instituto Geológico, Minero y Metalúrgico, Boletin 9, p. 1-47.

Jurassic-Cretaceous Biochronology and Paleogeography of North America,
edited by G.E.G. Westermann,
Geological Association of Canada Special Paper 27, 1984

THE ALLOCHTHONOUS JURASSIC-LOWER CRETACEOUS TERRANES OF THE CANADIAN CORDILLERA AND THEIR RELATION TO CORRELATIVE STRATA OF THE NORTH AMERICAN CRATON

H.W. Tipper
Geological Survey of Canada, 100 West Pender Street, Vancouver B.C. V6B 1R8

ABSTRACT

The Jurassic-Lower Cretaceous strata of the Canadian Cordillera are treated as parts of four distinct terranes roughly corresponding to Wrangellia, Stikinia, Quesnellia, and the Fernie Basin. The first three are allochthonous relative to the North American craton whereas the Fernie Basin can be linked physically to rocks on the stable craton. Each of the four terranes differs in varying degrees in stratigraphy, structural evolution, faunal succession, and geological history. The Lower Jurassic-early Middle Jurassic terranes are characterized by volcanic accumulations of oceanic or of arc-type origin. The late Middle Jurassic sequences are typically shale and suggest that the three terranes were juxtaposed by Callovian time. Clastic depositional sequences in Late Jurassic and Cretaceous time indicate that the three terranes were firmly locked together by Albian time although tectonic activity continued within and along the margins of this block into Tertiary time.

RÉSUMÉ

Les strates comprises entre le Jurassique et le Crétacé inférieure des Cordillères Canadiennes sont traitées comme fesant partie à quatre "terranes" correspondant grossierement au "Wrangellia", "Stikinia", "Quesnellia" et le Basin Fernie. Les trois premières sont exogènes relatif au craton Nord Americain alors que le Basin Fernie peut être physiquement lié aux roches du craton stable. Chacunes des quatres "terranes" différentes en plusieurs degrés d'après la stratigraphie, l'évolution structurale, la succession faunistique et l'histoire géologique. Les "terranes" comprises entre le Jurassique inférieur et le Jurassique moyen sont caractérisees par des accumulations volcaniques d'origine soit, oceanique ou de type d'arc insulaire. Les dernières séquences du Jurassique moyen sont Typiquement constituées di argiles suggérant ainsi que les trois "terranes" étaient juxtaposées au Callovien. Des séquences de déposition de type clastique vers la fin du Jurassique-Crétacé indiquent que les trois "terranes" étaient solidement fixées ensembles au Albien et ce malgre la persistance d'activité tectonique, à l'intérieur ainsi qu'au long des marges de ce block, jusqu'au Tertiaire.

INTRODUCTION

Investigations in Jurassic paleontology in the Canadian Cordillera during the years of geological reconnaissance can be credited to J.F. Whiteaves, P.S. Warren, Frank McLearn, Colin Crickmay and a few others. Through their efforts the presence of several standard stages was recognized and documented with ammonite faunas. However, their conclusions were based on detailed studies conducted in very small areas or on fragmentary collections submitted by other geologists. In the late 1940s Hans Frebold and J.A. Jeletzky began a new investigative phase in which fossils were collected in a far more systematic manner, leading to much greater precision in defining the biostratigraphic record. Expanded field mapping by the Geological Survey of Canada, by the British Columbia Department of Energy, Mines and Petroleum Resources, by oil companies, and mining companies produced a wealth of information and fossil material. The gradual increase in number of specialists in various fields in universities, government departments and private industry has produced and will produce better documented and more abundant data on the Jurassic Period.

EARLY MODELS OF THE CORDILLERA

When one looks at a map of the Canadian Cordillera for the first time much as the early geologists might have viewed the region after a few reconnaissance surveys, it is readily apparent that the plutonic rocks and spatially associated volcanics are most abundant in the western part. The eastern Cordillera is characterized by great thicknesses of sediment in the Rocky

Project 171:
Circum-Pacific Jurassic

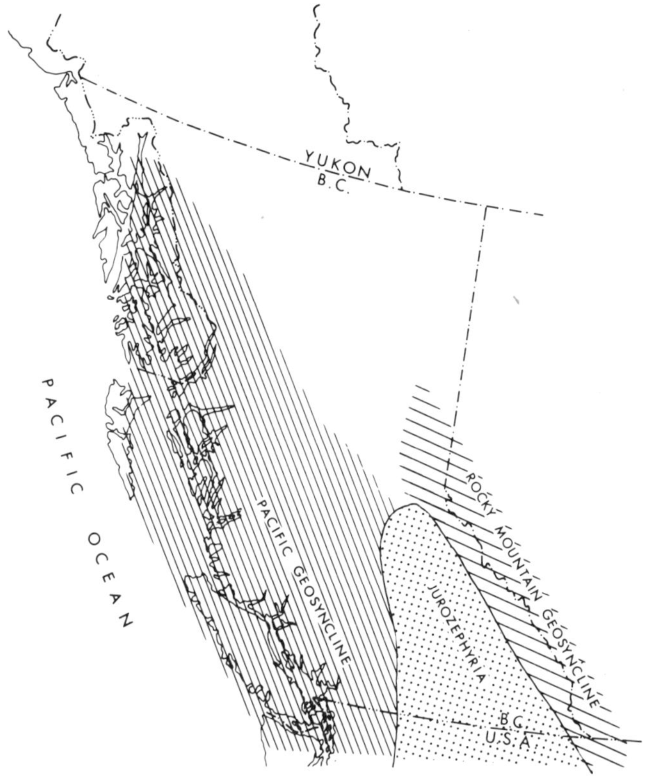

Figure 1. Crickmay's 1931 paleogeographic interpretation for Bajocian time in the Cordillera.

Mountains and Foothills. The dividing line between east and west lies roughly along the Rocky Mountain Trench in northern British Columbia and the Kootenay Arc in the south. Considering the layered rocks of the early Mesozoic era, this line then separates the Triassic-Jurassic volcanic piles in the west and the totally sedimentary sequences in the east.

This volcanic-sedimentary distinction was easily recognized by many geologists but was first clearly distinguished by Daly (1912, p. 7) and best put into diagrammatic form by Crickmay for the Jurassic (Fig. 1) when he described the Cordillera as two geosynclines – a volcanic Pacific Geosyncline and a sedimentary Rocky Mountain Geosyncline (Crickmay, 1931, p. 84). Marshall Kay (1951) defined these as a eugeosyncline and a miogeosyncline but the distinction remained essentially the same as Crickmay's geosynclines. That the eastern and western Cordillera differed in lithology and geological history was firmly established.

CRUSTAL TRANSPORT

In the last two decades concepts involving both latitudinal and longitudinal shifts of segments of the earth's crust have become prominent in geological thought. From many viewpoints it seemed abundantly clear that the western Cordillera was allochthonous relative to the North American continent, i.e., everything west of the northern Rocky Mountain Trench-Kootenay Arc formed elsewhere and was transported to its present site. The evidence for latitudinal shifts of terranes essentially falls into three categories, namely structural, faunal, and paleomagnetic. Many scientists have noted evidence for right-lateral transcurrent movement on major faults (Roddick, 1967; Tempelman-Kluit, 1970; Eisbacher, 1977; Gabrielse and Dodds, 1977; Tipper, 1969) based primarily on the demonstrable offset of structures or lithologic units by recognizable faults.

FAUNAL DISPLACEMENTS

Faunal displacements were pointed out by E.T. Tozer (1970, p. 635) who found that warm water Triassic faunas from the southern Yukon in the western Cordillera apparently now lie farther north than cooler water faunas of the same age from the Peace River in the eastern Cordillera. An even more dramatic example was described by Monger and Ross (1971) in which distribution of two Permian fusulinid faunas clearly substantiates latitudinal shifts of terranes. Comparison of one of these faunas with an Asiatic fauna argues strongly for a longitudinal shift as well. The occurrence of Permian strata with a North American fusulinid fauna presently lying west of a belt of Permian strata, the Cache Creek Group, with a distinct Asiatic fauna, clearly demonstrates the exotic nature of parts of some terranes.

More recently, studies in paleomagnetism (Irving *et al.*, 1980) are offering evidence in support of the shifting of the Cordilleran terranes relative to the craton.

ALLOCHTHONOUS TERRANES

The eugeosynclinal part of the Cordillera has been described by Monger and Irving (1980) as a "collage of terrains of varied

Figure 2. The four Early to Early Middle Jurassic terranes of the Canadian Cordillera (modified from Monger *et al.*, 1971). Jurassic outcrop areas of the allochthonous terranes shown in black.

origins that became attached to the continent at different times". Such a collage or mosaic includes rocks of a great range of ages and many origins transported in various manners for varying distances. Each fragment of the collage has its own peculiar history. This concept of fragments of the crust transported from their sites of origin has received fairly wide acceptance but arguments continue over details, such as the time and manner of transport and the amount of transport. An examination of the general lithostratigraphy and paleogeography of a few fragments of this collage, those of Jurassic and Cretaceous age, suggests an evolving Mesozoic history from independent shifting fragments to an amalgamated block attached to the continent.

The Early and Middle Jurassic layered rocks of the Canadian Cordillera can be assigned to four principal terranes that are each internally consistent on the basis of lithostratigraphy, paleontology, and underlying Triassic stratigraphy. The three westerly terranes (Fig. 2) namely Wrangellia, Stikinia, and Quesnellia are displaced terranes that were transported from their sites of origin to their present position. These comprise the allochthonous eugeosyncline. The Jurassic beds of the Alberta and Liard Troughs, the fourth terrane, was deposited on the continent or on strata physically linked to the continent, and are considered to be relatively in their original depositional position.

The blocks or terranes are fault bounded (Fig. 2). The faults may have been operative since the early Jurassic, or at least since the terranes began to shift which might have been shortly after or while the basins developed. The Kutcho-Pinchi fault follows the east side of Stikinia Terrane but is offset by the younger, possibly Tertiary, Finlay fault. The Yalakom-Pasayten fault forms the northeast side of Wrangellia and is offset by the younger Fraser-Straight Creek fault. The Finlay-Fraser-Straight Creek faults could be parts of a single Late Cretaceous-Tertiary fault system. The northern Rocky Mountain Trench fault separates the displaced terranes from the continent and may also be an old fault upon which later movement was localized. Without appealing to faunal displacement, there is ample evidence from the offset of structures and lithologic units across these faults to prove that lateral shifts have occurred and that right-lateral transcurrent movement is dominant in their early history. Verical movements in many areas can be ascribed to a younger tectonic history.

Faunal evidence has always been a factor when considering the allochthonous nature of the eugeosyncline. As already mentioned, Tozer cited an example of displaced faunas. Nichols and Silberling (1979) cited yet another example for the Triassic. The following example for the Jurassic has been previously reported by the writer (Tipper, 1981).

In Europe a north to south geographic zonation of late Pliensbachian faunas by climate has been described by Donovan and others (Donovan, 1967). Part of Figure 3 is simplified from Donovan. A boreal amaltheid fauna dominates northern Europe, a Tethyan hildoceratid fauna dominates in the Mediterranean region, and a mixed fauna lies between. Although the possibilities of other causes were considered, Donovan preferred a climatic control for this zonation. The same faunas are recognized in the North American Cordillera apparently with a similar north to south zonation. If the Pacific margin were sliced by right-lateral faults as shown in Figure 3, the resulting positions of the fossil localities suggest that the zones were offset and that the most westerly terrane, Wangellia, shifted north relative to Stikinia and Stikinia north relative to Quesnellia. The three combined terranes shifted north by a very large amount relative to the continent. Whether or not one accepts the validity of climatic zonation, it is obvious that Tethyan elements are shifted northward along the same structures by which Triassic faunas appear to be displaced.

TERRANE STRATIGRAPHY

The lithostratigraphy of the three displaced terranes offers some interesting comparisons (Fig. 4). The rock units can be grouped into four episodes of deposition identified by the predominant lithology, separated by hiastuses or unconformities, that are more or less reflected in each terrane. The episodes are as follows, from lowest to highest.

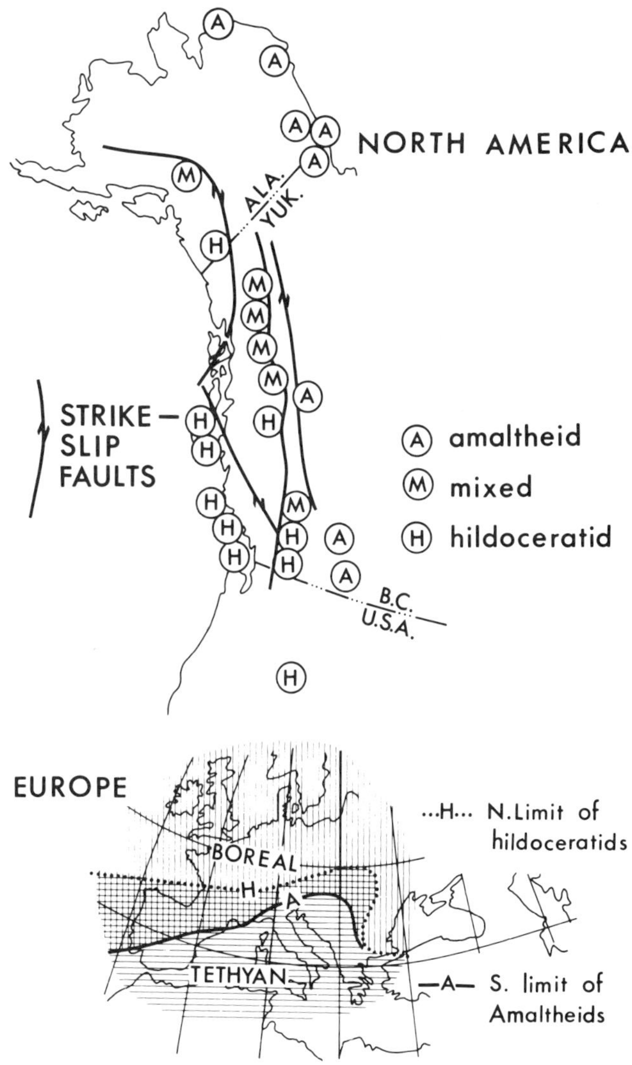

Figure 3. Offset of Late Pliensbachian faunas in the Canadian Cordillera after Tipper (1982) and distribution of similar faunas in Europe after Donovan (1967).

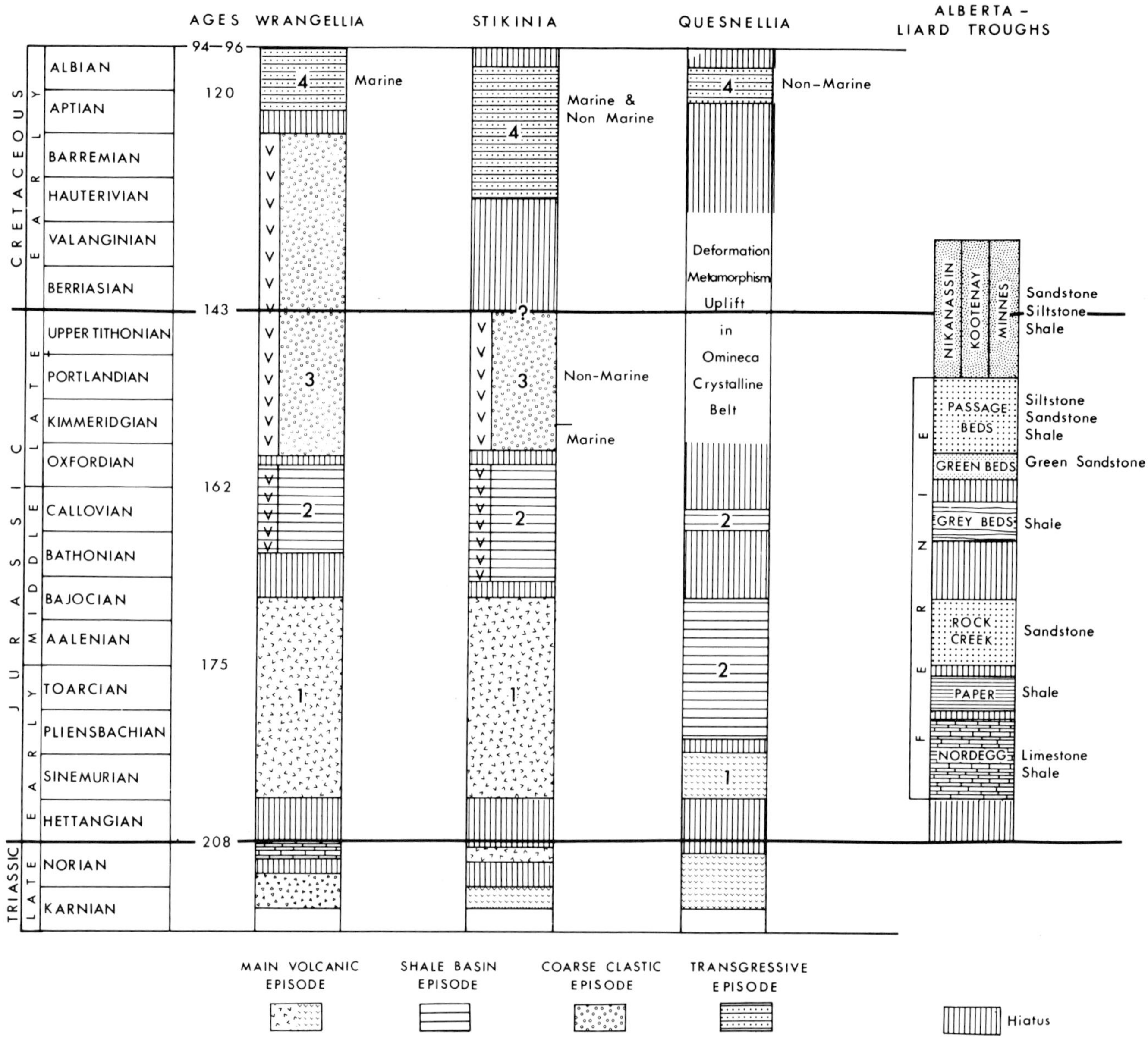

Figure 4. Jurassic and Lower Cretceous lithologic episode of the Mesozoic in the western Canadian Cordillera and stratigraphy of the Fernie Basin.

1). The volcanic episode had distinctive and different stratigraphy and volcanic lithology in each displaced terrane.

2). The succeeding shale episode featured black shales spread uniformly over wide areas.

3). The third episode is one of coarse clastic deposition in a basin or trough that was tectonically active along the margins.

4). The fourth episode involves a widespread marine transgression over much of the southwest quarter of British Columbia, culminating in Albian time, in which shale, sandstone, and basin margin conglomerate were deposited. A correlative non-marine, essentially conglomeratic unit was deposited east of the marine basin on part of the Quesnellia Terrane.

The major unconformities that define the episodes apparently apply to all four terranes regardless of the internal stratigraphy. If one considers the possibility that displacement of these terranes may have been distributed over hundreds of kilometres, or even thousands, it seems probable that the events that produced the hiatuses are of continental proportions. The duration or importance of these breaks may not be the same in all terranes. The presence of rocks of Hettangian age has not been proven unequivocally in any terrane south of the Yukon although the literature suggest otherwise. Late Bajocian has recently been indicated in the Fernie area (Hall and Stronach, 1981) but elsewhere is thought to be absent. A marked lithologic change in the mid-Oxfordian or the end of Callovian time seems to reflect the onset of tectonism and the production of coarse clastic basins. The last major interruption is below the Albian or Aptian.

Volcanic Episode. The distribution of Jurassic rocks within Wrangellia is typified by the Sinemurian-Pliensbachian stages (Fig. 5). Red, green, and grey feldspathic andesitic to rhyolitic

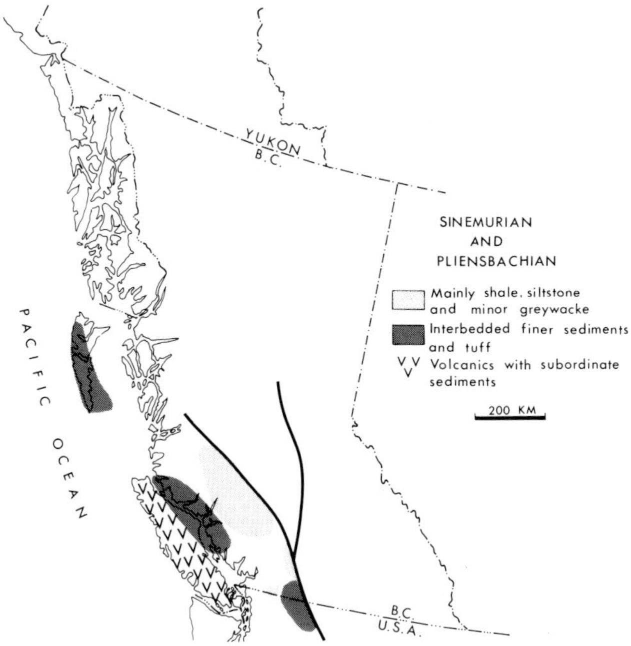

Figure 5. Distribution of Sinemurian-Pliensbachian lithologic units of Wrangellia.

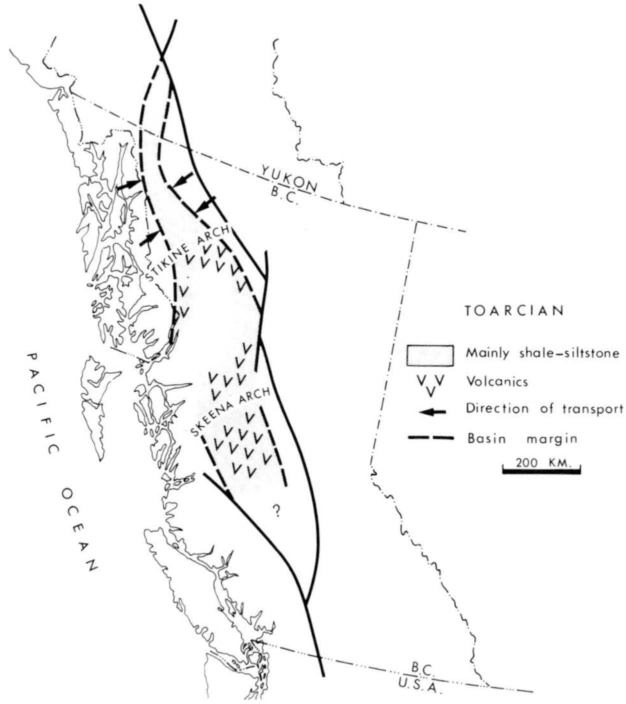

Figure 6. Distribution of Toarcian lithologic units of Stikinia.

pyroclastics outcrop along the west side of Vancouver Island and grade eastward into tuffaceous shale of a northwest trending belt from east of Vancouver Island to Queen Charlotte Islands. Eastward the exposures are almost entirely shale to the Yalakom Fault which truncates the basin along the northeast margin of Wrangellia (Sutherland Brown, 1968; Muller *et al.*, 1974; Tipper, 1978). Throughout this volcanic episode, volcanics abounded on the southwest margin of Wrangellia and shales on the northeast. In Toarcian to Bajocian time the distribution was probably similar except that the volcanic belt, as least in Bajocian time, had shifted eastward, possibly to the site of the present central Coast Mountains. There is no record of Toarcian to Bajocian rocks on Vancouver Island but this may be due to a lack of fossil evidence.

In Stikinia during Early Jurassic time (Fig. 6) the depositional area was a north-south trough known as the Whitehorse-Hazelton Trough (Tipper and Richards, 1976). Two east-west welts or arches across the trough developed during this time. On these arches, red, green, maroon, and grey pyroclastic andesitic to rhyolitic volcanics accumulated – the Hazelton volcanics on the southerly Skeena Arch and the Toodoggone volcanics on the northerly Stikine Arch. Figure 6, which represents Toarcian time, is typical of the paleogeographic configuration in Stikinia in Early Jurassic time. Sedimentary rocks, accumulated throughout the trough, interfingered with the volcanics. North of Stikine Arch mainly conglomerate, greywacke, and shale accumulated with rare thin volcanic sequences. There is increasing evidence to suggest that the two arches and associated volcanism enclosed a shale basin. South of the Skeena Arch the stratigraphy is poorly known because of Tertiary cover.

The Sinemurian volcanics (Fig. 7) of Quesnellia are characteristically augite porphyry basalt like the underlying Triassic

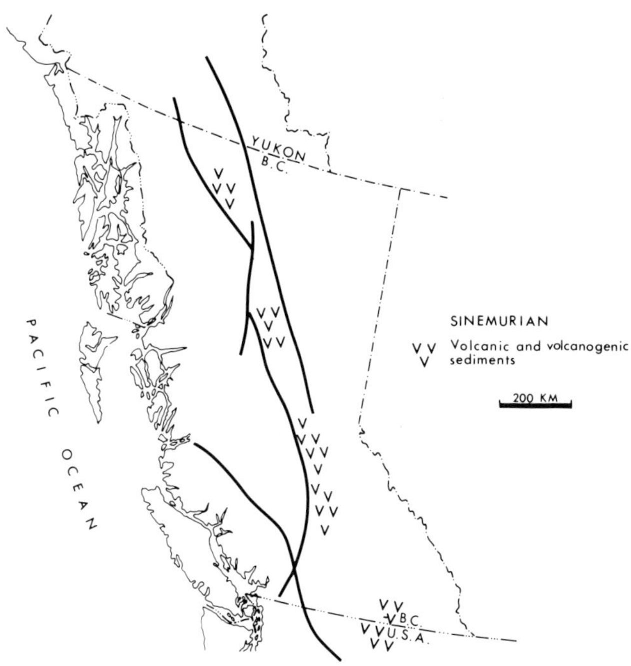

Figure 7. The distribution of Sinemurian volcanics of Quesnellia.

Nicola volcanics. They are known the length of the province from the Rossland volcanics in the south, to the Nicola, to the Quesnel River, and to the eastern belt of Takla volcanics in central and northern British Columbia. No varicoloured acid volcanics like those of Stikinia and Wrangellia are prominent in this terrane. The paleogeographic configuration of this basin is unknown.

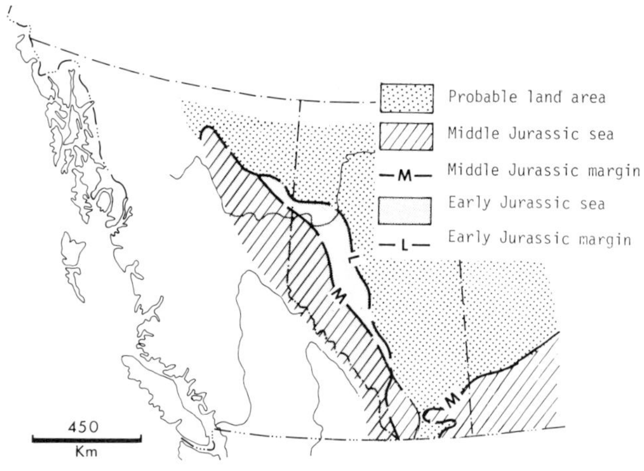

Figure 8. Lower and Early Middle Jurassic seas of the Alberta-Liard Troughs.

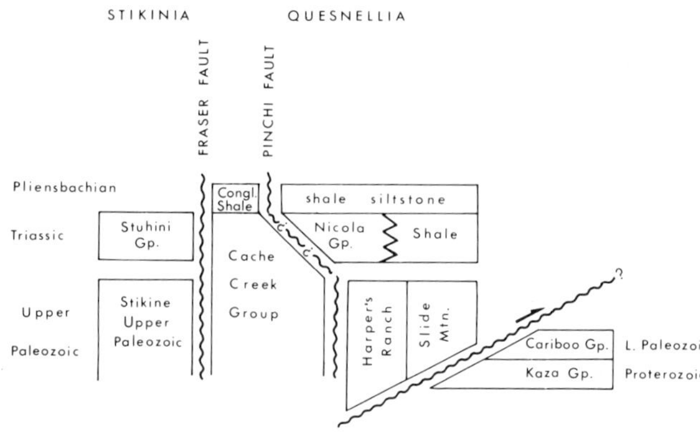

Figure 10. Contrasting rock units separated by the Pinchi and Fraser faults.

In the Alberta-Liard Troughs (Fig. 8) during the volcanic episode of the displaced terranes sedimentation was discontinuous in a relatively simple depositional site along the western margin of the North American craton. The source of the sediments was from the east and the western side of the basin is unknown. It may have possibly opened to the Pacific Ocean.

A comparison of the volcanic episodes of the three allochthonous terranes shows major dissimilarities in lithology and the underlying Triassic stratigraphy is even more distinctly dissimilar in each terrain. The Triassic rocks of Wrangellia (Fig. 4) are dominated by thick piles of Karmutsen oceanic pillow basalt, succeeded by marine sediments comprised of limestone, shale, and limy siltstone. The Triassic of Stikinia has thick augite porphyry basalts deposited in the Karnian and early Norian succeeded by vari-coloured fragmental volcanics including those of the Early Jurassic Hazelton volcanics. Interbedded with the later Triassic volcanics in many areas is a massive Late Norian grey limestone such as the Sinwa limestone of northwestern British Columbia. The Triassic of Quesnellia is typified by basaltic augite porphyry throughout much of the upper Triassic to late in the Norian. As previously mentioned this volcanic episode continued into the Jurassic.

Shale Episode. Following the volcanic episode the shale

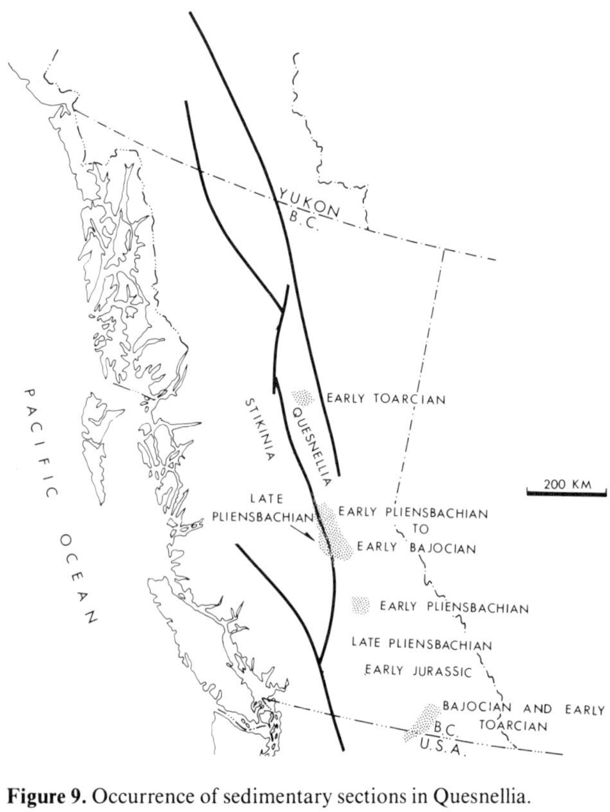

Figure 9. Occurrence of sedimentary sections in Quesnellia.

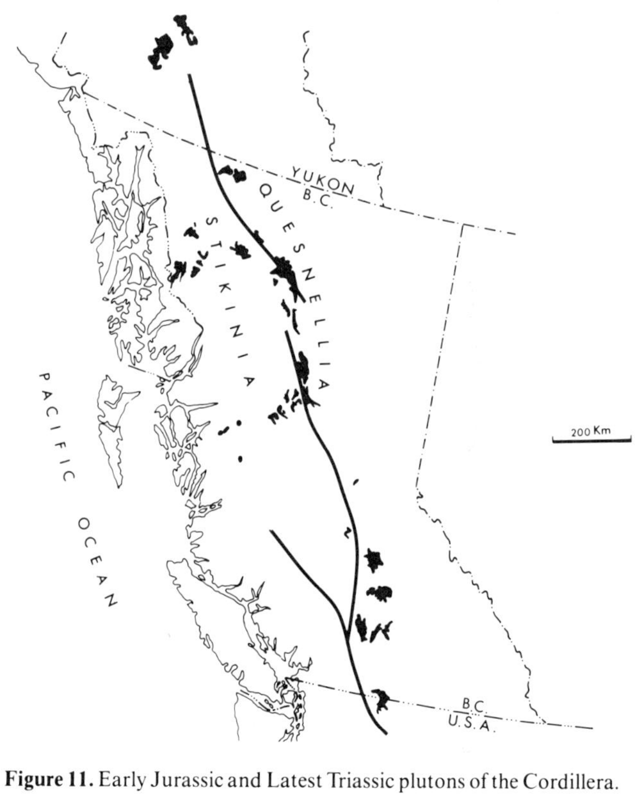

Figure 11. Early Jurassic and Latest Triassic plutons of the Cordillera.

basin episode links the allochthonous terranes together into a single plate. In Quesnellia, sedimentary rocks (Fig. 9), mostly shales, were deposited from Early Pliensbachian to Bajocian. Generally they are massive black to finely laminated dark grey shales and are preserved now as small fault blocks or erosional remnants scattered throughout the terrane. Near Quesnel, Late Pliensbachian sediments were deposited on both sides of the Pinchi Fault (Fig. 10) suggesting that the two terranes, Stikinia and Quesnellia were juxtaposed across the fault at this time. The stratigraphic sequences below the sediments on either side of the fault are totally dissimilar. Early Jurassic plutons of the Cordillera occur along the Pinchi-Kutcho faults (Fig. 11) and along the Stikine and Skeena Arches suggesting that these terranes were juxtaposed when the plutons were emplaced. For these reasons the amalgamation of Quesnellia and Stikinia is thought to have been accomplished during the Early Jurassic but they may have been juxtaposed even earlier by the end of the Triassic.

Many late Middle Jurassic shale formations throughout the Western Cordillera (Fig. 12) represent the shale episode that ties Wrangellia to the Stikinia-Quesnellia plate, thus providing evidence of amalgamation of the three displaced terranes. Typically the shales are massive black or grey and finely laminated and have yielded similar fossil faunas from each of the three terranes. In the Tyaughton Creek area, the Nechako area, and the Smithers area the shales grade westward into tuffaceous sediments; presumably there was limited volcanic activity in the site of the present Coast Mountains. If the shale areas could be shown conclusively to be parts of a single basin as is strongly suggested, then the time of amalgamation of the displaced terranes would be Bathonian-Callovian. The rising Omineca Crystalline Belt formed a land barrier on the northeast margin separating the shale area from the Alberta-Liard Troughs to the east. An uplifted area, but not necessarily a barrier, existed in the region now occupied by the Coast Mountains.

Coarse Clastic Episode. The shale episode, tectonically relatively quiescent, was followed by deposition of coarse clastic sediments (Fig. 13) indicating renewed tectonism throughout the displaced terranes. The Tyaughton Trough, formed along Yalakom fault, received detritus from both sides beginning early in the Late Jurassic; the fault remained active during deposition. The Bowser Basin received detritus from the adjacent arches and from the Omineca Crystalline Belt all of which were subjected to uplift and erosion from early Late Jurassic to late Early Cretaceous.

The Albian Transgression. Following the clastic depositional episode, many of the uplifted areas collapsed. The Skeena Arch was destroyed as a structure controlling sedimentation, the Tyaughton Trough became relatively inactive and the Coast Mountains area subsided. An Albian sea spread over central British Columbia and reached to the centre of the province (Fig. 14). The margin of the sea is marked by pebble conglomerate and coarse clastics, such as the Jackass Mountain Group, and formed an irregular northwest trending line through central British Columbia. The lower part of the Sustut Group and other non-marine Cretaceous basins in east-central British Columbia may have been the non-marine correlatives of the sediments of the marine transgression.

DISCUSSION

The three terranes by Albian time were firmly locked together and acted as one terrane. Young faults such as the Straight Creek-Fraser-Finlay disrupted the block but possibly did not follow old sutures. Tectonic activity continued on the margins of this block along Queen Charlotte Fault and the northern

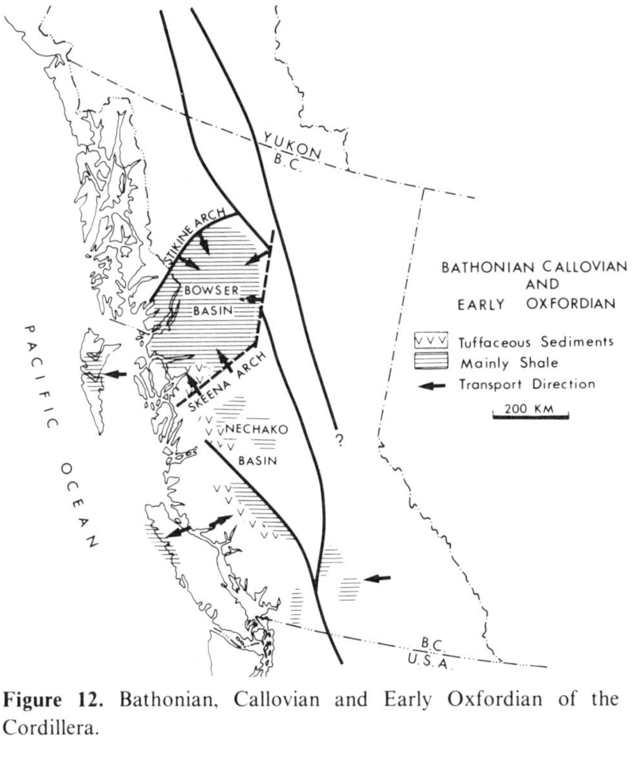

Figure 12. Bathonian, Callovian and Early Oxfordian of the Cordillera.

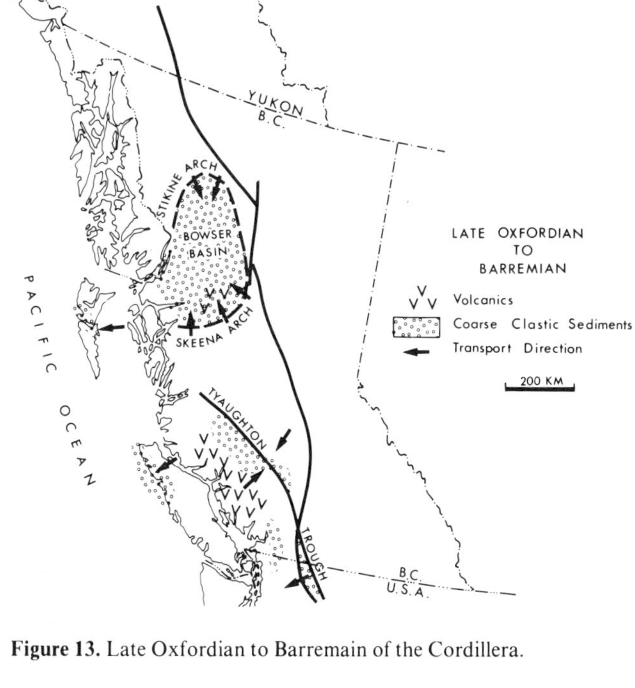

Figure 13. Late Oxfordian to Barremain of the Cordillera.

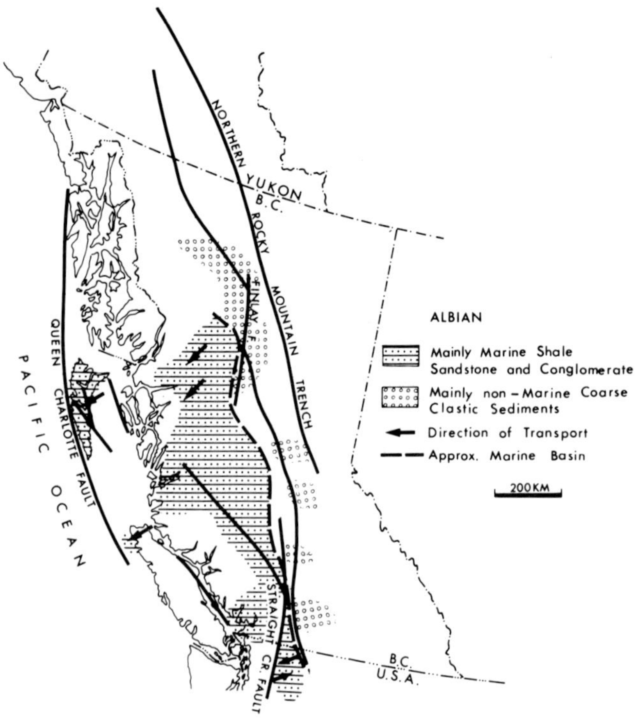

Figure 14. The Albian Transgression.

Rocky Mountain Trench Fault in Cretaceous, Tertiary, and younger time. The time of attachment of the amalgamated displaced terrains to the continental margin is not clear but probably the block was roughly in its present relative position by mid-Jurassic time. Certainly the Bajocian and Toarcian sediments of Quesnellia probably were deposited in a basin not far removed from the continent as the writer has found they contain detrital white micas, suggesting deposition near a metamorphic source; the Early Jurassic and Latest Triassic plutons (Fig. 11) in general are not a source of white micas.

A total lack of volcanic rocks distinguishes the Alberta-Liard Troughs lithostratigraphically from the allochthonous terranes. In Pliensbachian time the Fernie Group was clearly faunally distinct, as it may also have been at other times and its paleogeographic setting cannot be easily related to the western allochthonous terranes which themselves display dissimilarities in lithostratigraphy, in paleogeography, and, to a lesser extent, in paleontology.

The evidence presented here suggests that the three allochthonous terranes, Wrangellia, Stikinia and Quesnellia, may have been formed along the Pacific margin of the Americas, and were moved northwestward, relative to the Alberta-Liard Troughs along the margin by transcurrent displacements, and finally amalgamated into a coherent mass. There is no compelling argument to suggest that they were transported across the Pacific Ocean as has been proposed for some Triassic and Paleozoic terranes (Tozer, 1982; Coney *et al.*, 1981).

REFERENCES

Coney, P.J., Jones, D.A., and Monger, J.W.H., 1981, Cordilleran Suspect Terranes: Nature, v. 288, p. 329-333.

Crickmay, C.H., 1931, Jurassic History of North America: Its Bearing on the Development of Continental Structure: Proceedings American Philosophical Society, v. 70, no. 1, p. 15-102.

Daly, R.A., 1912, Geology of the North American Cordillera at the Forty-Ninth Parallel: Geological Survey of Canada Memoir 38.

Donovan, D.T., 1967, The Geographical Distribution of Lower Jurassic Ammonites in Europe and Adjacent areas: *in* Adams, C.G. and Ager, D.V., eds., Aspects of Tethyan Biogeography: Systematics Association, Publication No. 7, p. 111-134.

Eisbacher, G.H., 1977, Sedimentology of the Dezadeash Flysch and Its Implications for Strike-Slip Faulting along Denali Fault, Yukon Territory and Alaska: Canadian Journal of Earth Sciences, v. 13, p. 1495-1513.

Gabrielse, H., and Dodds, C.J., 1977, The Structural Significance of the Northern Rocky Mountain Trench and Related Lineaments in North-Central British Columbia: Geological Association of Canada, 1977 Annual Meeting Program with Abstracts, v. 2, p. 19.

Hall, R.L., and Stronach, N.J., 1981, First Record of Late Bajocian (Jurassic) Ammonites in the Fernie Formation, Alberta: Canadian Journal of Earth Sciences, v. 18, p. 919-925.

Irving, E., Monger, J.W.H., and Yole, R.W., 1980, New Paleomagnetic Evidence for Displaced Terranes in British Columbia: *in* Strangway, D.W., ed., The Continental Crust and Its Mineral Deposits: Geological Association of Canada Special Paper 20, p. 441-456.

Kay, G.M., 1951, North American Geosynclines: Geological Society of America Memoir 48.

Monger, J.W.H., and Irving, E., 1980, Northward Displacement of North-Central British Columbia: Nature, v. 285, p. 289-294.

Monger, J.W.H., and Ross, C.A., 1971, Distribution of Fusilinaceans in the western Canadian Cordillera: Canadian Journal of Earth Sciences, v. 8, p. 259-278.

Muller, J.E., Northcote, K.E., and Carlisle, D., 1974, Geology and Mineral Deposits of Alert-Cape Scott Map-Area, Vancouver Island, British Columbia: Geological Survey of Canada Paper 74-8.

Nichols, K.M., and Silberling, N.J., 1979, Early Triassic (Smithian) Ammonites of Paleoequatorial Affinity from the Chulitna Terrane, South-Central Alaska: United States Geological Survey Professional Paper 1121-B.

Roddick, J.A., 1967 Tintina Trench: Journal of Geology, v. 75, p. 23-32.

Sutherland Brown, A., 1968, Geology of the Queen Charlotte Islands: British Columbia Department of Mines and Petroleum Resources Bulletin 54.

Tempelman-Kluit, D.J., 1970, Stratigraphy and Structure of the 'Keno Hill Quartzite' in Tombstone River – Upper Klondike River Map-Areas, Yukon Territory: Geological Survey of Canada Bulletin 180, 102 p.

Tipper, H.W., 1969, Mesozoic and Cenozoic Geology of the Northeast Part of Mount Waddinton Map-Area (92N), Coast District, British Columbia: Geological Survey of Canada Paper 68-33, p. 72-73.

———, 1978, Taseko Lakes Map-Area: Geological Survey of Canada, Open File 534.

———, 1981, Offset of an Upper Pliensbachian Geographic Zonation in the North American Cordillera by Transcurrent Movement: Canadian Journal of Earth Sciences, v. 18, p. 1788-1792.

Tipper, H.W., and Richards, T.A., 1976, Jurassic Stratigraphy and History of North-Central British Columbia: Geological Survey of Canada Bulletin 270, 73 p.

Tozer, E.T., 1970, Marine Triassic Faunas: *in* Geology and Economic Minerals of Canada: Geological Survey of Canada, Economic Geology Report No. 1, p. 663-640.

———, 1982, Marine Triassic Faunas of North America: Their Significance for Assessing Plate and Terrane Movements: Geologische Rundschau, Band 71, Heft 3, p. 1077-1104.

Jurassic-Cretaceous Biochronology and Paleogeography of North America,
edited by G.E.G. Westermann,
Geological Association of Canada Special Paper 27, 1984

JURASSIC AMMONITE BIOGEOGRAPHY OF WESTERN NORTH AMERICA: THE TECTONIC IMPLICATIONS

D.G. Taylor
Department of Geology, Portland State University, Portland, Oregon 97207, U.S.A.
J.H. Callomon
Department of Chemistry, University College London, 20 Gordon Street, London, WC1H 0AJ, England
R. Hall
Department of Geology, University of Calgary, Calgary, Alberta T2N 1N4
P.L. Smith
Department of Geological Sciences, University of British Columbia,
6339 Stores Road, Vancouver, B.C. V6T 2B4
H.W. Tipper
Geological Survey of Canada, 100 West Pender Street, Vancouver, B.C. V6B 1R8
G.E.G. Westermann
Department of Geology, McMaster University, Hamilton, Ontario L8S 4M1

ABSTRACT

Three faunal realms and five provinces based primarily on ammonites are recognized for the Jurassic of western North America. The Western Interior Shoshonean, Cordilleran Athabascan, and Bering Provinces are new. Provinciality was modest until the Middle Jurassic and the late Early Bajocian is the last time in which truly pandemic ammonites were prevalent. Thereafter, the region was characterized by strong provincialism and, at times, marked southward spread of Boreal faunas.

The present-day distribution of the provinces indicates the following conditions regarding displacements of tectonic terranes. Contrary to certain paleomagnetic interpretations Wrangellia probably had a northern hemispheric position in the Triassic as did the Alaskan Peninsula throughout the Jurassic. The Alaskan Goodnews terrane likely had an Early Jurassic location significantly south of its present position. Late Pliensbachian evidence indicates northward movements of at least 500 km, 1800 km, and 2500 km for Quesnellia, Stikinia, and the Alaskan Peninsula and Wrangellia respectively. For later times there is no evidence for displacement of the first terrane, but there is evidence suggesting similar displacements for the others in the Bajocian. Post-Pliensbachian movement of at least 1000 km of the John Day terrane and several hundred kilometres for Oregonian Wrangellia, North Sierra terrane, and Sonomia are suggested. The western part of the John Day terrane did not attain its present location relative to North America until after the Bajocian, and the eastern part may have moved northward into position first and conceivably was in place then.

The Middle Jurassic faunas of the Mixteca terrane, southen Mexico, are mostly Andean, but also include species shared apparently only with the Mediterranean. A Jurassic position of this terrane closer to the central Andes (Peru) than at present seems probable, and direct access to a Hispanic Corridor (cratonic proto-Atlantic seaway) is probable.

RÉSUMÉ

Les ammonites permettent de distinguer trois domaines faunesques et cinq provinces du Jurassique de l'ouest de l'Amérique du Nord. Les provinces de: la partie septentrionale Shoshonéenne, de l'Athabasca Cordillerienne et de la région de Béring sont nouvelles. Le provincialisme était peu marqué jusqu'au Jurassique Moyen, et la fin du Bajocien Inférieur est la limite de la domination des ammonites pandémiques. Par la suite, cette région était caractérisée par un provincialisme très développé, et à certains moments, la faune boréale s'est développée vers le sud.

La distribution présente des provinces indiquent les conditions suivantes en regard des déplacements des terrains tectoniques. Contrairement à certaines interprétations paléomagnétiques, Wrangellia a probablement occupé une position hémisphérique septentrionale pendant le Triassique tout comme la péninsule de l'Alaska, pendant le Jurassique.

Le terrain "Alaskan Goodnews". avait probablement une

Project 171:
Circum-Pacific Jurassic

position significativement plus au sud de sa situation présente, au cours du Jurassique Inférieur. Les évidence de la fin du Pliensbachien indiquent des mouvements vers le nord de 500 km, 1,800 km et 2,500 km, pour, respectivement: Quesnellia, Stikinia, la péninsule de l'Alaska et Wrangellia. Il n'y a pas d'évidence pour des déplacements antérieurs des premier terrain, mais il y a des évidence qui suggèrent des déplacements similaires pour les trois derniers terrains pendant le Bajocien. Des mouvements Post-Pliensbachien de 1,000 km pour le terrain "John Day" et de plusieurs centaines de kilomètres pour "Oreginian Wrangellia" et, terrain "North Sierra" "Sonomia" sont suggérés. Relativement à l'Amérique du Nord, la partie ouest du terrain "John Day" n'occupe sa position présente qu'après le Bajocien. La partie est de ce terrain peut avoir bougée vers le nord pour se placer dans sa position présente.

Les faunes du Jurassique Moyen du terrain "Mixteca" situé au sud du Mexique, sont principalement Andéennes; cependant, les espèces incluses sont similaires aux espèces méditerranéennes seulement. Une position de ce terrain, près du corridor Hispanique, est aussi indiquée.

INTRODUCTION

Jurassic ammonites from western North America have become sufficiently well known in the last several years to make possible a meaningful review of their biogeography at the generic level. It has become clear, however, that with the rapidly accumulating information bearing on the tectonic evolution of the region, any such biogeographic study must be considered in the framework of theories relating to terrane movements. The picture that is emerging is a complex one, in which most of the Western Cordillera consists of a collage of more than 50 probably allochthonous terranes (Coney *et al.*, 1980; Coney, 1981). It is based, however, on three independent but supportive lines of evidence, i.e., structural, paleomagnetic and paleontologic. According to current theory, this collage had an accretionary history which continued throughout the Mesozoic and persisted into the Tertiary, the terranes being emplaced through convergence followed by right-lateral strike-slip motion. Many of these terranes are thought to have originated far south of their present locations, and some, such as Wrangellia, having come perhaps from south-equatorial regions (Irving and Yole, 1972; Symons, 1973; Hillhouse, 1977; Jones *et al.*, 1977, 1978; Yole and Irving, 1980; Irving *et al.*, 1980). Such a complex history would imply that these terranes have now juxtaposed faunas that were widely separated when living. Previous studies of Triassic (Tozer, 1970, 1982a; Nichols and Silberling, 1979) and Jurassic faunas (Tipper, 1981; Smith, 1983) have already pointed to major terrane displacements. This paper describes the patterns of distribution of Jurassic ammonites in western North America in terms of a set of provinces. The distribution of the provinces is then examined to assess to what extent it is consistent or otherwise with movements of terranes inferred from geologic and paleomagnetic data.

We wish to emphasize the importance of biogeography as a source of information that stands on its own for interpreting the tectonic history of the region. When lithostratigraphic correlations fail, the sense and magnitude of past terrane displacements usually cannot be reconstructed. In these cases paleomagnetic and faunal analyses become our only hope.

The basic procedures of biogeographic analysis most widely employed are simple but bear repeating. Faunas are collected place by place and then compared. Two situations can arise. First, two faunas from geographically distinct localities are the same (the case of *similar* faunas): all we can say positively is that they are geologically of the same age and faunal province. Second, and more interesting, two faunas from geographically distinct localities differ (the case of *dissimilar* faunas). Then any or all of three causes may be responsible. They may, (a) be of different ages, (b) have lived in different basinal settings, or (c) have belonged to different faunal provinces. After assessing (a) and (b) it is possible then to work with (c). In the case of (c) faunas of different provinces formerly widely separated may have been brought into juxtaposition through subsequent tectonic displacements from their places of entombment (or vice versa). Tectonic displacements might then show up as strike-slip discontinuities in former provincial boundaries much as they do in oceanic magnetic lineations. The sensitivity depends on the sharpness of the provincial boundaries. In the case of ammonites, these were at times surprisingly well defined.

Faunal analysis can thus, in favourable cases, provide the following indications that are often beyond the reach of paleomagnetic studies. It can (a) decide whether a terrane at a given time was in the northern or southern hemisphere, thereby placing constraints on the polarity of paleomagnetic determinations; (b) indicate displacements of as little as a few hundred kilometres; and (c) provide longitudinal constraints. However, only through the combined efforts of lithostratigraphic, paleomagnetic, and biogeographic analyses will a comprehensible history ultimately evolve from the collage that confronts us. This paper is a preliminary effort and it points to the bright future for biogeographic analysis when enough data accumulates to allow us to work consistently and quantitatively at the species level with numerous taxonomic groups.

Ammonoid identifications for the biogeographic maps (Figs. 3-19) are based on published and unpublished work. Identifications from the literature have been evaluated critically; determinations based on unfigured specimens were usually avoided unless one or more of us has examined the material.

Literature providing biogeographic data for the Early Jurassic (Figs. 3-8) includes coverage of faunas from (1) Nevada (Muller and Ferguson, 1939; Guex, 1980, 1981, 1982; Guex and Taylor, 1976), (2) Oregon and California (Crickmay, 1933; Imlay, 1968; Sanborn, 1960), (3) Northern Alaska (Imlay, 1955), (4) Alaska in general (Imlay, 1981; Imlay and Detterman, 1973), (5) British Columbia and the southern Yukon (Crickmay, 1928; Frebold, 1959, 1967, 1970; Frebold and Little, 1962; Frebold *et al.*, 1969; Lees, 1934; Sutherland Brown, 1968; Tipper and Richards, 1976), (6) Alberta and eastern-most British Columbia (Frebold, 1957, 1959, 1960, 1964a, 1966, 1969, 1976; Hall and Stronach, 1982; Tozer, 1982b), (7) Arctic Canada (Frebold, 1958, 1960; Frebold and Poulton, 1977), (8) Canada in general (Frebold, 1951, 1964b; Frebold and Tipper, 1970), (9) the United States in general (Imlay, 1980), and (10) the United States and Canada (Tipper, 1981; Smith, 1983).

The published information for the Aalenian and Bajocian

(Figs. 9-11) is based on faunas from (1) Oregon (Imlay, 1973; Smith, 1980; Taylor, 1980), (2) British Columbia, southern Yukon, and western Alberta (Frebold, 1957, 1964a; Frebold and Tipper, 1970; Frebold *et al.,* 1969; Hall and Stronach, 1981, 1982; Hall and Westermann, 1980; McLearn, 1928, 1929; Westermann, 1964b; Whiteaves, 1889), (3) Arctic Canada (Frebold, 1957, 1958, 1960), (4) western Canada in general (Frebold, 1964b), (5) Western Interior of the United States (Imlay, 1967), (6) southern Alaska (Imlay, 1961, 1964; Westermann, 1964a, 1969, 1979), (7) northern Alaska (Imlay, 1955), (8) Alaska in general (Imlay and Determan, 1973), and (9) the United States in general (Imlay, 1980).

Elsewhere in this volume, Callomon provides a comprehensive review of the late Middle and Late Jurassic ammonoid systematics, correlation, and provincialism in western Canada and the United States. He is also largely responsible for certain biogeographic maps (Figs. 11-19) and accompanying descriptions of distributions in the present paper. The reader is referred, therefore, to the accompanying paper by Callomon for more detailed discussions.

The references to ammonoids from Bajocian to Callovian in Mexico include Burckhardt (1927), Cantu Chapa (1979), Erben (1956b), Geyssant and Rangin (1979), Westermann (1983, 1984) and Westermann *et al.* (this volume). Westermann is also responsible for the unpublished Middle Jurassic data on Mexico. Late Jurassic molluscan faunas of Mexico are also treated in the accompanying papers by Alencaster (this volume) and Buitrón (this volume).

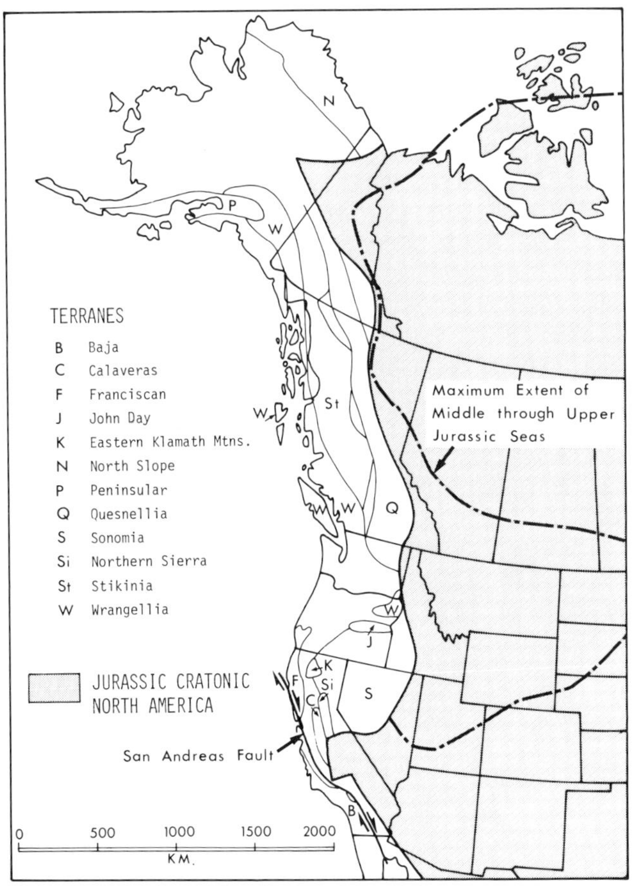

Figure 1. Tectonic terrane map for western North America. The most important terranes for reconstruction of Jurassic ammonite biogeographic patterns are depicted.

THE TERRANES

The extent of the potentially allochthonous tectonic collage (Fig. 1) is defined by those parts of North America whose Paleozoic stratal sequences differ clearly from the coeval miogeoclinal suites of the continental interior (Coney *et al.,* 1980). Each of the tectonostratigraphic terranes composing the collage is characterized by a stratal sequence different from that of its neighbors and by being commonly fault-bounded. The terranes of the conterminous United States on the map in Figure 1 follow those of Coney *et al.* (1980) except for an Oregon terrane. The southern part of the Blue Mountains terrane is separated and termed the John Day terrane, which is equivalent to the "southern coherent Mesozoic terrane" of Dickinson and Thayer (1978). Also, the Jurassic ammonite faunas from Nevada (Sunrise Formation) fall within the area denoted as Sonomia (Coney *et al.,* 1980), but it was originally defined by Speed (1979) as a terrane that accreted to North America in the Early Triassic. Consequently, the overlying Jurassic rocks do not pertain to Sonomia *s.s.* The term Sonomia, moreover, will not be used in forthcoming tectonostratigraphic terrane maps of the North American Cordillera being prepared by Silberling and others, because it is a concept that is now recognized as encompassing several discrete terranes (Silberling, pers. commun.). In the meantime, for lack of a better term, we discuss the Jurassic Nevadan ammonite faunas with reference to Sonomia. The terrane map of Canada follows that of Tipper (1981, this volume; note that the Alberta-Liard Troughs of Tipper, this volume, is equivalent to the Fernie Basin of Tipper, 1981), while that for Mexico (Fig. 21) follows Campa and Coney (1983). Only those terranes with ammonite occurrences important for this paper are depicted in Figure 1.

The definition and interpretation of the terranes based on geologic and paleomagnetic data may lead to complexities that cannot be ignored. Some terranes of independent history may have coalesced before finally docking on the craton. Such a development has been suggested for the Peninsular terrane and Wrangellia among others (Stone *et al.,* 1982). A few other terranes are certainly composite (Coney *et al.,* 1980), such as the Franciscan terrane, Sonomia, and the John Day terrane. The western part of the area encompassing the John Day Inlier (Dickinson and Thayer, 1978), which is the source of the rich Jurassic ammonite faunas from Oregon (Imlay, 1968, 1973), has a late Triassic and Jurassic lithologic sequence quite distinct from that in far eastern Oregon near Huntington (compare Dickinson and Thayer, 1978; Brooks and Vallier, 1978). At least one terrane, Wrangellia, fragmented as it amalgamated piecemeal with the North American continent. Some terranes with rock sequences similar to that of the craton may have been displaced with respect to the continental interior. For example, the composite terrane Sonomia yields, in part, rich Jurassic ammonite faunas from the Sunrise Formation, and has Lower Mesozoic stratigraphic sequences clearly indicating proximity with the craton; but this by itself does not exclude significant post-Jurassic displacement of up to several hundred kilometres by strike-slip movement adjacent to the craton. These "complexities" will be examined later, where faunal data provide clues to detailed terrane history.

AMMONITE BIOGEOGRAPHY

The faunal provinces recognized in the present work are based on the distributions of certain genera that are either confined to a restricted area or characteristic of one in which they are most widespread, diverse, or abundant. It is beyond the scope of this paper to work consistently down to the specific level, although characteristic species are discussed where their distribution is otherwise particularly illuminating. While the provincial definitions are based on ammonites, especially informative bivalve distributions are also mentioned.

The provinces are distributed among three Realms (Fig. 2): the Boreal, the Tethyan, and the East-Pacific of Westermann (1981). The Boreal Realm comprises the Boreal, Sub-Boreal, and new Bering provinces. The Boreal Realm is not further subdivided during the Early Jurassic, when provincialism was modest. But beginning with the Middle Jurassic it is valuable to distinguish more than one province within the realm. Therefore, in Figures 11-19 the term Boreal refers to the Boreal Province *s.s.* Boreal Realm faunas in North America are identified by their ties with typical Boreal successions in east Greenland, northern Europe and Siberia. The Sub-Boreal Province is carried over to North America from Europe (Cariou, 1973; Callomon, in Sykes and Callomon 1979, this volume). The Bering Province is characterized by a few Aalenian and Early Bajocian genera that are restricted to the north Pacific margin (Figs. 9, 10), from Japan to Oregon and the north slope of eastern Siberia and northwestern North America. These taxa occur both in eastern Asia and western North America and are found typically in terranes or on parts of the Jurassic craton that were at high latitudes during the Jurassic just as they are today. The Bering faunas are not referred to the Boreal Province proper, however, because they did not extend far into the Boreal Sea and are not known from coeval strata in Greenland or Europe. We stress that, as discussed in this paper, the Boreal Realm and its divisions excluding the Bering Province are always recognized by taxa that occur in northern Europe and/or Greenland as well as North America. These faunas in particular are indicative of comparatively high paleo-latitudes in the northern hemisphere.

REALM	REGION	PROVINCE
Boreal		Boreal
		Sub-Boreal
		Bering
Tethyan		Caribbean
East Pacific	North Cordilleran	Athabascan
		Shoshonean
		Andean
INFORMAL Pandemic "Pacific"		

Figure 2. Biogeographic terminology.

To the Tethyan Realm in western North America are assigned those taxa that are otherwise characteristic of its typical areas of development – the Jurassic Tethys and its continental margins in Europe, northern Africa, and southwestern and southern Asia. It must be clearly understood, however, that a Tethyan distribution *s.l.* does not necessarily imply a direct connection between the western Tethys and eastern Pacific via a proto-Altantic cratonic seaway (Hispanic Corridor of Smith, 1983). It merely expresses the observation that during the Early and Middle Jurassic the distribution of numerous genera extended circum-globally beyond the eastern portals of the Tethys, at about the same comparatively low paleo-latitudes as those of the Tethys itself. Inference of a Hispanic Corridor can, therefore, be made only if the taxa in question are absent in the eastern Tethys and/or western Pacific. The distinction between "Tethyan" *s.l.* and "pandemic" (see below) is therefore somewhat arbitrary. A late Jurassic Caribbean Province within the Tethyan Realm is shown in Figure 20. This provincial term is introduced only to show that taxa of Caribbean affinities occur in western North America, even though present-day occurrences are insufficient to demonstrate a continuous connection. These Caribbean distributions are discernible mainly at specific level only, for the genera as a whole are mostly found throughout the Tethys. The same holds true for some Oxfordian Submediterranean taxa mentioned in the text (not shown in the figures).

The East-Pacific Realm includes the provinces listed in Figure 2. The North-Cordilleran Province of Westermann and Riccardi (1976) is elevated to the status of Region, to incorporate two new provinces, the Athabascan and Shoshonean. The name Athabascan refers to a once widely distributed North American Indian linguistic group. The province is widely distributed in allochthonous Mesozoic North America and extends only rarely into the Western Interior region of the United States and the Canadian Rockies, where the Shoshonean Province lay.

The Shoshonean Province takes its name from the northern Shoshoni Indian tribe, which was the most important tribe in an area that covers approximately the region occupied by the great Jurassic epeiric sea of the Western Interior of the United States. The province is delimited geographically by the highly variable former extent of this sea (Fig. 1) and is first recognizable in the late Bajocian. The province is known from later time primarily from the Bathonian and Oxfordian but is poorly documented in the Callovian (Callomon, this volume) which appeared to have been a time of major regression.

The term Andean is used to denote the presence of faunas within the East-Pacific Realm primarily characteristic of South America. The northernmost extent of the Andean Province is in southern (and east-central?) Mexico.

The term Pacific is used here unqualified for Sinemurian and

Bajocian ammonites of the genera *Badouxia* and *Pseudotoites* (Figs. 4, 10) that are widely distributed around the periphery of that ocean, more widely than is implied by East-Pacific.

Pandemic describes those taxa that have a particularly wide distribution. As a minimum requirement this has to include Europe. The distribution usually marks a wider latitudinal spread than that of taxa referred to as Tethyan, but does not necessarily extend to the Boreal Province *s.s.*

These divisions into Realms and Provinces are necessarily rough and to some extent arbitrary. Provinces may overlap and their boundaries change with time. Thus, for instance, the genus *Pseudolioceras* began in the Toarcian as a pandemic inhabitant of the northern hemisphere. In the early Aalenian it retreated into the Boreal Realm, and in the latest Aalenian to earliest Bajocian it became restricted to the northern margins of the Pacific, now differentiated as the Bering Province. The southerly migration of the boundary of the Boreal Province, from the northern Yukon in the late Bajocian to Oregon and California in the Oxfordian, was already mentioned by Arkell (1956) who referred to it as the Boreal Spread.

Two suborders of ammonites in the Jurassic, the Lytoceratina and Phylloceratina, have traditionally been referred to as Tethyan, but this seems to have resulted largely from the historical accident of their first discovery in quantity in the Jurassic Tethys *s.s.* Elsewhere in this volume Callomon discusses the evidence indicating that the primary factor controlling the distribution of many of these ammonites may have been proximity to the open ocean and not environmental parameters related to latitude. Although we consequently de-emphasize the value of the phylloceratines and lytoceratines as Tethyan indicators, we do not always exclude them. In the Lower Jurassic, we count some of them among the Tethyan faunas when the evidence from Europe or elsewhere is not contradictory. Also, the Hettangian and Early Sinemurian lytoceratines (including *Pleuroacanthites* and its immediate derivatives; Guex, 1981) represent an early stage in the evolution of the group and were closely allied to the Ammonitina (Taylor, unpublished Early Sinemurian Data from Nevada and Oregon), which comprises the vast majority of the Jurassic ammonites. These early lytoceratines are probably reliable Tethyan indicators because they were abundant in the shelf seas of the Sunrise Formation in Nevada (Guex, 1980) and consequently do not seem to have been so strongly tied to waters close to an open marine environment, as was apparently the case with the later Lytoceratinae. The ammonites listed in the figures for which we signify special caution as Tethyan indicators, therefore, are the Juraphyllitidae (Figs. 3-5). We avoid basing interpretations of terrane displacements on the evidence of the juraphyllitids, or even exclusively on the early lytoceratines.

The Early Jurassic faunas from Mexico (Erben, 1956a) comprise pandemic and Tethyan taxa that provide, at generic level, little information for interpreting the microplate dis-

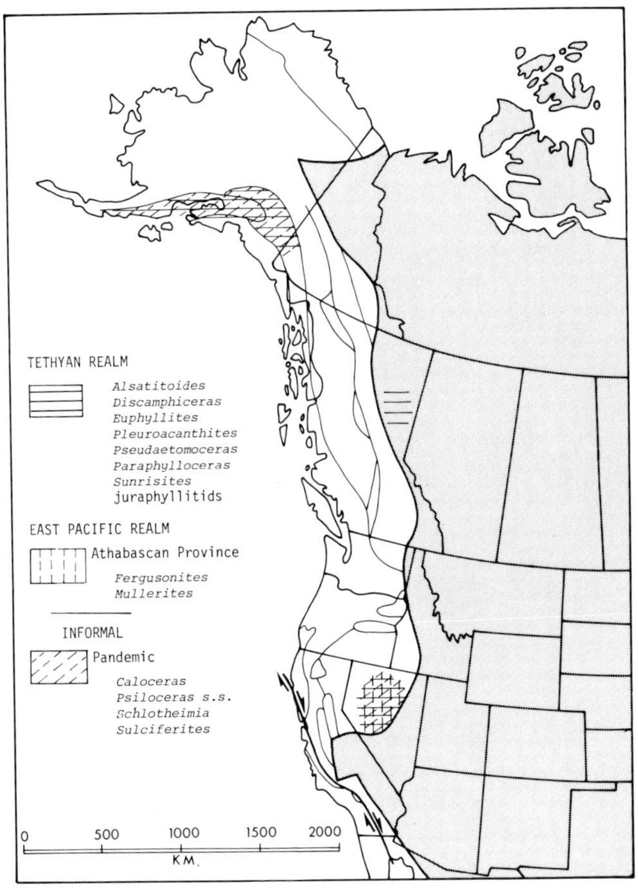

Figure 3. Hettangian.

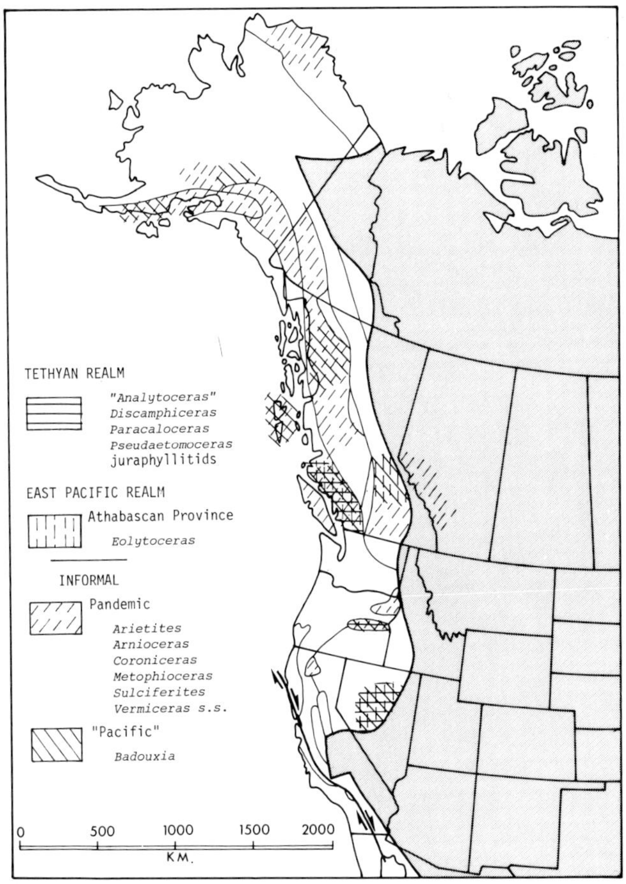

Figure 4. Early Sinemurian.

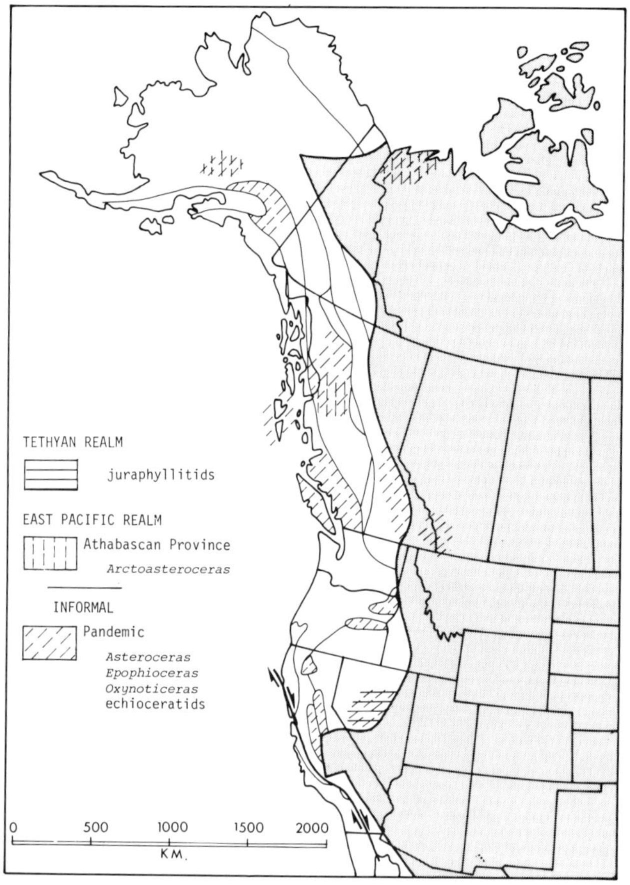

Figure 5. Late Sinemurian.

Figure 6. Early Pliensbachian.

placements discussed in this report. Coverage of Mexico therefore begins with the Bajocian Stage. This coverage is important, especially for those times in the Callovian and Oxfordian when the southward spread of Boreal faunas in North America was so extensive that Tethyan faunas ranged no further north than Mexico.

Hettangian

Hettangian ammonites from North America are known only from Sonomia, Wrangellia in southern Alaska, the Peninsular terrane, and on the craton in northeast British Columbia (Fig. 3). Previous Hettangian assignments based on the *Badouxia canadensis* fauna in Canada (Frebold, 1951, 1964a, 1967; see Frebold and Tipper, 1970 for summary of occurrences) are now considered to be Sinemurian (Guex and Taylor, 1976; Guex, 1980). In other cases, material from southeast British Columbia in Quesnellia (Frebold and Little, 1962), northern Alaska (Imlay, 1981), and the northern Yukon Territory (Frebold and Poulton, 1977) also assigned to the Hettangian is considered by us to be too poorly preserved for generic identification. The only report of a Hettangian ammonite from the North American craton is by Tozer (1982b). He reports on a specimen from northeastern British Columbia (Fernie Formation) which he refers to as *Psiloceras* (*Paraphylloceras*) *calliphyllum* (Neumayr). That species has Tethyan affinities.

By far the richest Hettangian ammonite faunas from North America are from the lower part of the Sunrise Formation from Sonomia in Nevada (Guex, 1980). All of the ammonites in Figure 3 characterizing the Tethyan Realm are known from that formation. The faunas from the Peninsular Terrane and Wrangellia in southern Alaska are also of Tethyan affinity, notably in that they yield the genus *Discamphiceras* (Imlay, 1981). The two Athabascan representatives *Fergusonites* and *Mullerites*, from the Sunrise Formation (Guex, 1980), also have affinities with ammonites of the Tethyan Realm.

Early Sinemurian

Early Sinemurian ammonites (Bucklandi and Semicostatum Zone equivalents) are much more widespread than Hettangian faunas in the Western Cordillera and even reached Alberta (Fig. 4). Probable pandemic arietitids from the Canadian Arctic figured by Frebold (1960) are not more closely determinable. Otherwise, occurrences of pandemic taxa are widespread in the United States and Canada. *Eolytoceras*, which occurs on Sonomia and southern Quesnellia, is restricted to the northern Cordillera and is, therefore, considered to be an Athabascan form. Its affinities are probably Tethyan by analogy with other coeval representatives of the family to which it belongs – the Analytoceratidae (*sensu* Wiedmann, 1970) which are Tethyan in distribution. Characteristic Tethyan ammonites are known as far north as central Stikinia, but the richest and most diverse

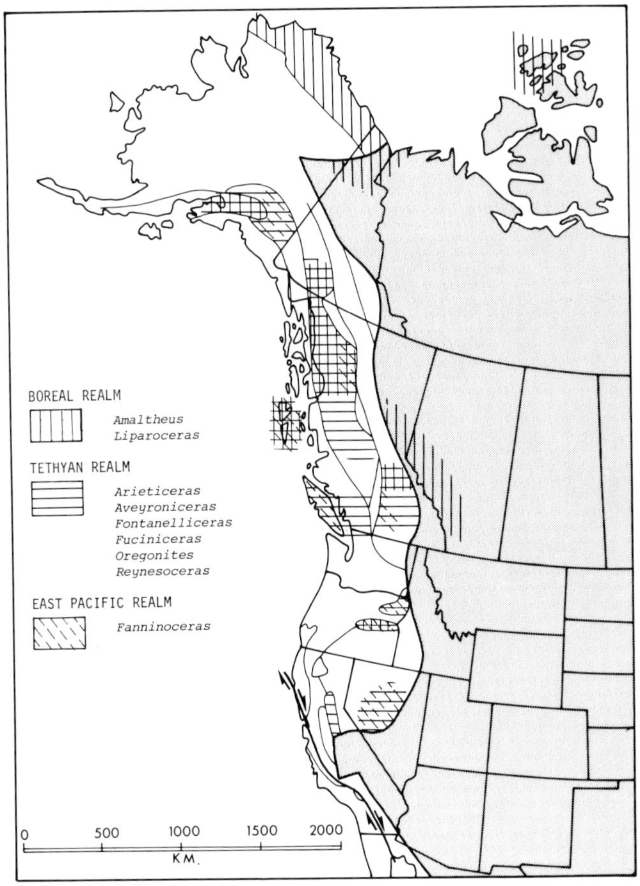

Figure 7. Late Pliensbachian.

Figure 8. Toarcian.

assemblages are from the John Day terrane and Sonomia. *Badouxia* has an East-Pacific or even Pacific distribution *s.l.;* besides North and South America (Hillebrandt, 1981b) it is known from eastern Siberia (Repin, 1977). The genus is widespread in western North America, from Sonomia to northernmost Alaska.

Weyla is a distinctive Tethyan pectinid bivalve found in the Mediterranean, eastern Africa (Saudi Arabia, Kenya, Madagascar) and the eastern Pacific (Damborenea and Manceñido, 1979). Occurrences in the Western Cordillera are often poorly dated but are most commonly Sinemurian and Pliensbachian in age. The genus occurs in conterminous United States in Sonomia, North Sierra, Eastern Klamaths, and the John Day terrane. The Canadian occurrences (Frebold and Tipper, 1970) are in southern Quesnellia, as far north in Stikinia as the southern Yukon Territory and in Wrangellia as far north as Alaska. The specimens from Wrangellia in Alaska are post-Sinemurian (MacKevett, 1970) and those from two localities in southwestern Alaska north of the Peninsular terrane are Sinemurian (Imlay and Detterman, 1973).

Late Sinemurian

Late Sinemurian ammonites are predominantly pandemic (Fig. 5). *Arnioceras* (ranging into Late Sinemurian and plotted only on the Early Sinemurian map, Fig. 4) and *Asteroceras* are not usually abundant, but they are widespread in both the United States and Canada. Also widespread in Canada but rare are oxynoticeratid ammonites tentatively identified as *Gleviceras* or *Oxynoticeras.* The echioceratids, which dominate the faunas in Nevada, are unknown in Oregon, but this almost certainly reflects collection failure since the relevant interval (within the Keller Creek Shale) is poorly exposed. The family is widespread farther north in Canada, particularly on the Queen Charlotte Islands and Harbledown Island east of Vancouver Island (Wrangellia), and in southern Alaska in the Talkeetna Mountains (Peninsular terrane). *Epophioceras,* recently recognized in Nevada, is now questionably recorded also from the Fernie Basin (Hall and Stronach, 1982). Rare *Juraphyllites* from Sonomia are the only late Sinemurian Tethyan ammonites. The Athabascan genus *Arctoasteroceras* is generally restricted to southern Alaska and northern Canada, but it is abundant at one locality in central Stikinia throughout a 7 m stratigraphic interval.

Early Pliensbachian

Early Pliensbachian pandemic and Tethyan taxa from the Western Cordillera occur as far north as southern Alaska (Fig. 6). *Crucilobiceras* (*Metaderoceras*) and *Dayiceras* are the most widespread Tethyan indicators, while most of the others are know primarily from the conterminous United States. *Phricodoceras,* which is widespread in Europe but most diverse and

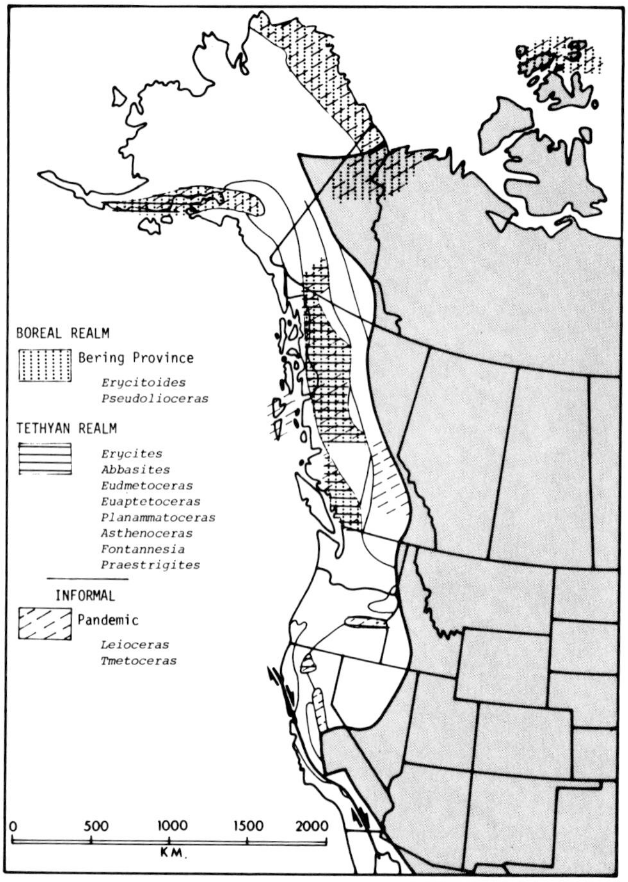

Figure 9. Aalenian.

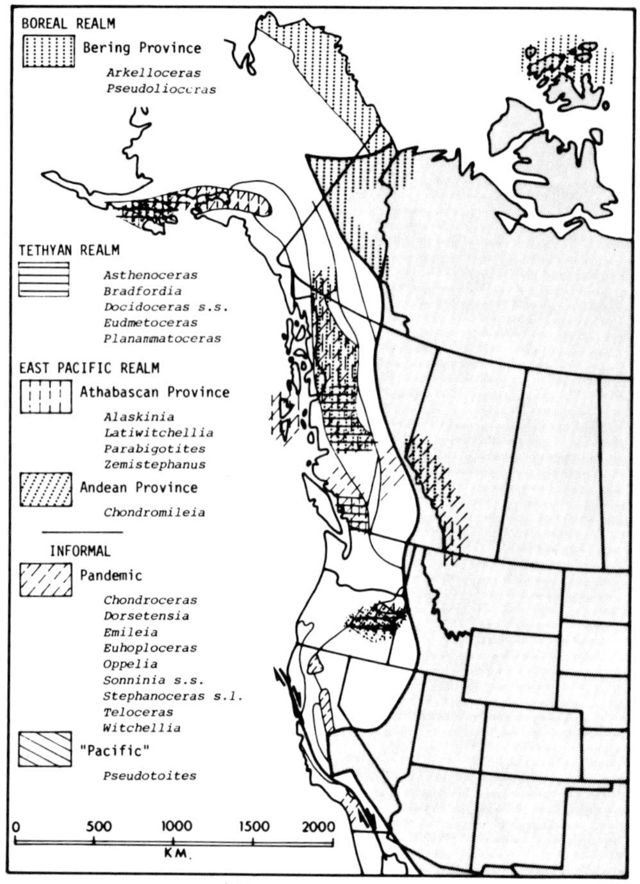

Figure 10. Early Bajocian.

long-ranging in the Mediterranean countries, is now known from rare specimens from Wrangellia in Oregon and the Queen Charlotte Islands, and possibly south of Fernie in British Columbia, where *Coeloceras* (*Reynesocoeloceras*) is also questionably present (Frebold, 1969).

Late Pliensbachian

Provincialism in the Early Jurassic was strongest during this substage, and North America reveals the same biogeographic separation of the Boreal amaltheids and Tethyan hildoceratids that is evident in Europe (Howarth, 1973; Tipper, 1981). Tethyan faunas contain many species representing genera such as *Arieticeras, Fuciniceras,* and *Fontanelliceras* (Fig. 7). Lytoceratids and phylloceratids are present in small numbers. Boreal faunas are low in diversity and dominated by *Amaltheus.* Tipper (1981) has divided the late Pliensbachian localities into those that yield amaltheids exclusively, those yielding hildoceratids exclusively, and those that are mixed. Specimens of the predominantly Boreal taxon *Liparoceras* (*Becheiceras*) are rare in North America, in that one example is known from northern Stikinia, three or four have been collected from the Mowich Group in the John Day terrane, and none are known elsewhere. *Amaltheus* is represented on the Peninsular terrane by two specimens from the Talkeetna Mountains (Imlay, 1981), and one each from Wide Bay (S.W. Muller manuscript, 1962: Report to the Richfield Oil Corporation; verified Callomon, 1981) and the Queen Charlotte Islands (Wrangellia). The Boreal pattern (Fig. 7) in the Peninsular terrane, in Wrangellia on the Queen Charlotte Islands, and in the John Day terrane, therefore pertain to rare specimens associated with otherwise characteristically Tethyan faunas. Faunas from Alaskan Wrangellia, southern Stikinia, southern-most Quesnellia, Oregonian Wrangellia, and Sonomia are purely Tethyan. Mixed assemblages include those from northern Stikinia and southern (but not southern-most) Quesnellia. Localities in the Alberta-Liard Troughs and those from the Arctic yield only amaltheids and thus are strictly Boreal.

Fanninoceras is the only known Early Jurassic ammonite restricted to the eastern Pacific, and ranges from Argentina (Blasco *et al.,* 1978; see also Hillebrandt, 1981a) to Alaska.

The rudist-like bivalve *Plicatostylus* is a highly distinctive genus that is restricted to the East-Pacific (Hillebrandt, 1981a). It has an Athabascan provincial distribution in North America, where it appears to be restricted to the Pliensbachian (circumstantial evidence indicates restriction to late Pliensbachian), and an Andean provincial distribution in South America, where it occurs only in Toarcian rocks in Chile and Argentina (Hillebrandt, 1973, 1981a). In North America *Plicatostylus* is restricted to the conterminous United States (Batten and Taylor, 1978), where it occurs in Sonomia and the North Sierra and John Day terranes.

Toarcian

Provincialism in Toarcian ammonites appears to have been as feeble in North America (Fig. 8) as in Europe (Howarth, 1973). North American ammonites of this stage are poorly known at present, although this will soon be rectified by studies in progress. Of particular importance is the diverse and abundant fauna from the Maude Formation of the Queen Charlotte Islands which is currently being studied by two of us (Smith and Tipper). Most of the Toarcian is present. In contrast, the United States faunas represent only the latest and earliest Toarcian. *Tiltoniceras propinquum* (Whiteaves), which is so well represented in Siberia (Dagis, 1971, 1974) and in the Queen Charlotte Islands (its type area), occurs as far south as Nevada, but is not known in Oregon. Other Early Toarcian harpoceratids are widespread in Canada in all terranes in reasonable abundance, but preservation is seldom good and the faunas have not been studied extensively. There is a general absence of lytoceratids and phylloceratids in Canada but such ammonites are known from the Upper Toarcian of Oregon where they occur with *Dumortieria* and hammatoceratids. *Dactylioceras commune* is represented in the Arctic Islands and possibly the Alberta-Liard Troughs but it is not recorded from western North America. In contrast, *D. canense* is known in the Queen Charlotte Islands and possibly east-central Oregon. Several indeterminate species of *Dactylioceras* are recognized from central and northern Stikinia and southern Quesnellia.

Aalenian

Early Aalenian faunas are of low diversity and generally poorly documented in western North America, but *Pseudolioceras mclintocki* s.s. and *Leioceras opalinum* in Arctic regions are of this age (Fig. 9). The pandemic genus *Leioceras* is absent from western North America south of the Arctic. The Graphoceratidae appear to be absent from the Pacific and America as a whole except for rare possible occurrences in the Andes of northern Chile (Hillebrandt and Westermann, unpublished). *Pseudolioceras* in Europe ranges from the late Early to latest Toarcian, and perhaps locally (N.W. Germany) into the basal Aalenian, lower Opalinum Zone (*Ps. beyrichi*). The earliest Aalenian representatives of this genus in North America may therefore be referred to as Boreal rather than Bering. Nevertheless, *Pseudolioceras* is depicted as a Bering representative, because of its strictly north Pacific distribution throughout all but the basal Aalenian and in the earliest Bajocian strata (Figs. 9, 10).

Late Aalenian ammonites, in contrast, are widely distributed in western North America and northeastern Siberia. The Bering form *Erycitoides* is especially widespread, and occurs as far south as the eastern part of the John Day terrane (Huntington area). *Pseudolioceras* is known as far south as central Stikinia. Tethyan faunal elements occur as far north as the Peninsular terrane, where there are diverse hammatoceratids. The fauna from the Snowshoe Formation from Oregon (western part of John Day terrane) is most diverse, in that it yields all of the Tethyan indicators except for *Erycites* (Imlay, 1973; Taylor, 1980).

Early Bajocian

The Early Bajocian sees the last of the Bering Province and the reappearance of the Athabascan Province (Fig. 10). Provincialism was modest in that most faunas were dominated by Tethyan and pandemic elements. The Arctic during this time was remarkable for its faunal poverty and yields only *Pseudolioceras* and *Arkelloceras*. Basal Bajocian faunas of the *Docidoceras widebayense* Zone and coeval strata are widespread, and are best represented in southern Alaska (Westermann, 1969) and Oregon (Imlay, 1973; Taylor, 1980). During this zonal interval the sole remaining representative of the Bering Province was *Pseudolioceras,* which is found as far south as central Stikinia but is not known from the Alberta-Liard Troughs. The Athabascan taxa during this time were *Witchellia* (*Latiwitchellia*) from the John Day terrane, and *Alaskinia* [*olim Alaskoceras*] from Alaska in the Peninsular Terrane at Wide Bay and the Talkeetna Mountains (=*Sonninia* cf. *nodata* of Imlay, 1964, p. B33, pl. 2, Figs. 1, 2). *Pseudotoites* of ambi-Pacific distribution is represented by a few specimens from the Peninsular and John Day terranes. Pandemics include genera such as *Euhoploceras, Oppelia,* and *Witchellia,* while *Docidoceras s.s., Eudmetoceras, Planammatoceras, Praeoppelia, Bradfordia,* and *Asthenoceras* could perhaps be counted as Tethyan indicators.

Higher in the lower Bajocian, in the *Parabigotites crassicostatus* Zone and coeval strata in northern Canada and Alaska, *Arkelloceras* replaces *Pseudolioceras* as the sole Bering representative. *Arkelloceras* occurs in Arctic Canada, northern Alaska and eastern Siberia, is known by one example from the Peninsular terrane, and is represented by isolated finds from central Stikinia and the Alberta-Liard Troughs. The Athabascan genus *Parabigotites* occurs in the Peninsular and John Day terranes. *Eudmetoceras* is also known from strata of this interval, which is higher than its range in Europe or elsewhere. It may, therefore, be regarded as an Athabascan form for this part of its range.

The late Early Bajocian (*Stephanoceras kirshneri* and *Chondroceras oblatum* Zones, equivalent roughly to the Humphriesianum Zone of Europe) is an important interval for world-wide correlation for it is the last, until the Cretaceous (Valanginian), in which truly pandemic ammonites are found: *Stephanoceras* (including *Stemmatoceras* and the microconch *Normannites*), *Sphaeroceras* and *Chondroceras.* These genera occur world-wide: Tethys and peri-Tethyan Europe as well as circum-Pacific. This includes the whole of the Andean Cordillera from Graham Land northwards, and almost all of the western North American Jurassic. One of the few major areas not populated by these genera was the Boreal Province, including Arctic Canada, which so far has yielded no ammonites of this age. There is some minor provincialism. The North American forms *Chondroceras* ex gr. *oblatum* (Whiteaves) differ slightly but consistently from their European counterparts, and could be distinguished as a subgenus *Defonticeras* McLearn (see Callomon, this volume). An Athabascan representative during this interval is *Zemistephanus,* which is widely distributed in the North American Jurassic. Conversely, Tethyan groups such as *Oppelia,* common in the Humphriesianum Zone of Europe, are missing or rare.

Immediately following the Early Bajocian, faunal provincialism in the ammonites became acute and world-wide. It was also marked by the largely cryptogenic appearances of major new taxa such as the whole of the superfamilies Perisphinctaceae and Spiroceratacea, families such as the Boreal Cardioceratidea, and subfamilies such as the Cadomitinae, the extra-Pacific Garantianinae, and the important Pacific Eurycephalitinae of the Sphaeroceratidae.

Early Late Bajocian

Beds of this age occur widely and can be securely correlated by means of the pandemic groups such as *Spiroceras, Lissoceras* and early *Leptosphinctes s. s.* (Fig. 11). The major new faunal realms (and provinces) are sharply delimited: the Boreal, with the first *Cranocephalites borealis* in the Arctic still strongly resembling *Chondroceras oblatum,* its probable ancestor; and the East-Pacific with *Megasphaeroceras,* the first of the Eurycephalitinae, extending from southern Alaska to the southern Andes (and ? Antarctic Peninsula). Slightly modified forms of *Megasphaeroceras* appeared to develop in the Western Interior, referred there to the genus *Eocephalites.* The boundary between Boreal and East-Pacific Realms is sharp, lying between southern and northern Alaska with no overlap.

In the Mixteca terrane of southern Mexico (Fig. 21), two early Late Bajocian ammonite faunas can be distinguished (Westermann, 1983 and 1984, with revision). Within the brief diachronous transgressive cycle of the Taberna Formation, however, only a single fauna appears to be present at any one locality. The probably older, *Duashnoceras* Association (Westermann, 1983) of Mixtepec, has yielded the endemic *D. floresi* (Burckhardt) and other, unnamed species (including *"Stephanoceras* aff. *psilacanthus"* of Burckhardt, 1927), some of which have affinities with the Chilean *Lupherites chongi* (Hillebrandt); the cosmopolitan *Leptosphinctes* spp., the Andean *Stephanoceres chilense* Hillebrandt with macroconch and microconch (*S. paucicostatum* Felix), and the Mediterranean *"Parastrenoceras"* [*Subcollina*] *lucretius* (d'Orbigny) *sensu* Sturani (1971) suggesting the lower Subfurcatum Zone (Banksi Subzone). The probably slightly younger *Parastrenoceras* Association, occuring mainly in the Tezoatlán area about 50 km farther north of Mixtepec, contains *P. mixteca* Ochoterena with micro-and macroconchs, *Oppelia subradiata erbeni* Westermann (with close affinity to the mainly European species), and abundant *Leptosphinctes tabernai* Westermann. *L. tabernai* and *P. mixteca* (under the microconch name *P. tlaxiacense*) probably occur rarely in the upper Subfurcatum Zone (Baculatum Subzone) and ?*Duashnoceras floresi, "P"* cf. *lucretius,* and *O. subradata erbeni* in the lower to middle Subfurcatum Zone (Polygyralis Subzone) of south-central Europe (Dietl, 1980, and pers. commun; Sturani, 1971). *Stephanoceras* spp. with Andean affinity (Caracoles) occur just below the *Duashnoceras* Association at Mixtepec and could be latest Early Bajocian. Surprisingly, the East-Pacific *Megasphaeroceras* has not been found. The *Duashnoceras* Association and the probably subjacent stephanoceratids thus consist of endemic and Andean elements, together with some Mediterranean elements, while the slightly later *Parastrenoceras* association strongly resembles Mediterranean and Submediterranean faunas except that *Parastrenoceras* is prevalent in Mexico and scarce in Europe.

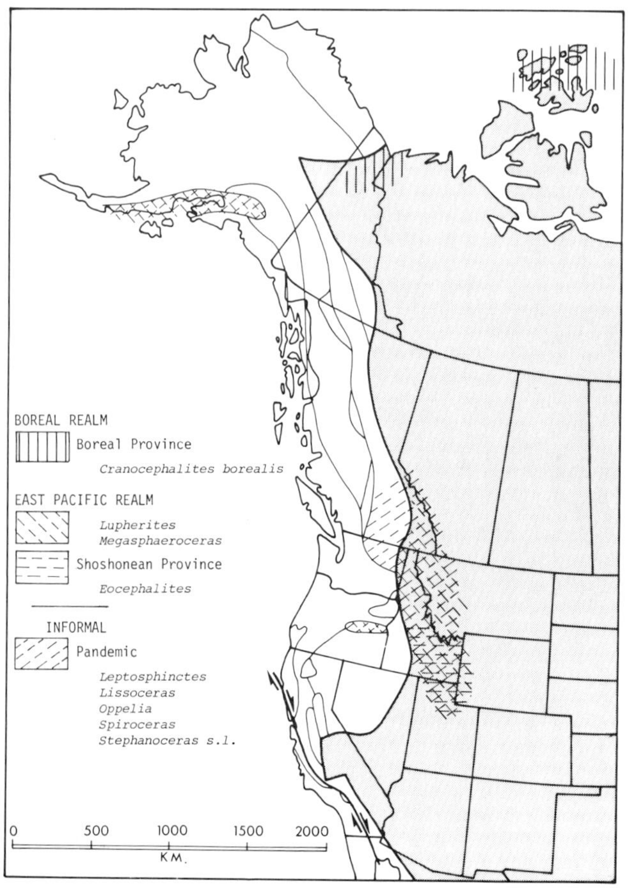

Figure 11. Early Late Bajocian

Latest Bajocian and Earliest Bathonian

Boreal Arctocephalitinae [*Cranocephalites pompeckji* group] extend in force as far south as the northern Yukon (Fig. 12). Athabascan faunas include the late stephanoceratid *Parareineckeia* found in the Alaskan Peninsula, Alaskan Wrangellia, Stikinia, the Alberta-Liard Troughs and possibly in the John Day terrane. A single specimen has been found in the Boreal northern Yukon, but unfortunately only as float.

Epizigzagiceras, probably a perisphinctid, is known so far only from Stikinia. *"C" costidensus* Imlay, an important form found in the Alaskan Peninsula and Alaskan Wrangellia and commonly referred to Boreal *Cranocephalites,* is more probably also a Pacific eurycephalitine closely homeomorphic with *Cranocephalites.* A closely similar form occurs in what is otherwise regarded as the typical Boreal Alaskan North Slope, *"C." ignekensis* Imlay, and it, too, is probably a eurycephalitine. More work is urgently needed in this area, possibly one of the few in which East-Pacific and Boreal Realms overlap in the Middle Jurassic. World-wide correlation of all these faunas depends on the identification of the genus *Cobbanites* Imlay as, at most, a subgenus of the well-known pandemic late *Leptosphinctes.* It probably belongs to the group of *L. martiusi*

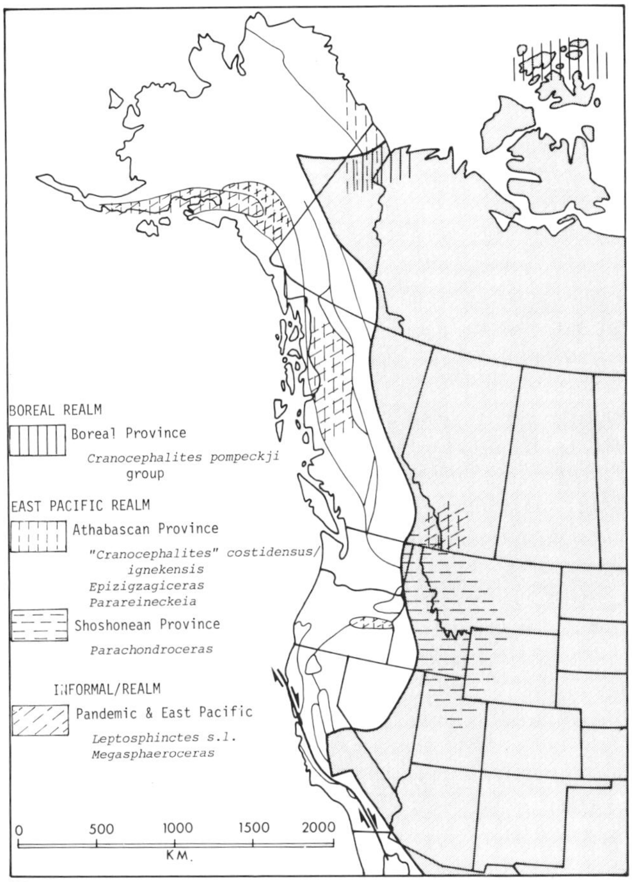

Figure 12. Latest Bajocian and earliest Bathonian.

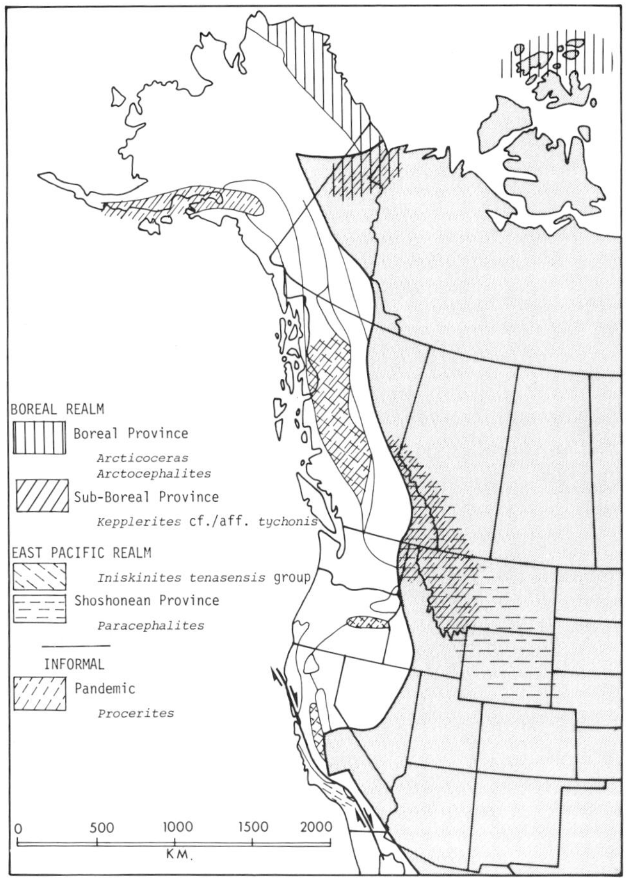

Figure 13. Early and Middle Bathonian.

(d'Orbigny), e.g., *L.* (*Vermisphinctes*) which is common in the Tethyan and European Garantiana and Parkinsoni Zones at the top of the Bajocian. East Pacific *Megasphaeroceras* persisted and a Shoshonean provincial fauna, represented by *Parachondroceras* (including *Sohlites*), developed in the Western Interior.

Early and Middle Bathonian

Fully differentiated Boreal Arctocephalitinae continued to occupy the Arctic in strength (Fig. 13). The Pacific Eurycephalitinae continued to evolve and the species found in Stikinia, the Alberta-Liard Troughs, and John Day at this level are, at present, conventionally placed already in *Iniskinites*. Their absence from the Alaskan Peninsula is probably only trivial, lost in some major non-sequence. The pandemic perisphinctid *Procerites* (widely recorded still as *Cobbanites*) occurs more or less sparsely in most areas and serves to indicate ages generally. The forms tend to be rather featureless and hence long-ranging, so that even in Europe their value for more precise stratigraphical correlation is very limited. They are, however, distinguishable from the earlier true *Leptosphinetes* (*Cobbanites*) and hence of value in distinguishing Bathonian from Bajocian.

Two further important events occurred in the Bathonian. The Western Interior saw the differentiation of an endemic, Shoshonean, group of Eurycephalitinae which may be grouped together under the name *Paracephalites*. They include forms that have been variously assigned in the past to *Arctocephalites* [*A. codyensis, A. henryi*] or *Cadoceras* [*C. muelleri, C. piperense*] but the clues to their true affinities lie in their microconchs, which are all closely similar to the unusually variocostate microconch of *Eurycephalites, Xenocephalites* under which name many of them have been correctly identified in the past. This interval also saw the sudden appearance of the first of a whole new family, the Kosmoceratidae. The earliest forms are barely distinguishable from what must certainly have been their ancestors, the Cadomitinae. The conventional use of the generic name *Kepplerites,* from the first forms upward derives from the impossibility of drawing a line between it and *Cadomites* at any level other than the one provided by the hiatus in the northern hemisphere below the first *Kepplerites* in the Bathonian. The family subsequently developed continuously in the northern hemisphere up to the top of the Callovian. Its distribution is consistently Sub-Boreal, overlapping with the Boreal Province (northern Yukon), the Athabascan Province of the East-Pacific Realm (Alaskan Peninsula to Foothills terrane of central California), and the Shoshonean Province. Whether the first appearances in all these areas were exactly isochronous remains to be established, but they could not have differed by very much. The earliest forms are wholly round-ventered in both dimorphs, as in *Cadomites*/*Polyplectites*. In the Late Bathonian, flat and sharp-edged venters developed proterogenetically (from the inner whorls outwards) and at about the Bathonian-

Callovian boundary the lateral and ventro-lateral tubercles characteristic of *Kosmoceras* itself began to appear.

Late Bathonian

The Boreal Province is revealed in the northern Yukon by early, abundant *Cadoceras* [*C. barnstoni*] (Fig. 14). The morphological break between *Arcticoceras* and *Cadoceras* seen there is smoothly bridged elsewhere in the Arctic by intermediate forms and is, therefore, due only to a local faunal nonsequence. The East-Pacific Realm is represented by *Iniskinites s. s.*, known from the John Day terrane of Oregon via Wrangellia of the Queen Charlotte Islands to the Alaskan Peninsula. Significantly, there is now some provincial overlap, for *Iniskinites* extended as far as the northern Yukon in some strength, and rare *Cadoceras* occurs in the Peninsula [*C. moffiti* Imlay]. Sub-Boreal *Kepplerites* also straddles the Boreal Province and East Pacific Realm. Pandemic perisphinctids occur occasionally almost everywhere as previously, but are again of little use for closer correlations. Similarly, scattered finds of Oppeliidae occur widely, but their significance remains to be evaluated largely because they are hard to identify. There is evidence that they may, in addition, have formed a provincially restricted distinct Boreal sub-group that has yet to be mapped. The Western Interior saw the development of yet another local group, this time of *Kepplerites: K. mclearni*, a spectacular giant

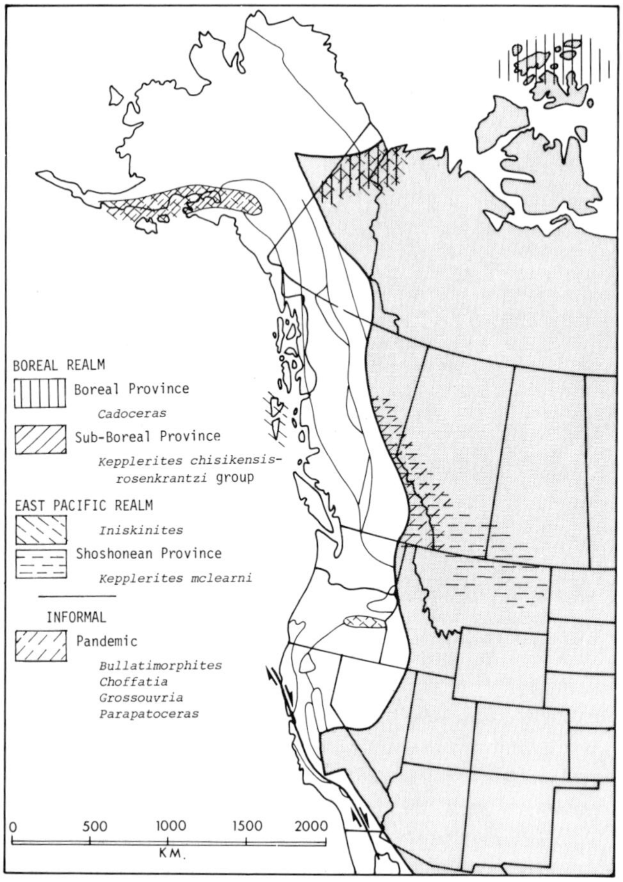

Figure 14. Late Bathonian.

homoeomorph of Early Bajocian *Stephanoceras*, the homoeomorphism being again resolved via the microconch, *Kosmoceras*-like *K. knechteli* Imlay. There are intriguing glimpses in Oregon of pandemic-Tethyan elements well known from Europe, the Andes and Mexico, such as *Bullatimorphites* and the heteromorph *Parapatoceras*.

The *Neuqueniceras* Association (latest Bathonian – earliest Callovian) from the Cualac area, Mixteca terrane, of Mexico (Fig. 21) has recently been re-examined by Westermann *et al.* (this volume). Of the 27 species he has identified, no fewer than half are identical with species from the central and southern Andes (South Peru to Neuquén), far fewer are the same as ones from the western Tethys, and the fewest are conspecific with ones from the North Cordilleran Province. *Neuqueniceras* [*Reineckeia neogaea* Burckhardt] was also recorded from the southernmost Sierra Madre terrane (Huayacocota Mountains) by Erben (1956b) and Cantu (1979 and earlier). Recent field work (Westermann), however, failed to duplicate these findings and the older collections have either been lost or are unavailable for examination. The well-core records of Cantu (1979 and earlier) of *Wagnericeras* and even *Kepplerites* from the Coahuila terrane also cannot be confirmed since the specimens are unavailable.

Prior to his recent field work, Westermann (1981, 1983) had tentatively followed Arkell (1956) and Imlay (this vol. and earlier) in placing *Stephanoceras floresi* as a new subgenus *Duashnoceras* in *Zigzagiceras* of the Zigzagiceratinae, implying the presence of Lower Bathonian for the Pacific area (see also Callomon, this volume). The much more compressed *"Stephanoceras* aff. *psilacanthum"* of Burckhardt (1927, pl. 12) was correctly considered as closely related and Andean *Stephanoceras* and *Lupherites chongi* were also compared.

Early Callovian

The Callovian saw a dramatic southward spread of Boreal *Cadoceras* (Fig. 15). The genus became dominant in the Alaskan Peninsula and common in at least northern Stikinia and Wrangellia of southwestern British Columbia, and it may have reached as far south as the Northern Sierra terrane of California. The East-Pacific Realm continues to be represented by the last of the Eurycephalitinae [*Eurycephalites, Lilloettia*], well known also from Guerrero in southern, Pacific Mexico (Mixteca terrane) and the Andean Cordillera. The Sub-Boreal Province is represented in the Wrangellian Queen Charlotte Islands, the type locality, and in the Alaskan Peninsula by *Kepplerites loganianus* (the type species of *Seymourites*, thought originally to be a Campanian kossmaticeratid) and species so similar to *K. keppleri* from the earliest Callovian of Europe, that they are the best indicators so far known for the position of the Bathonian-Callovian boundary in the whole of the Pacific. Poorly preserved material from the Northern Sierra terrane of California identified as *Reineckeia* (Crickmay, 1933) may indicate Tethyan affinities, as does the fauna found in accretionary melange on Cedros in Lower California (Geyssant and Rangin, 1979) which compares closely with species known from Mexico and the Peruvian Andes.

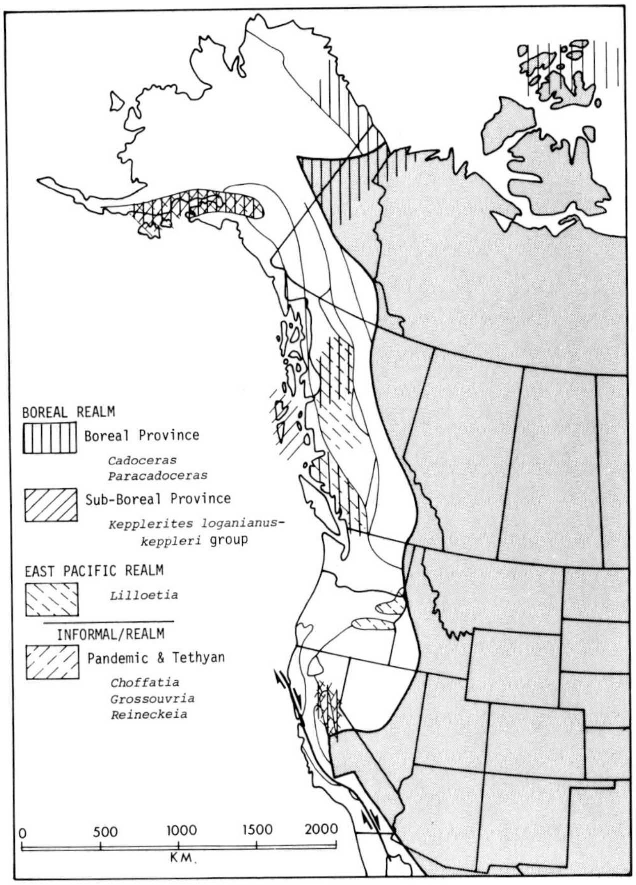

Figure 15. Earliest Callovian.

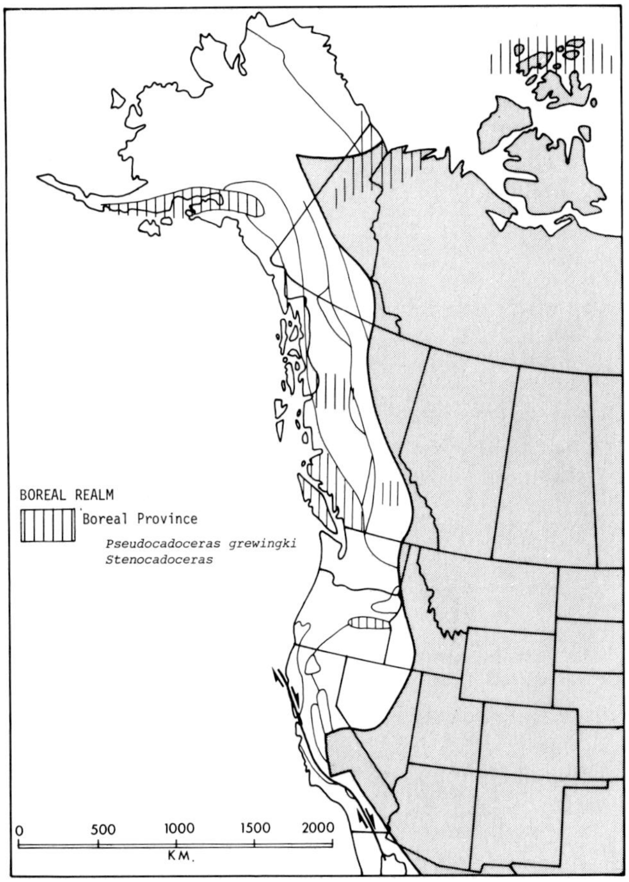

Figure 16. Middle Callovian.

Middle Callovian

Evidence is sparse north of Mexico and consists largely of Boreal *Cadoceras* (*Stenocadoceras*) and its microconchs *Pseudocadoceras* (Fig. 16). The southern Sierra Madre terrane of east-central Mexico yields typical Tethyan *Reineckeia* (*Reineckeia*) ex gr. *anceps* and *stuebeli* (Cantu, 1979; confirmed by new findings near Huauchinango by Westermann). Most of the Middle and Late Callovian ammonite records are spurious, however, owing to condensed facies with fragmentary fossils, except for a *Peltoceras* or early *Euaspidoceras* recently found at Mixtepec by Cardenas (1978). *Stephanoceras* (*Erymnoceras*) *mixtecorum* Burckhardt, 1927 is merely another *reineckeiid*, as are presumably the other specimens recorded as *Erymnoceras* by Erben (1956b).

Late Callovian

The evidence is even sparser for this interval than for the Middle Callovian (Fig. 17) and in many areas it is demonstrably lost in major non-sequences. What evidence there is reveals that the unambiguously Boreal Province continued to persist as far south as Montana: a single species of *Quenstedtoceras*, *Q. collieri* marked a modest attempt to recolonize the Western Interior which had been devoid of ammonites during the whole of the Callovian and as far back as the Late Bathonian.

Early and Middle Oxfordian

This marked a period of major transgressions (southern Alaska, Western Interior, and Mexico) and recolonization by Boreal ammonites [*Cardioceras*] as far south as Oregon (Fig. 18). The late Middle and earliest Late Oxfordian ammonites, widespread in Mexico consisting of *Perisphinctes* (*Dichotomosphinctes*), *Discosphinctes*, *Ochetoceras*, and *Euaspidoceras*, are predominantly Tethyan. Pandemic Tethyan or Sub-Mediterranean families are common in the Foothills terrane (California), but farther north never rise above being minor faunal constituents. A notable feature of southern Alaska is the co-occurrence in comparable proportions of the Cardioceratidae with Phylloceratidae and Lytoceratidae, to the exclusion of almost everything else.

Late Oxfordian and Early Kimmeridgian

Ammonites become scarce everywhere and are replaced by the bivalve *Buchia* which extends as far south as Mexico. Such ammonites as are found have the same distribution as in the Early and Middle Oxfordian: Boreal Cardioceratidae (*Amoeboceras*) ranged as far south as central California, there to overlap with Tethyan Perisphinctidae (Fig. 19).

Figure 17. Late Callovian.

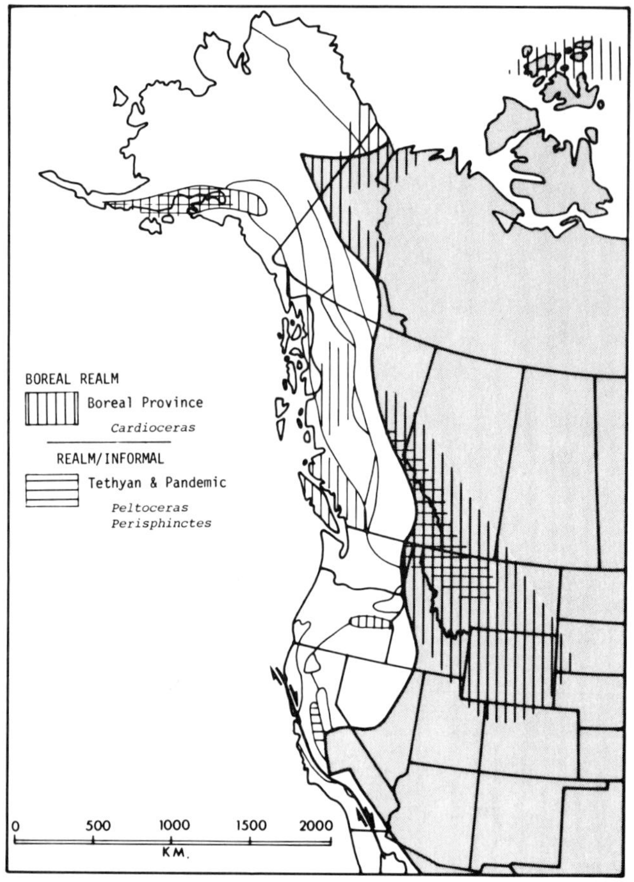

Figure 18. Early and Middle Oxfordian.

Late Tithonian and Berriasian

Tethyan/Andean/Caribbean faunas are found in a long narrow strip extending from northern California through southwestern Oregon (Franciscan/Elder Creek terrane; Imlay and Jones, 1970) as far as west Vancouver Island in Wrangellia (Jeletzky, 1965) (Fig. 20). They are wholly separated from the Boreal Volgian ammonite faunas found in the northern part of the Sverdrup Basin by intermediate regions yielding only *Buchia.* The famous allegedly Portlandian *Titanites occidentalis* Frebold (1957) from the Rockies at Fernie, British Columbia, has been reclassified (Callomon, this volume). Its age is wholly problematical; it could still be late Oxfordian.

TERRANE DISPLACEMENTS

It is necessary to assess the Jurassic faunal data for evidence concerning longitudinal and latitudinal displacements of terranes relative to the North American craton in order to gain the fullest appreciation of their history. A significant body of structural, paleomagnetic, and paleontologic data has accumulated bearing on latitudinal movements, while the evidence for or against significant longitudinal displacements is not as well documented. The one case for major longitudinal drift is based on Permian verbeekinid fusulinids (Yancey, 1975) and indicates significant eastward drift for the Cache Creek terrane of British Columbia (Monger *et al.,* 1982). It, however, yields no Jurassic ammonites. While there is no evidence for long distance longitudinal displacement of terranes as late as the Mesozoic, the Jurassic faunal data, nevertheless, provide clear indications of longitudinal positions. Athabascan, East-Pacific, Andean, and West-Tethyan faunal elements furnish longitudinal control by tying terranes to the eastern rather than western Pacific. Some West-Tethyan taxa (Westermann, 1981) occur in Europe, North Africa, and in the Americas, but are not known from the western Pacific or eastern Tethys. North-South displacements may be inferred from the observation of latitudinal juxtapositions of Boreal and Tethyan faunas. The faunas of the Boreal Realm tie terranes to the northern Pacific, and the faunas of the Athabascan province relate terranes to the northeastern Pacific when the ammonites have a foothold on the craton.

Evidence for Longitudinal Positions of Terranes

Northeast Pacific faunal ties between the craton and terranes are provided by some Athabascan ammonites. These appear as early as the Hettangian in the terranes, but the first found on the craton as well as on the terranes are Late Sinemurian in age: *Arctoasteroceras* (Fig. 5) occurs on the craton in Arctic Canada. Occurrences of the genus in Stikinia and in southern Alaska north of the Peninsular terrane suggest a northeastern Pacific position since the Sinemurian for the latter two areas. Athabas-

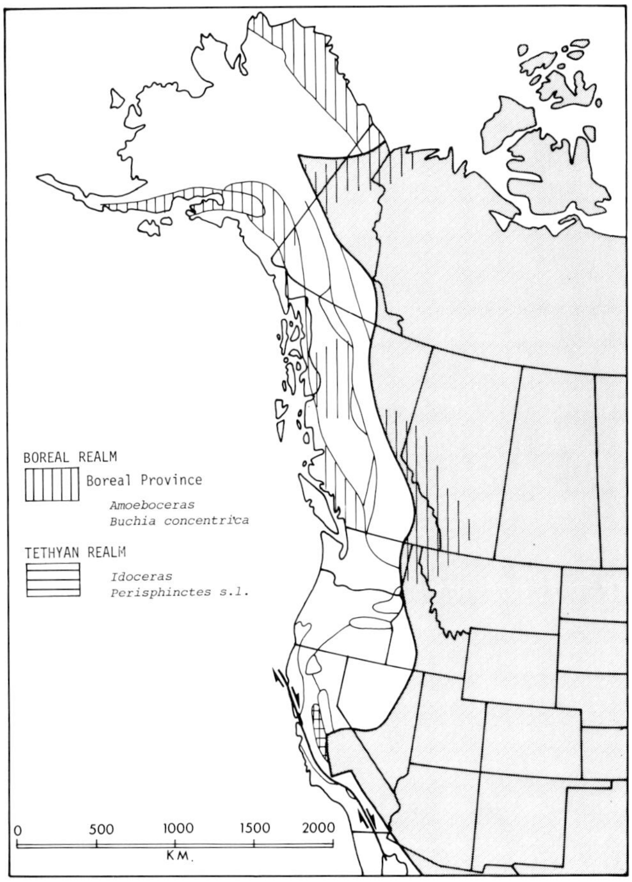

Figure 19. Late Oxfordian through Early Kimmeridgian.

Figure 20. Late Tithonian and Berriasian.

can faunas from the Aalenian through earliest Bathonian interval are well anchored to the craton and clearly tie the Peninsular terrane, most components of Wrangellia, Stikinia, and the John Day terrane, to the northeastern Pacific region.

The eastern Pacific Early Jurassic ammonite *Fanninoceras* (Fig. 7) and bivalve *Plicatostylus*, as well as the numerous taxa from the early Bajocian through early Callovian (see also Westermann, 1981; West-Tethyan genera), connect all of the Cordilleran terranes (except the younger Franciscan) to the western Americas. Neither of the above Early Jurassic taxa occurs on cratonic North America. They do occur on Sonomia which, as stated earlier, has a Mesozoic stratal sequence that unites the relevant part of this terrane with the craton. The Early Jurassic West-Tethyan and East-Pacific bivalve *Weyla* also allies the Goodnews terrane (see Coney *et al.*, 1980), Wrangellia, Stikinia, Quesnellia, John Day, North Sierra, Eastern Klamath, and Sonomia terranes to the Americas.

In southern Mexico, the Late Bajocian *Duashnoceras* and *Parastrenoceras* associations suggest a placement of the Mixteca terrane close to the opening of the Hispanic Corridor ("proto-Atlantic"), while the Late Bathonian *Neuqueniceras* Association allies it with the central and southern Andes. The wide-spread Callovian and Late Jurassic faunas which belong to the circum-global Tethyan Realm carry less information.

In summary, the faunal data tells us clearly that during the Jurassic all of the terranes mentioned above were allied with the eastern rather than the western Pacific.

Evidence for Latitudinal Positions and Displacements

The Athabascan genus *Arctoasteroceras* (Fig. 5) discussed above, and Boreal *Amaltheus* and *Liparoceras* (Fig. 7) demonstrate that the Peninsular terrane, Wrangellia, Stikinia, Quesnellia, and the John Day terrane were well in the northern hemisphere by the middle Early Jurassic. Similar evidence continues for the remainder of the Jurassic and the southward spread of the Boreal Realm from the Bathonian through later times takes in the Northern Sierra terrane as well. The remainder of the evidence for latitudinal displacement relates more specifically to the precise positions of terranes relative to one another and the craton.

Early Jurassic Tethyan ammonites are numerous, but Boreal faunas are poorly represented except for those of the late Pliensbachian. Nevertheless, the occurrence of Tethyan ammonites in the Hettangian and early Pliensbachian as far north as southern Alaska (Peninsular terrane and Wrangellia) is somewhat surprising, and not inconsistent with some northward displacement of these parts of Alaska. The Sinemurian occurrences of the Tethyan bivalve *Weyla* from the Goodnews terrane (Coney *et al.*, 1980) would suggest perhaps that parts of southwest Alaska north of the Alaskan Peninsula also may have been farther south than now.

By far the most interesting Early Jurassic distributive pattern is that of the late Pliensbachian when Boreal *Amaltheus* invaded North America in force (Tipper, 1981). Its strictly Boreal faunas are dominant on the craton as far south as southern Alberta, and coeval Tethyan faunas are found on terranes as far north as southern Alaska (Fig. 7). The evidence points to a progressively greater northward displacement of successively seaward Canadian and Alaskan terranes. If we regard the faunas from the John Day terrane, Wrangellia (Queen Charlotte Islands), and the Peninsular terrane to be Tethyan (since the Boreal ammonites occur as minor elements among abundant Tethyan forms), the northward limit of predominantly Tethyan faunas increases from terrane to terrane in a westerly direction. The corresponding displacements for Quesnellia, Stikinia, and the Peninsular terrane and Alaskan Wrangellia would then be at least 500 km, 1800 km, and 2500 km respectively.

If Stikinia were repositioned south of present-day Alberta, with its strictly Boreal ammonites, then the southerly limit in that terrane of mixed Tethyan and Boreal faunas would lie at the southern-most end of present-day Oregon. This would be south of those Pliensbachian faunas in Oregon that are characteristically Tethyan, indicating in turn a northward displacement of at least 400 to 500 kilometres of the John Day terrane and Oregonian Wrangellia since the Pliensbachian. Displacement of Cordilleran terranes relative to the craton also finds support in the distribution of *Plicatostylus*. The genus probably thrived only at low latitudes, as in the case of typical later Mesozoic rudists. *Plicatostylus* has been found no farther north than Oregon; yet this thick-shelled bivalve is so distinctive that if it had occurred in cratonic Canada, we would expect to have had reports of it from any of the shallow water facies there. By repositioning Stikinia, which lacks *Plicatostylus*, south of Alberta, the southern limit of its outcrops would extend to the latitude of present-day central Nevada. Displacements of about 1000 km for the John Day terrane and a few hundred kilometres for the North Sierra terrane and Sonomia would then be required to bring these *Plicatostylus*-bearing terranes south of the repositioned southern limit of the *Plicatostylus*-free Stikinia.

The faunal data for the early Middle Jurassic in the United States and Canada indicate essentially the same terrane relationships as in the Early Jurassic. There are a number of characteristically Tethyan late Aalenian and Early Bajocian genera, including *Praestrigites, Fontannesia, Abbasites, Asthenoceras,* and common hammatoceratids. Tethyan faunas yielding various combinations of these genera occur in the John Day terrane, Stikinia, and Peninsular terranes but are absent from the craton. Of the above taxa only *Asthenoceras* and the hammatoceratids are partly coeval with the rather meager (Early Bajocian) faunas known from the craton. Although the cratonic faunas are not as well documented for this interval as in the Pliensbachian, this pattern is at least consistent with a northward drift of the Peninsular terrane and Stikinia from south of present-day Alberta.

The Early Bajocian Andean subgenus *Emileia* (*Chondromileia*) and the Late Bajocian East-Pacific *Lupherites* occur in the John Day terrane (western part) and no farther north. An approximately 800 km or more northward displacement of the John Day terrane would then retain these Andean forms south of the repositioned Stikinia and Peninsular terrane.

The evidence for large-scale latitudinal displacements of terranes for much of the remainder of the Middle and Late Jurassic is tenuous. The faunal patterns mostly reveal a normal gradation of Boreal faunas in the north into Tethyan faunas in the south, which is consistent with interpretations in terms of even zero displacements (see also Jeletzky, this volume). The ammonite identifications are not firm enough to base any conclusions on the enigmatic apparently Early Callovian Boreal records in the North Sierra terrane (Fig. 15). Likewise, the Oxfordian *Perisphinctes* giving the Tethyan pattern in southern Alaska (Fig. 18) may have had a wide latitudinal range at this time (including Greenland) and should not be taken as evidence for terrane displacements.

In Mexico, the Late Bajocian and latest Bathonian-earliest Callovian ammonite faunas of the Mixteca terrane suggest (to Westermann) a paleoposition close to the Pacific opening of the Hispanic Corridor connecting with the Western Tethys, but not too far removed from the Andes of south Peru, Chile and Argentina.

Comparison of Faunal with Paleomagnetic and Structural Evidence

Paleomagnetic studies of the Middle and/or Late Triassic Nikoli Greenstone (Wrangellia) suggest paleolatitudes of about 15° from the equator in either the northern or southern hemispheres (Hillhouse, 1977; Jones *et al.*, 1977). Similar results were obtained from the Late Triassic Karmutsen basalt on Vancouver Island (Symons, 1971; Irving and Yole, 1972; Yole and Irving, 1980). Some authors prefer a southern hemisphere position (Yole and Irving, 1980; Irving *et al.*, 1980). Jones *et al.* (1978) also argue for such a location, considering data by Hillhouse (1977). These data and assumed reversed polarity and clockwise rotation give a northward displacement of 60° latitude relative to North America. A clockwise rotation is more consistent with the sense of rotation for tectonic blocks inferred generally for the region. On the other hand, Panuska and Stone (1981) and Stone *et al.* (1982) favor a northern hemispheric position for Wrangellian Alaska in the Triassic. They found a close correspondence in positions of the virtual geomagnetic poles for the terrane for the Triassic and Late Paleozoic, a time of known polarity. Muller (1977) also proposed a northern site and suggested that the extensive basaltic extrusions in the Wrangellian Triassic were related to extensional tectonics at a latitude corresponding to present-day "California".

None of the above arguments for the Triassic latitudinal position of Wrangellia is binding in itself. The Jurassic faunal data are not conclusive either, but are strongly suggestive. We place the components of Wrangellia north of Oregon into the northern hemisphere at a latitude greater than 30° by the Bajocian in order to draw the faunas of the Boreal Realm out of the tropical zone. More than 10 cm per year of northward movement would then be called for to bring Wrangellia from 15°S in the Carnian to at least 30°N in the Bajocian. While such a high rate of movement would have been possible, we prefer the slower rate that is implied by having the terrane already in the northern hemisphere in the Late Triassic.

The choice of hemispheres for the location of the Jurassic Alaskan Peninsula based on paleomagnetic data is also ambiguous, but the faunal evidence is conclusive. Palaeomagnetically a location approximately either 30° N or 30° S is possible, again depending upon the interpretation of magnetic polarity and sense of terrane rotation. The southern hemisphere Jurassic location favored by Panuska and Stone (1981) and Stone *et al.* (1982) would best satisfy the paleomagnetic data by calling for less complexity in north-south migrations of the terrane between the Triassic and Recent. The southern hemisphere Jurassic option reportedly calls for a southward migration of the terrane from the Triassic to the Jurassic and then a northward movement thereafter. The northern hemisphere Jurassic option apparently suggests a northward migration from the Triassic to the Jurassic, a southward migration from the Jurassic to the Cretaceous, and then a northward movement from the Cretaceous to the Recent. The Jurassic ammonites demonstrate quite clearly that the Peninsular terrane was in the northern hemisphere throughout the Jurassic. A model similar to the Alaskan Baja model seems preferable (Packer and Stone, 1974; Stone and Packer, 1977, 1979). Here the Alaskan Peninsula was close to the conterminous United States and at a latitude approximately equivalent to that of the present-day California-Oregon border.

Our faunal data largely agree with the suggestions of Monger and Irving (1980) and Irving *et al.* (1980) that Stikinia has moved northward about 1300 km since the Jurassic and that Symons' call (1973) for no displacement of the terrane results from a misinterpretation of his paleomagnetic data.

The time of accretion of the John Day terrane to North America, sometime between the Late Triassic and Middle Cretaceous, proposed by Dickinson and Thayer (1978), is based on geologic evidence. Our data indicate a post-Bajocian emplacement for the western part of this terrane.

There are marked Middle Jurassic faunal dissimilarities between the Alaskan North Slope and northern Yukon (Callomon, this volume) that could suggest that these two portions of North America may have been subsequently juxtaposed, perhaps by the combined displacements of the Tintina and Kaltag faults (Jones, 1980). Both areas have strong Boreal ties, however, and no major latitudinal displacement of the areas is indicated.

In Mexico, the Mojave-Sonora Megashear (Silver and Anderson, 1974) running from the northwestern corner of the country southeastward toward the Gulf of Mexico was active during the Late Triassic-Middle Jurassic opening of the Gulf by seafloor spreading. The half of Mexico southwest of the megashear would have moved some 1000 km southeastward. A counter-clockwise rotation of western and southwestern Mexico bounded by the megashear (or a fault zone in similar position) has also been proposed to prevent the overlap of Mexico with South America in the Pangea reconstruction; additional paleomagnetic evidence has suggested counter-clockwise rotation for at least several tectonic blocks (microplates?) along this fault zone, supporting the idea of a large-scale shear (Gose *et al.*, 1982 and earlier). Recently, Campa and Coney (1983 and earlier) have postulated that about 80% of the southern part of the North American Cordillera of Mexico is made up of suspect terranes resembling the Cordillera of Canada and United States. The eastern terranes, however, were accreted already by the late Paleozoic, but were cut in the Jurassic by the Mojave-Sonora and Torreon-Monterrey Megashears or Discontinuities trending through the Sierra Madre terrane (Fig. 21). Western and southwestern Mexico are largely made up of several distinct Jurassic-Cretaceous submarine volcanic arc terranes, mostly with unknown basement, that accreted against the North American craton during the Late Cretaceous-Early Tertiary. The combined effect of accretion and translation along the megashear(s) eliminate the overlap of Mexico with South America in the Pangea reconstruction.

The Mixteca terrane of the Sierra Madre Occidental (Fig. 21) is of particular interest because of its unique tectonostratigraphic history and fossil record. This is the only terrane in Mexico with well documented Late Bajocian-Early Callovian ammonite faunas which perhaps continue northward into doubtful parts of the Sierra Madre terrane. The faunas place the Mixteca terrane during the Middle Jurassic at the eastern Pacific margin in the vicinity of the Central Andes, where the northernmost related faunas are known from southern Peru (Westermann and Riccardi, 1980), and the western end of the Hispanic Corridor (proto-Atlantic seaway). The *Neuqueniceras* Association with abundant Phylloceratidae also supports an oceanic rather than cratonic environment. The paleontologic evidence, therefore, tentatively supports Campa and Coney's hypothesis of an allochthonous Mixteca terrane, rather than the more conservative view of a deep marine embayment (Oaxaca Embayment) with a somewhat restricted northwesterly opening (see Imlay, this volume; Alencaster, this volume). The abundance of terrigenous and associated continental deposits, however, indicates the proximity of a larger landmass.

The *Reineckeia* s.s. faunas of the Sierra Madre Oriental belongs to the common stock which appeared in the Tethys and Andes at about the same (mid-) Callovian time when the Central Atlantic ocean began to open by seafloor spreading. If *Reineckeia* s.s. evolved indeed in the Andean Province (including parts of Mexico) from Late Bathonian-Early Callovian *Neuqueniceras* and *Reineckeia* s.l., then eastward rather than westward faunal dispersal through the early Atlantic is indicated. This migration could have occurred either directly across the spreading Gulf where the hypersaline conditions (Louann Salt) would have disappeared by then at least in large areas, or indirectly through an extended Oaxaca Embayment opening into the Pacific (see Imlay, this volume; Alencaster, this volume) and thence through the proto-Caribbean area eastward, or directly through the Central Atlantic with the reineckeiids-bearing terranes of Mexico positioned offshore. The recent paleogeologic work of Longoria and Scott (both this volume) support the first alternative.

The problem of whether the southeastern end of the Sierra Madre terrane in eastern Mexico belonged to the Mixteca terrane would depend in part on the reliability of the earlier

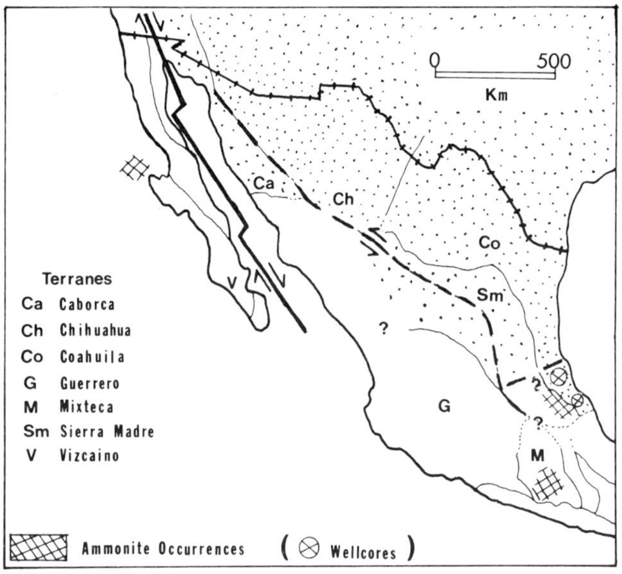

Figure 21. Late Bajocian-Callovian ammonites in Mexico relative to tectonic terranes.

records of Andean *Neuqueniceras* from poorly fossiliferous beds – this we were unable to confirm. The Trans-Mexico Volcanic Axis prevents structural recognition of plate boundaries in this area.

ACKNOWLEDGMENTS

The research for this paper by D. Taylor was conducted through a post-doctoral fellowship at McMaster University with G.E.G. Westermann. P. Smith acknowledges the support of NSERC. We thank J.W.H. Monger, B.C. Panuska, and N.J. Silberling for reviewing the manuscript, C. Burke, Portland State University for typing it, and T. Oliveric, Geological Survey of Canada, Vancouver, B.C. for preparing the figures.

REFERENCES

Arkell, W.J., 1956, Jurassic Geology of the World: London, Oliver and Boyd, 806 p.

Batten, B.L. and Taylor, D.G., 1978, Reassessment of the Age of Thompson Limestone, Lower Jurassic, Plumas County, California: Journal of Paleontology, v. 52, p. 208-209.

Blasco, G., Levy, R., and Nullo, F., 1978, Los amonites de la formacion Osta Arena (Liasico) y so Posicion Estratigráfica, Pampa de Agnia (Provincia del Chubut): Actas. VII Congr. Geol. Argent., Neuquen 1978, v. 2, p. 407-429.

Brooks, H.C., and Vallier, T.L. 1978, Mesozoic Rocks and Tectonic Evolution of Eastern Oregon and Western Idaho: in Howell, D.G. and McDougall, K.A. (eds.), Mesozoic Paleogeography of the western United States: Society of Economic Paleontologists and Mineralogists, Pacific Coast Paleogeography Symposium 2, p. 133-145.

Burckhardt, C., 1927, Cephalópodos del Jurasico Medio de Oaxaca y Guerrero: Instituto de Geologia de México Boletin 47, 108 p.

Campa, M.F. and Coney, P.J., 1983, Tectono-Stratigraphic Terranes and Mineral Resource Distributions in Mexico: Canadian Journal of Earth Sciences, v. 20, p. 240-251.

Cantu Chapa, A., 1979, Biostratigrafia de la Serie Husteca (Jurásico medio y superior) en el Subsuelo de Poza Rica, Veracruz: Instituto Mexicana de Petrol Revista, v. 11, p. 14-24.

Cardenas, V.J., 1978, Contributión al Conocimiento Geológico de la Mixteca Oaxaqueña: Thesis Instituto de Geologia México, p. 15-107.

Cariou, E., 1973, Ammonites of the Callovian and Oxfordian: in Hallam, A., Atlas of Biogeography: Elsevier, p. 287-295.

Coney, P.J., 1981, Accretionary Tectonics in Western North America: Arizona Geological Society Digest, v. 14, p. 23-37.

Coney, P.J., Jones, D.L., and Monger, H.W.H., 1980, Cordilleran Suspect Terranes: Nature, v. 288, p. 329-333.

Crickmay, C.H., 1928, The Stratigraphy of Parson Bay, British Columbia: University of California Publications, Department of Geological Sciences Bulletin, v. 18, p. 51-70.

———, 1933, Mount Jura investigation: Geological Society of America Bulletin, v. 44, p. 895-926.

Dagis, A.A., 1971, Genetic relationships between *Tiltoniceras* – *Eleganticeras-Harpoceras:* Paleontological Journal, Academy of Sciences, USSR, Moscow 1971, no. 2, p. 41-46 (in Russian).

———, 1974, Toarcian ammonites (Hildoceratidae) of North Siberia: Academy of Sciences, USSR, Siberian Branch, Transactions of the Institute of Geology and Geophysics, v. 99, 108 p. (in Russian).

Damborenea, S.E. and Mancenido, M.O., 1979, On the Paleogeographical Distribution of the Pectinid Genus *Weyla* (Bivalvia, Lower Jurassic): Palaeogeography, Palaeoclimatology, Palaeoecology, v. 27, p. 85-102.

Dickinson, W.R., and Thayer, T.P., 1978, Paleogeographic and Paleotectonic Implications of Mesozoic Stratigraphy and Structure in the John Day Inlier of central Oregon: in Howell, D.G. and McDougall, K.A. eds., Mesozoic paleogeography of the western United States: Society of Economic Paleontolgoists and Mineralogists, Pacific Coast Paleogeography Symposium, v. 2, p. 147-161.

Dietl, G., 1980, Die Ammoniten-Gattung *Leptosphinctes* aus dem südwestdeutschen Subfurcaten-Oolith (Bajocium, Mttl.Jura): Stuttgarter Beiträge zur Naturkunde, Serie B (Geologie und Paläontologie), 66, p. 49.

Erben, H.K., 1956a, El Jurásico Inferior de México y sus ammonitas: 20 Congreso Geológica International, 393 p.

———, 1956b, El Jurásico Medio y el Caloviano de México: 20 Congreso Geológico International, 140 p.

Frebold, H., 1951, Contributions to the Palaeontology and Stratigraphy of the Jurassic System in Canada: Geological Survey of Canada Bulletin 18, p. 1-14.

———, 1957, The Jurassic Fernie Group in the Canadian Rocky Mountains and Foothills: Geological Survey of Canada Memoir 287, 197 p.

———, 1958, Fauna, Age and Correlation of the Jurassic Rocks of Prince Patrick Island: Geological Survey of Canada Bulletin 41, 69 p.

———, 1959, Marine Jurassic Rocks in Nelson and Salmo Areas, Southern British Columbia: Geological Survey of Canada Bulletin 49, 31 p.

———, 1960, The Jurassic Faunas of the Canadian Arctic – Lower Jurassic and lowermost Middle Jurassic ammonites: Geological Survey of Canada Bulletin 59, 33 p.

———, 1964a, Lower Jurassic and Bajocian Ammonoid Faunas of Northwestern British Columbia and Southern Yukon: Geological Survey of Canada Bulletin 116, 31 p.

———, 1964b, Illustrations of Canadian Fossils: Jurassic of Western Canada and Arctic Canada: Geological Survey of Canada Paper 63-4, 106 p.

______, 1966, Upper Pliensbachian Beds in the Fernie Group of Alberta: Geological Survey of Canada Paper 66-27, 9 p.

______, 1967, Hettangian Ammonite Faunas of the Taseko Lakes Area, British Columbia: Geological Survey of Canada Bulletin 158, 35 p.

______, 1969, Subdivision and Facies of Lower Jurassic Rocks in the Southern Canadian Rocky Mountains and Foothills: Geological Association of Canada Proceedings, v. 20, p. 76-87.

______, 1970, Pliensbachian Ammonoids from British Columbia and Southern Yukon: Canadian Journal of Earth Sciences, v. 7, p. 435-456.

______, 1976, The Toarcian and Lower Middle Bajocian Beds and Ammonites in the Fernie Group of Southeastern British Columbia and Parts of Alberta: Geological Survey of Canada Paper 75-39, 33 p.

Frebold, H. and Little, H.W., 1962, Paleontology, Stratigraphy, and Structure of the Jurassic Rocks in Salmo map-area, British Columbia: Geological Survey of Canada Bulletin 81, 31 p.

Frebold, H. and Poulton, T.P., 1977, Hettangian (Lower Jurassic) Rocks and Fauna, Northern Yukon Territory: Canadian Journal of Earth Sciences, v. 14, p. 89-101.

Frebold, H. and Tipper, H.W., 1970, Status of the Jurassic in the Canadian Cordillera of British Columbia, Alberta, and Southern Yukon: Canadian Journal of Earth Sciences, v. 7, p. 1-21.

Frebold, H., Tipper, H.W., and Coates, J.A., 1969, Toarcian and Bajocian Rocks and Guide Ammonites from Southwestern British Columbia: Geological Survey of Canada Paper 67-10, 55 p.

Geyssant, J.R. and Rangin, C., 1979, Découverte D'Ammonites Calloviennes dans de Complexe à Blocs de l'ile de Cedros (Basse Californie, Mexique) et Implication pour l'age de la Série Ophiolitique Sous-jacente: Comptes Rendus des Séances de l'Academie des Séances de Paris, Séries D, p. 521-524.

Gose, W.A., Belcher, R.C., and Scott, G.R., 1982, Paleomagnetic Results from Northeastern Mexico: Evidence for Large Mesozoic Rotations: Geology, v. 10, p. 50-54.

Guex, J., 1980, Remarques Préliminaires sur la Distribution Stratigraphique des Ammonites Hettangiennes du New York Canyon (Gabbs Valley Range, Nevada): Laboratoires de Géologie, Mineralogie, Géophysique et du Musée Géologique de l'Université de Lausanne Bulletin, v. 250, p. 127-140.

______, 1981, Quelques Cas de Dimorphisme Chez les Ammonoides du Lias Inférieur: Laboratoires de Géologie, Minéralogie, Géophysique et du Musée géologique de l'Université de Lausanne Bulletin, 258, p. 239-248.

______, 1982, Relation Entre le Genre *Psiloceras* et les Phylloceratida au Voisinage de la Limite Trias-Jurassique: Labortoires de Géologie, Mineralogie, Géophysique et du Musée Géologique de l'Université de Lausanne Bulletin, v. 249, p. 1-6.

Guex, J. and Taylor, D.G., 1976, La Limite Hettangien-Sinémurien, des Préalpes Romandes au Nevada: Eclogae Geologicae Helvetiae, v. 69, p. 521-526.

Hall, R.L. and Stronach, N.J., 1981, First Record of late Bajocian (Jurassic) Ammonites in the Fernie Formation, Alberta: Canadian Journal of Earth Sciences, v. 18, p. 919-925.

______, 1982, A Guidebook to the Fernie Formation of Southern Alberta and British Columbia: Circum-Pacific Jurassic Research Group, First Field Conference, Calgary, Alberta, 48 p.

Hall, R.L. and Westermann, G.E.G., 1980, Lower Bajocian (Jurassic) Cephalopod Faunas from Western Canada and Proposed Assemblage Zones for the Lower Bajocian of North America: Palaeontographica Americana, v. 9, 93 p.

Hillebrandt, A. von, 1973, Neue Ergebnisse über den Jura in Chile und Argentinien: Münstersche Forschungen zur Geologie und Paläontologie, v. 31, p. 167-199.

______, 1981a, Kontinentalverschiebung und die Paläozoogeographischen Beziehungen des Südamerikanischen Lias: Geologische Rundschau, v. 70, p. 570-582.

______, 1981b, Faunas de Amonites del Liásico Inferior y Médio (Hettangiano hasta Pliensbachiano) de America del sur (excluyendo Argentina): in Volkheimer, W. and Musacchio, E.A. eds., Cuencas Sedimentarias del Jurásico y Cretácico de America del Sur: II Congreso Latinamericano de Paleontologia, Porte Alegre 1981, v. 2, p. 499-538.

Hillhouse, J., 1977, Paleomagnetism of the Triassic Nikolai Greenstone, South-Central Alaska: Canadian Journal of Earth Sciences, v. 14, p. 2578-2592.

Howarth, M.K., 1973, Lower Jurassic (Pliensbachian and Toarcian) ammonites: in Hallam, A., Atlas of Palaeobiogeography: Elseview, p. 275-282.

Imlay, R.W., 1955, Characteristic Jurassic Mollusks from Northern Alaska: United States Geological Survey Professional Paper 274-D, p. 69-96.

______, 1961, New Genera and Subgenera of Jurassic (Bajocian) Ammonites from Alaska: Journal of Paleontology, v. 35, p. 467-474.

______, 1964, Middle Bajocian Ammonites from the Cook Inlet Region, Alaska: United States Geological Survey Professional Paper 418-B, 61 p.

______, 1967, Twin Creek Limestone (Jurassic) in the Western Interior of the United States: United States Geological Survey Professional Paper 540, 105 p.

______, 1968, Lower Jurassic (Pliensbachian and Toarcian) Ammonites from Eastern Oregon and California: United States Geological Survey Professional Paper 593-C, p. C1-C51.

______, 1973, Middle Jurassic (Bajocian) Ammonites from Eastern Oregon: United States Geological Survey Professional Paper 756, 100 p.

______, 1980, Jurassic Paleobiogeography of the Conterminous United States in its Continental Setting: United States Geological Survey Professional Paper 1062, 134 p.

______, 1981, Early Jurassic Ammonites from Alaska: United States Geological Survey Professional Paper 1148, 49 p.

______, 1984, Jurassic Ammonite Succession in North America and Biogeographic Implications: This volume.

Imlay, R.W. and Dettermann, R.L., 1973, Jurassic Paleobiogeography of Alaska: United States Geological Survey Professional Paper 801, 34 p.

Imlay, R.W. and Jones, D.L., 1970, Ammonites from the *Buchia* Zones in Northwestern California and Southwestern Oregon: United States Geological Survey Professional Paper 647-B, 59 p.

Irving, E., Monger, J.W.H., and Yole, R.W., 1980, New Paleomagnetic Evidence for Displaced Terranes in British Columbia: in: Strangway, D.W., ed., The Continental Crust and Its Mineral Deposits: Geological Association of Canada Special Paper 20, p. 441-456.

Irving, E. and Yole, R.W., 1972, Paleomagnetism and the Kinematic History of Mafic and Ultramafic Rocks in Fold Mountain Belts: Earth Physics Branch Publication 42, No. 3, p. 87-95.

Jeletzky, J.A., 1965, Late Upper Jurassic and Early Lower Cretaceous Fossil Zones of the Canadian Western Cordillera, British Columbia: Geological Survey of Canada Bulletin 103, 70 p.

Jones, D.L., Silberling, N.J., and Hillhouse, J.W., 1977, Wrangellia – A Displaced Terrane in Northwestern North America: Canadian Journal of Earth Sciences, v. 14, p. 2565-2577.

______, 1978, Microplate Tectonics of Alaska – Significance for the Mesozoic History of the Pacific Coast of North America: in: Howell, D.G., and McDougall, K.A., eds., Mesozoic Paleogeography of the Western United States: Society of Economic

Paleontologists and Mineralogists, Pacific Coast Paleogeography Symposium, v. 2, p. 71-74.

Jones, P.B., 1980, Evidence from Canada and Alaska on Plate Tectonic Evolution of the Arctic Ocean Basin: Nature, 285, p. 215-217.

Lees, E.J., 1934, Geology of the Laberge Area, Yukon: Royal Canadian Institute Transactions, v. 20, p. 1-48.

MacKevett, Jr., E.M., 1970, Geologic Map of the McCarthy C-5 Quadrangle: Geologic Quadrangle Maps of the United States, Map GQ-844.

McLearn, F.H., 1928, New Jurassic Ammonoidea from the Fernie Formation, Alberta: Geological Survey of Canada Bulletin 49, p. 19-22.

———, 1929, Stratigraphic palaeontology: in Mesozoic Palaeontology of Blairmore Region, Alberta: National Museum of Canada Bulletin 58, p. 80-107.

Monger, J.W.H. and Irving, E., 1980, Northward Displacement of North-Central British Columbia: Nature, v. 285, p. 289-294.

Monger, J.W.H., Price, R.A., and Tempelman-Kluit, D.J., 1982, Tectonic Accretion and the Origin of the Two Major Metamorphic and Plutonic Welts in the Canadian Cordillera: Geology, v. 10, p. 70-75.

Muller, J.E., 1977, Evolution of the Pacific Margin, Vancouver Island and Adjacent Regions: Canadian Journal of Earth Sciences, v. 14, p. 2062-2085.

Muller, S.W. and Ferguson, H.G., 1939, Mesozoic Stratigraphy of the Hawthorne and Tonopah Quadrangles, Nevada: Geological Society of America Bulletin, v. 50, p. 1573-1624.

Nichols, K.M. and Silberling, N.J., 1979, Early Triassic (Smithian) Ammonites of Paleoequatorial Affinity from the Chulitna Terrane, South-Central Alaska: United States Geological Survey Professional Paper 1121-B.

Packer, D.R. and Stone, D.B., 1974, Paleomagnetism of Jurassic Rocks from Southern Alaska, and the Tectonic Implications: Canadian Journal of Earth Sciences, v. 11, p. 976-997.

Panuska, B.C. and Stone, D.B., 1981, Late Paleozoic Paleomagnetic Data from Wrangellia: Resolution of the Polarity Ambiguity: Nature, v. 293, p. 561-563.

Repin, Y.S., 1977, Early Jurassic Ammonites of the North Pacific Realm: in Krassilov, V.A., ed., Organic Evolution in the Circum-Pacific Belt, Chronological and Paleobiogeographic Boundaries: Academy of Sciences, USSR, Far-Eastern Scientific Centre, Institute of Biology and Pedology, p. 123-138.

Sanborn, A.F., 1960, Geology and Paleontology of the Southwest Quarter of the Big Bend Quadrangle, Shasta County, California: California Division of Mines, Special Report 63, p. 3-26.

Silver, L.T. and Anderson, T.H., 1974, Possible Left Lateral Early to Middle Mesozoic Disruption of the Southwestern North American Craton Margin: Geological Society of America, Abstracts with Programs, v. 6, p. 955.

Smith, P.L., 1980, Correlation of the Members of the Jurassic Snowshoe Formation in the Izee Basin of East-Central Oregon: Canadian Journal of Earth Sciences, v. 17, p. 1603-1608.

———, 1983, The Pliensbachian Ammonite *Dayiceras dayiceroides* and Early Jurassic Paleogeography: Canadian Journal of Earth Sciences, v. 20, p. 86-91.

Speed, R.C., 1979, Collided Paleozoic Microplate in the Western United States: Journal of Geology, v. 87, p. 279-292.

Stone, D.B. and Packer, D.R., 1977, Tectonic Implications of Alaska Peninsula Paleomagnetic Data: Tectonophysics, v. 37, p. 183-201.

———, 1979, Paleomagnetic Data from the Alaska Peninsula: Geological Society of America Bulletin, v. 90, p. 545-560.

Stone, D.B., Panuska, B.C., and Packer, D.R., 1982, Paleolatitudes Versus Time for Southern Alaska: Journal of Geophysical Research, v. 87, p. 3697-3707.

Sturani, C., 1971, Ammonites and Stratigraphy of the *"Posidonia alpina"* Beds of the Venetian Alps (Middle Jurassic, Mainly Bajocian): Memorie degli Instituti di Geologia e Mineralogia dell' Università di Padova, 27, 190 p.

Sutherland Brown, A., 1968, Geology of the Queen Charlotte Islands, British Columbia: British Columbia Department of Mines and Petroleum Resources Bulletin 54, 226 p.

Sykes, R.M. and Callomon, J.H., 1979, The *Amoeboceras* Zonation of the Boreal Upper Oxfordian: Palaeontology, v. 22, p. 839-903.

Symons, D.T.A., 1971, Paleomagnetic Notes on the Karmutsen Basalts, Vancouver Island, British Columbia: Geological Survey of Canada Paper 71-24, p. 10-24.

———, 1973, Paleomagnetic Results from the Jurassic Topley Intrusions near Endako, British Columbia: Canadian Journal of Earth Sciences, v. 10, p. 1099-1108.

Taylor, D.G., 1980, Jurassic (Bajocian) Ammonites from the Snowshoe Formation, Suplee Area, Oregon: Cordilleran Section, Geological Society of America 76th Annual Meeting (Abst.), Abstracts with Program, v. 12, p. 155.

Tipper, H.W., 1981, Offset of an Upper Pliensbachian Geographic Zonation in the North American Cordillera by Transcurrent Movement: Canadian Journal of Earth Sciences, v. 18, p. 1788-1792.

Tipper, H.W. and Richards, T.A., 1976, Jurassic Stratigraphy and History of North-Central British Columbia: Geological Survey of Canada Bulletin 270, 73 p.

Tozer, E.T., 1970, Marine Triassic faunas: in Geology and Economic Minerals of Canada: Geological Survey of Canada, Economic Geology Report No. 1, p. 663-640.

———, 1982a, Marine Triassic Faunas of North America: Their Significance for Assessing Plate and Terrane Movements: Geologische Rundschau, v. 71, p. 1077-1104.

———, 1982b, Late Triassic (Upper Norian) and Earliest Jurassic (Hettangian) Rocks and Ammonoid Faunas, Halfway River and Pine Pass Map Areas, British Columbia: in Current Research Part A, Geological Survey of Canada Paper 82-1A, p. 385-391.

Westermann, G.E.G., 1964a, The Ammonoid Fauna of the Kialagvik Formation at Wide Bay, Alaska Peninsula, Part I, Lower Bajocian (Aalenian): Bulletins of American Paleontology, v. 47, p. 325-503.

———, 1964b, Occurrence and Significance of the Arctic *Arkelloceras* in the Middle Bajocian of the Alberta foothills (Ammonitina, Jurassic): Journal of Paleontology, v. 38, p. 405-409.

———, 1969, the Ammonoid Fauna of the Kialigvik Formation at Wide Bay, Alaska Peninsula, Part II, *Sonninia Sowerbyi* Zone (Bajocian): Bulletins of American Paleontology, v. 57, 226 p.

———, 1979, *Alaskinia* nom.nov.for *Alaskoceras,* Westermann, 1969 non Miller and Kummel, 1945, Jurassic Ammonitina: Journal of Paleontology, v. 52, p. 604.

———, 1981, Ammonite Biochronology and Biogeography of the Circum-Pacific Middle Jurassic: in House, M.R., and Senior, J.R., The Ammonoidea: Systematic Association Special Volume No. 18, p. 459-498.

———, 1983, The Upper Bajocian and Lower Bathonian (Jurassic) Ammonite Faunas of Oaxaca, Mexico and West-Tethyan affinities: Paleontologia Mexicana, v. 46 (for 1981), 63 p.

———, 1984, The Late Bajocian *Duashnoceras* Association (Jurassic Ammonitina) of Mixtepec in Oaxaca, Mexico: Congreso Latinamericano de Paleontologia Memoria III, p. 192-199.

Westermann, G.E.G., and Riccardi, A.C., 1976, Middle Jurassic Ammonite Distribution and Affinities of the Andean Fauna:

Primer Congreso Geológico Chileno, 1, p. C23-C39.

_____, 1980, Jurásico Medio en Peru: Instituto de Géologia Mineralogia y Metalurgi Boletin 9, 47 p.

Westermann, G.E.G., and Riccardi, A.C., 1976, Middle Jurassic Ammonite Distribution and Affinities of the Andean Fauna: Primer Congreso Geológico Chileno, 1, p. C23-C39.

_____, 1980, Jurásico Medio en Peru: Instituto de Géologia Mineralogia y Metalurgi Boletin 9, 47 p.

Whiteaves, J.F., 1889, On some Cretaceous Fossils from British Columbia, and North West Territory and Manitoba: Geological Survey of Canada, Contributions to Canadian Paleontology, v. 1, p. 151-196.

Wiedmann, J., 1970, Über den Ursprung der Neoammonoideen. Das Problem einer Typogenese: Eclogae Geologicae Helvetiae, v. 63, p. 923-1020.

Yancey, T.E., 1975, Permian Marine Biotic Provinces in North America: Journal of Paleontology, v. 49, p. 758-766.

Yole, R.W., and Irving, E., 1980, Displacement of Vancouver Island: Paleomagnetic Evidence from the Karmutsen Formation: Canadian Journal of Earth Sciences, v. 17, p. 1210-1228.

Jurassic-Cretaceous Biochronology and Paleogeography of North America,
edited by G.E.G. Westermann,
Geological Association of Canada Special Paper 27, 1984

A REVIEW OF THE BIOSTRATIGRAPHY OF THE POST-LOWER BAJOCIAN JURASSIC AMMONITES OF WESTERN AND NORTHERN NORTH AMERICA

J.H. Callomon
University College London, Gower Street, London WC1E 6BT, U.K.

ABSTRACT

From the beginning of the Upper Bajocian, Jurassic North America straddled two strongly differentiated ammonite faunal realms, the East-Pacific Realm and the Boreal Realm. These in turn overlap so little with the Tethyan Realm, in which the primary chronostratigraphic ammonite zonations have been defined, that direct correlations between North America and the standard European successions continue to be almost impossible in most of the Middle Jurassic. This has led to much uncertainty, the identifications of ammonites and their assigned ages being frequently locked in circular argument. The taxonomy and biostratigraphy of the ammonites are, therefore, reviewed in the light of what has been learned in the last 30 years of their distribution and evolution both in North America and in the rest of the world. Four regions or domains are distinguished: A, the epicratonic Western Interior, from Wyoming to Alberta; B, the mobile Cordillera, from northern California to the Alaska Peninsula; C, epicratonic arctic Canada, including the Archipelago; and D, the Alaskan North Slope. Some 80 ammonite faunas are identified and arranged in four parallel time-ordered sequences, ranging from Upper Bajocian to the top of the Jurassic. Their correlations, among themselves and with standard successions in the Arctic and in Europe, are discussed and indicated as far as the evidence allows. The conclusions are: 1) The Middle Jurassic ammonites of region A formed a distinct Western Interior faunal province within the Pacific Realm. 2) With one exception; all the Middle Jurassic faunas of the Western Interior Province are of Upper Bajocian – Bathonian age, including the ones previously assigned to the Callovian: the Callovian is almost totally unrepresented. 3) The Upper Bajocian and Bathonian are similarly strongly represented in the Cordillera, region B, Lower Callovian being well-developed only in southern Alaska. 4) The succession in region C, arctic Canada, is closely similar to that of the rest of the Boreal Province as typified by East Greenland, with again an extensive (Boreal) Bathonian sequence and only intermittent evidence of Callovian. 5) Despite apparent faunal disparities, the succession in region D, the North Slope, has close affinities with that of adjacent arctic Canada, region C. 6) Comparison of ammonite distributions within and between domains A and B shows them to be consistent with possible lateral tectonic displacements of accretionary terranes in the Cordillera since the Middle Jurassic of between zero and a maximum of perhaps 2000 km northwards for the Alaska Peninsula relative to the continent. 7) Similarly, strike-slip displacements of northern Alaska relative to the Laurentian Shield are unlikely to have exceeded a few hundred km and could also have been zero.

RÉSUMÉ

A partir du début du Bajocien supérieur, l'Amérique du Nord se partageait entre deux domaines où les faunes d'ammonites étaient fortement distinctes; il s'agissait du Domaine Est Pacifique et du Domaine Boréal. Ceux-ci, à leur tour, ne se superposent que si peu au Domaine Téthysien, où les premières zonations chronostratigraphiques d'ammonites furent définies, que les corrélations directes entre les successions de référence en Europe et celle d'Amérique du Nord continuent à être presque impossibles pour la plus grande part du Jurassique moyen. Ceci a conduit à une grande incertitude, l'identification des ammonites et leurs âges supposeés étant fréquemment l'objet de démonstrations circulaires. La taxinomie et la biostratigraphie des ammonites sont donc révisées ici à la lumière de ce que les dernières 30 années nous ont appris sur leur distribution et leur évolution tant en Amérique du Nord que dans le reste du monde. Quatre régions ou domaines sont distingués: A, le "Western Interior" intracratonique s'étendent du Wyoming à l'Alberta; B, la Cordillère mobile, de la Californie Septentrionale à la Péninsule d'Alaska; C, le Canada arctique, intracratonique, comprenant l'Archipel; et D, le Versant Nord de l'Alaska. Quelques 80 faunes d'ammonites ont été identifiées et disposées en quatre séquences chronologiques parallèles qui s'étendent du Bajocien supérieur au sommet du Jurassique. Les corrélations, aussi bien entre elles qu'avec les successions de référence de l'Arctique et d'Europe, sont discutées et établies dans la mesure des preuves disponibles. Les conclusions suivantes sont tirées: 1) Les ammonites du Jurassique moyen de la région A constituent une province du Western Interior, bien distincte au sein du Domaine Pacifique. 2) A l'exception d'une seule, toutes les faunes du Jurassique moyen de la Province du Western Interior sont d'âge Bajocien supérieur – Bathonien, y compris celles précédemment attribuées au Callovien; cet étage n'est pratiquement pas repésenté. 3) De même, le Bajocien supérieur et le Bathonien sont largement représentés dans la Cordillère (région B), le Callovien inférieur n'étant bien développé qu'en Alaska méridional. La succession dans la région C, le Canada arctique, est très semblable à celle du reste de la Province Boréale comme elle a été établie dans l'Est du Groënland, avec encore une séquence très répandue du Bathonien boréal et, seulement, des indications sporadiques de Callovien. 5) Malgré ses apparentes particularités fauniques, la succession de la région D, le Versant Nord, a des affinités étroites avec celle du la région C, contigüe, du Canada arctique. 6) La comparaison des distributions d'ammonites au sein et entre les domaines A et B montre qu'il y a compatibilité avec des mouvements tectoniques horizontaux de zones d'accrétion dans la Cordillère à partir du Jurassique moyen; l'amplitude des mouvements, vers le Nord, de la Péninsule de l'Alaska par rapport au continent peut être évaluée entre zéro et, peut être 2000 km. 7) De même, les décrochements de l'Alaska septentrional par rapport au Bouclier Laurentien n'ont pas dû dépasser quelques centaines de km et ont pu aussi bien être nuls.

INTRODUCTION

Jurassic rocks have been recorded from almost innumerable localities in the western United States, Canada and Alaska, and listed and reviewed by Imlay in his masterly survey of 1980. Many ammonities have been found and described, but the majority came from scattered localities and were collected by field parties not primarily interested in ammonoid biostratigraphy. The number of places in which collections have been made carefully bed by bed in recorded sections is therefore very small. To work out the correct sequence of ammonite faunas and even the associations of genera and species occuring together in them has been very difficult, and is in many cases still uncertain. We have had to rely to a considerable extent on indirect correlations, mainly lithostratigraphical – the positions of finds within a formation, with luck at levels that could be located relative to the base or top of the formation. While this works quite well in the shallow-water epicratonic sediments east of the Rockies, in Wyoming, Montana and Alberta, in which relatively thin formations can be followed with little change over distances of hundreds of kilometres, it is often barely possible in the Cordilleran deposits whose facies are almost wholly volcanoclastic and often turbiditic, laid down in tectonically unstable basins. Fossils tend to be scattered throughout formations whose thicknesses can be enormous and rapidly varying, either in depositional wedges or through erosion. Lithologic correlations at the level of ammonite faunal horizons may therefore be reliable over distances no greater than a few kilometres. To collect bed by bed from such successions therefore may never be possible. Nevertheless, great progress has been made, and much of the ammonite sequence can now be pieced together with some confidence. The picture that emerges is as follows.

Correct identification of Lower Jurassic rocks presents no serious problems, for the ammonites they contain resemble those from the well-known standard successions of Europe quite closely – almost always to generic, sometimes even to specific level. Occurrences are scattered and successions locally highly incomplete, but somewhere or other almost every part of the Lias has been recognized. In the Aalenian (Lower Bajocian *olim* and *sensu* U.S. and Canadian Geological Surveys), more and more specifically Pacific elements enter the faunas, extending from the Andes to the Alaskan Peninsula, but broad correlations can be continued upwards as far as the top of the Lower Bajocian (Middle Bajocian *olim* and *sensu americano*), Humphriesianum Zone, whose *Stephanoceras*, ranging from the Andes through Oregon, Alberta, the Queen Charlotte Islands, the Talkeetnas to the Alaskan Peninsula, bear an astonishing resemblance to those from Dorset and the Swabian Jura. (For recent reviews see Westermann and Riccardi, 1976, 1979; Hall and Westermann, 1980; and Westermann, 1981). Thereafter, faunal similarities with Europe abruptly cease.

The next higher ammonites found both in South and North America came to be described during the 1920s to 1930s in the heyday of the Buckman-Spath schools of taxonomy, and almost every new ammonite found, no matter how scrappy, had to have its own new generic name: a) *Neuqueniceras* Stehn, 1924 (Andes), b) *Yakounites* McLearn, 1927 (Queen Charlotte Islands), c) *Yakounoceras* McLearn, 1927 (Queen Charlotte Islands), d) *Eurycephalites* Spath, 1928 (Andes), e) *Xenocephalites* Spath, 1928 (Andes), f) *Miccocephalites* Buckman, 1929 (Alberta), g) *Paracephalites* Buckman, 1929 (Alberta), h) *Metacephalites* Buckman, 1929 (Alberta), i) *Paracadoceras* Crickmay, 1930 (B.C.), and j) *Lilloettia* Crickmay, 1930 (B.C.).

Nevertheless, the affinities to known groups seemed fairly clear. The second and third [b) and c)], alias *Seymourites* Kilian and Reboul, 1909, were close to *Kepplerites* from the Lower Callovian of Europe; (a) appeared close to *Reineckeia* from the Middle Callovian of Europe; and the rest strongly resembled *Macrocephalites* from the Lower Callovian, together with *Reineckeia* the only non-American genera to be widely cited from South America. The ages of the beds containing these faunas were thus firmly ascribed to the Callovian; ergo, immediately succeeding as they did the Lower Bajocian, the whole of the Upper Bajocian and Bathonian had to be missing in North and South America. The situation in the Arctic appeared to be similar. There, Palaeozoic, Liassic or Aalenian rocks were followed by sequences containing *Cranocephalites*, *Arctocephalites*, *Arcticoceras* and *Kepplerites* also firmly ascribed to the Callovian. Such large and globally widespread non-sequences coincided with a Bathonian developed in northwest Europe often in highly condensed or even non-marine facies. Thus arose the doctrine of the great world-wide Bathonian regression, followed by the equally great world-wide Callovian transgression, summarized by Arkell (1965) and both still widely cited today, often to be found in curves of eustatic sea-level as a function of time.

Our knowledge of these Middle Jurassic faunas has been immensely enriched in the last 30 years almost entirely through the untiring efforts of Ralph Imlay and Hans Frebold. Many new faunas from many horizons have been described, yet the similarities with the classical Bathonian and Callovian faunas of Europe have become little closer. New forms now had to be given new generic names out of necessity, to emphasize the faunal differences within the broad similarities. The study of the ammonites was still in the analytical phase (Wright, 1981). It became clear that the main reason for the dissimilarity of American, Arctic and European ammonite successions in the Middle Jurassic was the development of strong faunal provincialism (e.g., Imlay, 1965). At the end of the Lower Bajocian, the differentiation that had begun mildly in the Aalenian became complete and the ammonites segregated into three almost non-overlapping faunal realms: the ancestral Tethyan Realm, with its adjacent epicontinental domains including the Northwest European Province with its rich ammonite faunas that form the primary world standard of Jurassic zonation; the Pacific Realm, including all the circum-Pacific domains from New Guinea via New Zealand through the western cordilleras of South and North America to the Sea of Okhotsk and Japan; and the Arctic or Boreal Realm, from the North Slope of Alaska through the Sverdrup Basin, north and east Greenland, Spitsbergen, the Petchora, through the breadth of north Siberia as far as Kolyma.

It seems an opportune moment to attempt another review of the post-Lower Bajocian ammonites of North America. We begin with a brief systematic review of the classification of the relevant ammonites at generic level and up. There has been progress in this field independently, for besides the recognition of faunal provincialism a lot has been learned in recent years about phyletic relationships among Middle and Upper Jurassic ammonites generally (Callomon, 1981a, b). Together with the recognition of dimorphism, this does much to simplify and bring order into what has grown into a daunting catalogue of generic names.

Next, the published occurences of ammonites are analyzed into distinguishable assemblages, each of which characterizes a *faunal horizon.* Attempts then to arrange these faunal horizons into time-ordered sequences reveals the presence in western North America, between central California and the Arctic Islands, of four distinct post-Lower Bajocian parallel faunal successions each characteristic of a separate *faunal domain.* Even the correlation of these domains still presents problems for the overlaps are at present tenuous – either because of lack of information, or because they no longer exist. Conversely, the possible reasons for this are interesting and topical, certainly in part due to faunal provincialism (primary), but perhaps also the results of subsequent macrotectonic events now manifested in disjointed terranes (secondary causes).

Finally, to avoid some of the circular arguments of the past, the correlation of the North American successions with the standard successions of Europe is left to the end. The basic difficulty remains. Despite many new discoveries, there are very few faunal elements between Lower Bajocian and Oxfordian these regions have in common. Neither the Bajocian-Bathonian nor the Bathonian-Callovian boundaries as defined in Europe can be directly recognized in North America, either by ammonites or as yet by any other fossil group. The Bathonian-Callovian boundary however can now be located fairly precisely through indirect correlations via the Arctic. Together with the discovery of what are now known to be undoubtedly Upper Bajocian and Bathonian ammonites in both South and North America, the previously held picture of the distribution of Bathonian and Callovian rocks will need considerable modification. It transpires that most of the formerly "Callovian" ammonites of North America are Upper Bajocian and Bathonian, and that Callovian proper is in turn very scantily represented in many areas, if at all.

SUMMARY OF AMMONITE SYSTEMATICS

The post-Bajocian ammonite faunas to be considered here, together with the numerous localities at which one or more of their elements have been found, are distinctly characteristic of four regions, or *domains.* It seems useful to introduce such a term to describe regions delimited by present-day geographical boundaries, chosen quite possibly wholly or in part for more or less arbitrary reasons of convenience, in contradistinction to faunal realms and provinces, tectonic terranes, and depositional basins, all of which may have contributed to the distribution of fossil faunas as found to-day but in ways usually very imperfectly known and yet to be determined. It corresponds roughly to the 'areas' described on a broader circum-Pacific scale by Westermann and Riccardi (1976). A domain in this sense, therefore, may be part of several faunal provinces and incorporate several terranes and basins. The four domains in North America to be distinguished here all lie north of the 37th parallel, and the Jurassic faunas of southern California, Mexico and the Gulf states, which fall into separate categories, will not be discussed. The domains are:

A) *Western Interior:* Utah – Colorado – Black Hills of Dakota – Wyoming – Alberta and eastern British Columbia: Rocky Mountains and Foothills as far as the Peace River (56°N). Includes numerous more or less ephemeral basins, including the Wind River, Bighorn, Williston and Fernie Basins.

B) *Cordillera:* Sierra Nevada of east-central California – east central Oregon – south, central and north-west British Columbia – Vancouver Island – Queen Charlotte Islands – Wrangell Mountains – Cook Inlet – Alaska Peninsula. Includes numerous disjoint tectonic units, terranes and their parts, and a mosaic of rapidly changing, unstable depositional basins.

C) *Arctic Canada:* northern Yukon Territory (Porcupine River around Old Crow, Richardson Mountains) – Arctic Islands (Prince Patrick to Ellesmere). (Continues to east Greenland – northern Russia – northern Siberia).

D) *Alaskan North Slope.*

Of these, the first three are of considerable extent. The epicratonic deposits of the Western Interior can be followed over more than 2000 km – the distance from the Sierra Nevada of California to the Alaska Peninsula is 3000 km and from the epicratonic deposits of the northern Yukon to those of the Petshora in northern Russia is 4000 km. The fourth region is small and is differentiated here only because its succession, as far as it is known, makes it uncertain whether its affinities are closest with those of B or C, and in the context of tectonic discussions this question could be vital.

Leaving aside minor differences, the uniformity of the faunal assemblages and their successions within each of these major regions is remarkable. In contrast, the differences between the regions range from modest to extreme. The problem under discussion is the possible origins of these similarities and differences. Three causes could be invoked:

1) Differences in ages. As indicated in the introduction, the successions known today are still very incomplete. It could have been the case, therefore, that highly intermittent regional sedimentary cycles in each of regions A and B, for instance, were responsible for the widespread extent and correlation of the faunas found within them, but that there was a mismatch of cycles between the regions so that the faunal correlation now found between them is correspondingly low on grounds of non-overlapping ages.

2) Differences arising from faunal provincialism, e.g., between the Pacific and Boreal Realms discussed in the introduction.

3) Differences across geographic boundaries observed today arising from plate-tectonic displacements, i.e., the juxtaposition of disparate fossil faunas by the subsequent movement of one of them from its original place of entombment.

Before assessing the possible importance and contributions of these factors it is necessary to summarize the faunas briefly. We take up the successions immediately above the last horizon to be identifiable world-wide, with ubiquitous *Stephanoceras* marking the top of the Lower Bajocian, Humphriesianum Zone, and begin with a short systematic review to make clear the interpretation of the generic names used. Ammonite taxa whose authors are not listed in the references may be found in the *Treatise* (Arkell, 1957). Dimorphs [M] and [m]: macro- and microconchs.

Superfamily STEPHANOCERATACEAE Neumayr, 1875

The main branches of this major group of Middle – Upper Jurassic ammonites are represented diagrammatically in Figure 1 (after Callomon, 1981a).

Family SPHAEROCERATIDAE Buckman, 1920

The derivation of this family from the Otoitidae and its subsequent blossoming world-wide in the Lower Bajocian is not of immediate relevance here. Of interest, however, is the phylogenetic splitting that began in the Humphriesianum Zone at the top of the Lower Bajocian. The family divided into at least three biogeographically separate branches, two of which are of major importance in North America. One branch, the Sphaeroceratinae *sensu stricto*, lingered on with little change in the northern Tethys up to about the Garantiana Zone of the Upper Bajocian (Sturani, 1971). Another branch, apparently restricted at first to the Eastern Pacific Province of the Pacific Realm, developed into the Eurycephalitinae. The third branch, probably also originating in the eastern Pacific, invaded the Arctic and then dominated the Boreal Province of the Boreal Realm as the Cardioceratidae up to the Kimmeridgian of the Upper Jurassic.

Subfamily SPHAEROCERATINAE Buckman, 1920

1a. *Sphaeroceras* BAYLE, 1878

Type species *Amm. brongniarti* J. Sowerby, 1817. Holotype, refigured by Arkell (1952, p.77, Text-Fig.20.2a,b), from the Bajocian of Normandy; microconch.

1b. *Chondroceras* MASCKE, 1907

Type species *Amm. gervillei* J. Sowerby, 1818. Holotype, refigured by Arkell (1952, p.77, Fig.20.3a,b) and Westermann (1956, Pl.1, Figs.1a,b), also from Normandy; microconch.

European forms of these two groups have been profusely illustrated by Westermann (1956) and Sturani (1971). Their general dimorphism was subsequently recognized by Westermann (1964c), and it was clearly illustrated by Sturani (1971, p.147, Fig.44), who could also study extensive collections from successive well-resolved stratigraphical horizons. By common consent, the differences between the two European genera were small. In *Sphaeroceras brongniarti* the coiling is globular and the umbilicus so minute as to be occluded. In *Chondroceras gervillei* the coiling is more open and less inflated. Even more evolute and planulate forms were grouped by Westermann into a third genus-group taxon, *Schmidtoceras*. The precise relations between these groups were not very clear. Westermann assumed that *Chondroceras* and *Sphaeroceras* differed in age, but Sturani showed that this was not so. Although the ages of the types of the type species from Normandy will probably never be determined absolutely precisely, topotypes occur together in the same bed in the lower Humphriesianum Zone, the Conglomérat de Bayeux (cf. Rioult, 1964). The extensive collections from northern Italy contain forms that are morphologically typical of all three genera ranging from the Romani Subzone (the lowest) of the Humphriesianum Zone, Lower Bajocian, through the whole of the Subfurcatum Zone of the Upper Bajocian. The generic taxa were therefore relegated to subgeneric status, but classification at specific level continued to be based on purely morphological criteria and *Sph. brongniarti* and *Ch. gervillei* were retained as distinct. Since then, large new collections have become available from the Romani Subzone of the Humphriesianum Zone of Sherborne, Dorset (Parsons, 1976, p.131, bed 4b). These contain both typical *Sph. brongniarti* and *Ch. gervillei* as well as intermediate forms, and it seems impossible to draw a dividing line between them. They give the impression of being but a single, highly variable biospecies in the which the typical morphology of the macroconchs is close to that of *Chondroceras wrighti* Buckman (1881; 1923, Pl.415) and the range of variability extends continuously from *Sphaeroceras brongniarti* as the one, involute end-member to *Schmidtoceras orbignyanum* (Wright, 1860) (cf. Westermann, 1956, p.74, Pl.5, Fig.6) as the other, evolute one. Similarly, although not precisely, the same ranges of variability appear then to be the rule at the higher levels also, giving the "form-genera" comparable parallel vertical ranges. On such an interpretation even the type species of *Sphaeroceras* and *Chondroceras* are synonymous, and the retention of *Chondroceras* and *Schmidtoceras* could be justified only in a purely morphological sense on grounds of convenience. Such ranges of variability seem to be typical of many other sphaeroconic relatives of *Sphaeroceras* also, e.g., *Cranocephalites borealis*, the first of the Cardioceratidae, and *Megasphaeroceras rotundum*, the first of the Eurycephalitinae.

2. *Sphaeroceras* (*Defonticeras*) MCLEARN, 1927 (including *Saxitoniceras* McLearn, 1927).

Type species *D. defontii*. This group has recently been revised by Hall and Westermann (1980). It is abundant in western North America,

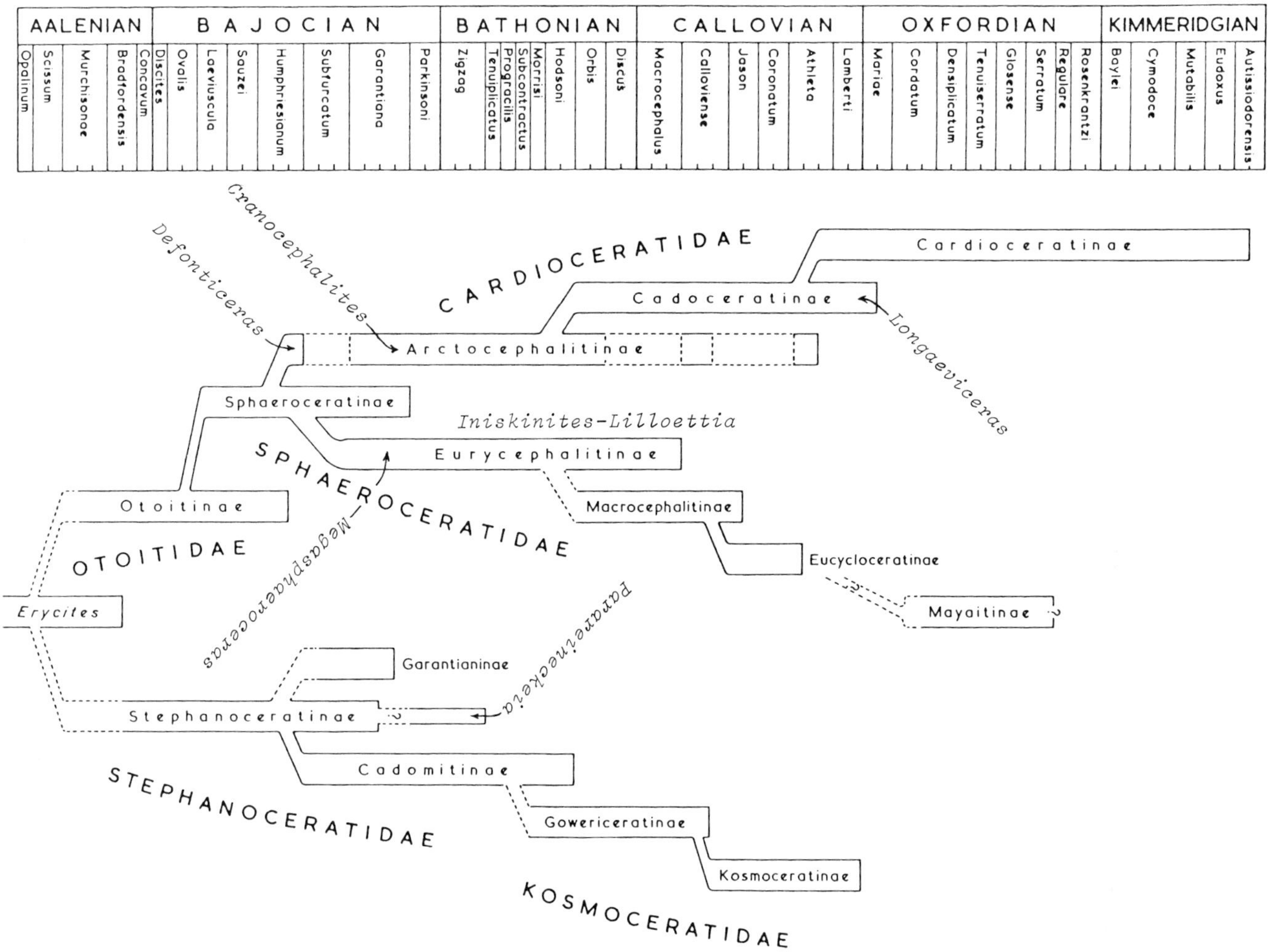

Figure 1. Phylogenetic classification of the Stephanocerataceae. The time-scale at the left is calibrated in units of European standard chrono-Subzones.

both in the Cordillera (region B) from Oregon to Alaska and in the Rockies of the Western Interior (region A); but it appears to be unknown elsewhere. A number of slightly different faunas can be recognized, reflecting small differences in age, but the range of the group appears to extend little beyond the equivalents of the Humphriesianum Zone as indicated by the associated stephanoceratids and other rare elements. One of these faunas, that of *Spaeroceras oblatum* (Whiteaves), has been chosen as index of an Oblatum Zone in a zonation of the Bajocian of North America (Westermann and Riccardi, 1979, p.112; Hall and Westermann, 1980, p.19), equivalent to the upper part of the Humphriesianum Zone.

Morphologically, the American assemblages differ from the European ones of the same age in being larger, having less strongly modified adult peristomes, and having somewhat more strongly varicostate macroconchs with characteristicly coarse but subdued ribbing on the body chamber. These differences are small but quite consistent; the ranges of variability barely overlap. They hardly merit separation at generic rank and were simply incorporated as separate species in a single comprehensive genus *Chondroceras* by Hall and Westermann. If the problem of variability in the European assemblages is resolved in the manner discussed above, however, the appropriate generic taxon would be *Sphaeroceras*. As the European and American assemblages differ in the same consistent fashion over a range of levels, equivalent to at least two ammonite Subzones, the distinctions appear to be more than geographically subspecific in a biological sense, i.e., already phylogenetic albeit at an early stage. The appropriate taxonomic distinction seems to be at subgeneric level, therefore, retaining *Defonticeras* as a geographical subgenus of *Sphaeroceras* in a North Cordilleran Province of the East Pacific Realm. (Isolated reports of *Chondroceras* from the Lena Basin of Siberia, e.g., *Ch. sphaericum* and *Ch. custodium* Tuchkov, 1972, are most probably based on misidentifications of *Cranocephalites* spp. of the *pompeckji* group).

Subfamily EURYCEPHALITINAE Thierry, 1976 [including Paracephalitinae Tintant and Mouterde, 1982]

Successors of *Sphaeroceras* in the Pacific Realm: the dividing line is to some degree arbitrary. The systematic relations between the numerous American "genera" still need to be fully evaluated so that it is necessary to retain a lot of separate names for the time being. Their relation to South American *Eurycephalites* itself has to await a similar revision of the succession in the Andes (A.C. Riccardi, in prep.). Their close affinities are revealed e.g., by the fact that many of them share the same microconch "genus", the highly characteristic *Xenocephalites* Spath, 1928 – the micronconch *sensu stricto* of *Eurycephalites*. Homoeomorphisms are common so that generic identifications without stratigraphical evidence can be hazardous and misleading.

The following names, listed in order of priority, are available. Interpretation is based strictly on type specimens of type species.

1. *Eurycephalites* SPATH, 1928 [M]
Type species *Macrocephalites vergarensis* Burckhardt, 1903, from Mendoza. Assigned by Imlay to the Bathonian because it resembled the European *Morrisiceras:* it was subsequently removed to the Callovian during the period in which Bathonian was regarded as unrepresented in the Andes (Arkell, 1956; Westermann and Riccardi, 1976). More recently the genus has been cited as ranging from late Bajocian to early Callovian (Westermann, 1981), but pending a full systematic and stratigraphical revision the use of the name for North American fauna seems of questionable value.

2. *Xenocephalites* SPATH, 1928 [m]
Type species *Macrocephalites neuquenensis* Stehn, 1924. The microconch of *Eurycephalites* in South America; but also used extensively in North America for forms that are microconchs of other "genera", including *Lilloettia, Paracephalites* and *Iniskinites partim.*

3. *Paracephalites* BUCKMAN, 1929 [M] [incl. *Metacephalites* Buckman, 1929 *Miccocephalites* Buckman, 1929 (see revision by Frebold, 1963), and *Oligocadoceras* Meledina, 1977 [m]]
Includes the group of "*Cadoceras*" *muelleri* Imlay, *piperense* Imlay, *tetonense* Imlay (all [M]) and *shoshonense* Imlay ([m], type species of *Oligocadoceras,* = *Xenocephalites*?)

4. *Lilloettia* CRICKMAY, 1930 [M] incl. *Buckmaniceras* Crickmay, 1930
Type species *L. lilloetensis,* from an isolated horizon of uncertain stratigraphical position in southern British Columbia. The use and meaning of this generic name also has oscillated among its close homoeomorphs, as synonym of *Arctocephalites* (Spath, 1933, p.878), or subgenus of *Macrocephalites* (Arkell, 1956, p.541) and, more recently, as subgenus of *Eurycephalites* (Westermann and Riccardi, 1976; Westermann, 1981, and this volume). Its supposed age has ranged from Middle Bathonian (Westermann, 1981) to Middle Callovian (e.g., Imlay, 1953b, Frebold and Tipper, 1967), but the age now proposed for it is early Lower Callovian. The microconchs include typical *Xenocephalites.*

5. *Megasphaeroceras* IMLAY, 1961a [M]
Type species *M. rotundum,* from southern Alaska. The microconch appears plausibly to be *Sphaeroceras talkeetnanum* Imlay (1962b). As a pair their morphological affinities are closest with Tethyan *Sphaeroceras* s.s. as most probable ancestors, rather than with North Pacific *Sph.* (*Defonticeras*). They have the fine, sharp, dense ribbing and, in the case of the mircoconch, the strongly collared and hooded peristome of *Sphaeroceras* s.s.. They differ principally in the much larger size. This contrasts with the coarse but subdued ribbing of adult *Defonticeras.* They could, therefore, still be included in *Sphaeroceras* of the Sphaeroceratinae as a subgenus in the same way as *Defonticeras* has been. *Megasphaeroceras* appear sharply early in the Upper Bajocian along the whole of the eastern border of the Pacific, from southern Alaska through Oregon (Imlay, 1973) to Peru (Westermann, 1981), Chile (Hillebrandt, 1970), possibly Mendoza (Hall and Westermann, 1980) and Antarctica (Quilty, 1970), – a distribution that characterizes the East-Pacific Realm of Westermann (1981). It is succeeded by numerous morphologically similar groups that constitute the Eurycephalitinae, initially confined to the same area, and is most probably their root. Because of this affinity and its complete biogeographical segregation from the coeval *Sphaeroceras* s.s. of the Tethys and Arctocephalitinae of the Boreal Province, it seems most appropriate to draw the dividing line at the Lower-Upper Bajocian boundary and to include *Megasphaeroceras* already in the Eurycephalitinae as their first member.

6. *Warrenoceras* FREBOLD, 1963 [M]
The involute, smooth compressed forms originally recorded as *Arcticoceras* (*henryi* Meek and Hayden, 1865); microconchs again typical *Xenocephalites* ("*Arcticoceras*" *crassicostatum, loveanum* Imlay, 1953a).

7. *Imlayoceras* FREBOLD, 1963 [M]
Used so far only for a single isolated fauna from the Miette-Rock Lake area of the Rocky Mountains, Alberta (53°-53½°N). Possibly only an inflated *Lilloettia* or a coarse-ribbed *Iniskinites.*

8. *Eocephalites* IMLAY, 1967 [M]
Questionably distinct from *Megasphaeroceras;* early forms, probably still Upper Bajocian.

9. *Parachondroceras* IMLAY, 1967 [M] [incl. *Sohlites* Imlay, 1967 [M]]
Local fauna in the Western Interior, probably still of Upper Bajocian age (see fauna A2).

10. *Iniskinites* IMLAY, 1975 [M]
The densely-ribbed forms with minute umbilici, often sphaeroconic, originally recorded as *Kheraiceras.*

11. *Chinitnites* IMLAY, 1975 [M]
The microconchs of *Iniskinites;* closely related to, if not synonymous with, *Xenocephalites.*

12. *Talkeetnites* IMLAY, 1980b [M]
Probably merely the inflated end-members of the variational range of a group, the most typical variants of which are still usually recorded as *Cranocephalites* Spath, although the resemblance is probably homoeomorphic ("*Cr.*" *costidensus* Imlay; see below, faunas B4 and D1).

13. *Tuxednites* IMLAY, 1980b [m]
The microconch of *Talkeetnites;* also closely related to *Xenocephalites.*

Family CARDIOCERATIDAE Siemiradzki, 1891
Subfamily ARCTOCEPHALITINAE Meledina, 1968

1. *Cranocephalites* SPATH, 1932 [M] and [m] [incl. *Boreiocephalites* Meledina, 1967, and *Pachycephalites* Meledina, 1973]
The first of the Arctocephalitinae – *Cranocephalites borealis* (Spath) – appeared throughout the whole of the Boreal Province as close to simultaneously as one can tell, and these early forms bear such a strong morphological similarity to *Sphaeroceras* (*Defonticeras*) that a close and direct derivation seems highly likely (see Fig. 2, and fauna C1 below). Thereafter the distribution and development of the subfamily appears to have been purely Boreal, with the possible exception of one group occurring in the Cordillera and on the Alaskan North Slope, that of "*Cranocephalites*" *costidensus* Imlay and "*Cr.*" *ignekensis* Imlay (see discussion below, faunas B4 and D1).

2. *Arctocephalites* SPATH, 1928 [M] and [m]
Boreal successors of *Cranocephalites.*

3. *Arcticoceras* SPATH, 1924 [M] [including *Costacadoceras* Rawson, 1982 [m].
Boreal successors of *Arctocephalites.* Shares its microconch genus *Costacadoceras* with those of *Arctocephalites* and early *Cadoceras.*

Subfamily CADOCERATINAE Hyatt, 1900

The transition from *Arcticoceras* to *Cadoceras* took place in the Boreal Province, where it is gradual – the dividing-line being to some extent arbitrary. The diagnostic character of *Cadoceras* is a sharp edge on the umbilical margin of at least the adult macroconch body-chamber. This immediately distinguishes it from many forms that have been assigned to it in the past, e.g., "*Cadoceras*" *muelleri,*

Figure 2. The morphological affinities of the Sphaeroceratinae and the Arctocephalitinae. A,a: *Sphaeroceras (Defonticeras) oblatum* (Whiteaves), Oblatum Zone (equivalent to the upper Humphriesianum Zone), Canada. B,b: *Cranocephalites borealis* (Spath), Borealis Zone, East Greenland. (*A*: macroconch; R.L. Hall collection, Fernie Formation, Ribbon Creek, Alberta. McMaster J1836b, figured Hall and Westermann, 1980, Plate 14, Figures 3a,b. *a*: microconch; R.L. Hall collection, Yakoun Formation, Queen Charlotte Is., B.C. McMaster J1794a, figured Hall and Westermann, 1980, Figure 8. *B*: macroconch; more than usually strongly ribbed variant retaining ribbing on the mature bodychamber, Vardekloft Formation, Jameson Land, East Greeland. JHC collection 1368. *b*: microconch, typical; same locality and collection, JHC 1206a).

pipersense, etc., now placed in the Eurycephalitinae. There are numerous generic names available, and as their relations to each other are well understood, representing in many cases the successive evolutionary stages of an evolving lineage, they can be incorporated in classification at subgeneric level.

1. *Cadoceras* FISCHER, 1882

1a. *C. (Catacadoceras)* BODYLEVSKY, 1960 [M]

The earliest forms, including the American *C. barnstoni* (Meek 1859), and *C. variabile* Spath, 1932, from the Boreal Upper Bathonian in east Greenland.

1b. *C. (Paracadoceras)* CRICKMAY, 1930 [M] [including *Streptocadoceras* Meledina, 1977]

The status of this name continues to be highly unsatisfactory. The holotype of the type species (*P. harveyi* Crickmay, 1930, Pl.16, Figs.1,2), although complete, appears to be a juvenile; the peristome is distorted, and what looks like a terminal constriction indicating maturity is not real. Neither do the last visible sutures show the approximation and simplification otherwise so common in the Cadoceratinae. *C. (P.) harveyi* could therefore be the immature stages of any of several forms widely distributed in the Cordillera, including many of the ones described by Imlay from Alaska (1953, *PP 249*-B), e.g., *C. glabrum* (cf. Pl.37, Fig.2), *multiforme, chisikense,* etc. This is close to the interpretation followed by Imlay himself. Amplification of the generic name, which unfortunately is the oldest available for macroconchs after *Cadoceras* itself, must therefore be based on the subjective choice of one such species showing the adult stages so necessary for closer characterisation of the Cadoceratinae. *Cadoceras glabrum* Imlay seems as good as any. The characteristic features of the subgenus as thus interpreted are fairly evolute inner whorls with rounded venter becoming inflated and depressed only gradually, developing the sharp umbilical edge of the genus only on the middle whorls, and retaining a very wide and open umbilicus to the end. The differences between *C. (Paracadoceras)* and *Cadoceras sensu stricto* are only ones of degree, and no sharp line can be drawn between them. On the whole, *C. (Paracadoceras)* is more serpenticone and relatively small when adult, whereas *Cadoceras* s.s. tends to become cadicone and large. Density and persistence of ribbing can be quite variable in either. Such an interpretation of *Paracadoceras* fully encompasses *Cadoceras subtenuicostatum* Voronets, 1962, the type species of *Streptocadoceras* Meledina, 1977. Both in Europe (*Cadoceras breve* Blake, 1905) and in east Greenland, *C. (Paracadoceras)* occurs in the lowest part of the Lower Callovian and is succeeded by *Cadoceras* s.s. in the Calloviense Zone, upper Lower Callovian.

1c. *C. (Cadoceras)* FISCHER, 1882 [M] [including *Bryocadoceras* Meledina, 1977]

1d. *C. (Stenocadoceras)* IMLAY, 1953b [M]

Inner whorls compressed, involute, subcarinate, heralding the transition to *Longaeviceras* and hence the Cardioceratinae, the umbilicus widening and developing sharp edges only in the middle and outer whorls. Middle Callovian.

1e. *C. (Rondiceras)* TROITSKAYA, 1956 [M]

Similar development to that of *Stenocadoceras,* from which it may be doubtfully distinct; used so far only for some species found on the Russian Platform.

1f. *C. (Pseudocadoceras)* BUCKMAN, 1918 [m] [including *Novocadoceras* SAZONOV, 1965 [m].]

The microconchs of all the subgenera of *Cadoceras;* the type species, *Ps. boreale,* from the Koenigi Subzone of the Lower Callovian, is unusually involute and subcarinate. The dividing line between *Pseudocadoceras* and its predecessors, *Costacadoceras* Rawson, 1982 (type species *C. bluethgeni,* [m] of *Arcticoceras ishmae*) is arbitrary.

2. *Longaeviceras* BUCKMAN, 1918 [M] and [m]
Retaining the sharp umbilical edge of *Cadoceras* but developing the style of ribbing and incipient carina of the later Cardioceratinae. Upper Callovian in Europe and the Arctic.

Subfamily CARDIOCERATINAE Siemiradzki, 1891

The morphological variability of the family reaches its peak in *Cardioceras.* Past classifications have almost invariably been "vertical", in terms of morphological form-genera or subgenera of varying, overlapping ranges apparently evolving in parallel. These include *Vertebriceras* (most strongly ribbed), *Subvertebriceras, Cawtoniceras, Maltoniceras* (with vestigial and strongly differentiated secondary ribbing), *Scoticardioceras* (smooth and discoidal), *Pachycardioceras* and *Goliathiceras* (smooth and inflated). Precise ages are hard to pin down on the strength of only one or two specimens, but given sufficient material at any one level the range of forms found together usually spans two or more of these (sub)genera continuously in a way that can be highly diagnostic stratigraphically. As the classification of the American representatives of the subfamily raises no special problems, however, it need not be considered further in detail here.

Family STEPHANOCERATIDAE Neumayr, 1875
Subfamily STEPHANOCERATINAE Neumayr, 1875

1. *Parareineckeia* IMLAY, 1962a [M] & [m]
A North American Cordilleran local group, probably derived from the otherwise quite ordinary Stephanoceratinae that lived on in the Pacific after they had vacated the western Tethys and Europe, where they were replaced by Cadomitinae. *Steph.* (*Normannites*) (alias *Dettermanites* Imlay, 1961a) occurs widely in the Cordilleran Upper Bajocian *Megasphaeroceras rotundum* Zone. *Parareineckeia* [m] differs little from *Normannites* (cf. Imlay 1980b, *PP 1091*, Pl.10, Fig. 1). Westermann (1981, p.488), in contrast, considers it to belong to the Cadomitinae.

2. *Lupherites* IMLAY, 1973 [m] and *Domeykoceras* HILLEBRANDT, 1977 [M]
The systematic position of these forms has been discussed by Westermann and Riccardi (1979, p.174). *Lupherites* was described originally from Oregon but its occurrence together with *Domeykoceras* in the Andes suggests strongly that the two are merely a dimorphic pair. Their horizon must lie close to the Lower-Upper Bajocian boundary, for they occur both in Oregon and in the Andes in close proximity to the *Leptosphinctes-Spiroceras-Megasphaeroceras* fauna, at present taken as the lowest in the Upper Bajocian (see fauna B1, below). But no absolute association appears to have been recorded, so that their age could still be topmost Lower Bajocian.

Family KOSMOCERATIDAE Haug, 1887
Subfamily GOWERICERATINAE Buckman, 1926
[incl. Keppleritinae TINTANT, 1963]

In East Greenland the earliest forms appear suddenly in the Boreal Upper Bathonian, but in North America they may appear even somewhat earlier. The earliest forms in both regions are small, very finely and densely ribbed, and round-whorled at all stages of growth, with no trace of ventral flattening either in macro- or microconchs. Adult bodychambers are relatively inflated and short, little more than half a whorl, and the early forms could therefore be equally well retained in the Cadomitinae of the Stephanoceratidae, from which they must certainly have been derived. No sharp dividing-line can be drawn on purely morphological grounds. The division appears to have been a biogeographical one, for at the same time as the early Kosmoceratidae colonized North America and East Greenland, there to evolve continuously into and throughout the Callovian, the original Cadomitinae persisted independently in the Tethys and its margins until well into the Upper Bathonian. At specific level coeval *Kepplerites* and *Cadomites* are quite distinct and there appears to have been no intercommunication.

Although regarded classically as members of the Boreal Realm *par excellence,* the Kosmoceratidae never in fact populated the truly northerly reaches of the Boreal Province proper. Thus, the Gowericeratinae flourished in western North America, southern Alaska and East Greenland; but they became rare even in the northern Yukon, Spitsbergen and the Petshora and are unknown from the Boreal Bathonian of Siberia. Later, in the Middle and Upper Callovian, the Kosmoceratinae at times dominate the ammonite faunas of the Sub-Boreal Province of north-west Europe and the Russian Platform, but became minor elements in East Greenland and are absent now both from North America and the whole of the Arctic, their place there being taken by *Longaeviceras* of the Cadoceratinae. The southern margins of the distribution of the Kosmoceratidae are often very sharp. Isolated reports of their occurrence very much further south should therefore be treated with great caution, particularly in view of the ease with which they could be confused with more or less homoeomorphic members of other branches of the Stephanoceratidae.

1. *Kepperites* NEUMAYR and UHLIG, 1892 [M]
From their sudden first appearance in the Bathonian of North America or the Boreal Upper Bathonian of East Greenland these early Kosmoceratidae changed very little until nearly the top of the Lower Callovian. What changes did take place, e.g., the transition from round whorl-section to tabulate venter bordered by tubercles with concomitant coarsening of the secondary ribbing, occurred so gradually that no sharp boundaries can be drawn. Most of the many names of the past can therefore be abandoned, or retained at most at subgeneric level.

1a. *K.* (*Kepplerites*) [M] [=*Seymourites* (Kilian Reboul, 1909; *Cereriiceras* Buckman, 1922; *Galilaeiceras* Buckman, 1922; *Galilaeanus* Buckman, 1922; *Galilaeites* Buckman, 1922; *Yakounites* McLearn, 1927; *Yakounoceras* McLearn, 1927].

1b. *K.* (*Gowericeras*) Buckman, 1921 [M] [=*Toricelliceras* Buckman, 1922]

1c. *K.* (*Toricellites*) BUCKMAN, 1922 [m]
Used for the microconchs of the whole genus. The earliest members are still round-whorled. Tabulate sections appeared in the Upper Bathonian, with flat venters which the secondary ribs cross undiminished (e.g., "*Cosmoceras*" *vigorosum* Imlay, *alaskanum* Imlay, see below); and ventro-lateral tubercles bordering a ventral furrow which may be smooth, *Kosmoceras*-like, appeared in the Lower Callovian.

Younger members of the family need not be considered further here.

Superfamily PERISPHINCTACEAE
Family PERISPHINCTIDAE Steinmann, 1890
Subfamily LEPTOSPHINCTINAE Arkell, 1950

1. *Cobbanites* IMLAY, 1962a [M] and [m], = *Leptosphinctes* (*Vermisphinctes*) Buckman, 1920
The type species, *C. talkeetnanus* Imlay, is morphologically so close to well-known European forms from the Garantiana Zone of the late Upper Bajocian (e.g., *L.* (*V.*) *meseres* (Buckman, 1927; pl.446A,B)

– itself probably a synonym of *L.* (*V.*) *pseudomartinsi* (Siemiradski, 1899)), that its age and affinities can hardly be in doubt. The name *Cobbanites* has come since to be applied to almost all other Perisphinctidae that have been found in the North American Middle Jurassic, but most of these can also be confidently reassigned to previously well-known genera, e.g., *Procerites* or *Choffatia* (see below under Zigzagiceratinae and faunas A3-A6, B7). It appears that the Middle Jurassic Perisphinctidae of North America are all representatives of rather general, pandemic groups and developed few if any local faunas. Conversely, they make useful indicators of age within the somewhat broad ranges characteristic of perisphinctids.

Subfamily ZIGZAGICERATINAE Buckman, 1920 [incl. Siemiradzkiinae Westermann, 1958, and Gracilisphinctinae Besnosov, 1982]

1. *Epizigzagiceras* FREBOLD, 1973 [M] (in Frebold and Tipper, 1973).
This genus appears to have all the characteristics of the Zigzagiceratinae, particularly those of the earliest forms: compare *E. evolutum* Frebold, 1973, Pl. 1, Fig. 2, 4 with *Zigzagiceras* (*Franchia*) *arkelli* Sturani, 1967, Pl. 18, Fig. 2,3, from the basal subzone of the Bathonian of south-east France.

The only other forms from North America that have been assigned to Zigzagiceratinae close to *Zigzagiceras* itself are those described by Burckhardt (1927) from Oaxaca in Mexico as Middle Bajocian *Stephanoceras floresi* and *St.* aff. *brodiae* (Pl.12, Figs.10-16, Figs.18-20). They were re-identified as Lower Bathonian *Zigzagiceras* by Arkell (1956, p.564) and this identification was recently upheld (Imlay, 1980; Westermann, 1981) until the present revision by Westermann (*in* Taylor *et al.*, this volume) based on new field work. "*S.*" *floresi* and relatives are very early Upper Bajocian stephanoceratids as also classified by Sturani (1967, p.50). The material as a whole bears a very close resemblance to the early Upper Bajocian *Stephanoceras* from southern Alaska described recently by Imlay (1982a, *PP. 1189*, e.g., Pl.5, Figs.6,8; Pl.4, Fig.4). The perisphinctid nuclei (Burckhardt 1927, Pl.11, Figs.5-8) identified by Arkell as *Siemiradzkia* or *Planisphinctes* could equally well be *Leptosphinctes* (cf. Imlay 1982a, Pl.4, Figs.5-7). The forms developing the extremes of coronate ribbing found in *Zigzagiceras* are untypical of the Zigzagiceratinae as a whole and are otherwise known only from a quite restricted area of sub-Mediterranean (but not Tethyan) Europe. Reassignment of the Mexican forms to *Stephanoceras* s.s. would remove what otherwise is a serious anomaly, with its strong implications of a direct marine link between Oaxaca and the western Tethys via the proto-Atlantic in the Lower Bathonian (Westermann, 1981, p.476). In the more typical representatives of the Zigzagiceratinae such as *Procerites*, coronate ribbing rarely persists beyond the rudimentary state on the inner whorls. It is an unfortunate historical accident that the atypical *Zigzagiceras* should have given its name to a whole subfamily otherwise lying directly on the smooth evolutionary path of the Perisphinctidae between the more typical Leptosphinctinae and Pseudoperisphinctinae.

2. *Procerites* SIEMIRADZKI, 1898.
Some at least of the perisphinctids from North America seem to be safely assignable to this genus, e.g., *P. engleri* Frebold, 1957 (Pl.39; Pl.40, Fig.1) – the latter finding its closest match with *P.* (*Gracilisphinctes*) of the Middle Bathonian of Europe. The microconch is *P.* (*Siemiradzkia*) Hyatt. (Frebold, 1957, Pl.40, Fig.1).

3. *Choffatia* SIEMIRADZKI, 1898.
This genus, too, appears to be represented in North America, e.g., *Ch. irregularis* Imlay, 1953 (*PP 249-B*, Pl.54) from Alaska. In Europe the transition from *Procerites* to *Choffatia* is gradual. *Choffatia* appears as a minor element in the Middle Bathonian, and *Procerites* persists barely into the early Upper Bathonian.

Superfamily HAPLOCERATACEAE Zittel, 1984

Members of the family Oppeliidae are found occasionally in all three regions A-C in north America but are too rare and poorly understood to have much stratigraphical value at present. Lissoceratidae are moderately common in the Upper Bajocian of the Cordillera, but the same remarks apply.

Superfamily PHYLLOCERATACEAE Zittel, 1884 Family PHYLLOCERATIDAE Zittel, 1884

Members of this family occur widely in north America but beyond formal descriptions, their value for stratigraphical and palaeogeographical purposes tends to be dismissed as minimal indicating little more than the former extent of the Tethyan Realm latitudinally bounded roughly by the tropics. One of the discoveries of recent times, however, has been their barely diminished occurrence in parts of the world far removed from the tropics even during the Jurassic – sometimes associated with truly Boreal ammonites such as the Arctocephalitinae or Cadoceratinae in the northern hemisphere. From this and other evidence it seems more probable that the phylloceratids differed from most other ammonites in being pelagic ocean-dwellers rather than shelf-dwellers, and that the significance of their distribution today is as indicators of proximity to former oceans rather than of low palaeolatitudes. Their common occurrence in, for example, the Jurassic of Oaxaca, parts of the north American Cordillera and in southern Alaska reflected the presence nearby of the Pacific Ocean, then as now. Away from the oceanic margins their frequency diminishes rapidly, so that although they are still relatively common in the northern Yukon they have become rarities in north-eastern Siberia (Voronets, 1962) and the Western Interior of the U.S., and are practically unknown in the rest of the Arctic including East Greenland. Their rarity in the epicontinental deposits of, for example, Caracoles, Mendoza, and Neuquen in the Andean Cordillera, now barely 200 km from the ocean, immediately suggests that these areas may have been considerably further from the Pacific during the Jurassic. A lot of continental crust appears to have been lost since. It seems that a general re-appraisal of the world-wide distribution of the whole suborder Phylloceratina could be palaeogeographically illuminating.

THE AMMONITE SUCCESSIONS IN NORTH AMERICA

The more important faunas that can be recognized are reviewed briefly. No attempt is made to use them to construct chronostratigraphical scales of zones; rather, they may be regarded as characterising faunal horizons of greater or lesser importance whose relations to each other remain in most cases to be clarified. Nevertheless, it is convenient to label each fauna/horizon with the name of one characteristic species as an index. Important accessory elements will be indicated and brief notes will draw attention to other points of interest. The faunal successions are summarized in Figures 3-5.

Region A. Western Interior (Dakota-Wyoming-Utah-Montana-Alberta)

Fauna A1. Eocephalites primus IMLAY (1967)
\+ *Megasphaeroceras* cf. and aff. *rotundum*
Spiroceras sp
Stephanoceratinae indet.

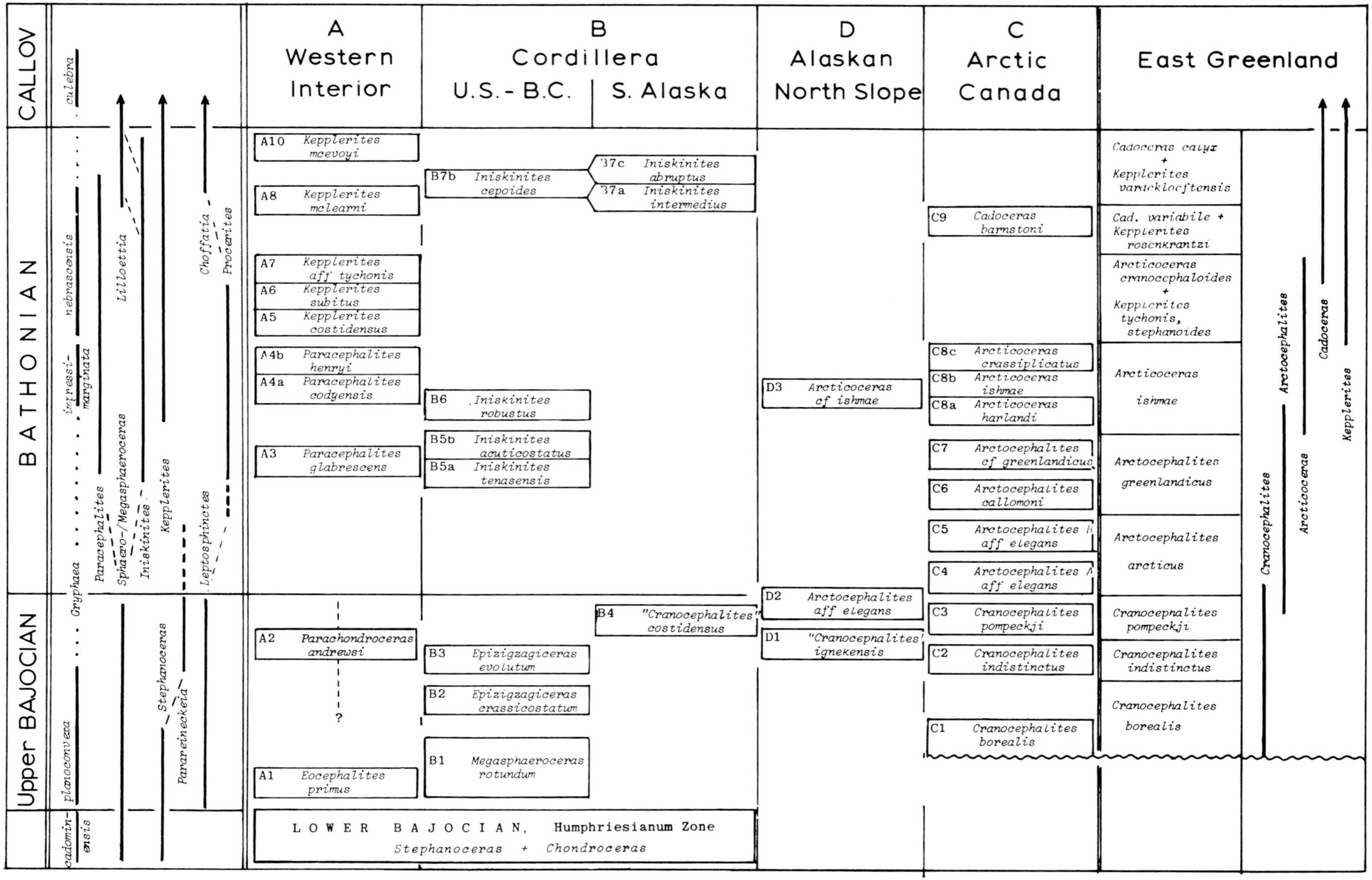

Figure 3. The successions of ammonite faunas in North America: Upper Bajocian – Bathonian. Note that the boxes represent no more than time-ordered sequences in any column; neither vertical widths nor spacings are proportional to time-durations. Boxes in adjacent columns shown as partly or wholly overlapping are meant to suggest that their faunas probably differ little in age, although their precise order is usually not known. Correlations are discussed further in the text.

Utah-Idaho-Wyoming; Upper Bajocian; close if not exact equivalents of faunas with *Megasphaeroceras* in the Cordillera. A fauna of about this age has also recently been found in the foothills of south Alberta (Hall and Stronach, 1981).

Fauna A2. Parachondroceras andrewsi IMLAY (1967)
Utah-Wyoming-Idaho-Montana; possibly still Upper Bajocian, but no independent indicators for the fauna is known only from the Western Interior. *Parachondroceras* and *Sohlites* seem generically doubtfully distinct, but there is evidence to suggest that the forms may extend over some range stratigraphically; further horizons may, therefore, be recognizable one day.

The precise age of this fauna is conjectural. Imlay still favoured Upper Bajocian, largely because of the gradational transition between the Rich Member of the Twin Creek Limestone containing it and the underlying Sliderock Member with fauna A1. However, such evidence is not compelling, and fauna A2 could be younger, perhaps already Bathonian.

Fauna A3. Paracephalites glabrescens BUCKMAN (Frebold 1963; Imlay 1962) (= *P. sawtoothensis* Imlay, 1962)
+ *Xenocephalites saypoensis* Imlay [m]
Procerites ("*Cobbanites*") cf. or aff. *engleri* Frebold.

Montana, Sawtooth Formation and Alberta, *Corbula munda* Beds (lowest member of the Grey Beds), Fernie Group. This is the first of five more or less successive faunas that are closely related, characterized by large, smooth *Paracephalites* with its associated proportionally large *Xenocephalites* micronconchs (*Oligocadoceras* Meledina). They give the impression of spanning only a short interval of time, separated from the last fauna (A2) known below them, by a considerable gap. Age presumably Lower Bathonian.

Fauna A4(a) Paracephalites (*Warrenoceras*) *codyensis* IMLAY (1953; Frebold 1957, 1963) (incl. *P. rierdonensis* Imlay, *imlayi* Frebold)
+ *Xenocephalites crassicostatum* Imlay [m]
Kepplerites sp. aff. *tychonis*
Procerites ("*Cobbanites*") *engleri* Frebold [M] and [m]

Montana, lower Rierdon Formation, and Alberta, middle Grey Beds of the Fernie, especially the Gryphaea Bed of Blairmore.

Fauna A4(b) Paracephalites (*Warrenoceras*) *henryi* (MEEK AND HAYDEN, 1865) (Imlay 1953; 1967, p.60)
South Dakota-Wyoming. Variants of fauna 4a are so close to the true *P. henryi* that they have been recorded as such by Frebold from the *Gryphaea* Bed.

Fauna A5. Kepplerites costidensus (IMLAY) (1953)
+ *Paracephalites muelleri, tetonensis, piperensis* (Imlay) [M]
Xenocephalites shoshonensis (Imlay) [m]
Procerites sp. [M]
P. (*Siemiradzkia*) *warmdonensis* (Imlay) [m]

These early, finely and densely-ribbed, small, round-whorled species of *Kepplerites* bear a strong resemblance to the first *Kepplerites* to occur in East Greenland, *K. stephanoides* sp. MS, in the lower Cranocephaloides Zone, which marks the beginning of the Boreal Upper Bathonian there. Montana, basal Rierdon Formation.

Fauna A6. Kepplerites ["*Gowericeras*"] *subitus* (IMLAY) (1948, 1953)
+ *K.* (*Toricellites*?) *costicrassum* (Imlay) [m] *partim*
Paracephalites muelleri, tetonensis etc. [M]
Xenocephalites shoshonensis [m]
Procerites [M] and [m] ("*Grossouvria*")

This fauna of *Kepperites* is quite distinct from the finely-ribbed forms found below and above, unlike any known from elsewhere, and seems to be a local one. *Gowericeras* has strongly tabulate and tuberculate inner whorls, and is probably much younger. Montana, Rierdon Formation.

Fauna A7. Kepplerites aff. *tychonis* (RAVN) (Imlay 1953)
+ *Xenocephalites* sp.

Another small, densely ribbed species resembling the forms in A5, and the same remarks apply. Montana, middle Rierdon Formation, and Alberta, upper Grey Beds of the Fernie, 20 ft above the *Grypaea* Bed (Frebold 1963, Pl.9, Fig.3)

If the resemblance of the *Kepplerites* of North America to those of east Greenland is not coincidental and indicates affinity, the ages of faunas A4-7, and perhaps A3 as well, are equivalent to early Upper Boreal Bathonian. Similarly, the Proceritids would indicate a correspondence to Middle-early Upper Bathonian of Europe.

Fauna A8. Kepplerites mclearni IMLAY [M] (1948, 1953) (incl. *K. rockymontanus, planiventralis* (Imlay)
+ *K.* (*Toricellites*) *knechteli* (Imlay [m] (incl. *K.* (*T.*) *zortmanensis, vigorosus* (Imlay))
+ *Lilloettia* sp. [M]
Xenocephalites phillipsi Imlay [m]

These singular giant *Kepplerites* with their long bodychambers, occupying the last 1¼ whorls, would be readily mistaken if found in isolation with their almost perfect homoeomorphs, *Stephanoceras* ex gr. *rhytum* of the Lower Bajocian: only their compressed tabulate-ventered *Kosmoceras*-like micronconchs reveal their true nature. Only known in North America from the upper Rierdon Formation of Montana and its equivalents in Jasper Park and sub-surface in Saskatchewan (Frebold, 1963), they give the impression of being yet another local fauna.

It seems once again more than coincidence that at a single precisely analogous level in Greenland a closely similar assemblage of giant *Kepplerites* occurs: *K. antiquus* Spath and *K. peramplus* Spath (1932) with hitherto undescribed microconchs very similar to those from America (cf. Callomon and Birkelund, 1980, Pl.1 Fig.4). The position of this fauna in Greenland is in the lower Calyx Zone, the highest Zone of the Boreal Upper Bathonian.

Fauna A9. Lilloettia? ["*Imlayoceras*"] *miettensis* (FREBOLD) (1963)
Found so far only in Alberta near Jasper Park, but allegedly above fauna A8 (Frebold, 1963, p.31)

Fauna A10? Kepplerites mcevoyi (McLEARN) (1928, Pl.4)
Only the holotype is known, from an unknown level in the Foothills of Ribbon Creek, southern Alberta (see Frebold, 1957, p.64). The generally fine ribbing and rounded venter suggests an age predating *K. keppleri* and its supposed American equivalents, *K. loganianum* of fauna B8(a) (q.v., below), and hence still Bathonian. It has close matches in the Kepplerit-ids of the *Cadoceras barnstoni* fauna of the Yukon, C9, and in the Calyx Zone of east Greenland (cf. Spath, 1932, Pl.25,

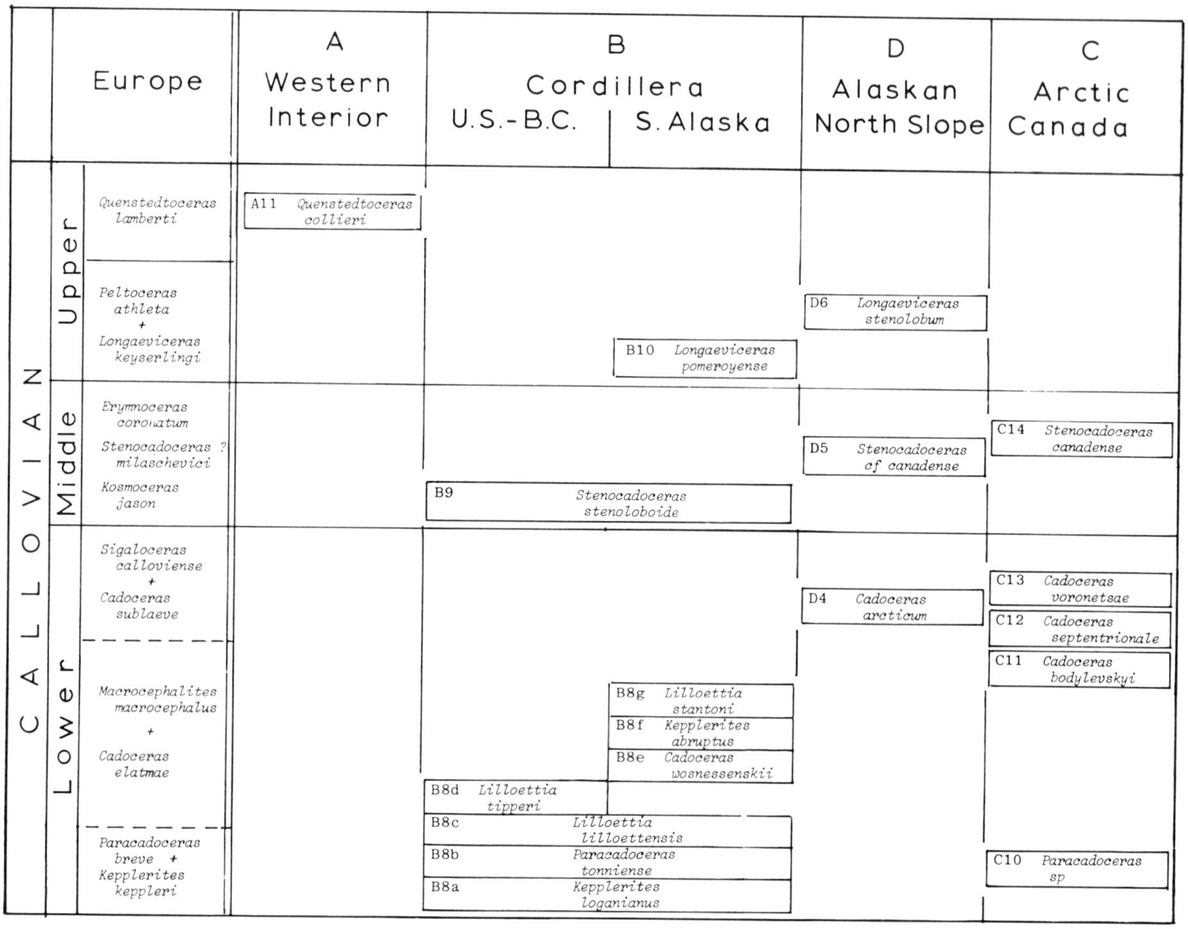

Figure 4. Callovian (see captions and note for Fig. 1).

Fig.1), both of which are still placed in the Upper Boreal Bathonian. Its level relative to faunas A7 and A8 is not yet known.

– non-sequence

Fauna A11. Quenstedtoceras collieri (Reeside, 1919)
+ *Peltoceras* sp. (Imlay 1982, Pl.2, Figs. 3-6, ?7,8, as *Quenst.* cf. *omphaloides*)
Perisphinctes (*Prososphinctes*) sp. (Imlay, 1982b, Pl.26, Figs.8-10)

Montana, basal Swift Formation; Upper Callovian, Lamberti Zone, probably lower part.

– Oxfordian:

Faunas A12 – A16: Quenstedtoceras and *Cardioceras* spp.

The rich faunas of Cardioceratidae from the Western Interior described by Reeside (1919) and most recently by Imlay (1982b) have been discussed by Arkell (1956, p.549-50) who pointed out their remarkable resemblance to analogous faunas of northern Europe. Equally remarkable is the monogeneric character of the faunas: unlike those of Europe, the Cardioceratidae occur almost without accessory elements of any other family, e.g., Perisphinctidae or Oppeliidae. They have therefore a purely Boreal stamp, and this appears to apply wherever they are found in the whole of North America. The shells occur at several levels in the Redwater Shales of the Sundance Formation of Wyoming and its correlatives from the Canadian border in Montana to northern Utah (see Imlay, 1982b). It is rare however that they are sufficiently abundant at any locality to establish the range of variability of an assemblage, much less to work out the succession. Descriptions have had to be primarily morphological, and the current classification is very much "vertical", expressed in range-charts of subgenera of *Cardioceras*. Stratigraphically Reeside's material was almost wholly unlocalized and unfortunately has to form the basis for the majority by far of the specific names available and in use today. It seems possible now to sort it into a rather few assemblages guided by the range of variability to be expected by analogy with the well-known faunas of Europe. The variability of cardioceratid biospecies is large, highly characteristic and almost unchanging from the Upper Callovian to the Middle Oxfordian: at every level one can find a morphologically continuous spectrum extending from compressed involute forms tending to be carinate (*Lamberticeras – Scoticardicoeras*), to highly inflated, more evolute, round-whorled forms (*Eboraciceras – Pavloviceras – Goliathiceras*). Equally characteristic is the form of the dimorphism. In what follows, the names of species refer only to the type specimens in the first instance, and to what extent these assemblages really are homogeneous in age remains to be seen. The four that are distinguished here correspond roughly to separate Zones or Subzones in Europe. By today's standards this is still very coarse, compared for instance with the twenty or so faunal horizons of cardoceratids that can now be recognized in the Lower to Middle Oxfordian of Europe. However, it is sufficient to provide a first frame of reference. The American faunas are as follows.

Fauna A12. Cardioceras cordiforme (MEEK HAYDEN, 1859) [M] (incl. *Quenstedtoceras hoveyi, subtimidum, suspectum, tumidum* Reeside, *Card. albaniense, auroraensis, bellefourchense, crassum, incertum, russeli* Reeside, all [M])
+ *Cardoceras latum* Reeside [m].

Lower Oxfordian, Mariae Zone, Scarbugense Subzone. *C. cordiforme*, as exemplified by the lectotype, and the oldest name available, happens to be the compress end-member of the series. The transition from "*Quenstedtoceras*" to "*Cardioceras*" within the series is almost exactly as found in the Mariae Zone of England (e.g., Arkell 1939, Woodham: *Qu. omphaloide* – *Qu. mariae* – *Qu. woodhamense* – *Card. scarburgense* all imperceptibly intergrading). The variability in America is more striking because of the presence of abundant well-preserved macroconchs, which in Europe are rare. Several of the American forms are very close to *C. scarburgense*. Arkell regarded some of them as still belonging to *Qu.* (*Eboraciceras*) of the Lamberti Zone, Upper Callovian, but the aspect of the assemblage as a whole favours the slightly younger age. Wyoming, Montana, Alberta.

Fauna A13. Cardoceras martini REESIDE, 1919 ([m?], Alaska)
(a) *Cardioceras reesidei* MARIE, 1938 (= *C. cordiforme*, Reeside 1919 *partim*, Pl.8, Figs.4,5, holotype [m]; Figs.6,7 [m]) (incl. *C. whitfieldi* Reeside ?)
(b) *Cardioceras mountjoyi* Frebold, 1959 [M]

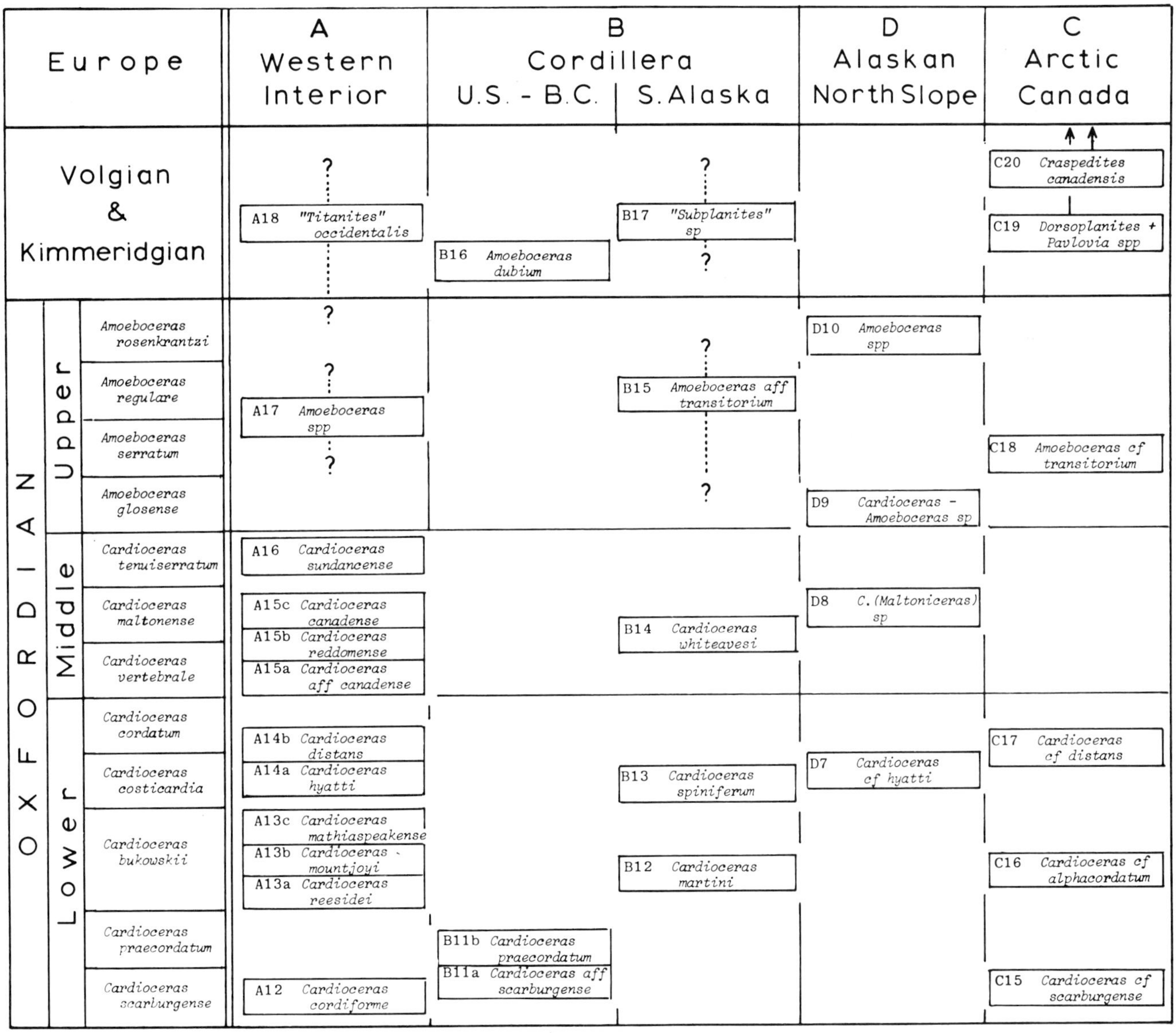

Figure 5. Upper Jurassic (see captions and note for Fig. 1).

(c) *Cardoceras mathiaspeakense* Imlay, 1982b [M]
Lower Oxfordian, Cordatum Zone, Bukowskii Subzone. The resemblance of these forms to those of the Bukowskii Subzone of Europe so profusely illustrated by Maire (1938) and Arkell (1946) is very close. Some variants, especially among the microconchs, are indistinguishable. Whether *C. reesidei* and *C. mountjoyi* are of precisely the same age remains to be seen. In Europe such a pairing of micro- and macroconchs would be quite normal. The macroconchs of *C. mountjoyi* (from British Columbia) and of *C. martini* (from Alaska) do appear to be slightly different, although both can still be accommodated within the Bukowskii Subzone. *C. mathiaspeakense* gives the impression of being a little younger, and could already belong to the Costicardia Subzone. *Scarburgiceras*-like forms range well up into the Cordatum Zone in Europe. Wyoming (Reeside, 1919; Imlay, 1982b), Rocky Mountains of Alberta and British Columbia, in the Green Beds of the Fernie Group (Frebold *et al.*, 1959). Also in the Cordillera (q.v.): eastern Oregon, southern and central British Columbia, southern Alaska. Arctic Islands (Axel Heiberg Is., Frebold, 1961).

Fauna A14. Cardoceras distans (WHITFIELD, 1880) [m] (incl. *C. stantoni, hyatti, haresi* Reeside [m], and ? *C. wyomingense, crookense* Reeside [m]
+ *Cardoceras schucherti* Reeside, *minnekahtense* Imlay [M] *Grossouvria* cf. *trina* (Buckman) (Imlay 1982b)

Lower Oxfordian, Cordatum Zone, Costicardia and ?Cordatum Subzones; possibly divisible, 14(a), *C. hyatti,* and 14(b), *C. distans.* The microconchs again span a range from compressed (*C. hyatti:* cf. *C. persecans, anacanthum* Buckman in Europe) to depressed and coarsely-ribbed (*C. distans, haresi:* cf. *C.* (*Vertebriceras*) *quadrarium* Buckman in Europe). The macroconchs seem little known so far, but the characteristic style of ribbing of *C. schucherti* is the same as that of *C. studleyense* Arkell, 1946, also from the Costicardia Subzone. Wyoming and ?Montana, southern Alaska.

Fauna A15. Cardioceras (*Cawtoniceras*) canadense WHITEAVES, 1903 [m]
+ *C.* (*Vertebriceras*) *whiteavesi* Reeside ? [m]
C. (*Maltoniceras*) *plattense* Reeside, *bighornense* Imlay, *reddomense* Imlay [M]

Middle Oxfordian, Densiplicatum Zone. The holotype of *C. canadense,* from the Rockies near Fernie, B.C. (refigured by Frebold 1957, Pl.34, Fig.2a,b; 1964b, Pl.47, Fig.3a,b) resembles *C.* (*Cawtoniceras*) *cawtonense* from England so closely that its age can hardly be in doubt. None of the other specimens ascribed to this species by Reeside and Imlay from Montana and Wyoming exactly matches either the holotype or English material, and there may be small differences in age. Late Middle Oxfordian is suggested by the already strong differentiation of primary and secondary ribbing and a crenulate keel. The dense residual tertiary ribbing joining the secondary ventro-lateral tubercles to the crenulations of the keel is peculiar to the American faunas and at best rare in Europe.

Two slightly different faunas in sequence can be distinguished in at least one area – that of the Pryor Mountains of Montana (Imlay, 1982, P.3, Fig.2, extreme right-hand column = locality 17 of Fig.1 and Table 2):

A15(a) (the lower): Imlay, 1982, Pl.24, Figs.1,2,5-8,11,12; Pl.26, Figs.20-26, 29-31: all [m]; Pl.4, Figs.9,10,14, Pl.23, Fig.10, Pl.24, Figs.15-18,21: all [M] – described variously as *C. sundancense, reddomense, canadense* and *Goliathiceras tumidum.*

A15(b) (the upper): Imlay, 1982, Pl.24, Figs.3,4,13,14, Pl.26, Figs.32,33: all [m]; Pl.20, Figs.6,7 (type of *C. bighornense*), 1,3, Pl.23, Fig.16, Pl.24, Figs.9,10 (type of *C. reddomense*): all [M].

But the nearest approach to the true

A15(c): C. canadense sensu stricto

seems to lie in some of the forms described as *C. whiteavesi* (Imlay, 1982, Pl.22, Figs.14-21) from other localities also in Montana.

These assemblages still span the full range of variability typical of *Cardioceras* as found elsewhere up to about the middle of the Densiplicatum Zone, including inflated *Goliathiceras* and ultra-coarse *Vertebriceras* (Pl.26, Figs.32,33) in A15(a) and A15(b). These two may therefore still belong to the late Vertebrale Subzone. The true *C. canadense,* however, seems to fit securely into the higher Maltonense Subzone. These zonal names refer to the recently-established zonation of the Boreal Oxfordian (Sykes and Callomon, 1979) – *alias* Plicatilis Zone, Vertebrale and Antecedens Subzones.

Fauna A16. Cardioceras sundancense REESIDE, 1919 [M] (incl. *C. stillwelli, americanum* Reeside)
+ *C. canadense* Reeside and Imlay *partim, non* Whiteaves [m] (Reeside, 1919, Pl.17, Figs.5-11; Imlay, 1982, Pl.26, Figs.14-19)

Middle Oxfordian, Tenuiserratum Zone. The differentiation of the keel is now virtually complete, as in *Amoeboceras:* inner whorls are characteristically smooth with well-spaced and wholly differentiated primary ribbing reappearing on the outer whorl of some microconchs as in *C. blakei* and *C. tenuiserratum* (see Sykes and Callomon, 1979), and the range of variability has shrunk to mainly compressed forms. The homogeneity of this fauna is supported by the fact that all the types of the species listed above came from a single locality (Reeside's no.22, Imlay's no.63) in north-east Wyoming, as did the microconchs cited, and that this locality did not seem to yield any of the other faunas. Many forms figured by Imlay under the names given above other than the types, and from other localities, belong to other faunas.

Fauna A17. Amoeboceras spp. (with *Buchia concentrica*)
Said to be widespread in the Passage Beds of Alberta and adjacent British Columbia, but not yet well described (Frebold and Tipper, 1970). These Passage Beds are the highest fully marine deposits below the coal-bearing Kootenay Formation and its equivalents.

Fauna A18. "*Titanites*" *occidentalis* FREBOLD, 1957 (see also Arkell, 1956, Pl.26; Westermann, 1966; Frebold and Tipper, 1970).

This "fauna" is based on what must be the single most famous ammonite in Canada, lying in Ammonite Gully off Coal Creek, west of Fernie, B.C., where it was discovered by Newmarch

(1953) during mapping. Its fame derives from two sources. The first is its spectacular size, 147cm (5 ft) across. The second is its position, on the top surface of the Moose Mountain Member of the Morrisey Formation (Gibson, 1979; the Basal Sandstone of the Kootenay Formation in the older literature). This is the last fully marine horizon underlying the coal-bearing Mist Mountain Formation, the principal coal-measures of the Blairmore coal-fields of southern Alberta and adjacent British Columbia, and thence thick non-marine Neocomian, etc. (dated by fossil flora), up to Albian, with ammonites, of the Alberta Group. Its dating is crucial to the question of the time of the onset in the area of the major non-marine cycle that has its equivalents in the Morrison Formation further south. There is no independent evidence of its age. Below it, the Green Beds of the Lower Oxfordian (fauna A13) are followed by the Passage Beds (100 m +) which were probably the source of the type of *Cardioceras canadense* Whiteaves (fauna A15), found not more than a mile or two from "*T.*" *occidentalis*. The Passage Beds also yield sporadic *Amoeboceras* and *Buchia concentrica* (fauna A17), but so far none appear to have been reported from the immediate vicinity. The bed with "*T.*" *occidentalis* contains no other macrofauna, nor has it yielded any microfauna. Besides the holotype, only one other fragment of a similarly huge ammonite has been found, 25 km to the east at more or less the same level (Hamblin, 1978; Hamblin and Walker, 1979). All that can be said independently therefore is that these big ammonites are post-Middle Oxfordian and probably pre-Valanginian in age. Any further conclusions must rest on the identifications of the ammonites themselves.

All previous opinions have been essentially in agreement. The ammonites are Perisphinctids, very large, and hence by analogy with what was then known of the classical successions of Europe, probably *Titanites sensu lato* of Portlandian/Middle Volgian/Upper Tithonian age. Since then a great deal more has been learned about *Titanites* itself, however, and the grounds for equating the Canadian form with it have largely disappeared. The difficulties are twofold: morphological and biogeographical.

The ammonite in Ammonite Gully is only partially preserved, mostly as an external mould of the outer 1¼ whorls that cannot be removed or easily photographed. While the coarse ribbing on the bodychamber could be matched perhaps by some members of the Dorsoplanitinae or Pavloviinae (of which *Titanites* is a member), a cast made by Westermann (1966) and photographed under appropriate oblique lighting brought out enough of the next inner whorl to show that the shell was strongly variocostate, i.e., dense fine ribbing on the inner whorls, with about 60 ribs per whorl, suddenly changes at the commencement of the adult bodychamber to coarse strong ribbing with less than 30 ribs per whorl. Such variocostation is unknown in the "Portland giants" – the Pavloviinae. Something approaching it can be found in some Dorsoplanitinae of the Middle Volgian, or Pectinatitinae of the Lower Volgian, but none of these attain a maximum size more than about a quarter that of the Canadian species.

Biogeographically, it has also become quite clear that the late Pavloviinae became more and more restricted in their distribution so that the last of them, the "Portland giants", were confined to an ever-shrinking Anglo-Saxon Basin (later to become the non-marine Purbeckian) extending no further north than central east Greenland, if that, and no further east than the Moscow Basin on the Russian Platform, at most. All possible connecting-routes to British Columbia via the north were definitely populated by quite different faunas of Dorsoplanitinae, found i.a. in the Canadian Arctic; and all possible routes to the south were blocked by yet other faunas, largely Ataxioceratidae and their decendents, found in the sub-Mediterranean Tithonian i.a. of the Caribbean, Mexico and southern California. Negative evidence includes the apparently total absence of the various forms of *Buchia* that are otherwise ubiquitous in western North America, Alaska, and the Yukon in beds of Portlandian/Middle-Upper Volgian/late Tithonian age.

It seems time, therefore, to abandon the assignment of the Canadian species to *Titanites, sensu lato* or otherwise, and with it the implied time-correlation. The only alternative that comes to mind is that the species represents yet another local or Pacific fauna of perisphinctids whose closer affinities to others are still unknown – certainly not a member of the Pavloviinae, and probably not of the Dorsoplanitinae, but perhaps an independent descendent of some Oxfordian Perisphinctinae. Its age is presumably not later than Hauterivian, the youngest stage to have yielded Perisphinctids anywhere, including the Cordillera of North America. It could still be just Upper Oxfordian or Lower Kimmeridgian, in which case the onset of non-marine sedimentation in the Fernie Basin would have been roughly sychronous with the base of the Morrison Formation. It could even be "Portlandian" as has always been assumed – but not for the reasons given.

The paucity of Upper Jurassic perisphinctids in the northern Pacific as a whole continues to be a mystery, in great contrast to their abundance almost everywhere else, including the Boreal Province (Dorsoplanitinae), the Tethys and its margins, and even the Andes in the south Pacific. There is just no general framework within which isolated finds such as the one at Fernie can be discussed. Clues are sparse. Perhaps two specimens recently described by Imlay may belong to the same group: "*Subplanites* sp." [M] (Imlay, 1981, Pl.8, Fig.15), and "*Aulacosphinctoides* sp." [m] (Imlay, 1981, Pl.7, Fig.1-3), both from the same locality on the Alaska Peninsula and associated with *Buchia mosquensis* and *B. rugosa* indicating roughly a Lower Volgian age. (For the ranges of *Buchia* and the approximate equivalence between Lower Volgian and Lower Tithonian, see Imlay, 1980a, Figs.15,16).

Region B. North American Cordillera

Fauna B1. Megasphaeroceras rotundum IMLAY [M] (1962)
+ *Sphaeroceras talkeetnanum* Imlay [m]
Normannites vigorosus (Imlay) [m]
Leptosphinctes cliffensis and *delicatus* Imlay
Oppelia kellumi Imlay
Lissoceras bakeri Imlay
Spiroceras sp.

South Alaska, Cook Inlet. Unmistakably Upper Bajocian, probably upper Subfurcatum or lower Garantiana Zone in the European standard zonation. The *Leptosphinctes* could probably be matched exactly with European forms. Equivalence of this fauna to that of the Subfurcatum Zone was already clearly pointed out by Imlay (1954a, p.7); made the basis of a Rotundum Zone by Hall and Westermann (1980).

Fauna B2. Epizigzagiceras crassicostatum FREBOLD (Frebold and Tipper, 1973)
+ *Megasphaeroceras* aff. *rotundum* Imlay
Leptosphinctes sp. (as "*Cobbanites talkeetnanus*")
Central British Columbia; Smithers Formation, lower fauna.

Fauna B3. Epizigzagiceras evolutum FREBOLD (Frebold and Tipper, 1973)
+ *Megasphaeroceras* ["*Morrisiceras*"] *dubium* Frebold
?*Megasphaeroceras* ["*Arctocephalites*?"] *multicostatum* (Frebold)
"*Cranocephalites*" aff. *costidensus* Imlay
Leptosphinctes sp. [M] and [m] cf. *subdivisus* (Buckman) (as "*Cobbanites talkeetnanus*")
Parareineckeia cf. *shelikofana* Imlay
Central British Columbia; Smithers Formation, upper fauna. Faunas B2 and B3 found in sequence, presumed to follow B1. Alaska Peninsula, unlocalized (the type of *Parareineckeia shelikofana)*

Fauna B4. "*Cranocephalites*" [*Iniskinites*?] *costidensus* IMLAY [M] (1962, 1980)
+ "*Cranocaphlites*" *globosus, alaskanus* Imlay [M]
Talkeetnites cadiformis Imlay [M]
Tuxednites alticostatus Imlay [m]
Parareineckeia hickersonensis Imlay
– *nelchinensis* Imlay
– sp. nov.? (*Cadomites* cf. *deslongchampsi* in Imlay, 1980b)
Cobbanites talkeetnanus Imlay [M] and [m]
Oppelia sp.
Southern Alaska: Cook Inlet, Talkeetnas, Wrangells, widespread and abundant. Of all the forms found in the Cordillera, those described here as "*Cranocephalites*" *costidensus* and *alaskanus* are certainly the ones bearing the closest resemblance to any of the true Boreal Arctocephalitinae. Nevertheless, there are some important differences between "*C.*" *costidensus* and *C. pompeckji* Spath – the species it most closely resembles: the ribbing in the former is denser, finer and more flexuous, and the apertural constriction is always deeper and narrower, most pronounced on the lower whorl-side in *costidensus* as in *Sphaeroceras* and not as in *Cranocephalites* s.s. Taken in isolation, such differences need hardly be more than specific; but in the context of two lineages of recently common origin now evolving separately in adjacent faunal provinces the possibility of transitory convergent homoeomorphism should not be ruled out. "*Cranocephalites*" *costidensus* [M] (and its microconch *Tuxednites alticostatus*) would certainly fit into the Pacific succession from *Megasphaeroceras* to *Iniskinites* as well as, or even better than, it would fit into the Boreal Arctocephalitinae as a temporary emigrant. Which of these interpretations is correct remains to be seen, but in the meantime it is important not to assume a close time-correlation between the *costidensus* fauna of the northern Pacific and the Pompeckji Zone of the Arctic that an assignment of the former to *Cranocephalites* might imply.

Fauna B5(a). Iniskinites tenasensis FREBOLD (1978)
+ *Kepplerites* sp. (? – cited, but not described: Frebold, 1978)
Central British Columbia; Ashman Formation, lower fauna (section XX, beds 1-5, in Tipper and Richards, 1976), in sequence above B3.

Fauna B5(b). Iniskinites acuticostatus IMLAY (1981a) [M]
+ *Xenocephalites vicarius* Imlay [m]
Iniskinites acuticostatus seems hard to distinguish from *I. tenasensis.* Eastern Oregon, upper part of the upper member of the Snowshoe Formation (450 m). According to Imlay (1981a, p.7), the top member of this formation is divisible into three parts: basal, middle (ca. 250 m) and upper (ca. 200 m). The basal part yielded a fragment of *Leptosphinctes* sp. [?, indet.] and hence was assigned to the basal Upper Bajocian. The middle part was unfossiliferous; and the species listed above came from the upper part. Numerous additional forms from these beds were assigned to *Kepplerites, Toricelliceras, Parareineckeia, Cobbanites, Bullatimorphites* and *Choffatia.* On the strength of these the whole of the upper part of the Snowshoe Formation was assigned to late Bathonian – early Callovian. Some of them appear however to have been misidentified. Thus, the type of *Toricelliceras izeeense* Imlay (1981a, Pl.3, Fig.6) has a lappet and is a large microconch; this, together with the style of tuberculation of the inner whorls, points strongly to a *Stephanoceras* (*Normannites*) aff. *vigorosum* (Imlay), early Upper Bajocian. Its macroconch may well be the very large "*Kepplerites* cf. *plenus*" on Plate 2, Figure 9 (shown reduced x 0.73), whose inner whorls are hard to distinguish from those of *Kepp. mclearni* of fauna A8 but whose outer whorl, apparently still at least partly septate, is quite different and again typical of *Stephanoceras. Parareineckeia* cf. *hickersonensis* (Imlay, 1964 Pl.4, Figs.1-6) and *Toricelliceras izeeense* pars (1981, Pl.3, Figs.7-10) both strongly resemble Upper Bajocian *Stephanoceras* from Alaska (Imlay, 1982a, Pl.4, Fig.4, Pl.5, Figs.6,8). *Cobbanites ochocoensis* Imlay seems at least in part (1981, Pl.5, Fig.6) close to *Caumontisphinctes* (cf. *C. aplous diniensis* Pavia, 1973, Pl.21, Fig.5) or *Leptosphinctes* (cf. *L. schmiereri* (Bentz) Dietl, 1980, Pl.2, Fig.1) from the Subfurcatum Zone. Some of the uppermost beds of the Snowshoe Formation appear therefore to be still of Upper Bajocian age. This is not to deny that Bathonian and Callovian may also be represented: the identification of the most interesting *Bullatimorphites* (*Treptoceras* [Enay, 1960, non Flowers, 1942]) sp. [m] is not in doubt and its closest resemblance is to European forms of about Middle Bathonian age. But attempts to subdivide the faunas of the upper Snowshoe formation further must await more stratigraphical evidence.

Fauna 5(a) was collected from a single well-defined succession of beds spanning only 19 m and seems to be nearly homogeneous. It apparently marks the first appearance of *Kepplerites* in the Cordillera.

Fauna B6. Iniskinites robustus FREBOLD (1978) [M]
+ *Xenocephalites vicarius, hebetus* Imlay [m]
"*Lilloettia*" [*Iniskinites*] aff. *lilloetensis* Crickmay

Kepplerites cf. *tychonis* Ravn (as "aff. *ingrahami*, cf. *chisikensis*" Frebold; *snugharborensis* Imlay, 1981, Oregon, here?)

Central British Columbia, Ashman Formation, upper fauna (Section XX, beds 8-11, 25 m). This fauna, if present at all, is not conspicuous in Oregon. There may be traces of it in southern Alaska ("*Cranocephalites*" *pompeckji*, Imlay, 1962, Pl.1, Figs. 11-13). The *Kepplerites* are in the same style – relatively small, evolute, very densely and finely ribbed – as the first ones to appear in the succession both in the Western Interior (*K. costidensus*, fauna A5) and in East Greenland (*K. tychonis* var. *fasciculata* Spath, 1932, and *K. tychonis* Ravn *s.s.*). In Greenland they mark the lowest Zone of the Boreal Upper Bathonian.

Fauna B7. Iniskinites intermedius (IMLAY) (1953, 1975)
Iniskinites s.str. ranges over a number of horizons.

Fauna B7(a). Iniskinites intermedius (IMLAY) [M]
+ *I. magniformis, martini* Imlay [M]
Chinitnites chinitnaensis, parviformis Imlay [m]
Kepplerites chisikensis Imlay [M]
K. cf. *abruptus* (McLearn) [M]
K. (*Toricellites*) *alaskanus* (Imlay) [m]
Cadoceras moffiti Imlay [M]
Choffatia irregularis Imlay [M]

Southern Alaska: Talkeetnas and Cook Inlet (e.g., Imlay, 1975, p.5, first and last columns, basal 100 ft of Tonnie Silstone Member). Characterized by the spectacularly globuliform, finely ribbed spheroconic *Iniskinites*. The original identification as *Kheraiceras* gave support to the assignment of these beds to the Lower Callovian. The two genera are however easily distinguished, i.a., by the form of their dimorphism: only the macroconchs are homoeomorphic, the microconchs (*Chinitnites* and *Kh.* (*Bomburites*)) bearing little resemblance to each other. The sphaeroceratid ancestry of *Iniskinites* is also clearly revealed in the form of its peristome, marked by deep lateral constrictions unfolding vertrally into a projected hood resembling the peak of a cap exactly as in *Sphaeroceras* itself (see Imlay, 1975, Pl.3, Figs.8,9).

The precise ranges of the various genera and species described from this part of the succession appear to be very uncertain and further separate faunal horizons may come to be distinguished. The Keppleritids seem to be characteristic, however, consisting still of densely and finely-ribbed forms resembling the true *K. tychonis* but compressed and involute in both macro- and microconchs. Such forms occur in Greenland in the upper Variabile Zone and lower Calyx Zone of the Upper Boreal Bathonian. This fauna apparently also sees the first appearance of *Cadoceras* in the coarsely-ribbed forms with open, round-shouldered umbilici of *C. moffiti* (Imlay, 1953, Pl.42, Figs.3,4,9) whose position at these levels seems to be established in the Talkeetnas (Imlay, 1975, p.9, second column from the right, USGS Mesozoic locs. 24151, 24147, 24224). In East Greenland such early forms of *Cadoceras* also occur in the lower part of the Calyx Zone. Imlay also records *C. catostoma* Pompeckj, but the principal occurrence of the forms ascribed by him to this species is higher and is further discussed below (fauna B8(f)).

Fauna B7(b). Iniskinites cepoides (WHITEAVES, 1884) (Frebold, 1979)

Queen Charlotte Islands, top member of the Yakoun Formation. Apparently found in isolation without any other groups. Exact stratigraphic position conjectural, placed here on grounds of morphological resemblance to *Iniskinites* of fauna B7(a).

Fauna B7(c). Iniskinites abruptus (IMLAY) (1953, 1975) [M]
+ *Kepplerites chisikensis* Imlay 1975 [M] (holotype, Pl.1, Figs.1,5)
K. alticostatus Imlay [M] (incl. *K.* cf. *abruptus* (McLearn) – Imlay, 1953, Pl.52, Fig.6; *K.* cf. *ingrahami* (McLearn) – Imlay, 1953, Pl.49, Fig.21)
K. (*Toricellites*) cf. *alaskanus* (Imlay) (Imlay, 1953, Pl.48, Figs.12,13) [m]

Cook Inlet, localities 9, 77, 78; Peninsula, Wide Bay, locality 125; Talkeetnas, localities 18, 32, 34. The Keppleritids differ little from the ones found below and above, but the *Iniskinites* is distinct and unmistakable. The association, apparently always free from *Cadoceras*, seems to be so widespread as to be characteristic of a distinct horizon. The evidence from Cook Inlet indicates that it lies higher than that with fauna B7(a) (see Imlay, 1975, p.5, right-hand column, 350-750 ft above base).

Fauna B8. Cadoceras comma IMLAY, 1953 (1975)

This index is used here for the faunas of what is roughly the "*Cadoceras catostoma* Zone" of Imlay (1975), best known from the famous "*Cadoceras*-beds" of the Chinitna Formation in southern Alaska. The reasons for abandoning *C. catostoma* Pompeckj, 1900, as index are twofold. Firstly, the species as defined by its type series is only barely interpretable. It was based on two syntypes of which only one is at all identifiable (Pompeckj 1900b, Pl.5, Fig.1), a complete microconch only 23 mm in diameter and here designated lectotype. Such microconchs of *Cadoceras* change little from their first appearance in the Upper Bathonian up to the top of the Lower Callovian; and whether the various "species" named by Pompeckj (*C. catostoma, grewingki, nanseni, petelini* and *schmidti*), as represented by their microconch type series, really are distinct and of different ages, or whether they are merely variants of a single species, may never be determined. The type localities are not well enough localized to be able to recollect topotypes. Secondly, the material described by Imlay (1953, Pl.34) consists of a mixture of forms also from poorly known horizons, so that even if one of them were chosen as basis for re-interpretation of the name, e.g., the macroconch from Iniskin Bay shown on Pl.34, Figs.12,14, little would have been gained for stratigraphical purposes. Neither would a precisely known level have been characterized, nor would an identifiable fauna have been labelled. The name *catostoma* seems best left as a *nomen dubium*, and the labelling of faunal horizons restricted to well-defined nominal species.

Stratigraphic uncertainties reach their peak here. The range-charts of the rich Callovian faunas in southern Alaska given by Imlay (1953, 1975) indicate long and largely overlapping ranges of many species of *Cadoceras, Lilloettia* and *Kepplerites*. Yet the figured material includes forms that resemble species found elsewhere in isolation or in succession over ranges of strata representing considerable intervals, extending over several ammonite zones. The range-charts for Alaska must inevitably depend very much on the systematic classification adopted for the material available; and as the bulk of this consists of small lots made by various collectors at many localities, the classification has to be primarily morphological. What follows is an attempt to identify what might be separable faunas on the strength of the material that has been figured, and to suggest some possible ages and correlations.

Fauna B8(a). Kepplerites loganianus (WHITEAVES, 1884) (McLearn, 1929; Imlay 1953, 1975, *PP836)*(incl. *K. plenus, multus, gitinsi, abruptus, ingrahami, torrensi* (McLearn) [M]
K. (*Toricellites*) *newcombii* (Whiteaves, 1900) [m]
K. (*Toricellites*) *penderi* (McLearn) [m]

Queen Charlotte Islands, top member of the Yakoun Formation; apparently monogeneric, free of *Cadoceras.*

This fauna is very close to the Keppleritids of the zone of *Cadoceras apertum* sp.nov. MS in East Greenland and of the immediately overlying beds, both in the macro- and microconchs. One of the former is *Kepp. traillensis* Donovan, 1953. Compared with the earlier forms of *Kepplerites,* the ribbing on the inner whorls has become stronger and coarser, and the innermost whorls of the macroconchs have flattened venters. The microconchs are also more evolute, strongly and coarsely ribbed. Both the American and Greenland faunas in turn are very close to, if not quite identical with, the fauna of *Kepplerites keppleri* (Oppel) in northwest Europe: there are individual variants in all three faunas that are morphologically indistinguishable. The *K. keppleri* fauna was first described from Württemberg in southern Germany (Oppel, 1862; Buckman, 1922, Pls.289 [M], 310,318-9 [m]; Quenstedt, 1886, Pl.77, Figs.1-5 [M], Pl.76, Fig.9 [m]) where it occurs at the base of the "Macrocephalenoolith" (Dietl, 1981). This condensed ironshot oolite is regionally the lowest bed of Callovian age, but its time-duration within the Callovian is highly variable. Its upper boundary is demonstrably diachronous, and the precise age of its base relativeto the base of the Callovian Stage as defined in England (Callomon 1964) has not yet been established independently. Fortunately the level at which the *K. keppleri* fauna occurs in England is known rather more precisely (Callomon, 1959). Two stratigraphically not closely localized specimens have been figured ("*Cerericeras*" *cereale* Buckman 1922; Arkell, 1954, p.118, Text-fig.42), but several others in the Oxford University Museum came from Swan Inn quarry, Long Handborough, 12 km northwest of Oxford, Upper Cornbrash (Douglas and Arkell, 1928, p.129, bed 4), *Ornithella siddingtonensis* brachiopod zone. This is the lowest brachiopod zone in the Macrocephalus Subzone of the Macrocephalus Zone in England, and hence of the Callovian Stage by definition. *K. keppleri* appears to occur at only a single horizon in the Macrocephalus Zone both in England and Swabia, so it seems safe to assume that these occurrences reflect but a single brief and closely isochronous southerly invasion of northwest Europe by what was at the time still a strictly Boreal group. Conversely, these occurrences provide the crucial key to the correlation of the standard northwest European, Boreal and northwest Pacific ammonite provinces at the base of the Callovian, and make it possible to locate the Bathonian-Callovian boundary there as well as can be done, to within half a standard ammonite Zone, say. Fauna B8(a) may therefore be taken to mark the base of the Callovian in the Cordillera. In the Queen Charlottes it occurs immediately above fauna B7(b). In Greenland its analogues occur some distance above the analogues of fauna B7(c). Its position elsewhere in the Cordillera remains somewhat uncertain, however. It has also been recorded from southern Alaska, e.g., localities 85-87 on the east shore of Iniskin Bay (Imlay, 1953b), but the assemblage from this area seems slightly different and is assigned a separate horizon, discussed further below (fauna B8(f)).

Fauna B8(b). Cadoceras (*Paracadoceras*) *tonniense* IMLAY, 1953 [M]
\+ *C. glabrum* Imlay, partim, including
? *C.* (*Paracadoceras*) *harveyi, brooksi* Crickmay
C. wosnessenskii (Grewingk) *sensu* Imlay partim

This appears to be the lowest fauna in which *Cadoceras* predominates. The problems of classifying the Alaskan representatives of the genus have already been discussed in the systematic introduction. Although the range-charts show many species extending over long intervals of strata in parallel, the faunas as a whole give the impression of changing little, implying that they represent only a short time-interval – perhaps the equivalent of no more than a single standard ammonite zone altogether. Most of the collections are small lots from scattered localities, but when occasionally a single locality did yield more extensive collections (e.g., USGS Mesozoic loc.2921, no.87 in Imlay, 1953) the impression is very strongly one of all the "species" blending into each other in a single homogeneous assemblage. The characteristic features almost all the forms have in common are open, conical umbilici, only moderate inflation, and small size as *Cadoceras* goes. In Alaska the *C. tonniense* fauna ("*Paracadoceras*") appears to be well localized in the Iniskin Peninsula near Tonnie Peak, at about 450 ft above the base of the Chinitna Formation (Imlay, 1975, p.5, second column from the right).

The type and only specimen of *Paracadoceras harveyi* came from Crickmay's "Zone of *Cadoceras brooksi*" near Harrison Lake in southern British Columbia, which yielded in addition *C. catostoma* (*teste* Crickmay and Frebold), *C.* (*Pseudocadoceras*) *schmidti* and *grewingki* Pompeckj (see Frebold and Tipper, 1967). The position of this horizon relative to others could not however be firmly established, and there seems no point in separating "a horizon of *C. harveyi*/*brooksi*" in any other than the most local sense.

In East Greenland *Paracadoceras* occurs commonly in beds immediately above the levels with *Kepp. traillensis* and *K.* cf. *keppleri* and then ranges upwards as a minor element in the Zone of *Cadoceras nordenskjoeldi* sp.nov. MS. *Paracadoceras* occurs also very rarely in the Cornbrash of southern England (*Cadoceras breve* Blake, 1905, Pl.5, Fig.1), in beds taken to be Upper Cornbrash (*teste* Blake) and hence of Macrocephalus Zone and Subzone age. The only other known specimen, in the British Museum (C.7331), is from an old collection without locality, but its matrix points strongly to the Upper Cornbrash of the Malmsbury area of Wiltshire.

All the evidence strongly and consistently indicates an early Lower Callovian age for this fauna, equivalent to the Macrocephalus Zone of Europe and probably also still the Macrocephalus Subzone.

Fauna B8(c). Lilloettia lilloetensis CRICKMAY, 1930 (incl. *L.* ("*Buckmaniceras*") *buckmani* and *L. mertonyarwoodi* Crickmay – all [M]
Xenocephalites vicarius Imlay [m]

Southern Alaska, Tonnie Siltstone Member of the Chinitna Formation; southern British Columbia, "Zone of *L. lilloetensis*" of Crickmay; eastern Oregon, Trowbridge Shale (Imlay, 1964, 1981). The three "species" of *Lilloettia* appear to represent no more than the partial range of variability of a single biospecific assemblage; *L. milleri* Imlay may be the spheroconic end-member.

The precise age of *Lilloettia* was for a long time uncertain, for it had been recorded from many scattered localities at few of which, if any, it was associated or found in succession with other genera. In Alaska it is recorded as ranging over almost the whole of the *Cadoceras*-beds, but its precise levels within these beds seems now securely established in at least two places on the Iniskin Peninsula: USGS Mesozoic loc. 21322 (no. 38 in Imlay, 1953, and second column from the right in Imlay, 1975, p.5); and USGS Mesozoic locs. 21340, 22556 (nos. 35 and 36, and third column from the right). In both places it succeeds *Cadoceras* but is not associated with it. At the first locality it thus directly overlies fauna B8(b), which puts it in the same relation to *Paracadoceras* as Crickmay thought it did at Harrison Lake, although he had not proof. At the second locality the *Cadoceras* does not seem to be quite abundant enough to tell whether it belongs to fauna B8(b) or to a higher one, B8(f), described below. Either way, the early Lower Callovian age of these species of *Lilloettia* now seems firmly established.

This is of great importance in the wider context of the Pacific Realm as a whole, for very closely related, possibly identical faunas occur further south, e.g., in the material from Cualac, Province of Guerrero, Mexico, described by Burckhardt and discussed in a preliminary revision elsewhere (Westermann *et al.*, this volume)

Macrocephalites boesei Burckhardt [M] (1927, Pl.15, Figs. 1-4, = *Lilloettia* cf. buckmani (Crickmay, 1930))
Macrocephalites aff. *macrocephalus* (Schlotheim) [M] (1927, Pl.16, Figs.1-3, = *Lilloettia* cf. lilloetensis Crickmay)
Macrophalites nikitini Burckhardt [m] (1927, Pl.16, Figs. 4-9, = *Xenocephalites* aff. *vicarius* Imlay, 1953)

The close homoeomorphism between macroconchs of *Lilloettia* and of *Macrocephalites* is neatly resolved by means of the microconchs. It should be possible to follow these correlation further, into the Andes (Riccardi, in progress) and hence finally to settle the question of the Bathonian-Callovian boundary there.

Fauna B8(d). Lilloettia tipperi FREBOLD (1967) [M] (in Frebold and Tipper, 1967)

Southern British Columbia: isolated horizon and locality. Listed separately here to emphasize the point that *Lilloettia* ranges over several horizons. The material is also sufficiently plentiful and well preserved to show that at some horizons, at least, the biospecific variability of *Lilloettia* can be quite small.

Fauna B8(e). Cadoceras wosnessenskii (GREWINGK, 1850) *sensu* Imlay 1953 [M]
+ *Cadoceras catostoma* Pompeckj, 1900, *sensu* Imlay [M]
Cadoceras glabrum Imlay [M]
Cadoceras comma Imlay *partim* [M]

This label is applied to the two collections from the east shore of Iniskin Bay at USGS locs. 2921 and 3029 (nos. 87 and 86 in Imlay, 1953, respectively) which, to judge from the admittedly scanty indications, came from the lower part of the Paveloff Siltsone Member of the Chinitna Formation (Imlay, 1975, p.5, right-hand column). Although there may be some mixing of horizons, much of the fauna gives the impression of homogeneity. If the correlations given by Imlay are correct, this upper *Cadoceras* fauna lies considerably higher than the lower *Cadoceras* fauna B8(b) and above the *Lilloettia* fauna B8(c).

It includes good macroconchs described under two of the old Russian names revied by Imlay. One of these, *C. catostoma,* has already been discussed above. The case for the other one, *C. wosnessenskii,* is more hopeful. The type series seemed to consist of the single specimen figured by Grewingk (1850, Pl.4, Figs.1(a)-(c)). It is the inner part of a phragmocone of a macroconch that does resemble the forms described by Imlay under the name quite closely, as far as it goes, although it could have belonged probably equally well to several other of Imlay's species. If the name is to be revived it should be firmly tied to a reference specimen, a possible neotype, should the holotype be lost. The holotype came from the Shelikoff Formation near Katmai Bay on the Alaskan Peninsula (not from Chignik Bay as stated by Imlay). No new material seems to have been collected in this area and hence there are no topotypes. For the next best candidate there is not much to choose between the fine specimens described by Imlay from Wide Bay, 100 km south-west, or Iniskin Bay, 200 km north-east, and hence the specimen suggested here is Imlay's (1953, *PP 249*-B) Plate 40, Figures 9 and 12. This closely matches Fig. 1 on the same plate from Wide Bay, and this in turn seems to have come from the same level and locality as the type of *C. glabrum* (Imlay, 1953, Pl.37, Figs.5,6,8,9). In this way a picture of the fauna can be built up. It still resembles the lower *Cadoceras* fauna B8(b) very closely and the time-gap between them must have been very short.

One or both of these *Cadoceras* faunas are also present in northern Siberia (e.g., *C. anabaranse* Bodylevsky, 1960, Pl.4, Fig.3; Meledina, 1977, Pl.21, Fig.1, Pl.22, Fig.1, Pl.23, Fig.1, Pl.24, Fig.1; *C.* cf. *multiforme* Imlay – Bodylevsky 1960, Pl.8; Meledina, 1977, Pl.19, Fig.1), but it has so far been described neither from the Alaskan North Slope, nor the Arctic Islands, nor East Greenland. In Greenland the reason may be that at the time the area was occupied by a peculair local fauna, *C. nordenskjoeldi* sp. nov. MS (cf. Madsen, 1904, Pl.10, Fig.2), between the Apertum Zone of the basal Lower Callovian and the Septentrionale – Calloviense Zones of the late Lower Callovian.

Fauna B8(f). Kepplerites cf. *abruptus* (McLEARN)

Southern Alaska, localities 85-87 on the east shore of Iniskin

Bay, figured by Imlay under various of McLearn's names (1953, Pls.50-51, and including *Kepp.* (*Toricellites*) cf. *newcombii,* Pl.53, Figs.4,5 [m]). Those from locality 85 at least (USGS Mesozoic loc. 20763) were confidently placed into the upper part of the Chinitna Formation by Imlay (1975, P.5, right-hand column) and said to be associated with *Cadoceras,* including *C. wosnessenskii, C. doroschini* and *C. kialagvikense,* suggestive of fauna B8(e). Morphologically these forms differ little from the ones both from the Queen Charlotte Islands (B8(a)) and from some of the others described by Imlay from beds with *Iniskinites* in Alaska. There is, perhaps, a more persistent tendency for the venters to become tabulate as far as the bodychamber (cf. Imlay, 1953, Pl.50, Figs.1,3) which is rare in the true *Kepp. keppleri* and, according to McLearn, in the *Kepp. loganianus* fauna. It may well be that faunas B8(a) and B8(f) merely reflect temporary predominances in the fluctuating abundance of *Kepplerites,* ranging with little change in Alaska from the latest *Iniskinites* beds throughout the *Cadoceras* beds as Imlay's range-charts suggest. This would support the impression given by *Cadoceras* of a relatively short time interval.

Fauna B8(g). Lilloettia stantoni IMLAY
+ *Lilloettia buckmani* (Crickmay)

There is consistent evidence of a higher assemblage of *Lilloettia* differing somewhat from that in fauna B8(c) at the very top of the *Cadoceras* beds with the *C. wosnessenskii – comma* group (e.g., Imlay, 1975, p.7; p.5, right-hand column, USGS Mesozoic loc. 22434; third column, loc. 22556). Also eastern Oregon, Trowbridge Shale ? (Imlay, 1981).

Fauna B9. Cadoceras (*Stenocadoceras*) *stenoloboide* POMPECKJ, 1900 [M] (Imlay 1953, 1975) (incl. *C.* (*S.*) *multicostatum, bowserense, straitum?* Imlay [M]
+ *C.* (*Pseudocadoceras*) *petelini* Pompeckj [m] (Imlay, 1953, Pl.44, Figs.6-8,12; Pl.47, Figs.7,8; Pl.48, Figs.1-6; Pl.49, Figs.1,2,8 – under various names)

Southern Alaska: Peninsula, Cook Inlet, Talkeetnas (Imlay, 1953, 1975); central and southern British Columbia (Frebold and Tipper, 1967). Canadian Arctic: northern Yukon, Babbage River and Aklavik Range (Frebold, 1964). Northern Siberia: eastern Taimyr (Meledina, 1977). *Pseudocadoceras* spp. recorded from a number or other localities, including Vancouver Island, eastern Oregon, and possibly as far south as the Sierra Nevada of northern California (Imlay, 1961, Pl.2, Figs.6,7 only) may belong to this horizon. The use of the name *stenoloboide* to label this fauna again follows Imlay's interpretation, for the type series (4 syntypes) seemed to consist mainly of microconchs and is also barely interpretable.

The age of fauna appears to be Middle Callovian although there is no direct proof. *Stenocadoceras* [M] is known neither in Europe nor in East Greenland. Cadoceratinae are relatively common in the lower part of the Middle Callovian of Britain (Jason Zone), but the macroconchs are still typical *Cadoceras* s.s. Higher, in the Coronatum Zone, they are exemplified by the group of *C.* ("*Rondiceras*") *milaschewici* (Nikitin) [M] (see illustrations by Sazonov, 1957) whose inner whorls do resemble *Stenocadoceras* but whose outer whorls revert to the style of ordinary *Cadoceras.* The Alaskan microconchs, however, closely resemble those from the Middle Callovian of Europe, e.g., *Pseudocadoceras concinnum, laminatum* Buckman (1927, Pls.727,735) from around the Middle-Upper Callovian boundary; but they also resemble some earlier forms, including *Pseudocadoceras boreale* Buckman itself (1919, Pl. 121B), from the Koenigi Subzone of the Calloviense Zone (not Upper Callovian, as Buckman thought), and *Ps. mundum* (Sazonov, 1957, Pl.6, Figs.2-4) from the Elatmae Zone (pre-Koenigi) of the Russian Platform. The main argument in favour of a Middle Callovian age is indirect, based on the morphologically intermediate position of *Stenocadoceras* [M] and [m] between the undoubtedly Lower Callovian faunas B8(a)-(f) below, and the Upper Callovian fauna B10 above.

The correctness of this assignment would, however, raise another problem. As argued above, all the *Cadoceras* in the horizons below, B8, indicate an early Lower Callovian age, rising no higher than the Macrocephalus Zone. Of the large cadicone *Cadoceras* ex gr. *sublaeve* typical of the late Lower Callovian, Calloviense Zone, there is no sign. Yet such forms are present not far away, in the northern Yukon and the Arctic Islands (*Cad. arcticum, voronetsae, septentrionale* var., Frebold, 1964) and on the North Slope. The implications would be a considerable non-sequence in southern Alaska between the *Cad. comma* (B8) and *Stenocadoceras* (B9) horizons in the Chinitna and Shelikoff Formations. This would not in itself be at all surprising, considering the other non-sequences that can be recognized. The difficulty would lie in the reports of admixture or overlap of the two faunas. These should be carefully re-examined. If confirmed, it would become necessary to consider the possibility that *Stenocadoceras* is also still Lower Callovian, perhaps a local genus.

Fauna B10. Longaeviceras pomeroyense IMLAY [M] (1953, Pl.43, Figs.1-3)
+ *Pseudocadoceras crassicostatum* Imlay *partim* [m] (1953, Pl.49, Figs.19,20)
Pseudocadoceras chinitnense Imlay [m] (Pl.48, Figs. 7-10)

Southern Alaska, Cook Inlet. These forms are unmistakably of early Upper Callovian age, lower Athleta Zone. They occur near the top of the Chinitna Formation.

– regional non-sequence –

Oxfordian

Fauna B11(a). Cardioceras aff. *scarburgense* (YOUNG and BIRD) (Frebold and Tipper, 1975) (= *Quenstedtoceras* (*Lamberticeras*) *henrici, intermissum* and sp. aff., Frebold and Tipper)

Central British Columbia: Smithers area, GSC locality 85376. Although a general resemblance to *Quenstedtoceras* of the Lamberti Zone cannot be denied, the sharpness of the ribbing and rather involute coiling points strongly to a somewhat later age, probably in the Mariae Zone of the Lower Oxfordian.

Fauna B11(b). Cardioceras cf. *praecordatum* DOUVILLÉ Frebold and Tipper 1975, Pl.2, Figs.8,9,12, GSC locality 90821, central British Columbia; top Mariae Zone, Praecordatum Subzone.

Fauna B12. Cardioceras martini REESIDE [M] + [m] (Reeside 1919; Imlay 1964, 1981)

Very close to, if not identical with, *C. reesidei* Maire, fauna A13. Eastern Oregon (Imlay 1964); central British Columbia (Frebold and Tipper 1975, Pl.2, Figs.7,10; GSC locality 90824); southern Alaska (Imlay). Cordatum Zone, Bukowskii Subzone.

Fauna B13. Cardioceras spiniferum REESIDE [M] (Reeside 1919; Imlay 1981)
+ *C. alaskense* Reeside, cf. *stella* Arkell – Imlay [M]
C. distans (Whitfield) – Imlay, cf. *lillooetense* Reeside – Imlay [m]

Southern Alaska, Lower Oxfordian, Cordatum Zone, upper Bukowskii or lower Costicardia Subzone. In Europe, where many successive faunas can be collected even within a single Subzone, each one differs slightly in aspect from the others as the *Cardioceras* lineage evolved and the boundaries between Subzones are arbitrary and not always easy to recognize. Individual morphs can range over several faunas or even Subzones. It is not surprising therefore that Imlay records similar overlaps in Alaska (1981, p.14-15, Tables 2,3).

The precise position of *Cardioceras lillooetense* Reeside as exemplified by its type from southern British Columbia requires confirmation, but it could well belong here.

Fauna B14. Cardioceras whiteavesi REESIDE [m] (Imlay, 1981)
+ *Perisphinctes* cf. *muehlbachi* Hyatt ? [m] (Imlay, 1981, Pl.7, Fig.9)

Southern Alaska. The *Cardioceras* does strongly resemble the type which came from Wyoming, fauna A15, assigned there to the Middle Oxfordian, Densiplicatum Zone. The *Perisphinctes* came from the same area, and it is not unlikely that it may have come from beds of the same age. It is certainly very close to *Per. antecedens* Salfeld or *Per. buckmani* Arkell from the Plicatilis – Tenuiserratum Zones of Europe, and it could be significant that this horizon, about the only one at which a more southerly faunal element ever penetrated into the northerly territory of the Cardioceratinae in America, has the same age as those zones in the Oxfordian of Europe in which the Perisphinctidae similarly achieved their maximum northerly extent.

Fauna B15. Amoeboceras aff. *transitorium* SPATH (Imlay, 1981, Pl.12, Figs.2-15) + *Buchia concentrica*

Alaska, Wrangell Mountains. This fauna is not easy to place. The macroconchs (Imlay, 1982, Pl.12, Figs. 12,15) resemble *Amoeboceras schulginae* Mesezhnikov, 1967 (Pl.3, Fig.1) in the straightness and coarseness of the primary ribs with their ventrolateral tubercles; but the general resemblance of the inner whorls and microconchs (Figs.5-11) seems closer to *A. glosense* and *transitorium* (see Sykes and Callomon, 1979). The most likely ages are therefore either Upper Oxfordian, Glosense Zones (*A. transitorium*) or Regulare Zone (*A. schulginae*).

Fauna B16. Amoeboceras dubium (HYATT, 1894) [m] (Reeside, 1919, Imlay, 1961)

Northern California. These seem to be the only indisputably Lower Kimmeridgian Boreal ammonites so far described. They are close to *A. elegans* Spath (1935) and their age is Eudoxus Zone.

Fauna B17. "*Subplanites* [M]/*Aulacosphinctoides* [m] spp." IMLAY (1981)
+ *Buchia* spp.

Southern Alaska. These forms are entirely strange and new. They were accompanied only by *Buchia mosquensis* and *B. rugosa,* indicating about a Lower Volgian age (Middle Kimmeridgian *sensu anglico*).

Higher Jurassic strata appear to be widely represented in the Cordillera but have yielded no ammonites, only *Buchia.* A number of ammonites from various localities have not been considered here, because they are not closely identifiable or because they are stratigraphically not very significiant at present. This includes numerous Phyllocertidae and Lytoceratidae, some Oppeliidae, and some Perisphinctaceae of southerly affinities found in California and southwestern Oregon, such as the Upper Tithonian faunas described by Imlay and Jones (1970, p.B5-11).

Region C. Arctic Canada: Northern Yukon (excluding British Mountains) – Richardson Mountains – Arctic Archipelago

Due largely to the intensive exploration for hydrocarbons in the last 30 years, the Jurassic stratigraphy of this region has become rather well known. Emerging in the Keele Range near the Alaskan border at about 67°N, the outcrop sweeps in a broad band northeastwards via Prince Patrick Island into the Sverdrup Basin as far as Axel Heiberg and Ellesmere Islands (80°N) before being lost into the Arctic Ocean (cf. Poulton *et al.,* 1982, Fig.3; Poulton, 1982, Fig.3). It is bounded to the northwest by a fault belt that runs northeastwards from the Brooks Range into the Beaufort Sea at the mouth of the Mackenzie Delta and is said to be associated with the Kaltag Fault of western Alaska, dividing it from the Brooks-Mackenzie Basin occupying the North Slope and British Mountains which are discussed separately as region D below. The overall distribution of ammonites in region C is so uniform that its evolution as a single coherent pericratonic basinal entity can be in little doubt (see Poulton *et al.,* 1982). The major features of the ammonite faunas are almost identical with those of East Greenland. There are lesser differences, in the form of clearly discernible faunal gradients, that are of great significance. It is useful, therefore, to distinguish three subregions, from southwest to northeast:

C(i): Keele Range – Old Crow – upper Porcupine River. This areas contains what must be the finest single section through the Middle Jurassic in the whole of the Arctic at Salmon Cache Canyon on the Porcupine River, at the "Big Bend" where its flow changes direction from northwards to westwards (67½°N, 138°W) 60 km east of Old Crow settlement (cf. Poulton, 1978b, Pl.II). Some 70 beds can be distinguished in succession (cf. Poulton and Callomon, 1976, Poulton, 1978a,b for preliminary descriptions) of which at least 35 have yielded ammonites that are in course of being described (Poulton, 1983).

C(ii): Richardson Mountains. Jurassic rocks are seen in many fine exposures from the Mackenzie Delta westwards to the White Mountains and their surroundings. Over 30 detailed sections have been recently described (Poulton *et al.,* 1982). They reveal very rapid lateral changes of thickness and facies, reflecting synsedimentary tectonic movements; yet the ammonites that occur tend always to belong to the same rather limited set of faunas. This suggests that the limitations were imposed by sporadic faunal colonization rather than by gaps in the sedimentary record, and that with some exceptions, the succession of faunas as we now know it may be regionally fairly complete.

C(iii): Sverdrup Basin, from Prince Patrick to Ellesmere Islands. Ammonites have been found at widely scattered localities and only rarely has any one locality yielded much of a faunal succession. The evidence for relative dating, therefore, is sometimes indirect.

The early Middle Jurassic is represented by all but one member of what appears to be a standard Boreal – North Pacific succession defining a distinct, Bering Province (see Taylor *et al.*, this volume) shared by regions B, C and D and extending without modification into eastern Siberia, from the Chukhotsk Peninsula via Kolyma to the southern shores of the Sea of Okhotsk and the Bureya basin (cf. Saks, 1976).

Be-1: *Pseudolioceras mclintocki* (HAUGHTON, 1858)
Prince Patrick, Mackenzie King and Melville Islands (Frebold, 1958, 1960); northern Yukon (Poulton, 1978b); northern Alaska (Imlay, 1976); Okhotsk: Tugur and Bureya (Sey and Kalacheva, 1980).

Be-2: *Leioceras opalinum* (REINECKE)
Prince Patrick and Melville Islands (Frebold, 1958, 1960); northern Yukon (Poulton, 1978); Spitzbergen (Wierzbowski *et al.*, 1981); northern Siberia, Lena basin (Bidzhiyev, 1965).

Be-3: (a) *Erycitoides howelli* (WHITE, 1889) [M]
+ *E.* (*Kialagvikites*) *kialagvikensis* (White, 1889) [m]
(b) *Pseudolioceras whiteavesi* (White, 1889)
(c) *Pseudolioceras* [*"Tugurites"*] *tugurense* (Kalacheva and Sey, 1970; see 1980)
(d) *Pseudolioceras spitsbergenense* Frebold, 1975
(e) = *Pseudolioceras* [*"Canavarella"*] *crassifalcatum* (Imlay, 1976) ? (*PP 854*)
(f) *Tmetoceras* spp.
Ellesmere Island [(d), Frebold, 1975]; northern Yukon [(a), (b)?, (e)?, Poulton, 1978b]; northern Alaska [(a), (e), Imlay, 1955]; southern Alaska [(a), (b), (d), (f), Westermann, 1964a, the basis in part of his *E. howelli* Zone]; Okhotsk: Tugur and Bureya [(a), (b), (c), Sey and Kalacheva, 1980]; Japan [(b), as *Graphoceras*, (f), Sato, 1954, 1957, 1958].

Be-4: *Pseudolioceras fastigatum* WESTERMANN, 1969
Southern Alaska (Westermann, 1969); Okhotsk: Bureya (Sey and Kalacheva, 1980).

Be-5: *Arkelloceras tozeri* FREBOLD, 1958 [M]
+ *A. mclearni* FREBOLD, 1958 [m]
Axel Heiberg, Melville and Prince Patrick Islands (Frebold, 1958, 1961, 1964); northern Yukon (Poulton, 1978b); southern Alberta (Westermann, 1964b). Southern Alaska (Imlay, 1964); northern Alaska, Canning River (Imlay, 1955). Siberia: Chukotsk Peninsula, Anadyr (Efimova *et al.*, 1968, p.80); Kolyma Plateau (Efimova *et al.*, 1968, pl.79); Bureya basin (Sey and Kalacheva, 1980).

There is no sign of the rich accessory faunal elements found in the Alaska Peninsula. The abundant *Stephanoceras* faunas of the south, of the Humphriesianum Zone and early Upper Bajocian, are similarly absent, with the possible exception of a single specimen from the Big Bend section on the Porcupine. Instead, there follows abruptly everywhere the first of the purely Boreal faunas, unknown further south in regions A and B, that then held sway during the rest of the Jurassic. Yet some Pacific elements from region B did penetrate a little distance into the Arctic as far as Salmon Cache Canyon. These include *Phylloceras*, relatively common there at several horizons; *Iniskinites*, not rare in the Upper Bathonian; *Choffatia;* and even a single specimen, unfortunately not in place, of the very local North Pacific *Parareineckeia*. Yet hardly any of these seems to have got as far even as the Richardsons [C(ii)] and with the exception perhaps of very rare *Phylloceras* in Prince Patrick Island, none reached the Sverdrup Basin [C(iii)]. Old Crow was thus very much at the gateway between Pacific Ocean and Boreal Sea during the Jurassic as far as the ammonites were concerned.

Fauna C1. Cranocephalites borealis (SPATH) (incl. *C. warreni* Frebold, 1961).
Salmon Cache Canyon; Richardsons, widespread; East Greenland; northern Siberia. As in Greenland, the duration of this fauna seems to have been rather short and all the material that has been collected probably can be safely regarded as effectively isochronous. The ranges of variability of the Canadian and Greenland assemblages then largely overlap but do not quite coincide. The small differences could well represent geographical subspeciation. If due to differences of age, this could only have been very slight. No other genera have been found so far in association and the question of precise age and correlation with Europe and the rest of North America remains open. The similarities in both dimorphs with the Pacific Sphaeroceratinae of the Humphriesianum Zone have already been noted in the systematic introduction above (see Fig. 2) suggesting a close phyletic and temporal relation, and the best guess for the age of *C. borealis* remains early Upper Bajocian as suggested previously (Callomon, 1959).

Fauna C2. Cranocephalites indistinctus CALLOMON, 1959.
Richardsons (Poulton *et al.* 1982, p.84); Prince Patrick Island (Frebold 1958, Pl.8); East Greenland. In Greenland there are at least two slightly different assemblages in the Indistinctus Zone.

Fauna C3. Cranocephalites pompeckji (MADSEN, 1904) and spp. aff.
Richardsons (Poulton *et al.*, 1982); Prince Patrick Island (Frebold, 1958); East Greenland; Novaya Zemlya (Sokolov, 1913); northern Siberia (Voronets, 1962). In East Greenland the Pompeckji Zone is subdivisible, encompassing at least six distinct faunas; and some of the forms from Siberia (e.g., *C. nordvikensis* Voronets) are different again. Close correlation of the Canadian material is not yet warranted either by the quality or quantity of material available.

Fauna C4. Arctocephalites sp. *A.* aff. *elegans* SPATH, 1932.
Small, compressed, fine-ribbed species of *Arctocephalites* that resemble *A. elegans* Spath (1932, Pl.10, Fig.4) in these respects were recorded as such from various localities by Frebold (1961, Salmon Cache Canyon; 1964a, Axel Heiberg Is.). New collections *in situ* both at Salmon Cache Canyon and in East Green-

land show that the resemblance is close only between the peripheral variants of what are quite distinct assemblages although the ages are probably not very different. The Greenland fauna from the lower Arcticus Zone includes *A. nudus* Spath, *A. ornatus* Spath, *A. sphaericus* Spath as well as the Franz Josef Land species *A. koettlitzi* (Pompeckj, 1900a), *A. ellipticus* Spath, *A. pilaeformis* Spath and the true *A. arcticus* (Newton and Teall, 1897), all in the same metre thick concretionary bed. The assemblages from the Porcupine are being fully described by Poulton (1983). Porcupine and Richardsons, various localities (Poulton *et al.* 1982); ?Axel Heiberg Island.

Fauna C5. Arctocephalites sp. nov. *B*
At Salmon Cache Canyon the beds with fauna C4 are succeeded by horizons yielding even smaller forms of this genus, the macroconchs, with minute umbilici, barely exceeding 50 mm in diameter. Fragmentary material may be hard to distinguish from the small *Phylloceras* or *Partschiceras* that occurs in the same beds (cf. Frebold *et al.*, 1967, Pl.2, Figs.1,2 ?). Also on the Old Crow River near its junction with the Porcupine. These forms are also being described by Poulton. In East Greenland, higher parts of the Arcticus Zone yield similarly fine-ribbed smoothish forms with minute umbilici, but they are not as small as those of fauna C5.

Fauna C6. Arctocephalites callomoni FREBOLD, 1964a.
Axel Heiberg Island, in a well-defined bed; and Salmon Cache Canyon on the Porcupine. ? Northern Siberia (*A. voronetsae, kigilakhensis* Meledina, 1973, Pls.13,14,16). Occasional variants resembling these species occur in East Greenland in the Greenlandicus Zone, but the assemblage as a whole is distinct.

Fauna C7. Arctocephalites cf. or aff. *greenlandicus* SPATH, 1932
Porcupine and Richardsons, widespread. In East Greenland the Greenlandicus Zone is divisible into at least three Subzones. It therefore represents a considerable time-span and although the general correspondence to the faunas from the Yukon is clear, an exact match may not be possible. There are also a number of levels at Salmon Cache Canyon yielding faunas of this general aspect. They include inflated variants comparable with *A. crassus* (Madsen) and *A. freboldi* Spath found in East Greenland mainly in the higher parts of the Zone.

Fauna C8. Arcticoceras ishmae (KEYSERLING, 1846).
In East Greenland the Ishmae Zone (formerly Kochi Zone: see Callomon, 1972) is clearly divisible into three Subzones: of *A. harlandi* Rawson, 1982 (type from Spitsbergen) (lowest); *A. ishmae* (lectotype from the Petchora); and *A. crassiplicatum sp. nov. MS* (type from Jameson Land (highest). Copious new collections from the Ishmae Subzone match the lectotype of *A. ishmae* and the topotypes in the British Museum (cf. Spath, 1932, Pl.15, Fig.7) perfectly, and show that the poorly-preserved holotype of *A. kochi* Spath came from this assemblage. In *A. harlandi* the ribs are straight and rectiradiate, in *A. ishmae* somewhat curved and moderately projected on a more sharply curved venter, and in *A. crasiplicatum* the coarse and ventrally strongly projected ribs persist almost undiminished to the adult peristome on whorls of almost carinate section. All three Subzones can be recognized in the Porcupine – Old Crow area of the northern Yukon but seem to have been cut out at least in the eastern Richardsons.

Fauna C8(a). Arcticoceras harlandi RAWSON, 1982
Salmon Cache Canyon, Porcupine (cf. Frebold, 1961, Pl.12, Pl.16, Fig.1).

Fauna C8(b). Arcticoceras ishmae (KEYSERLING, 1846)
+ *Iniskinites* sp. nov.
Oppelia (? *Oxycerites*) sp. cf. and aff. *undata* and *jugata* Ershova and Meledina, 1968
Choffatia sp.
Salmon Cache Canyon. It seems significant that this is precisely the same horizon as the only one to have yielded Oppelids in East Greenland (Birkelund, Hakansson and Surlyk, 1971). The forms from northern Siberia appear to be slightly older, said to be from beds with *Arctocephalites. A. ishmae* has a wide distribution, including northern Russia (Keyserling, 1846), Novaya Zemlya and northern Siberia (Meledina, 1973).

Fauna C8(c). Arcticoceras cf. *crassiplicatum* sp. nov. MS
A spot locality north of the Porcupine, between Old Crow and Ramparts House: D.K. Norris collection, 1981 (1409 NC; I am indebted to Dr. Norris for showing me this material).

Fauna C9. Cadoceras barnstoni (MEEK, 1859)
+ *Iniskinites* aff. *varicostatus* (Imlay, 1953)
Kepplerites cf. or aff. *rosenkrantzi* Spath
Porcupine: Salmon Cache Canyon (almost certainly the source of the type of *C. barnstoni,* refigured by Frebold, 1964a), abundant; Axel Heiberg Island, "lower *Cadoceras* beds" *partim* ? (Frebold, 1964a, Pl.10, Fig.2); Ellef Ringnes Island ? (Frebold, 1964a, Pl.11, Fig.2). Individual variants are close to *C. variabile* Spath, although the assemblage as a whole is distinct. Both *Cadoceras* and *Kepplerites* are so close to those of the Variabile Zone of East Greenland that a very similar age cannot be in doubt – it is still Boreal Upper Bathonian. In Greenland there are several further faunas between Ishmae and Variabile Zones indicating a considerable time gap between faunas C8 and C9, but the absence, so far, of any evidence of these faunas or their equivalents in the Yukon may well be merely collection-failure. At Salmon Cache Canyon there is a considerable thickness of poorly exposed shales in the interval in question.

Fauna C10. Cadoceras (*Paracadoceras* ?) sp. indet.
A single layer of concretions at about the middle of some 50 m of very poorly exposed beds in the upper part of the Salmon Cache Canyon section, between faunas C9 and C11, contains rather badly preserved crushed specimens of what appeared to be *Paracadoceras* (pers. observation, 1975). If correct, it is unfortunate that it would be just this sparsely-documented part of the north Canadian succession in which the equivalents of the rich *Cadoceras*-beds of southern Alaska should be sought.

Fauna C11. Cadoceras bodylevskyi FREBOLD, 1964
The type material came from the "lower *Cadoceras* beds" of Axel Heiberg Island (Frebold, 1964a). A closely similar, if not identical, assemblage occurs in great profusion in the top beds at Salmon Cache Canyon and analogous levels at other localities in the vicinity.

Fauna C12. Cadoceras septentrionale FREBOLD, 1964.
Axel Heiberg Island, "upper *Cadoceras* beds" (Frebold, 1964a). These forms retain vestigial primary bullae on the umbilical margin to fairly late stages, reminiscent of earlier species, but begin to assume the globular shape and depressed whorl section characteristic of the later Lower Callovian forms like *C. sublaeve.* Extreme variants can become almost spindle-shaped. What appears to be an identical assemblage occurs in East Greenland between *Paracadoceras* and *Cadoceras sublaeve* and forms the basis there of the Septentrionale Zone which is assumed to lie immediately below the Koenigi Subzone of the Calloviense Zone.

Fauna C13. Cadoceras voronetsae FREBOLD, 1964 (incl. *Cad. septentrionale* var. *latidorsata* Frebold *partim,* 1964a, Pl.7, Fig.1, Pl.8, Fig.1; *Cad. arcticum* Frebold, 1964a?)
Large spherocones with smooth steep-walled conical umbilici as in the group of *C. sublaeve.* Richardsons, not closely localized and not in sequence; Cornwall Island ? (type of *C. arcticum*). The relative sequence of faunas C12 and C13 is conjectural.

Fauna C14. Cadoceras (Stenocadoceras) canadense FREBOLD, 1964a.
Richardson Mountains, not closely localized. The finer and ventrally more strongly projected ribbing gives the impression that this species may be somewhat younger than those of the *Stenocadoceras stenoloboide/multicostatum* assemblage (B9) of southern Alaska, intermediate between it and early *Longaeviceras,* e.g., *L. filarum* Meledina (1977, Pl.24, Fig.3). This suggests a late Middle Callovian age.

Faunas C15 - C18: Oxfordian, *Cardioceras* and *Amoeboceras* spp.
Small collections have been made from widely scattered localities. The material is rarely well preserved and almost never found in faunal succession. As a whole it indicates that almost all the Oxfordian is represented at one place or another and that the ammonites are essentially the same as those found elsewhere in North America.

Fauna C15. Cardioceras cf. or aff. *scarburgense* (YOUNG and BIRD).
Richardsons, Berry Creek (see Poulton *et al.* 1982, Fig.1; new collection, locality Pu-2-81). ? Tombstone area, central Yukon (recorded as *Cadoceras* ? sp. in Frebold *et al.,* 1967, Pl.2, Figs.3-6). Lower Oxfordian, Mariae Zone, Scarburgense Subzone.

Fauna C16. Cardioceras cf. or aff. *alphacordatum* SPATH or *mountjoyi* FREBOLD
Richardsons, various localities (Poulton *et al.,* 1982, p.33,37,91) including the one cited previously, in sequence above fauna C15. Axel Heiberg Island (Frebold, 1961, Pl.18, Fig.2). Lower Oxfordian, upper Mariae or lower Cordatum Zones.

Fauna C17. Cardioceras cf. or aff. *distans* (WHITFIELD)
Tombstone area, central Yukon (Frebold *et al.,* 1967, Pl.2, Figs.8,9, Pl.3, Fig.4; Poulton and Templeman-Kluit, 1982). Lower Oxfordian, Cordatum Zone.

Fauna C18. Amoeboceras sp. cf. or aff. *transitorium* SPATH
Richardsons, Martin Creek, near base of Husky Formation (recorded as *Cardioceras* sp. in Poulton, 1978b, p.464). Enough of the diagnostic serrated keel and lateral tubercles are preserved to make the identification as *Amoeboceras (Prionodoceras)* firm (see Sykes and Callomon, 1979). Early Upper Oxfordian, Serratum Zone. The specimen is important because it dates the onset of the *Buchia* facies that spread over so much of North America and displaced the otherwise ubiquitous ammonites during the Upper Jurassic.

Fauna C19. Dorsoplanites aff. *maximus* SPATH + *Pavlovia* (? *Paravirgatites*) sp.
Ellesmere Island (Frebold, 1961), Middle Volgian. The general resemblance to the well-known Volgian faunas of East Greenland (see Callomon and Birkelund, 1983, for a recent review), the Barents Shelf and northern Siberia (Zakharov and Mesezhnikov, 1974) is obvious but a more precise match remains elusive. It is not clear whether the two specimens are strictly associated and of precisely the same age. Their closest matches in Greenland would be of somewhat different ages. The reluctance of these otherwise prolific ammonites to penetrate further than the northern end of the Sverdrup Basin during the late Jurassic and early Cretaceous remains mysterious (see also discussion below, under fauna A18).

Fauna C20. Craspedites canadensis JELETZKY, 1966
Ellesmere Island, a prolific fauna from a single concretion at an isolated level in the Deer Bay Formation; Upper Volgian.

The relatively rich succession of ammonite faunas in the Neocomian is summarized by Jeletzky (1973), but with rare exceptions it, too, is confined to the northern Sverdrup Basin.

Region D. Alaskan North Slope – British Mountains

The Jurassic rocks of the Brooks-Mackenzie Basin north of the "disturbed belt" attributed to the Kaltag Fault are poorly exposed and ammonites are known from only a few horizons. The Aalenian – Lower Bajocian succession has already been described in the previous section and is essentially the same as in regions B and C; faunas Be1, Be3 and Be5 have been recognized. They are succeeded by the following horizons.

Fauna D1. "Cranocephalites" [Iniskinites ?] ignekensis IMLAY (1976)
The identity of this fauna is subjet to the same doubts as is that of "*Cranocephalites*" *costidensus,* B4, from southern Alaska. It does not fit very readily into *Cranocephalites sensu stricto,* having equally close affinities with the Pacific genus *Iniskinites,* as discussed above under fauna B4, to which it bears a very close resemblance. True *Cranocephalites* occurs not far away, in Prince Patrick Island (fauna C3; Frebold, 1958, Pl.7).

Fauna D2. Arctocephalites aff. *elegans* SPATH – IMLAY (1976)
Undoubtedly this is a member of the true Boreal Arctocephalitinae close to the forms from the northern Yukon described above under fauna C4. The exact stratigraphical relations between faunas D1 and D2 still seem somewhat uncertain. At some localities D2 lies above D1, as it should if both are correctly identified. At others they appear to occur together

(Imlay, 1976), which would be simply explicable only if D1 were an *Iniskinites* rather than a *Cranocephalites*. Boreal Bathonian, Arcticus Zone.

Fauna D3. ? Arcticoceras cf. *ishmae* (KEYSERLING)
Canning River (Imlay, 1976 Pl.3, Figs.22,?17,18. For the remaining material figured as *A. ishmae,* see fauna D6 below).

Fauna D4. Cadoceras cf. *arcticum* FREBOLD, 1964
Babbage River (eastern boundary of the British Mountains, northern Yukon; Frebold, 1964a). Aichillik River (western end of British Mountains, north-eastern Alaska; U.S.G.S. Mesozoic loc.31274, Detterman coll. 1976). The specimens from the Babbage River are paratypes of *C. arcticum,* although the holotype came from Cornwall Island. The scanty material most closely resembles the late Cadoceratids of the Calloviense Zone (cf. fauna C13), but it is not well-enough preserved for closer comparisons.

Fauna D5. Cadoceras (*Stenocadoceras*) cf. or aff. *canadense* FREBOLD, 1964.
Babbage River (Frebold, 1964a); cf. faunas B9, C14. Middle Callovian?

Fauna D6. Longaeviceras cf. *stenolobum* (KEYSERLING, 1846)
(= *Pseudocadoceras grewingki,* Imlay, 1955, Pl.12, Fig.11; *Arcticocers ishmae* and sp. juv., Imlay, 1976, Pl.3, Figs.1-4,7-16,19-21; *Cardioceras dettermani* Imlay *partim* (not the holotype), 1981, Pl.10, Figs.1-4,8-11, = 1976, pp.3,5 as *Card.* (*Scarburgiceras*) sp.)
+ *Grossouvria* sp.
Longaeviceras has a wide distribution in the Boreal Realm, and many of the Alaskan forms can be matched elsewhere. Imlay's macroconch (1976, Pl.3, Figs.7,8) is close to the true *L. stenolobum* itself, as represented by Sokolov's (1912) Plate 1, Figure 4 from the Petchora [neither "the" type specimen, as claimed by Sokolov, nor even a syntype, according to Bodylevsky (1960, p.78) who renamed it *Cad. lahuseni,* but probably a topo- or chorotype]. This probably includes *L. nikitini* (Sokolov) (1912, Pl.1, Fig.3), and Newton and Teall's *Cad. modiolare* from Cape Flora, Franz Josef Land (1897, Pl.39, Fig.10, Institute of Geological Sciences Collection, London, *For 964;* cf. Clark, 1982). Imlay's Figures 9 and 10 show the onset of the characteristic shallow conical umbilicus to be seen i.a. in *L. bodylevskii* Meledina (1977, Pl.30, Fig.2), from Anabar Bay in northern Siberia. The Alaskan microconchs include some (Pl.3, Figs. 1,12,16) that match Pompeckj's *Cad. nanseni* (1900a, Pl.2), also from Cape Flora; and others (Pl.3, Figs.14,20) that are close to *L. holtedahli* (Salfeld and Frebold, 1924, Pl.1, Figs.3,4), from Novaya Zemlya. In the Boreal Province proper (northern Alaska; northern Siberia: Lena, Olenek, Anabar; northern Russia and Barents Shelf: Petchora, Novaya Zemlya, Franz Josef Land, Spitsbergen) *Longaeviceras* dominates and occurs free of *Kosmoceras.* A little further south on the borders of the Sub-Boreal Province (e.g., in East Greenland), *Longaeviceras* and *Kosmoceras* are found in comparable numbers; while in the Sub-Boreal Province proper [e.g., England (*L. longaevum* and *placenta* (Leckenby, 1859) in Buckman, 1919-20, Pls.121, 148) and the Moscow Basin (Nikitin, 1881)], *Longaeviceras* occurs only as a minor constituent of the faunas, which now includes abundant more southerly elements, such as Perisphinctidae and Peltoceratinae. Upper Callovian, Athleta Zone, probably middle part, a little younger than fauna B10 of southern Alaska; *Longaeviceras keyserlingi* Zone of Soviet authors.

Fauna D7. Cardioceras cf. *hyatti* REESIDE *sensu* IMLAY, 1982
British Mountains, Babbage River (Frebold *et al.,* 1967, Pl.3, Figs.2,3). The scanty material seems to match some of the forms figured by Imlay (1982, Pl.15), from fauna A14, q.v. Lower Oxfordian, Cordatum Zone.

Fauna D8. Cardioceras (*Maltoniceras*) sp.
British Mountains, Babbage River (Frebold, *et al.,* 1967, Pl.2, Fig.10, Pl.2, Fig.1). Middle Oxfordian, upper Densiplicatum Zone.

Fauna D9. Cardioceras – Amoeboceras sp. nov. + *Buchia concentrica*
Canning River (Imlay, 1976 p.4, Text-fig.1; 1981, Pl.12, Figs.16-21). Hard to place closely; the macroconch (Pl.12, Figs.19-21) still resembles late *Card.* (*Maltoniceras*) in the crenulation of the keel (cf. Arkell, 1941, Pl.51, Figs.3b,7b) but *Amoeboceras* (*Prionodoceras*) in ribbing and coiling, whereas the microconch (Fig.16) resembles *Amoeboceras* s.s in its keel but *Card.* (*Vertebriceras*) in ribbing (cf. de Riaz, 1898, Pl.16, Fig.8; Gygi and Marchand, 1982, Pl.13, Fig.2) of the Blakei Subzone of the Tenuiserratum Zone. Hence the age is probably earliest Upper Oxfordian, basal Glosense Zone.

Fauna D10. Amoeboceras spp. + *Buchia concentrica*
Canning River (Imlay, 1955, Pl.12, Figs.2-6; 1981, Pl.12, Fig.1) and Babbage River (Frebold *et al.,* 1967, Pl.3, Figs.5-7). These barely determinable fragments suggest a late Upper Oxfordian age, perhaps Rosenkrantzi Zone.

Higher Jurassic beds are indicated by further *Buchia* spp., but ammonites first reappear in the Albian (see Imlay, 1961, *PP 335*).

CORRELATION

The ammonite faunas and their supposed successions in regions A to D are shown in Figures 3, 4 and 5. It must be stressed that these tables represent no more than time-ordered sequences and that vertical spacing are not proportional to time-intervals. Neither are the thicknesses of the boxes representing the faunas meant to suggest equal time-durations. The effective time-durations of the faunas as catalogued in this review must have varied enormously, from the instantaneous to periods comparable to that represented by the whole of an average standard ammonite Zone. Boxes in adjacent columns shown as partly or wholly overlapping are meant to suggest that their faunas probably differ but little in age, although their precise order is usually not known.

Upper Bajocian – Bathonian (Figure 3)

The relevant part of the standard Boreal Middle Jurassic ammonite zonation is indicated as a reference scale on the right. Standard European Stages are named on the left, but their ammonite Zones are not included because correlation of the American faunas with them at this level of precision is still quite impossible. Ranges of some important genera are also shown.

The boundary between Bajocian and Bathonian has been drawn to lie above faunas B2 – B4 with *Leptosphinctes* ("*Cobbanites*" s.s.) and below fauna A4 with (fairly late) *Procerites*. If the interpretation of *"Cranocephalites" costidensus* (B4) and *"C." ignekensis* (D1) suggested in the text is correct, the proximity of the latter to the earliest *Arctocephalites* of D2, which in turn correlates closely with C4, would indicate that the Bajocian – Bathonian boundary in the standard Boreal succession lies somewhere around the Pompeckji – Arcticus boundary. This would be new. Hitherto, except for a suggestion that the Borealis Zone might still be Upper Bajocian, no attempt has been made to indicate this boundary more closely. There has just been no evidence, and everything from the *Cranocephalites indistinctus* Zone to the *Cadoceras calyx* Zone inclusive below the beds with *Kepplerites keppleri*, has been simply referred to as "Boreal Bathonian". The closeness of the resemblance between *Cranocephalites borealis* (C1) and *Chondroceras oblatum* of the North American Oblatum Zone, correlated with the European Humphriesianum Zone, that has become apparent in recent times strengthens the idea that all the true *Cranocephalites* beds of the Arctic may still be Upper Bajocian.

Faunas A4 and B6 are shown close together because both appear to mark the lowest appearance in their regions of *Kepplerites* with forms that are closely similar. Faunas A5 and A7 are correlated with the Boreal Cranocephaloides Zone because of the characteristic small, fine-ribbed *Kepplerites* they contain. In East Greenland *Kepplerites* first appears suddenly and in quantity in this Zone. Hence, if the correlation is correct, it must occur in North America already somewhat earlier. Assuming that the time-interval between A4 and A5 is short, this correlation leaves a considerable gap between A4 or B6 and B4, filled in the north by faunas C5 – C8 whose close correlation with the standard Boreal succession is in no doubt. The Cranocephaloides Zone has been taken arbitrarily to mark the base of the Boreal Upper Bathonian, but that does not imply that it also correlates with the base of the European Upper Bathonian. The *Procerites* of faunas A4 – A7 still resemble rather more those of the European Middle Bathonian, although the genus does range up into the lowest, Hodsoni Zone of the Upper Bathonian there (cf. Torrens in Cope, 1980).

Faunas A8 and B7 are correlated with the Peramplus Subzone of the Calyx Zone of the Boreal succession, again on the strength of the *Kepplerites* they contain, particularly the microconchs (*K. knechteli, alaskanus*). The single *Kepplerites* of A10 resembles those of the upper Calyx Zone. Nothing can be said about the precise relative age of fauna A9, and hence it has been omitted.

Looking at Figure 3 as a whole, probably the most striking feature of the correlations proposed is the inclusion of all of the long and well-known "Callovian" faunas A3 to A8 of the Western Interior still in the Bathonian. In fact, with the exception of the single fauna A11 from Montana at the very top, the Callovian appears not to be represented by ammonites in the Western Interior at all. It may seem ironic that this marks a return to the interpretation that Imlay would clearly have preferred originally on the strength of the mollusca other than the ammonites (Imlay, 1956). These include *Gryphaea*, and the ranges of the important species are also indicated in Figure 3 (cf. Hallam and Gould, 1975).

Callovian (Figure 4)

Standard Boreal and European zonal scales are indicated on the right and left. The richness of the faunas from southern Alaska distorts the table and weighs it heavily in favour of its lower part.

The position of the Bathonian – Callovian boundary is based on the correlation via East Greenland of *Kepplerites loganianus* with *K. keppleri*, and hence with the stratotype of the base of the Callovian in England. The Macrocephalus Zone is thus as defined stratigraphically (Callomon, 1964) and not by the supposed first appearance of *Macrocephalites* in Europe (Thierry, 1978). Most of the faunas fall into the equivalents of this Zone and probably its lowest Subzone at that, and they reveal the fact that the thick volcanoclastic turbiditic sequences of southern Alaska were deposited in short bursts of rapid sedimentation separated by major non-sequences that are inconspicuous.

Oxfordian – Volgian (Figure 5)

This figure calls for few comments. The standard Boreal Oxfordian ammonite zonation shown at the left applies to the whole of North America north of the 37th parallel, with the exception of the small corner in south-western Oregon, probably displaced tectonically from the south, that bears Tithonian faunas. Assignments of individual faunas, and their uncertainties, are discussed in the text.

CONCLUSIONS

This review has attempted to re-examine the Middle and Upper Jurassic ammonite faunas of North America in the light of what has become known in the last 25 years of ammonoid biostratigraphy more widely, particularly in the Arctic, and of what has been learned of the distribution, evolution, and classification of the Jurassic ammonites as a whole. The principal conclusions are as follows.

Faunal Provincialism

During the early Middle Jurassic, up to the Humphriesianum Zone of the Bajocian, the ammonites of North America divide into two broad faunal domains. Regions A and B, Western Interior and Cordillera, were inhabited mainly by pandemic Tethyan faunas with only minor locally provincial elements (e.g., *Zemistephanus, Parabigotites, Chondroceras oblatum*). Regions C and D, Arctic Canada and Alaska, belonged to a clearly differentiated Boreal Realm (*Pseudolioceras*) and North Pacific or Bering Province (*Erycitoides, Arkelloceras*). The provinces overlapped strongly in southern Alaska, so that correlation presents no problems.

At the beginning of the Upper Bajocian, provincialism suddenly became acute. Regions A and B were now firmly part of a well-differentiated Pacific Realm, in which Eurycephalitinae of the Sphaeroceratidae dominated, with strong faunal affinities between, for example, the Cordilleras of both North and South America. Region C (Arctic Canada) was part of an equally well-differentiated Boreal Realm characterized mainly by Arctocephalitinae and Cadoceratinae of the Cardioceratidae; and region D appeared to be one of overlap between Pacific and Boreal Realms. The ammonites of Pacific, Boreal and Tethyan Realms became so mutually exclusive in the Upper Bajocian and Bathonian that correlation at zonal level is still not possible. Each realm, therefore, has to retain its own independent ammonite zonation up to about the top of the Lower Callovian.

A more restricted provincialism is also discernible in these realms during this time. The shallow epicontinental basins of region A, far distant from the Pacific Ocean, appeared to develop their own endemic faunas of Eurycephalitinae, forming a separate Middle Jurassic Western Interior or Shoshonean Province (see Taylor *et al.*, this volume). Some other genera, e.g., *Parareineckeia, Epizigzagiceras* and *Iniskinites,* extending over all of region B, mark a North Cordilleran, or Athabascan Province (Taylor *et al.,* this volume). A Sub-Boreal Province is characterized by *Kepplerites* of the Kosmoceratidae. It overlaps with the North Cordilleran and Western Interior Provinces of North America, the Boreal Province of North America and East Greenland, and thence the Submediterranean Province of Europe, thereby providing one of the few links that make even rough interprovincial correlations possible at the Bathonian – Callovian boundary.

In the Upper Jurassic, the ammonite faunas of the whole of regions A to D, mainly Cardioceratidae, become homogeneously representative of the Boreal Realm and Province. They stand in sharp contrast to the Submediterranean and Tethyan faunas of Mexico. The boundary between the realms probably oscillated somewhat with time and lay mainly in southern California, but its positions are rarely known more closely, lost or heavily obscured by subsequent tectonic deformations.

Regional Allochthony and Tectonic Displacements

Current discussions of the growth of western North America by the accretion of terranes pose three questions on which the present review may have a bearing. They are: what light, if any, can the distribution of ammonites as found today throw on: 1) possible movements since the Middle Jurassic of terranes within the Cordilleran domain, region B, relative to each other? 2) movements between terranes in the Cordillera, B, and the cratonic domains of the Western Interior, A, and Arctic Canada, C? and 3) movements of the Alaskan Brooks Range and North Slope, D, relative to the craton, C?

An upper limit to movements within the Cordillera can be put forward from a comparison of the faunas of the Alaska Peninsula with those of British Columbia and Oregon. There are striking similarities that have been pointed out above, but there are also differences. The most significant lies in the distribution of *Cadoceras*. This Boreal Lower Callovian genus dominated in Alaska and at other localities as far south as Vancouver Island and southern British Columbia, but not further south. This southern limit appears to have been the Boreal provincial boundary, which suggests that had the Alaskan Peninsular terrane moved north, it could probably not have come from much further south than about the 49th parallel, putting an upper limit on the displacement of perhaps 2000 km (1300 miles). On the other hand, the evidence is equally consistent with a movement of zero.

Evidence for or against lateral displacement between Cordilleran terranes and the cratonic Western Interior is tenuous. The Sub-Boreal *Kepplerites* found in the northern parts of the Western Interior have rough equivalents in east-central Oregon and southern British Columbia, and in one case as far south as north-central California. Similarly, the Boreal Cardioceratidae of the Western Interior go right across at the same latitude into Oregon. Conversely, with the exception of the Tithonian faunas found in northern California and southwestern Oregon (and, in the Neocomian, as far north as Vancouver Island), Tethyan ammonites are not found north of California. This strongly suggests that the Boreal – Tethyan faunal boundary lay at about the same latitude in the Cordillera as it did in the Western Interior, i.e., that with the exceptions mentioned above, right lateral movements cannot have exceeded a few hundred kilometres. They could again equally well have been zero. The presence of strongly Boreal faunas (B16) as far south as Mariposa County in central California could even be construed as evidence in favour of a left-lateral (southward) displacement, although associated sub-Mediterranean forms would favour provincial overlap as a preferred explanation. Similarly, the proximity of purely Boreal Callovian and Tethyan Neocomian faunas on the west coast of Vancouver Island (Jeletzky, 1964) would seem to cancel each other out as possible indicators of serious left- and right-lateral movements respectively. In summary, Middle and Upper Jurassic ammonites as found today are consistent with lateral Cordilleran tectonic displacements of from zero to perhaps some hundreds of kilometres, with 2000 km for the Alaskan Peninsula as absolute maximum.

With regard to the North Slope terrane, D, its Jurassic ammonites are typical of the Boreal Province throughout and its affinities close with those of the adjacent region C. Comparing columns C and D in Figures 3 to 5, the number of mismatches greatly exceeds the coincidences, which could be thought to suggest that the two regions may have been much further apart during the Jurassic than they are now. However, the differences could be attributed equally satisfactorily to a combination of facies-differences and collection-failure. The Jurassic sediments of region C yielding abundant ammonites are predominantly of near-shore shallow-water facies, whereas those of D (sparsely fossiliferous Kingak Shale) are of more distal basinal type; and the distances between them today are consistent with the facies-gradients that are observed in region C. The presence of a strong residual Pacific influence (*Arkelloceras,* and *"Cranocephalites" ignekensis* of fauna D4) points, moreover, to a former position not much further from the Pacific than what it is at present. The evidence as a whole cannot rule out relative displacements of up to at most a few hundred kilometres, but is also again consistent with displacements of zero.

The catalogue of faunas put together in this review leaves many smaller assemblages from widely scattered localities unassigned for lack of evidence. The positions of some of those that have been assigned are open to question and will doubtlessly have to be revised. Yet others will come to be discovered and will have to be slotted into the gaps. Whatever the permanence of the present scheme may turn out to be, it will have fulfilled its purpose if, by pointing to the many gaps and uncertainties, it stimulates further efforts to make the state of our knowledge more complete.

ACKNOWLEDGEMENTS

This review owes much to those who have stimulated my interest in the Jurassic of North America, both in the field, in the collections, and through innumerable discussions. I am grateful particularly to H. Frebold, J.A. Jeletzky, T.P. Poulton, F.G. Young and the Geological Survey of Canada for making it possible for me to see the Jurassic of the Canadian Rockies, northern Yukon and the Richardson Mountains; and to R.W. Imlay, R.L. Detterman and the U.S. Geological Survey for showing me something of the Western Interior and the Alaskan Peninsula. Financial assistance from N.E.R.C. (U.K.) is also gratefully acknowledged.

REFERENCES

Arkell, W.J., 1939, The Ammonite Succession at the Woodham Brick Company's Pit, Akeman Street Station, Buckinghamshire, and Its Bearing on the Classification of the Oxford Clay: Quarterly Journal of the Geological Society, London, v.95, p.135-222.

———, 1941, 1946, A Monograph of the Ammonites of the English Corallian Beds: Monograph of the Palaeontographical Society, London, Pt.vii, 217-238, Pt.xi, p.297-332.

———, 1952, 1954, Monograph of the English Bathonian Ammonites: Monograph of the Palaeontographical Society, London, Pt.iii, p.73-102; pt.iv, p.103-128.

———, 1956, Jurassic Geology of the World: Oliver and Boyd, Edinburgh and London, 806 p.

———, 1957, in Moore, R.C., ed., Treatise on Invertebrate Paleontology, pt.L, Mollusca 4: Kansas, University of Kansas Press, 490 p.

Besnosov, N.V., 1982, On the systematics of the Perisphinctidae (Ammonoidea): Paleontologicheskyi Zhurnal, 1982, p.54-64 (in Russian).

Bidzhiyev, R.A., 1965, Zonal Analysis of the Jurassic Deposits of the Northern Pre-Verkhnoyan Regional Depression: Geologiya i Geofizica, v.4, p.49-57, in Russian.

Birkelund, T., Hakansson, E., and Surlyk, F., 1971, New Finds of Bathonian, Callovian and Oxfordian Ammonites in Northern Jameson Land, East Greenland: Geological Society of Denmark Bulletin 20, p.240-259.

Blake, J.F., 1905, A Monograph of the Fauna of the Cornbrash: Monograph of the Palaeontographical Society, London, Pt.i, p.1-100.

Bodylevsky, V.I., 1960: Callovian ammonites from northern Siberia: Zapisky Leningradskogo ... gornogo Institut, v.37, vyp.2, p.49-82, in Russian.

Buckman, S.S., 1881, A Descriptive Catalogue of Some of the Species of Ammonites from the Inferior Oolite of Dorset: Quarterly Journal of the Geological Society, v.37, p.588-608.

———, 1918-1927, Yorkshire Type Ammonites, v.2, and Type Ammonites, v.3-7, Plates 68-790: London, published by the author and Wheldon and Wesley; reprinted 1972-1976 as Historia Naturalis Classica, v.93, J. Cramer Verlag, Lehre, and Wheldon and Wesley, Codicote, Herts., U.K.

———, 1929, Jurassic Ammonoidea: National Museum Canada Bulletin 58, p.1-27.

Burckhardt, C., 1903, Beiträge zur Kenntniss der Jura- und der Kreide-Formation der Cordillere: Palaeontographica, v. 50, Pt.i, p.1-72.

———, 1927, Cefalopodes del Jurasico Medio de Oaxaca y Guerrero: Instituto Geologico de Mexico, Bol.47, 108p.

Callomon, J.H., 1959, The Ammonite Zones of the Middle Jurassic Beds of East Greenland: Geological Magazine, v.96, p.505-513.

———, 1964, Notes on the Callovian and Oxfordian Stages: Institut Grand-Ducal Luxembourg, Sect. Sciences Naturelles, Physiques et Mathematiques, Comptes Rendues et Mémoires du Colloque Jurassique, Luxembourg, 1962, p.269-291.

———, 1972, An Annotated Map of the Permian and Mesozoic Formations of East Greenland: Jurassic System: Meddelelser om Gronland, v.168, no.3, p.15-20.

———, 1981a, Classification of the Jurassic Ammonitina: *in* House, M.R., and Senior, J.R., eds., The Ammonoidea: Systematics Association Special Publication 18, Academic Press, London and New York, p.101-155.

———, 1981b, Dimorphism in Ammonoids: *in* House, M.R., and Senior, J.R., eds., The Ammoidea: Systematics Association Special Publication 18, London and New York, Academic Press, p.257-273.

Callomon, J.H. and Birkelund, T., 1980, The Jurassic Transgression and the Mid-Late Jurassic Succession in Milne Land, Central East Greenland: Geological Magazine, v.117, p.211-226.

———, 1983, The ammonite Zones of the Boreal Volgian (Upper Jurassic) in East Greenland: *in* Embry, A.F., and Balkwill, H.R. eds., Arctic Geology and Geophysics: Canadian Society of Petroleum Geologists, Memoir 8, p.349-369.

Clark, R.D., 1982, Type, Figured and Cited Jurassic Cephalopoda in the Collection of the Institute of Geological Sciences: Institute of Geological Sciences, London, Report 82/9, 104p.

Cope, J.C.W., ed., 1980, A Correlation of Jurassic Rocks in the British Isles: Pt.II: Middle and Upper Jurassic: Geological Society of London, Special Report 15, 109p.

Crickmay, C.H., 1930, Fossils from Harrison Lake Area, British Columbia: National Museum Canada, Bull.63, Geol. Ser. No.51, p.33-66.

———, 1933, Mount Jura Investigation: Geological Society of America Bulletin 44, p.895-926.

Dietl, G., 1980, Die Ammoniten-Gattung *Leptosphinctes* aus dem südwestdeutschen Subfurcaten-Oolith (Bajocium, Mittl. Jura): Stuttgarter Beiträge Naturkunde, v.66B, 49p.

———, 1981, Über *Macrocephalites* (Ammonoidea) aus dem Aspidoides-Oolith und die Bathonium/Callovium-Grenzschichten der Zollernalb (SW-Deutschland): Stuttgarter Beiträge Naturkunde v.68B, 15p.

Donovan, D.T., 1953, The Jurassic and Cretaceous Stratigraphy and Palaeontology of Traill 0, East Greenland: Meddelelser om Gronland, v.111, no.4, 150p.

Douglas, J.A., and Arkell, W.J., 1928, The Stratigraphic Distribution of the Cornbrash: I, The Southwestern Area: Quarterly Journal of the Geological Society, London, v.84, p.117-178.

Efimova, A.F., Kinasov, V.P., Paraketsov, K.V., Polubotko, I.V., Repin, Y.S., and Dagis, A.S., 1968, Field Atlas of the Jurassic Faunas and Floras of Northeastern U.S.S.R.: Magadan, Ministry of Geology of the R.S.F.S.R., 382p.

Enay, R., 1959, Note sur quelques Tulitidés (Ammonitina) du Bathonien: Bulletin de la Société Géologique de France, 7e série, v.1, p.252-289.

Ershova, E.S., and Meledina, S.V., 1968, Late Bathonian Oppeliidae of Northern Siberia: *in* Saks, V.N., ed., Mesozoic Marine Faunas of the Northern and Far-eastern U.S.S.R. and Their Stratigraphic significance: Trudy, Instituta Geologii i Geofiziki Akad. Nauk S.S.S.R., Sibirskoe otdelenie, v.48, p.42-50, in Russian.

Flower, R.H., 1942, Antarctic Cephalopod Faunule from the Cynthiana of Kentucky: Bulletins of American Paleontology, v.27, p.5-41.

Frebold, H., 1957, The Jurassic Fernie Group in the Canadian Rocky Mountains and Foothills: Geological Survey of Canada Memoir 287, 197 p.

______, 1958, Fauna, Age and Correlation of the Jurassic Rocks of Prince Patrick Island: Geological Survey of Canada Bulletin 41, 69 p.

______, 1959, Systematic Description of Fossils: *in* Frebold, H., Mountjoy, E., and Reed, R. The Oxfordian Beds of the Jurassic Fernie Group, Alberta and British Columbia: Geological Survey of Canada Bulletin 53, p.20-29.

______, 1960, The Jurassic Faunas of the Canadian Arctic – Lower Jurassic and Lowermost Middle Jurassic Ammonites: Geological Survey of Canada Bulletin 59, 33 p.

______, 1961, The Jurassic Faunas of the Canadian Arctic – Middle and Upper Jurassic Ammonites: Geological Survey of Canada Bulletin 74, 43 p.

______, 1963, Ammonite Faunas of the Upper Middle Jurassic Beds of the Fernie Group in Western Canada: Geological Survey of Canada Bulletin 93, 33 p.

______, 1964a, The Jurassic Faunas of the Canadian Arctic – Cadoceratinae: Geological Survey of Canada Bulletin 119, 27 p.

______, 1964b, Illustrations of Canadian fossils – Jurassic of Western and Arctic Canada: Geological Survey of Canada Paper 64-3, 106 p.

______, 1967, Systematic Descriptions: in Frebold, H., and Tipper, H.W., Middle Callovian Sedimentary Rocks and Guide Ammonites from Southwestern British Columbia: Geological Survey of Canada Paper 67-21, p.9-16.

______, 1975, The Jurassic Faunas of the Canadian Arctic – Lower Jurassic Ammonites, Biostratigraphy and Correlations: Geological Survey of Canada Bulletin 243, 24 p.

______, 1978, Ammonites from the late Bathonian "*Iniskinites* fauna" of central British Columbia: Geological Survey of Canada Bulletin 307, 26 p.

______, 1979, Occurrence of the Upper Bathonian Ammonite Genus *Iniskinites* in the Queen Charlotte Islands, British Columbia: Geological Survey of Canada, Current Research C, Paper 79-1C, p.63-66.

Frebold, H., Mountjoy, E., and Reed, R., 1959, The Oxfordian Beds of the Jurassic Fernie Group, Alberta and British Columbia: Geological Survey of Canada Bulletin 53, 47 p.

Frebold, H., Mountjoy, E.W., and Templeman-Kluit, D.J., 1967, New Occurrences of Jurassic Rocks and Fossils in Central and Northern Yukon: Geological Survey of Canada, Paper 67-12, 35 p.

Frebold, H., and Tipper, H.W., 1967, Middle Callovian Sedimentary Rocks and Guide Ammonites from Southwestern British Columbia: Geological Survey of Canada Paper 67-21, 29 p.

______, 1970, Status of the Jurassic in the Canadian Cordillera of British Columbia, Alberta and Southern Yukon: Canadian Journal of Earth Sciences, v.7, p.1-21.

______, 1973, Upper Bajocian – Lower Bathonian Ammonite Faunas and Stratigraphy of Smithers Area, British Columbia: Canadian Journal of Earth Sciences, v.10, p.1109-1131.

______, 1975, Upper Callovian and Lower Oxfordian Ammonites from Southeastern Bowser Basin, British Columbia: Canadian Journal of Earth Sciences, v.12, p.145-157.

Gibson, D.W., 1979, The Morrisey and Mist Mountain Formations – Newly Defined Lithostratigraphic Units of the Jura-Cretaceous Kootenay Group, Alberta and British Columbia: Bulletin of Canadian Petroleum Geology, v.27, p.183-208.

Grewingk, C., 1850, Beitrag zur Kenntniss der orographischen und geognostischen Beschaffenheit der Nord-West-Küste Amerikas mit den anliegenden Inseln: Verhandlungen der Russisch-kaiserlichen mineralogischen Gesellschaft zu St. Petersburg, Jahrgang 1848 and 1849, p.76-424.

Gygi, R.A., and Marchand, D., 1982, Les Faunes de Cardioceratinae (Ammonoidea) du Callovien Terminal et de l'Oxfordien Inférieur et Moyen (Jurassique) de la Suisse Septentrionale: Stratigraphie, Paléoécologie, Taxonomie Préliminaire: Géobios, v.15, p.517-571.

Hall, R.L., and Stronach, N.J., 1981, First Record of Late Bajocian (Jurassic) Ammonites in the Fernie Formation, Alberta: Canadian Journal of Earth Sciences, v.18, p.919-925.

Hall, R.L., and Westermann, G.E.G., 1980, Lower Bajocian (Jurassic) Cephalopod Faunas from Western Canada and Proposed Assemblage Zones from the Lower Bajocian of North America: Palaeontographica Americana, v.9, no.52, 93 p.

Hallam, A., and Gould, S.J., 1975, The Evolution of British and American Middle and Upper Jurassic *Gryphaea:* a Biometric Study: Proceedings of the Royal Society of London, B, v.189, p.511-542.

Hamblin, A.P., 1978, Sedimentology of a Prograding Shallow Marine Slope and Shelf Sequence, Upper Jurassic Fernie-Kootenay Transition, Southern Front Ranges: M.Sc. Thesis, McMaster University, Hamilton, 196 p.

Hamblin, A.P., and Walker, R.G., 1979, Storm-Dominated Shallow Marine Deposits: the Fernie-Kootenay (Jurassic) Transition, Southern Rocky Mountains: Canadian Journal of Earth Sciences, v.16, p.1673-1690.

Haughton, S., 1858, Geological Notes and Illustrations in M'Clintock's Reminiscences of Arctic Ice Travel: Journal of the Royal Dublin Society, v.1, p.183-250.

Hillebrandt, A. von, 1970, Zur Biostratigraphie und Ammoniten-Fauna des südamerkanischen Jura (insbes. Chile): Neues Jahrbuch für Geologie und Paläontologie, Abhandlungen, v.136, p.166-211.

______, 1977, Ammoniten aus dem Bajocien (Jura) von Chile (Südamerika): Mitteilungen aus der Bayerischen Staatssamlung für Paläontologie und historischen Geologie, v.17, p.35-69.

Hyatt, A., Trias and Jura in the Western States: Geological Society of America Bulletin 5, p.395-434.

Imlay, R.W., 1948, Characteristic Marine Jurassic Fossils from the Western Interior of the United States: United States Geological Survey Professional Paper 214B, p.13-33.

______, 1953a, Callovian (Jurassic) Ammonites from the United States and Alaska, Part I: Western Interior United States: United States Geological Survey Professional Paper 249-A, 39 p.

______, 1953b, Callovian (Jurassic) Ammonites from the United States and Alaska, Part II: Alaska Peninsula and Cook Inlet: United States Geological Survey Professional paper 249-B, p.41-108.

______, 1955, Characteristic Jurassic Molluscs from Northern Alaska: United States Geological Survey Professional Paper 274-D, p.69-96.

______, 1956, Review of Arkell, W.J., 1956, Jurassic Geology of the World: Journal of Paleontology, v.30, p.1376-1377.

______, 1961a, New Genera and Subgenera of Jurassic (Bajocian) Ammonites from Alaska: Journal of Paleontology, v.35, p. 467-474.

______, 1961b, Characteristic Lower Cretaceous Megafossils from

Southern Alaska: United States Geological Survey Professional Paper 335, 74 p.
———, 1961c, Late Jurassic Ammonites from the Western Sierra Nevadas, California: United States Geological Survey Professional Paper 374-D, 30 p.
———, 1962a, Jurassic (Bathonian or Early Callovian) Ammonites from Alaska and Montana: United States Geological Survey Professional Paper 374-C, 32 p.
———, 1962b, Late Bajocian Ammonites from the Cook Inlet Region, Alaska: United States Geological Survey Professional Paper 418-A, 15 p.
———, 1964a, Middle Bajocian Ammonites from the Cook Inlet Region, Alaska: United States Geological Survey Professional Paper 418-B, 54 p.
———, 1964b, Upper Jurassic Molluscs from Eastern Oregon and Western Idaho: United States Geological Survey Professional Paper 483-D, 24 p.
———, 1965, Jurassic marine faunal differentiation in North America: Journal of Paleontology, v.39, p.1023-1038.
———, 1967, Twin Creek Limestone (Jurassic) in the Western Interior of the United States: United States Geological Survey Professional Paper 540, 105 p.
———, 1973, Middle Jurassic (Bajocian) Ammonites from Eastern Oregon: United States Geological Survey Professional Paper 756, 100 p.
———, 1975, Stratigraphic Distribution and Zonation of Jurassic (Callovian) Ammonites in Southern Alaska: United States Geological Survey Professional Paper 836, 28 p.
———, 1976, Middle Jurassic (Bajocian and Bathonian) Ammonites from North Alaska: United States Geological Survey Professional Paper 854, 22 p.
———, 1980a, Jurassic Paleobiogeography of the Conterminous United States in its Continental Setting: United States Geological Survey Professional Paper 1062, 134 p.
———, 1980b, Middle Jurassic (Bathonian) Ammonites from Southern Alaska: United States Geological Survey Professional Paper 1091, 42 p.
———, 1981a, Jurassic (Bathonian and Callovian) Ammonites in Eastern Oregon and Western Idaho: United States Geological Survey Professional Paper 1142, 25 p.
———, 1981b, Late Jurassic Ammonites from Alaska: United States Geological Survey Professional Paper 1190, 40 p.
———, 1982a, Late Bajocian Ammonites from Southern Alaska: United States Geological Survey Professional Paper 1189, 19 p.
———, 1982b, Jurassic (Oxfordian and Late Callovian) Ammonites from the Western Interior Region of the United States: United States Geological Survey Professional Paper 1232, 44 p.
Imlay, R.W., and Jones, D.L., 1970, Ammonites from the *Buchia* Zone in Northwestern California and Southwestern Oregon: United States Geological Survey Professional Paper 647-B, 59 p.
Jeletzky, J.A., 1964, Illustrations of Canadian Fossils – Early Lower Cretaceous (Berriasian and Valanginian) of the Canadian Western Cordillera, British Columbia: Geological Survey of Canada Paper 64-6, 18 p.
———, 1966, Upper Volgian (latest Jurassic) Ammonites and Buchias of Arctic Canada: Geological Survey of Canada Bulletin 128, 52 p.
———, 1973, Biochronology of the marine Boreal latest Jurassic, Berriasian and Valanginian of Canada: in Casey, R., and Rawson, P.F. eds., The Boreal Lower Cretaceous: Liverpool, Geological Journal Special Issue No. 5, p.41-80.
Keyserling, A. von, 1846, Wissenschaftliche Beobachtungen auf einer Reise in das Petschoraland im Jahre 1843: St. Petersburg, Carl Kray, 465 p.
Kilian, W., and Reboul, P., 1909, Les Céphalopodes Néocrétacés des Îles Seymour et Snow Hill: Wissenschaftliche Ergebnisse der Schwedischen Südpolar-Expedition, 1901-1903, v.3, Lief.6, 75 p.
Leckenby, J., 1859, On the Kelloway Rock of the Yorkshire Coast: Quarterly Journal of the Geological Society, London, v.15, p.4-15.
Madsen, V., 1904, On Jurassic Fossils from East-Greenland: Meddelelser om Gronland, v.29, p.157-211.
Maire, V., 1938, Contribution a la Connaissance des Cardioceratidés: Mémoires de la Société Géologique de France, Nouvelle Série, Mém. 34, 132 p.
McLearn, F.H., 1927, Some Canadian Jurassic Faunas: Transactions of the Royal Society of Canada, Sect. IV, Ser.3, v.21, p.61-73.
———, 1928, New Jurassic Ammonoidea from the Fernie Formation, Alberta: Geological Survey of Canada Bulletin 49, Geological Series No. 48, p.19-22.
McLearn, F.H., 1929, Contributions to the Stratigraphy and Palaeontology of Skidegate Inlet, Queen Charlotte Islands, B.C.: National Museum of Canada Bulletin 58, p.1-27.
Meek, F.B., 1859, Remarks on the Cretaceous fossils collected by Professor Henry Y. Hind, on the Assiniboine and Saskatchewan Exploring Expedition, with Description of Some New Species, Northwest Territory: *in* Hind, H.Y., Reports of Progress, Together with a Preliminary and General Report on the Assiniboine and Saskatchewan Exploring Expedition: Toronto.
Meek, F.B., and Hayden, F.V., 1859, Description of New Organic Remains Collected in Nebraska Territory in the Year 1857, by Dr F.V. Hayden, Geologist to the Exploring Expedition under the Command of Lieut. G.K. Warren, Top. Engr. U.S. Army, Together with Some Remarks on the Geology of the Black Hills and Portions of the Surrounding Country: Academy of Natural Sciences, Philadelphia, Proceedings for 1858, p.41-59.
———, 1865, Palaeontology of the upper Missouri: a Report upon Collections Made Principally by the Expedition under the Command of Lieut. G.K. Warren, U.S. Top. Engineers, in 1855 and 1856. Invertebrates: Smithsonian Contributions to Knowledge, v.14, art.5, 135 p.
Meledina, S.V., 1967, A New Genus of Middle Jurassic Arctic Ammonites: *in* Saks, V.N., ed., Problems of Paleontologic Substantiation of Detailed Mesozoic Stratigraphy of Siberia and the Far East of the U.S.S.R.: Leningrad, Academy of Sciences of the U.S.S.R., Siberian Branch, Institute of Geology and Geophysics, p.103-109 (in Russian).
———, 1968, Middle Jurassic Boreal ammonites of Siberia and their stratigraphical significance: Doklady Akademiya Nauk S.S.S.R., v. 183, p. 416-419; and Doklady Earth Science Sections, American Geological Institute, v.183, p.54-57.
———, 1973, The Ammoniand Zonal Stratigraphy of the Bajocian – Bathonian of Siberia: Trudy, Institut Geologii i Geofiziki, Akad. Nauk S.S.S.R., Sibirskoe otdel., v. 153, 152 p. (in Russian).
———, 1977, The ammonites and Zonal Stratigraphy of the Callovian of Siberia: Trudy, Institut Geologii i Geofiziki, Akad. Nauk S.S.S.R., Sibirskoe otdel., v.356, 290 p. (in Russian).
Mesezhnikov, M.S., 1967, A New Ammonite Zone of the Upper Oxfordian and the Position of the Oxfordian-Kimmeridgian Boundary in Northern Siberia: *in* Saks, V.N., ed., Problems of Paleontologic Substantiation of Detailed Mesozoic Stratigraphy of Siberia and the Far East of the U.S.S.R.: Leningrad, Academy of Sciences of the U.S.S.R., Siberian Branch, Insitute of Geology and Geophysics, p.110-113, in Russian.
Newmarch, C.B., 1953, Geology of the Crowsnest Coal Basin with Special Reference to the Fernie Area: British Columbia Department of Mines Bulletin 53, 107 p.
Newton, E.T., and Teall, J.H.H., 1897, Notes on a Collection of

Rocks and Fossils from Franz Josef Land, Made by the Jackson-Harmsworth Expedition During 1894-1896: Quarterly Journal of the Geological Society, London, v.53, p.477-519.

Nikitin, S.N., 1881, Der Jura der Umgegend von Elatma: Mémoires de la Société Impériale des Naturalistes de Moscou, Nouvelle Série, v.14, p.85-133.

Oppel, A., 1862, Über jurassische Cephalopoden: Palaeontologische Mittheilungen des Museum des königlichen Bayerischen Staates, v.3, p.127-162.

Parsons, C.F., 1976, A Stratigraphic Revision of the *Humphriesianum/Subfurcatum* Zone Rocks (Bajocian Stage, Middle Jurassic) of Southern England: Newsletter of Stratigraphy, v.5, p.114-142.

Pavia, G., 1973, Ammoniti del Baiociano Superiore di Digne (Francia SE, Dip. Basses-Alpes): Società Paleontologica Italiano, Boll. 10, p.75-142.

Pompeckj, J.F., 1900a, The Jurassic Fauna of Cape Flora, Franz Josef Land: *in* Nansen, F., ed., The Norwegian North Polar Expedition 1893-1896: Scientific Results, v.1, pt.ii, p.33-148.

______, 1900b, Jura-Fossilien aus Alaska: Verhandlungen der kaiserlichen Russischen Mineralogischen Gesellschaft St. Petersburg, Ser.2, v.38, p.239-281.

Poulton, T.P., 1978a, Internal Correlations and Thickness Trends, Jurassic Bug Creek Formation, Northeastern Yukon and Adjacent Northwest Territories: Geological Survey of Canada Paper 78-1B, p.27-30.

______, 1978b, Pre-Late Oxfordian Jurassic Biostratigraphy of Northern Yukon and Adjacent Northwest Territories: *in* Stelck, C.R., and Chatterton, B.D.E., eds., Western and Arctic Canadian Biostratigraphy: Geological Association of Canada Special Paper 18, p.445-471.

______, 1982, Paleogeographic and Tectonic Implications of the Lower and Middle Jurassic Facies Patterns in Northern Yukon Territory and Adjacent Northwest Territories: Embry, A.F., and Balkwill, H.R., eds., Arctic Geology and Geophysics: Canadian Society of Petroleum Geologists, Memoir 8, p.13-27.

______, 1983, Boreal Bathonian and Lower Callovian Ammonites, Zonation and Correlation, Salmon Cache Canyon, Northern Yukon: Geological Survey of Canada Bulletin, in press.

Poulton, T.P., and Callomon, J.H., 1976, Major Features of the Lower and Middle Jurassic Stratigraphy of Northern Richardson Mountains, Northeastern Yukon Territory and Northwestern District of Mackenzie: Geological Survey of Canada Paper 76-1B, p.345-52.

Poulton, T.P., Leskiw, K., and Audretsch, A., 1982, Stratigraphy and Microfossils of the Jurassic Bug Creek Group of Northern Richardson Mountains, Northern Yukon and Adjacent Northwest Territories: Geological Survey of Canada Bulletin 325, 137 p.

Poulton, T.P., and Templeman-Kluit, D.J., 1982, Recent Discoveries of Jurassic Fossils in the Lower Schist Division of Central Yukon: Geological Survey of Canada Paper 82-1C, p.91-94.

Quenstedt, F.A., 1886-7, Die Ammoniten des schwäbischen Jura, 2. Band: Der Braune Jura, mit Atlas: Stuttgart, Schweizerbart, p.441-815; reprinted 1973.

Quilty, P.G., 1970, Jurassic Ammonites from Ellsworth Land, Antarctica: Journal of Paleontology, v.44, p.110-116.

Rawson, P.F., 1982, New Arctocephalitinae (Ammonoidea) from the Middle Jurassic of Kong Karls Land, Svalbard: Geological Magazine, v.119, p.95-100.

Reeside, J.B., 1919, Some American Jurassic Ammonites of the Genera *Quenstedticeras, Cardioceras,* and *Amoeboceras,* Family Cardioceratidae: United States Geological Survey Professional Paper 118, 38 p.

Riaz, A. de, 1898, Description des ammonites des couches à *Peltoceras transversarium* (Oxfordien supérieux) de Trept (Isère): Paris, Masson, 69 p.

Rioult, M., 1964, Le stratotype du Bajocien: Colloque du Jurassique, Luxembourg, 1962: Volume des Comptes Rendus et Memoires, Institut Grand-Ducal, Section des Sciences Naturelles, Physiques et Mathematiques, p.239-258.

Saks, V.N., 1976, ed., Stratigraphy of the Jurassic System of the North of the U.S.S.R.: Moscow, Nauka, 436 p. (in Russian).

Salfeld, H., and Frebold, H., 1924, Reports on the Scientific Results of the Norwegian expedition to Nowaya Zemlya, 1921; No.23, Jura- und Kreidefossilien von Nowaya Zemlya: Oslo, Broggers, 12 p.

Sato, T., 1954, Découverte de *Tmetoceras* dans le Plateau de Kitikama au Nord du Japon: Japanese Journal of Geology and Geography, v.24, p.115-121.

______, 1957, Biostratigraphie de la Série de Shizukawa (Jurassique inférieur) du Japon Septentrional: Journal of the Faculty of Science, University of Tokyo, v.10, p.313-350.

______, 1958, Supplement à la Faune de la Série de Shizukawa (Jurassique inférieur) du Japon Septentrional: Japanese Journal of Geology and Geography, v.29, p.153-159.

Sazonov, N.T., 1957, Jurassic deposits of the central region of the Russian Platform: Leningrad, VNIGNI, 154 p. (in Russian).

______, 1965, New data on Callovian, Oxfordian and Kimmeridgian ammonites: *in* Sazonova, N.T., and Shutskoy, E.K., eds., Mesozoic and Cainozoic faunas of the European part of the U.S.S.R. and Middle Asia: Moscow, VNIGNI, p.1-99, in Russian.

Sey, I.I., and Kalacheva, E.D., 1980, Biostratigraphy of the Lower and Middle Jurassic Deposits of the Far East: Leningrad, Trudy VNIGI, new series, v.285, 187 p., in Russian, with English summary.

Siemiradzki, J. von, 1898-9, Monographische Beschreibung der Ammonitengattung *Perisphinctes:* Palaeontographica, v.45, p. 69-352.

Sokolov, D.N., 1912, Zur Ammonitenfauna des Petschoraschen Jura: Mémoires du Comité Géologique de St. Petersbourg, Nouvelle Série, livr.76, p.1-49 (in Russian), 50-65 (in German).

______, 1913, Sur les Fossiles des Blocs Erratiques de Novaia Zemlia: Travaux du Musée géologique de St. Petersbourg, v.7, p.59-92.

Spath, L.F., 1924, On the Blake Collection of Ammonites from Kachh, India: Palaeontologia Indica, New Series, v.9, no.1, 39 p.

______, 1928, 1933, Revision of the Jurassic cephalopod fauna of Kachh (Cutch), India: Palaeontologia Indica, New Series, v.9, no.2, pt.iii, p.163-278; pt.vi, p.659-945.

______, 1935, The Invertebrate Faunas of the Bathonian-Callovian deposits of Jameson Land (East Greenland): Meddelelser om Gronland, v.87, no.7, 158 p.

______, 1935, The Upper Jurassic Invertebrate Faunas of Cape Leslie, Milne Land. I. Oxfordian and Lower Kimmeridgian: Meddeleser om Gronland, v.99, no.2, 82 p.

Stehn, E., 1924, Beiträge zur Kenntniss des Bathonium und Callovium in Südamerika: Neues Jahrbuch für Mineralogie, Geologie und Paläontologie, Beilage-Band 49, p.52-158.

Sturani, C., 1967, Ammonites and Stratigraphy of the Bathonian in the Digne – Barrème Area (Southeastern France, Dépt. Basses-Alpes): Società Paleontologica Italiano, Boll. 5, 55 p.

______, 1971, Ammonites and Stratigraphy of the *Posidonia alpina* Beds of the Venetian Alps: Memorie degli Instituti di Geologia e Mineralogia dell' Universita di Padova, v.28, 190 p.

Sykes, R.M., and Callomon, J.H., 1979, The *Amoeboceras* Zonation of the Boreal Upper Oxfordian: Palaeontology, v.22, p.839-903.

Thierry, J., 1976, Essai de Taxonomie et de Phylogenie de la Famille des Macrocephalitidae (Ammonitina, Stephanocerataceae; Jurassique moyen): 4e Réunion Annuelle des Sciences de la terre, Société Géologique de France, Paris, 1 p.

———, 1978, Le Genre Macrocephalites au Callovien Inferieur (Ammonites, Jurassique moyen): Mémoires Géologiques de l'Université de Dijon, v.4, 492 p.

Tintant, H., 1963, Les Kosmoceratidés du Callovien Inférieur et Moyen d'Europe Occidentale; Essai de Paléontologie Quantitative: Publications de l'Université de Dijon, v.29, 500 p. and Atlas.

Tintant, H., and Mouterde, R., 1981, Classification et Phylogenie chez les Ammonites Jurassiques: Martinell, J., ed., International Symposium on Concepts and Methodology in Paleontology, Barcelona 1981, Contributed Papers, p.85-101.

Tipper, H.W., and Richards, T.A., 1976, Jurassic Stratigraphy and History of North-Central British Columbia: Geological Survey of Canada Bulletin 270, 73 p.

Troitskaya, E.A., 1956, The genus *Rondiceras* Troitskaya gen. nov.: *in* Kiparisova, L.D., Markovski, B.P. and Radchenko, G.I., eds.: Contributions to Paleontology, VSEGEI, Moscow, New Series, Paleontology, v.12, p.79-80 (in Russian).

Tuchkov, I.I., 1972, New Data on the Stratigraphy and the Upper Boundary of the Jurassic in the Boreal and Arctic regions:: Izvestiya Akad. Nauk SSSR, ser. geologicheskaya, 1972, no.2, p.111-125, in Russian.

Voronets, N.S., 1962, Stratigraphy and Cephalopodean Molluscs of the Jurassic and Lower Cretaceous Deposits of the Lena-Anabar region: Moscow, Trudy NIIGA, v.110, 237 p., in Russian.

Westermann, G.E.G., 1956, Monographie der Bajocien-Gattungen *Sphaeroceras* und *Chondroceras* (Ammonoidea): Geologisches Jahrbuch, Hannover, Beihefte, v.24, 125 p.

———, 1958, Ammoniten-Fauna und Stratigraphie des Bathonien NW-Deutschlands: Geologisches Jahrbuch, Hannover, Beihefte, v.32, 103 p.

———, 1964a, The Ammonite Fauna of the Kialagvik Formation at Wide Bay, Alaska Peninsula, Part I: Lower Bajocian (Aalenian): Bulletins of American Paleontology, v.47, no.216, p.329-503.

———, 1964b, Occurrence and Significance of the Arctic *Arkelloceras* in the Middle Bajocian of the Alberta foothills (Ammonitina, Jurassic): Journal of Paleontology, v.38, p.405-409.

———, 1964c, Sexual-Dimorphismus bei Ammonoideen und seine Bedeutung für die Taxionomie der Otoitidae (einschliesslich Sphaeroceratinae; Ammonitina, M.-Jura): Palaeontographica, v.A125, p.33-73.

———, 1966, The Holotype (Plastotype) of ?*Titanites occidentalis* Frebold from the Kootenay Sandstone (Upper Jurassic) of southern British Columbia: Canadian Journal of Earth Sciences, v.3, p.623-625.

———, 1969, The Ammonite Fauna of the Kialagvik Formation at Wide Bay, Alaska Peninsula, Part II: *Sonninia sowerby* Zone (Bajocian): Bulletins of American Paleontology, v.57, no.255, 226 p.

———, 1981, Ammonite Biochronology and Biogeography of the Circum-Pacific Jurassic: *in* House, M.R. and Senior, J.R., eds., The Ammonoidea: Systematics Association Special Publication 18, London and New York, Academic Press, p.459-498.

Westermann, G.E.G., and Riccardi, A.C., 1976, Middle Jurassic Ammonite Distribution and the Affinities of the andean Faunas: Primer Congresso geologico Chileno, Santiago de Chile, v.1, p.C23-C39.

———, 1979, Middle Jurassic Ammonoid Fauna and Biochronology of the Argentine-Chilean Andes, Part II: Bajocian Stephanocerataceae: Palaeontographica, v.A164, p.85-188.

White, C.A., 1889, Mesozoic molluscs from the Southern Coast of the Alaskan Peninsula: United States Geological Survey Bulletin 51, p.64-70.

Whiteaves, J.F., 1900, On Some Additional or Imperfectly Understood Fossils from the Cretaceous Rocks of the Queen Charlotte Islands, With a Revised List of Species from these Rocks: Geological Survey of Canada, Mesozoic Fossils, IV, p.263-307.

———, 1903, Description of a Species of *Cardioceras* from the Crowsnest Coalfields: Ottawa Naturalist, v.17, p.65-67.

Whitfield, R.P., 1880, Report on the geology and Resources of the Black Hills of Dakota: United States Geographic and Geologic Survey, Rocky Mountains Region, p.380-381.

Wierzbowski, A., Kulicki, C., and Pugaczewska, H., 1982, Fauna and Stratigraphy of the Uppermost Triassic and the Toarcian and Aalenian Deposits in the Sassenfjorden, Spitsbergen: Acta Palaeontologia Polonica, v.26, p.195-241.

Wright, C.W., 1981, Cretaceous Ammonoidea: *in* House, M.R. and Senior, J.R., eds.: The Ammonoidea, Systematics Association Special Volume No. 18: Academic Press, Ch.6, p.157-174.

Zakharov, V.A., and Mesezhnikov, M.S., 1974, The Volgian Stage in the sub-polar Urals: Novosibirsk, Institut Geologii i Geofiziki Akad. Nauk S.S.S.R., Sibirsk otdel., Trudy, v.196, 216 p., in Russian.

Jurassic-Cretaceous Biochronology and Paleogeography of North America,
edited by G.E.G. Westermann,
Geological Association of Canada Special Paper 27, 1984

JURASSIC-CRETACEOUS BOUNDARY BEDS OF WESTERN AND ARCTIC CANADA AND THE PROBLEM OF THE TITHONIAN-BERRIASIAN STAGES IN THE BOREAL REALM

J.A. Jeletzky
Geological Survey of Canada, 601 Booth Street, Ottawa, Ontario K1A 0E8

ABSTRACT

The Buchia terebratuloides *s.l. and* B. okensis *s.s. Zones of western British Columbia correspond respectively to the* B. *aff.* B. okensis *Zone and the basal part of* B. uncitoides *s.l. Zone of California. In Mexico they correspond respectively to the* Substeueroceras-Berriasella *complex (that also contains* Durangites *and* Kossmatia*) and the* Subthurmannia tenochi *Zone. These Mexican faunas are, in turn, respectively coeval with the* Substeueroceras koeneni *Zone and the* Argentiniceras noduliferum *Zone of western South America. Therefore, the* B. terebratuloides *s.l. Zone corresponds to the topmost Tithonian (i.e., the uppermost* Paraulacosphinctes transitorius *Zone or the upper part of regional* Durangites *Zone) of South Western Europe. The next younger* B. okensis *s.s. Zone corresponds to the Lower Berriasian (i.e.,* Pseudosubplanites grandis *Zone and lower* Tirnovella occitanica *Zone) of that region.*

The Buchia terebratuloides *s.l. Zone persists into high Boreal basins of Arctic Canada. In the Sverdrup Basin this zone contains* Buchia unschensis *s.s.,* B. terebratuloides *s.l. and* B. *n. sp. aff.* okensis *in association with* Praetollia (Praetollia) fedorovi (=P. antiqua *Jeletzky 1973),* P. maynci aberrans *and* Pseudocraspedites anglicus. *This uppermost Tithonian zone corresponds to the allegedly basal Berriasian* Chetaites sibiricus *and* Praetollia fedorovi *Zones of North Siberia and the* Praetollia maynci *Zone of East Greenland. It appears to be absent in England. The* Buchia terebratuloides *s.l. Zone may possibly include the basal beds of allegedly Late Berriasian* Hectoroceras kochi *Zone.*

Buchia okensis *s.s. Zone also persists into Arctic Canada. In the Sverdrup Basin it overlies gradationally the Boreal facies of* B. terebratuloides *s.l. Zone and contains* Pseudocraspedites *aff.* anglicus (=Subcraspedites *aff.* suprasubditus *Jeletzky 1973) and* Praetollia? suritiformis *Klimova 1972 (*=Subcraspedites *aff.* hoeli *Jeletzky 1973). This basal Berriasian zone represents the bulk (?or all) of the Boreal* Hectoroceras kochi *Zone of northern Eurasia and, possibly, the basal part of the next younger* Surites analogus *Zone. The bulk of the* B. okensis *s.s. Zone is represented by a hiatus in the type Riasanian sections of Central Russia. Therefore, the Tithonian-Berriasian boundary either coincides approximately with the base of the interregional Boreal* H. kochi *Zone or runs through its basal part, leaving the regional Boreal* Praetollia maynci, Praetollia fedorovi, Chetaites sibiricus *and* Buchia unschensis *s.s. Zones in the Tithonian.*

The basal Riasanian occurrences of Riasanites *ex gr.* rjasanensis *and* "Euthymiceras" *(a new genus) ex gr.* transfigurabilis *correspond to the upper part of* Chetaites chetae *Zone of North Siberia. Therefore, they correspond to the upper Tithonian of the Tethyan Realm. Hence, the basal Riasanian occurrences of these ammonites are considerably older than their oldest known occurrences in the uppermost Lower Berriasian Dalmasiceras dalmasi Subzone of Mangyshlak and North Caucasus and the still younger Berriasian beds of Crimea.* Riasanites *and* "Euthymiceras" *ex gr.* transfigurabilis *must, therefore, be Boreal endemics. They were restricted to the Central Russian Basin throughout the late Late Tithonian and Early Berriasian and then migrated into the marginal Tethyan basins of southern U.S.S.R. in Mid-Berriasian. They also migrated westward into the central Polish Basin in the Late Berriasian. These data discredit all preceding attempts to correlate the bulk of the Upper Volgian with the Lower (or Lower and Middle) Berriasian.*

All latest Jurassic and earliest Cretaceous marine macroinvertebrate faunas of western North America and northern Eurasia are pronouncedly climatically zoned. The geographical distribution of the high Boreal, marginal Boreal, marginal Tethyan (=North Pacific) and high Tethyan faunas of that time and their regular north-south and east-west intergradation indicate that all latest Jurassic and earliest Cretaceous marine basins of western North America were already situated at their approximate present-day latitudes and longitudes. The "allochtonous" geosynclinal basins (terranes) of western British Columbia and Alaska at this time were accreted to the North American Craton because of near-total identity of their shallow water faunas to those of adjacent intracratonic basins.

RÉSUMÉ

Les zones à Buchia terebratuloides *s.l. et* Buchia okensis *s.s. de l'ouest de la Colombie-Britannique correspondent respectivement à la zone à* Buchia *aff.* B. okensis *et à la partie basale de la zone à* Buchia uncitoides *s.l. de la Californie. Au Mexique, elles correspondent respectivement au complexe à* Substeueroceras-Berriasella *(qui contient aussi* Durangites *et* Kossmatia*) et à la zone à* Subthurmannia tenochi. *Ces faunes mexicaines sont à leur tour, respectivement comtemporaines de la zone à* Substeueroceras koeneni *et de la zone à* Argentiniceras noduliferum *de l'ouest de l'Amérique du Sud. Par conséquent, la zone à* Buchia terebratuloides *s.l. correspond au sommet du Tithonique (soit au sommet de la zone à* Paraulacosphinctes transitorius *ou à la partie supérieure de la zone régionale à* Durangites*) du sud-ouest de l'Europe. La zone suivante dans l'ordre ascendant, la zone à* Buchia okensis *s.s., correspond au Berriasien inférieur (soit à la zone à* Pseudosubplanites grandis *et à la zone inférieure à* Tirnovella occitanica*) dans cette région.*

La zone à Buchia terebratuloides *s.l. se prolonge dans les bassins boréaux les plus septentrionaux de l'Arctique canadien. Dans le bassin de Sverdrup, cette zone contient* Buchia unschensis *s.s.,* B. terebratuloides *s.l. et* B. *n. sp. aff.* okensis, *en association avec* Praetollia (Praetollia) fedorovi (=P. antiqua *Jeletzky 1973),* P. maynci aberrans *et* Pseudocraspedites anglicus. *Cette zone la plus haute du Tithonique correspond aux zones, prétendument situées à la partie basale du Berriasien, à* Chetaites sibiricus *et* Praetoilla fedorovi *du nord de la Sibérie, et à la zone à* Praetollia maynci *de l'est du Groenland. Elle semble absente en Angleterre. La zone à* Buchia terebratuloides *s.l. contient peut-être des couches basales de la zone à* Hectoroceras kochi *prétendument situées, à la Berriasien supérieure.*

La zone à Buchia okensis *s.s. se prolonge aussi dans l'Arctique du Canada. Dans le bassin de Sverdrup, elle recouvre graduellement le facies boréal de la zone à* Buchia terebratuloides *s.l., et contient* Pseudocraspedites *aff.* anglicus *(=*Subcraspedites *aff.* suprasubditus *Jeletzky 1973) et* Praetollia? suritiformis *Klimova 1972 (=*Subcraspedites *aff.* hoeli *Jeletzky 1973). Cette zone basale du Berriasien représente la majeure partie (?ou la totalité) de la zone boréale à* Hectoroceras kochi *du nord de l'Eurasie et, peut-être, la partie basale de la zone suivante, dans l'ordre ascendant, à* Surites analogus. *La majeure partie de la zone à* Buchia okensis *s.s. est représentée par une lacune dans les sections-types du Riasanien de la Russie centrale. Par conséquent, la limite entre le Tithonique et le Berriasien coincide approximativement avec la base de la zone interrégionale boréale à* Hectoroceras kochi, *ou traverse sa partie basale, ce qui laisse les zones boréales régionales à* Praetollia maynci, Praetollia fedorovi, Chetaites sibiricus *et* Buchia unschensis *s.s. dans le Tithonique.*

Dans la partie basale du Riasanien, la présence de Riasanites *ex gr.* rjasanensis *et* "Euthymiceras" *(nouveau genre) ex gr.* transfigurabilis *correspond à la partie supérieure de la zone à* Chetaites chetae *du nord de la Sibérie. Par conséquent, cette couche correspond à la partie inférieure du Tithonique supérieur du domaine de la Téthys. De ce fait, les ammonites que l'on rencontre à la base du Riasanien sont beaucoup plus anciennes que leur apparances les plus anciennes déjà connues dans la partie la plus haute de la sous-zone à* Dalmasiceras dalmasi *du Berriasien inférieur de Mangyshlak et du nord du Caucase, et dans les couches berriasiennes encore plus récentes de Crimée. Donc,* Riasanites *et* "Euthymiceras" *ex gr.* transfigurabilis *doivent être des espèces endémiques de la région boréale. Elles étaient limitées à la partie centrale du bassin Russe pendant toute la fin du Tithonique supérieur et le Berriasien inférieur, avant de migrer dans les bassins marginaux de la Téthys dans le sud de l'U.R.S.S., à l'époque du Berriasien moyen. Elles ont aussi migré vers l'ouest jusque dans le centre du bassin Polonais au Berriasien supérieur. Ces données rendent invalides toutes les tentatives déjà faites pour établir une corrélation entre la majeure partie du Volgien supérieur et la Berriasien inférieur (ou inférieur et moyen).*

Toutes les faunes de macroinvertébrés marins, du Jurassique le plus récent au Crétacé le plus ancien, dans l'ouest de l'Amérique du Nord et le nord de l'Eurasie, présentent une répartition des zones climatiques bien définies. La répartition géographique des faunes provenant de la zone boréale haute, la zone boréale marginale, la zone marginale de la Téthys (=Pacifique Nord) et la zone de la Haute-Téthys à cette époque, et leur intergradation régulière du nord au sud et de l'est à l'ouest, indiquent que tous les bassins marins du Jurassique le plus récent et du Crétacé le plus ancien dans l'ouest de l'Amérique du Nord se situaient approximativement déjà à leurs latitudes et longitudes respectives actuelles. Les bassins géosynclinaux «allochtones» (terrains) de l'ouest de la Colombie-Britannique et de l'Alaska devaient à cette époque déjà être rattachés au Craton nord-américain, en raison de la similarité presque totale de leurs faunes d'eau peu profonde avec celles des bassins intracratoniques adjacents.

INTRODUCTION

This paper is a much expanded and amended version of the talk presented in August, 1982, at the Jurassic-Cretaceous biochronology and biogeography symposium of the Third North American Paleontological Convention (Jeletzky, 1982). However, the subject is so vast and involved that this paper remains a summary only. Therefore, the documentation and illustration of the conclusions advanced in this paper had to be severely restricted, but in spite of this restriction, the vastness and complexity of the subject necessitated a segregation of principal conclusions and the most important items of supporting evidence from the bulk of supporting evidence. To facilitate the use of this paper by a general reader, the former items are set in normal type while the rest of the text is set in small type.

The first objective of this paper is to demonstrate the key role of the fossil-rich, marine uppermost Jurassic and basal Cretaceous rocks of western and Arctic Canada for the reliable correlation of the Tethyan and Boreal facies of these beds on the global scale. This subject was already discussed by Jeletzky (1971, 1974) in a summary form and touched upon in connection with the discussion of other, related problems (Jeletzky, 1965; Jeletzky, in Jeletzky and Tipper, 1968, p. 11, 12; Jeletzky, 1973). However, there was no opportunity so far to cover the complete scope of the problem. Furthermore, many conclusions presented in the above papers are either partly or entirely outdated by the wealth of new information that became available since their compilation.

The second objective of this paper is to produce an up-to-date, concise summary of the stratigraphy and fossil zonation of the Jurassic-Cretaceous boundary beds of western British Columbia and Sverdrup Basin that would facilitate further geological study and mapping of the Jurassic and Lower Cretaceous rocks of these basins.

The detailed interrealm correlations attempted in this paper were facilitated greatly by the descriptions of the uppermost Jurassic and basal Cretaceous *Buchia* and ammonite faunas of northern California (Jones *et al.*, 1969; Imlay and Jones, 1970) and the recent monographic description of the Late Jurassic and early Early Cretaceous *Buchia* faunas of northern U.S.S.R. (Zakharov, 1981).

Recent descriptions of ammonite and calpionellid faunas and biochronology of the Jurassic-Cretaceous boundary beds in southeastern France (Le Hégarat, 1971, 1973; Remane, 1964), southern Spain (e.g., Barthel *et al.*, 1966; Enay and Géyssant, 1975; Allemann, Gruen and Wiedmann, 1975; Oloriz and Tavera, 1979a, 1979b; Hoedemaeker, 1979, 1982) and the Crimean-Caucasian region (e.g., Luppov *et al.*, 1975, 1979; Sakharov, 1975, 1976, etc.) were found to be equally important. These and other recent publications referred to elsewhere in the paper clarified a number of important to outright critical paleontological and biochronological misconceptions that had previously hindered the recognition of correct intercontinental correlations of the ammonite and *Buchia* faunas of the uppermost Jurassic and basal Cretaceous beds of western and Arctic Canada as well as those of the classical Tethyan and Volgian fossil zones with one another.

BIOCHRONOLOGIC DEFINITIONS: THE JURASSIC-CRETACEOUS BOUNDARY BEDS AND THE BOUNDARY

The Jurassic-Cretaceous boundary beds are defined as the Berriasian standard stage and the Upper Tithonian standard substage for the purposes of this paper. Their limits and zonal contents favored by the writer are elucidated by Figures 5 and 10. Some of the numerous other interpretations currently used are elucidated by Figures 1 and 2. It should also be mentioned that the writer follows Enay (1972) in subdividing the Tithonian stage into the Lower and Upper Tithonian substages. The Middle Tithonian of authors is amalgamated with the Lower Tithonian. As pointed out recently by Enay and Géyssant (1975, p. 53): "La partie inférieure du Tithonique est caracterisée par l'abondance et la variété des Perisphinctidae, des Aspidoceratidae et des Oppeliidae avec l'évolution propre des Streblitinae (*Neochetoceras, Semiformiceras)*. Dans la partie supérieure du Tithonique, les Perisphinctidae se raréfient pour se limiter pratiquement au genre *Paraulacosphinctes*; les Streblitinae ne sont plus représentes que par *Substreblites* et, sans doute, *Cyrtosiceras; Semiformiceras* et *Neochetoceras* ont disparu. Par contre, apparaissent et se dévelopent les Himalayitinae, les Berriasellinae et les Olcostephanidae, ainsi que les Calpionelles vrais parmis les microfaunes. Il apparait donc, dans l'ensemble, une division assez nette du Tithonique en deux parties."

Exactly the same conclusion was more recently reached by Imlay (1980, p. 32): "For correlations between the continents, preference is given herein to a twofold division of the Tithonian because the faunal boundary between the two divisions is much sharper than are the boundaries within the threefold division."

In my opinion the twofold division of the Tithonian is also advantageous in that it parallels closely the classical twofold division of its Boreal equivalents into the lower and upper Volgian stages. I have used this twofold division in all publications to date and see the recently proposed switch to a unified Volgian stage with a threefold division into substages (Gerassimov and Michailov, 1966) as a retrograde step.

The still controversial problem of the delimitation of the international Tithonian and Berriasian stages within their type areas, which happen to be situated in the northern part of the Tethyan Realm, shall not be discussed in this paper. Nor shall the equally controversial problem of the placement of the Jurassic-Cretaceous boundary be discussed. These problems have been effectively settled by the majority opinion of specialists that have attended the Lyon-Neuchâtel Colloquium in 1973 (see its Proceedings Volume, Mémoires du B.R.G.M. No. 86, 1975, p. 380-393). Accordingly, the boundary between the Upper Tithonian substage and the Berriasian stage (and consequently that between the Jurassic and Cretaceous systems) is placed at the base of the so called combined Grandis-Jacobi Zone as it is developed in southeastern France (i.e., the standard), including the Chomerac and d'Aizy-sur-Noyarey Beds (see Le Hégarat, 1971, 1973). The only deviation from this majority decision consists in the renaming of the Grandis-Jacobi Zone as Grandis Zone, necessitated by a subsequent discreditation of *Berriasella jacobi* Mazenot as the zonal fossil of this basal Berriasian Zone (see below for further details).

This somewhat cavalier treatment of the above problems in the absence of any official decision of the international stratigraphic bodies reflects my (Jeletzky, 1975a) feeling that none of the previously proposed variants of the delimitation of the Jurassic and Cretaceous Systems is based on a well defined biochronological event that permits an easy recognition of the Jurassic-Cretaceous boundary on the global scale. This is also true of the latest proposal of Hoedemaeker (1982, p. 56-60, enclosure 4) to place this boundary between the Grandis and Occitanica Zones. However well defined in the southwestern Europe (and possibly in the Balcanian and Crimean-Caucasian regions of the Mediterranean Province), this boundary still appears to be based on a provincial biochronological event that cannot be easily recognized outside of the Mediterranean Faunal Province.

None of the hitherto proposed Jurassic-Cretaceous boundaries, including the one favored here (Fig. 5), are "natural" geological boundaries comparable, for example, to the Permian-Triassic, Triassic-Jurassic or Cretaceous-Tertiary (i.e., the Maastrichtian-Danian) boundaries. Because of the by now well established absence of a comparable world-wide renewal of the biota anywhere within (or, for that matter, anywhere near) the Jurassic-Cretaceous boundary beds as defined in this paper, the delimitation of the Jurassic and Cretaceous Systems is not a matter of principle but that of a formal agreement among specialists ratified by an appropriate international commission.

FAUNAL CRITERIA FOR THE DELIMITATION OF GRANDIS AND TRANSITORIUS ZONES

The Status of Jacobi Zone (Subzone) of Authors

The sections in southeastern France, on which the Grandis-Jacobi Zone of authors is based, are underlain by limestones that did not yield any diagnostic macrofossils. They were only tentatively correlated with the latest Tithonian Transitorius Zone because of their stratigraphic position (Le Hégarat, 1971; Le Hégarat, 1973, p. 289, Table 29). This "gap is our knowledge" of the macrofaunas occurring at the Jurassic-Cretaceous boundary in the type-area of the Berriasian stage was presenting problems ever since its recognition (Mazenot, 1939, p. 264). However, it was largely closed following the description of biochronology of the Upper Tithonian-Berriasian sections in Southern Spain where the equivalents of these unfossiliferous beds proved to be fossiliferous. It is primarily these recent studies in Southern Spain that necessitate a review of the faunal criteria for the delimitation of the Grandis and Transitorius Zones and led the writer to the complete discard of the Jacobi Zone.

Barthel and Remane (in Barthel *et al.*, 1966) were apparently the first to discover the association of such alleged index fossils of the Jacobi Subzone as *Berriasella moreti, B.* cf. *jacobi* and *B. delphinensis* in their Bed B42 with *Paraulacosphinctes transitorius* and the calpionellids of Remane's (1964) Subzone A2. However, this discovery did not gain the appropriate recognition until the publication of Hoedemaeker's (1979, 1982) field and office studies of these and other sections in Southern Spain.

Hoedemaeker's (1982) review of presently available data about the time range of *Berriasella jacobi* (and of the related forms *B. delphinensis, B. richteri* and *B. moreti* also) clearly indicates that it is a boundary breaking, long ranging species that appears already in the upper part of the regional *Durangites* Zone of Enay and Géyssant (1975) in association with *Paraulacosphinctes transitorius*, other diagnostic late Upper Tithonian ammonites, and calpionellids of Remane's (1964) Subzones A2 and A3. These results reveal that the previously generally accepted essential equivalence of time ranges of *Berriasella jacobi* and *Pseudosubplanites grandis* and their association with the calpionellid fauna of the Zone B are an artefact simulated by the so far unfossiliferous character of local equivalents of the upper part of the Transitorius Zone of southeastern France. The above discussed Spanish occurrences of *B. jacobi* cannot be a different faunal facies of its "zone" exposed at Chomerac and Aizy. The ammonite fauna of the latter is devoid of any diagnostic Late Tithonian elements that are associated with *B. jacobi* in Southern Spain with an apparent (but far from certain) exception of *Proniceras* (Le Hégarat, 1971). Furthermore, these French sections contain the calpionellid fauna of the Zone B of Remane (1964), as pointed out by Le Hégarat (1971, 1973, p. 293, Table 30) and Hoedemaeker (1982, p. 10-12). Hoedemaeker's (1982) analysis also invalidates the claim of Zeiss (in Westermann, 1982, p. 146) that "the vertical ranges of *Pa. transitorius* and *M. micracanthum* are apparently different in the Mediterranean and Submediterranean areas, as evident also from their different calpionellid faunas (Oloriz and Tavera, 1981, Sapunov, 1977, Makarjeva, 1979)." These apparent discrepancies of their time ranges between the Bulgarian and Czechoslovakian sections on the one hand and those of Southern Spain on the other are simulated, instead, by the previously unrecognized common downward extension of time ranges of *Berriasella jacobi, B. delphinenses, B. chaperi* and *B. moreti* into the upper parts of the *Durangites* and Transitorius Zones. Furthermore, the claims of Oloriz and Tavera (1981) concerning the upward extension of the principal Stramberg ammonite fauna into the Jacobi Zone of authors are most unconvincing. They reflect this biochronological unreliability of *Berriasella jacobi, B. delphinensis* and *B. chaperi*, which evidently range down into the upper part of the Transitorius Zone in Stramberg Limestone just as they do in Southern Spain (compare Hoedemaeker, 1982, p. 10, 12). This is clearly indicated by the studies of Housa (1965, 1975) and Le Hégarat (1971), the results of which are fully valid in the writer's opinion.

The above data demonstrate that *Berriasella jacobi* cannot be used as either the zonal or subzonal fossil of Grandis Zone as defined in southeastern France and Southern Spain. Instead it joins the ranks of other notoriously unreliable berriasellids (e.g. *B. chaperi, B. moreti, B. delphinensis*) long believed to be good zonal fossils (e.g., Mazenot, 1939; Enay, 1964) but found more recently to be long-ranging and boundary-breaking (i.e., ranging across the Jurassic-Cretaceous boundary accepted in this paper; see above for further details). The confusion created by numerous attempts to use *B. chaperi* and *B. delphinensis* as zonal indices is well illustrated by Yegoyan's (1975, Fig. 1) correlation table. *B. jacobi* is, therefore, discarded as a zonal index of the basal Beriasian Grandis-Jacobi Zone of authors (Figs. 5, 10) and this zone is renamed the Grandis Zone after *Pseudosubplanites grandis* (Mazenot). It is defined principally by the abundance (the "acme") of *Pseudosubplanites*, most of the species of which (e.g., *P. grandis, P. ponticus* and *P. euxinus*) are restricted to it. This Grandis Zone is also characterized by the calpionellid fauna diagnostic of the lower subzone of the Zone B of Remane (1964) and by an apparently total absence of *Paraulacosphinctes, Durangites,* and *Micracanthoceras*, that are restricted to the underlying *Durangites* Zone according to Hoedemaeker (1982, p. 12). Contrary to Hoedemaeker's (*loc. cit.*) claim, *Aulacosphinctes* does range up into the Grandis Zone.

According to Hoedemaeker (1982, p. 12), *Himalayites, Corongoceras, Parapallasieras, Aspidoceras, Cyrtosiceras, Proniceras, Protancyloceras*, and *Neolissoceras* range above the calpionellid Subzones A2 and A3 of Remane (1964) and, hence, across the here proposed Tithonian-Berriasian boundary. However, all reliable records of *Corongoceras* in southern Spain and elsewhere appear to be restricted to the Late Tithonian as defined above. Furthermore, *Himalayites* s.s. (i.e., exclusive of *H.? nieri* Pictet), *Proniceras, Aspidoceras, Aulacosphinctes*, and *Parapallasiceras* mostly do not range above the uppermost Tithonian beds containing *Berriasella jacobi* and *Berriasella* ex gr. *chaperi-richteri-delphinense* in association with calpionellids diagnostic of the Subzones A2 and A3. Only the last, rare representatives of these ammonite genera ascend into the basal Berriasian Grandis Zone as defined herein. Therefore, they are considered here to be basically Late Tithonian index fossils (Figs. 5, 10).

All beds in Southern Spain and elsewhere (e.g., in Bulgaria; see Sapunov, 1977, p. 56-58, Figs. 2, 3 and the Stramberg Limestone proper) that contain *Berriasella jacobi* and *Berriasella* ex gr. *chaperi-richteri-delphinense* associated with diagnostic ammonites of the regional *Durangites* Zone, diagnostic ammonites of the equivalent upper part of the Transitorius Zone and/or calpionellids of the Subzones A2 and A3 are excluded from the basal Berriasian Grandis Zone and placed into the interregional UpperTithonian Transitorius Zone. This zone spans the greater upper part of the Upper Tithonian stage in the Submediterranean and Mediterranean Belts of the European Tethys alike (Figs. 5, 10). *"Malbosiceras" chaperi* Subzone of the Transitorius Zone proposed by Sapunov (1977, p. 57, 58, Figs. 2, 3) is considered to be just as invalid as the Jacobi Zone.

Contrary to Oloriz and Tavera (1979b, p. 503, Fig. 3), Messezhnikov (in Vakhrameev *et al.*, 1982, Tables 17, 20) etc., the South Spanish *Durangites* Zone (better Subzone; Sapunov, 1977) is treated here as an equivalent of the uppermost part of the Transitorius Zone. Except for its rare presence in the lower Upper Tithonian Micracanthum Zone in southern Spain (Enay and Géyssant, 1975), *Durangites* appears to be restricted to the upper Upper Tithonian where it coexists with *Paraulacosphinctes* ex gr. *transitorius-senex* and calpionellids of the Subzones A2 and A3 (Barthel *et al.*, 1966; Enay *et al.*, 1971, Enay and Géyssant, 1975; Hoedemaeker, 1982, p. 10-12). In southern Spain this *Durangites* Subzone is directly and apparently gradationally overlain by the *Berriasella jacobi* fauna (Enay and Géyssant, 1975) that did not yet yield *Durangites* but appears to be, nevertheless, referable to the latest Tithonian as defined above because of the scarcity of *Pseudosubplanites*.

Geographic Restriction and Biochronological Value of Durangites

So far as known, *Durangites* Burckhardt 1912 is a strictly Late Tithonian genus in Europe (see above) and the Americas (see p. 201-203 for further details). Enay *et al.*, (1971) had once claimed its presence in the uppermost Lower Tithonian Zitteli Zone of Turkey but this claim was withdrawn almost immediately (Enay, 1972). However, it is not generally appreciated that the time range of *Durangites* and the frequency of its occurrence in different parts of its range vary strongly from one region to another. This fact and its rather restricted geographical range limit strongly its usefulness as a Late Tithonian index fossil.

As already mentioned, *Durangites* is not restricted to its subzone in southern Spain but is also present, albeit rarely, in the next older Micracanthum Zone. It is also present in the basal Upper Tithonian in Turkey. Therefore, *Durangites*, actually ranges right through the Upper Tithonian in Europe while being much more common in its upper part. As shown below (see p. 201-203), it also ranges right through the Upper Tithonian in Mexico. However, in Mexico, *Durangites* is much more common in the lower Upper Tithonian (i.e., the regional *Kossmatia-Durangites-Corongoceras* Subzone) than in the upper Upper Tithonian (i.e., the regional *Substeueroceras-Berriasella* Subzone and its equivalents). These biochronologically discrepant maxima of frequency of *Durangites* may have been caused by a temporary presence of "sieve-like" geographical barriers that inhibited but not prohibited its transcontinental migration. However, it seems more likely that these frequency maxima were caused by a combination of a gradual cooling of the waters in Mexican basins during Late Tithonian time with a persistence of warm water temperatures in southern Spain. This could have induced a migration of the bulk of the late Late Tithonian *Durangites* populations from its original evolutionary center in Mexico to southern Spain. This hypothesis is favored because of the well established restriction of *Durangites* to the central zone of the Tethyan climatic belt of Europe and the Americas.

In Europe *Durangites* appears to be restricted to southern Spain and Turkey while being totally unknown in the Submediterranean Belt (e.g., in Southeastern France, Stramberg, Balkan Peninsula, Crimean-Caucasian Region, etc.). This paleogeographical restriction of *Durangites* in Europe is mirrored in the Americas where it is all but restricted to Mexico and adjacent Central American countries (Imlay, 1980, p. 34-36, Figs. 15, 16), that form the core of the Tethyan climatic belt. Further north, *Durangites* is unknown even in California since the writer does not accept the validity of its only older record there (see p. 189 for further details).

South of the Central American core of the Tethyan Realm *Durangites* is known from the Puente Ingo near Lima, Peru only. There the *"Kosmoceras" limense* of Lisson (1907) appears to be a *Durangites*, as already noted by Enay (1973). This locality has also yielded other Late Tithonian ammonites, such as *Micracanthoceras noduliferum* (Lisson, 1907) and *Protancyloceras sabaudianum* (Pictet-Loriol), which likewise were re-interpreted generically by Enay (1973).

South of Lima *Durangites* is totally unknown in the equivalent *Corongoceras alternans* and *Substeueroceras koeneni* Zones of Chile and Argentina, in spite of an extremely close affinity of the rest of their ammonites with those of Mexico and other Central American countries.

Absence of true Substeueroceras from the European Berriasian

All verifiable claims of *Substeueroceras* in the Berriasian of the Mediterranean and Submediterranean Belts of the European Tethys are invalid. This conclusion is most important in view of recent proposals (see p. 201-203) to correlate the Koeneni Zone of Central and Southern America with the Grandis and/or Occitanica Zones of the type Berriasian.

I disagree with Hoedemaeker's (1979, p. 245, Footnote 5, Fig. 1, Table 2) conclusion that *"Pseudargentiniceras" flandrini* Le Hégarat 1973 (Pl. 26, Figs. 7, 10; Pl. 51, Fig. 3; Pl. 52, Fig. 2) and *"Pseudargentiniceras" beneckei* Jacob (Le Hégarat, 1973,

Pl. 26, Fig. 4; Pl. 51, Fig. 2) from the basal Berriasian (Grandis Zone) belong to the genus *Substeueroceras* Spath 1925. Their inner whorls are closely similar to those of the typical representatives of *Substeueroceras* ex gr. *koeneni* from Western South America, Mexico and California in their shape, proportions, and ribbing habit. However, *"Pseudargentiniceras" beneckei* (Mazenot, 1939, Pl. XXXII, Figs. 8-13) differs in its regularly rounded and strongly furrowed venter that contrasts strongly with distinctly flattened, nonfurrowed venter of true *Substeueroceras*. The holotype of *"Pseudargentiniceras" flandrini* (Le Hégarat, 1973, Pl. 52, Fig. 2) develops umbilical bullae and numerous *Fauriella*-like bidichotomous rib bundles starting from these bullae on its adult ultimate whorl. Furthermore, the sculpture becomes strongly reduced at first and then all but lost on its living chamber. These features of its ultimate whorl contrasts strongly with those of the true representatives of *Substeueroceras* which retain the fine flexuous single primaries, that lack umbilical bullae and either bifurcate or trifurcate at various heights, to the oral end of the living chamber. Furthermore, their adult living chamber remains very sculptured throughout. Therefore, the similarity of whorl shape and proportions and that of the sculpture of the inner whorls of these French berriasellids to *Substeuroceras* must be a matter of homoemorphy only.

The type species of *Pseudargentiniceras* – *P. abscissum* (Oppel, 1868; see Zittel, 1868, Pl. 19, Fig. 1b, 4b) also develops umbilical bullae and at least some bidichotomous rib bundles on the adult ultimate whorl. Therefore, it is not a *Substeueroceras*, in spite of the fact that it retains pronounced ribbing to the oral end of its living chamber. However, this distinction, in combination with a pronounced coarsening and sparcening of the ribbing habit on the adult living chamber of *P. abscissum*, appears to be incompatible with the reference of *"Pseudargentiniceras" flandrini* and *"P." beneckei* to the genus *"Pseudargentiniceras"* Nikolov 1966 based on *P. abscissum*. They may be more properly assigned to the genus *Fauriella* Nikolov 1966 as its earliest known, most primitive representatives.

The "*Neocomites suprajurensis* Mazenot, *N. allobrogensis* Mazenot, *Tirnovella davidi* Le Hégarat" and *"Neocomites neocomiensis* variété plate à côtes fines et fasciculées in Sayn, 1907 ... ", which were assigned questionably to *Substeueroceras* by Hoedemaeker (*loc. cit.*), differ very strongly from its American respresentatives in the extremely high and slender whorl cross-section and even more importantly in most of their secondaries arising out of tiny (e.g., in *"Neocomites allobrogensis"*; see Mazenot, 1939, Pl. XXXIII, Fig. 4b) or large (e.g., in *"N. suprajurensis"*; see Mazenot, 1939, Pl. XXXIII, Fig. 5b) umbilical bullae. Furthermore, the holotype of *"N. suprajurensis"* becomes almost smooth on the oral half of its ultimate whorl (the presumably adult living chamber). All these forms are only superficially similar to the true American *Substeueroceras*.

It is impossible to appraise the only remaining, unfigured record of *Substeueroceras* (Oloriz and Tavera, 1979b, p. 503). However, this ammonite was found in the *Durangites* Zone of southern Spain. If correctly identified, it would support the late Late Tithonian age of the *Substeueroceras-Berriasella* Complex and the Koeneni Zone documented below (p. 201-203).

CONFLICTING CORRELATIONS BETWEEN THE BOREAL AND TETHYAN BOUNDARY BEDS

There are two principal concepts of correlation of the Boreal and Tethyan facies of the Jurassic-Cretaceous boundary beds. One is adhered to by most of the recent Soviet workers (see Saks *et al.*, 1979 and Saks *et al.*, 1972 for further details) and, with some modifications, by Jeletzky (1965, 1968, 1971a, 1973, 1974, 1979). According to these workers, the boundary between the Upper Volgian stage (or substage) and the Ryazanian Beds (or the boreal facies of the Berriasian) in the Boreal Realm corresponds approximately with that between the Upper Tithonian and Berriasian stages (or substages) in the Tethyan Realm so long as the Berriasian is interpreted to include either the Grandis-Jacobi Zone or the Grandis Zone exclusive of the Jacobi Zone.

In the opinion of a majority of these workers, the type Upper Volgian stage and the type Riasanian Beds are separated by a prolonged hiatus corresponding to the uppermost Tithonian and the basal Berriasian beds of the Tethyan Realm. This hiatus is filled up by the *Chetaites chetae, Chetaites sibiricus* and, according to some workers (e.g., Jeletzky, 1965, 1968, 1973, 1979; Sazonova, 1971, 1977; Sazonova and Sazonov, 1979a), by the *Hectoroceras kochi* Zones in the continuous sections of the Jurassic-Cretaceous boundary beds in North Siberia (see Saks *et al.*, 1972, Table 5 for further details). These interpretations are supported by the presence of *Virgatosphinctes* spp. in the *Chetaites chetae* Zone, which is accordingly interpreted as the equivalent of the latest Tithonian *Virgatosphinctes transitorius* Zone.

According to the above workers, the time relationships between the Boreal and Tethyan facies of the boundary beds are sufficiently obvious to employ the Upper Tithonian and Berriasian stages in the Boreal Realm and to abandon their regional equivalents (i.e., the Upper Volgian and Ryazanian stages). All workers of this group, except Jeletzky (loc. cit.) and Sazonova (*loc. cit.)*, place the Jurassic-Cretaceous boundary in the Boreal Realm between the *Chetaites chetae* and *Chetaites sibiricus* Zones. Of the latter two workers, Jeletzky (1979, Fig. 8 and Figs. 9, 10 in this article) places this boundary either between the *Chetaites sibiricus* and *Hectoroceras kochi* Zones or, more likely, within the basal part of the *Hectoroceras kochi* Zone while Sazonova (1971, Table opp. p. 14 and in the text) is inclined to place this boundary underneath the *Chetaites chetae* Zone. The most widespread variant of this concept is elucidated by Figure 1. However, Sazonova and Sazonov (1979b) have later adopted the opposite point of view of Casey (1962, etc.), Zeiss (1979) and Enay (1972) discussed below. Furthermore, Messezhnikov (in Vakhrameev *et al.*, 1982, p. 137, 138) has also indicated his readiness to give up his previously held ideas.

The above discussed ideas are opposed by Casey (1962, 1963, 1973) Enay (1972, 1973) and Zeiss (1968, p. 49, 139, Footnote; 1974; 1977; 1979 (p. 18, Correlation Table), and in Westermann, 1982, p. 146). These workers (see Fig. 2) claim that only the

MEDITERRANEAN PART OF THE TETHYAN REALM — STAGES	(SOUTHEASTERN FRANCE)
VALANGINIAN (BASAL PART)	*THURMANNICERAS PERTRANSIENS*
BERRIASIAN	*FAURIELLA BOISSIERI*
	TIRNOVELLA OCCITANICA
	BERRIASELLA GRANDIS
UPPER TITHONIAN	*BERRIASELLA JACOBI*
	PARAULACOSPHINCTES TRANSITORIUS
	PSEUDOVIRGATITES SCRUPPOSUS
MIDDLE TITHONIAN (PART)	*PSEUDOVIRGATITES PUSCHI*

SOVIET PART OF THE BOREAL REALM — STAGES		CENTRAL RUSSIA	NORTH SIBERIA
VALANGINIAN (BASAL PART)		*PSEUDOGARNIERIA UNDULATOPLICATILIS*	*NEOTOLLIA KLIMOVSKIENSIS*
BERRIASIAN		BEDS WITH *PEREGRINOCERAS* AFF. *ALBIDUM*	*BOJARKIA MESEZHNIKOWI*
		SURITES TZIKWINIANUS	
		BEDS WITH *RIASANITES, EUTHYMICERAS* AND *SURITES*	*SURITES ANALOGUS*
		HECTOROCERAS KOCHI	*HECTOROCERAS KOCHI*
		BEDS WITH *RIASANITES* AND *GARNIERICERAS* ?	*CHETAITES SIBIRICUS*
VOLGIAN	UPPER	? *CRASPEDITES NODIGER*	*CHETAITES CHETAE*
			CRASPEDITES TAIMYRENSIS
		CRASPEDITES SUBDITUS	*CRASPEDITES OKENSIS*
		KACHPURITES FULGENS	
	MIDDLE	*EPIVIRGATITES NIKITINI*	*EPIVIRGATITES VARIABILIS*
		VIRGATITES VIRGATUS	*TAIMYROPHINCTES EXCENTRICUS*
			DORSOPLANITES MAXIMUS
			DORSOPLANITES ILOVAISKII
		DORSOPLANITES PANDERI	*PAVLOVIA IATRIENSIS*
	LOWER (PART)	*SUBPLANITES PSEUDOSCYTHICUS*	*PECTINAITES PECTINATUS*

GSC

Figure 1. Interrealm correlation of Jurassic-Cretaceous boundary beds according to Saks *et al.* (1979).

EUROPEAN PART OF TETHYAN REALM (MAINLY SOUTHEASTERN FRANCE AND SPAIN) — STAGES		
VALANGINIAN (BASAL ONLY)	*THURMANNICERAS PERTRANSIENS, PLATYLENTICERAS* SPP.	
BERRIASIAN	*FAURIELLA BOISSIERI*	
	DALMASICERAS DALMASI	
	TIRNOVELLA OCCITANICA	
	BERRIASELLA GRANDIS	
UPPER TITHONIAN	*BERRIASELLA JACOBI*	
	PARAULACOSPHINCTES TRANSITORIUS ↓	*DURANGITES* ↑
	MICRACANTHOCERAS MICRACANTHUM	
MIDDLE TITHONIAN (PART)	*MICRACANTHOCERAS PONTI*	

EURASIAN PART OF THE BOREAL REALM — STAGES	CENTRAL RUSSIA	NORTH SIBERIA
VALANGINIAN (BASAL ONLY)	*PSEUDOGARNIERIA UNDULATOPLICATILIS*	*NEOTOLLIA KLIMOVSKIENSIS*
RIASANIAN	*SURITES SPASSKENSIS*	*BOJARKIA MESEZHNIKOWI*
	RIASANITES RJASANENSIS	*HECTOROCERAS KOCHI*
	HIATUS	*CHETAITES SIBIRICUS*
UPPER VOLGIAN	*CRASPEDITES NODIGER*	*CHETAITES CHETAE*
		CRASPEDITES TAIMYRENSIS
	CRASPEDITES SUBDITUS	*CRASPEDITES ORIGINALIS*
		CRASPEDITES OKENSIS
	KACHPURITES FULGENS	*"VIRGATOSPHINCTES" EXOTICUS*
MIDDLE VOLGIAN	*LOMONOSOVELLA BLAKEI*	*EPIVIRGATITES VARIABILIS, EPIVIRGATITES VOGULICUS*
	EPIVIRGATITES NIKITINI	
	VIRGATITES ROSANOVI	*TAIMYROSPHINCTES EXCENTRICUS*
	ZARAISKITES ZARAJSKENSIS	*DORSOPLANITES MAXIMUS*
		DORSOPLANITES ILOVAISKYI
	ZARAISKITES SCYTHICUS	*PAVLOVIA IATRIENSIS*
LOWER VOLGIAN (PART)	*ILOWAISKYA TENUICOSTATA*	*PECTINAITES LIDERI*

GSC

Figure 2. Interrealm correlation of Jurassic-Cretaceous boundary beds according to Zeiss (1979).

Lower Volgian stage *sensu* Nikitin (1884) or the Lower and Middle Volgian of Gerassimov and Mikhailov (1966) and more recent Soviet workers (e.g., Saks *et al.*, 1972; Saks, Messezhnikov and Shulgina, 1979) are correlative with the Upper Tithonian of the Tethyan Realm. The bulk of the Upper Volgian stage (= the Upper Volgian substage of Gerassimov and Mikhailov, 1966 and more recent Soviet workers) is believed to correspond either to the Grandis Zone alone or to this zone and the Occitanica Zone combined of the type Berriasian. This conclusion is based on the restriction of *Riasanites rjasanensis* and allies and of *"Euthymiceras"* ex gr. *transfigurabilis* to the high Berriasian Boissieri Zone in the Transcaspian, North

Caucasian, Crimean and Polish sections and the assumption that all of the *Riasanites rjasanensis* Zone of the type Ryasanian is of that particular age. The previously mentioned recent Soviet workers and the writer counter these arguments by invoking the already mentioned existence of a prolonged hiatus between the type Riasanian Beds and the underlying *Craspedites nodiger* Zone. The validity of these counterarguments, including the correctness of the identification and dating of "*Virgatosphinctes* spp." in the upper Volgian beds (including Chetae Zone) of Northern Siberia, is, in turn, being disputed by the above mentioned adherents of the second point of view.

PALEOBIOGEOGRAPHY AND PALEOCLIMATOLOGY

The still raging controversy about the correlation of the Boreal and Tethyan facies of the Jurassic-Cretaceous boundary beds is rooted in the long recognized fact (at least since Waagen, 1864, p. 98) that the latest Jurassic and earliest Cretaceous time was a climax of differentiation of the Boreal and Tethyan faunal realms. These sublatitudinally oriented, faunally contrasting belts of marine rocks obviously register the existence of well differentiated paleoclimatic belts (e.g., Arkell, 1956, p. 606-615; Saks *et al.*, 1973; Kaufmann, 1973, 1979; Hölder, 1979, p. 339), in spite of objections of some workers (e.g., Hallam, 1969). The faunal contrast of Boreal and Tethyan basins of the latest Jurassic and earliest Cretaceous was not limited to an almost total dissimilarity of their most diagnostic invertebrate groups (e.g., ammonites, belemnites, inoceramids and pteriid pelecypods) on specific, generic and even family level. These Boreal faunas had extremely low diversity as compared with the Tethyan faunas, although all their taxa were very abundant. Furthermore, they were devoid of all demonstrably warm water faunal elements (e.g., hermotypic corals, trigoniid and diceratid pelecypods, other thick-shelled, commonly bioherm-building molluscs, large benthonic foraminifers) that were restricted to the low lattitudinal Tethyan basins. Finally, these Boreal basins all but backed limy sediments that were, in contrast, extremely abundant in the low lattitudinal basins of the Tethyan Realm.

The marine connections between the latest Jurassic and earliest Cretaceous marine basins of the Boreal and Tethyan Realms were strongly restricted (Fig. 3). These connecting seaways were definitely more restricted than those of the immediately preceding Early Volgian (e.g., Saks *et al.*, 1973) and immediately following Valanginian (e.g., Saks *et al.*, 1973). This scarcity of connecting seaways must have been contributing to the faunal dissimilarity of the Boreal and Tethyan marine basins of the latest Jurassic and earliest Cretaceous. However, the decisive influence of the temperature gradient upon the differentiation of these Boreal and Tethyan faunas is attested by the fact that their mixing within these connecting seaways was either rather limited or almost nonexistent. It is the resulting scarcity of mixed Boreal and Tethyan faunas that continues to be the principal obstacle to the reliable correlation of the Boreal and Tethyan facies of the Jurassic-Cretaceous boundary beds.

The idea that the alleged total dissimilarity of Late Volgian ammonite faunas from their Tethyan counterparts was caused by a complete isolation of the Late Volgian basins of Central and Arctic Russia from all other contemporary basins (e.g., Arkell, 1956, p. 612; Callomon, 1961) was discredited by subsequent research (e.g., Jeletzky, 1966; Shulgina in Saks *et al.*, 1969 and Casey, 1973) that demonstrated the presence of elements of these faunas all over the Boreal Realm. However, these ammonites do not penetrate significantly into any of the connecting seaways, the existence of which is evident because of considerable southward penetrations of contemporary Boreal belemnites and buchias.

This far reaching faunal differentiation of the Boreal and Tethyan Realms was accompanied by a considerable differentiation of invertebrate faunas of individual marine basins (or groups of basins) within the Boreal Realm. These latest Jurassic to earliest Cretaceous basins, usually designated as Boreal faunal provinces, are shown in Figure 3. They include: 1) The Boreal-Atlantic Province; 2) The Central Russian Province; 3) The North Uralian Province; 4) The North Siberian Province; 5) The North American Boreal Province of Jeletzky (1971b, p. 12-16), which is the same as the Chukchi-Canadian Province of Saks *et al.*, (1972, p. 273, Fig. 17).

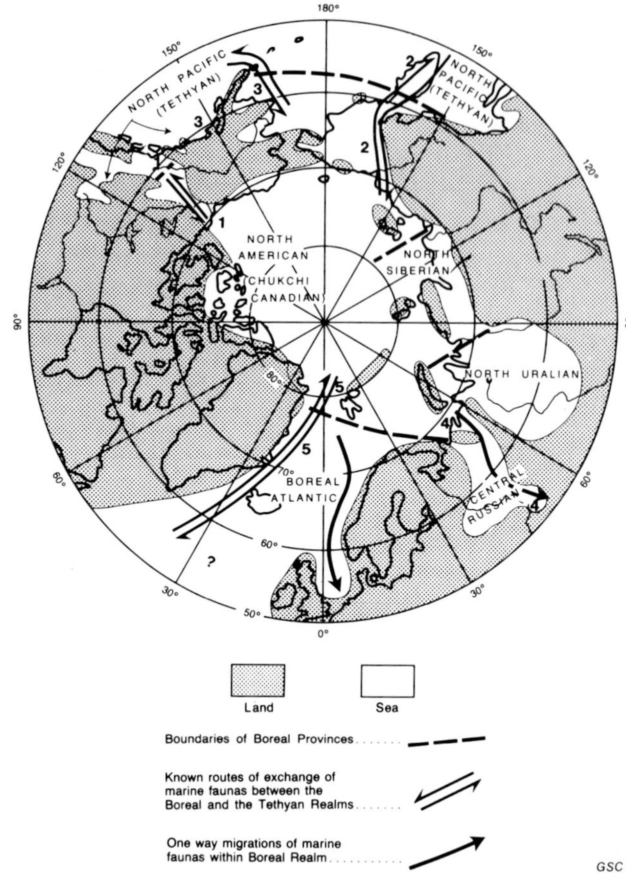

Figure 3. Late Volgian (= Late Tithonian) paleobiogeography of the Boreal Realm (data from Saks *et al.*, 1973; Jeletzky, 1971b, 1973, etc., modified). 1) Midwestern strait of North America (= Dawson Strait). 2) Western Pacific Seaway (or Western Shelf of Paleopacific Ocean). 3) Eastern Pacific Seaway (or North American Shelf of Paleopacific Ocean). 4) Pechora-Caspian Seaway (its marine connection with the Tethyan basins of the Crimean-Caucasian regions was closed in the Late Volgian). 5) East Greenland Seaway.

The North Pacific faunal province (or a realm) of Jeletzky (1965, p. 61, Fig. 3, 1970, p. 649, Table XI-8) that extends along the Pacific Slope of North America from southeastern Alaska to southern California and along the Pacific Slope of Asia from Kamchatka to Japan and farther south, is included among the Boreal provinces by most of the recent Soviet workers (e.g., Saks *et al.*, 1972, 1973). However, the original treatment of this province as the northernmost part of the Tethyan Realm is maintained in this paper because of the predominantly Tethyan affinities of its latest Jurassic and earliest Cretaceous ammonite faunas (Fig. 3).

The latest Jurassic and earliest Cretaceous Boreal basins were connected with the low lattitudinal Tethyan basins by the following five marine straits numbered 1 to 5 in Figure 3.

1) The Western North American Strait connecting the North Pacific geosynclinal marine troughs of western British Columbia with the Boreal marine basins of the Canadian Arctic mainland and the Sverdrup Basin (see Jeletzky, 1971b, p. 34-37, Fig. 5 for further details).
2) The Western Pacific seaway extending along the Pacific slope of Siberia from the Anadyr'-Koriak region through the Soviet Far Eastern Province to the Japanese Islands and apparently farther south to the Taivan Island beneath the Pacific Ocean. This seaway must have ultimately opened into the Himalayan-Indonesian Geosyncline.
3) The Eastern Pacific seaway that extended along the western shore of Alaska from the Cape Lisbourne area to its southeastern base and then into the geosynclinal troughs of western British Columbia. This seaway(s) then must have extended southward into the Mexican and South American geosynclinal basins because of the Tethyan affinities of its ammonite faunas in western British Columbia discussed below. This seaway also must have joined the Midwestern North American Strait somewhere in the northwestern corner of British Columbia because of the Boreal affinities of its *Buchia* and belemnite faunas discussed by Jeletzky (1971b).

Because of an inferred existence of the Paleopacific Ocean throughout the Phanerozoic Eon at least (Khudoley, 1974; Khain, 1981), the Eastern and Western Pacific Seaways are interpreted as the eastern and western shelves of the latest Jurassic and earliest Cretaceous generations of that ocean (Fig. 3). A wide Bering Shelf must have connected these two shelves in the north because of very close affinities of their shallow water Jurassic and Cretaceous marine faunas with one another and with those of Northern Siberia and the North American Arctic.[1]

NOTE: I adhere to the hypothesis of the Expanding Earth in the sense of W.S. Carey (e.g., The necessity for Earth expansion; The Expanding Earth, A Symposium. Earth Resources Foundation, University of Sydney, *1981*, p. 375-396). Active (i.e. convection cells driven), large scale horizontal movements of lithospheric plates and their subduction postulated by the hypothesis of orthodox plate tectonics are considered to be improbable and rejected (compare, Jeletzky, 1978, p. 15 and p. 242-244 below).

4) The Pechora-Caspian Seaway that connected the Boreal basin of the European Arctic with the Central Russian Boreal Basin and, at certain times only, with the epicontinental Polish Basin and the geosynclinal Tethyan seas of the Crimea and Caucasus. It must be stressed emphatically that, in the writer's opinion, the marine connections of the Central Russian Basin with the Polish Basin and the Crimean-Caucasian Basins only existed in the Early Volgian and Mid- to Late-Berriasian (Fig. 12b) times. In contrast, this Boreal Basin was connected with the Boreal Basin of the European Arctic throughout the Late Volgian and Early Berriasian (= early Riasianian) time (Figs. 3, 12a). The reasons for these conclusions and their implications are discussed below (p. 241-242).
5) The East Greenland Seaway that was situated between Greenland and the Scandinavian Peninsula. This strait lapped only slightly onto the eastern margin of Greenland in the west. The recent discovery of the shallow water marine uppermost Jurassic and basal Cretaceous rocks near the northern end of Greenland (Hakansson *et al.*, 1981) suggests strongly that this seaway was connected not only with the Boreal Basin of the European Arctic but with the Sverdrup Basin as well. In the east, East Greenland Seaway apparently onlapped the Scandinavian Peninsula in the Andoy area only (Zacharov, Surlyk, and Dalland, 1981).

This paper is concerned immediately with the following of the above discussed connecting seaways: 1) The Midwestern Strait of North America, that was named the Dawson Strait by Jeletzky (1970), including the Sverdrup Basin situated at its northern end but excluding the Peace River Basin of northeastern British Columbia connected to it by the Vanderhoof Strait (Fig. 4); 2) The Eastern Pacific Seaway (or the North American Shelf of the Pacific Seaway); and 3) The Pechora-Caspian Seaway in the European Part of U.S.S.R.

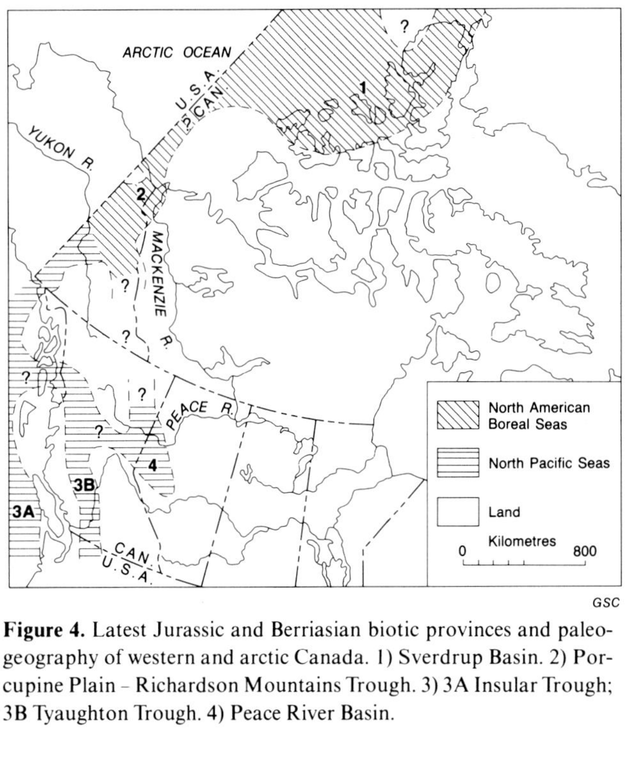

Figure 4. Latest Jurassic and Berriasian biotic provinces and paleogeography of western and arctic Canada. 1) Sverdrup Basin. 2) Porcupine Plain – Richardson Mountains Trough. 3) 3A Insular Trough; 3B Tyaughton Trough. 4) Peace River Basin.

This selection was dictated by the fact that the uppermost Jurassic and basal Cretaceous deposits of these three seaways have recently yielded a wealth of new data bearing on the paleontological correlation of the Jurassic-Cretaceous boundary beds of adjacent Tethyan and Boreal basins. This new evidence bears directly on the subject of this paper.

Of the other above mentioned connecting seaways, the West Pacific Seaway (or the Western Shelf of the Pacific Ocean) has so far only yielded poorly preserved and scarce latest Jurassic and basal Cretaceous ammonites and buchias (e.g., Matsumoto, 1978; Sato, 1962; Takarashi, 1973; Khudoley, 1959, 1963, 1970, 1974; Yefimova *et al.*, 1968; Paraketsov, 1980) that are difficult to interpret biochronologically and paleobiogeographically. The same is true of the ammonite and *Buchia* faunas of the East Greenland Seaway. Furthermore, the Tethyan influences within the latter seaway appear to be restricted to such fossil groups as brachiopods, gastropods and pelecypods other than *Buchia*. No ammonites of Tethyan affinities were so far encountered anywhere within the East Greenland Seaway (Casey, 1973; Surlyk, 1973, 1977, 1978, 1979), except for the still doubtful *Virgatosphinctes*? ex gr. *tenuicostatus* Shulgina.

UPPERMOST JURASSIC AND BASAL CRETACEOUS STRATIGRAPHY AND BIOCHRONOLOGY OF WESTERN BRITISH COLUMBIA

General and Historical Remarks

In western British Columbia the Jurassic-Cretaceous boundary beds are only known to occur in the so called Tyaughton and Insular Troughs (Fig. 4). These late Mid-Jurassic (Callovian and later) successor troughs were originally defined by Jeletzky and Tipper (1968, p. 3, 5, 74-77, Figs. 7, 8, 9) and the reader is referred to this publication for further stratigraphic and structural details. An unwieldy name "Vancouver Island-Queen Charlotte Islands Trough" was used for the Insular Trough in that publication. It was later replaced by the term Insular Trough (Jeletzky, 1971b, p. 36; 1976, p. 95) that was originally proposed for the same structural element by Sutherland Brown (1966).

The earliest detailed paleontological-stratigraphical studies of the Jurassic-Cretaceous boundary beds of western British Columbia dealt with the incomplete sequences exposed on the western flank of Tyaughton Trough in Harrison Lake area (Crickmay, 1930a, 1930b, 1962) and on the Vancouver Island within the Insular Trough (Jeletzky, 1950, 1965).

Later a detailed study of the Late Jurassic and Neocomian stratigraphy and biochronology of the Taseko map-area (Jeletzky, in Jeletzky and Tipper, 1968) and the adjacent parts of the Pemberton map-area (Jeletzky, 1967, 1971c) demonstrated the presence of a complete, entirely marine sequence of the Jurassic-Cretaceous boundary beds that contained a succession of rich *Buchia* faunas that matched almost exactly that known to occur in northern Eurasia. However, some of these *Buchia* faunas were found to contain Tethyan ammonites of Mexican and South American affinities. These rocks and their faunas have completely filled up the latest Jurassic hiatus previously encountered in the Insular Trough and in the Harrison Lake area of the Tyaughton Trough.

The Berriasian *Buchia* faunas of the Taseko Lakes map area are the same as those of the western Vancouver Island and the Harrison Lake area. However, some of them have yielded Berriasian ammonites that are new for the entire North America (Jeletzky, 1971a, 1973).

Shallow water to nonmarine facies of the Jurassic-Cretaceous boundary beds also were found in the Mount Waddington map-area at the northwestern flank of the Tyaughton Trough (Jeletzky, 1968b and in Tipper, 1969, p. 38-40).

Finally, the presence of the Jurassic-Cretaceous boundary beds is indicated in the Manning Park area on the southeastern flank of the Tyaughton Trough (Coates, 1974).

Buchia cf. blanfordiana Zone

Stratigraphy and Geographical Distribution. In western British Columbia *Buchia* cf. *blanfordiana* Zone is known in both the Insular and the Tyaughton Troughs.

In the Insular Trough the zone is known only in the Kyuquot-Esperanza area of western Vancouver Island where it is represented by argillaceous rocks. It overlies gradationally the beds characterized by buchias previously identified as early representatives of *Buchia piochii* var. *russiensis* and its allies (now assigned to *Buchia russiensis* s.l.) and is overlain disconformably and apparently regionally unconformably by the basal Berriasian *Buchia okensis* s.s. Zone (Jeletzky, 1965, p. 17, Fig. 4). The reader is referred to Jeletzky (1950, 1965, p. 17-20, Figs. 2, 4) for further details of the thickness, lithology and stratigraphic relationships of *Buchia* cf. *blanfordiana* Zone in that area.

Gritty and pebbly greywacke containing a poorly preserved *Buchia* fauna, that is either late Late Tithonian (i.e., early *Buchia terebratuloides* s.l. time) or late Early Tithonian (i.e., *Buchia* cf. *blanfordiana* time) in age, was recently discovered on a small island south of White Point on the west side of Graham Island, Queen Charlotte Islands (GSC loc. C-081990; J.A. Jeletzky, 1982, unpublished intradepartmental fossil report Km-2-1982-JAJ). This suggests that the marine sediments of *Buchia* cf. *blanfordiana* Zone were originally deposited in some other areas of the Insular Trough.

The previously inferred (Jeletzky, 1965, p. 17, explanation of Pl. III, Fig. 8) presence of *Buchia* cf. *blanfordiana* Zone in the Lillooet and Bridge River areas of the Tyaughton Trough was confirmed by a detailed stratigraphical-biochronological study of the Jurassic and Cretaceous rocks in the Taseko Lakes map-area (Jeletzky, in Jeletzky and Tipper, 1968, p. 12). In that area this zone comprises the topmost part of the lithologically diversified Mid(?)-Kimmeridgian to Portlandian s.str. unit of the Relay Mountain Group (Jeletzky, in Jeletzky and Tipper, 1968, p. 17-21, Fig. 2). The zone is represented predominantly by fine to coarse grained greywacke with minor to major intercalations of largely sandy siltstone and a minor gritty to conglomerate interbeds. This shallow marine, arenaceous facies of *Buchia* cf. *blanfordiana* Zone forms two narrow northwest-trending belts flanking the Relay Mountain Massif. They were presumably deposited on the flanks of the Tyaughton Trough that was very strongly reduced because of a regression throughout *Buchia* cf. *blanfordiana* time. This is indicated by these belts being separated from one another by a narrow northwest-trending belt where *Buchia* cf. *blanfordiana* Zone appears to be represented exclusively to almost exclusively by greenish grey, rust-weathering, more or less sandy siltstone. This belt, which is centered at the Relay Mountain Massif and appears to be only 6.5 to 8 km wide across the strike, is believed to comprise the strongly restricted axial part of the Tyaughton Trough (Jeletzky, in Jeletzky and Tipper, 1968, p. 17-20, Figs. 2, 14).

Buchia cf. *blanfordiana* Zone is at least 95 m thick in its marginal arenaceous facies belts (e.g., the combination of Sections 13 and 14 on the northeastern flank; see Jeletzky, in Jeletzky and Tipper, 1968, p. 148-149, 166-168). However, its complete thickness remains uncertain as all the longest sections of this facies studied (i.e., Sections 3, 8, 13 and 14) are faulted. Both contacts of this facies are gradational in all suitably exposed sections of both belts.

No complete sections of the argillaceous mid-basin facies of *Buchia* cf. *blanfordiana* Zone are known in the Taseko Lakes map-area. However,it seems to be strongly reduced in comparison with the adjacent arenaceous facies as its fossiliferous siltstone is only about 7.5 m thick in the apparently undisturbed Section 16 (Jeletzky, in Jeletzky and Tiper, 1968, p. 181) where its normal contact with the overlying *Buchia piochii* s.s. Zone is gradational. The underlying unfossiliferous siltstone is undisturbed and appears to grade imperceptibly into the *Buchia* cf. *blanfordiana*-bearing rocks above and into the *Buchia concentrica*-bearing rocks below. Therefore, this only 21.5 m thick siltstone is believed to represent a strongly reduced facies of the next older *Buchia russiensis* s.l. (formerly *Buchia* n.sp. aff. *piochii*) and *Buchia mosquensis* s.l. Zones.

In the Manning Park area of the Tyaughton Trough *Buchia* cf. *blanfordiana* (Stoliczka, 1856) was so far only found as a minor faunal admixture in two fossil collections dominated by *Buchia russiensis* s.l. (formerly *B.* n. sp. aff. *piochii*; see Jeletzky in Coates, 1974, p. 84, 103, 122). Neither of these sections was visited by the writer and the stratigraphic position of these two fossil lots in the sequence of *Buchia*-bearing Dewdney Creek Formation remains obscure. They are believed to represent the very base of the transitional beds between the *Buchia russiensis* s.l. and *Buchia* cf. *blanfordiana* Zones where the first representatives of the latter species begin to appear (e.g., unit 25 of Section 13; Jeletzky, in Jeletzky and Tipper, 1968, p. 150). It is not known whether the remainder of *Buchia* cf. *blanfordiana* Zone is concealed, unfossiliferous or absent by a subsequent erosion.

Marine sediments of *Buchia* cf. *blanfordiana* Zone are not known to outcrop anywhere else in the Tyaughton Trough. However, one could expect their presence in the very poorly known marine Upper Jurassic rocks outcropping east of the Taseko Lakes area in the vicinity of Lilloet.

In the Mount Waddington map-area the non-fossiliferous or plant-bearing rocks of *Buchia* cf. *blanfordiana* Zone are assumed to be present in the lower part of the Berriasian to Portlandian s.s. carbonaceous and plant-bearing arenites that separate the later lower Berriasian marine rocks (i.e., the upper part of *Buchia okensis* s.s. Zone) from the underlying mid-Portlandian (i.e., *Buchia russiensis* s.l. Zone) shallow marine rocks (see Jeletzky, in Tipper, 1969, p. 40, 41).

In the Bowser Basin, that occupies the northwestern part of British Columbia's mainland, the *Buchia* cf. *blanfordiana* time is apparently only represented by a hiatus. According to Tipper and Richards (1976, p. 40, Fig. 12), the region was uplifted and was undergoing erosion through the Kimmeridgian to Hauterivian time.

Paleontology and Biochronology. Geographical Range. The *Buchia* cf. *blanfordiana* Zone is a regional biochronological unit as beds characterized by the prevalence of this fossil appear to be restricted to the western British Columbia and northwestern part of the State of Washington, U.S.A. (Jeletzky, 1965; Jeletzky, in Jeletzky and Tipper, 1968). Elsewhere, *Buchia* cf. *blanfordiana* was only noted occasionally in California (Jones *et al.*, 1969), on the western flank of the Porcupine Plain-Richardson Mountains Trough and questionably in northeastern British Columbia (Jeletzky, unpublished intradepartmental fossil reports). In these regions *Buchia* cf. *blanfordiana* was encountered only as a minor faunal admixture in beds characterized by the prevalence of other *Buchia* forms that were previously assigned to *Buchia piochii* var. *russiensis* (Pavlow), *Buchia p.* var. *mniovnikensis* (Pavlow) and *Buchia* aff. *fischeriana* (d'Orbigny).

Affinities of Buchia cf. *blanfordiana.* The restricted geographical range of *Buchia* cf. *blanfordiana* and its intergradation with the latest known, thick and wide forms of *B.* ex gr. *mosquensis* reproduced by Anderson (1945, Pl. 12, Fig. 3) as *Aucella* cf. *mosquensis* in the basal beds of its zone in the Taseko Lakes map-area (Jeletzky, in Jeletzky and Tipper, 1968, p. 12, 150) indicates its being an endemic, presumably dead end offshoot of that Early to mid-Lower Volgian species group. The previously favored idea of Indo-Pacific affinities of *B.* cf. *blanfordiana* (Jeletzky, 1965, p. 20, 59, Fig. 3) that is reflected in its name is abandoned herewith and the species is now considered to be a new species belonging to the *Buchia mosquensis* (von Buch) species group. This idea has already been advanced by Konovalov (1977).

Status of Buchia cf. blanfordiana and its Beds. Jones *et al.* (1969, p. A10, Pl. 2, Figs. 38, 40-42) have pointed out that some of the large, thick and markedly coiled *Buchia* specimens from the regional *Buchia elderensis* Subzone of *Buchia piochii* s.l. Zone (here included in *Buchia russiensis* s.l. see below) are identical with the Canadian *Buchia* cf. *blanfordiana*. However, they proposed to treat them and the Canadian form as an extreme variant of *B. elderensis*. The writer agrees with the Jones *et al.* identification, at least where the Californian specimen reproduced in their Plate 2, figures 38, 40-42 is concerned. Furthermore, he has himself collected other, morphologically similar representatives of *B.* cf. *blanfordiana* from the upper part of the regional *Buchia elderensis* Subzone (e.g., GSC loc. 77481 and 77482). However, he disagrees where other similarly large and thick *Buchia* specimens reproduced from the same subzone by Jones *et al.* (1969, Pl. 2, Figs. 39, 43-45) are concerned. These specimens lack entirely the pronounced right-hand twist of the beak part of the left valve and the strong swelling of the left beak itself that are diagnostic of *B.* cf. *blanfordiana*. They are unusually large and sturdy representatives of *Buchia russiensis* s.l. that are particularly similar to, and probably subspecifically identical with, the thickest representatives of *Buchia taimyrensis* Zakharov 1981 (e.g., Pl. XVIII, Fig. 4; Pl. XIX, Fig. 1). The latter is considered to be but an extreme variant of *Buchia russiensis* s.l. by the writer.

Buchia cf. *blanfordiana* characterizes a well defined sequence of beds that intervenes between the *Buchia russiensis* s.l. Zone (formerly *Buchia* n. sp. aff. *piochii* Zone) on the one hand and the *Buchia piochii* s.s. Zone on the other in the Taseko Lakes map-area (Jeletzky, in Jeletzky and Tipper, 1968, p. 12, 13, Table 1) and, apparently (except for the missing upper contact and the next younger zones) on the west coast of Vancouver Island (Jeletzky, 1965, p. 17, 18, Figs. 2, 4). Within this interval it occurs either alone or dominates the *Buchia* populations almost to the exclusion of *Buchia* ex gr. *russiensis-piochii*. These preceding and subsequent *Buchia* forms only are prevalent in the basal and topmost parts of *Buchia* cf. *blanfordiana* Beds. Therefore, *Buchia* cf. *blanfordiana* is a well defined, morphologically distinctive species in western British Columbia. It will be formally named elsewhere. Its beds represent a regional fossil zone designated herein as *Buchia* cf. *blanfordiana* Zone. This full zonal status of *Buchia* cf. *blanfordiana* Beds in our region discredits the conclusion of Jones *et al.* (1969, p. A10) that its previously discussed, rare Californian representatives are only extremely large and heavily built representatives of *Buchia elderensis* (*recte B. russiensis* s.l.). These rare specimens are instead isolated representatives of *B.* cf. *blanfordiana* that migrated southward way beyond its normal geographical range. As such they provide a valuable indication about the correspondence of the Canadian *Buchia* cf. *blanfordiana* Zone to part or all of the regional Californian Zone of *Buchia elderensis (recte Buchia russiensis* s.l. Zone). This problem shall be discussed in a greater detail elsewhere in this paper.

Age and Correlation. In spite of regional restriction of *Buchia* cf. *blanfordiana* Zone, its approximate age in terms of the Boreal standard zones and stages is indicated by its stratigraphic position between the *Buchia russiensis* s.l. Zone below and *Buchia piochii* s.s. Zone above.

It is because of the gradational superposition of *Buchia* cf. *blanfordiana* Zone on beds bearing *Buchia russiensis* s.l. and its allies (which were then treated as variants of *Buchia piochii* s.l.) that Jeletzky (1965, p. 17) has originally suggested that this zone, "is either of the latest Portlandian (s. str.) or of Upper Tithonian (Purbeckian) age in terms of the international standard stages." Later Jeletzky (1965, p. 18) notes also that he "favors latest Portlandian (s. str.) age of *Buchia* cf. *blanfordiana* zone and assumes tentatively that it corresponds only to the lower part of *Buchia piochii* zone (i.e. to the *Buchia elderensis* Subzone of the current Californian usage; writer's comment) of the adjacent regions of North America (Fig. 1). It would seem that in the Canadian western Cordillera the upper half of *Buchia piochii* zone and all of *Buchia fischeriana* zone are missing and are represented only by the erosional gap between *Buchia* cf. *blanfordiana* and *Buchia okensis* zones (Fig. 1)."

A better defined stratigraphic position of *Buchia* cf. *blanfordiana* Zone in the Taseko Lake map-area (Jeletzky, in Jeletzky and Tipper, 1968, p. 12, Table 1) and the subsequent discovery of rare *Buchia* cf. *blanfordiana* in the regional *Buchia elderensis* Subzone of northern California (Jones *et al.*, 1969, p. A10, Pl. 2, Figs. 38, 40-42, and collecting personally in 1966) permit its more definitive dating and correlation.

A re-appraisal of the age of the lower part of the *Buchia piochii* s.l. Zone in California that was designated as its *Buchia elderensis* Subzone by Jones *et al.* (1969, p. A9-A11, Fig. 2) and a re-assessment of the taxonomic status of its name fossil are indispensable in order to determine more closely the lower and upper age limits of *Buchia* cf. *blanfordiana* Zone.

The bulk of *Buchia* forms figured by Anderson (1945, Pl. 3, Figs. 8, 9, 15, 21; Pl. 4, Figs. 1-4, 12) and Jones *et al.* (1969, Pl. 2, Figs. 17-33) from the lower part of the Knoxville Beds representing the *Buchia elderensis* Subzone is regarded conspecific with the Canadian forms previously assigned by Jeletzky (1965, Pl. I, Figs. 3, 4, 5, 6, 9; 1966, Pl. VIII, Figs. 6, 10) to *Buchia piochii* (s.l.) var. *russiensis* (Pavlow 1907) and *B. p.* var. *mniovnikensis* (Pavlow 1907). These latter forms are, in turn, conspecific with the Central Russian and North Siberian forms described and figured by Pavlow (1907, Pl. II, Fig. 6; Pl. III, Figs. 2, 3, 14, 15; Pl. IV, Figs. 21, 22) and Gerassimov (1955, Pl. XIV, Figs. 1-3) under the names *Aucella mniovnikensis Pavlow, Aucella russiensis* Pavlow, *Aucella gabbi* Pavlow and *Aucella subovalis* Pavlow and by Zakharov (1981, Pls. XIV-XVII; Pls. XVIII, Figs. 1-8; Pl. XIX, Figs. 1-7) as *Buchia russiensis* and *Buchia taimyrensis*.

As already pointed out by Zakharov (1981, p. 89), all these "species" of Pavlow (1907) are but varieties of one and the same polytypic *Buchia* species that characterize the middle and upper parts of the Lower Volgian (or the lower part of the Middle Volgian of current Soviet usage; see Gerassimov and Mikhailov, 1966) throughout the northern Eurasia (Zakharov, 1981, p. 222-224, Table 3) and in the Arctic Canada (Jeletzky, 1966). In northern California the beds containing this species (i.e., *Buchia elderensis* of Jones *et al.*, 1969, p. A11) are "probably not older than middle Tithonian as indicated by an absence of *Buchia rugosa (Fischer) and B. mosquensis* (von Buch), which are common in North America and the Boreal region in beds of middle Kimmeridgian to early middle Tithonian ages (Imlay, 1955, p. 75, 85)." In the Taseko lakes map-area of Western British Columbia *Buchia russiensis* s.l. (previously designated "*Buchia* n. sp. aff. *piochii*" by Jeletzky, in Jeletzky and Tipper, 1968, p. 12, Table 1) characterizes beds that are presumed to be of the same age as the "*Buchia elderensis*" Beds in northern California and *Buchia russiensis* Zone in northern Eurasia because of their stratigraphic position between *Buchia mosquensis* Zone below and *Buchia* cf. *blanfordiana* Zone above (see Jeletzky, in Jeletzky and Tipper, 1968, p. 12, Table 1). These "*Buchia* n. sp. aff. *piochii* Beds" did not yield any diagnostic ammonites in western British Columbia.

I now propose to use the name *Buchia russiensis* (Pavlow, 1907) s.l. for the forms discussed here and to place into its synonymy all other above listed species names of Pavlow (1907), as well as the names *Aucella elderensis* Anderson 1945, *Buchia* n. sp. aff. *piochii* Jeletzky 1968 and *Buchia taimyrensis* Zakharov 1981.

The beds characterized by the prevalence of *Buchia russiensis* s.l. in western British Columbia (formerly *Buchia* n. sp. aff. *piochii* Zone) and northwestern California (formerly *Buchia elderensis* Subzone) are now accorded a full zonal status (Figs. 5, 10) because of their equivalence with the interregional *Buchia russiensis* Zone of Northern Eurasia. The intercalation of the regional *Buchia* cf. *blanfordiana* Zone between the *Buchia piochii* s.s. Zone and the equivalents of the California Subzone of *Buchia elderensis* in western British Columbia is another reason for the complete separation of this "subzone" from *Buchia piochii* "subzone" proper, which is also elevated to the full zonal status. Furthermore, the "*Buchia elderensis* Subzone" must be renamed *Buchia russiensis* Zone because of the priority of the latter specific name.

The above proposed synonymy of *Buchia elderensis* with *Buchia russiensis* s.l. and the rare occurrence of *Buchia* cf. *blanfordiana* in "*Buchia elderensis*" Subzone of California should not be interpreted as an indication of a complete equivalence of the regional *Buchia* cf. *blanfordiana* Zone of western British Columbia with the *Buchia russiensis* Zone of northern California, Arctic Canada and northern Eurasia. This idea is contradicted by the presence of a considerable thickness of beds characterized by *Buchia russiensis* s.l., late forms of *Buchia mosquensis* s.l., and forms transitional between *Buchia mosquensis* s.l. and *B.* cf. *blanfordiana* immediately below the *Buchia* cf. *blanfordiana* Zone in the Taseko Lakes map-area and on the west coast of Vancouver Island. In combination with the presence of rare representatives of *Buchia* cf. *blanfordiana* in the upper part of *Buchia russiensis* Zone in California (e.g., GSC loc. 77481 and 77482) this indicates the correspondence of the Canadian zone to the upper part of *Buchia russiensis* s.l. Zone only. Therefore, *Buchia* cf. *blanfordiana* Zone is assigned the latest Lower (or the latest Middle in terms of Gerassimov and Michailov's 1966 subdivision) Volgian age and is correlative approximately with the *Epivirgatites variabilis* and *Taimyrosphinctes excentricus* Zones of northern Siberia and their latest Lower Volgian equivalents in Central Russian basin (Figs. 9, 10).

The proposed correlation of *Buchia* cf. *blanfordiana* Zone with part or all of the *Virgatites virgatus* and *Epivirgatites nikitini* Zones of Central Russian Basin suggests its approximate contemporaneity with the West European Tethyan Zone of *Micracanthoceras ponti* (Fig. 5). This rather tenuous correlation can be checked by comparing the scarce to relatively scarce ammonites of the Canadian and Californian sections of the *Buchia* cf. *blanfordiana*Zone with the Portlandian s.s. ammonite faunas of northwestern Europe and the rich, high Tethyan ammonite faunas of the Mexican Tithonian.

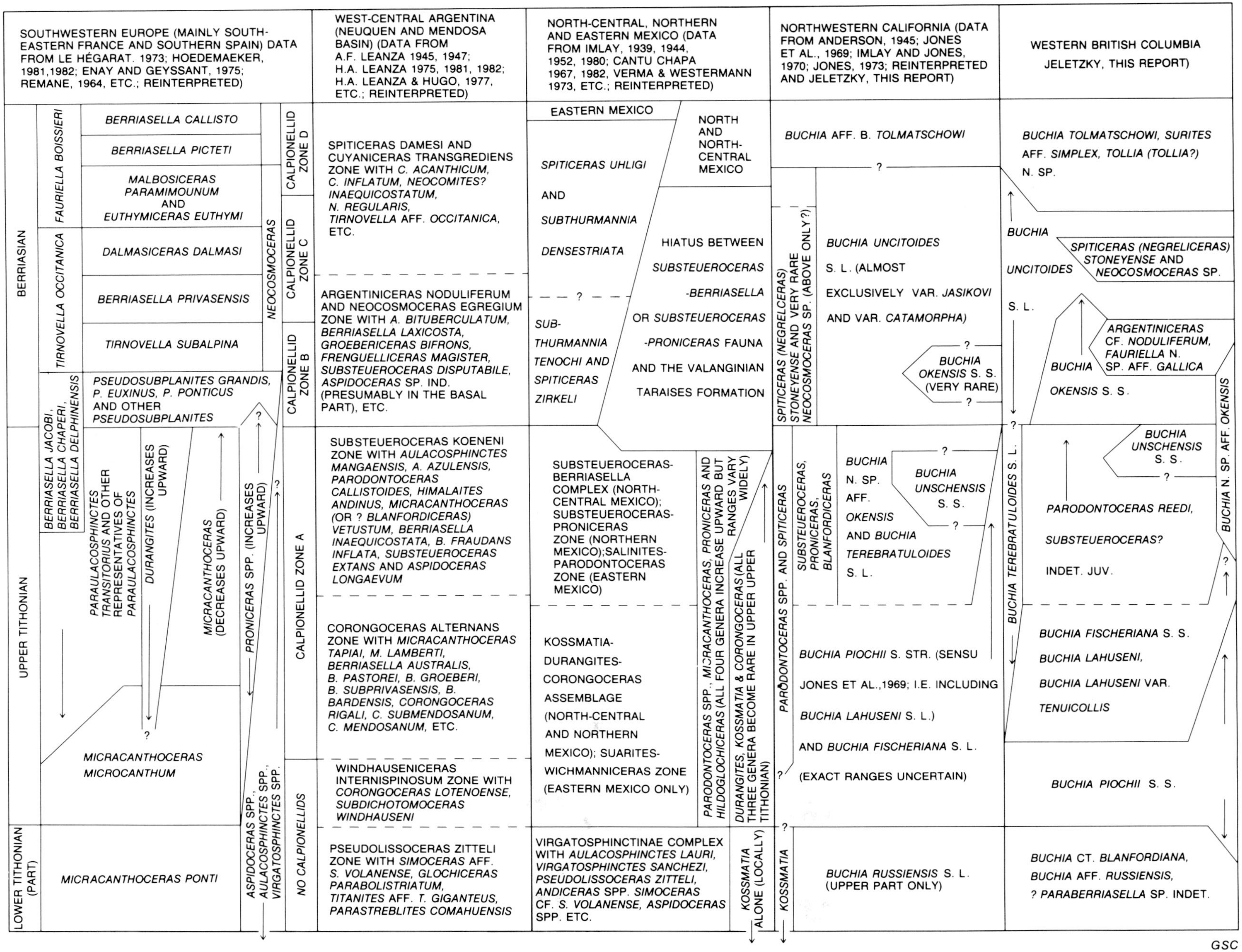

Figure 5. Proposed Correlation of Jurassic-Cretaceous Boundary Beds of Western British Columbia with those of northwestern California, Mexico, Western South America and Southwestern Europe.

In Canada the *Buchia* cf. *blanfordiana* Zone so far has yielded only a couple of identifiable ammonites on the west coast of Vancouver Island. These were originally identified by Jeletzky (1965, p. 18, Pl. 11, Figs. 5, 8, 9) as: "a poor berriasellid ammonite resembling *Substeueroceras stantoni* Anderson 1945 and a poor oppelioid ammonite resembling *Gymnodiscoceras* Spath 1925". This tentative comparison of the first of these ammonites was unfortunately misinterpreted as its positive identification by some Soviet (e.g., Shulgina in Saks *et al.*, 1972, Table 5) and West European (e.g., Enay, 1973) workers. A clarification is, therefore, in order.

The inferred latest Lower Volgian age of *Buchia* cf. *blanfordiana* Zone and its stratigraphic position directly beneath the *Buchia piochii* s.s. Zone are against the occurrence of either *Substeueroceras* s.s. or its close ally *Parodontoceras* in this zone. As explained on p. 197-198, *Substeueroceras* does not range down even into the next younger *Buchia piochii* s.s. Zone in California. According to Jones *et al.* (1969) and Imlay and Jones (1970, p. B12) *Parodontoceras* does range down into *Buchia piochii* s.s. Zone. However, it is not known to occur in the next older *Buchia elderensis* Zone, the upper part of which is correlative with the Canadian *Buchia* cf. *blanfordiana* Zone (see p. 186 for further details). Furthermore, the stiff and straight course of primary and secondary ribs on the flanks of the Vancouver Island ammonite (Jeletzky, 1965, Pl. II, Fig. 5), the subsequently established absence of any weakening, let alone interruption of secondaries in its venter and an absence of constrictions are against its assignment to either one of the above mentioned Late Tithonian genera or to the Berriasellinae in general. The stratigraphically more plausible reference to this ammonite to *Kossmatia* is ruled out by an exactly to almost exactly transversal course of its secondaries across but slightly deformed oralmost part of its venter and by the absence of their weakening there. An additional preparation of this specimen has shown that the cross-section of its but little deformed oralmost part is much higher than wide, has subparallel, straight flanks and a narrow, low arched venter. In combination with moderately evolute whorl proportions, very fine and closely spaced, almost invariably dichotomous ribbing habit lacking any flexuosity of secondaries and primaries and the previously discussed character of secondaries on the venter, these features strongly suggest the assignment of this specimen (i.e., GSC No. 16580; see Jeletzky, 1965, Pl. II, Fig. 5) to the subgenus *Paraberriasella* Donze 1948 (compare Arkell in Arkell *et al.*, 1957, p. L330, Fig. 428). It resembles so closely *Pectinatites (Paraberriasella) blondeti* Donze 1948, that the writer compares it tentively with that West European genus and species. Even better preserved ammonites apparently referable to *?Pectinatites (Paraberriasella)* ex gr. *blondeti* Donze have been found in the Manning Park area of the Tyaughton Trough (Jeletzky, in Coates, 1974, p. 126; this paper, Pl. IV, Fig. 14). Unfortunately, they were found on the float associated with diagnostic fossils of several other latest Jurassic and early Early Cretaceous ages.

The genus *Pectinatites* is considered to be diagnostic of the Late Kimmeridgian (= early Early Tithonian) in England and Greenland and of the equivalent early Early Volgian in North Siberia (Saks *et al.*, 1979, Correlation Table I and Figs. 2, 9 of this paper). Therefore, it should not range up into the *Buchia* cf. *blanfordiana* Zone that is correlative with the latest Lower Volgian and, apparently, with the latest Lower Tithonian *sensu* Enay (1972) on other grounds. However, the genus *Pectinatites* could conceivably range into considerably younger Volgian (or Tithonian) beds in western North America than it does in northern Eurasia. Whatever is the exact nature of the ammonite concerned, it is an Early rather than a Late Volgian type and so confirms in a general way the latest Early rather than earliest Late Volgian (or latest Early Tithonian) age of the *Buchia* cf. *blanfordiana* Zone that is being proposed on other grounds.

The tentative comparison of the above mentioned oppelioid ammonite with *Gymnodiscoceras* was also misinterpreted as its definitive generic identification by Enay (1973, p. 305). However, Enay (*loc. cit.*) concluded that: "The latter ammonite (i.e., the *Gymnodiscoceras*-like specimen of Jeletzky, 1965, Pl. II, Figs. 8, 9; writer's remark) is badly preserved and may be the same as *Substreblites* from southern Europe." This tentative re-assignment, which the writer is ready to second, would only suggest a general Tithonian, rather than a Late Tithonian, age (*sensu* Enay, 1973, p. 300) of the *Buchia* cf. *blanfordiana* Zone. As already mentioned, this assignment is in a general agreement with other evidence that indicates the latest Early Tithonian age of this zone.

The records of *Kossmatia* spp. and *?Durangites* in the *Buchia russiensis* s.l. Zone of northwestern California (Anderson, 1945, p. 983, Pl. 2, Fig. 4; Jones *et al.*, 1969, p. A10, A11; Imlay and Jones, 1970, p. B11) seem to contradict the latest Early Volgian (or latest Early Tithonian) dating of *Buchia* cf. *blanfordiana* Zone based on its *Buchia* fauna because of an allegedly Late Tithonian age of these ammonites (e.g., Enay, 1973, p. 299, Table 1).

According to Imlay and Jones (1970, p. B11, B12), *Kossmatia* was only found in the middle part of the *Buchia russiensis* s.l. Zone (i.e., the regional *Buchia elderensis* Zone) in northwestern California. The previously discussed latest Early Volgian (or latest Early Tithonian) assignment of this *Buchia* Zone in California agrees well with the recorded restriction of the Late Tithonian *Proniceras* and *Parodontoceras* to the next younger *Buchia piochii* s.s. Zone in all hitherto studied Californian sections (see Jones *et al.*, 1969, p. A10, A11, Imlay and Jones, 1970, p. B11, B12, Fig. 2 and below in this paper for further details). Therefore, contrary to Enay (1973, p. 299, Table 1), all these Californian records of *Kossmatia* appear to be of a late, but not the latest (as it does not appear to range up into the upper third of *Buchia russiensis* Zone), Early Tithonian age *sensu* Enay (1972, 1973). Hence, there does not appear to be any contradiction between the *Buchia* and *Kossmatia* ages of *Buchia russiensis* s.l. Zone (and hence of the Canadian *Buchia* cf. *blanfordiana* Zone) in northwestern California. However, this very restricted time range of *Kossmatia* in northwestern California does not apply to the adjacent Mexican basins. There the genus is reliably recorded to range from at least the lower Lower Volgian (i.e., Lower Tithonian) beds with *Buchia mosquensis* (von Buch) through the lower Upper Tithonian beds containing the *Kossmatia-Durangites-Corongoceras* fauna (that also includes locally *Micracanthoceras, Proniceras*, and *Parodontoceras*) into the upper Upper Tithonian beds containing *Substeueroceras-Berriasella* spp. fauna (that also includes less common *Durangites, Proniceras, Micracanthoceras* and *Parodontoceras*).

Although well documented, this extremely long time range of *Kossmatia* in Mexico apparently was never adequately summarized in the literature. The following remarks are intended to remedy this situation. The association of *Kossmatia* with *Buchia mosquensis* (von Buch) at San Pedro del Gallo, Durango was originally recorded by Burckhardt (1912, p. 206, 221). As pointed out by Imlay (1980, p. 34), who had re-studied the specimens concerned (pers. commun., 1968), this "shows that some of the beds concerned are at least as old as the *Zaraiskites albani* zone of northwest Europe and the equivalent part of the *Dorsoplanites panderi* zone in Russia. ... "

The time range of *Kossmatia* appears to be different in the more southerly San Luis Potosi area where it is not known to occur in the Lower Tithonian beds characterized by various Virgatosphinctinae (Verma and Westermann, 1973, p. 153, 154, Table 3). Instead *Kossmatia* spp. are common in the next younger, regional *Kossmatia-Durangites-Corongoceras* assemblage (Verma and Westermann, *loc-cit.*, p. 153, 155, Tables 3, 6). The beds containing this assemblage are obviously younger than the *Buchia russiensis* s.l. Zone of California (*i.e.* the *Buchia russsiensis* s.l. and *Buchia* cf. *blanfordiana* Zones of western British Columbia combined) because of the presence of such diagnostic late Tithonian ammonites as *Corongoceras mendosanum*, *C. cordobai* and *Micracanthoceras acanthellum*. Furthermore, the presence of the earliest known representatives of *Parodontoceras* (*i.e.* "*Substeueroceras*" *catorcense* Verma and Westermann, 1973) indicates its correlation with the Californian *Buchia piochii* s.s. Zone. The same is, finally, indicated by the first appearance in these beds and a frequent presence there of the Late Tithonian *Durangites* spp. as this genus is restricted to the Mediterranean *Micracanthoceras microcanthum* and *Durangites* Zones in southern Spain (Enay and Géyssant, 1975, p. 44, 45, Fig. 2; this paper, Figs. 5, 10). Furthermore, *Kossmatia* ascends into the youngest Tithonian ammonite assemblage observed in the San Luis Potosi area, which is correctly correlated with the Koeneni Zone of South America and the upper part of the European Transitorius Zone by Verma and Westermann (1973, p. 155, 158, Table 4) see p. 201-203 for further details).

Kossmatia is also known to range well up into the Upper Tithonian elsewhere in Mexico as it is associated with the calpionellid microfossils in the Sierra Cruillas, Tamaulipas (Cantú Chapa, 1982, p. 315; Bonet, 1956,p. 468, locs 44, 45 and 48). These examples are deemed to be sufficient to demonstrate that the time range of *Kossmatia* in Mexico is greatly different from that in the northwestern California and so has no bearing on the age of the *Buchia russiensis* s.l. Zone in the latter region.

The above data about the time ranges of *Kossmatia* in California and Mexico are incompatible with the recent conclusion of Enay (1972, 1973, p. 299, Table 1) that it is almost exclusively a Late Tithonian genus. Enay (1972, p. 372) states: "De la même façon, *Kossmatia* que j'ai considéré partout comme d'âge tithonique supérieur à été placé récemment dans le Kimméridgien (Arkell, 1957, p. L323; Fleming et Kear, 1960; Stevens, 1965, 1967, 1968). Les seule espèce réellement connue du Tithonique inférieur (avec *Semiformiceras* en Ardèche et en Espagne) est *K. richteri*, si bien que les formes mexicaines et himalayennes du Tithonique supérieur auraient pu évoluer et se différencier à partir de cette espèce méditérranéenne, dans deux voies et deux domaines différents. Cette conclusion est à l'opposé de celle de Stevens (1967) pour qui, *Kossmatia*, apparue dans le Kimméridgien en Nouvelle-Zélande, aurait étendu son domaine géographique au Tithonique inférieur, en particulier vers l'Europe méditerranéenne."

These conclusions of Enay are untenable. There is also no valid reason to claim (Enay, 1972, 1973) that the *Kossmatia*-bearing Ohauan Stage of New Zealand is an Upper Tithonian deposit in contradiction to much better founded conclusions of Arkell (1956, p. 454, 455 and in Arkell *et al.*, 1957, p. L 323) and Stevens (1965, 1976, 1968), which are also supported by all other faunal (i.e., belemnites, *Malayomaorica*, etc.) and stratigraphical evidence available. However, the opinion of Stevens (1967) that New Zealand is the center of origin of *Kossmatia* does not appear to be tenable either. The Kimmeridgian records of the European *K. richteri* appear to be at least as old as the records of *Kossmatia* spp. in the Ohauan Stage of New Zealand. Furthermore, the oldest known Mexican records of *Kossmatia* in association with *Buchia mosquensis* are either similarly old or but slightly younger than its Kimmeridgian records elsewhere (depending on whether the Mid- to Late Kimmeridgian or the Early Portlandian s.s. representatives of *B. mosquensis* are involved). Therefore, it appears to be most feasible to consider the European *K. richteri* to be the progenitor of both the Indo-Pacific (including the New Zealandian) and Mexican-Californian *Kossmatia* stocks via migration in opposite directions along the Tethys in the Late Kimmeridgian and a subsequent differentiation (still in the Kimmeridgian) of the migrants in the Mexican and Indo-Pacific basins. This hypothesis only differs from the above cited hypothesis of Enay (*loc. cit.*) dealing with the evolution of *Kossmatia* in the earlier (Kimmeridgian) date of the migrations and subsequent differentiation of the Indo-Pacific and Mexican stocks of *Kossmatia*.

The data presently available about the regional variations of time ranges of *Kossmatia* within the Tethyan Realm are suggestive of its gradual retreat southward into the high Tethyan basins of Mexico and southern Europe in the course of the Early and Late Tithonian. This retreat may be reasonably ascribed to the gradual deterioration of the climatic conditions during that time interval that is amply indicated by other evidence discussed earlier in this paper.

The *Titanites occidentalis* Frebold 1957 found in the basal Kootenay sandstone in the Fernie Basin (eastern British Columbia) is generally considered to be of the Late Portlandian s.s. age (Arkell, 1956, p. 540; Frebold, 1957, p. 66, 67, Plates XLII, XLIII and XLIV; Frebold and Tipper, 1970, p. 14, Table 1). Therefore, it appears to be appreciably older (mid-Early Tithonian *sensu* Enay, 1973, p. 299, Table 1) than either the Fisher Creek example of the *Buchia piochii* s.s. fauna containing *Proniceras* sp. indet. juven. or its Californian examples containing *Parodontoceras* (see p. 191). However, to the best of the writer's knowledge, no buchias of any kind have been found either at this *Titanites occidentalis* locality or elsewhere in the basal sandstone of the Kootenay Formation. It is important to stress this point as Frebold and Tipper's (1970, p. 17, Rocky Mountains graph of the table) placement of *Titanites occidentalis* in the same slot with *Buchia piochii* was already misinterpreted (e.g., Zeiss, 1979, p. 17) in the sense that these two fossils are actually associated and hence contemporary. This has resulted, in turn, in an erroneous dating of *Buchia piochii* Zone in terms of the Boreal zonal standard discussed later in this paper (see p. 192). Actually *Titanites occidentalis* appears to be contemporary either with the upper part of the *Buchia mosquensis* s.s. (i.e., lower Portlandian s.s. only) Zone or with some part of the overlying *Buchia russiensis* (i.e., mid- to early Late Portlandian s.s.) Zone, provided always that it is a true *Titanites* and not a North American homoeomorph of that European genus, as suggested by Callomon (this volume).

I discount the recorded presence (e.g., Anderson, 1945, p. 983, Pl. 2, Fig. 4; Imlay and Jones, 1970, p. B11) of *Durangites* aff. *D. vulgaris* high in the *Buchia russiensis* s.l. Zone of northwestern California. The figures of the poorly preserved solitary specimen, that was so identified by Anderson (1945) and is now lost, do not permit its generic identification in the writer's opinion. No description of this specimen was provided by Anderson (1945) and judging by its photographs alone it could even more likely be a *Kossmatia*. This is suggested, in particular, by the forward-deflected, somewhat shevron-like course of the ribs on its venter on both sides of their prominent interruption combined with an apparent absence of ventro-lateral nodes there (see Anderson, 1945, Pl. 2, Fig. 4). These features are definitely suggestive of a *Kossmatia* instead of a *Durangites*. Furthermore, the occurrence of this ammonite in the demonstrably uppermost Lower Tithonian *Buchia russiensis* s.l. Zone militates against its being a *Durangites*. As pointed out below in the discussion of age and correlation of *Buchia terebratuloides* s.l. Zone, *Durangites* is a strictly Late Tithonian genus.

Buchia piochii s.s. Zone

Stratigraphy and Geographical Distribution. The marine sediments of *Buchia piochii* s.s. Zone are only known in the Taseko Lake map-area (Jeletzky, in Jeletzky and Tiper, 1968,p. 12, Table 1, Fig. 13) where they are widespread. In this area the zone was found to persist from argillaceous mid-basinal facies into partly to entirely arenaceous to conglomeratic marginal facies of the southwestern flank. This indicates a great facies tolerance (from at least outer neritic depths to either innermost neritic or/and upper littoral depths) of *Buchia piochii* s.s. and so enhances its biochronological value.

The argillaceous, mid-basin facies of *Buchia piochii* s.s. Zone comprises the lower part of the Latest Portlandian s.s. to Late Berriasian shale unit of Relay Mountain Group (Jeletzky, in Jeletzky and Tipper, 1968, p. 21, 22), and may or may not extend downward into the uppermost part of the Variegated mid-Kimmeridgian to Portlandian s.s. unit in all suitably exposed sections studied. The zone is about 48.5 m thick in the Section 14 where it is completely exposed and essentially undisturbed by faulting (Jeletzky, in Jeletzky and Tipper, 1968, p. 163-166, Fig. 13). In this section the basal 13 m of *Buchia piochii* s.s. Zone, are represented by black, fine to very fine-grained greywacke. Wherever the contacts of the mid-basinal facies of *Buchia piochii* Zone are exposed, it grades upward into the *Buchia fischeriana* Zone and downward into the *Buchia* cf.*blanfordiana* Zone. The thickness of this mid-basin facies of *Buchia piochii* s.s. Zone appears to be in order of 50 to 60 m throughout the area, judging by its apparently unfaulted Section 16 (Jeletzky, in Jeletzky and Tipper, 1968, p. 181, 182, Fig. 13). There, the *Buchia piochii* s.s. Zone is represented exclusively by siltstone and shale. Unfortunately, the scarcity of fossils in the upper part of the zone precludes an exact placement of its upper boundary in this section.

The southwestern marginal facies of *Buchia piochii* s.s. Zone is well developed in the headwaters of Tyaughton Creek and in the vicinity of Elbow Mountain where it comprises the basal part of the Latest Portlandian s.s. – (?)Early Valanginian arenaceous unit of the Relay Mountain Group. In this facies, which is exemplified by the most nearly complete Section 3 (see Jeletzky, in Jeletzky and Tipper, 1968, p. 106-108, Fig. 13), the zone is at least 67 m thick. Furthermore, the units 18 to 20, about 8 m thick, of that section lacking diagnostic fossils may also form part of that zone. Another, indeterminate thickness of this facies of *Buchia piochii* s.s. Zone is missing at the fault contact that separates its topmost preserved unit 27 from the directly superimposed unit 28 that represents the *Buchia okensis* s.s. Zone. This contact of the zone is either faulted or covered in all its other sections studied. The lower contact of the here discussed facies of *Buchia piochii* s.s. Zone appears to be gradational in the Section 3 regardless of whether it is placed at the base of the bed 21 or at the base of the bed 18 and the same appears to be true of the other unpublished sections of that facies.

No rocks of *Buchia piochii* s.s. Zone have been observed anywhere on the northeastern flank of the Tyaughton Trough in the Taseko Lakes map-area. However, there does not seem to be any valid reason to doubt the extension of its mid-basinal, largely or entirely argillaceous facies into the upper Yalakom River area and farther east into the Camelsfoot Range – considering the presence of fault slices of the argillaceous facies of the next older and next younger *Buchia* Zones (Jeletzky, in Jeletzky and Tipper, 1968, p. 74, Figs. 7-9; Jeletzky, 1967, p. 67; 1971c, p. 226, and elsewhere in the report) in these areas. It is inferred that the Tyaughton Trough extended at least into the Ashcroft and Princeton map-areas (probably somewhat beyond the Frazer River in the Lillooet-Pavillion area) in *Buchia piochii* s.s. time.

In the Mount Waddington map-area, the equivalents of *Buchia piochii* s.s. Zone are only known in a nonmarine facies. There, they must form part of the Berriasian to Portlandian s.s. carbonaceous and plant-bearing arenites that separate the later Lower Berriasian marine rocks from the underiying Mid-Portlandian s.s. (i.e., late, but not the latest, Lower Volgian) marine rocks (Jeletzky, in Tipper, 1969, p. 40, 41). It is not known whether the apparent absence of *Buchia piochii* s.s. Zone elsewhere in the Tyaughton Trough (e.g., in the Harrison Lake area) is caused by its nondeposition, subsequent erosion or the scarcity of detailed stratigraphical work.

In Bowser Basin the *Buchia piochii* s.s. time, like the rest of the Late Kimmeridgian to Early Hauterivian time, is apparently only represented by a hiatus. According to Tipper and Richards (1976, p. 40, Fig. 12) "Kimmeridgian to Early Cretaceous (Hauterivian?) time was apparently a period of regional uplift, faulting and erosion that ended with the deposition of the Skeena and Sustut Groups on this deformed terrane (the late Bowser Lake Group is meant; writer's remark)".

In the Insular Trough, finally, the *Buchia piochii* s.s. time, like the rest of the Late Tithonian (see Jeletzky, 1965, 1976, 1978, Fig. 4; Sutherland-Brown, 1968), appears to be only represented by a hiatus. However, the previously mentioned (see p. 184), far from proven presence of rocks of the preceding *Buchia* cf. *blanfordiana* Zone in Queen Charlotte Islands suggests that the *Buchia piochii* s.s. Zone may also be present in that area of the trough.

Paleontology and Biochronology. Concept of Species and Synonomy. As already mentioned in the discussion of the status of *Buchia russiensis* s.l., the previously held opinion (Jeletzky, in Jeletzky and Tipper, 1968, p. 12) that *Buchia piochii* (Gabb 1865) s.s. is a senior synonym of *Buchia gabbi* (Pavlow 1907) and *Buchia russiensis* (Pavlow 1907) even at the infraspecific level, is abandoned herewith. My study of its holotype (figured by Anderson, 1945, Pl. 3, Fig. 1 and Imlay, 1959, Pl. 17, Fig. 10) and the publication of an extensive series of specimens of *Buchia russiensis* (Pavlow 1907) by Zakharov (1981, Pls. XIV-XVIII) have confirmed the opinion of Imlay and Jones (1970, p. B12) that typical representatives of *Buchia piochii* s.s. from California are specifically distinct from the Canadian *Buchia* forms figured as its varieties by Jeletzky (1965, 1966). This is true of *B. piochii* f. typ. reproduced by Jones *et al.* (1969, Pl. 3, Figs. 23-29) and Anderson (1945, Pl. 3, Figs. 1-5) from the earlier Late Tithonian (or earlier Late Volgian) of California and the still unfigured typical representatives of that species from the equivalent beds of the Taseko Lakes map-area in western British Columbia (Jeletzky, in Jeletzky and Tipper, 1968, p. 12). These forms differ from typical representatives of *Buchia russiensis* (Pavlow) from the upper Lower Volgian (= late Mid-Tithonian) rocks of Central Russia (Gerassimov, 1955, Pl. XIV, Figs. 1-3) and Northern Siberia (Zakharov, 1981, Pls. XIV-XVII) in a considerably more attenuated and relatively longer beak part of the left valve. When viewed from the outside, the flanks of this beak part are distinctly concave in contrast with the distinctly convex to approximately straight flanks of the beak part of *Buchia russiensis* (= *B. gabbi* Pavlow 1907). Furthermore, the shells of the North American *B. piochii* s.s. are, as a rule, appreciably higher, more narrow and considerably smaller than those of the Eurasian *B. russiensis* (= *B. gabbi*). There may be some morphological overlap between the two species as exemplified by the extremely slender and elongated specimens reproduced by Gerassimov (1955, Pl. XIV, Fig. 3) and Zakharov (1981, Pl. XVII, Fig. 2a-2v). However, the bulk of their representatives appear to be morphologically distinguishable when their left valves are available. Their isolated right valves do not seem to be reliably distinguishable where such extreme homoeomorphs are concerned.

As the rocks characterized by *Buchia piochii* s.s. are younger than those characterized by *Buchia russiensis* s.l. (formerly *B. elderensis*) in northwestern California and as the *Buchia* cf. *blanfordiana* Zone is inserted between the approximate equivalents of these beds in western British Columbia, the above morphological distinctions between the two forms are judged to be ample for their full specific separation. *Buchia piochii* s.s. appears to be a direct descendant of *Buchia russiensis* s.l. that arose out of its late forms in the latest Early Tithonian (or latest Early Volgian) time somewhere in western North America. This is clearly indicated by their intergradation in the topmost beds of *Buchia* cf. *blanfordiana* Zone in the Taseko Lakes map-area and in the presumably contemporary topmost beds of *Buchia russiensis* s.l. Zone in California.

As already noted by Zakharov (1981, p. 94), *Buchia piochii* s.s. forms part of the Late Tithonian species group of *Buchia lahuseni* (Pavlow 1907). It does intergrade with *B. lahuseni* f. typ. and its var. *tenuicollis* (Pavlow 1907), *Buchia stantoni* (Pavlow 1907) and *Buchia krotovi* (Pavlow 1907) in the upper part of its stratigraphic range at least. This is clearly indicated by a series of Californian representatives of these two species reproduced by Imlay (1959, Pl. 17, Figs. 7-10, 12-29) and Jones *et al.*, (1969, Pl. 3, Figs. 20-29) under the name of *B. piochii* (see p. 188 for further details). The same appears to be true of western British Columbia (Jeletzky, in Jeletzky and Tipper, 1968, p. 11, 12) where such intergradation appears to be restricted to the lower part of the regional *Buchia fischeriana* Zone, that is inferred to correspond to the upper part of the *Buchia piochii* s.s. Zone of northwest California. This intergradation and the prevalence of *Buchia piochii* s.s. over all related *Buchia* forms throughout its zone indicates its being an independent evolutionary stage of the *Buchia russiensis* s.l.-*Buchia* ex gr. *lahuseni* plexus rather than a morphological variant of the latter species. It appears that this intermediate evolutionary stage remained confined to western North America and the Canadian Arctic (see below) while the older and younger members of the plexus, including *Buchia lahuseni* itself, spread over the most part of the Boreal Realm. Additional work is needed to identify whether or not *B. piochii* s.s. is conspecific with the apparently contemporary Californian and British Columbian *Buchia* forms described as *Aucella knoxvillensis, Aucella gracilis Pavlow 1907, Aucella* aff. *A. bononiensis* Pavlow 1907 and *Aucella stantoni* Pavlow 1907 by Anderson (1945, Pl. 3, Figs. 12-14, 17-20, 22; Pl. 4, Figs. 6, 11).

Age and Correlation. The gradational superposition of uppermost Lower Volgian (= uppermost Lower Tithonian) rocks by the *Buchia piochii* s.s. Zone in western British Columbia and California indicates that the lower boundary of that zone either coincides with the Lower-Upper Volgian (or Lower-Upper Tithonian) boundary or is situated in its close proximity.

No identifiable ammonites have been found in the so far studied sections of *Buchia piochii* s.s. Zone in the Taseko Lakes map-area. However, this zone has yielded some diagnostic ammonites in northeastern British Columbia, a region that is generally speaking beyond the scope of this paper. There, on Fisher Creek in Peace River Foothills *Notostephanus*? sp. (previously recorded as *Notostephanus* cf. *kurdistanense* Spath by Warren and Stelck, 1958, p. 58 and Jeletzky, 1968, p. 4) and *Proniceras* sp. indet. juven. were found associated with typical *Buchia piochii* s.s. fauna in a loose boulder. The sandy siltstone lithology suggests that this boulder was derived from the Fernie-Minnes Transition Beds rather than from the overying lower Monteith Formation. These diagnostic ammonites and buchias are reproduced in Plate IV, Figs. 1-5, 7, 8.

Though small and specifically indeterminate, this Fisher Creek *Proniceras* sp. indicates a Late Tithonian age for this particular occurrence of *Buchia piochii* s.s. fauna. As pointed out below, this genus is not known to descend into the Lower Tithonian. The presence of *B.* aff. *lahuseni* var. *krotovi* and *B.* aff. *terebratuloides* s.l. in this fauna suggests that this loose boulder is derived from the upper part of *Buchia piochii* s.s. Zone. The total absence of *B. fischeriana* appears to contradict its placement into that mid-Upper Tithonian zone.

The association of this *Proniceras* sp. indet. juven. with an only questionably identified example of *Notostephanus* suggests that this fauna is not younger than the early late Tithonian. This association could even be interpreted as suggestive of this fauna representing the uppermost Lower Tithonian (*sensu* Enay, 1973, p. 297, 299, Table 1) *Pseudolissoceras zitteli* Zone of the Mediterranean standard instead of its basal Upper Tithonian *Micracanthoceras micracanthum* Zone. However, the only fragmentary specimen of *Notostephanus*? is too poorly preserved to be certain of its generic nature. Furthermore, the latest Early Tithonian age of the Anatolian and Kurdistanian beds containing *Notostephanus* in association with *Pseudolissoceras* and *Proniceras* (Spath, 1950; Enay *et al.*, 1971) is rather doubtful. Enay (1972) has reassigned these beds to the early Late Tithonian almost immediately after the appearance of his earlier publication where he held them to be the latest Early Tithonian. Furthermore, as pointed out astutely by Imlay (1980, p. 32) "the association of *Pseudolissoceras* with typical late Tithonian ammonite *Proniceras* in Kurdistan (Spath, 1950, p. 125) suggests that *Pseudolissoceras* may range higher than recorded in Europe."

As already mentioned, the *Titanites occidentalis* Frebold 1957 found in the Fernie Basin of eastern British Columbia is older than the *Buchia piochii* s.s. Zone as defined herein.

The early Late Tithonian dating of *Buchia piochii* s.s. Zone in western Canada, based on the Fisher Creek example discussed above, agrees well with the age of its approximate equivalents in northwestern California where *Buchia piochii* s.s. is associated with *Buchia fischeriana* and a considerably more variegated and better preserved fauna of Tethyan ammonites.

According to Imlay and Jones (1970, p. B11, B12) "the upper 4,000-5,000 feet of the *Buchia piochii* Zone (*i.e.* of their *Buchia piochii* s.s. Subzone that is largely but not entirely equivalent to the Canadian *Buchia piochii* s.s. Zone; writer's remark) is definitely middle to late Tithonian in age as shown by the stratigraphic position between occurrences of *Kossmatia* below and *Substeueroceras* above and by the presence of ammonites *Parodontoceras* and *Spiticeras*. Of these, *Parodontoceras* (Mesozoic loc. 1084), found several thousand feet below the top of the zone, is excellent evidence for an age not older than late middle Tithonian and *Spiticeras* (Mesozoic loc. 1085), found a few hundred feet below the top of the zone, is evidence of an age not older than late Tithonian. Also, the presence of an ammonite resembling *Groebericeras* (Mesozoic loc. M2024), found about 2,100 feet below the top of the zone, favors an age not older than late Tithonian because that genus has been found only in beds of Berriasian age in Argentina (Leanza, 1945, p. 82)." I agree essentially with these conclusions, except that I contend that *Parodontoceras* does not range beneath the Late Tithonian *sensu* Enay (1972). Furthermore, I believe that *Groebericeras? baileyi* of Imlay and Jones (1970, p. B35, Pl. 5; Pl. 6, Figs. 1, 2, 5, 6, 8, 11; Pl. 15) from the *Buchia piochii* s.s. Zone is unrelated to *Groebericeras*. Its innermost exposed whorls exhibit pronounced shevron-like bends of the secondaries in the mid-venter (Imlay and Jones, 1970, p. B35, Pl. 6, Figs. 9, 10) that resemble closely those of halfgrown to adult shells of Californian and Mexican *Kossmatia* species (compare Imlay and Jones, 1970, Pl. 3, Figs. 1-7; Verma and Westermann, 1973, Pl. 39, Figs. 7a-7c; Pl. 40, Figs. 2-4; Pl. 47, Figs. 16, 26, 36; Pl. 48, Fig. 1). Therefore, *"Groebericeras? baileyi"* Imlay and Jones 1970 belongs to a new perisphinctid genus derived directly from American representatives of *Kossmatia*. This unnamed genus, that may well be an ancestor of the latest Tithonian to Berriasian *Protothurmannia* Crickmay 1932 (compare Crickmay, 1932, Pl. II, Fig. 2 and *Protothurmannia* n. sp. A, Pl. 2, Fig. 1), had developed umbilical bullae, bundled bidichotomous to trichotomous ribbing habit and the high-oval *Berriasella*-like whorl shape on its intermediate whorls. This makes it similar to *Protothurmannia* – the innermost whorls of which are poorly known but may be *Kossmatia*-like. The degeneration and coarsening of this *Fauriella*-like ribbing habit on the outer whorls of *"Groebericeras? bayleì"* (Imlay and Jones, 1970, p. B35, Pl. 15) also allies it to *Protothurmannia* (compare Crickmay, 1932, Pl. 1, Fig. 5; this paper, Pl. III, Fig. 1). So interpreted, *"Groebericeras? bayleì"* is suggestive of the early Late Tithonian instead of the latest Tithonian or still younger age of the beds (i.e., the middle part of the *Buchia piochii* s.s. Zone) that have yielded it. Furthermore, this Californian zone underlies directly the *Buchia terebratuloides* s.l. Zone (i.e., the *Buchia* aff. *B. okensis* Zone of Jones *et al.*, 1969, p. A12) which is definitely upper Upper Tithonian (see p. 201-203).

The *Buchia piochii-Buchia fischeriana* Zone of northwestern California is correlative with the early Late Tithonian *Kossmatia-Durangites-Corongoceras* assemblage of San Luis Potosi area, Mexico (Verma and Westermann, 1973, p. 153, 155, Tables 5, 6) because of the presence there of the oldest known representatives of *Parodontoceras* (synonymized with *Substeueroceras* by Verma and Westermann, *loc. cit.*). Furthermore, the *Buchia piochii-Buchia fischeriana* Zone and the *Kossmatia-Durangites-Corongoceras* assemblages are both overlain directly by the upper Tithonian beds containing *Substeueroceras* (s.s.) ex gr. *koeneni* and other diagnostic ammonites (see p. 201-203 for further details). Elsewhere in Mexico the *Buchia piochii* s.s. Zone must correspond approximately to the *Suarites* (or *Suarites bituberculatum*) Zone of Cantú Chapa (1967, p. 20; 1982, p. 315). The largely endemic fauna of that zone includes *Wichmanniceras hernandense* Cantú Chapa and cf. *Corongoceras filicostatum* Imlay that indicate its correlation with the *Kossmatia-Durangites-Corongoceras* complex of San Luis Potosi area. Furthermore, this zone is overlain immediately and apparently conformably by the upper Upper Tithonian *Salinites-Parodontoceras* aff. *callistoides* Zone that contains *Proniceras, Kossmatia, Durangites, Protancyloceras* and sometimes *Substeueroceras* (Cantú Chapa, 1967, p. 18; 1982, p. 315, Figs. 3, 4, 6). This Zone is obviously correlative with the *Substeueroceras-Berriasella* complex of the San Luis Potosi area (see p. 202 for further details).

Unlike the Californian *Buchia piochii-B. fischeriana* Zone, the Canadian *Buchia piochii* s.s. Zone is correlative with the lower part of the Mexican *Kossmatia-Durangites-Corongoceras* complex only. This is indicated by the intercalation of the *Buchia fischeriana* Zone between the *Buchia piochii* Zone and the upper Upper Tithonian *Buchia terebratuloides* s.l. Zone (see Fig. 5 and p. 195 for further details).

The above correlations of the *Buchia piochii-Buchia fischeriana* Zone of northwestern California and the largely but not entirely correlative *B. piochii* s.s. Zone of western Canada indicate that the former is equivalent to all of the *Windhauseniceras internispinosum* and *Corongoceras alternans* Zones of western South America while the latter corresponds approximately to the *Windhauseniceras internispinosum* Zone only. These correlations are indicated by the fact that these two South American zones combined contain essentially the same ammonite fauna and occupy the same stratigraphic position as the Mexican *Kossmatia-Durangites-Corongoceras* complex and its equivalents (Fig. 5).

These new data about the age and correlation of the Californian *Buchia piochii-B. fischeriana* Zone and the previously presented information (see p. 188-189) about the great variability of time ranges of *Kossmatia* in adjacent regions of North America invalidate the following, rather far-reaching conclusions of Zeiss (1979, p. 17; writer's translation from Russian) concerning the intercontinental correlation of the Tithonian and Volgian ammonite and *Buchia* zones. According to Zeiss, "in the western part of the U.S.A. *Buchia piochii* is largely confined to beds that carry *Kossmatia*. This ammonite occurs, in turn, in Middle and South America in the zone of *Corongoceras alternans* where also occur *Micracanthoceras* and *Durangites*. This permits a reliable correlation with the both above mentioned upper Tithonian zones of Southwest Europe [the zones of *Micracanthoceras micracanthum* and *Durangites*; writer's comment]. On the other hand, in Canada *Buchia piochii* was found associated with *Titanites*, a genus that characterizes the upper part of the middle Volgian. This species was also recorded on the Russian Platform, where it occurs somewhat scantily in the middle and upper parts of the middle Volgian. However, it occurs much more commonly in these beds than it does in the upper Volgian (Gerassimov, 1964). Therefore, the species *Buchia piochii* may be utilized for the purpose of a direct correlation of the Boreal middle Volgian deposits with the basal part of the upper Tithonian (*Kossmatia*-bearing beds). This correlation was proposed, in part, by Imlay and Jones (1970). It was also utilized by V.N. Saks *et al.* (The Stratigraphy of the Jurassic.................., 1976)".

This correlation reflects the then existing confusion about the taxonomic interpretation and age of *Buchia piochii* in California and Canada, the erroneous synonymization of this early Late Tithonian North American species with the late Early Tithonian *Buchia russiensis* (Pavlow) (Jeletzky, in Jeletzky and Tipper, 1968, p. 12) and the incorrect assumption of Frebold and Tipper (1970, p. 14, 17) that the Middle Volgian *Titanites* is associated with *Buchia piochii* in the Fernie Basin. Following the clarification of these misconceptions earlier in this paper, it is inadmissible to correlate the Boreal Middle Volgian (or late Lower Volgian, Figs. 5, 10) even with the lower part of the Upper Tithonian in southwestern Europe. The latter (i.e., the *Micracanthoceras micracanthum* Zone and the lower part of the *Paraulacosphinctes*-bearing beds; Fig. 5) are considerably younger and must correspond instead to the basal Upper Volgian (*Kachpurites fulgens* Zone). The *Durangites* Zone and its equivalents are younger yet. Furthermore, it is incorrect to utilize the genus *Kossmatia* as an intercontinental or even broadly regional index of these lower Upper Tithonian beds.

Buchia fisheriana Zone

Stratigraphy and Geographical Distribution. In western British Columbia the rocks of *Buchia fischeriana* Zone are only known to occur in the Tyaughton Trough. Their main outcrop-area there is in the Taseko Lakes map-area where they comprise the middle part of the mid-basinal Latest Portlandian s.s. to Late Berriasian shale unit of the Relay Mountain Group (Jeletzky, in Jeletzky and Tipper, 1968, p. 21, 22). This unit forms a northwest-trending belt centred at the Relay and Cardtable Mountains and extending through the investigated part of the area. Within it, *Buchia fischeriana* Zone is represented mostly by moderately hard to friable, pure to silty shale that is medium grey when fresh and weathers light to brownish grey. This shale includes minor interbeds and pods (up to 0.6 m thick) of moderately to strongly glauconitic shale and siltstone and similarly thick calcareous siltstone that grades locally into strongly argillaceous, cryptocrystalline limestone. No arenaceous interbeds have been noted in any of the sections studied. Some small and large, irregularly rounded concretions of very hard, commonly calcareous shale and siltstone occur locally. Marine fossils consisting almost exclusively of buchias but including rare to very rare other pelecypods, rare brachiopods, fairly common belemnites and rare to very rare phylloceratid and lytoceratid ammonites occur scattered in most beds. However, they may form coquinoid accumulations in some, mostly silty and limy interbeds or in concretions. An outer shelf, generally speaking low to very low energy, open marine environment is indicated by the composition and preservation (mostly double-valved *Buchia* shells with closed to only slightly displaced or gaping valves) of the fauna. The relatively rare instances of coquinoid accumulation of dismembered but mostly undamaged *Buchia* valves were apparently caused by episodes of bottom current activities.

Buchia fischeriana Zone is approximately 90 m thick in this mid-basin facies, as exemplified by the units 10 to 6 of the Section 14 (Jeletzky, in Jeletzky and Tipper, 1968, p. 160-164). This representative section of the zone is complete, unfaulted and exposes its both contacts, which are lithologically and paleontologically gradational.

No exposures of *Buchia fischeriana* Zone have been observed in either of the marginal, mostly arenaceous facies belts of the Jurassic-Cretaceous boundary beds that flank their mid-basin belt in the Taseko Lakes map-area. This appears to be caused exclusively by beds of this zone being either faulted out or covered in all studied sections of that facies (compare Section No. 8; Jeletzky, in Jeletzky and Tipper, 1968, p. 129). No information about the width of the marine Tyaughton Trough during the time of *Buchia fischeriana* is, therefore, available for the Taseko Lake and Pemberton map-areas.

In the Manning Park area *Buchia fischeriana* (d'Orbigny) var. *trigonoides* (Lahusen) was so far only found once. This solitary (Pl. IV, Fig. 15) specimen was collected on the float at the base of high bluffs towering above Thunder Lake (Jeletzky, in Coates, 1974, p. 126). It was associated with *?Pectinatites (Paraberriasella)* ex gr. *blondeti* (Pl. IV, Fig. 14) and diagnostic fossils of several other latest Jurassic and early Early Cretaceous ages. No information about the lithology and age of the rocks outcropping in those bluffs is available to the best of the writer's knowledge.

In the Mount Waddington map-area the non-fossiliferous, presumably non-marine rocks of *Buchia fischeriana* Zone are assumed to be present in the middle part of the already mentioned nonmarine unit that separates the later Lower Berriasian marine rocks from the underlying Mid-Portlandian marine rocks (see Jeletzky in Tipper, 1969, p. 40, 41). It is not known whether the apparent absence of *Buchia fischeriana* Zone elsewhere in the Tyaughton Trough is caused by its non-deposition, subsequent erosion or the scarcity of detailed stratigraphic work. In the Bowser Basin *Buchia fischeriana* Zone is unknown and presumed to be absent for the same reason as the older latest Jurassic zones (see p. 184, 190 for further details).

Paleontology and Biochronology. General Remarks. I am no longer certain that the term *Buchia fischeriana* Zone, introduced for a faunally distinctive rock sequence that separates *Buchia piochii* s.s. and *Buchia terebratuloides* s.l. Zones in the Taseko Lakes map-area (Jeletzky, in Jeletzky and Tipper, 1968, p. 11), is appropriate. This extension of *Buchia fischeriana* Zone from the high Boreal basins of Canada into the North Pacific basins may have been unfortunate as its name fossil was later found to be a lesser to outright minor faunal element in a number of sections of that rock sequence in the Taseko Lakes map-area (see below for further details). This situation reflects *B. fischeriana* being an immigrant from the high Boreal basins that had only settled on the Pacific Slope temporarily (see p. 194-195 for further details). Furthermore, the very existence of separate *Buchia fischeriana* Zone seems to be doubtful in California where this species is reported to be associated with *Buchia piochii* s.s. throughout its time range (Jones *et al.*, 1969, p. A9; Imlay and Jones, 1970, p. B11, B12, Fig. 2). Though the succession of *Buchia* species concerned appears to be different in the Taseko Lakes map-area where *B. fischeriana* is associated not with *B. piochii* s.s. but with the younger *B.* ex gr. *lahuseni*, its fitness as a zonal index obviously is questionable even there. It may have been better to select *B. lahuseni* (Pavlow 1907) as the index fossil of the zone concerned as it is its frequent, commonly prevalent faunal element. Furthermore, *B. lahuseni* has the advantage of being endemic to the Pacific slope of North America where it forms part of plexus consisting of *B. russiensis* s.l., *B. piochii* s.s. and *B. lahuseni* itself (see p. 189, 192 for further details). However, it does not seem to be practical to change the name of the zone at present when the writer's study of taxonomy and time ranges of *B.* ex gr. *russiensis-piochii-lahuseni* remains unfinished and the time relationships of this species group with *B. fischeriana* in northwestern California are questionable. For these reasons *Buchia fischeriana* Zone continues to be used in this paper.

Concept of Buchia fischeriana and its Interprovincial Morphological Distinctions. The species concept of *B. fischeriana* adhered to in this paper is the same as that employed in a previous study of its high Boreal representatives (Jeletzky, 1966). The same broad interpretation of *B. fischeriana* was adhered to by Zakharov (1981).

The recent proposal of Konovalov (1977, p. 142) to treat *B. trigonoides* (Lahusen, 1888) as an "independent, easily diagnosed species" is erroneous. The status of any species is determined by the nature of its type material. In our case both cotypes of *B. trigonoides* (see Lahusen, 1888, Pl. II, Figs. 21-24) are medium-sized, closely to moderately closely and fairly finely ribbed, thin forms that fall well within the range of infraspecific variation of North Siberian (Zakharov, 1981, Pl. XXI and PL. XXIII, Figs. 1-3), North Canadian (Jeletzky, 1966, Pl. VII, Fig. 2; Pl. VIII, Figs. 1-6, 8, 9) and Californian (Imlay, 1959, Pl. 17, Figs. 2-7; Jones *et al.*, 1969, Pl. 1, Figs. 3, 7-9) population samples of *B. fischeriana*. The much larger, pronouncedly triangular, much more distantly and coarsely ribbed buchias (e.g., Anderson, 1945, Pl. 13, Fig. 4; Jones *et al.*, 1969, Pl. 1, Figs. 18-22, 25-35), that are representative of *B. trigonoides* in Konovalov's (*loc. cit.*) opinion, are not conspecific with its type material. They belong instead to a younger, morphologically well defined *Buchia* n. sp. aff. *okensis* that is largely restricted to *Buchia terebratuloides* s.l. Zone (see p. 196-197 for further details).

The Californian and British Columbian *Buchia* forms assigned herein to *Buchia fischeriana* (d'Orbigny 1845) are conspecific with this East European species. However, all hitherto studied samples (e.g., Anderson, 1945, Pl. 10, Figure 2; Imlay, 1959, Pl. 17, Figs. 2-6; Jones *et al.*, 1969, Pl. 1, Figs. 3, 7-9) differ appreciably from North Canadian (e.g., Jeletzky, 1966, Pl. VII, Fig. 2; Pl. VIII, Figs. 2-6, 8, 9) and North Siberian (e.g., Zakharov, 1981, Pls. XXI-XXII, Pl. XXIII, Figs. 1-3) samples in the apparent absence of relatively large to very large, either strongly elongated and *Buchia volgensis*-like (e.g. Zakharov, 1981, Pl. XXII, Figs. 1-3), or broadly rounded to broadly oval (e.g., Jeletzky, 1966, Pl. VII, Fig. 2; Pl. VIII, Figs. 2-3) forms. Even the smaller, broadly rounded to broadly oval forms (e.g., Jeletzky, 1966, Pl. VIII, Figs. 5, 6) common in the high Boreal samples are either absent or rare in the British Columbian and Californian samples. The latter consist largely or entirely of small to medium-sized, elongated forms similar to those figured by Jeletzky (1966, Pl. VIII, Figs. 8, 9) and Zakharov (1981, Pl. XXI, Figs. 2, 3, 4, 7; Pl. XXIII, Figs. 1-3). Many of them are indistinguishable from the type material of *B. fischeriana* var. *trigonoides* (Lahusen, 1888, Pl. II, Figs. 21-24) while other subtriangular specimens are stronger elongated and so morphologically transitional to *Buchia russiensis* s.l.

The above discussed morphological distinctions of the British Columbian and Californian representatives of *B. fischeriana* are evidently provincial in nature as our populations lived in the marginal North Pacific Province of the Tethyan Realm while the Canadian Arctic and North Siberian populations inhabited high Boreal basins. They are ascribed to the adaptation of the British Columbian and Californian populations of *B. fischeriana* to warmer water temperatures because they are paralleled by similar decreases in size and other morphological changes (e.g., decrease or loss of ornamentation) of younger *Buchia* species (e.g., the almost total replacement of *Buchia okensis* s.s. and *B. uncitoides* var. *spasskensoides* by *Buchia uncitoides* var. *jasikovi* and var. *catamorpha* between western British Columbia and northwestern California (see p. 207, 209 for further details).

These morphological distinctions of "northern" and "southern" populations of *B. fischeriana* appear to be phylogenetically significant as well. Its British Columbian and Californian populations exhibit a definite morphological trend toward the next younger *Buchia* form that was described as *Buchia* n. sp. aff. *okensis* by Jeletzky (1971a, p. 3, Fig. 2) and as *Buchia* aff. *B. okensis* by Jones *et al.* (1969, p. A12, Pl. 1, Figs. 18-22, 25-35). As pointed out below (see p. 196-197), these two forms are connected by transitions in the Taseko Lake map-area and, apparently, in northwestern California. This endemic Pacific Coast lineage then leads to *Buchia okensis* s.s. The high Boreal populations of *Buchia fischeriana*, in particular the North Siberian ones (see Zakharov, 1981, Pl. XXII, Figs. 1-3), exhibit an equally definite morphological trend toward the next younger *Buchia volgensis* (Lahusen 1888). This Eurasian species, that is totally unknown in western and Arctic North America, except for extremely rare examples in the basal Berriasian or Sverdrup Basin (see p. 223, for further details), appears to be a direct descendant of these *Buchia volgensis*-like variants of *B. fischeriana* that had evolved out of them somewhere in the Eurasian Arctic. These vicarious relationships of the *Buchia volgensis* and *Buchia okensis* s.s. lineages account for the almost total absence of the former species in North America and a much more prolific and morphologically variable representation of *Buchia okensis* s.s. there (see p. 204-205).

Other Faunal Elements. So far as is known, the *Buchia fischeriana* fauna of all sections studied in the Taseko Lakes map-area is dominated by its name fossil, *Buchia lahuseni* (Pavlow 1907) f. typ. and its var. *tenuicollis*, and some other insufficiently understood, thick but slender-beaked, and strongly elongated *Buchia* forms allied either to *B. lahuseni* s.s. or to *B. piochii* s.s. (see Pl. IV, Figs. 9, 11, 13). As already mentioned, the ratios of these *Buchia* are highly variable and *B. lahuseni* et var. may be prevalent over *B. fischeriana* in many beds.

Buchia piochii s.s. appears to be rare or absent, except in the basal beds of the zone. Judging by these data and personal observations in some Californian sections in 1966, Anderson's (1945), Imlay's (1959) and Jones *et al.* (1969, p. A8, A9) observations of a common occurrence of *B. piochii* s.s. in association with *B. fischeriana* pertain either to the superficially similar *B. lahuseni* or other closely related late Upper Tithonian *Buchia* forms. Though similarly elongated and slender-beaked, *B. lahuseni* differs from *B. piochii* s.s. in a considerably greater thickness of its distinctly more narrow valves, which have considerably more steeply sloping flanks (compare Gerassimov, 1955, p. 90, 91, Pl. XIV, Figs. 4-8). Furthermore, the beak parts of the left valve of *B. lahuseni* and its var. *tenuicollis* tend to be distinctly right-handedly incurved while that of *B. piochii* s.s. is, as a rule, straight to slightly right-handedly incurved. Imlay's (1959, Pl. 17, Figs. 7-10, 12-20) specimens from USGS Mes. Loc. 1225 appear to be *B. piochii* s.s. while those from the USGS Mes. Loc. 2275 (Imlay, 1959, Pl. 17, Figs. 21-29) appear to be *B. lahuseni*. Unfortunately, no cross-sections or lateral views of these specimens are provided, which makes their definitive identification difficult. As already mentioned, the typical representatives of these two closely allied species are connected by a host of morphologically transitional forms, which are especially numerous in beds transitional between their zones. Further study of large collections of Californian *Buchia* ex gr. *piochii-lahuseni* collected bed by bed in undisturbed sections is needed to confirm these tentative conclusions.

In the sections studied in the Taseko Lakes map-area, *Buchia terebratuloides* s.l. and *B.* n. sp. aff. *okensis* are, as a rule, either rare or absent in the lower and middle parts of the *Buchia fischeriana* Zone. However, their numbers and the numbers of forms transitional between *B. fischeriana* var. *trigonoides* and *B.* n. sp. aff. *okensis*, characteristically increase rapidly in its upper part until the latter two forms become common to prevalent in its topmost beds (e.g., in Bed 10 of Section 14; see Jeletzky, in Jeletzky and Tipper, 1968, p. 160). In contrast, *B.* n. sp. aff. *okensis* becomes common only higher within the *Buchia terebratuloides* s.l. Zone.

The ratio of *Buchia terebratuloides* s.l., including its variants *B.t. subinflata, B.t. subuncitoides* and *B.t. occidentalis*, also increases rapidly in the topmost beds of the zone. The ratio of *B. fischeriana*, exclusive of its var. *trigonoides*, decreases rapidly in these beds.

Age and Correlation. In western British Columbia, northwestern California, and probably in northeastern British Columbia, *Buchia fischeriana* appears to be an immigrant from the high Boreal basins of Arctic Canada and northern Eurasia. This is indicated by: 1) the apparent absence of *Buchia* faunas that consist exclusively or almost exclusively of *B. fischeriana* and by an apparent absence of morphologically transitional forms between it and *B. lahuseni* s.l. in any of the personally studied sections; 2) the lesser morphological variability of *B. fischeriana* on the Pacific Slope as compared with that in the high Boreal basins of Arctic Canada and northern Eurasia. Such decrease of infraspecific variability is typical of peripheral populations; 3) the relative rarity of *B. fischeriana* in comparison with the contemporary *Buchia* ex gr. *lahuseni*, which are known to be endemic to western British Columbia and northwestern California; and 4) the considerably later appearance in western British Columbia in comparison with the Sverdrup Basin (see p. 225, 229 and Fig. 9) where *B. fischeriana* ranges down either to the base of the Upper Volgian stage ?or somewhat deeper. In northern Eurasia and European Russia *B. fischeriana* appears already in the late Lower Volgian stage (beginning with the *Dorsoplanites maximus* Zone) and ranges up to the very top of the Upper Volgian stage (Zakharov, 1981, p. 104, 105, 214, 215, Table 3).

In western British Columbia, in contrast, *B. fischeriana* first appears high in the regional *Buchia piochii* s.s. Zone, which is demonstrably early Late Tithonian or early late Volgian (see p. 191-192). The above described regional time ranges clearly indicate a gradual late Early Tithonian migration of *B. fischeriana* from northern Eurasia into the Canadian Arctic and then an early Late Tithonian migration into the North Pacific Province.

The exact timing of immigration of *B. fischeriana* into the

North Pacific Province of western North America is somewhat obscured by its recorded time range in northwestern California. According to Jones *et al.* (1969, p. A9-A11) and Imlay and Jones (1970, p. B11, B12, Fig. 2), *B. piochii* s.s. and *B. fischeriana* have the same stratigraphic ranges in that region. As already mentioned, I am inclined to question this conclusion, but do not have enough data to demonstrate that the Californian *Buchia piochii* s.s. Zone corresponds approximately to the *Buchia piochii* s.s. and *Buchia fischeriana* Zones of western British Columbia combined (Fig. 5) and that *B. fischeriana* is associated exclusively or largely with *B.* ex gr. *lahuseni* in the upper part of that zone. This problem requires further study. However, it is not critical as far as the time ranges and migrations of *B. fischeriana* is concerned, as the early Late Tithonian *Parodontoceras, etc.* fauna ranges right through the beds characterized by the association of *B. piochii* s.s. and *B. fischeriana* in California. This first appearance of *B. fischeriana* would still be considerably later than its first appearance in northern Eurasia if one admits the validity of Jones *et al.* (1969) and Imlay and Jones (1970) conclusions.

Buchia fischeriana apparently disappears appreciably earlier in western British Columbia and northwestern California than in northern Eurasia. It becomes very rare in the upper part of its zone in the Taseko Lakes map-area and the same is probably true of the topmost part of its recorded range in California. Furthermore, *B. fischeriana* only questionably ascends into the next younger, uppermost Tithonian (see below for documentation) *Buchia terebratuloides* s.l. Zone in these two regions. Its occasional presence in this zone may be simulated by morphologically extreme juvenile specimens of *B.* n. sp. aff. *okensis* bearing particularly closely spaced and fine concentric ribs. Such *B. fischeriana*-like juvenile forms occur occasionally even in the still younger *Buchia okensis* s.s. Zone (*e.g.*, Jeletzky, 1965, Pl. IV, Fig. 5a). They can only be identified correctly when their later, coarsely and sparcely ribbed growth stages are available. I suggest that the true *B. fischeriana* does not range up into the *Buchia terebratuloides* s.l. Zone in western North America, except for its basal beds where it is largely or entirely represented by its var. *trigonoides* (Lahusen).

The above data on time range of *B. fischeriana* in various parts of its geographical range, the stratigraphic position of its zone in the Taseko Lakes map-area between the *Buchia piochii* s.s. and *Buchia terebratuloides* s.l., and its association with the *Parodontoceras, etc.* ammonite fauna in northwestern California, indicate the correspondence of its regional zone to some middle part of the Upper Volgian stage of northern Eurasia (Figs. 9, 10). Closer correlation of this zone is impossible in the absence of Boreal ammonites in either western British Columbia or northwestern California.

Because of the association of *B. fischeriana* in northwestern California with the *Parodontoceras, etc.* fauna and because of other evidence about the age of this ammonite presented in the discussion of the age and correlation of *Buchia piochii* s.s. Zone, the *Buchia fischeriana* Zone of our region must correspond to some upper part of the same Mexican, western South American and Mediterranean ammonite zones as the regional *Buchia piochii* s.s. Zone. It is definitely older than the Koeneni Zone of Mexico and western South America or the *Durangites* Zone of southern Spain (Fig. 5), because of its stratigraphic position immediately below the *Buchia terebratuloides* s.l. Zone (see p. 201-203).

Buchia terebratuloides s.l. Zone

Stratigraphy and Geographical Distribution. The *Buchia terebratuloides* s.l. Zone is unknown in the Insular Trough. On the west coast of Vancouver Island *Buchia terebratuloides* time is represented by the prolonged hiatus that separates the latest Lower Tithonian *Buchia* cf. *blanfordiana* Zone from the Early Berriasian *Buchia okensis* Zone (Jeletzky, 1965, p. 17, 18, 20-23, 64-66, Fig. 4).

The only possible exception is the previously mentioned (p. 184) shallow water gritty and pebbly greywacke discovered recently on Graham Island, Queen Charlotte Islands. If its poorly preserved *Buchia* fauna really represents the lower part of *Buchia terebratuloides* s.l. Zone, it would suggest that this zone was also deposited elsewhere in the Insular Trough but was destroyed by erosion during or after a brief tectonic phase (Diablan orogeny of Anderson, 1945) that preceded the deposition of the *Buchia okensis* rocks.

In the Tyaughton Trough, rocks of the *Buchia terebratuloides* s.l. Zone are known in the Taseko Lakes map-area and in the adjacent part of the Pemberton map-area (Jeletzky, in Jeletzky and Tipper, 1968, p. 10, 11, Table 1; Jeletzky, 1971a, p. 4, Fig. 4). There its mid-basin facies consists of 75 m or argillaceous rocks that form part of the Latest Portlandian to Late Berriasian shale of the Relay Mountain Group (Jeletzky, in Jeletzky and Tipper, 1968, p. 21, 22). All studied sections of this facies are incomplete. However, its normal stratigraphic contacts with the underlying *Buchia fischeriana* Zone and the overlying *Buchia okensis* s.s. Zone have been observed in the Sections 15 and 14 respectively (Jeletzky, in Jeletzky and Tipper, 1968, p. 159, 169). The above mentioned, estimated thickness of the argillaceous facies of the zone is based on the combination of these two sections with some other sections.

The belt of the mid-basin argillaceous facies of *Buchia terebratuloides* s.l. Zone is flanked from the southwest by that of its arenaceous facies centered at the northern flank of Elbow Mountain. This shallow water facies comprises the middle part of the Latest Portlandian s.s. to (?)Early Valanginian arenaceous unit of the Relay Mountain Group (Jeletzky, in Jeletzky and Tipper, 1968, p. 24, 25), which was deposited on the southwestern flank of the Tyaughton Trough and evidently was derived from a southwestern source area. All hitherto studied sections of the arenaceous facies of *Buchia terebratuloides* s.l. Zone are incomplete, less than 20 m thick and faulted. The most important of them was studied by H.W. Tipper (pers. commun.) in the proximity of the Section 8 (see Jeletzky, in Jeletzky and Tipper, 1968, p. 128, 129) on the southern spur of a sharp, east-west-trending ridge situated about 4 km southeast of Elbow Mountain. There, the fine-grained greywacke of the arenaceous facies underlies, apparently gradationally, the lithologically similar greywacke of *Buchia okensis* s.s. Zone. The latter appears to represent the downward extension of the unit 7 of Section 8 (see Jeletzky, in Jeletzky and Tipper, 1968, p. 129). The base of this section of *Buchia terebratuloides* s.l. Zone, which is some 15 m thick, was not observed.

Isolated outcrops of the argillaceous facies of *Buchia terebratuloides* s.l. Zone occurs on the northeastern flank of Tyaughton Trough in the Camelsfoot Range. However, they are either only known from collections of fauna (e.g., GSC loc. 15995) made by older workers (Leech, 1946) in sections that were not found by the writer or represent thin fault slivers of shale and siltstone in severely disturbed or largely tectonically brecciated sections of the eastern side of the Yalakom Fault Zone. So far as known, these isolated outcrops are represented by the same siltstones and shales as the previously described mid-basin facies of *Buchia terebratuloides* s.l. Zone. It is suggested accordingly that the northeastern shoreline of the Tyaughton Trough was situated farther eastward in *Buchia terebratuloides* s.l. time than it was in the preceding Early and early Late Tithonian (compare Jeletzky, in Jeletzky and Tipper, 1968, p. 19, 20, Fig. 2). It may well have been situated somewhere on the eastern side of Fraser River in the Lillooet-Pavillion area.

No marine rocks of the *Buchia terebratuloides* s.l. Zone were found in the Mount Waddington map-area, on the southwestern flank of Tyaughton Trough. In that area this zone appears to be represented by the plant-bearing, carbonaceous to coaly clastics that form the upper part of the nonmarine unit intercalated between the conglomeratic rocks containing *Buchia uncitoides* (Pavlow) and small, presumably late forms of *Buchia okensis* and marine arkoses of the *Buchia russiensis* s.l. Zone (formerly *Buchia* n. sp. aff. *piochii* Zone) (see Jeletzky in Tipper, 1969, p. 40, 41).

No marine or nonmarine rocks referable to the *Buchia terebratuloides* s.l. Zone are known anywhere else in western British Columbia, with the possible exception of Graham Island, Queen Charlotte Islands (see p. 184). It is not certain whether their absence was caused by nondeposition or by a subsequent erosion. However, the erosionally disconformable contact of the *Buchia okensis* Zone with the underlying, predominantly argillaceous *Buchia* cf. *blanfordiana* rocks observed on the west coast of Vancouver Island (Jeletzky, 1965, p. 17, 18, Fig. 4) seems to favor the latter interpretation.

Paleontology and Biochronology. Concept of Species. The broad interpretation of *Buchia terebratuloides* (Lahusen 1888) proposed and explained earlier (Jeletzky, in Jeletzky and Tipper, 1968, p. 10) was somewhat expanded in a later paper (Jeletzky, 1973, p. 47, Pl. 4, figs. 2, 3; Pl. 5, fig. 2) by the inclusion of *Buchia obliqua* (Tullberg 1881). Essentially the same broad species concept is adhered to by Jones *et al.* (1969). This concept, which does not differ materially from Lahusen's (1888) original concept, is adhered to in this paper. The nomenclatorially incorrect use of the junior synonym – *Buchia terebratuloides* (Lahusen 1888) – is aimed at the avoidance of confusing changes of zonal nomenclature that would have been introduced by the use of the very little known senior subjective synonym *Buchia obliqua* (Tullberg 1881). The specific separation of *B. terebratuloides* s.l. and *B. obliqua* proposed by Zakharov (1981, p. 100) is unacceptable. In northern Canada these forms have exactly the same stratigraphic ranges and appear to integrade at all levels (see p. 229 for further details). Furthermore, the same appears to be true of western British Columbia and northwestern California, although *B. terebratuloides* var. *obliqua* appears to be a rare form in these regions.

Geographical Range, Facies Tolerance and Zonal Value. *Buchia terebratuloides* s.l., as defined above, is an extremely widespread species that ranges within the Boreal climatic belt right around the globe. Furthermore, it ranges south into the North Pacific Faunal Province of the Tethyan Realm where it occurs at least as far south as northwestern California. *B. terebratuloides* s.l. occurs without any significant morphological changes in every possible kind of marine environment from outer neritic, or possibly upper bathyal, pure clay shale to littoral coarse pebble conglomerate. Even the range of infraspecific variation of *B. terebratuloides* s.l. apparently remains the same in population samples obtained from all these contrasting facies in the Taseko Lakes map-area. Its north-south range, that extends from the Ellesmere Island in the proximity of the pole to northwestern California, attests to its outstanding climatic tolerance, as I do not accept large scale north-south movements of "allochtonous suspect terranes" in western North America and Alaska after the Middle Jurassic (p. 242-244). The only paleobiogeographically significant morphological distinctions of *B. terebratuloides* s.l. samples obtained in the high Boreal and North Pacific Provinces of North America consist in a more common occurrence of *B. terebratuloides* var. *obliqua* in the former and *B.t.* var. *subinflata* in the latter. These "varieties" are probably geographical subspecies of *B. terebratuloides* s.l.

Considering its rather short (Late Volgian only) and, so far as we know, constant time range throughout its realm, *B. terebratuloides* s.l. is just as valuable and reliable an index fossil as *Buchia okensis* s.s. (compare Jeletzky, 1965, p. 23) and most of the latest Jurassic and earliest Cretaceous zonal ammonites.

Stratigraphic Range in Western Canada. So far as known, *Buchia terebratuloides* s.l. is common to abundant throughout the thickness of its zone in the Tyaughton Trough. As already mentioned (see p. 192), it does range down at least into the upper part of *Buchia fischeriana* Zone. Furthermore, closely allied and possibly conspecific forms appear rarely in the upper part of *Buchia piochii* s.s. Zone (e.g., Pl. IV, Fig. 5). However, it is not known to range up even into the basal beds of the overlying *Buchia okensis* s.s. Zone, in spite of their gradational contact in the Taseko Lake map-area sections (*e.g.* in the unit 1 of Section 15; Jeletzky, in Jeletzky and Tipper, 1968, p. 170). *Buchia terebratuloides* s.l. is not known to occur in the *Buchia okensis* Zone on the west coast of Vancouver Island as attested by the representative sample of *Buchia* fauna from its lowermost preserved beds figured by Jeletzky (1965, Pl. IV). However, the equivalent of the lowermost part of the complete sections of *Buchia okensis* Zone, such as occur in the Taseko Lakes map-area, may be absent on the west coast of Vancouver Island.

Buchia terebratuloides s.l. versus Buchia aff. okensis Zone. The Californian Zone of *Buchia* aff. *okensis* is synonymized with *Buchia terebratuloides* s.l. Zone for the following reasons. In Western British Columbia *Buchia* n. sp. aff. *okensis* first becomes common in the middle part of *Buchia terebratuloides* s.l. Zone and ranges up into the next younger *Buchia okensis* s.s. Zone (Fig. 5). Furthermore, it is morphologically similar and so easier to confuse with *B. okensis* s.s. than *B. terebratuloides* s.l. and is abundant only in the upper part of *Buchia terebratuloides* s.l. Zone. *Buchia terebratuloides* s.l. is, finally, an extremely wide ranging fossil, the paleogeographic range of which far exceeds the ranges of other *Buchia* species occurring in its zone (e.g., *B. unschensis* s.s., *B.* n. sp. aff. *okensis* and *B. lahuseni*). The paleozoogeographical and biochronological importance of the *Buchia terebratuloides* var. *subinflata* fauna in California was already stressed by Bodylevsky (1936, p. 130, 131) and his conclusion was fully confirmed by subsequent research, including that of the writer. The term *Buchia terebratuloides* s.l. Zone (Jeletzky, in Jeletzky and Tipper, 1968, p. 10) has priority over the *Buchia* aff. *okensis* Zone (Jones *et al.*, 1969, p. B12) and the term *Buchia* aff. *okensis* Zone is unacceptable as a formal zonal term because it is based on a fossil described in open nomenclature.

Other Diagnostic Faunal Elements. Other than its name fossil, the fauna of *Buchia terebratuloides* s.l. Zone of the Taseko Lake map-area contains numerous representative of a large *Buchia*, the Californian representatives of which were originally (and erroneously; p. 193) identified as *Aucella trigonoides* Lahusen 1888 (Anderson, 1945, p. 971, Pl. 13, Fig. 4). They were later assigned to a new informally named species *Buchia* aff. *okensis* (Pavlow 1907) by Jones *et al.* (1969, p. A10, Pl. 1, Figs. 18-22, 25-35). In the writer's opinion, other *Buchia* forms described and figured by Jones *et al.* (loc. cit., Pl. 1, Figs. 1, 2, 4-6, 10-17, 23, 24) as *Buchia trigonoides* (Lahusen) also do not belong to this variant of *B. fischeriana* (d'Orbigny) as defined earlier in this paper (p. 193). Most of them (e.g., Jones *et al.*, 1969, Pl. 1, Figs. 1, 2, 11-17, 23, 24) are transitional forms between *B.* aff. *okensis* and *B. fischeriana* var. *trigonoides* (e.g., their Pl.1, Figs. 3, 7-9) or belong to *B. terebratuloides* s.l. (e.g., their Pl. 1, Figs. 4-6).

In the Taseko Lakes map-area, *B.* aff. *okensis* of Jones *et al.* (1969) was originally designated by Jeletzky (in Jeletzky and Tipper, 1968, p. 10, 11, Table 1) as "early forms of *Buchia okensis* s. lato". Later it was recognized as an independent species diagnostic of *Buchia terebratuloides* s.l. Zone and designated informally as *B.* n. sp. aff. *okensis* (Pavlow 1907) (Jeletzky, 1971a, p. 4, Fig. 2). I was unaware at that time of a similarly sounding name introduced somewhat earlier for the same form by Jones *et al.* (1969). The informal name *B.* n. sp. aff. *okensis* was subsequently used in all my publications dealing with the stratigraphy and faunas of the Jurassic-Cretaceous boundary beds (e.g., Jeletzky, 1973, p. 50, Fig. 3). A formal description is planned in another publication.

The morphology of *B.* n. sp. aff. *okensis* was adequately described by Jones *et al.* (1969, p. A12 and Table 2 on p. A10, A11). Additional comments are provided below in the description of *B. okensis* s.s. and of its morphological distinctions from allied *Buchia* species (p. 202-205).

As already pointed out (p. 193), *B.* n. sp. aff. *okensis* and forms transitional between it and *B. fischeriana* var. *trigonoides* appear already in the *Buchia fischeriana* Zone where they are connected with that species by a host of morphologically intermediate forms. The ratio of *B.* n. sp. aff. *okensis* and these transitional forms rapidly increases upward but its typical representatives only become common (or even prevalent locally) un the upper two-thirds of *Buchia terebratuloides* s.l. Zone. As explained in the description of the fauna of *Buchia okensis* Zone below (p. 204-205, figs. 5, 10), *B.* n. sp. aff. *okensis* ranges up well into this basal Berriasian zone. It is for that reason that Jones *et al.* (1969) choice of *B.* n. sp. aff. *okensis* as a name fossil of the *Buchia terebratuloides* s.l. Zone is unfortunate. However, this was not obvious to these workers because of a nearly total absence of *Buchia okensis* s.s. and its variants in northwestern California.

Other than *B. terebratuloides* s.l. and *B.* n. sp. aff. *okensis*, this zone contains relatively rare representatives of *Buchia* ex gr. *lahuseni* (Pavlow) and poorly understood *Buchia* forms that appear to be transitional between *B. terebratuloides* var. *subuncitoides* (Bodylevsky) and *Buchia uncitoides* s.l. The former form, also poorly understood, appears to be restricted to the lower (?basal only) part of the zone while the latter are probably restricted to its upper part. Additional study is needed to decide whether or not some of these *Buchia* aff. *uncitoides* s.l. may already belong to its varieties *catamorpha* Crickmay and/or *jasikovi* Pavlow. Regardless of the exact character of these specimens, *B. terebratuloides* s.l. and *B. uncitoides* s.l. appear to be connected by transitional forms in the uppermost part of *Buchia terebratuloides* s.l. Zone.

The rare occurrence of *B. unschensis* s.s. (Plate I, Fig. 9) in the upper part of the *Buchia terebratuloides* s.l. Zone is most important in strengthening its correlation with the high Boreal facies of the *Buchia terebratuloides* s.l. Zone, which commonly contains a significant to high ratio of *B. unschensis* s.s. or is dominated by it almost to the exclusion of all other *Buchia* forms (Jeletzky, 1966).

In addition to numerous representatives of *B.* n. sp. aff. *okensis*, the upper part of our zone has yielded some forms transitional to *B. okensis* s.s. (Plate I, Fig. 6) and a few forms indistinguishable from that basal Cretaceous species (Plate III, Fig. 2). The presence of these forms in the Taseko Lakes map-area explains and verifies the records of *Buchia okensis* s.s. (e.g., Pavlow, 1907, p. 40, Pl.1, Fig. 10; Anderson, 1945, p. 940, 942) in the topmost beds of the so called Knoxville Series of northwestern California, which have greatly puzzled Imlay (1959). However, this inferred presence of very rare, typical representatives of *B. okensis* s.s. in the *Buchia terebratuloides* s.l. Zone of northwestern California has no bearing on its equally rare presence in the overlying basal Cretaceous beds of that region recorded by Jones *et al.* (1969, p. A13).

The rare appearance of *Buchia okensis* s.s. in the uppermost part of the *Buchia terebratuloides* s.l. Zone of the Pacific Slope of North America does not discredit it as the first-rate diagnostic fossil of the next younger *Buchia okensis* s.s. Zone. It is a natural result of the ancestor-descendant relationship of *B.* n. sp. aff. *okensis* and *B. okensis* s.s. that is discussed in detail in the next section of this paper. It stresses, once more, the unreliability of solitary *Buchia* specimens for exact age determinations (Jeletzky, 1965, p. 14, 15). A reliable differentiation of the *Buchia okensis* s.s. Zone from the *Buchia terebratuloides* s.l. Zone depends, therefore, on the strong prevalence of typical and giant *B. okensis* forms in the former zone and the absence of *B. terebratuloides* s.l. and *B. lahuseni* s.l. there. *B. unschensis* s.s. is also unknown in the *Buchia okensis* s.s. Zone in western British Columbia. However, it is difficult to use this diagnostic species of *Buchia terebratuloides* s.l. Zone in this region (and also in northwestern California) where it occurs only as a rare immigrant from the high Boreal basins of northern Canada and northern Alaska.

B. n. sp. aff. *okensis* appears to be an endemic form on the Pacific Slope of North America that only migrated northward from there into the Arctic regions of North America. The reasons for this conclusion are explained below in the description of the geographical range of *B. okensis* s.s. However, some of the most recent citations of *B. okensis* (Zakharov 1981, p. 225, 227) in the basal Riasanian beds on Oka River seem better referable to *B.* n. sp. aff. *okensis*. This is suggested in particular by the fact that these specimens are reported to be associated with numerous *B. fischeriana* and to occur in beds that underlie those containing *Hectoroceras kochi* and numerous *B. unschensis* s.s. (p. 236-237).

Rare juvenile to halfgrown berriasellids occurring in the uppermost beds of the argillaceous facies of the *Buchia terebratuloides* s.l. Zone apparently belong to the genus *Substeueroceras* Spath 1923 (Plate I, Figs. 4, 5). More numerous and much better preserved berriasellids are found in the uppermost beds of its arenaceous facies. They belong to *Parodontoceras reedi* (Anderson 1945) (Plate I, Figs. 1-3) and to other generically indeterminate (?) berriasellid ammonites. The specimens identified here as *P. reedi* were originally assigned to *P. callistoides* (Behrendsen) (Jeletzky, 1971a, p. 3, Footnote). This may well be correct as *P. reedi* likely is but a variant or geographic subspecies of *P. callistoides* according to Imlay and Jones (1970, p. A12). However, the Taseko Lakes specimens are indistinguishable from the Californian form as revised by Imlay and Jones (1970) and so are re-assigned to it.

PLATE I

Equivalents of Buchia terebratuloides s.l. Zone in northwestern California. As pointed out by Jeletzky (in Jeletzky and Tipper, 1968, p. 10, 11; 1971a, p. 4, Fig. 2; 1973, p. 47, 50, 52, 54, Fig. 3), the *Buchia terebratuloides* s.l. fauna of the Taseko Lakes map-area can be confidently correlated with those Tithonian rocks of northwestern California that have yielded that species in association with *Buchia trigonoides* (Anderson 1945 *non* Lahusen 1888) and a much more diversified fauna of Tethyan ammonites.

These Californian beds were already assigned correctly to the Late Tithonian by Anderson (1945) and Imlay (1959) in spite of the recorded presence of *B. okensis* (e.g., Pavlow, 1907, Pl. 1, Fig. 10) and *B. subokensis*. However, the zonal value of their *Buchia* fauna was not recognized by these workers who amalgamated it with the *Buchia fischeriana* Zone. Only Jones *et al.* (1969, p. A12) recognized the zonal independence of this fauna and errected a *Buchia* aff. *okensis* Zone for it, following a somewhat complicated discussion and joint field work with me in northwestern California in 1966 (Jones *et al.*, 1969, p. A2, A3). In spite of the almost complete absence of *B. okensis* s.s. in northwestern California, Jones *et al.* (1969, p. A12) noted that their *Buchia* aff. *okensis* s.s. Zone differs from the *Buchia okensis* s.s. Zone of western British Columbia and is older according to its latest Tithonian ammonites. These ammonites were later re-described and biochronologically re-evaluated by Imlay and Jones (1970). Jones *et al.* (1969) and Imlay and Jones (1970), however, neither mentioned nor discussed the paper of Jeletzky and Tipper (1968) in the text of their reports, although it is included in the list of references. They must have been unaware of my paleontological-stratigraphical results in the Jurassic-Cretaceous transition beds of the Taseko Lakes map-area, which were published prior to the publication of their final reports. This appears to be the reason why Jones *et al.* (1969, p. A12) have designated these latest Tithonian beds of northwestern California as the *Buchia* aff. *okensis* Zone instead of using *Buchia terebratuloides* s.l. that was designated as their zonal index in the Taseko Lakes map-area. This is unfortunate as *Buchia* aff. *okensis* is obviously a less diagnostic fossil of these beds that is *Buchia terebratuloides* s.l. (p. 196-197).

EXPLANATION FOR PLATE I

All photographs are natural size.

Figure 1. *Parodontoceras reedi* (Anderson 1945). Hypotype, GSC Cat. 73669. GSC loc. 94250 (=56484). Photograph of rubber mold of intermediate whorls preserved as an external cast. Note numerous and prominent constructions diagnostic of the genus. Latest Portlandian s.s. to (?)Early Valanginian arenaceous unit of the Relay Mountain Group, upper part of *Buchia terebratuloides* s.l. Zone, uppermost Tithonian. Taseko Lakes map-area (92-0), B.C. Ridge between heads of the Graveyard and Tyaughton Creeks, outcrop at ridge's crest at its S end, Lat. 51°07′34″N.; Long. 123°04′57″W. Collected by T.P. Poulton, GSC.

Figure 2a, b. *Parodontoceras reedi* (Anderson 1945). Hypotype, GSC Cat. 73670. The same locality, *etc.* as for Figure 1. An almost undeformed fragment of intermediate whorl. 2a) lateral view; 2b) ventral view. Recorded as from: "Taseko Lakes map-area (92-0), B.C. 228.5 metres S of the locality 2Fb-20K-TD, on ridge SE of Elbow Mountain. Collected by H.W. Tipper, GSC.

Figures 3a-c. *Parodontoceras reedi* (Anderson 1945). Hypotype. GSC Cat. 73671. The same locality, *etc.* as for Figure 1. A slightly deformed fragment of an adult? living chamber. Note the oblique orientation of the umbilical wall and the broadly rounded umbilical shoulder in Figure 3b which are diagnostic of the genus. 3a) lateral view; 3b) cross-section; 3c) ventral view.

Figure 4. *Substeueroceras*? sp. indet. juven. GSC Cat. 73672. GSC loc. 62562. Uppermost Tithonian, upper part of *Buchia terebratuloides* s.l. Zone. Relay Mountain Group. Latest Portlandian s.s. to Late Berriasian shale unit, Taseko Lakes map-area (92-0), southwestern British Columbia. On the crest of northeastern spur of the Summit 8,222′ of the 4 mile topo map at the point about 320 m northeast of its top; Lat. 51°09′N.; Long. 122°57′W. Collected by J.A. Jeletzky, 1964. Lateral view of moderately deformed specimen. Note the pronouncedly flexuous secondary ribs that branch at different levels and form trifurcating bundles.

Figure 5. *Substeueroceras*? sp. indet. juv. GSC Cat. 73673. Same locality, etc. as for Figure 4. The ribbing habit is identical to that of this specimen.

Figures 6a-d. *Buchia* n. sp. aff. *okensis* (Pavlow 1907). GSC Cat. 73674. Same locality, *etc.* as for Figure 1. 6a) lateral view of the exterior of the left valve; 6b) lateral view of the exterior of the right valve; 6c) beak part and hinge margin of the left valve; 6d) anterior view of both valves but the right valve presents a pronouncedly oblique view. Large, rather wide, but still pronouncedly triangular-shaped, shell transitional to *B. okensis* s.s.

Figures 7a-c. *Buchia* n. sp. aff. *okensis* (Pavlow 1907). GSC Cat. Nos. 73675 and 73676. Same locality, etc. as for Figure 1. Two typical left valves in the same piece of rock. 7a) lateral view of the exterior of the larger completely exposed valve (GSC Cat. No. 73675); 7b) anterior view of both valves. The smaller, more coarsely ribbed (GSC Cat. No. 73676) valve is partly pushed inside the larger, completely exposed valve (GSC Cat. No. 73675); 7c) lateral view of the exterior of the incompletely exposed, fragmentary left valve (GSC Cat. No. 73676). These two left valves are typical of *B.* n. sp. aff. *okensis* (Pavlow 1907). The valve shown in Figure 7a is morphologically closer to *B. fischeriana* (d'Orbigny) var. *trigonoides* (Lahusen 1888) while that shown in Figure 7c is a more advanced form (note its more prominent, heavier and more widely spaced ribbing and the markedly left-handedly incurved beak) that retains the strongly elongated, pronouncedly triangular shape diagnostic of the species.

Figures 8a-d. *Buchia terebratuloides* (Lahusen 1888) var. *subinflata* (Pavlow 1907). GSC Cat. 73677. Same locality, etc. as for Figure 1. 8a) lateral view of the exterior of left valve; 8b) lateral view of the exterior of right valve; 8c) anterior view of both valves. Much of the left beak is broken off; 8d) beak parts and hinge margins of both valves viewed from above. A very typical, large (presumably adult) bivalved representative of *B. t.* var. *subinflata*.

Figures 9a-d. *Buchia unschensis* (Pavlow 1907) s.s. GSC Cat. 73678. Same locality, *etc.* as for Figure 1. 9a) lateral view of the exterior of left valve. The posterior wing is broken off. The convexly rounded, short appearance of the anterior and anterior-lower margins is an artefact caused by loss of their fringe parts. The complete margins would have been just as long and straight as those of the Sverdrup Basin representatives of *B. unschensis* s.s. (e.g., Pl. V, Fig. 2a, Pl. VI, Figs. 2a, 7a; Pl. VIII, Figs. 1a, 4a, 5a); 9b) lateral view of the exterior of right valve that is partly squeezed into the left valve. The angular, posterior wing is still visible; 9c) beak parts and hinge margins of both valves. Much of the left beak is broken off; 9d) posterior view of both valves. Much of the left beak is broken off.

The studies of Jones *et al.* (1969), Imlay and Jones (1970) and Jeletzky (in Jeletzky and Tipper, 1968) revealed the complete identity of all *Buchia* species of these two faunas. To these now we can add the rare *B. unschensis* s.s. (e.g., at California Academy of Science loc. 28037) and *Parodontoceras reedi* (Anderson). The extremely close affinities of these two faunas and their almost identical positions in the closely similar zonal sequences attest to their complete equivalence. Such faunal differences as do exist between these faunas are attributable to the difference in their latitudinal position.

The sections of the *Buchia terebratuloides* s.l. Zone in the Taseko Lakes map-area are uniquely important for detailed zonal correlation of the Jurassic-Cretaceous boundary beds of the Pacific Slope of North America with their high Boreal equivalents in Arctic North America and northern Eurasia on the one hand and with their high Tethyan equivalents in Mexico and western South America on the other. These sections remain the only ones known in which the interregional *Buchia terebratuloides* s.l. fauna is directly and gradationally overlain by the equally important interregional *Buchia okensis* s.s. fauna. The latter fauna is not clearly developed in northwestern California where the much less diagnostic *Buchia uncitoides* s.l. fauna overlies directly the *Buchia terebratuloides* s.l. fauna (Fig. 5). Furthermore, the *Buchia terebratuloides* s.l. and *Buchia okensis* s.s. Zones of the Taseko Lakes map-area are represented by richly fossiliferous (mostly *Buchia*), open marine rocks throughout, and contain enough Tethyan ammonites for the recognition of their zonal equivalents among the *Buchia*-devoid high Tethyan ammonite faunas of Mexico and western South America.

Boreal and Tethyan Equivalents of Buchia terebratuloides *s.l. Zone Outside Western British Columbia and Northwestern California.* The contemporaneity of the *Buchia terebratuloides* s.l. Zone of western British Columbia with the so called *Buchia* aff. *okensis* Zone (=*Buchia terebratuloides* s.l. Zone) of northwestern California is critical for its exact dating in terms of the high Tethyan ammonite zones of the Jurassic-Cretaceous boundary beds of Mexico and western South America. Even more critical is the fact that these ammonite zones are, in turn, comparable with the Mediterranean ammonite zones of southwestern Europe. However, the correlation of the zone with the Boreal ammonite and *Buchia* zones of the classical regions of northern Eurasia depends on the Taseko Lake map-area where the *Buchia terebratuloides* s.l. Zone overlies gradationally the interregional Boreal Zone of *Buchia fischeriana* and is overlain gradationally by the interregional Boreal Zone of *Buchia okensis* s.s.

As stressed by all American and European workers since Anderson (1945), the fairly rich ammonite fauna of the Californian *Buchia terebratuloides* s.l. Zone is purely Tethyan. It is but a considerably depauperated faunal facies of the so called *Substeueroceras-Berriasella* assemblage of northern Mexico (e.g., Verma and Westermann, 1973) and the *Substeueroceras koeneni* Zone of western South America (e.g., A. Leanza, 1945, 1947; H. Leanza, 1981a, 1981b; Widemann, 1980; Biro-Bagoczky, 1980).

The contemporaneity of the *Substeueroceras-Parodontoceras* ammonite fauna of the Californian *Buchia terebratuloides* s.l. Zone with the *Substeueroceras-Berriasella* complex of northern Mexico and the Koeneni Zone of western South America has been excellently documented by Imlay and Jones (1970, p. B11, B12, Fig. 2). This fauna includes *Substeueroceras, Parodontoceras, Blanfordiceras, Proniceras* and *Spiticeras (Spiticeras)* of Mexican and South American affinities. The presence of *Substeueroceras stantoni* (Anderson) closely allied to *S. subquadratum* described by Imlay (1939) from the highest Jurassic beds in northern Mexico, of *S.* cf. *kellumi* Imlay closely allied to that Mexican species, and of *Parodontoceras reedi* (Anderson) closely allied to and probably conspecific with *P. callistoides* (Behrendsen), is considered to be particularly diagnostic by these workers.

This extremely close affinity and obvious contemporaneity of the Californian *Substeueroceras-Parodontoceras* fauna with that of the youngest Jurassic rocks of Mexico (the *Substeueroceras-Berriasella* complex of the San Lois Potosi area) and the Koeneni Zone was re-confirmed by Verma and Westermann (1973, p. 155, 157, Table 4). These workers established the presence of the Californian *Parodontoceras storrsi* (Stanton) and of a new subspecies of the South American *Substeueroceras koeneni* in their *Substeueroceras-Berriasella* assemblage.

The generic and specific affinities of these three ammonite faunas are so close that their full zonal equivalence is beyond a reasonable doubt (Fig. 5). Further proof is provided by correspondence, which was discussed previously, of the ammonite fauna of the underlying Californian *Buchia piochii* s.s. – *Buchia fischeriana* Zone with the next older *Kossmatia-Durangites-Corongoceras* Complex of Mexico and the *Windhauseniceras internispinosum* and *Corongoceras alternans* Zones of Argentina (see p. 191-192). This correlation is accepted by most of the recent American (e.g., Cantú Chapa, 1967, 1971, 1982; Imlay, 1980, p. 34-37, Figs. 15, 16; A. Leanza, 1945, 1947; H. Leanza, 1980, 1981a, 1981b; H. Leanza and Hugo, 1977; Biro-Bagoczky, 1980) and European (e.g., Enay, 1972, 1973; Widemann, 1968, 1980) workers. However, it is disputed by Zeiss (1979, and in Westermann, 1982, p. 142, 143) and others (p. 181, Fig. 2).

Zeiss (1979, p. 17, 18, corr. chart) correlates the *Buchia* aff. *okensis* Zone of California (= *Buchia terebratuloides* s.l. Zone; Fig. 5) and equivalent *Buchia* beds of western and Arctic Canada with the upper part of the *Proniceras* and *Substeueroceras* Zone of Mexico, the upper part of the *Substeueroceras koeneni* Zone of South America, and the Occitanica Zone of the type Berriasian. Furthermore, he correlates this *Buchia* Zone and its alleged Mexican, South American and Southwest European equivalents with the uppermost Volgian *Craspedites nodiger* Zone of the Central Russia (Fig. 2). The next oldest *Buchia piochii-B. fischeriana* Zone of California and its Canadian equivalents are correlated with the lower part of the *Proniceras* and *Substeueroceras* Zone in Mexico, the lower part of the *Substeueroceras koeneni* Zone in South America, and the Jacobi and Grandis Zones combined of the type Berriasian (p. 192). As with the previously discussed correlations (p. 186, 188-189) of the *Buchia russiensis* s.l. and *Kossmatia*-bearing beds and *Buchia piochii-B. fischeriana* Zone of California, the above discussed conclusions of Zeiss (1979) concerning the *Buchia terebratuloides* s.l. Zone of California were outdated by the monograph of Zakharov (1981), the principal results of whom are summarized in the North Siberian graph of Figure 10.

As stressed by Zakharov (1981, p. 104, 105, 224, 225, Table 3), *Buchia fischeriana* is a largely earlier Late Volgian species. It is not known to range up into the *Buchia okensis* s.s. Zone that roughly corresponds to the restricted *Hectoroceras kochi* Zone (the problem of the occurrence of *B. fischeriana* in the basal Riasanian Beds is discussed on p. 236-237). *Buchia terebratuloides* s.l. (i.e. including *B. obliqua*) is also a Late Volgian species that ranges right through this stage and is not known to ascend into the *Hectoroceras kochi* Zone anywhere in northern U.S.S.R. (Zakharov, 1981, p. 100, 109, 224-226, Table 3). Finally, *Buchia volgensis* (Lahusen) and *Buchia okensis* s.s. are strictly Early Cretaceous species that make their first appearances in the restricted *Hectoroceras kochi* Zone throughout the northern part of U.S.S.R. and also in its more southerly regions (Zakharov, 1981, p. 121, 130, 226-229, Table 3). Contrary to the claims of earlier Soviet workers (Saks *et al.*, 1972) utilized by Zeiss (1979), neither of them occurs in the underlying latest Volgian rocks of northern Siberia and other regions of U.S.S.R. It is only the immediate predecessor of *B. okensis* s.s. – *Buchia* n. sp. aff. *okensis* that may occur in the uppermost part of the Volgian stage in northern Siberia and Central Russia (p. 236-237). Even the *Buchia* aff. *okensis* Zone (= *Buchia terebratuloides* s.l. Zone) cannot correspond to the restricted North Siberian Zone of *Hectoroceras kochi* (except for its basal beds) as the latter is characterized by the *Buchia okensis* s.s. fauna. The *Buchia okensis* s.s. Zone is not known in northwestern California. However, it is typically developed in the Taseko Lakes map-area, western British Columbia (p. 203-204), where it overlies gradationally the *Buchia terebratuloides* s.l. Zone. The latter contains common *B.* n. sp. aff. *okensis* and rare *B. unschensis* s.s. just as its Californian equivalent does. *Buchia fischeriana* occurs in a zone of its own in the Taseko Lakes map-area that underlies directly and gradationally the *Buchia terebratuloides* s.l. Zone (Fig. 5). Apart from somewhat different zonal nomenclature, the sequence of these three *Buchia* Zones in that area virtually duplicates the sequence of topmost Jurassic and basal Cretaceous *Buchia* Zones established by Zakharov (1981) in northern Siberia. These data are deemed to be sufficient to demonstrate the Late Volgian age of both the *Buchia piochii* s.s. – *B. fischeriana* and *Buchia terebratuloides* s.l. Zones of northwestern California in their entirety.

I also reject the proposal of Zeiss (1979) to correlate the *Buchia piochii* – *B. fischeriana* Zone with the lower part of the *Substeuero-ceras-Berriasella* complex in Mexico (Verma and Westermann, 1973), and the Mediterranean Grandis/Jacobi Zone, while correlating the superposed *Buchia* aff. *okensis* Zone (= *Buchia terebratuloides* s.l. Zone) with the upper part of this complex and the Mediterranean Occitanica Zone.

As indicated earlier in the discussion of age and correlation of the *Buchia piochii* s.s. and *Buchia fischeriana* Zones of western British Columbia, these two zones combined are correlative with the *Kossmatia-Durangites-Corongoceras* complex of northern Mexico rather than with the next younger *Substeueroceras-Berriasella* complex of that country. Furthermore, the same is true of the approximately equivalent *Buchia piochii* s.s. – *B. fischeriana* Zone of northern California (Fig. 5), because the fine to very fine and densely ribbed *Substeueroceras* ex gr. *koeneni*, that characterizes the Mexican *Substeueroceras-Berriasella* complex in its entirety, is restricted to the next younger *Buchia terebratuloides* s.l. Zone (Imlay and Jones, 1970, B11). Therefore, this Californian zone (and hence its Canadian equivalent) alone is correlated with the *Substeueroceras-Berriasella* complex of northern Mexico and the equivalent *Substeueroceras koeneni* Zone of western South America. Furthermore, there is no indication in Verma and Westermann's (1973) paper that any of the *Substeuer-oceras* species found in the *Substeueroceras-Berriasella* complex are stratigraphically restricted within the complex. Nor is there any indication that any of its *Berriasella* species are so restricted. Therefore, I follow Verma and Westermann (1973, p. 158) in treating this complex as a unit and an exact equivalent of the South American *Substeueroc-eras koeneni* Zone. I see no reason to subdivide either of these two units into subzones as Imlay and Jones (1970) and Zeiss (1979) propose.

As already pointed out (p. 188, 189), the proposal of Zeiss (1979) to correlate the Mexican *Substeueroceras-Berriasella* complex and the *Substeueroceras koeneni* Zone with the combined Jacobi, Grandis and Occitanica Zones of the West European standard was decisively influenced by the apparent restriction of *Durangites* to the next older *Kossmatia-Durangites-Corongoceras* complex of Verma and Westermann (1973, p. 153, 155, Tables 5, 6). The assumption that *Durangites* is restricted to beds underlying *Substeueroceras* in Mexico and elsewhere in the Gulf Province is rather widespread among Mexican and American workers (compare Imlay, 1980, p. 34, 35, Fig. 15). Following the recent discovery of *Durangites* fauna in Southern Spain (p. 179-180), where it occurs in the uppermost Tithonian beds corresponding to the upper part of the Transitorius Zone of the Submediterranean Belt, this assumption convinced Zeiss (1979) to correlate the entire *Substeueroceras-Berriasella* assemblage of Verma and Westermann (1973) or, for that matter; the *Substeueroceras-Proniceras* Zone of Imlay (1939, 1980) with the Jacobi, Grandis and Occitanica Zones of the type Berriasian. His correlation of the bulk of the Upper Volgian stage with the Jacobi, Grandis and Occitanica Zones of the type Berriasian followed naturally.

It is unfortunate for these intercontinental and interrealm correlations of Zeiss (1979, and in Westermann, 1982, p. 142, 143) that the generalizations of the Upper Tithonian zonal sequence of ammonite genera in Mexico and other Middle American countries (e.g., Verma and Westermann, 1973; Imlay, 1939, 1980, etc.) on which they are based conflict with the actually observed zonal sequence there. Firstly, Verma and Westermann's (1973) zonal scheme ignores the common to rare occurrence of *Durangites* spp. and *Kossmatia* spp. in the *Substeueroceras-Berriasella* assemblage of the San Luis Potosi area. Secondly, Cantu Chapa (1967, 1982) has recorded these diagnostic Tithonian genera in the equivalent *Salinites-Parodontoceras* (or *Parodontoceras* aff. *callistoides*) Zone of Eastern Mexico.

The ascendance of *Durangites* spp. and *Kossmatia* spp. into the *Substeueroceras-Berriasella* complex is not apparent from the perusal of its general description provided by Verma and Westermann (1973, p. 153, 155, Table 4) and a recent critical paper of Cantu Chapa (1982, p. 315, Figs. 4, 6, 7) was published in the Petroleum Geology Journal. This necessitates a brief summary of the specific evidence available therein.

Kossmatia bifurcata and *Substeueroceras koeneni tabulatum* were collected *in place* at Verma and Westermann's (1973, p. 132) locality 19C of the Los Alamitos section. Furthermore, adjacent localities 19A, 19D and 20A, all of which represent: "talus, the source beds [of which; writer's remark] belong almost certainly to the El Verde Member ... ", have yielded *Berriasella subprivasensis, Substeueroceras koeneni tabulatum* and *S. catorcense* in association with *Kossmatia bifurcata, Durangites vulgaris, D. heilprini, D.* aff. *D. heilprini* and *D. alencasteri.*

Elsewhere in the area (Verma and Westermann, 1973, p. 134), the locality 22 of El Verde Member in Los Alamitos valley has yielded: "above, *Berriasella* cf. *B. picteti, B.* aff. *B. zacatecana, Kossmatia bifurcata, K. exceptionalis, Substeueroceras alticostatum, S. catorcense* and, somewhat lower *Corongoceras*? n. sp. indet., *Durangites heilprini, D. vulgaris, D.* n. sp. aff. *rarifurcatus, D. alencasteri* n. sp., and *Micracanthoceras acantellum*". Finally, the locality 25 of the El Verde Member situated north of La Leona has yielded *Kossmatia, Durangites* and *Corongoceras* associated with *Substeueroceras* spp. and *Berriasella* spp. Verma and Westermann (1973, p. 135) state that: "from the undisturbed part (of this section; writer's remark) at about 10 to 15 m below the contact with the Taraises Formation were collected

Aspidoceras cf. *A. andinum, Berriasella* cf. *B. picteti, B. subprivasensis, B.* aff. *B. zacatecana, Kossmatia bifurcata, K. exceptionalis, K. flexicostata, K. purisma n. sp., Substeueroceras alticostatum, S. catorcense* n. sp., *S. storrsi, Corongoceras cordobai* n. sp. and *Durangites heilprini*".

These data demonstrate that the *Kossmatia-Durangites-Corongoceras* and the *Substeueroceras-Berriasella* complexes of Verma and Westermann (1973) are much more closely allied than is apparent from their generalized description. They are, actually, one diversified ammonite complex where *Kossmatia, Durangites* and *Corongoceras* are more common in the lower part while *Substeueroceras* and *Berriasella* (especially the latter) are more common in the upper part. The complexes should, therefore, rank at the most as faunal subzones of one Upper Tithonian faunal zone. The correlation of any part of this zone with the basal Berriasian, let alone with its Occitanica Zone, is prohibited by the presence of *Durangites* and *Kossmatia* in its upper subzone. As already pointed out in the discussion of the Tithonian/Berriasian boundary in southwestern Europe (p. 179-180), *Durangites* is a strictly Late Tithonian genus. *Kossmatia* does range from the Late Kimmeridgian into the Late Tithonian but definitely not into the latest Tithonian, let alone the basal Berriasian. In southern Spain *Kossmatia richteri* was only found in the upper Lower Tithonian (i.e., *Semiformiceras semiforme* and *Semiformiceras fallauxi* Zones) by Enay and Géyssant (1975, p. 44, Fig. 2). However, a solitary specimen of *Kossmatia* sp. was found in the upper Upper Tithonian *Durangites* Zone by Oloriz and Tavera (1979b, p. 503, Fig. 3).

Therefore, the two assemblages of Verma and Westermann (1973) discussed above are correlative with the *Micracanthoceras micracanthum* and *Durangites* Zones of the southern Spain combined. This intercontinental correlation indicates the contemporaneity of these two complexes combined with the *Paraulacosphinctes transitorius* Zone of the Submediterranean Belt of the European Tethys, which was shown to be coeval with the two above mentioned Late Tithonian zones of southern Spain combined (Figs. 5, 10).

Another indication of the late Late Tithonian age of these two ammonite complexes is provided by a recent analysis of ammonite faunas of the oil well cores in eastern Mexico where Cantú Chapa (1982, p. 315) has found the following two Upper Tithonian ammonite zones:

"a zone with *Salinites-Parodontoceras* at the top; and a zone with *Suarites* at the base. In the first zone there are also *Proniceras, Kossmatia, Durangites,* and *Protancyloceras,* but in small numbers. Just below ["above" obviously a *lapsus calami;* writer's remark] this upper zone, *Suarites* and very rare *Acevedites* and *Haploceras* have been found (Plate 1)."

The ammonite association of the upper zone is clearly coeval with the *Substeueroceras-Berriasella* assemblage of the San Luis Potosi area, even more so, as it includes *Substeueroceras* sp. in at least one instance. This critical find, which is not listed in the above cited faunal summary description of the zone, is recorded in the Camotal 1 well where *Substeueroceras* sp. was found in association with *Proniceras* sp. and *Parodontoceras* sp., just below the lithologically defined position of the Jurassic-Cretaceous boundary (Cantú Chapa, 1982, p. 316, Fig. 3). *Durangites* sp. was found at about the same level below that boundary in two other wells (Cantú Chapa, 1982, p. 318, Figs. 6, 7).

The above evidence demonstrates that the beds carrying *Substeueroceras* in these Mexican outcrop areas are not younger than the topmost beds of the *Paraulacosphinctes transitorius* Zone that are correlative with it. The ascendance of *Kossmatia* spp. into these beds makes me doubt that the so-called *Substeueroceras-Berriasella* complex and the *Salinites-Parodontoceras* Zone reach as high as the top of the European *Durangites* Zone.

Verma and Westermann's (1973, p. 158) and Imlay's (1980, p. 34, 35, Figs. 15, 16) conclusion that the *Substeueroceras-Berriasella* assemblage of the San Luis Potosi area is correlative with the Argentinian Zone of *Substeueroceras koeneni* and "is clearly uppermost Tithonian (Table 4) in age" is obviously valid. However, their proposal to correlate this assemblage with the *Berriasella chaperi* Zone of southeastern France is outdated as *Berriasella chaperi* was more recently discredited as a zonal fossil. Like *Berriasella jacobi* and *Berriasella delphinensis,* it was found to range right through the basal Berriasian (i.e., the lower part of the *Pseudosubplanites grandis* Zone) and the topmost part of the *Paraulacosphinctes transitorius* Zone (p. 178-179).

The bulk of the above conclusions concerning the latest Tithonian age of the *Substeueroceras-Berriasella* complex and the *Salinites-Parodontoceras* Zone (or *Parodontoceras* aff. *callistoides* Zone) of northern and eastern Mexico are applicable to the *Substeueroceras koeneni* Zone of western South America because of the extremely close affinity of their *Substeueroceras* and *Parodontoceras* species. Furthermore, the Mexican zones and the Koeneni Zone overlie immediately and apparently gradationally another almost identical and definitely older Late Tithonian fauna – the *Kossmatia-Durangites-Corongoceras* complex in Mexico and the *Corongoceras alternans* and *Windhauseniceras internispinosum* Zones in Argentina (Fig. 5). As shown earlier (p. 179-180), the total absence of *Durangites* in these Argentinian zones has no biochronological significance, but is caused by climate.

In addition to this "external" evidence (Jeletzky, 1973, p. 43) of the latest Tithonian age of the *Substeueroceras koeneni* Zone, there is good "internal" evidence confirming the age. This is the combined presence of *Aspidoceras longaevum* A. Leanza, *Aulacosphinctes mangaensis* (Steuer), *A. azulensis* A.F. Leanza, *Micracanthoceras* (= *Blanfordiceras,* H.A. Leanza, 1981a) *vetustum* (Steuer) and *Himalayites andinus* H. Leanza (H.A. Leanza, 1981a). This association clearly indicates correspondence with the upper Upper Tithonian (i.e., the regional *Durangites* Zone) of southern Spain, to which these genera are largely restricted (p. 179).

The recent citation of "*Virgatosphinctes*" aff. *transitorius* (Oppel) from the *Corongoceras alternans* and *Substeueroceras koeneni* Zones (H.A. Leanza and C.A. Hugo, 1977, p. 250, 256) would provide yet another argument in favor of the correspondence of these two Argentinian zones with the *Paraulacosphinctes transitorius* Zone of the European Tethys. However, this record is not mentioned in the later papers of H. Leanza (e.g., 1981a, 1981b) available to me, which suggests that it has been discarded.

Regardless of whether one accepts a broader concept and an essentially trans-Tethyan occurrence of such perisphinctid genera as *Virgatosphinctes, Aulacosphinctes* and *Aulacosphinctoides* that is favored by Imlay (1939, 1942, 1980), Imlay and Jones (1970), Arkell (1956, and in Arkell *et al.,* 1957), Verma and Westermann (1973), A.F.

Leanza (1945, 1947), H.A. Leanza (1980) and others, or attempts to split these genera into a number of morphologically restricted and geographically limited taxa (e.g., Enay, 1973), these taxa remain restricted largely to the Early Tithonian *sensu* Enay (1971, 1972, 1973). They are relatively rare already in the Late Tithonian and their very rare, last representatives barely cross the Tithonian/Berriasian boundary as defined above (p. 179). Therefore, the presence of two *Aulacosphinctes* species (e.g., *A. mangaensis* and *A. azulensis*) in the Koeneni Zone of western Argentina is perhaps the best "internal" evidence for its age i.e., no younger than late Late Tithonian. These two species also tend to support the previous suggestion that the Koeneni Zone may be older than the latest Tithonian of the European Tethys where *Berriasella jacobi* and other berriasellids (Fig. 5) mix with *Durangites*, *Paraulacosphinctes* ex gr. *transitorius* and other characteristic late Late Tithonian ammonites.

The recorded presence of *Riasanites* in the Koeneni Zone of Argentina has been used to support its Berriasian age (Zeiss, 1979), but I am convinced that this genus ranges downward into the uppermost Tithonian beds of the type Riasanian (p. 237, 240 and Fig. 10). True *Riasanites*, however, is unknown in Argentina. *"Riasanites" rjasanensoides* of Krantz (1926, p. 441) belongs elsewhere because of the common irregular spacing of primaries, forming closely spaced pairs separated from adjacent pairs by wide interspaces (Krantz, 1926, Pl. XVII, Fig. 1). This feature is quite foreign to the true *Riasanites*, including *Tauricoceras* (Kvantaliani and Lyssenko, 1979a, Pl. I, Figs. 1a, 2a). Furthermore, one of the primaries of these pairs commonly remains single and intercalated secondaries are absent on the early whorls of the Argentinian form. These morphological distinctions and the occurrence of *"R." rjasanensoides* in the Southern Hemisphere indicate that it is a homoeomorph of the true *Riasanites*. It is interesting that similar pairing of primaries is a common feature in other late Late Tithonian ammonites in Argentina (e.g., *Berriasella australis* A.F. Leanza, 1945, Pl. II, Fig. 10; *Micracanthoceras lamberti* A.F. Leanza, 1945, Pl. III, Fig. 9; *Berriasella*? *pastorei* A.F. Leanza 1945, Pl. III, Fig. 12; *Corongoceras rigali* A.F. Leanza, 1945, Pl. IV, Fig. 3, etc.). A number of these forms also resemble *"R." rjasanensoides* in the shape and proportions of the whorl, the moderately evolute umbilicus and their simple septal suture. Therefore, it appears to be more feasible to consider *"R." rjasanensoides* as a relative of some such Argentinian forms than to attach it to the only superficially similar and geographically remote *Riasanites*.

The Late Tithonian age of the Koeneni Zone is also supported by its stratigraphic position. It is overlain directly and apparently gradationally (A.F. Leanza, 1945; H.A. Leanza and Hugo, 1977; H.A. Leanza, 1981a, 1981b; Biro-Bagoczky, 1980) by the *Argentiniceras noduliferum* Zone. The latter is correlative with the Grandis Zone, as used in this paper, and at least the lower part of the Occitanica Zone for reasons presented in the discussion of age and correlation of *Buchia okensis* s.s. Zone (p. 210, 211). The Koeneni Zone is underlain immediately and apparently gradationally by the *Corongoceras alternans* Zone – the himalayitid ammonites of which are extremely closely related to those of the *Micracanthoceras micracanthum* Zone of the European Tethys. As pointed out by A.F. Leanza (1945, p. 45), his *Micracanthoceras lamberti* resembles closely *"Berriasella" spinulosa* Gerth (1925, Pl. VI, Figs. 1, 2, 2a) that is hardly distinct from *M. micracanthum* (Oppel) according to Enay (1972, p. 375). Since *M. micracanthum* is diagnostic of the lower Upper Tithonian zone of that name in southern Spain (Enay and Géyssant, 1975, p. 44, 45), correlation of the Argentinian *Corongoceras alternans* Zone with the lower rather than the upper part of the *Paraulacosphinctes transitorius* Zone is indicated. Furthermore, *M. lamberti* (compare A.F. Leanza, 1945, Pl. III, Figs. 9, 10 with Mazenot, 1939, Pl. XXXVII, Figs. 7a-7c) is so similar to the early whorls of *M. micracanthum* (Oppel) that it may be considered conspecific with this Stramberg and southern Spain species. A.F. Leanza (1945, p. 45) also stresses the close affinity of these two species, although he does not synonymize them because of the slightly higher bifurcation points in the Argentinian species and other unspecified distinctions. At any rate the abundance of *Micracanthoceras* spp. in the *Corongoceras alternans* Zone and their close affinities with *M. micracanthum* clearly indicate its correlation with that West European zone.

As already pointed out by Enay (1972, p. 376), the fauna described by Lisson (1907, p. 56-58, Pl. X, Figs. 1-6; Pl. XI, Figs. 1a, 1b) from the Jurassic-Cretaceous boundary beds of coastal Peru near Lima includes *Micracanthoceras* (i.e., *"Ancyloceras" nodosum*) that was found at the same locality (and presumably in association) with a *Durangites* (*"Kosmoceras" limense*) of Mexican affinities and a *Protancyloceras* (his *"Ancyloceras" sabaudianum*) of Middle American and European affinities. This fauna provides a tie between the *Corongoceras alternans* fauna of western Argentina and Chile on the one hand and the *Kossmatia-Durangites-Micracanthoceras* fauna of northern Mexico (Verma and Westermann, 1973; Cantú Chapa, 1967) on the other. The paleogeographical significance of the Lisson's (1907) fauna and its obviously Late Tithonian (apparently mid-Late Tithonian) age were discussed earlier (p. 180).

Buchia okensis s.s. Zone

Stratigraphy and Geographical Distribution. No additional information about the stratigraphy and geographical distribution of *Buchia okensis* s.s. Zone on the west coast of Vancouver Island has become available since Jeletzky's (1965) paper on the Canadian Western Cordillera; nor has that zone been discovered elsewhere in the Insular Trough. Similarly, no additional information has become available from the Harrison Lake area since Crickmay (1962) and Jeletzky (1965).

The marine sediments of *Buchia okensis* s.s. Zone are fairly widespread in the Taseko Lakes map-area where it outcrops in the central zone of the trough and on its southwestern flank. In this central zone it is represented by the argillaceous facies, which comprises the middle part of the Latest Portlandian s.s. to Late Berriasian shale unit of the Relay Mountain Group (Jeletzky, in Jeletzky and Tipper, 1968, p. 21, 22). This facies of *Buchia okensis* s.s. Zone is estimated to be about 70 m thick in the Section 15 where it is completely exposed and appears to be almost unfaulted (Jeletzky, in Jeletzky and Tipper, 1968, p. 169, 170). However, the basal 16 m of shale assigned to this zone in this section are unfossiliferous and so conceivably could belong to the underlying *Buchia terebratuloides* s.l. Zone, in part at least. The mid-basin facies of *Buchia okensis* s.s. Zone is represented exclusively by friable, pure to silty shale, dark grey to dark brown when fresh, weathering medium to ash grey, brownish tinged and flaky to earthy. The shale is rich in large and small, variously shaped concretions and up to 60 cm thick bands of hard to very hard, light brownish grey weathering clay ironstone and impure limestone that are richly fossiliferous for the most part. This fauna consists almost exclusively of buchias with a minor to considerable admixture of ammonites in some beds. This faunal composition and the prevalence of bivalved shells, with the valves mostly only slightly displaced in relation to one another among the buchias, indicate the deposition of this shale in a low energy, outermost shelf (outer neritic or ?upper bathyal) environment.

All other sections of the mid-basinal facies of *Buchia okensis* s.s. Zone are either incomplete (e.g., Section 12; Jeletzky, in Jeletzky and Tipper, 1968, p. 146) or too badly disturbed (or both) to be measured.

The argillaceous facies of the *Buchia okensis* s.s. Zone grades into the underlying *Buchia terebratuloides* s.l. Zone and the overlying

Buchia uncitoides s.l. Zone. The contact relationships observed in Sections 12 and 15 are typical of all of the appropriately exposed sections studied.

The southwestern marginal facies of *Buchia okensis* s.s. Zone is best exposed in the headwaters of Tyaughton Creek and in the vicinity of Elbow Mountain where it comprises the middle part of the Latest Portlandian s.s. (?)-Early Valanginian arenaceous unit of the Relay Mountain Group. No complete section of this facies of *Buchia okensis* s.s. Zone was found so far, its upper part being cut off invariably by the splays of the Elbow Mountain thrust zone. The longest Section 3 of this facies was measured on the northern spur of Elbow Mountain that overlooks the Big Creek from the east (Jeletzky, in Jeletzky and Tipper, 1968, p. 105, 106). This incomplete section is only about 13.5 m thick. Its upper 8 m consist of sandy, dense and hard siltstone, which is blue-grey when fresh and weathers dull-grey with a rusty tinge and specks and is blocky and splintery. The siltstone contains some interbeds of very fine-grained, silty greywacke near the base and grades downward into the greywacke of the lower part of *Buchia okensis* s.s. Zone. The siltstone is separated from the overlying siltstone of the *Buchia uncitoides* s.l. Zone by about a 4.5 m wide shear zone – the inferred significance of which will be discussed below.

The lower part of *Buchia okensis* s.s. Zone in Section No. 3 consists of an about 5.5. m thick bed of hard (silicified) greywacke which is dark-green when fresh and fine to medium grained. This greywacke is massive when fresh and weathers rust-coloured and is splintery. At its base, a 0.5 to 0.6 m wide interval of powdery, arenaceous and apparently pebbly rock (greywacke?) occurs and it includes somes patches of fine pebble conglomerate. This interval is probably a brecciated (mylonitized?) fault plane because a mixed *Buchia piochii* s.s.-*Buchia* cf. *terebratuloides* s.l. fauna, diagnostic of either the lower part of *Buchia terebratuloides* s.l. Zone or of the still older upper part of *Buchia fischeriana* s.l. Zone, occurs only about 5 m below. Furthermore, no rocks similar to the richly fossiliferous, arenaceous to conglomeratic facies of the topmost part of *Buchia terebratuloides* s.l. Zone present in adjacent section, occur either within or immediately below this interval.

The sheared contact of this section of the *Buchia okensis* s.s. Zone with the overlying *Buchia uncitoides* s.l. Zone (see units 29 to 31 of Section 3; Jeletzky, in Jeletzky and Tipper, 1968, p. 104, 105) does not seem to indicate any significant loss of strata. This is suggested by the presence of a mixed fauna of *Buchia okensis* s.s. and *Buchia uncitoides* s.l. characteristic of the upper part of the *Buchia okensis* s.s. Zone, immediately below this contact. Furthermore, the *Buchia uncitoides* s.l. fauna apparently devoid of *Buchia okensis* s.s. occurs in both the sheared interval and the overlying 14.5 m (approximately) thick *Buchia uncitoides* s.l. beds. Finally, a poorly preserved berriasellid fauna apparently equivalent to the *Argentiniceras* cf. *noduliferum*-*Fauriella* n. sp. aff. *gallica* fauna that occurs elsewhere in the upper but not uppermost part of the argillaceous facies of *Buchia okensis* s.l. fauna, is present closely below the siltstone unit that carries the mixed *Buchia okensis* s.s.-*Buchia uncitoides* s.l. fauna. Section No. 3 appears, therefore, to include an either complete or an almost complete sequence of the upper and middle parts of the arenaceous facies of *Buchia okensis* s.s. Zone, but it is not known how much of its lower part is missing.

The basal beds of this facies of *Buchia okensis* s.s. Zone and its normal lower contact are exposed in a short but important adjacent section studied and collected from by H.W. Tipper and described under the *Buchia terebratuloides* s.l. Zone (p. 195).

The only fossiliferous section of *Buchia okensis* s.s. Zone known to exist in the northeastern part of the Mount Waddington map area (Jeletzky, 1968b; and in Tipper, 1969, p. 39, 40) consists of about 5 m of greywacke that is bluish-grey when fresh and weathers brownish-orange with dark specks, and is mostly fine grained but locally pebbly. It is interbedded with 7.5-25 cm beds of pebble conglomerate that includes granitic pebbles. Small, presumably late forms of *Buchia okensis* s.s. and rare *Buchia uncitoides* s.l. were collected in the topmost part of this bed 20. In combination with the presence of a rich *Buchia uncitoides* s.l. fauna that appears to be devoid of *B. okensis* s.s. in the immediately overlying bed 21 of the section, this suggests that the bulk of *Buchia okensis* s.s. Zone is represented by the upper part of the unfossiliferous, partly or entirely nonmarine (presmably beach to deltaic) arkoses and pebble conglomerates that underlie bed 20. This unfossiliferous sequence overlies conformably and apparently gradationally the shallow marine Upper Portlandian s.s. (or uppermost Lower Tithonian) bed containing the *Buchia russiensis* s.l. fauna (Jeletzky, in Tipper, 1969, p. 41). Therefore, it is impossible to segregate the nonmarine equivalents of the *Buchia okensis* s.s. Zone from those of the *Buchia terebratuloides* s.l., *Buchia fischeriana* s.l., and *Buchia piochii* s.s. Zones and to estimate the thickness of the former.

The inferred prevalence of nonmarine rocks in the *Buchia okensis* s.s. Zone and adjacent uppermost Jurassic and basal Cretaceous beds of northeastern Mount Waddington map-area clearly indicates its deposition at the western margin of Tyaughton Trough and derivation from an adjacent western source area.

Paleontology and Biochronology. Concept of Species. The broad interpretation of *Buchia okensis* s.s. used by Jeletzky (1965, and in Jeletzky and Tipper, 1968) was considerably restricted (Jeletzky, 1971a, p. 4, Fig. 2) by the exclusion of the latest Tithonian *B.* n. sp. aff. *okensis* that is its direct ancestor. The same step was taken slightly earlier by Jones *et al.* (1969) who recognized specific independence of the latest Tithonian *B.* aff. *okensis* (p. 196-197). Zakharov's (1981, p. 116-121, 226, 227) interpretation of *Buchia okensis* is almost identical with that of Jeletzky (1965) in that he does not recongize the specific independence of *B.* n. sp. aff. *okensis* and consequently is not clear about its time range (see below). Zakharov (1981, p. 120) may well be right in excluding the type specimen of *B. spasskensis* (Pavlow 1907) from *B. okensis* s.s. However, I had in mind the typical late forms of *B. okensis* f. typ. described and figured under the name *Aucella spasskensis* by Gerassimov (1955, Pl. XIII, Figs. 9, 10).

Geographical Range, Facies Tolerance and Subspecies. Jeletzky (1971a, 1973, 1974) has confirmed earlier conclusions (Jeletzky, 1965, p. 23) about the exceptionally wide geographical range, very restricted and uniform time range and truly remarkable facies tolerance of *Buchia okensis* s.s. Following the full specific separation of *B.* n. sp. aff. *okensis* (= *B.* aff. *B. okensis* of Jones *et al.*, 1969), *B. okensis* is an early Berriasian index fossil wherever it is known in the Boreal Realm, except for its extremely rare earliest representatives in the uppermost beds of the underlying *Buchia terebratuloides* s.l. Zone and the somewhat more common latest representatives in the lower part of the next younger *Buchia uncitoides* s.l. Zone and its equivalents in Arctic Canada and Northern Eurasia. *Buchia okensis* s.s., however, does not penetrate as far south into the North Pacific Province as either its immediate predecessor *B. terebratuloides* s.l. or its companion and successor *B. uncitoides* s.l. (p. 207-208).

The description and figures of North Siberian representatives of *Buchia okensis* provided by Zakharov (1981, p. 116-121, Pls. XXXI-XXXV) confirm Jeletzky's (1973, p. 54) earlier conclusion that the most diagnostic Early Berriasian variants of *B. okensis*, such as the large to giant representatives of its typical form and *B. okensis* var. *canadiana*, are entirely or largely restricted to the Canadian (Jeletzky,

1964) and American (i.e., Northern Alaska; Imlay, 1961) parts of the North American Boreal Province and to adjacent areas of the North Pacific Province (e.g., Western British Columbia and the State of Washington, Jeletzky, 1965). Like the Eurasian representatives of *Buchia okensis* figured by various earlier workers (see Jeletzky, 1973, p. 54), those figured by Zakharov (1981) represent mostly small to medium-sized, relatively closely and less prominently ribbed variants approaching *Buchia* n. sp. aff. *okensis*, *Buchia okensis* var. *subokensis* Pavlow, and late forms of the species typified by its holotype (Pavlow 1907, Pl. I, Fig. 11) and by that of *B. okensis* var. *cascadensis* Crickmay 1930. In my opinion, only the specimens reproduced in Pl. XXXI, Fig. 2, and Pl. XXXIV, Figs. 1-3 of Zakharov (1981) belong to those wide, posteriorly extended, extremely coarsely ribbed *B. okensis* forms that dominate its Lower Beriasian populations in Western British Columbia (e.g., Jeletzky, 1965, Pl. IV, Figs. 1, 4, 6, 10, 12, 18, 19, 20, 21; Pl. V, Figs. 2, 6, 7, 10; Pl. VI, Figs. 1, 3, 4, 5, 6; Pl. VIII, Fig. 1; Pl. XXII, Fig. 5) and Arctic Canada (e.g., Jeletzky, 1964, Pl. I, Figs. 1, 3). Furthermore, Zakharov (1981, p. 120, 121) stresses that these specimens represent his fourth form of *B. okensis* that is rare in his collections and but little known in the Eurasian literature. All other North Siberian representatives of *Buchia okensis* figured and described by Zakharov (1981) and stated to form the bulk of its North Siberian populations, either closely resemble the largest and most advanced representatives of *Buchia* n. sp. aff. *okensis* figured by Jones *et al.* (1969, Pl. I, Figs. 25-35) and in this paper (e.g. Pl. I, Figs. 6, 7; Pl. III, Figs. 3, 4), or are comparable to *B. okensis* var. *subokensis* (Pavlow 1907) as interpreted by Jones *et al.* (1969) and Jeletzky (1965). This *B.* n. sp. aff. *okensis*-like morphology of most North Siberian representatives of *B. okensis* s.s. is presumed to be responsible for Zakharov's (1981, p. 116, 227) decision to amalgamate *B.* aff. *okensis* of Jones *et al.* (1969) with *B. okensis* s.s. in spite of a statement to the contrary made by Jones *et al.* (1969, p. A12). However, this amalgamation of the two species and of their zones in western and Arctic North America apparently did not result in a similar amalgamation of *B.* n. sp. aff. *okensis* and *B. okensis* s.s. and their zones in northern Siberia. I suggest that *B.* n. sp. aff. *okensis* proper may not occur there, nor almost anywhere else in northern Eurasia. Zakharov (1981, p. 226, 227) stresses that the North Siberian beds containing *B. okensis* are confined between those dominated by the *B. unschensis* and *B. terebratuloides* fauna below and those dominated by the *B. jasikovi* (mostly = *B. uncitoides* s.l. of Jeletzky, 1965) and *B. volgensis* fauna above. Furthermore, Zakharov (1981) stresses that *B. unschensis* but rarely ascends into the basal part of the *Buchia okensis* Zone in North Siberia. Finally, the earliest Berriasian age (in the sense of the writer but not Zakharov, who places the *Chetaites sibiricus* Zone into the basal Berriasian; see Fig. 1) of the *Buchia okensis* s.s. Zone in northern Siberia is indicated by the association of this species with *Hectoroceras kochi* in beds representing the bulk of its zone. These beds directly and gradationally overlie the basal beds of the *Hectoroceras kochi* Zone and the *Chetaites sibiricus* Zone where *Chetaites sibiricus* and sometimes *Hectoroceras kochi* are associated with *B. unschensis*, *B. terebratuloides* s.l. and/or *B. fischeriana* (Zakharov, 1981, p. 225, 227). *B. okensis* is stated to be absent from these beds, except for the previously mentioned, exceptional occurrences in the basal part of the type Riasanian (p. 197, 237).

I conclude that the differences between the North Siberian and North American *B. okensis* s.s. indicate that *B. okensis* var. *canadiana* Jeletzky (1965, p. 21, 23, 24) is not a mere morphological extreme variant but is an almost exclusively North American geographical subspecies. The same appears to be true of the previously described and figured (Jeletzky, 1964, 1965) large to giant, early representatives of its typical form. Outside of North Siberia, I only know typical, giant representatives of *B. okensis canadiana* from East Greenland (Surlyk, 1973, p. 91-93, Pl. 1, Fig. 1). The recent discovery of marine Jurassic-Cretaceous boundary beds containing *B. okensis* fauna near the northern end of Greenland (Hakansson *et al.*, 1981) explains this penetration of the North American subspecies into the European Arctic.

As already mentioned (p. 197), the Zone of *Buchia* n. sp. aff. *okensis* proper appears to be unknown outside of Western and Arctic North America. So far as known, it appears to be replaced by the upper part of the *Buchia unschensis* s.s. Zone throughout northern Eurasia, with the possible exception of the basal Riasanian beds in their type-area.

The *B. okensis* s.s. populations contained in the arenaceous and argillaceous facies of its zone do not provide an alternative explanation of the above discussed morphological distinctions of the North American and Eurasian representatives of the species. The samples of *B. okensis* s.s. obtained in Canada from arenaceous and argillaceous (commonly from pure shales) rocks were found to be dominated by *B. o. canadiana* and its giant forms. This was observed, in particular, in the northern Richardson Mountains and the Sverdrup Basin (see Jeletzky, 1964, Pl. I, Fig. 1; Pl II, Fig. 2).

Diagnostic Features and Composition of Fauna in Western British Columbia. As already pointed out by Jeletzky (in Jeletzky and Tipper, 1968, p. 10), the still unfigured *B. okensis* s.s. fauna of the Taseko Lakes map-area does not differ from its previously described and exhaustively figured Vancouver Island counterpart (Jeletzky, 1965, p. 20-27, Pls. IV-VIII, Pl. X, Figs. 1, 7; Pl. XXII, Figs. 4, 7). This fauna also includes variable ratios of *B. uncitoides* s.l. that, generally speaking, increase upward in section until *Buchia uncitoides* s.l. becomes either equally common or prevalent in the topmost beds of the zone. Another common feature of *Buchia okensis* faunas in these two areas is the presence of variable but always subordinate ratios of *Buchia* n. sp. aff. *okensis* (Pavlow). This ancestral form is particularly common in the lower part of this zone on Vancouver Island (Jeletzky, 1965, Pl. IV, Figs. 2, 8, 9, 11, 15, 23, and 24). These specimens are indistinguishable from the representatives of *Buchia* n. sp. aff. *okensis* (Pavlow) prevalent in the upper part of *Buchia terebratuloides* Zone in the Taseko Lakes map-area (Pl. I, Figs. 6, 7; Pl. III, Figs. 3, 4).

Judging from specimens described and published by Crickmay (1930a) and the material collected by other workers, the *Buchia okensis* s.s. fauna of the Harrison Lake sections does not differ from that collected on western Vancouver Island.

The *Buchia* fauna from the basal beds (e.g., GSC loc. 56479) of the *Buchia okensis* s.s. Zone in the Taseko Lakes map-area is exactly like that reproduced in the plate IV of Jeletzky (1965). The two are easily distinguishable from the next older *Buchia* fauna of the upper part of *Buchia terebratuloides* s.l. Zone because of the prevalence of broad, heavily and sparcely ribbed forms of *Buchia okensis* s.s. (e.g., Jeletzky, 1965, Pl. IV, Figs. 1, 3, 4, 18, 19, 20, 21, 22) with broad to extremely broad and well defined posterior wing. As already mentioned (p. 197), such typical representatives of *Buchia okensis* s.s. and those of its var. *canadiana* (Crickmay) are either absent or extremely rare here in the immediate underlying upper part of *Buchia terebratuloides* s.l. Zone (e.g., Pl. III, Fig. 2). Another distinguishing feature of the *Buchia* fauna of the lower part of *Buchia okensis* s.s. Zone in the two areas discussed here is the total or near total absence of forms referable to or even comparable with *Buchia terebratuloides* s.l.

The rich *Buchia* fauna of the argillaceous facies of *Buchia okensis* Zone in the Taseko Lakes map-area only differs from that of its arenaceous facies and from that collected on Vancouver Island (arenaceous facies) in a distinctly greater but nevertheless subordinate ratio of shells either referable to or comparable with *Buchia* n. sp. aff. *okensis* (Pavlow).

Though the *Buchia* fauna of *Buchia okensis* s.s. Zone appears

PLATE II

to be identical in its Insular Trough and Tyaughton Trough sections, their ammonite faunas differ strongly. Only *Lytoceras* (s.l.) sp. indet. was found so far in the Vancouver Island sections (Jeletzky, 1965, p. 21). No ammonites have been found in the Harrison Lake sections. However, the sections in the Taseko Lakes map-area have yielded a fairly diversified but unfortunately poorly preserved fauna of berriasellid ammonites. The early Berriasian ammonites identified, include *Argentiniceras* cf. *noduliferum* (Steuer 1897) of west-central Argentinian affinities (Pl. II, Figs. 3, 6) previously identified as *Berriasella (Mazenoticeras)* aff. *broussei* Mazenot 1939 (Jeletzky, in Jeletzky and Tipper, 1968, p. 170), and *Fauriella* n. sp. aff. *gallica* (Mazenot, 1939) (figured by Jeletzky, in Douglas *et al.*, 1970, Pl. XXIII, Fig. 13) of European Tethyan affinities. This fauna includes *Protothurmannia* n. sp. A (Pl II, Fig. 1) and *Protothurmannia* n. sp. B (Pl. II, Figs. 2, 4; Pl. III, Fig. 1). These two ammonites resemble *Fauriella* Nikolov 1966 and *Subthurmannia* Spath 1939 in the small umbillical bullae and predominantly bidichotomous mode of ribbing in their intermediate growth stages, but their ultimate and, in part, penultimate whorls are ornamented by blunt bullae which are separated from coarse secondary ribs covering upper flank and venter by an almost smooth band. Furthermore, their poorly known early whorls seem to be circular. Because of the close similarity of their adult ornament to that of the latest Tithonian *Protothurmannia rezanoffiana* (Crickmay 1932, p. 2, Pl. I, Fig. 5), these ammonites are tentatively assigned to *Protothurmannia* (Crickmay 1932). I tentatively derive *Protothurmannia* from "*Groebericeras*?" *baylei* Imlay and Jones 1970 and interpret it as an endemic North American stock that is only homoeomorph to *Fauriella* and *Subthurmannia* (see p. 191-192).

Equivalents of Buchia okensis s.s. Zone in northwestern California. According to Jones et al. (1969, p. A12, A13), the *Buchia terebratuloides* s.l. Zone (i.e., *Buchia* aff. *B. okensis* Zone) is here directly overlain by the *Buchia uncitoides* s.l. Zone. However, the equivalence of the lower part of the latter zone to the *Buchia okensis* s.s. Zone of western British Columbia is indicated by the apparently transitional character of the contact between the two zones and by the rare occurrence of *B. okensis* s.s. in the lower part of *Buchia uncitoides* s.l. Zone. As pointed out by Jones *et al.* (1969, p. A13): "a few specimens of

EXPLANATION FOR PLATE II

All photographs are natural size except where indicated otherwise.

Figure 1. *Protothurmannia* n. sp. A. GSC Cat. 73679. GSC loc. 56693. Late Early Berriasian (equivalent of Lower Occitanica Zone), upper part of *Buchia okensis* s.s. Zone. Relay Mountain Group, Uppermost Portlandian s.s. to Upper Berriasian shale unit. Taseko Lakes map-area (92-0), southwestern British Columbia. Collected on western slope of a steep north-south trending ridge overlooking the headwaters of an unnamed east-west trending creek at the point about 2.8 km WNW of the top of Relay Mountain and about 1 km WNW of the Summit 8,222′ of the 4 mile topo map. Lat. 51°9′N.; Long. 122°5′W. Collected at the level of 19.8-22.8 m below top of the bed 1 of Section 12 (see Jeletzky, in Jeletzky and Tipper, 1968, p. 146). Lateral view of fragmentary ?intermediate whorl. Photograph of a rubber mold of a natural external cast.

Figures 2a-c. *Protothurmannia* n. sp. B. GSC Cat. 73680. Same locality, *etc.* as for Figure 1. 2a) lateral view of fragmentary but apparently undeformed intermediate whorl. The sculpture is pronounced at the whorl diameter where that of *Protothurmannia* n. sp. A. is already strongly reduced and disorganized; 2b) cross-section of the whorl shown in Figure 2a; 2c) ventral view of the best preserved segment of the whorl where the forward bent secondaries are completely interrupted in the mid-venter.

Figures 3a, b. *Argentiniceras* cf. *noduliferum* (Steuer 1897). GSC Cat. 73681. GSC loc. 62601. Late Early Berriasian (equivalent of Lower Occitanica Zone), upper part of *Buchia okensis* s.s. Zone. Relay Mountain Group, Uppermost Portlandian to Upper Berriasian shale unit. Taseko Lake map-area (92-0), southwestern British Columbia, southeastern slope of the Summit 8,222′ of the 4 mile topo map at the point about 0.8 km toward 110° from its top at the altitude of about 6,700 feet and about 90.5 m southeast of a prominent conical auxiliary hill occurring on that slope. Collected at the level 15.8 to 18.9 m below top of the unit 1 of Section 15 (see Jeletzky, in Jeletzky and Tipper, 1968, p. 170). 3a) lateral view of the complete specimen. 3b) ventral view of the innermost whorl visible in 3a. This undeformed whorl is regularly rounded and equidimentional, which is diagnostic of *Argentiniceras*. Note the forward bends of secondary ribs and their interruption on the venter.

Figures 4a-c. *Protothurmannia* n. sp. B. GSC Cat. 73682. The same locality, *etc.* as for Figure 3 but collected from the fresh, locally derived float. 4a) lateral view of undeformed fragment of an apparently adult penultimate whorl (compare Pl. III, Fig. 1). This fully septate fragment is still pronouncedly and regularly ribbed, unlike *Protothurmannia* n. sp. A (compare Figure 1); 4b) cross-section of the apical end of the fragment; 4c) best preserved segment of the venter. The secondaries are weakened but not completely interrupted in the mid-venter.

Figures 5a-c. *Buchia terebratuloides* (Lahusen 1888) var. *obliqua* (Tullberg, 1881). GSC Cat. 73683. GSC loc. 56484 (=92450). Uppermost Upper Tithonian, upper part of *Buchia terebratuloides* s.s. Zone. Taseko Lakes map-area (92-0), southwestern British Columbia. Ridge between heads of the Graveyard and Tyaughton Creeks, outcrop at ridge's crest at its S end, Lat. 51°07′34″N.; Long. 123°04′57″W. Collected by H.W. Tipper, Geol. Survey of Canada, 1964. 5a) lateral view of the exterior; 5b) posterior view; 5c) beak part and the hinge line viewed from above. Typical, late juvenile representative of the variant.

Figure 6. *Argentiniceras* cf. *noduliferum* (Steuer 1897). GSC Cat. 73684. The same locality, etc. as for Figures 3 and 4 but collected from the fresh, locally derived float. Lateral view of a fragment of presumably adult living chamber bearing a characteristic ribbing habit including umbilical bullae and lateral nodes, from which the primaries bifurcate (compare oralmost whorl of the specimen reproduced in A.F. Leanza, 1945, Pl. XXIII, Fig. 1).

Figure 7. *Neocosmoceras* sp. indet. GSC Cat. 16610. GSC loc. 20661. Ventral view of the oralmost whorl fragment of the specimen reproduced in Jeletzky, 1965, Pl. XI, Fig. 4, x 2. Mid-Berriasian (*Buchia uncitoides* s.l. Zone). West coast of Vancouver Island. Southern rocky fringe of Grassy Island. One Tree Formation, about 828 feet above its base and 322 feet above the Jurassic-Cretaceous contact.

PLATE III

B. okensis (Pavlow) were found at one locality (USGS Mesozoic loc. M4066) in the *Buchia uncitoides* Zone".

These relationships indicate that *B. okensis* s.s. is almost replaced by *B. uncitoides* s.l. in the equivalent beds of northwestern California. This southward disappearance of *B. okensis* s.s. between western British Columbia and northwestern California (Fig. 5) is believed to be caused by a gradual southward increase of the water temperature of the Berriasian seas of the Pacific Slope of North America. This temperature differential may have favored smaller but thicker and more elongated forms (e.g., *Buchia* n. sp. aff. *okensis* and *Buchia uncitoides* s.l.) over larger but much broader and flatter forms (e.g., *Buchia okensis* f. typ. and its giant variants). As already pointed out by Jeletzky (1965, p. 25, 26), *Buchia uncitoides* s.l. occurs throughout the *Buchia okensis* Zone in western British Columbia, although it is less numerous than the latter species in most beds of that zone. *B. uncitoides* and *B. okensis* f. typ. intergrade throughout the Zone. Therefore, they appear to be extreme variants of the same polytypic species rather than independent species. The nonpenetration of *B. okensis* f. typ. into northern California indicates that this form and *B. uncitoides* s.l. were geographical subspecies rather than morphological variants during *Buchia okensis* time (*see* Jeletzky, 1965, p. 57-63, Fig. 3).

The Ancestry and Centre of Origin of Buchia okensis. Jeletzky (1965, p. 25, Fig. 3) was undecided whether *Buchia okensis* (Pavlow 1907) arose out of the ancestral *Buchia fischeriana-piochii-terebratuloides* group in the western Cordillera of North America or in some area of northern Eurasia. However, he favored the former region as its evolutionary center. The subsequent discovery of closely similar and ancestral *B.* n. sp. aff. *okensis* of Jeletzky, 1971 (=*B.* aff. *B. okensis* of Jones *et al.*, 1969) in the next older, latest Jurassic beds of the Tyaughton Lake map-area and northwestern California and the presence of an uninterrupted series of transitional forms

EXPLANATION FOR PLATE III

All photographs are natural size.

Figure 1. *Protothurmannia* n. sp. B. GSC Cat. 73685. GSC loc. 56700. Late Early Berriasian (equivalent of Lower Occitanica Zone) upper part of *Buchia okensis* s.s. Zone. Relay Mountain Group, Uppermost Portlandian s.s. to Upper Berriasian shale unit. Taseko Lakes map-area (92-0), southwestern British Columbia. Collected on the western slope of a steep north-south-trending ridge overlooking the headwaters of an unnamed east-west-trending creek at the point about 2.8 km WNW of the top of Relay Mountain and about 1 km WNW of the Summit 8,222′ of the 4 mile topo map. Lat. 51°09′N.; Long. 122°57′W. Collected at the level 19.8-22.8 m below top of the bed 1 of Section 12 (see Jeletzky, in Jeletzky and Tipper, 1969, p. 146). Lateral view of strongly deformed but complete adult representative. The lower part of the mouth border is preserved. The living chamber comprises about one-half whorl. The sculpture is pronouncedly weakened and disorganized as in *Protothurmannia* n. sp. A. but at a considerably larger size.

Figures 2a-d. *Buchia okensis* (Pavlow 1907) s.s. GSC Cat. 73686. GSC loc. 56484 (=94250). Uppermost Tithonian, upper part of *Buchia terebratuloides* s.l. Zone. Taseko Lakes map-area (92-0), southwestern British Columbia. Ridge between heads of the Graveyard and Tyaughton Creeks, outcrop at ridge's crest at is S end, Lat. 51°07′34″N.; Long. 123°04′57″W. Collected by H.W. Tipper, Geol. Survey of Canada, 1964. 2a) lateral view of the exterior of left valve. The fringe parts of the anterior and lower-anterior margins are lost; 2b) lateral view of the right valve; 2c) anterior view of both valves. The left valve was originally thicker than it appears to be now because of the loss of the fringe part of its anterior rim. The right valve is squeezed into the left valve; 2d) beak parts and hinge margins of both valves viewed from above. This is a typical, fairly large representative of *B. okensis* s.s. indistinguishable from specimens occurring in the overlying *Buchia okensis* s.s. Zone.

Figures 3a-d. *Buchia* n. sp. aff. *okensis* (Pavlow 1907). GSC Cat. 73687. The same locality, etc. as for Figure 2. 3a) lateral view of the exterior of the left valve; some of the lower anterior margin is absent. 3b) lateral view of the exterior of the right valve that is somewhat squeezed into the left valve; 3c) anterior view of both valves; 3d) beak parts and hinge margins of both valves (that of the right valve is hidden inside of the left valve) viewed from above. A small-sized adult of *B.* n. sp. aff. *okensis* that combines a characteristically elongated, subtriangular shape, right-handedly incurved left beak and anteriorly convex crest line (all of these features are reminiscent of its *B. fischeriana* var. *trigonoides* ancestry; compare Figures 5-7) with unusually prominent and sparce concentric ribs that are already similar to those of the younger *B. okensis* s.s. This combination of features is characteristic of the small variant of *B.* n. sp. aff. *okensis.*

Figures 4a-d. *Buchia* n. sp. aff. *okensis* (Pavlow 1907). GSC Cat. 73688. The same locality, etc. as for Figure 2. 4a) lateral view of the exterior of the left valve; 4b) lateral view of the exterior of the right valve; 4c) beak parts and hinge margins of both valves viewed from above; 4d) anterior view of both valves. A still smaller, possibly halfgrown representative of the morphological variant of *B.* n. sp. aff. *okensis* reproduced in Figure 3. The coarsest and sparcest ribs always occur in these smaller forms of the species while its larger forms are more finely and closely ribbed.

Figures 5a-c. *Buchia* ex aff. *okensis* (Pavlow 1907). GSC Cat. 73689. The same locality, etc. as for Figure 2. 5a) lateral view of the exterior; 5b) posterior view; 5c) beak part and hinge margin viewed from above. A small left valve that is morphologically transitional between *B. fischeriana* var. *trigonoides* (Lahusen 1888) and small variants of *B.* n. sp. aff. *okensis* (Pavlow 1907) but closer to the latter form.

Figures 6a-b. *Buchia fischeriana* (d'Orbigny) cf. var. *trigonoides* (Lahusen 1888). GSC Cat. 73690. The same locality, etc. as for Figure 2. 6a) lateral view of the exterior; 6b) anterior view. This somewhat defective (no beak part) left valve appears to be indistinguishable from *B. f.* var. *trigonoides* (Lahusen 1888). Such forms are extremely rare in the upper part of *Buchia terebratuloides* s.l. Zone.

Figures 7a, b. *Buchia* aff. *fischeriana* (d'Orbigny 1845) s.l. GSC Cat. 73691. The same locality, etc. as for Figure 2. 7a) lateral view of the exterior; 7b) posterior view. The left valve belongs to a *Buchia* form transitional between *B. fischeriana* (d'Orbigny) var. *trigonoides* (Lahusen 1888) and morphologically primitive form of *B.* n. sp. aff. *okensis* (Pavlow 1907) reproduced in Pl. I, Fig. 7. Such forms are common in the lower part of *Buchia terebratuloides* s.l. Zone but become rare in its upper part where *B.* n. sp. aff. *okensis* is common.

Figures 8a-c. *Buchia terebratuloides* (Lahusen 1888) var. *subinflata* (Pavlow 1907). GSC Cat. 73692. The same locality, etc. as for Figure 2. 8a) lateral view of the exterior; 8b) anterior view; 8c) beak part and hinge margin viewed from above. A large and typical right valve of the variant.

connecting *B.* n. sp. aff. *okensis* with *B. okensis* in the latest Jurassic beds of the Tyaughton Trough (see Jeletzky, in Jeletzky and Tipper, 1968, p. 10, 11, Table 1; this paper Pl. I, Figs. 6, 7; Pl. III, Figs. 2-5) indicate that their transmutation has taken place in the geosynclinal troughs of western British Columbia. The all but complete absence of *Buchia okensis* s.s. in the Berriasian rocks of California, where *Buchia* n. sp. aff. *okensis* was discovered slightly earlier by Jones *et al.*, (1969, p. A12, A13), confirms this conclusion and indicates that the former species was prevented from penetrating into the marginal Tethyan basins of California and ?Oregon by relatively higher water temperatures. This and the common presence of the most advanced evolutionary phase of the *Buchia* n. sp. aff. *okensis* fauna in California (Jones *et al.*, 1969) suggests that this appreciable warming of the Californian basins occurred at the end of the Jurassic.

The evolution of *Buchia* n. sp. aff. *okensis* into *Buchia okensis* s.s. could hardly have taken place in any Boreal basin of western and Arctic Canada that was situated north of the geosynclinal troughs of western British Columbia. As already pointed out by Jeletzky (1965, p. 25) and confirmed by later data discussed below, the typical and giant forms of *Buchia okensis* s.s. appear suddenly in the basal Berriasian beds of these basins. Though *Buchia* n. sp. aff. *okensis* occurs locally in the underlying uppermost beds of the next older *Buchia terebratuloides* s.l. Zone of these basins, (i.e., in its *Praetollia fedorovi* Beds; see Pl. VI, Figs. 5, 6; Pl. VIII, Figs. 3, 8, 9, 12) and in their basal Berriasian beds (i.e., in the *Buchia okensis* s.s. and *Praetollia (Pseudocraspedites)* aff. *anglicus* Zone), it does not appear to be connected there by an uninterrupted series of transitional forms with the *Buchia okensis* s.s.

Age and Correlation. The Canadian *Buchia okensis* s.s. Zone corresponds essentially to the interprovincial *Hectoroceras kochi* Zone of the Boreal facies of northern Siberia and the Central Russian Basin in a restricted sense (i.e., exclusive of *Chetaites sibiricus* Zone). This problem is discussed in the section dealing with the Sverdrup Basin.

The diagnostic early Berriasian ammonites occurring in the Taseko Lakes sections of the *Buchia okensis* s.s. Zone are totally unknown from northwestern California. It is, nevertheless, possible to correlate the lower part of *Buchia uncitoides* s.l. Zone of northwestern California with the Canadian *Buchia okensis* s.s. Zone. First it occupies a stratigraphic position between the latest Tithonian *Buchia terebratuloides* s.l. Zone below and the upper part of the local *Buchia uncitoides* s.l. Zone above, that is directly correlative with the middle part of that zone on Vancouver Island. Second, *B. okensis* s.s. occurs very rarely in the lower part of *Buchia uncitoides* s.l. zone (*see* p. 207).

The *Buchia okensis* s.s. Zone of the Taseko Lakes map-area contains the western South American ammonite genus *Argentiniceras* and, most likely, the index species of the *Argentiniceras noduliferum* and *Neocosmoceras egregium* Zone of that region (A.F. Leanza, 1945, 1947; H.A. Leanza, 1981a, 1981b; H.A. Leanza and Hugo, 1977; Biro-Bagoczky, 1980). This permits a direct correlation of the Boreal *Buchia okensis* s.s. Zone with this Tethyan ammonite zone. This *A. noduliferum* and *N. egregium* Zone is equivalent to the Grandis Zone (or Grandis-Jacobi Zone) of southwestern Europe and placed into the basal Berriasian by all American and European workers, except for Zeiss (1979, 1982) who correlates this zone with the middle part of the Boissieri Zone of the upper Berriasian (Fig. 2). This correlation is untenable in view of the uppermost Tithonian age of the next older, *Substeueroceras koeneni* Zone as demonstrated earlier in this paper. The basal Berriasian age of some part at least of the *A. noduliferum* and *N. egregium* Zone is also indicated by the recorded presence of *Aspidoceras* sp. indet. in its Malargue section (Bodenbender, 1892, p. 39; A. Leanza, 1945, p. 10). According to A.F. Leanza, who has revised this fauna, *Aspidoceras* sp. is associated with *"Perisphinctes" noduliferous* Steuer, *"P." loncochensis* Steuer, *"Holcostephanus" bodenbenderi* Steuer, *"Odontoceras" theodorii* Steuer, *etc.* in unit III of the Malargue section. This unit overlies the unfossiliferous unit II that is, in turn, underlain by the upper Upper Tithonian unit I that contains the *Substeueroceras koeneni* fauna. As pointed out above (p. 178-179, Figs. 5, 10), the genus *Aspidoceras* does not range above the lower part of the Grandis Zone as defined here. Therefore, and because of the presence of *S. koeneni* in unit I, unit III of the Malargue section is coeval with the Grandis Zone of the type Berriasian. However, the association of *Argentiniceras noduliferum* and other characteristic ammonites of that Argentinian zone with *Neocosmoceras egregium* in its other sections (e.g. A.F. Leanza, 1945, p. 52, 89) indicates that it also includes equivalents of the next younger Occitanica Zone of the type Berriasian. It is well established that the genus *Neocosmoceras* makes its first appearance in the *Tirnovella subalpina* Subzone of the Occitanica Zone throughout the Mediterranean Province (Le Hégarat, 1973; Luppov *et al.*, 1975, 1979; Sacharov, 1975, 1976). Although I do not know of any stratigraphic data supporting this interpretation, *Neocosmoceras egregium* is presumed to occur in the upper part of the *A. noduliferum* and *N. egregium* Zone while the above mentioned *Aspidoceras* sp. is assumed to be derived from the basal beds of that zone. The presence of *Tirnovella* cf. *occitanica* Pictet (see A.F. Leanza, 1945, p. 62, 63, Corr. Table, Pl. XI, Figs. 3, 4) in the next younger *Spiticeras damesi* and *Cuyaniceras transgrediens* Zone indicates that the *A. noduliferum* and *N. egregium* Zone includes the lower part of the Occitanica Zone only (Fig. 5).

The presence of *Fauriella* n. sp. aff. *gallica* (Mazenot) in the *Argentiniceras* cf. *noduliferum* fauna of the Taseko Lakes map-area confirms the correlation of this fauna, and hence of the upper part of the *Buchia okensis* s.s. Zone in western British Columbia, with the type Berriasian. According to Le Hégarat (1973, p. 146-149, Table 10), the genus *Fauriella* is restricted to the Berriasian as interpreted here. Furthermore, only *F.* aff. *carpathica* and *F. floquinensis* make their first, rare appearance in the lower part of the Grandis Zone. *F. gallica* is restricted to the Dalmasi Subzone of the Occitanica Zone and the Paramimounum and Picteti Subzones of the Boissieri Zone. Therefore, the presence of *Fauriella* allied to *F. gallica* in the upper part of our *Buchia okensis* s.s. Zone tends to confirm the early Mid-Berriasian age (i.e., the basal part of the Occitanica Zone) inferred on other grounds for the *Argentiniceras* cf. *noduliferum* fauna of western British Columbia and the correlative *A. noduliferum-N. egregium* fauna (its upper part only?) of midwestern Argentina.

Because of the absence of any diagnostic ammonites in the lower and middle parts of the *Buchia okensis* s.s. Zone in the Taseko Lakes map-area, one might claim that these beds are still latest Tithonian.

However, the restriction of *Parodontoceras reedi* and *Substeueroceras*? sp. indet. juv. to the underlying uppermost beds of the *Buchia terebratuloides* s.l. Zone, the uniformity of *Buchia okensis* s.s. fauna throughout its zone and, last but not least, the presence of *Fauriella* n. sp. aff. *gallica* in the upper part of this zone, all militate against this idea. These data suggest instead that the *Argentiniceras* cf. *noduliferum-Fauriella* n. sp. aff. *gallica* fauna represents the upper part of the Argentinian zone only while the underlying part of the *Buchia okensis* s.s. Zone corresponds to its lower part. I correlate the lower and middle part of the *Buchia okensis* s.s. Zone with the Grandis Zone of southwestern Europe (Fig. 5).

The representatives of *Protothurmannia* associated with *Argentiniceras* cf. *noduliferum* are nondiagnostic at present, as the only other representative of *Protothurmania* (i.e., *P. rezanoffiana* Crickmay 1932) was reportedly collected in the uppermost Tithonian of west-central California (Crickmay, 1932, p. 4). This suggests that *Protothurmannia* crosses the Jurassic-Cretaceous boundary as defined in this paper.

In Northern California the *Buchia uncitoides*-bearing equivalents of the *Buchia okensis* s.s. Zone as discussed above, have yielded *Spiticeras (Negreliceras) stoneyense* Imlay and Jones. This is confirmed by Imlay and Jones (1970, p. B12, B13, B35) statement that this ammonite is restricted to the lower two-thirds of the *Buchia uncitoides* s.l. Zone, including USGS locality M4066 where this *Buchia* species is associated with rare *B. okensis* (Pavlow). Since the subgenus *Negreliceras* Djhanelidze 1922 is restricted to the Berriasian, its presence in the equivalents of the *Buchia okensis* s.s. Zone provides further indication of its earliest Cretaceous age.

There is no faunal basis for a direct zonal correlation of the *Buchia okensis* s.s. Zone with the Berriasian ammonite zones of Mexico. I correlate this Canadian zone with the *Subthurmannia tenochi* and *Spiticeras zirkeli* Zone of northeastern Mexico (Fig. 5) because of their similar stratigraphic position.

Buchia uncitoides s.l. Zone

Stratigraphy and Geographical Distribution. In western British Columbia, rocks of this zone occur in the same areas as the underlying *Buchia okensis* s.s. Zone on which the zone rests gradationally.

There has been no additional information about the distribution of the *Buchia uncitoides* s.l. Zone on the west coast of Vancouver Island since Jeletzky (1965), nor was that zone discovered later anywhere else in the Insular Trough. Similarly, no additional information has become available from the Harrison Lake area since Crickmay (1962) and Jeletzky (1965). The more recent discovery (Roddick, 1965, p. 41, 42) of the *Buchia uncitoides* s.l. fauna west of Harrison Lake in a limestone interbed of conglomeratic to arenaceous, slightly to markedly metamorphosed rocks in the Fire Lake Group, is interesting only in delimiting more precisely the western margin of the Tyaughton Trough in that area.

In the mid-basin zone of the Tyaughton Trough in the Taseko Lakes map-area the lower part of this zone comprises the topmost part of the Latest Portlandian s.s. to Late Berriasian shale member of the Relay Mountain Group (Jeletzky, in Jeletzky and Tipper, 1968, p. 21, 22). The *Buchia uncitoides* s.l.-bearing part of that member is 17 to 19 m thick in the two best yet incomplete sections no. 12 and 15 (Jeletzky, in Jeletzky and Tipper, 1968, p. 146, 169) that are believed to be entirely complementary. Section no. 12 exhibits the gradational contact with the overlying upper part of the zone, while the section no. 15 has the gradational contact with the underlying shale of the *Buchia okensis* s.s. Zone. No complete sections of the lower part of *Buchia uncitoides* s.l. Zone were found but the combination of several incomplete sections suggest that it is at least 45 m thick. The shale of this part of the zone is dark grey when fresh, weathers light grey to ash grey or brownish grey, mostly pure, friable, and apparently noncalcareous except near the top. It includes rows of 12 to 60 cm thick concretions and bands of lithologically similar but harder to very hard, commonly calcareous shale, and some similarly thick pods and interbeds of white-weathering coquinoid shale near the top.

The upper part of this mid-basin facies of *Buchia uncitoides* s.l. Zone comprises the lower part of the overlying White-weathering coquinoid member of the Relay Mountain Group (Jeletzky, in Jeletzky and Tipper, 1968, p. 22, 23). This member consists of calcareous light grey siltstone interbedded with impure coquina limestone that are replete in *Buchia uncitoides* s.l. associated with very rare *Acroteuthis*-like belemnites and *Spiticeras?* (s.l.) sp. These white-weathering rocks are characteristically devoid of arenaceous admixture. This upper part of the *Buchia uncitoides* s.l. Zone (and the White-weathering Coquinoid Member in general; see Jeletzky, in Jeletzky and Tipper, 1968, p. 23) was apparently deposited during a period of shallowing of the mid-basinal belt of the Tyaughton Trough, to which it is restricted. The upper part of *Buchia uncitoides* s.l. Zone represented by the White-weathering Coquinoid Member is only about 5 m thick in the section no. 12 where it is completely exposed (Jeletzky, in Jeletzky and Tipper, 1968, p. 145). However, its thickness increases to about 16.5 m in the equally complete section no. 16 (Jeletzky, in Jeletzky and Tipper, 1968 p. 179, 180). These rapid thickness changes of the upper part of the zone suggest a lenticular nature of the member and a minor diachronism of its lower boundary. The contact of *Buchia uncitoides* s.l. and *Buchia tolmatschowi* Zones within the White-weathering Coquinoid Member is gradational both lithologically and faunally.

This mid-basinal facies is flanked at the southwest by its arenaceous facies, which evidently was deposited on the southwestern flank of the trough near its southwestern shoreline. This facies occurs in the upper, but apparently not uppermost, part of the Upper Portlandian s.s. to (?) Lower Valanginian arenaceous unit of the Relay Mountain Group (Jeletzky, in Jeletzky and Tipper, 1968, p. 24, 25) and forms a northwest-trending belt that extends at least from the headwaters of Tyaughton Creek to the north shoulder of Elbow Mountain. All sections are incomplete and disturbed by splays of the Elbow Mountain thrust zone. The *Buchia uncitoides* s.l. Zone is about 19 m thick in the longest and most representative Section no. 3 (Jeletzky, in Jeletzky and Tipper, 1968, p. 104, 105), which may be nearly complete, judging by the morphology of *B. uncitoides* s.l. in its basal and topmost parts. However, both contacts of the zone are faulted in that section.

The arenaceous facies is largely represented by fine grained greywacke in Section no. 4 and other less extensive sections (*e.g.* Sections 4 and 8; Jeletzky, *loc. cit.*, p. 117, 128, 129) but it includes lesser to minor interbeds of sandy siltstone. Coarser arenites and conglomeratic rocks are absent. The greywacke is medium to dull blue or greenish grey when fresh, weathers brown to rust coloured and speckled, hard and dense, massive to irregularly and conchoidally bedded in most sections. However, it may be blackish green to dark green or greenish orange, moderately hard and at least somewhat porous in other sections, which were less affected by shearing and faulting. The siltstone is mostly blue-grey when fresh, weathers dull grey, rusty speckled and tinged, hard and dense (siliceous). However, it too includes dark grey to brown coloured, moderately hard and dense varieties in the sections that were less affected by shearing and faulting. The lower contact of the facies was not observed. Its contact with the overlying Volcanic member appears to be gradational. The mostly rich, shallow marine macrofauna of the arenaceous facies contains numerous, large and heavily built pectinids (*Mclearnia, etc.*) and numerous, sometimes

PLATE IV

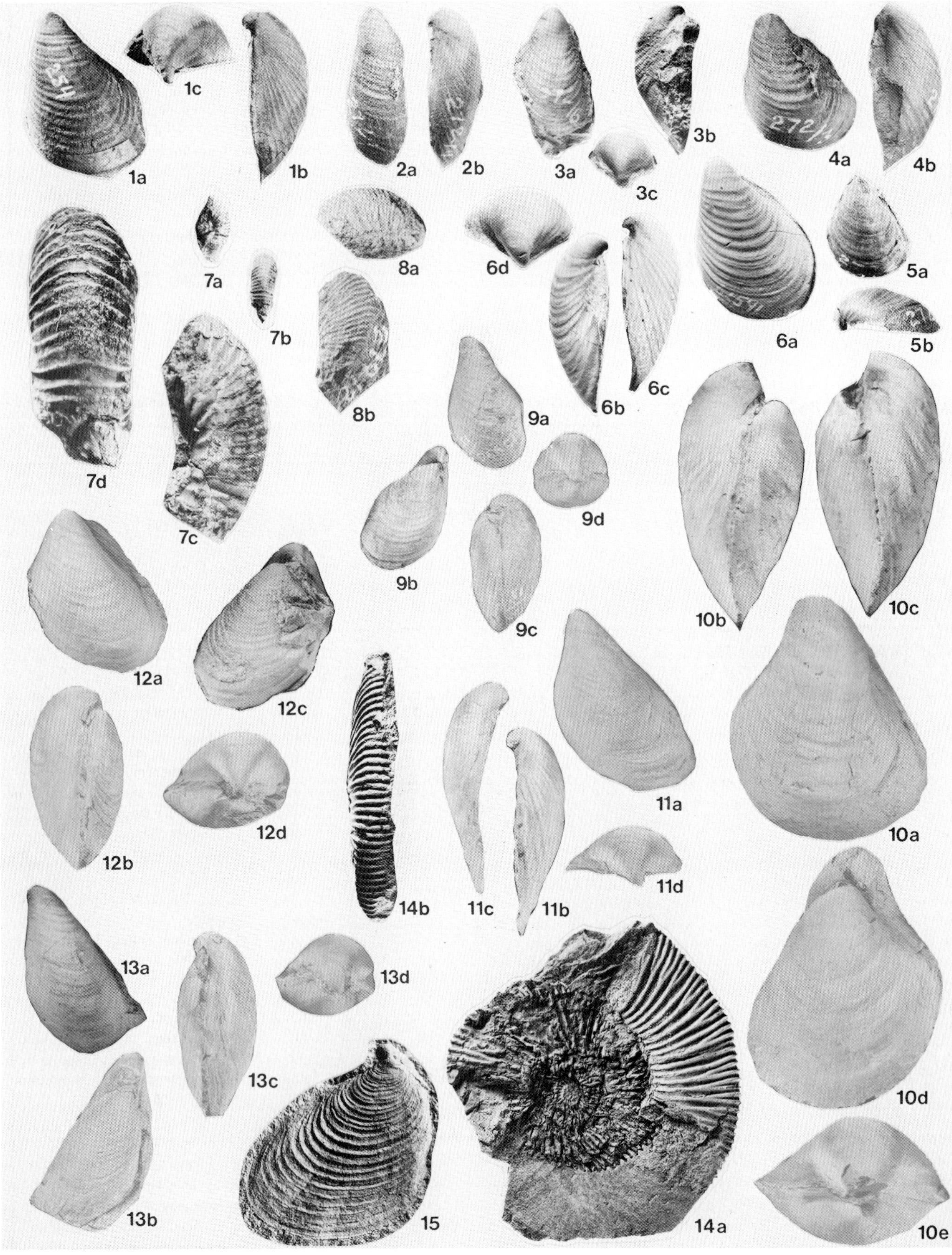
1a
1b
1c
2a
2b
3a
3b
3c
4a
4b
5a
5b
6a
6b
6c
6d
7a
7b
7c
7d
8a
8b
9a
9b
9c
9d
10a
10b
10c
10d
10e
11a
11b
11c
11d
12a
12b
12c
12d
13a
13b
13c
13d
14a
14b
15

EXPLANATION FOR PLATE IV

All photographs are natural size, except where otherwise indicated.

Figures 1a-c. *Buchia piochii* (Gabb 1865) s.s. Univ. of Alberta Cat. No. 7193, loc. No. 254. Reportedly from the lower part of the Lower Bullhead Formation (=?Fernie-Minnes Transition Beds). Martin Creek on Pine River, northeastern British Columbia. Collected by C.R. Stelck. Introduced for comparison with the fragmentary *B. piochii* s.s. shown in Figure 2. 1a) lateral view of the exterior; 1b) anterior view; 1c) beak part and hinge margin viewed from above.

Figures 2a, b. *Buchia piochii* (Gabb 1865) s.s. Univ. of Alberta. Cat. No. 7194, loc. No. 272. Loose, presumably locally derived boulder in creek's bed. Apparently derived from the Fernie-Minnes Transition Beds. Fisher Creek, northeastern British Columbia. Fragmentary left valve similar to that reproduced in Figure 1. 2a) lateral view of the exterior; 2b) anterior view.

Figures 3a-c. *Buchia* aff. *B. lahuseni* (Pavlow 1907) var. *krotovi* (Pavlow 1907). Univ. of Alberta Cat. 7199. The same repository, locality and collector as for Figure 2. 3a) lateral view of the exterior; 3b) anterior view; 3c) beak part and hinge margin viewed from above.

Figures 4a-c. *Buchia* cf. *B. russiensis* (Pavlow 1907) var. *taimyrensis* Zakharov 1981. Univ. of Alberta Cat. 7195. The same repository, locality and collector as for Figure 2. 4a) lateral view of the exterior; 4b) posterior view; 4c) beak part and hinge margin viewed from above.

Figures 5a, b. *Buchia* aff. *B. terebratuloides* (Lahusen 1888) s.l. Univ. of Alberta Cat. No. 7196. The same repository, locality and collector as for Figure 2. 5a) lateral view of the exterior; 5b) anterior view.

Figures 6a-d. *Buchia* aff. *B. piochii* (Gabb 1865) s.s. GSC Cat. 73693. GSC loc. 62591. Late Late Tithonian, basal part of *Buchia terebratuloides* s.l. Zone. Relay Mountain Group, Uppermost Portlandian s.s. to Late Berriasian shale unit. Taseko Lakes map-area, southwestern British Columbia. Southeastern slope of summit 8,222′ of the 4 mile topo map, about 1,2 km toward 110° from its top and at the altitude of about 6,700 feet; Lat. 51°09′N., Long. 122°57′W. From the basal 3.8 m of the unit 1 of the Section 15 (see Jeletzky, in Jeletzky and Tipper, 1968, p. 170, 171 for further details). The typical shape and proportions of *B. piochii* s.s. is combined with a considerably more prominent concentric ribbing habit resembling that of *B. uncitoides* s.l. Illustrates the morphology of the youngest known representatives of *B.* ex gr. *B. piochii* (Gabb 1865). 6a) lateral view of the exterior; 6b) posterior view; 6c) anterior view; 6d) beak part and hinge margin viewed from above.

Figures 7a-d. *Proniceras* sp. indet. juv. Univ. of Alberta Cat. 7197. The same repository, locality and collector as for Figure 2. 7a) lateral view of the fragment, x 1; 7b) ventral view, x 1; 7c) same view as in 7a but x 4 to show details of characteristic ornament; 7d) same view as in 7b but x 4 to show the characteristic ornament, including a pronounced constriction.

Figures 8a, b. *Notostephanus*? sp. indet. Univ. of Alberta Cat. 7198. The same repository, locality and collector as for Figure 2. 8a) lateral view, x 2. Note the considerable height of the whorl combined with characteristic bundling of secondary ribs near the umbilical shoulder; 8b) somewhat oblique ventral view, x 2. Note distinct forward bends of very fine and closely spaced secondary ribs.

Figures 9a-d. *Buchia* ex gr. *lahuseni* (Pavlow 1907). GSC Cat. 73694. GSC loc. No. 15995. Late Late Tihonian, either from the upper part of *Buchia fischeriana* Zone or from the lower part of *Buchia terebratuloides* s.l. Zone. Relay Mountain Group, presumably from the Uppermost Portlandian s.s. to Late Berriasian shale unit. Pemberton map-area, southwestern British Columbia, collected high on the eastern canyon wall of Yalakom River (i.e., on the western slope of Camelsfoot Range) about 9.6 km up from its mouth according to the late Dr. H.M.A. Rice (pers. commun. 1951). This specimen combines the thickness and short left beak of *B. lahuseni* var. *krotovi* (Pavlow 1907) with the strongly attenuated shape and a left-handed twist of its left beak. Elucidates the morphology of the still insufficiently understood *Buchia* ex gr. *lahuseni* (Pavlow 1907) that are commonly prevalent in the *Buchia fischeriana* Zone of western British Columbia. 9a) lateral view of the exterior of the left valve; 9b) the same view of the right valve; 9c) anterior view of both valves; 9d) beak parts and hinge margins of both valves viewed from above.

Figures 10a-e. *Buchia terebratuloides* (Lahusen 1888) var. *occidentalis* (Anderson 1945). GSC Cat. 73695. Locality and collector as for Figure 9. 10a) lateral view of the exterior of the left valve; 10b) posterior view of both valves; 10c) anterior view of both valves; 10d) lateral view of the exterior of the right valve; 10e) beak parts and hinge margins of both valves viewed from above. Much of the left beak is broken off.

Figures 11a-d. *Buchia* aff. *B. piochii (Gabb 1865) s.s.* GSC Cat. 73696. Locality and collector as for Figure 9. 11a) lateral view of the exterior; 11b) anterior view; 11c) posterior view; 11d) beak part and hinge margin. This specimen presents another combination of features characteristic of the still insufficiently understood *Buchia* ex gr. *piochii-lahuseni* (Pavlow 1907) from the *Buchia fischeriana* Zone of western British Columbia. It combines the thinnes and slenderness of *B. piochii* s.s. with a strongly protruding anterior margin and a markedly leftward twisted left beak (Fig. 11d). It is similar to *B. hyatti* (Pavlow 1907).

Figures 12a-d. *Buchia lahuseni* (Pavlow 1907) var. *krotovi* (Pavlow 1907) GSC No. 73697. Locality and collector as for Figure 9. 12a) lateral view of the exterior of the left valve; 12b) posterior view of both valves; 12c) lateral view of the exterior of the right valve; 12d) beak parts and hinge margins of both valves. This specimen is more similar to specimens assigned to *B. krotovi* by Anderson (1945, Pl. 6, Fig. 7) than to its type specimens (Pavlow, 1907, Pl. II, Figs. 31-33).

Figures 13a-d. *Buchia* ex gr. *lahuseni* (Pavlow 1907). GSC Cat. 73698. Locality and collector as for Figure 9. 13a) lateral view of the exterior of the left valve; 13b) lateral view of the exterior of the right valve that is partly squeezed into the left valve; 13c) anterior view. The shell appears to be thinner than it actually is because the right valve is partly squeezed into the left valve. Undeformed shell would be as thick as that reproduced in Figure 9c, which it matches closely in most other respects; 13d) beak parts and hinge margins of both valves. Here again the thickness of the undeformed shell would match closely that of the specimen shown in Figure 9d. Such forms were figured as *B. tenuicollis* by Anderson (1945, Pl. 13, Fig. 8).

Figures 14a, b. ?*Pectinatites (Paraberriasella)* ex gr. *blondeti* (Donze 1948). Hypotype. GCS Cat. 73722. GSC loc. 68706. Almost complete but mostly flattened (except for the undeformed living chamber) specimen. Presumably derived from mid- to upper Lower Tithonian (?*Buchia* cf. *blanfordiana* Zone) rocks. Provincial Manning Park, B.C., Lat. 49°01′30″N.; Long. 120°57′30″W. North side of Thunder Lake. Float specimen from talus below high cliffs of Upper Jurassic and Lower Cretaceous rocks towering above the lake. 14a) lateral view; 14b) ventral view of undeformed oralmost part of the specimen (living chamber). Introduced for comparison with the apparently conspecific ammonite from *Buchia* cf. *blanfordiana* Zone figured by Jeletzky (1965, Pl. II, Fig. 5).

Figure 15. *Buchia fischeriana* (d'Orbigny 1945) var. *trigonoides* (Lahusen 1888). GSC Cat. 73723. GSC loc. 68706. Lateral view of the exterior of right valve. Photograph of rubber mold of complete right valve and beak of left valve preserved as an external cast. This float specimen was presumably derived from the mid-Upper Tithonian (*Buchia fischeriana* Zone). The same locality as for the ammonite reproduced in Figures 14a and 14b.

waterworn belemnites, in addition to *Buchia uncitoides* s.l. No ammonites were seen.

The former tentative assignment of the arenaceous rocks outcropping on the northeastern flank of the Tyaughton Trough east of Yalakom River to the "Late Berriasian(?) to (?)mid Valanginian arenaceous unit" of the Relay Mountain Group (Jeletzky, in Jeletzky and Tipper, 1968, p. 25) is withdrawn herewith. A subsequent more detailed study of buchias occurring at GSC loc. 15995, collected by the late Dr. H.M.A. Rice has shown that they belong either to an early phase of the *Buchia terebratuloides* s.l. Zone or to a late, *Buchia* ex gr. *lahuseni*-dominated phase of the *Buchia fischeriana* Zone. The most important elements of this poorly understood fauna include *Buchia* aff. *piochii* s.s., *B.* ex gr. *lahuseni* and *B. terebratuloides* var. *occidentalis* (Pl. IV, Figs. 9-13).

Considering the regressive character of the *Buchia uncitoides* s.l. Zone in the mid-basin and southwestern marginal zones of the trough, its northeastern marginal zone may have been situated west of the Yalakom fault zone.

In the northeastern part of Mount Waddington map-area situated on the northwestern flank of Tyaughton Trough the *Buchia uncitoides* s.l. Zone is represented by part of or (?)all of an about 208 m thick unit of almost unfossiliferous, presumably largely nonmarine rocks (Jeletzky, 1968b; and Jeletzky in Tipper, 1969, p. 38, 39). This unit consists of fine to coarse grained arkose and greywacke with considerable interbeds and pods of pebble conglomerate and grit and minor interbeds of carbonaceous siltstone with poor plant remains. The unit is overlain, apparently conformably, by a thick unit consisting of sandstone and coquinoid limestone containing a typical *Buchia tolmatschowi* (Sokolov) fauna in the basal part. The basal bed of the lower unit is a coquina limestone replete in *Buchia uncitoides* s.l. As it is underlain by a greywacke bed containing a mixed fauna of *B. uncitoides* s.l. and small, presumably late forms of *B. okensis* s.s., the overlying nonfossiliferous beds 22 to 36 of this section (Jeletzky, in Tipper, 1969, p. 38, 39) must correspond, partly to (?)entirely, to the bulk of marine facies of the *Buchia uncitoides* s.l. Zone in the Taseko Lakes map-area.

Paleontology and Biochronology. Concept of Species. Buchia uncitoides s.l. in the sense of Jeletzky (1965, p. 28-33) and Jones *et al.* (1969, p. A12, A13, Table 3) corresponds to *B. jasikovi* and *B. uncitoides* of Zakharov (1981) combined. *B. jasikovi* (Pavlow), as used by Zakharov (1981, p. 121-125, 227, 228; Pl. XXXVI, Figs. 1-5; Pl. XXXVII, Figs. 1-4; Text – fig. 21), represents the more slender, feebly sculptured to almost smooth variants of *B. uncitoides sensu* Jeletzky 1965, which were named *B. acutistriata* and *B. catamorpha* by Crickmay (1930).

Zakharov (1981) did not figure any *Buchia uncitoides* but his comments about its morphology indicate that he placed there the more strongly and sparsely ribbed,larger forms locally associated with *B. jasikovi sensu* Zakharov 1981 in Northern Eurasia. Therefore, *B. uncitoides sensu* Zakharov 1981 corresponds to those extreme Canadian variants of *B. uncitoides* s.l. named *B. spasskensoides* by Crickmay (1930). These variants mostly correspond to *B. uncitoides* s.s. Pavlow (1907) and subsequent Russian and Soviet workers that preceded Zakharov (1981). However, as demonstrated by Jeletzky (1965), they are invariably associated with *B. jasikovi sensu* Zakharov 1981 in western British Columbia and connected by a host of intermediate forms. They are but extreme forms of the same polytypic species *Buchia uncitoides* (Pavlow 1907) s.l. in that region. However, the range of morphological variation of *B. uncitoides* s.l. in Oregon and Northern California differs from that in western British Columbia. As demonstrated by Jones *et al.* (1969, Pl. 3, Figs. 1-19) and confirmed by my field research, the North Californian populations of *B. uncitoides* are almost devoid of those larger, more coarsely and distantly ribbed forms that correspond to *B. uncitoides* s.s. and *B. u.* var. *spasskensoides* Crickmay 1930. Furthermore, the same appears to be true of the North Siberian sections studied by Zakharov (1981, p. 124). In the case of the Oregonian and North Californian populations, this scarcity or absence of its larger, more coarsely and distantly ribbed variants parallels the disappearance of *B. okensis* in Early Berriasian between western British Columbia and northern California as discussed previously (p. 207-208). As with *B. okensis*, the cause of this southward disappearance of *B. uncitoides* s.s. (or *B. uncitoides* var. *spasskensoides*) is believed to be climatical. Apparently it reflects a better adaptation of smaller, more densely ribbed, relatively thicker forms (i.e., *B. jasikovi*) to higher water temperatures. Conversely, larger, more coasely ribbed, relatively flatter *Buchia* forms (e.g., *B. okensis* s.s. and *B. uncitoides* s.s.) were better adapted to lower temperatures. There are not enough data to judge whether or not the geographical restriction of *B. jasikovi* and *B. uncitoides* s.s. in northern Eurasia exhibits a similar, climate-related pattern, though they appear to have different geographical ranges there and do not always coexist (Zakharov, 1981, p. 192, 193). At any rate, it seems best to consider them as true geographical subspecies, rather than mere varieties of the polytypic species *Buchia uncitoides* s.l. The name *B. uncitoides jasikovi* (Pavlow 1907), a senior synonym of *B. acutistriata* (Crickmay 1930) and *B. catamorpha* (Crickmay 1930), therefore, will be used for the smaller, finer ribbed forms. Its larger, coarsely ribbed, flatter forms shall be named *B. uncitoides* f. typ., while the extremely coarsely and sparsely ribbed forms merging into *B. okensis* s.s. shall be named *B. u. spasskensoides*. I do not see any reason to treat the first two forms as independent species, considering their apparently identical time ranges and complete intergradation in those populations where they coexist (see Jeletzky, 1965, p. 27-35).

Other Diagnostic Faunal Elements in Western British Columbia. Relatively to very rare representatives of *B. okensis* s.s. range up at least into the lower part of *Buchia uncitoides* s.l. Zone in western British Columbia (Jeletzky, 1965, p. 34). It is for that reason that Crickmay (1930) has amalgamated these two species in his *"Aucella canadiana"* Zone. As will be shown below, *B. okensis* s.s. also ranges into the equivalents of *B. uncitoides* s.l. Zone (*i.e. Buchia jasikovi* Zone of Zakharov, 1981, p. 227, 228, Table 3) in northern Eurasia. Otherwise, only rare, early representatives of *B. keyserlingi* var. *visiginensis* (Sokolov) and var. *sibirica* (Sokolov) were encountered in the topmost part of *B. uncitoides* s.l. Zone on the west coast of Vancouver Island (Jeletzky, 1965, Pl. X, Fig. 1; Pl. XI, Figs. 1, 5; Pl. XII, Figs. 1, 2). To my best knowledge *B. keyserlingi* was never authentically recorded elsewhere in similarly old rocks and no explanation of this anomaly can be offered.

No buchias either referable to, or comparable with, *Buchia volgensis* (Lahusen) have been seen from any sections of the *B. uncitoides* s.l. Zone studied on the Pacific Slope of North America. *B. uncitoides* s.l. occurs already in the *B. okensis* s.s. Zone in British Columbia (Jeletzky, 1965, p. 28-32, 60-62, Fig. 3) where it freely intergrades with the latter species.

It also intergrades freely with *B. tolmatschowi* in the boundary beds separating their zones and transmutates into the latter species. In the Taseko Lakes map-area the *B. uncitoides* s.l. Zone has yielded only *Spiticeras*? (s.l.) sp. (Jeletzky, in Jeletzky and Tipper, 1968, p. 146). To the best of my knowledge no ammonites have been found in this zone in the Harrison Lake area. The diagnostic, exclusively Tethyan ammonites found in the upper, but not the uppermost, part of the *Buchia uncitoides* s.l. Zone on the west coast of Vancouver Island (Jeletzky, 1965, p. 33) were originally identified as *Spiticeras (Spiticeras)* cf. *scriptus* (Strachey), *S. (S.)* cf. *mojsvari* Uhlig, *S. (S.)* sp. indet. juv., ?*S. (Groebericeras)* sp. indet., *Berriasella (Protacanthodiscus)* sp. nov. aff. *micheicus* (Bogoslovsky) and *Neocomites* (s.l.) sp. nov. indet. The correctness of the first two identifications, and that of the identification of ?*S. (Groebericeras)* sp. indet. are maintained herein. The *Neocomites* (s.l.) sp. nov. indet. is still believed to be more similar to that genus than to any other. However, it could belong to a new neocomitid genus. The forms assigned to *S. (S.)* sp. indet. juv. (Jeletzky, 1965, Pl. XI, Figs. 2, 3) are now assigned to the subgenus *Negreliceras* Djanelidze 1922. Furthermore, they appear to be conspecific with *Spiticeras (Negreliceras) stoneyense* Imlay and Jones 1970 from the *Buchia uncitoides* Zone of northwestern California. These Vancouver Island specimens are particularly similar to the more coarsely and sparcely ribbed variant of *S. (N.) stoneyense* (see Imlay and Jones, 1970, Pl. 4, Figs. 7, 13, 16 and 22).

The small fragmentary specimen previously assigned to *Berriasella (Protacanthodiscus)* sp. nov. aff. *micheicus* (Bogoslovsky) (see Jeletzky, 1965, Pl. XI, Fig. 4) is now assigned to *Neocosmoceras* Blanchet 1922. The previously unfigured venter of its last whorl is reproduced on Plate II, Figure 7. It is smooth and bears prominent, widely spaced ventro-lateral tubercles. These features and the sturdy, only slightly higher than wide cross-section of the whorl ally this Vancouver Island specimen with *Neocosmoceras euchrense* Imlay and Jones (1970, Pl. 14, Figs. 24-26) and *N.? crossi* (Anderson) (see Imlay and Jones, 1970, Pl. 14, Figs. 20-22). This specimen differs much more from the European *Neocosmoceras* species that are characterized by a markedly more compressed, rectangular instead of subtrapezoidal cross-section (see Le Hégarat, 1973, Pl. 44, Figs. 1, 2; Pl. 45, Figs. 1-2, 4-6; Pl. 20, Figs. 2, 4, 5).

Age and Correlation. Zakharov's (1981, p. 125, 227, 288, Table 3) re-appraisal of the time ranges of Berriasian *Buchia* forms of northern Siberia and other regions of northern U.S.S.R., invalidates the older range data of *B. uncitoides* s.l. (i.e., *B. jasikovi* and *B. uncitoides* s.s. of Zakharov, 1981, p. 227 combined) that appeared to contradict its exclusively later Berriasian age on the west coast of Vancouver Island, in the Taseko Lakes map-area, and in northwestern California (Jeletzky, 1965, p. 34, 35 and in Jeletzky and Tipper, 1968, p. 9, 10, Table I; Jones *et al.,* 1969, p. A12, A13; Imlay and Jones, 1970, p. B13, Fig. 2). According to Zakharov (1981), *B. uncitoides* s.l. is restricted to the middle part of the Berriasian stage throughout the northern part of U.S.S.R. There its zone is confined between the *B. okensis* s.s. Zone below and *B. tolmatschowi* Zone above, just as on the west coast of Vancouver Island (but not northwestern California). In terms of the Boreal ammonite zones of northern Siberia, the *B. uncitoides* s.l. Zone corresponds to the topmost part of the *Hectoroceras kochi* Zone and the lowermost part of the *Surites analogus* Zone. This confirms Jeletzky's (1965) conclusions, except for his correlation of the next younger *Buchia tolmatschowi* Zone with the basal Valanginian which is amended here.

In northwestern California (Jones *et al.,* 1969, p. A12, A13; Imlay and Jones, 1970, p. B12, B13, Fig. 2), the *B. uncitoides* s.l. Zone extends deeper than it does in either western British Columbia or northern Siberia. However, this does not reflect any differences in the time ranges of *B. uncitoides* s.l. in western British Columbia and northern California, but rather the almost total absence of *B. okensis* s.s. in the latter region.

The presence of *Neocosmoceras* sp. in the upper, but not the uppermost part of *B. uncitoides* Zone on Vancouver Island indicates that this part, beginning about 51 m above its base, is not older than the basal part of the Occitanica Zone. As the upper part of *B. okensis* s.s. Zone corresponds to the basal part of the Occitanica Zone (p. 210), this occurrence of *Neocosmoceras* sp. suggests a correlation of the upper part of *B. uncitoides* s.l. Zone with the upper part of this type Berriasian Zone (Fig. 5).

The presence of *Spiticeras* (*Negreliceras*) *stoneyense* (Imlay and Jones) at a slightly lower level than the *Neocosmoceras* sp. indicates only a general Berriasian age of that part of the *B. uncitoides* s.l. Zone. The subgenus *Negreliceras* apparently ranges right through the Berriasian (Imlay and Jones, 1970, p. B12, B13) and the species *S. (N.) stoneyense* is known to range down into the equivalents of the *B. okensis* s.s. Zone in northwestern California.

Enay (1973, p. 305) has suggested that the Vancouver Island *Spiticeras* spp. figured by Jeletzky (1965a, Pl. XI, Figs. 2A-2C, 3; Pl. XII, Figs. 4A-4C, 5) from the upper part of the *B. uncitoides* s.l. Zone resemble more closely the Upper Tithonian *Proniceras* of Mexican affinities. This generic assignment and the inferred Late Tithonian dating of the *B. uncitoides* s.l. Zone are erroneous; its Mid-Berriasian age is well established.

The presence of *Neocosmoceras* sp. high in the *Buchia uncitoides* s.l. Zone and the related stratigraphical evidence mentioned previously indicate the approximate correlation of this Canadian zone with the lower part of the *Spiticeras damesi* and *Cuyaniceras transgrediens* Zone of Argentina. This correlation is also supported by the occurrence of *Tirnovella* cf. *occitanica* in the Argentinian zone (A.F. Leanza, 1945, p. 62, 63, Corr. Table, Pl. XI, Figs. 3, 4). The association of *Neocosmoceras euchrense* Imlay and Jones (1970, p. B12, B13, B52, Fig. 2) with abundant *B. uncitoides* s.l. in the Riddle Formation of Oregon suggests that these particular beds are correlative with the upper part of the *B. uncitoides* s.l. Zone on Vancouver Island. This rather obvious correspondence of the middle part of the Californian *B. uncitoides* s.l. Zone to the bulk of the Canadian *B. uncitoides* s.l. Zone does not indicate even an approximate synchrony of their upper boundaries. Because of strong stratal and paleontological reduction of the *Buchia tolmatschowi* Zone in northwestern California, it is likely that *B. uncitoides* s.l. ranges higher there than in western British Columbia.

As with the *B. okensis* s.s. Zone, there is no faunal basis for a direct correlation of the Canadian *B. uncitoides* s.l. Zone with the Tethyan ammonite faunas of the Mexican Berriasian. However, its approximate Mid-Berriasian age suggests its approximate correlation with the lower part of the Mexican *Spiticeras uhligi* and *Subthurmannia densistriata* Zone (Fig. 5). The Californian *B. uncitoides* s.l. Zone should, in contrast, correspond to the bulk of the Mexican Berriasian, with the exception of its topmost beds.

Buchia tolmatschowi Zone

Stratigraphy and Geographical Distribution. No new information about the *B. tolmatschowi* Zone on the west coast of Vancouver Island has become available since Jeletzky (1965), nor were its rocks discovered anywhere else in the Insular Trough. Similarly, no new information about the unfossiliferous, largely volcanic facies of this zone outcropping on the southwestern flank of Tyaughton Trough in the Harrison Lake area has become available since Crickmay (1962) and Jeletzky (1965, p. 64, 65, Fig. 4). In the mid-basin zone of

Tyaughton Trough in the Taseko Lakes map-area, this zone comprises the upper, but not the uppermost, part of the White-weathering Coquinoid Member of the Relay Mountain Group (Jeletzky, in Jeletzky and Tipper, 1968, p. 23, 179, 180). As exemplified by its complete Section no. 16, this mid-basin facies is about 14.5 m thick and represented by exactly the same rocks as the underlying upper part of *B. uncitoides* s.l. Zone (p. 211, 214). Both contacts of the zone are gradational in all sections studied.

In the extreme western part of Taseko Lakes map-area in the vicinity of Elbow Mountain, the *B. tolmatschowi* Zone appears to be represented by the nonfossiliferous, probably nonmarine Volcanic member that comprises the topmost part of the Latest Portlandian s.s. to ?Early Valanginian arenaceous unit of the Relay Mountain Group. This southwestern marginal facies is typified by units 10 and 11 of Section no. 8 (Jeletzky, in Jeletzky and Tipper, 1968, p. 24, 168). The Volcanic member begins with a more than 1.3 m thick bed of pebble conglomerate that overlies gradationally fossiliferous greywacke and siltstone of the upper part of *B. uncitoides* s.l. Zone but its bulk consists of more than 25 m of green, porphyritic flows of (?)basic lava interbedded with similarly coloured volcanic tuff and tuffaceous greywacke. The contact with the underlying conglomerate appears to be conformable. The top of the member was not seen and its stratigraphic relationship with the late Valanginian-Hauterivian marine rocks outcropping nearby are obscure. However, it appears to be the northeastern pinchout wedge of the lower part of Brockenback Hill Formation of the Harrison Lake area (Crickmay, 1930b, 1962) that consists of about 760 m of unfossiliferous pyroclastics and arkoses. This part of the formation is unfossiliferous but is inferred to include ?nonmarine equivalents of *B. tolmatschowi* Zone because of the stratigraphic position between the fossiliferous rocks of *B. uncitoides* s.l. and *B. crasssicollis* s.s. Zones (Jeletzky, 1965, p. 64, Fig. 4).

Only some 15 m of the uppermost part of *B. tolmatschowi* Zone outcrops on the eastern slopes of Mount Sheba, some 20 km southwest of the Section no. 8. The beds are commonly pebbly and gritty arenaceous coquina limestone replete in a mixed fauna of *B. tolmatschowi* and *B. pacifica* (Jeletzky, 1967, p. 66). These littoral rocks appear to form part of the same southwestern marginal zone of the Tyaughton Trough as the Volcanic member. They grade upward into the conglomeratic rocks of the *B. pacifica* Zone, but their base is not exposed. The presumably older Volcanic member may or may not be concealed beneath them.

The approximately 208 m thick unit of arkoses and pebble conglomerates outcropping in the northeastern part of Mount Waddington map-area (Jeletzky, in Tipper, 1969, p. 38-40, Beds 22 to 36 inclusive) described earlier may possibly include beach to nonmarine (?alluvial) equivalents of the lower part of *B. tolmatschowi* Zone in its upper part. However, the *B. tolmatschowi*-bearing basal 13 m of the overlying unit 37 could conceivably represent all of that zone in a littoral facies. In any case, these northwesternmost exposures known in the Tyaughton Trough appear to represent an extension of its above described western marginal facies outcropping in the Taseko Lakes map-area. No deposits of *B. tolmatschowi* Zone are known on the eastern flank of Tyaughton Trough.

Paleontology and Biochronology. Concept of Species. All modern workers dealing with *Buchia tolmatschowi* (Sokolov 1908) (e.g., Jones *et al.*, 1969; Konovalov, 1977; Zakharov, 1979, 1981) interpreted it in the same sense as Jeletzky (1965, p. 35-43; Pl. XIII, Figs. 1, 3, 4; Pl. XIV, Fig. 1; Pl. XV, Figs. 4, 7, 10, 11).

Geographical Range and Phylogeny. Buchia tolmatschowi is a member of the so called morphologically progressive *Buchia* lineage that consisted of *B. uncitoides* s.l., *B. tolmatschowi* and *B. pacifica* and was almost restricted to the marginal Tethyan North Pacific Province (Jeletzky, 1965, p. 61-63, Fig. 4). This lineage was inferred to have evolved independently from the morphologically conservative lineage (e.g., *B. volgensis* s.s., *B. keyserlingi*, etc.) that was largely restricted to the Boreal Realm in North America and Northern Eurasia. This conclusion was criticized by Zakharov (1981, p. 11, 190-193) because of the presence of *B. tolmatschowi* in the northern part of Central Siberia and in Northern Urals. However, these scattered occurrences evidently represent but localized and temporary invasions into regions situated beyond its normal geographical range. As pointed out by Zakharov (1981, p. 131-134, 228), *B. tolmatschowi* only occurs in these regions as a minor faunal admixture in *Buchia* populations dominated by the endemic *B. volgensis* s.s. This contrasts with western British Columbia where *B. tolmatschowi* occurs in masses and forms monospecific associations throughout its zone. Furthermore, it intergrades with *B. uncitoides* s.l. below and *B. pacifica* above (Jeletzky, 1965, p. 37), which also occur in masses and either form monospecific populations or dominate the *Buchia* populations throughout their respective zones. South of this evolutionary centre, *B. tolmatschowi* was temporarily invading northwestern California as is evidenced by a strong attenuation of beds that contain it and the rarity of the typical representatives there (Jones *et al.*, 1969, p. A13). North of western British Columbia *B. tolmatschowi* was recently found to occur locally and rarely in the Sverdrup Basin (e.g., at GSC loc. C-22245 on Amund Ringnes Island) in beds of the Deer Bay Formation that are characterized by *Buchia* ex gr. *keyserlingi*. Like the North Siberian and North Uralian occurrences, these rare occurrences indicate temporary invasions of *B. tolmatschowi* into regions situated beyond its normal habitat; they do not discredit Jeletzky's (1965) conclusions about its phylogenetic relationships, evolutionary centre, and normal geographical range.

Age and Correlation. Unlike the next-older, *Buchia uncitoides* s.l. Zone, the ammonite fauna of the *B. tolmatschowi* Zone is almost entirely Boreal in character. So far, this fauna is found only on Vancouver Island where it is restricted to the uppermost beds of its zone and the overlap-beds with the *B. pacifica* Zone (Jeletzky, 1965, p. 38-41). The fauna is dominated by *Tollia*- and *Surites*-like craspeditids, that were assigned previously to *T. (T.) mutabilis* (Stanton) s.l. (Jeletzky, 1965, Pl. XV, Figs. 2, 6 8, 9) and *T. (T.) paucicostata* (Donovan) (Jeletzky, 1965, Pl. XII, Figs. 2, 5-7, 9; Pl. XIV, Figs. 4-6). However, Imlay and Jones (1970, p. B15, B16) have concluded correctly that "*T. (T.) mutabilis*" should have been left specifically unassigned while "*T. (T.) paucicostata*" is not conspecific with this East Greenland species. They all appear to be a new endemic species closely allied to those early *Tollia* currently separated as *Bojarkia* Shulgina 1969 by Saks and Shulgina (1969) and other Soviet workers, a genus synonymized with *Tollia* by Jeletzky (1979, p. 46, 48). Saks and Shulgina (in Saks *et al.*, 1972, p. 114) reached essentially the same conclusions as Imlay and Jones (1970) and re-interpreted the bulk of these Canadian forms as *Tollia* sp. However, I take strong exception to their assignment of some *Tollia* specimens (Pl. XIII, Fig. 9; Pl. XIV, Fig. 4) to "*Subcraspedites* (?*Borealites*) sp.". These larger specimens represent the adults of the same early *Tollia* (i.e. "*Bojarkia*") form as the smaller specimens associated with them.

Saks and Shulgina (in Saks *et al.*, 1972, p. 114) have re-identified the ammonite identified as "*Tollia* (*Tollia*?) aff. *simplex* (Bogoslovsky 1902)" by Jeletzky (1965, Pl. XIII, Fig. 8) as *Surites* (*Surites*) cf. *tzikwinianus* (Bogoslovsky 1902). This is feasible if one interprets the holotype of *Olcostephanus simplex* Bogoslovsky 1902 as a representative of the Early Valanginian genus *Temnoptychites* Pavlow 1914. However, Jeletzky, (1979, p. 28) follows Sazonova (1971, p. 41-43; Pl. II, Fig. 4; Pl. V, Fig. 3; Pl. X, Fig. 3; Pl. XIX, Fig. 3; Pl. XX, Fig. 3; Pl. XXIV, Fig. 7) in interpreting this holotype as a juvenile representative of a Riasanian *Surites* s.s. that is closely allied to *Surites tzikwinianus*. Therefore, this Vancouver Island ammonite may represent equally well a *Surites* form closely allied to *Surites (Surites) simplex* (Bogoslovsky) *sensu* Sazonova (1971). This close affinity and possible specific identity of the Canadian ammontie with the Late Riasanian (i.e., Zone of *Surites tzikwinianus;* Fig. 5) *Surites* species indicates that, regardless of its exact specific character, Saks and Shulgina (in Saks *et al.*, 1972, p. 114) are right in assigning Late Berriasian age to it, and hence to the entire Canadian *Buchia tolmatschowi* Zone. This conclusion was subsequently confirmed by Zakharov (1979, p. 127, Table I; 1981, p. 185, 228, 229, Table 3) who clarified the time range of *Buchia tolmatschowi* in Northern Eurasia. Contrary to the previous literature data reviewed by Jeletzky (1965, p. 41-43), Zakharov has found that in its Northern Siberian key sections *B. tolmatschowi* appears somewhat above the base of the *Surites analogus* Zone and disappears somewhat below the top of the "*Bojarkia*" *mesezhnikovi* Zone. This time range is entirely within the Boreal facies of the Berriasian as interpreted by all Soviet workers (e.g., Saks *et al.*, 1972; Saks and Shulgina, 1974; Saks, Mesezhinikow and Shulgina, 1979, p. 99, 100, Table 2). However, it includes at least the lower part of the North Siberian "*Bojarkia*" *mesezhnikovi* Zone that Jeletzky (1979, p. 56, fig. 8) has treated as equivalent of the Central Russian *Pseudogarnieria undulatoplicatilis* Zone and hence as the basal Valanginian unit. The latter interpretation would permit maintenance of the previously favored (Jeletzky, 1965, p. 41-43, Figs. 1-2) basal Valanginian dating of the *B. tolmatschowi* Zone, but I now question its validity. The close affinity and a possible specific identity of "*Tollia* (*Tollia*?) aff. *simplex*" with *Surites* (*Surites*) *tzikwinianus*, that was found to be diagnostic of the second youngest zone of the type Riasanian (Gerassimov, 1962, 1977; Mesezhnikov *et al.*, 1979, p. 78, 79), and the apparently erosionally disconformable superposition of the lower, but not basal (i.e., *P. undulatoplicatilis* Zone), Valanginian beds with *Temnoptychites* spp. on this *Surites tzikwinianus* Zone (Saks *et al.*, 1979, p. 97, Table 2), indicate instead a latest Berriasian age of the "*Bojarkia*" *mesezhnikovi* Zone (Fig. 5).

The thicker lower part of *B. tolmatschowi* Zone can be correlated with the bulk of the North Siberian *Surites analogus* Zone, which is characterized by the presence of *B. tolmatschowi* (Zakharov, 1981, p. 228, 229, Table 3). However, it may also correspond to the basal part of the North Siberian Zone of "*Bojarkia*" *mesezhnikovi* because of its intergradation with the upper part of the zone that contains *Surites* (*Surites*) cf. *simplex sensu* Sazonova 1971 on the west coast of Vancouver Island.

Diagnostic Tethyan ammonites have not been found either in the *B. tolmatschowi* Zone of western British Columbia or in the strongly attenuated (about 30.5 m thick) beds that carry *B. tolmatschowi* in northwestern California (Jones *et al.*, 1969, p. 13A; and pers. observations, 1966). Therefore, this regional *Buchia* Zone, which is all but unknown in North America except for the Pacific Slope, cannot be correlated directly with the Tethyan ammonite zones of Mexico, western South America and southwestern Europe. Nor is the only juvenile neocomitid ammonite found in the basal beds of the overlying *B. pacifica* Zone on Vancouver Island (Jeletzky, 1965, Pl. XVI, Fig. 5) a reliable zonal index. This ammonite was identified as *Neocomites* cf. *rota* (Spath 1925) by Jeletzky (1965), but as pointed out by Imlay and Jones (1970, p. B15), is too small to be assigned definitively to any neocomitid genus.

It is, nevertheless, possible to infer the approximate position of the Canadian *Buchia tolmatschowi* Zone in the sequence of standard Berriasian ammonite zones of southwestern Europe. The next older, *B. uncitoides* s.l. Zone is approximately correlative with the lower part of the Argentinian *Spiticeras damesi* and *Cuyaniceras transgrediens* Zone and the upper part of the Occitanica Zone of southwestern Europe (Fig. 5). This suggests strongly that the base of the Canadian *B. tolmatschowi* Zone is either coeval with the upper boundary of the Occitanica Zone or somewhat younger.

The upper age limit of *B. tolmatschowi* Zone can only be approximated in northwestern California where, stratally and faunally abbreviated, it is gradationally overlain by the ammonite-bearing facies of *B. pacifica* Zone. According to Imlay and Jones (1970, p. B14): "The age of the *Buchia pacifica* zone in California on the basis of ammonites is probably entirely early to middle Valanginian. Its lower part, about 150 feet above its base, has furnished the ammonite *Thurmanniceras,* whose presence suggests an age not older than the Valanginian (Busnardo, Le Hégarat and Magne, 1965, p. 27, 32)." This conclusion agrees well with Zakharov's (1981, p. 228, 229, Table 3) that the top of *Buchia tolmatschowi* Zone in Northern Siberia is situated closely below the Berriasian-Valanginian boundary. The position of this boundary appears to be reliably established in the Boreal Realm following the discovery of diagnostic fossils of the *Pseudogarnieria undulatoplicatilis* Zone in England and Northwest Germany (Casey, 1973; Casey *et al.*, 1977).

These data suggest strongly that the *Buchia tolmatschowi* Zone of western British Columbia spans most or all of the Boissieri Zone of southwestern Europe and so is entirely Late Berriasian in age (Fig. 5). In Argentina, and elsewhere in western South America, this zone appears to correspond approximately to the upper part of the regional *Spiticeras damesi* and *Cuyaniceras transgrediens* Zone.

The description and interregional correlation of the next younger, Lower Valanginian *Buchia pacifica* Zone are beyond the scope of this paper. The reader is referred to Jeletzky (1965), Imlay and Jones (1970) and Zakharov (1981, p. 229, 230, Table 3) for the most recent accounts of this zone.

UPPERMOST JURASSIC AND BASAL CRETACEOUS STRATIGRAPHY AND BIOCHRONOLOGY OF THE SVERDRUP BASIN

General and Historical Remarks

The stratigraphy and biochronology of the Jurassic-Cretaceous boundary beds of the Sverdrup Basin were decribed fairly recently in considerable detail (Jeletzky, 1964, 1966, 1971a, 1973, 1974). These papers also contain illustrations and descriptions of most of the then known diagnostic ammonites and buchias of these beds. The following account is, therefore, restricted to the description of the stratigraphic setting, biochronological evaluation, and illustration of rather numerous new ammonite and *Buchia* faunas discovered in these beds since the compilation of these papers. It also attempts a re-evaluation of some of the previously reached conclusions in the light of these new discoveries.

As pointed out earlier (Jeletzky, 1973, p. 43, 44), the Jurassic-Cretaceous boundary beds of the Sverdrup Basin are unique among the high Boreal basins of North America in that their *Buchia* faunas are commonly associated with well preserved high Boreal ammonites. All of the previously published ammonites are either specifically identical with, or closely related to, the contemporary ammonites of northern Siberia and the European Arctic. Some of the more recently discovered ammonites discussed and figured below have the same affinities. However, some others are either specifically identical with, or closely related to, the contemporary ammonites of Central Russia and/or Eastern England. Most of the latest Jurassic and earliest Cretaceous *Buchia* faunas of the Sverdrup Basin are identical with the Late Volgian and Berriasian *Buchia* faunas of northern Eurasia, as revised recently by Zakharov (1979, 1981). At the same time these *Buchia* faunas of the Sverdrup Basin are largely the same as the latest Jurassic and earliest Cretaceous *Buchia* faunas of western British Columbia and northwestern California, which were reliably dated in terms of the standard ammonite zones and stages of the Tethyan Realm earlier in this paper. Below, the results of this dating of the British Columbian and Californian *Buchia* zones (Fig. 5) are utilized to date the latest Jurassic and earliest Cretaceous *Buchia* and ammonite faunas of the Sverdrup Basin in terms of the standard Tethyan zones and stages. This makes it possible, in turn, to recognize the approximate positions of these Tethyan zones and stages in the sequence of high Boreal zones and stages of North Siberian and Central Russian basins. It is hoped that these long-ranging correlations via the North Pacific and Boreal basins of Canada will contribute to the settlement of the previously discussed (Figs. 1, 2) longlasting controversy about the detailed time relationships of the latest Jurassic and earliest Cretaceous provincial zones and stages (substages) of Tethyan and Boreal Realms.

Stratigraphy of Principal Fossiliferous Sections

The following six fossiliferous sections or isolated localities of the Jurassic-Cretaceous boundary beds are sufficiently important for the understanding of the regional sequence of their ammonite and *Buchia* faunas to be described in some detail. Their approximate geographical locations are shown in Figure 6. The ages and mutual correlations (i.e. internal *sensu* Jeletzky, 1973, p. 43) of their diagnostic ammonite and *Buchia* faunas are shown in Figure 7.

Rollrock Lake Section. This most important section is situated north of the Rollrock Lake, northern Ellesmere Island (Figs. 6, 8). The base of this section, which was discovered and originally described by Wilson, (1976, p. 450, Fig. 91.1), is at the Lat. 81°35′05″N.; Long. 75°50′00″W. and its top is at the Lat. 85°35′25″N.; Long. 75°55′00″W. (NTS-34 D). The following description utilizes much unpublished stratigraphical data kindly communicated by Mr. D.G. Wilson and Dr. A.F. Embry.

The Upper Jurassic to lower Lower Cretaceous argillaceous unit of the Rollrock Lake section was assigned to the Deer Bay Formation by

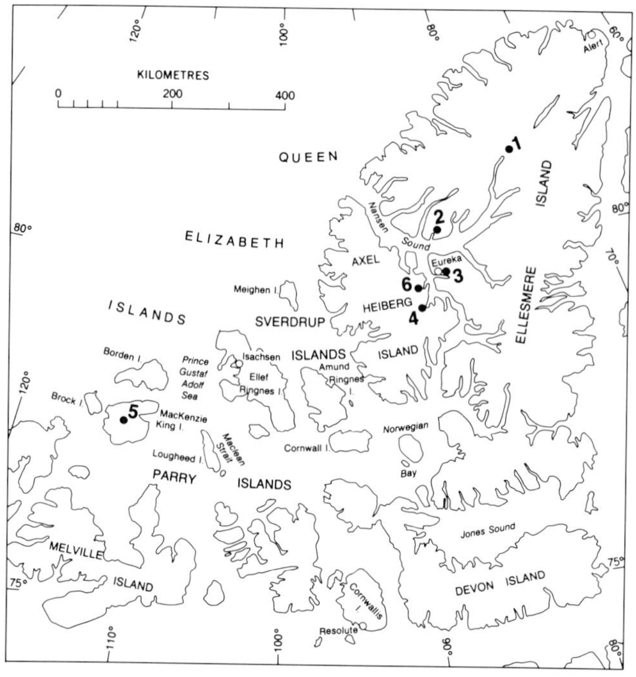

Figure 6. Key map showing the principal fossiliferous sections and fossil localities of the Jurassic-Cretaceous boundary beds discussed in the text. 1) Rollrock section. 2) Greely Fiord sections. 3) Slidre Fiord (Camp Creek) section. 4) Buchanan Lake (Awingak River) section. 5) Mackenzie King Island fossil localities. 6) Gibbs Fiord locality.

Wilson (1976, p. 450). However, Dr. A.F. Embry, whose usage is followed here, treates it as an unnamed J-K shale unit that deserves another formal formational name; he states (written commun., February 5, 1980): "The unit is underlain by the Jaeger Fm. (Bajocian) and overlain by the Isachsen Fm. It is about 580 m thick in the area and consists mainly of medium to dark grey, silty shales. Fine grained sandstone units up to 15 m thick occur in the lower part of the unit. Formal nomenclature has not yet been applied to the unit. Present data indicate that it is stratigraphically equivalent to the Upper Savik, Awingak and Deer Bay Formations of the Eureka area." According to Dr. A.F. Embry, the Rollrock Lake section of this J-K shale unit begins about one-third of the way up through its thickness and goes into the Isachsen Formation (Fig. 8).

The oldest diagnostic fauna of the Rollrock section, at the 99 m level above its base (GSC loc. C-55234), has yielded large forms of *Buchia fischeriana* (d'Orbigny) identical with those described and figured by Jeletzky (1966, p. 25, Pl. VIII, Figs. 2, 3, 5) from the Eureka Weather Station area. It also includes *Laugeites* sp. indet. resembling *Laugeites*? sp.indet. and *Dorsoplanites* cf. *gracilis* Spath apparently conspecific with the forms described and figured by Jeletzky (1966, p. 21, 23, Pl. VIII, Figs. 7, 10). These two ammonites are associated with *Buchia fischeriana* (D'Orbigny) in the Slidre Fiord and Greely Fiord sections respectively (Jeletzky, 1966, p. 22, 43).

Another diagnostic fauna was found 111 m above the base (GSC loc. C-55236). This fauna includes *Buchia richardsonensis,* originally described and figured by Jeletzky (1966, p. 34, 35, Pl. VIII, Figs. 1, 4, 8, 9) from the Greely and Slidre Fiords sections of the Deer Bay Formation, and *Dorsoplanites* (s.l.) sp. indet. This fauna appears to be essentially a biofacies of the *Buchia fischeriana* s.l. fauna of GSC loc. C-55234, although *B. richardsonensis* is known to range above and below that zone (Jeletzky, 1966, p. 34, 35, 43).

The fauna found 171 m above the base (GSC loc. C-55248) includes

STANDARD STAGES AND SUBSTAGES	NORTHERN ELLESMERE ISLAND (ROLLROCK LAKE AREA)	WESTERN ELLESMERE ISLAND (GREELY FIORD AREA)	WESTERN ELLESMERE ISLAND (SLIDRE FIORD AREA)	AXEL HEIBERG ISLAND (BUCHANAN LAKE AREA)	MACKENZIE KING ISLAND
BERRIASIAN (PART)	NO DIAGNOSTIC FOSSILS KNOWN	NO DIAGNOSTIC FOSSILS KNOWN	NO DIAGNOSTIC FOSSILS KNOWN	NO DIAGNOSTIC FOSSILS KNOWN *PSEUDOCRASPEDITES* AFF. *ANGLICUS, BUCHIA OKENSIS* AND VAR. *CANADIANA*	NO DIAGNOSTIC FOSSILS KNOWN *PSEUDOCRASPEDITES* AFF. *ANGLICUS BUCHIA OKENSIS* S. S. *PARTSCHICERAS* SP. INDET
UPPER TITHONIAN (= UPPER VOLGIAN)	*CRASPEDITES* S. S. AND *SUBCRASPEDITES* S. S. — *BUCHIA UNSCHENSIS* S. S. *BUCHIA TEREBRATULOIDES* S. L. *B. FISCHERIANA* (SMALL TO MEDIUM SIZED), *B. RICHARDSONENSIS* AND *B. PIOCHII* S. S. (VERY RARE) LARGE *BUCHIA FISCHERIANA B. RICHARDSONENSIS* AND *DORSOPLANITID* AMMONITES	LARGE *BUCHIA FISCHERIANA* AND *DORSOPLANITID* AMMONITES;	*CRASPEDITES (TAIMYROCERAS?) CANADENSIS* AND *BUCHIA UNSCHENSIS* S. S. (MASS OCCURENCE). *B.* AFF. *TEREBRATULOIDES* VAR. *SUBINFLATA* (RARE) NO DIAGNOSTIC FOSSILS KNOWN LARGE *BUCHIA FISCHERIANA* AND *DORSOPLANITID* AMMONITES	*PRAETOLLIA FEDOROVI, BUCHIA TEREBRATULOIDES* S. L., *B.* N. SP. AFF. *OKENSIS, B. UNSCHENSIS* S. S., *B. OKENSIS* S. S. (VERY RARE) NO DIAGNOSTIC FOSSILS KNOWN LARGE *BUCHIA FISCHERIANA* AND POOR *DORSOPLANITID* AMMONITES	*PSEUDOCRASPEDITES ANGLICUS, BUCHIA TEREBRATULOIDES* S. L., *B.* N. SP. AFF. *OKENSIS, B. UNSCHENSIS* S. S., *B. OKENSIS* S. S. (VERY RARE) NO DIAGNOSTIC FOSSILS KNOWN
LOWER TITHONIAN (= LOWER VOLGIAN) (PART)	NO DIAGNOSTIC FOSSILS KNOWN	NO DIAGNOSTIC FOSSILS KNOWN	*BUCHIA RUSSIENSIS* S. L. (INCLUDING *B.R.* VAR. *MNIOVNIKENSIS, BUCHIA* AFF. *FISCHERIANA*		*BUCHIA RUSSIENSIS* S. L. (INCLUDING *B.R.* VAR. *MNIOVNIKENSIS) BUCHIA* AFF. *FISCHERIANA*

GSC

Figure 7. Principal Late Tithonian (= Late Volgian) Marine Faunas of the Sverdrup Basin and their Internal Correlation.

Craspedites (*Subcraspedites*) cf. *sowerbyi* Spath 1952 (Pl. V, Fig. 1) and *C.* (*Craspedites*) n. sp. indet. aff. *subditus* (Trautschold) (Pl. V, Figs. 5a, 5b; Pl. VI, Fig. 1) associated with *Buchia terebratuloides* (Lahusen 1888) s.l. and *B.* cf. *richardsonensis* Jeletzky 1966.

The level 174 m above the base (GSC loc. C-55249) has yielded small to medium-sized forms of *Buchia fischeriana* (d'Orbigny) s.l., similar to those described by Jeletzky (1966, p. 25, Pl. VIII, Figs. 8, 9) from northwestern Richardson Mountains. A couple of shells referable to *B. piochii* s.s. were found also.

Figure 8. Rollrock Lake section of Upper Jurassic and lower Lower Cretaceous argillaceous rocks (unit J-K of Dr. A.F. Emery; see text for details). View due NE from a station near the northern shore of Rollrock Lake. The prominent light coloured bluffs at the section's top are built of sandstones of the Isachsen Formation while the darker coloured, gentler slopes below them are underlain by J-K shales (courtesy of Dr. A.F. Emery, GSC Cat. no. 203333-G).

The latest Jurassic craspeditids were also found at 190.5 m above base (GSC loc. C-55251). This fauna has yielded *Craspedites* (*Subcraspedites*) n. sp. indet. aff. *praeplicomphalus* Swinnerton (Pl. V, Fig. 7; Pl. VI, Fig. 4), *C.* (*Craspedites*) n. sp. indet. aff. *subditus* (Trautschold) (Pl. V, Fig. 5) and typical representatives of *Buchia unschensis* (Pavlow 1907) *sensu* Jeletzky 1966 (Pl. V, Fig. 6).

The occurrence of *Craspedites* s.s. and *Subcraspedites* s.s. fauna at levels 171 and 190.5 m indicates a considerable thickening of the earlier Late Tithonian part of J-K argillaceous unit in comparison with the corresponding part of the Deer Bay Formation in its Slidre Fiord section. In the Rollrock Lake section, GSC loc. C-55234 representing the oldest occurrence of the large *Buchia fischeriana* associated with dorsoplanitid ammonites is separated from GSC loc. C-55248 with the oldest occurrence of *Craspedites* s.s.-*C.* (*Subcraspedites*) s.s. fauna by 72 m of argillaceous rocks. In the Slidre Fiord section the equivalent *Buchia fischeriana,* etc. fauna is separated from the *Craspedites* (*Taimyroceras*?) *canadensis* fauna, which is younger than the *Craspedites* s.s. and *C.* (*Subcraspedites*) fauna of the Rollrock Lake section (see below for further details), by only 6 to 8 m of argillaceous rocks. No diagnostic latest Jurassic and earliest Cretaceous fossils have been found higher in the Rollrock Lake section (Fig. 7).

Greely Fiord Section. The section of the Deer Bay Formation measured by R. Thorsteinsson in 1961 southeast of the syncline situated 3.7 km east of the mouth of a major unnamed river entering a prominent bay that joins Greely Fiord 19.2 km east of Hare Fiord, has only yielded the widespread fauna of large *Buchia fischeriana* (d'Orbigny) s.l., *B. richardsonensis* Jeletzky, *Dorsoplanites* n. sp. aff. *crassus* Spath?, *D.* cf. *gracilis* Spath and *Laugeites*? sp. indet. (GSC loc. 47882, 47881 and 47883; see Jeletzky, 1966, p. 20-23; Pl. VIII, Figs. 10, 11). The regional correlation of this fauna found in the basal part of the formation was already discussed under the Rollrock Lake section (Fig. 7).

Slidre Fiord Section. The section of upper sandstone member of the Awingak Formation studied by Dr. J.C. Troelsen in 1952 on the

PLATE V

northern coast of Slidre Fiord, 1.9 km northwest of the Eureka Weather Station, has yielded an important late (possibly latest) Early Tithonian (=late Early Volgian) *Buchia russiensis*, s.l. fauna (GSC loc. 22776). Its buchias were originally figured and discussed by Jeletzky (1966, p. 34, Pl. VIII, Figs. 2, 3, 10) as *Buchia piochii* (Gabb) var. *mniovnikensis* (Pavlow) and *B.* aff. *fischeriana* (d'Orbigny) (see p. 186, 190, 191 for further details).

The fauna of large *B. fischeriana* s.l. and dorsoplanitid ammonites that occurs in the Rollrock Lake and Greely Fiord sections also occurs 30 to 35 m above the base of the next younger, Deer Bay Formation in its section studied but not measured by R. Thorsteinsson along the Camp (=Reptile) Creek 1.6 to 4 km north of the Eureka Weather Station, Slidre Fiord. The GSC loc. 28714 of this fauna includes *B. fischeriana* (d'Orbigny), *Dorsoplanites* sp. indet. ex gr. *panderi* (Michalski), *Dorsoplanites* sp. indet., and *Pavlovia*? sp. indet. (Frebold, 1961, p. 23, 24; Pl. XVII, Fig. 2; Pl. XX, Fig. 1; Pl. XXI, Fig. 1). Its GSC loc. 28714 and 53398 have yielded large *B. fischeriana* (d'Orbigny) s.l., *Dorsoplanites* sp. indet. ex gr. *D. panderi* (Michalski) and *Laugeites*? sp. indet. (Jeletzky, 1966, p. 20-23, 25-30; Pl. VIII, Figs. 1-5, 7).

Numerous *Craspedites* (*Taimyroceras*?) *canadensis* Jeletzky 1966, equally numerous *Buchia unschensis* (Pavlow 1907) and rare *B.* aff. *terebratuloides* (Lahusen) var. *subinflata* (Pavlow 1907) were found about 7.5 m higher in that section. This fauna (GSC loc. 28712), which occurs as a lenticular body of calcareous coquina in shale, was described and figured by Jeletzky (1966, p. 6-20, 35-42, Pls. I-VI). This fauna is considered to be younger than the *Craspedites* s.s.-*Subcraspedites* s.s. fauna of the Rollrock Lake section for reasons presented below.

No younger late late Tithonian or Berriasian ammonite and *Buchia* faunas have been found in this Camp (=Reptile) Creek section (Fig. 7).

Buchanan Lake Section. The widespread fauna of large *B. fischeriana* s.l. and of dorsoplanitid ammonites reappears in the section of the Deer Bay Shale originally measured by Souther (in Fortier *et al.*, 1963, p. 438) on Awingak River near Buchanan Lake, eastern Axel Heiberg Island (Figs. 6, 7). This fauna (GSC loc. 26156) was found in an 8 m bed commencing 31 m above the formation's base and identified originally as "*Aucella*" aff. "*A.*" *fischeri* d'Orbigny (Pavlow) fauna (Frebold in Fortier *et al.*, 1963, p. 438). Its dorsoplanitids are generically indeterminate.

Neither the *Craspedites* s.s.-*Subcraspedites* s.s. nor the *T.*? *cana-*

EXPLANATION FOR PLATE V

All photographs are natural size.

Figure 1. *Craspedites (Subcraspedites)* cf. *sowerbyi* Spath 1952. GSC Cat. 64448. GCS loc. C-55248. Mid Upper Tithonian stage (=mid Upper Volgian stage). *Craspedites* s.s.-*Subcraspedites* s.s. beds, Unnamed Jurassic-Cretaceous argillaceous unit, collected 170 m above base of its Rollrock Lake section (see text and Figs. 7, 8 for further details). N.W.T., District of Franklin, Northern Ellesmere Island, Tanquary Fiord-Ekblaw Lake area, Lat. 81°33′30″N.; Long. 76°05′W. The outlined external suture lines are typical of the genus *Craspedites* s.l. in their subradial to descendant orientation, feeble denticulation and presence of only three lateral lobes. Except for the suture line, this specimen would be difficult to differentiate from *Praetollia (Pseudocraspedites)* ex gr. *anglicus* (Shulgina 1972) (compare Pl. VII, Figs. 1a, 1c, 1k).

Figures 2a-d. *Buchia unschensis* (Pavlow 1907) s.s. GSC Cat. 73699. GSC loc. 70387. Uppermost Tithonian (=lowermost Riasanian or uppermost Upper Volgian), upper part of *Buchia terebratuloides* s.l. Zone (a faunal phase of its *Praetollia fedorovi* (=*antiqua*) beds). Mould Bay Formation, presumably from an argillaceous interbed in its Middle arenaceous member. Canadian Arctic Archipelago, N.W.T., District of Franklin. Mackenzie King Island, Section 65 MA-1 situated 24 to 32 km NE of McConnell Island. Collected by Petropar Canada Ltd., 1965. 2a) lateral view of the exterior; 2b) posterior view; 2c) anterior view (somewhat oblique); 2d) beak part and hinge margin viewed from above. A typical, juvenile representative closely matching that reproduced by Jeletzky (1966, Pl. VI, Fig. 3).

Figure 3. *Praetollia (Praetollia) maynci* Spath 1952 var. *aberrans* Spath 1952. GSC Cat. 73700. GSC loc. C-33897. Uppermost Tithonian (=lowermost Riasanian or uppermost Volgian), upper part of *Buchia terebratuloides* s.l. Zone (a faunal phase of its *Praetollia fedorovi* (=*antiqua*) beds.). Deer Bay Formation. Axel Heiberg Island, Gibbs Fiord, Lat. 79°52′N.; Long. 87°37′W. Collected by H.R. Balkwill, 1974. An almost flattened specimen with a completely preserved mouth border. Most of the ultimate whorl is occupied by the body chamber.

Figures 4a-c. *Buchia terebratuloides* (Lahusen 1888) var. *obliqua* (Tullberg 1881). GSC Cat. 73701. The same locality, etc. as for Figure 2. A typical, halfgrown representative of the variant. 4a) lateral view of the exterior; 4b) anterior view; 4c) beak part and hinge margin viewed from above.

Figures 5a, b. *Craspedites (Craspedites)* n. sp. aff. *subditus* (Trautschold 1877) GSC Cat. 64450. GSC loc. C-55251. The same locality, etc. as for Figure 1 but collected at 190.5 m level. 5a) lateral view of the oralmost preserved whorl that exhibits external suture lines typical of the genus *Craspedites* s.l. near the oral end (outlined in white). As with *Craspedites (Subcraspedites)* cf. *sowerbyi* shown in Figure 1, reliable differentiation of this specimen from *Praetollia* Spath 1952 (=*Boreal-ites* Klimova 1969) depends on the much more primitive character of its suture line (compare Jeletzky, 1979, p. 4-7, fig. 1L); 5b) lateral view of the other flank that exposes the preceding whorl. Note that the typical dichotomous ribbing habit is replaced by an alternating trichotomous and dichotomous ribbing habit on the next whorl shown in Figure 5a.

Figures 6a-c. *Buchia unschensis* (Pavlow 1907) s.s. GSC Cat. 73702. The same locality, etc. as for the specimen reproduced in figure 2. A typical, juvenile specimen matching closely those reproduced by Jeletzky (1966, Pl. I, Figs. 2, 4). 6a) lateral view of the exterior; 6b) anterior view; 6c) beak part and hinge margin viewed from above.

Figure 7. *Craspedites (Subcraspedites)* n. sp. aff. *praeplicomphalus* (Swinnerton 1936). GSC Cat. 64451. The same locality, etc. and stratigraphic level as for Figure 5. Lateral view of a fragment of adult living chamber superimposed on the flank of fully septate penultimate whorl reproduced in Pl. VI, Fig. 4a. Note a well preserved adult left valve of *Buchia unschensis* s.s. (GSC Cat. 64452) protruding from the oral (upper) end of the fragment. Its large, angular posterior ear is outlined in white. Another poorly preserved valve of *B. unschensis* s.s. protrudes from the apical end of the fragment.

PLATE VI

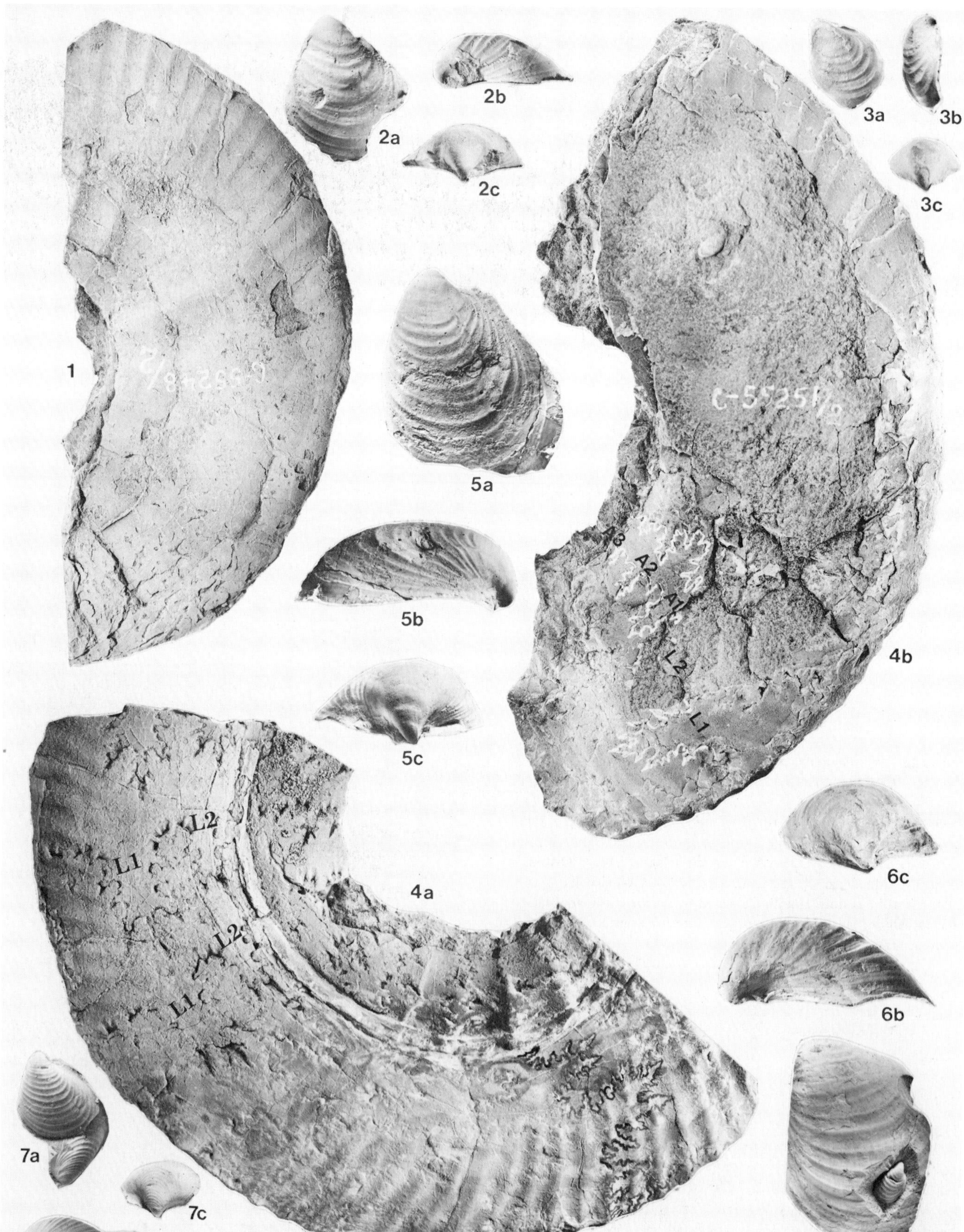

densis fauna was found in this section. However, one of them may be represented by the GSC loc. 26172 reportedly collected (Souther, in Fortier *et al.*, 1963, p. 438): "Within 100-foot stratum commencing 216 feet above the base of the formation." This fauna includes (Frebold, in Fortier *et al.*, 1963, p. 438) "Ammonites genus et sp. indet., *Camptonectes* cf. *C. praecinctus* Spath.." which cannot be identified any closer in my opinion.

The next younger fauna (GSC loc. 26171), at the 95 m level, was originally assigned a general Valanginian age (Frebold in Fortier *et al.*, 1963, p. 438, 439). It was reassigned later to the basal Berriasian Zone of *Buchia okensis and* Tollia (*Subcraspedites*) aff. *suprasubditus* (see below)

R. Thorsteinsson and E.T. Tozer's study and measurement of the Buchanan Lake section in 1962 provided additional fossils that were assigned to the following levels (above base): 1) 128 m – GSC loc. 52488; 2) 150 m – GSC loc. 52490; 3) 166 m – GSC loc. 52486; and 4) 168.5 m – GSC loc. 52487.

According to the collectors (labels), localities 2 to 4 inclusive correspond to the GSC loc. 26171 of Souther (in Fortier *et al.*, 1963) collected at the 95 m level. Despite this discrepancy in measurement, the writer is satisfied that the fossils from the localities 52486 and 52487 are the same fauna as that from the locality 26171. The ammonites of the next older locality 52490 have a transitional character between those of the GSC locs. 52486 and 52487 and those of the lowermost locality 52488. However, its *Buchia* fauna is the same as that of the two overlying localities. The fauna collected in 1976 by Dr. E. Kemper (GSC loc. 91330) at 204 m below the base of Isachsen Formation is identical to that of GSC loc. 26171, 52486, 52487.

The fauna of the GSC loc. 52488 was described as the *Praetollia antiqua* and *Buchia terebratuloides* s.l. fauna by Jeletzky (1973, p. 47, 50, 74-76; Pl. 4, Figs. 1-3; Pl. 5, Figs. 1-2; Pl. 7, Fig. 1; Text-Fig. 3) and assigned to the latest Jurassic. However, this diagnostic ammonite is redetermined here as *Praetollia* (*Praetollia*) *fedorovi* (Klimova 1969). The only other ammonite found is *Praetollia* (*Ronkinites*) n. sp. aff. *rossica* (Shulgina 1972). In addition to abundant *Buchia terebratuloides* s.l. figured by Jeletzky (1973) and here (Pl. VIII, Fig. 7), this fauna includes abundant *B.* n. sp. aff. *okensis* (Pl. VIII, Fig. 5), abundant *B. unschensis* s.s. (Pl. VIII, Figs. 1, 2, 3, 6) and very rare *Buchia okensis* s.s. collected by Dr. E. Kemper at the GSC loc. 91329.

The bed 52488 (=91329) underlies conformably and gradationally the basal Berriasian beds 52490, 52486 and 52487. Therefore, and for other reasons presented below (p. 233-234, Figs. 7, 9), this *Praetollia fedorovi* bed is considered to be younger than the *Craspedites* (*Taimyroceras*?) *canadensis* beds.

EXPLANATION FOR PLATE VI

All photographs are natural size.

Figure 1. *Craspedites (Craspedites)* n. sp. aff. *subditus* (Trautschold 1877). GSC Cat. 64449. GSC loc. C-55248. Mid Upper Tithonian (=mid Upper Volgian stage), *Craspedites* s.s.-*Subcraspedites* s.s. beds. Unnamed Jurassic-Cretaceous argillaceous unit, collected at 171 m level above the base of its Rollrock Lake section (see text; Figs. 7, 8 for further details), N.W.T., District of Franklin, Northern Ellesmere Island, Tanquary Fiord-Ekblaw Lake area, Lat. 81°33′30″N.; Long. 76°05′W. Lateral view of a fragmentary living chamber of the form reproduced in Plate V, Figure 5.

Figures 2a-c. *Buchia unschensis* (Pavlow 1907) s.s. GSC Cat. 73703. GSC loc. 70387. Uppermost Tithonian (=lowermost Riasanian or uppermost Volgian), upper part of *Buchia terebratuloides* s.l. Zone (a faunal phase of its *Praetollia fedorovi* (=*antiqua*) beds). Mould Bay Formation, presumably from an argillaceous interbed in its Middle arenaceous member. Canadian Arctic Archipelago, N.W.T., District of Franklin. Mackenzie King Island, Section 65-MA-1 situated 24 to 32 km NE of Mcconnell Island. Collected by Petropar Canada Ltd., 1965. 2a) lateral view of the exterior; 2b) anterior view; 2c) beak part and hinge margin viewed from above. A typical halfgrown left valve.

Figures 3a-c. *Buchia* n. sp. aff. *okensis* (Pavlow 1907). GSC Cat. 73704. The same locality, etc. as for Figure 2. 3a) lateral view of the outside; 3b) posterior view; 3c) beak part and hinge line (concealed behind the beak) viewed from above. A small, presumably juvenile specimen that combines the shape of *B. terebratuloides* s.l. with prominent and heavy concentric ribs characteristic of *B.* n. sp. aff. *okensis* (compare specimen reproduced in Pl. III, Fig. 4).

Figures 4a, b. *Craspedites (Subcraspedites)* n. sp. indet. aff. *praeplicomphalus* (Swinnerton 1936). GSC Cat. 64451. GSC lco. 55251. The same locality, etc. as for Figure 1 but collected at 190.5 m level. 4a) lateral view of penultimate whorl of the specimen produced in Plate V, Figure 7. The adumbilical third of the flank is pushed under its lateral two-thirds and is invisible, except for the umbilical seam. Therefore, only the first (marked L1) and second (marked L2) lateral lobes are visible on the flank (see Fig. 4b for the complete width of the whorl and a complete external suture line). Note pronouncedly bullate primaries of the early part of the preceding whorl visible in the umbilicus and their gradual adoral weakening. 4b) lateral view of the other, worn out flank that exhibits the whole width of the whorl and an almost complete external suture line (outlined in white). Once again (as in specimens shown in Plate V, Figs. 1, 5), this suture line is typical of the genus *Craspedites* s.l. The second lateral lobe (L2) is almost entirely abraded but the first lateral lobe (L1) and the three, distinctly suspensive auxiliary lobes (marked A1 to A3) are clearly visible.

Figures 5a-c. *Buchia* n. sp. aff. *okensis* (Pavlow 1907). GSC Cat. 73705. The same locality, etc. as for Figure 2. 5a) lateral view of the exterior; 5b) anterior view; 5c) beak part and hinge margin viewed from above. A medium sized adult that differs from similarly elongated representatives of *B. unschensis* s.s. in a slight right handed curvature of its beak, irregular spacing of ribs and gradual outward widening of their interspaces.

Figures 6a-c. *Buchia* cf. n. sp. aff. *okensis* (Pavlow 1907). GSC Cat. 73706. The same locality, etc. as for Figure 2. 6a) lateral view of the exterior; 6b) anterior view; 6c) beak part and hinge line from above. A fragment of large left valve exhibiting a mixture of morphological features. Its straight anterior margin (Fig. 6a) and pronouncedly left handedly incurved beak (Fig. 6c) ally it with *B. unschensis* s.s. However, the coarse, elevated and widely spaced concentric ribs are those of *B.* n. sp. aff. *okensis* and the restoration of the lacking lower and lower posterior parts of the valve would give it a triangular, longer than wide shape typical of that species.

Figures 7a-c. *Buchia unschensis* (Pavlow 1907) s.s. GSC Cat. 73707. The same locality, etc. as for Figure 2. 7a) lateral view of the exterior. In addition to the left valve, it offers an oblique view of the exterior of a right valve possibly belonging to the same specimen; 7b) anterior view; 7c) beak part and hinge margin (concealed behind the beak). Left valve of a very typical, late juvenile shell that is indistinguishable from specimens from the *Craspedites (Taimyroceras?) canadensis* beds figured by Jeletzky (e.g., 1966, Pl. VI, Fig. 3).

PLATE VII

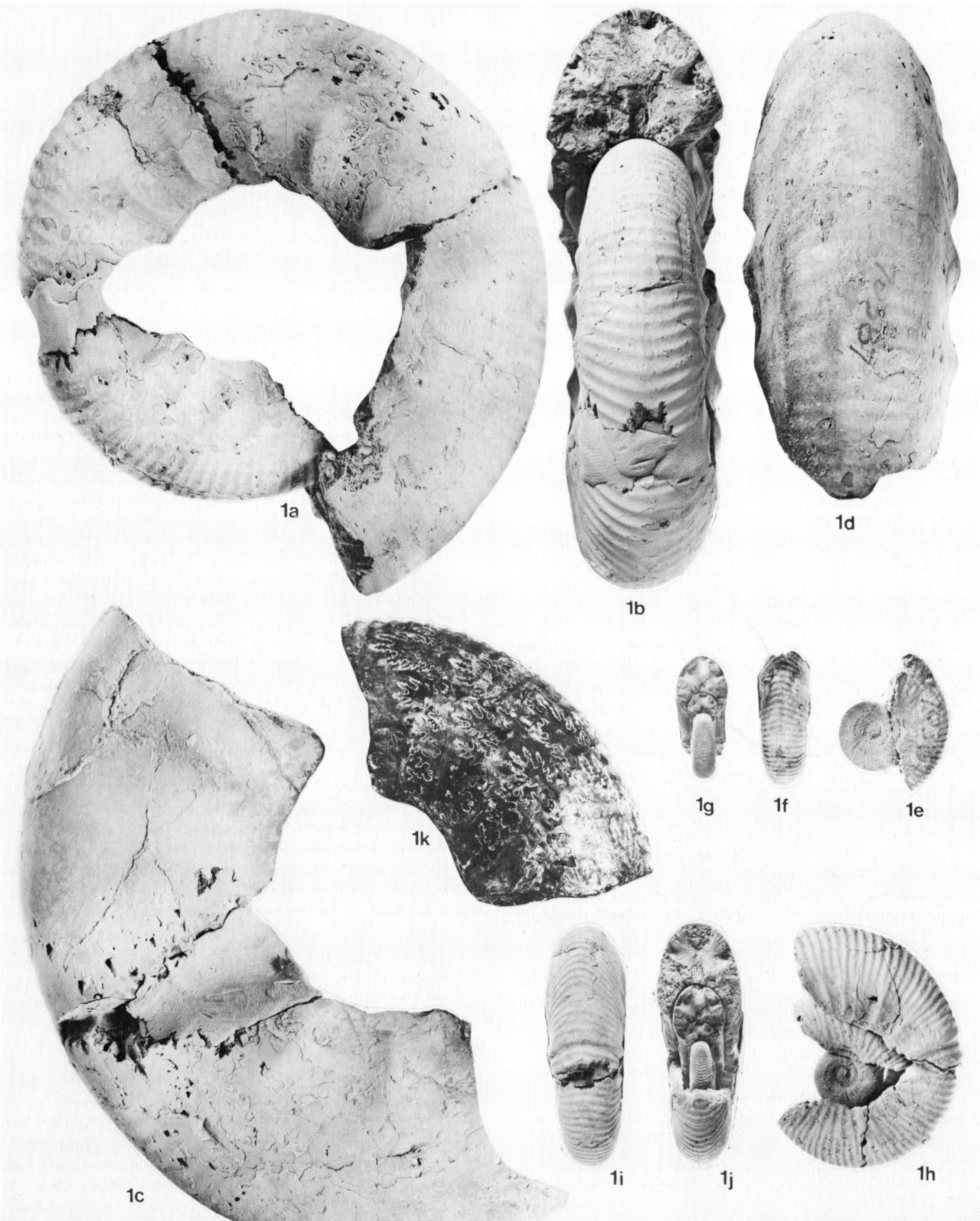

The basal Berriasian fauna of the GSC locs. 52490, 52486, 52487 and 91330 was described as *Buchia okensis* s.s. and *Tollia* (*Craspedites*) (or *Subcraspedites*) aff. *suprasubditus* by Jeletzky (1964, p. 30, 32, 34; Pl. 1, Fig. 1; Pl. II, Figs. 1, 2, 4, 5; Pl. III, Figs. 1, 2; 1973, p. 52-54, 56; Figs. 2, 3; Pl. 6, Fig. 1), but its diagnostic ammonite is reassigned herein to *Praetollia* (*Pseudocraspedites*) aff. *anglicus* (Shulgina 1972) (p. 234-235 and Figs. 9, 10). Later Dr. E. Kemper collected very rare representatives of *Buchia volgensis* (Lahusen 1888) at GSC loc. 91330 (Pl. VIII, Figs. 14a-14c) which, to the best of my knowledge, are the only authentic representatives of this Eurasian species found in North America.

The *Buchia okensis* s.s-*Praetollia* (*Pseudocraspedites*) aff. *anglicus* fauna is the youngest fauna found in the Jurassic-Cretaceous boundary beds of the Buchanan Lake section.

Uppermost Jurassic Rocks on the Mackenzie King Island. The survey of the Mackenzie King Island by geologists of the Petropar Canada, Ltd. produced an ammonite referable to the diagnostic North Siberian species "*Subcraspedites*" *anglicus* Shulgina 1972 (GSC loc. 70387). Here it is re-assigned to *Praetollia* (*Pseudocraspedites*). This ammonite, (Pl. VII, Fig. 1) is associated with common *B. terebratuloides* s.l. (Pl. V, Fig. 4), *B.* n. sp. aff. *okensis* (Pl. VI, Figs. 3, 5, 6) and *B. unschensis* s.s. (Pl. VI, Figs. 2, 7) and very rare *B. okensis* s.s. (like that figured in Pl. VIII, Fig. 13). Because of the complete identity of these buchias to those of the *Praetollia fedorovi* and *Buchia terebratuloides* s.l. fauna of the Buchanan Lake section, this Mackenzie King fauna (GSC loc. 70387) is tentatively considered to be its biofacies and dated latest Tithonian (Figs. 7, 9).

This *Praetollia* (*Pseudocraspedites*) *anglicus* and *Buchia terebratuloides* s.l. fauna was said to be collected in the Mould Bay Formation. No further details are available. This fauna probably was found in the same thin argillaceous interbed of its middle member as the slightly younger "*Craspedites* (*Subcraspedites*)" cf. "*C.*" *suprasubditus* (Bogoslovsky), *Phylloceras* (*Partschiceras*) sp. and *Buchia okensis* s.s. of Tozer and Thorsteinsson (1964, p. 145). The latter fauna is the same as the *Buchia okensis* s.s. and *Praetollia* (*Pseudocraspedites*) aff. *anglicus* fauna of the Buchanan Lake section discussed earlier (Fig. 7).

Uppermost Jurassic Rocks at Gibbs Fiord, Axel Heiberg Island. When mapping the Deer Bay Formation of Gibbs Fiord, Dr. H. Balkwill (formerly with GSC and presently with Petrocanada Co.) found at the Lat. 70°52′N.; Long. 87°37′W (GSC loc. C-33897) an ammonite that is assigned here to *Praetollia* (*Praetollia*) *maynci* var. *aberrans* Spath 1952 (Pl. V, Fig. 3). No further stratigraphic information is available. The associated *Buchia* fauna consists of common *B. terebratuloides* s.l. (Pl. VIII, Fig. 7), *B.* n. sp. aff. *okensis* (Pl. VIII, Figs. 8, 9, 12) and *B. unschensis* s.s. (Pl. VIII, Fig. 10), and very rare *B. okensis* s.s. (Pl. VIII, Fig. 13). It is identical with *Buchia* faunas of GSC loc. 52488 (=91329) in the Buchanan Lake section and GSC loc. 70387 (i.e., *Praetollia* (*Pseudocraspedites*) *anglicus* fauna) on the Mackenzie King Island. Therefore, the *Praetollia maynci* var. *aberrans* fauna is believed to be yet other biofacies of the *Praetollia fedorovi* and *Buchia terebratuloides* s.l. fauna. Like the other two biofacies, it is dated as latest Tithonian.

Buchia fischeriana Zone

Fauna of large Buchia fischeriana s.l. and Dorsoplanitid Ammonites. The oldest probably Upper Tithonian (=Upper Volgian) Sverdrup Basin fauna is the widespread fauna characterized by an abundance of large *B. fischeriana* associated with *Laugeites*? sp., *Dorsoplanites* cf. *gracilis* Spath and other mostly poorly preserved dorsoplanitid ammonites. The large *B. fischeriana* either occurs alone forming monospecific associations, or is associated with variable ratios of *B. richardsonensis* Jeletzky, and *B. russiensis* s.l. Neither *B. unschensis* s.s. nor *B. terebratuloides* s.l. were found in this fauna.

This *B. fischeriana* fauna was previously placed in the lower part of the Upper Volgian (Jeletzky, 1966, p. 20, 22, 30, 34, 43) because of its stratigraphic superposition on beds containing buchias, then identified as varieties of *B. piochii* (Gabb). Furthermore, the early Late Volgian age of this fauna appeared to be indicated by the occurrence of the *Craspedites* (*Taimyroceras*?) *canadensis* fauna only about 8.5 m above it. Finally, the large *B. fischeriana* forms of this fauna were interpreted as advanced forms of that species primarily because of their stratigraphic proximity to the *C.* (*T.*?) *canadensis* fauna. Therefore, the fauna of large *B. fischeriana* and dorsoplanitid ammonites was tentatively correlated with either the upper part of the *Kachpurites fulgens* Zone or the lower part of the *Craspedites subditus* Zone of the Central Russian standard. Jeletzky (1966, p. 22), however, pointed out that latest Early Volgian (*Epivirgatites nikitini* Zone) cannot be ruled out.

The recent discovery of the *Craspedites* s.s.-*Subcraspedites* s.s. fauna above the fauna of large *B. fischeriana* and dorsoplanitid ammonites has invalidated these earlier considerations (Fig. 9). This fauna is older than the *Craspedites* (*Taimyroceras*?) *canadensis* fauna and correlative with the *Craspedites subditus* Zone and, possibly, with part or ?all of the underlying *Kachpurites fulgens* Zone of the Central Russian standard. The buchias occurring in the immediately and gradationally underlying beds and previously identified as *B. piochii* var. *russiensis* and *B. p.* var *mniovnikensis* are neither conspecific nor contemporary with the early Late Tithonian *B. piochii* s.s., but are conspecific with the widespread Eurasian *B. russiensis* s.l. of the middle and upper Lower Volgian. As shown earlier in the paper, this *B. russiensis* s.l. is late Early Tithonian in age and, apparently, does not range even into the earliest Late Tithonian (Figs. 5, 10). Rare representatives of true *B. piochii* s.s. were found instead to be associated with *Craspedites* s.s. and *Subscraspedites* s.s. in the next younger beds (Fig. 7).

These recent data indicate that the fauna of large *Buchia fischeriana* and dorsoplanitid ammonites are not younger than the Lower/Upper Volgian boundary beds (Fig. 9). Considering

---

EXPLANATION FOR PLATE VII

Figures 1a-k. *Praetollia (Pseudocraspedites) anglicus* Shulgina 1972 (incl. of "*Subcraspedites*" *stenomphalus* Shulgina 1972 *non* Sowerby 1823). GSC No. 73708. GSC loc. 70387. Canadian Arctic Archipelago, N.W.T., District of Franklin. Mackenzie King Island, Section 65-MA-1 situated 24 to 32 km NE of McConnell Island. Collected by Petropar Canada Ltd. 1965. Mould Bay Formation. Upper part of *Buchia terebratuloides* s.l. Zone (=*Chetaites subiricus* Zone of Northern Siberia). Uppermost Tithonian (=lowermost Riasanian or uppermost Volgian). 1a) lateral view of penultimate whorl; 1b) ventral view and cross-section of the same; 1c) lateral view of the oralmost preserved whorl fragment that includes the beginning of ?adult body chamber; 1d) ventral view of the same; 1e-g) lateral view, ventral view and the cross-section of earliest accessible whorls; 1h-j) lateral view, ventral view and the cross-section of early intermediate whorl. The earlier whorls shown in Figures 1e-g are visible inside of this whorl in Figure 1j; Figure 1k) three suture lines of the penultimate whorl. These sutures are also visible in the middle of the upper part of Figure 1a. All figures x 1.

PLATE VIII

1a
1b
1c
2a
2b
2c
3a
3b
4a
4b
4c
5a
5b
5c
6a
6b
6c
7a
7b
7c
8a
8b
8c
9a
9b
9c
10a
10b
10c
10d
10e
11a
11b
12a
12b
12c
13a
13b
13c
14a
14b

EXPLANATION FOR PLATE VIII

All photographs are natural size

Figures 1a-c. *Buchia unschensis* (Pavlow 1907) s.s. GSC Cat. 73709. GSC loc. 91329 (=52488). Uppermost Tithonian (=lowermost Riasanian or uppermost Volgian), upper part of *Buchia terebratuloides* s.l. Zone (its *Praetollia fedorovi* (=*antiqua*) beds). Deer Bay Formation. Canadian Arctic Archipelago, N.W.T., District of Franklin, Axel Heiberg Island, measured section on Awinghak River situated 6.4 km southwest of Buchanan Lake. Collected 128 metres above base of the Deer Bay Formation (see text, Figs. 6, 7 for further details) by Dr. E. Kemper, 1976. 1a) lateral view of the exterior; 1b) anterior view; 1c) beak part and hinge margin viewed from above. A very typical medium-sized left valve, except in its beak being markedly right handedly incurved and the crest line being posteriorly convex like that of *B. fischeriana* or *B.* ex gr. *okensis*.

Figures 2a-c. *Buchia unschensis* (Pavlow 1907) s.s. GSC Cat. 73710. The same locality, etc. as for Figure 1. 2a) lateral view of the exterior; 2b) posterior view; 2c) beak part and hinge margin viewed from above. A typical, halfgrown left valve.

Figures 3a, b. *Buchia* n. sp. aff. *okensis* (Pavlow 1907). GSC Cat. 73711. The same locality, etc. as for Figure 1. 3a) lateral view of the exterior; 3b) anterior view (somewhat oblique). Fragment of a large, rather typical left valve matching that reproduced in Plate I, Figure 6.

Figures 4a-c. *Buchia unschensis* (Pavlow 1907) s.s. GSC Cat. 73712. The same locality, etc. as for figure 1. 4a) lateral view of the exterior; 4b) posterior view; 4c) beak part and hinge margin viewed from above. A broad asymmetrically triangular halfgrown left valve.

Figures 5a-c. *Buchia unschensis* (Pavlow 1907) s.s. GSC Cat. 73713. The same locality, etc. as for Figure 1. 5a) lateral view of the exterior; 5b) anterior view; 5c) beak part and hinge margin viewed from above. A typical, late juvenile left valve.

Figures 6a-c. *Buchia* ex aff. *unschensis* (Pavlow 1907) s.s. GSC Cat. 73714. The same locality, etc. as for Figure 1. 6a) lateral view of the exterior; 6b) anterior view; 6c) beak part and hinge margin viewed from above. A large left valve of a *Buchia* form that matches the variant of *B. unschensis* s.s. described and figured by *Jeletzky* (1966, p. 37, 38, Pl. VI, Figs. 7, 8) from *Craspedites (Taimyroceras?) canadensis* beds in its shape and large angular posterior wing. However, it has a fine, closely spaced and faint ribbing resembling that of *B. terebratuloides* s.l. from the *Praetollia fedorovi* beds (Jeletzky, 1973, Pl. IV, Figs. 2, 3; Pl. V, fig. 2) and may conceivably be an extreme variant of this species. Zakharov (1981, p. 110, Pl. XXX, figs. 2, 4, 5) treats morphologically similar forms as "smooth forms of *B. unschensis* s.s.").

Figures 7a-c. *Buchia terebratuloides* (Lahusen 1888) var. *subuncitoides* (Bodylevsky 1936). GSC Cat. 73715. The same locality, etc. as in Figure 1. 7a) lateral view of the exterior; 7b) posterior view; 7c) beak part and hinge margin (concealed behind the beak) viewed from above. A very large but incomplete left valve resembling closely that figured by Jeletzky (1973, Pl. 4, Fig. 2) from the same locality.

Figures 8a-c. *Buchia* n. sp. aff. *okensis* (Pavlow 1907). GSC Cat. 73716. GSC loc. C-33897. Uppermost Tithonian (=lowermost Riasanian or uppermost Volgian), upper part of *Buchia terebratuloides* s.l. Zone (a faunal phase of its *Praetollia fedorovi* (=*antiqua*) beds). Deer Bay Formation. Axel Heiberg Island, Gibbs Fiord, Lat. 79°52'.; Long. 87°37'W. Collected by H.R. Balkwill, 1974. 8a) lateral view of the exterior; 8b) posterior view; 8c) beak part and hinge line viewed from above. Left valve of a small, presumably juvenile shell that matches closely those reproduced in Plate VI, Figure 3 and Plate III, Figure 4.

Figures 9a-c. *Buchia* n. sp. aff. *okensis* (Pavlow 1907). GSC Cat. 73717. The same locality, etc. as in Figure 8. 9a) lateral view of the exterior; 9b) anterior view; 9c) beak part and hinge margin viewed from above. A very typical, somewhat deformed adult left valve. The concentric ribbing is prominent, coarse and sparce enough to place this specimen in *B. okensis* s.s. However, its much longer than wide, pronouncedly triangular shape and very small posterior wing are typical of *B.* n. sp. aff. *okensis*. The valve matches very closely those of the specimens from western British Columbia reproduced in Plate I, Figure 7 and Plate III, Figure 3.

Figures 10a-e. *Buchia unschensis* (Pavlow 1907) s.s. GSC Cat. 73718. The same locality, etc. as in Figure 8. 10a) lateral view of the exterior of the left valve; 10b) lateral view of the exterior of the right valve; 10c) posterior view; 10d) anterior view; 10e) beak parts and hinge margins of both valves viewed from above. A bivalve, late juvenile shell that is typical of the species, except in an almost complete absence of a posterior ear. This results in its looking like a small *B. keyserlingi* (Lahusen).

Figures 11a, b. *Buchia* cf. *B.* n. sp. aff. *okensis* (Pavlow 1907). GSC Cat. 73719. The same locality, etc. as in Figure 8. 11a) lateral view of the exterior; 11b) posterior view. A large right valve probably belonging to a more closely and irregularly ribbed variant of *B.* n. sp. aff. *okensis*. However, it could also belong to a large and somewhat prominently ribbed representative of *B. terebratuloides* var. *occidentalis* (Anderson) or var. *subuncitoides* (Bodylevsky).

Figures 12a-c. *Buchia* n. sp. aff. *okensis* (Pavlow 1907) var. GSC Cat. 73720. The same locality, etc. as in Figure 8. 12a) lateral view of the exterior; 12b) anterior view; 12c) beak part and hinge margin viewed from above. A large adult, excellently preserved (mostly shell-covered) left valve that has a sculpture of *B. terebratuloides* s.l. in its early and middle growth stages and only acquires prominent, coarse and widely spaced concentric ribs characteristic of *B.* ex gr. *okensis* late in its ontogeny. The shape and proportions are typical of *B.* n. sp. aff. *okensis*.

Figures 13a-c. *Buchia okensis* (Pavlow 1907) s.s. GSC Cat. 73721. The same locality, etc. as in Figure 8. 13a) lateral view of the exterior; 13b) posterior view; 13c) beak part and hinge margin viewed from above. A typical, medium-sized left valve of *B. okensis* s.s. combining prominent, heavy and widely spaced concentric ribs approaching those of *B. okensis* var. *canadiana* Crickmay 1930 with anteriorly convex crest line and wider than high proportions of the shell. However, the beak itself is somewhat left-handedly incurved (Fig. 13c).

Figures 14a, b. *Buchia volgensis* (Lahusen 1888) s.s. GSC Cat. 73724. GSC loc. 91330. Basal Berriasian (*Buchia okensis* s.s. and *Praetollia (Pseudocraspedites)* aff. *anglicus* Zone). Canadian Arctic Archipelago, District of Franklin, Axel Heiberg Island, Avinghak River Section, about 7 km southwest of Buchanan Lake, 204 m level below the base of Isachsen Formation. This fossil locality is the same as the GSC locs. 26171, 52486 and 52487 (see text for further details.) 14a) lateral view of the exterior; 14b) anterior view.

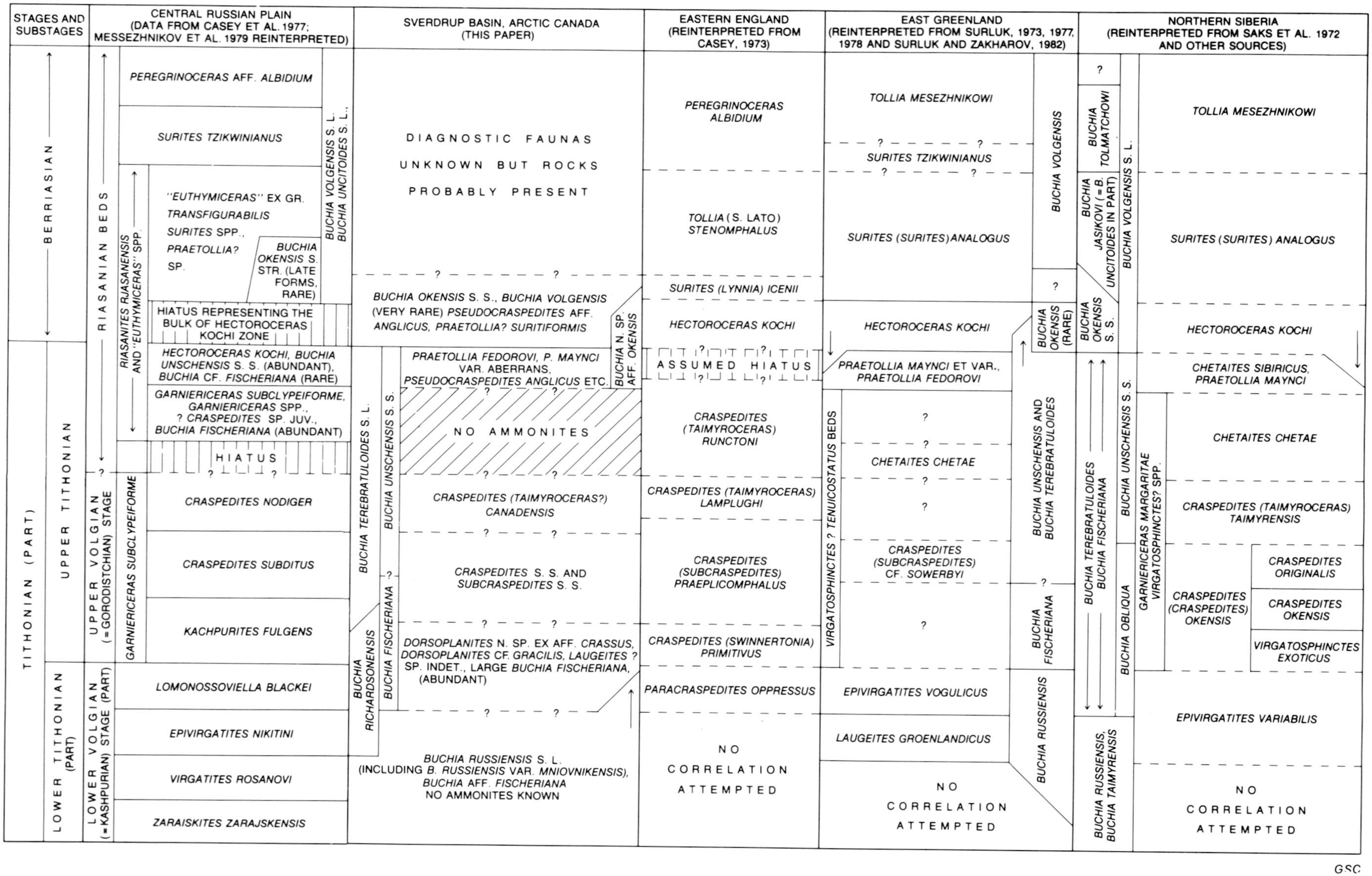

Figure 9. External Correlation and Age of the Jurassic-Cretaceous Boundary beds of the Sverdrup Basin, Canadian Arctic Archipelago within the Boreal Realm.

that the *Dorsoplanites gracilis* Spath and allied *Dorsoplanites* forms appear to be confined to the upper part of the Lower Volgian (e.g., Hakkansen *et al.*, 1981), one may feel inclined to assign a latest Early Volgian age to this fauna and to correlate it with the *Epirvirgatites variabilis* Zone of northern Siberia and the *Epivirgatites nikitini* and *Lomonossovella blakei* Zones of Central Russia. This conclusion appears to be supported by the rare finds of the so far strictly early Late Tithonian *B. piochii* s.s. in the immediately overlying beds of the Rollrock section (p. 219, Fig. 7). On the whole, however, the data available are not sufficient for such a refined correlation of this Canadian fauna, which is left to straddle the Lower/Upper Volgian (and accordingly the Lower/Upper Tithonian) boundary for the time being (Figs. 9, 10). Because of this new dating of the fauna concerned and the presence of locally numerous smaller forms of *B. fischeriana* in the overlying beds with *Craspedites* s.s.-*Subcraspedites* s.s., the large forms of *B. fischeriana* described above are confined to the lower part of *Buchia fischeriana* Zone as it is developed elsewhere in the Boreal Realm (i.e., Zakharov, 1981) and appear to be primitive rather than advanced forms of the species. It would appear that it is mostly the younger (i.e., advanced) smaller forms of *Buchia fischeriana,* which occur in the next younger *Craspedites* s.s.-*Subcraspedites* s.s. beds, that did penetrate into western British Columbia and northwestern California where they mixed with the endemic early Late Tithonian *Buchia* ex gr. *piochii-lahuseni.* Hence, the *Buchia fischeriana* Zone of those regions is largely (or even entirely?) younger than the restricted *Buchia fischeriana* Zone of the Sverdrup Basin.

Buchia terebratuloides s.l. Zone

Stratigraphical and Time Range in Sverdrup Basin. As shown by the sections (p. 217-223), *Buchia terebratuloides* s.l. ranges from the base of the *Craspedites* s.s.-*Subscraspedites* s.s. beds through the *Taimyroceras? canadensis* beds to the top of the *Praetollia fedorovi* beds (Fig. 7). These beds comprise its range zone in the Sverdrup Basin. *B. terebratuloides* s.l. is not known to range either down into the beds with large *Buchia fischeriana* and dorsoplanitid ammonites or up into the *Buchia okensis* s.s. Zone. However, *B. terebratuloides* s.l. is not equally common in all parts of its zone in the Sverdrup Basin. While it is dominant locally as, for example, in the basal part of the *Craspedites* s.s.-*Subcraspedites* s.s. beds in the Rollrock Lake section or in the *Praetollia fedorovi* beds in the Buchanan Lake section (Fig. 7), it is only a minor element of the *Buchia* fauna in the *Craspedites* (*Taimyroceras*?) *canadensis* beds in the Slidre Fiord (Camp Creek) section (Jeletzky, 1966, p. 39, 40). However, *B. terebratuloides* s.l. is known to be common in the equivalent *Taimyroceras taimyrensis* Zone in northern Siberia (Zakharov, 1981, p. 106, 107). So far as it is known, all morphological forms of *B. terebratuloides* s.l. (i.e., vars. *typica, obliqua, subinflata* and *occidentalis*) range throughout the above discussed stratigraphical interval. However, in contrast with western British Columbia (p. 194), *B. t.* var. *obliqua* is more common to prevalent while *B. t.* var. *subinflata* (Pavlow 1907) is rare in the Sverdrup Basin.

Other Diagnostic Buchia Forms. Beside its name fossil, the *Buchia terebratuloides* s.l. Zone of Sverdrup Basin contains *Buchia* n. sp. aff. *okensis* (Pavlow), *B. fischeriana* (d'Orbigny), *B. unschensis* (Pavlow) s.s. and *B. okensis* (Pavlow) s.s.

B. n. sp. aff. *okensis* is restricted to the uppermost part of this zone that is designated below as *Praetollia fedorovi* beds. It is common in its Buchanan Lake section and other biofaces of these beds which have yielded *Praetollia* (*Pseudocraspedites*) *anglicus* and *Praetollia maynci* var. *aberrans.* It is associated with the oldest known, very rare representatives of *B. okensis* s.s. at all of its principal localities. *B.* n. sp. aff. *okensis* ranges up into the next younger *Buchia okensis* s.s. Zone where it may be fairly to rather common, but always less so than its name fossil. However, it is not known to range down even into the *Taimyroceras? canadensis* beds.

Buchia unschensis s.s. ranges through the bulk of the *B. terebratuloides* s.l. Zone in the Sverdrup Basin, but is completely unknown in the lower part of its *Craspedites* s.s.-*Subcraspedites* s.s. beds in the Rollrock Lake section (p. 219, Figs. 7, 9). Since no other sections containing these ammonites are known, this absence may be an accident of collecting. The abundance ratios of *Buchia unschensis* s.s. in relation to those of other buchias vary very strongly from one locality to another, without any apparent stratigraphical regularity. However, the data now available are scarce.

Buchia fischeriana appears to be restricted to the lower part of *B. terebratuloides* s.l. Zone (e.g., the lower part of its *Craspedites* s.s.-*Subcraspedites* s.s. beds in the Rollrock Lake section). Here again, however, a number of fossil collections containing *B. fischeriana* in association with *B. terebratuloides* s.l. and/or *B. unschensis* s.s. lack ammonites and are of an uncertain stratigraphic position within the zone. Some of them could conceivably be from its higher beds. Nevertheless, the absence of *B. fischeriana* in the *Taimyroceras? canadensis* bed and in all biofacies of the *Praetollia fedorovi* beds suggests strongly that it does not range into that upper part of *Buchia terebratuloides* s.l. Zone in the Sverdrup Basin.

Buchia terebratuloides s.l. Zone versus Buchia unschensis s.s. Zone. Zakharov (1979, p. 126, corr. table; 1981, p. 116, 224-226, Fig. 3) has recently erected the *Buchia unschensis* Zone for the upper part of the Upper Volgian of northern Eurasia. The scope of this zone approximates that of the *Buchia terebratuloides* s.l. Zone proposed previously by Jeletzky (in Jeletzky and Tipper, 1968, p. 10, Table 1) for the uppermost Jurassic in western British Columbia and extended into the Sverdrup Basin in this paper (except possibly for the basal part of the latter). The following relative merits of *B. terebratuloides* s.l. and *B. unschensis* s.s. as zonal indices have convinced me to reject the *Buchia unschensis* s.s. Zone.

When originally introducing the term "*Craspedites* (*Taimyroceras*?) *canadensis* n. sp. and *Buchia unschensis* emend. fauna" for the latest Volgian fauna then known to exist in the Sverdrup Basin, Jeletzky (1966, p. 39, 40, 43) did not intend to use *B. unschensis* s.s. as a zonal fossil. He (1966, p. 40) states: "To our knowledge at least, *Buchia unschensis* is thus restricted to the upper part of the upper Volgian stage and does not occur either in the underlying or in the overlying rocks. Because of its rather erratic distribution, however, this probably facies-bound *Buchia* species can hardly be considered a reliable zonal index fossil despite its rather wide geographical range." These conclusions, which agreed with the earlier conclusions of Sokolov (1926, p. 34), were confirmed more recently by Messenzhikov *et al.*, (1979, p. 79, 80) and Zakharov (1979, p. 125, 126).

Zakharov (1979, p. 126; 1982, p. 114-116, 224-226, Table 3) has shown that *B. unschensis* s.s. is widespread in Northern Eurasia where it ranges from the *Craspedites okensis* Zone to the basal part of *Hectoroceras kochi* Zone. However, the base of its zone was drawn at the base of the *Taimyroceras taimyrensis* Zone while its top was placed at the base of *Buchia okensis* Zone, which according to Zakharov coincides with the boundary between the *Chetaites sibiricus* and *Hectoroceras kochi* Zones. This time range roughly corresponds to the time ranges combined of his *B. obliqua* and *B. terebratuloides* (i.e., *B. terebratuloids* sensu Jeletzky; see Figs. 9, 10). In East Greenland *B. unschensis* s.s. and *B. terebratuloides* s.l. apparently range up into the basal part of the *Hectoroceras kochi* Zone where they are associated with *B. okensis* var. *canadiana* (Suryk and Zakharov, 1982, p. 749).

I agree with Zakharov (1979, 1981, p. 226) that the time range of *B. unschensis* s.s. in the Sverdrup Basin is wider than it appeared formerly (Jeletzky, 1966) and about the same as in northern Eurasia. This species appears now to have about the same time range as *B. terebratuloides* s.l. over most (?or all) of the Boreal Realm and to be a much more common fossil there than previously believed. Nevertheless, Zakharov's selection of *B. unschensis* s.s. as a zonal index for these uppermost Volgian beds is not acceptable. In spite of the above developments, the geographical range of *B. unschensis* s.s. remains much more restricted than that of *B. terebratuloides* s.l. Furthermore, its distribution within the marginal belt of the Boreal Realm and the adjacent parts of the Tethyan Belt (e.g., North Pacific Province of North America) remains more erratic than that of *B. terebratuloides* s.l. For example, *B. unschensis* s.s. is a very rare, erratically present fossil in the Porcupine Plain-Richardson Mountains Trough where it was only encountered a couple of times in the uppermost Jurassic beds, during intensive search. It is absent in the only locality in Northwestern Yukon where *Praetollia* (*Praetollia*) *fedorovi* (=*antiqua*) was found associated with an abundant *B. terebratuloies* s.l. fauna (Jeletzky, 1973, p. 50). As mentioned earlier in this paper (p. 197, 198), *B. unschensis* s.s. occurs even more rarely and sporadically in western British Columbia and northwestern California where *B. terebratuloides* s.l. remains the most common and diagnostic fossil of its zone. Finally, the recent research (Zakharov, 1981, p. 225, 226; Messezhnikov *et al.*, 1979, p. 79, 80) confirmed the erratic, obviously climatically controlled occurrence of *B. unschensis* s.s. in the southern regions of the Boreal Realm of U.S.S.R. Therefore, *B. terebratuloides* s.l. is rated as a much superior biochronological marker than *B. unschensis* s.s. and is retained as the index fossil of this *Buchia* Zone.

Time Range in Terms of Tethyan Zones and Stages. The dating of the *Buchia terebratuloides* s.l. Zone of the Sverdrup Basin in terms of the Tethyan standard zones and stages depends entirely on the comparison of its *Buchia* fauna and the *Buchia* faunas of adjacent Sverdrup Basin zones with the *Buchia* faunas of western British Columbia, the Tethyan ages of which were worked out in detail above (Fig. 5).

In addition to the equally abundant occurrence of *Buchia terebratuloides* s.l. in these two regions and the rare presence of *B. unschensis* s.s. in the *B. terebratuloides* Zone of western British Columbia, an essential contemporaneity of the *B. terebratuloides* Zone of the Sverdrup Basin with its counterpart in western British Columbia is indicated by an almost total restriction of *B. okensis* s.s. to beds that gradationally overlie this zone in both of these regions (see Jeletzky, 1973). Furthermore, *Buchia* n. sp. aff. *okensis* has about the same time range in the two regions, from the upper part of *B. terebratuloides* s.l. Zone well up into the *B. okensis* s.s. Zone (Figs. 5, 10). Finally, the beds containing *B. terebratuloides* s.l. in the Sverdrup Basin are underlain conformably and gradationally, first by beds containing abundant *B. fischeriana* and rare *B. piochii* s.s. (in the upper part), and then by beds containing *B. russiensis* s.l. These species recur in approximately the same sequence and stratigraphic position in western British Columbia (Fig. 10). This overall similarity of zonal sequences of adjacent *Buchia* Zones in these two regions (Fig. 5, 10) attests that the *B. terebratuloides* s.l. Zone of the Sverdrup Basin is entirely uppermost Tithonian.

Ammonite Faunas of *Buchia terebratuloides* s.l. Zone

The biochronologically important ammonite faunas of this zone are discussed below in ascending order (Figs. 9, 10). These beds cannot yet be treated either as regional or as provincial (i.e., Boreal) biozones because of a scarcity of material and are designated as informal fossil beds. However, most or ?all of these regional fossil beds are expected to become the basis of a new regional zonal standard valid for all of the Canadian Arctic basins, at least.

Taxonomic Status of Diagnostic Craspedititds. For reasons presented elsewhere (Jeletzky, 1979, p. 3-9), only the following two full genera of round-ventered Late Tithonian (=Late Volgian) to basal Berriasian craspeditids are considered as valid.

1) *Craspedites* Pavlow 1892, with subgenera *Craspedites* s.s. (=*Trautscholdiceras* Sazonova 1977), *Swinnertonia* Shulgina *1972 emend.* Casey 1973, *Subcraspedites* Spath 1924 *sensu* Casey 1973, and *Taimyroceras* Bodylevsky 1960 (=*Volgidiscus* Casey 1973 =*Runctonia* Casey 1973). This genus, as interpreted herein, is distinguished by an adolescent to adult external septal suture with only three auxiliary lobes in addition to two lateral lobes and mostly subradially oriented (see Pl. V, Figs. 1, 5a, 5c; Pl. VI, Figs. 4a, 4b and caption). This suture may be either slightly ascendant or slightly descendant, especially in the adumbilical part, but not pronouncedly ascendant. *Craspedites* s.l. appears to be restricted to the Late Tithonian (i.e., Late Volgian) older than the *Chetaites sibiricus* Zone.

2) *Praetollia* Spath 1952, with subgenera *Praetollia* s.s. (=*Borealites* Klimova 1969), *Ronkinites* Shulgina 1972, and *Pseudocraspedites* Casey, Messezhnikov and Shulgina 1977. This genus is distinguished by an adolescent to adult external septal suture with four to five auxiliary lobes in addition to two lateral lobes, always pronouncedly (i.e., from 20° to 40°) ascendant, and straight to almost straight through all or most of its extent. *Praetollia* is a direct descendant of *Craspedites* s.l. (probably through *Taimyroceras*). It appears to be restricted to the uppermost Tithonian (i.e., *Chetaites sibiricus* Zone and its equivalents) and the basal Berriasian (i.e., *Hectoroceras kochi* Zone proper and its equivalents). *Praetollia* s.l. appears to merge imperceptibly into *Surites* Sazonov 1951 via some poorly understood *Pseudocraspedites*-like forms (e.g., *Praetollia* (*Praetollia*?) *suritiformis* Klimova 1972) of the restricted *Hectoroceras kochi* Zone.

All morphologic features (e.g., ribbing habit, shape and

proportions of the whorl, etc.) used by Shulgina (in Saks *et al.*, 1972), Casey (1973), Klimova (1969; in Saks *et al.*, 1972) and Casey *et al.*, (1977) to support the full generic status of all other round-ventered craspeditid taxa, are extremely variable and hence trivial and low-ranking (Jeletzky, 1979, p. 2-9). The mostly specific to infraspecific (only rarely subgeneric) value of these features is further decreased by the fact that "genera" and even "subgenera" based on them tend to occur in the same or almost the same beds and that their morphologies freely intergrade horizontally and vertically. Finally, all of these morphological features occurring in the subgenera of *Praetollia* s.l. recur repeatedly in most (or all?) subgenera of the ancestral *Craspedites* s.l. This contrasts with the external septal suture that apparently evolved irreversibly throughout the known evolutionary history of the *Craspedites* s.l. and *Praetollia* s.l.

Craspedites s.s.-Subcraspedites s.s. Beds. The ammonites of the lower locality C-55248 at the 171 m level of the Rollrock Lake section are so similar to those of its upper locality C-55251 at the 190.5 m level that the two are treated below as the same fauna.

This discovery of the *Craspedites* s.s. and *Subcraspedites* s.s. forms, either conspecific with or closely allied to the Central Russian and East English species, permits for the first time a direct zonal correlation of the provincial Upper Volgian ammonite and *Buchia* faunas of Arctic Canada with the standard ammonite zones of these faunal provinces of northern Europe (Fig. 9). This provides an important control of the earlier correlations (Jeletzky 1971a, 1973, p. 43), which utilized the North Siberian ammonite and *Buchia* faunas as an indispensable steeping stone.

The close similarity (?specific identity) of *Craspedites* (*Subcraspedites*) cf. *sowerbyi* Spath (Pl. V, Fig. 1) from the Rollrock Lake section with typical English representatives of the species, indicates the equivalence of the *Craspedites* s.s.-*Subcraspedites* s.s. beds to part or all of the English *Craspedites praeplicomphalus* Zone (see Casey, 1973, p. 220-222, Table 2). The presence of *Craspedites* (*Subcraspedites*) n. sp. aff. *praeplicomphalus* (Swinnerton) in this Canadian fauna supports the correlation which is proposed (Fig. 9). The presence of *Craspedites* (*Craspedites*) n. sp. aff. *subditus* (Trautschold) (Pl. V, Fig. 5; Pl. VI, Fig. 1) in the Canadian fauna links it with the *Craspedites subditus* s.s. Zone of the Central Russian Basin. Furthermore, the Canadian fauna may include some (?upper only) part of the underlying *Kachpurites fulgens* Zone of Central Russia because of a close similarity of its *C.* (*S.*) n. sp. aff. *praeplicomphalus* to the *Craspedites* (*Subcraspedites*) sp. described and figured by Gerassimov (1969, p. 93, Pl. XXX, Fig. 3) from that Central Russian zone.

The proposed correlations of the *Craspedites* s.s.-*Subcraspedites* s.s. beds invalidate Casey's (1973, p. 219, 220, Table 2) correlation of the *Craspedites praeplicomphalus* Zone with the *Craspedites nodiger* Zone of Central Russia and the *Taimyroceras taimyrensis* Zone of Northern Siberia. This correlation was already questioned by Casey *et al.* (1977, p. 19), who noted (writer's translation from Russian): "However, *Craspedites arcticus* Shulgina from the Okensis Zone (Originalis Subzone) on Bojarka River, basin of Khatanga River (Shulgina, 1969) may be a *Subcraspedites* s. str. as it was established during a joint study of the material. This indicates that the lower part of Praeplicomphalus Zone, where *Subcraspedites* s. str. make their first appearance, may correspond also to the upper part of the Okensis Zone." Casey (1973) correlated the *Craspedites praeplicomphalus* Zone with the *Craspedites nodiger* and *Taimyroceras taimyrensis* Zones because of the *C.* (*C.*) *nodiger*-like morphology of the *Craspedites* (*Craspedites*) *plicomphalus* (J. Sowerby 1822) present in that East English zone. However, this evidence is inconclusive because of the long known intergradation of *C.* (*C.*) *nodiger* with its predecessor *C.* (*C.*) *okensis*. One may, therefore, expect to encounter the first, rare *C.* (*C.*) nodiger-like forms in the next older, *Craspedites subditus* Zone and its English equivalents. The other evidence presented above indicates the correctness of this explanation (Fig. 9).

The pre-Nodiger age of the *Craspedites praeplicomphalus* Zone and the Canadian *Craspedites* s.s.-*Subcraspedites* s.s. beds is also supported by their inferred older age in comparison with the *Taimyroceras canadensis* beds documented below.

Buchia richardsonensis and *B. terebratuloides* s.l. occur in the lower part of the *Craspedites* s.s.-*Subcraspedites* s.s. beds together with numerous, small to medium-sized *B. fischeriana* and rare *B. piochii* s.s. However, the upper part of these beds has yielded only *B. unschensis* s.s. This suggests that the lower part of the *Craspedites* s.s.-*Subcraspedites* s.s. beds corresponds to the upper part of *Buchia piochii* s.s. Zone and to part or all of the *Buchia fischeriana* Zone of western British Columbia. This part of the *Craspedites* s.s.-*Subcraspedites* s.s. beds is also correlative with the beds containing a mixed *Buchia piochii* s.s. and *B. fischeriana* fauna in northwestern California (Fig. 5). Only their upper part with an apparently monospecific fauna of *B. unschensis* s.s., appears to be coeval with the basal part of the *B. terebratuloides* s.l. Zone of western British Columbia, considering that this species ranges somewhat below its proper zone in that region (Fig. 10). These data confirm the earlier conclusion that the *Craspedites* s.s.-*Subcraspedites* s.s. beds and the equivalent *Craspedites praeplicomphalus* and *Craspedites subditus* Zones are early Late Tithonian rather than late Early Tithonian.

Taimyroceras? canadensis Beds. The *Craspedites* s.s.-*Subcraspedites* s.s. and *Craspedites* (*Taimyroceras*?) *canadensis* faunas were not yet found in superposition in the same continuous section. However, the former fauna is definitely not younger than the English *Craspedites praeplicomphalus* Zone, the equivalent Central Russian *Craspedites subditus* Zone and the equivalent *Craspedites originalis* Subzone of the North Siberian Okensis Zone (Figs. 9, 10). The *Taimyroceras? canadensis* fauna is, in contrast, definitely coeval with the next younger, *Taimyroceras taimyrensis* Zone of Northern Siberia because of the recorded presence of *C.* (*T.*?) *canadensis canadensis*, C. (*T.*?) *canadensis pseudosubditus* and *C.* (*T.*?) *canadensis* aff. *pseudosubditus* in the Urdyuk-Khaia Point (Paks Peninsula) section of that zone (Bassov *et al.*, in Saks *et al.*, 1972, p. 35).

In Central Russia, the *Taimyroceras? canadensis* beds are coeval with the *Craspedites nodiger* Zone because of a close morphological similarity of the Canadian species with *C.* (*C.*) *nodiger* (Jeletzky, 1966, p. 19, 20). Furthermore, the *Taimyroceras taimyrensis* Zone of northern Siberia contains *C.* (*C.*) *pseudonodiger* Shulgina. As noted by Shulgina (in Saks *et al.*, 1969, p. 136, footnote, 144-148), this form occurs also in the *Craspedites nodiger* Zone of Central Russia in association with

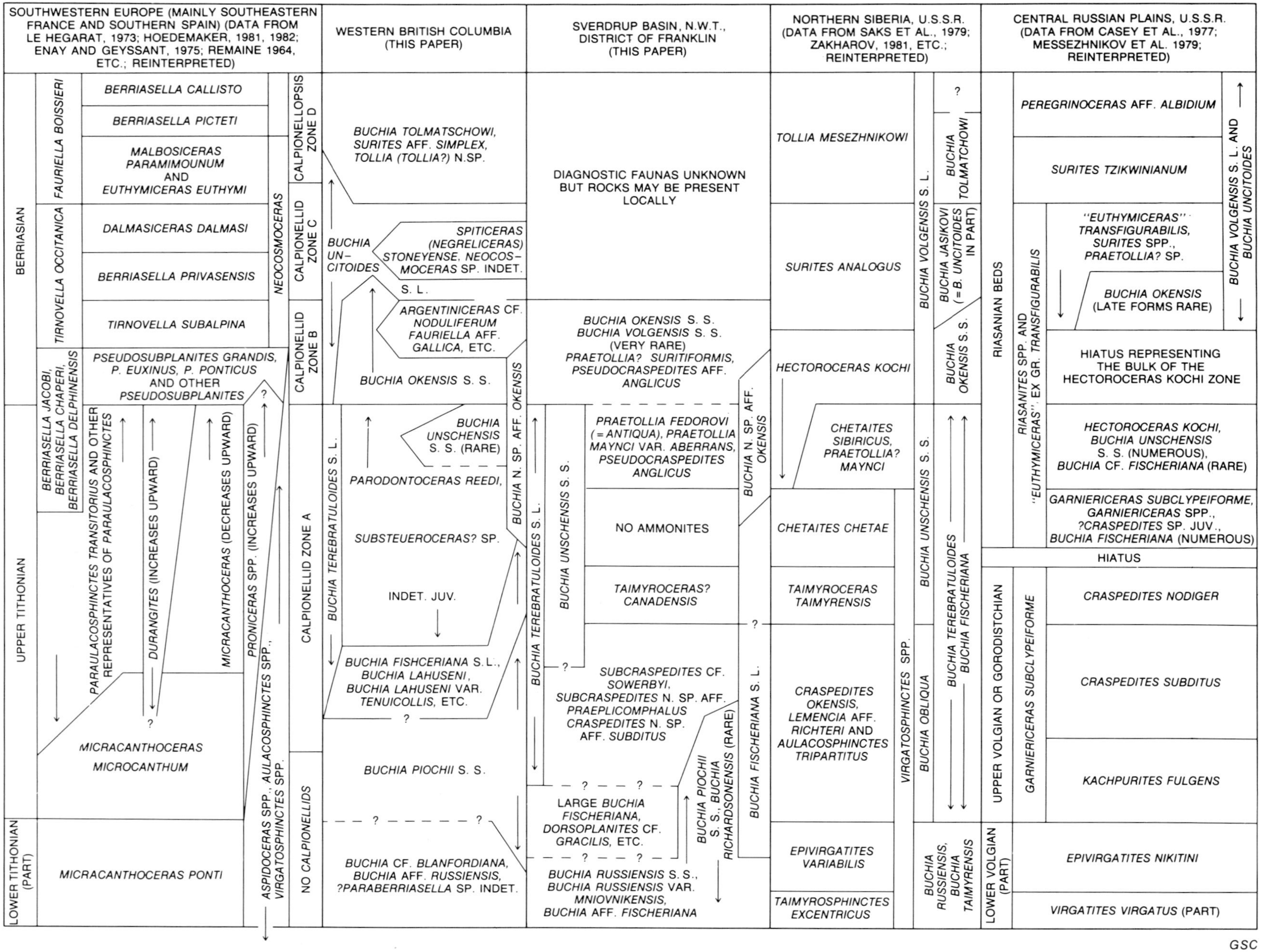

Figure 10. Proposed correlation of Jurassic-Cretaceous Boundary Beds of Western and Arctic Canada with the standard zones and stages of the Boreal and Tethyan Realm.

C. (*C.*) *nodiger*. The very close affinity of *C.* (*C.*) *nodiger pseudonodiger* Shulgina 1969 (especially of the specimen reproduced in Pl. XXXI, Fig. 2; Pl. XXXII, Fig. 1) with *C.* (*Taimyroceras*?) *canadensis eurekae* Jeletzky (1966, Pl. IV, Fig. 3) provides support for the here proposed correlation of the *Taimyroceras? canadensis* beds with the *Craspedites nodiger* Zone.

In England the *Taimyroceras? canadensis* beds correspond approximately to the "*Craspedites* (*Volgidiscus*)" *lamplughi* Zone of Casey (1973, p. 209, 210, 220, Table 2). This is indicated by the superposition of that zone on the *Craspedites praeplicomphalus* Zone and supported by Jeletzky's (1979, p. 7, footnote) conclusion that "*C.*(*V.*)" *lamplughi* is a *Taimyroceras* closely allied to *C.*(*T.*) *canadensis pseudosubditus*, and his placement of the "*Runctonia*" *runctoni* Zone into the uppermost Jurassic and its correlation with the *Chetaites chetae* Zone (Figs. 9, 10).

Because of their inferred stratigraphic position between the *Craspedites* s.s-*Subcraspedites* s.s. beds and the latest Jurassic *Praetollia fedorovi* beds representing the topmost *Buchia terebratuloides* s.l. Zone, the *Taimyroceras? canadensis* beds correspond to some middle part of the *Buchia terebratuloides* s.l. Zone of western British Columbia and northwestern California. Hence, they are mid-Upper Tithonian (Fig. 10).

Praetollia fedorovi (=Praetollia antiqua) Beds. The interpretation of the *Praetollia fedorovi* beds proposed here is much expanded and emended in comparison with that of their prototype – the *Praetollia antiqua* and *Buchia terebratuloides* s.l. beds – introduced by Jeletzky (1973, p. 47, 50, 52, Figs. 2, 3). These beds now include the *Praetollia* (*Pseudocraspedites*) *anglicus* and *Praetollia* (*Praetollia*) *maynci* var. *aberrans* faunules in addition to the *Praetollia* (*Praetollia*) *fedorovi* (formerly *Praetollia antiqua*) fauna proper. As pointed out in the stratigraphic section and Figure 7, these three ammonite faunas occur at rather distant geographical localities and are not known to have any ammonite species in common. Furthermore, their stratigraphic relationships with adjacent, diagnostic ammonite faunas are only known for the *Praetollia fedorovi* faunule proper (Buchanan Lake section). However, an approximate zonal contemporaneity of these three ammonite faunas is inferred because they all include the same, highly characteristic *Buchia* fauna, i.e., abundant *B. terebratuloides* s.l., *B. unschensis* s.s. and *B.* n. sp. aff. *okensis*, and very rare *B. okensis* s.s. (Pl. V, Figs. 2, 4, 6; Pl. VI, Figs. 2, 3, 5-7; Pl. VIII, Figs. 1-13). The same *Buchia* fauna occurs in the topmost beds of the *B. terebratuloides* s.l. Zone in the Taseko Lakes map-area (Pl. I, Figs. 6-9; Pl. III, Figs. 2-8), in western British Columbia, where this fauna is consistently restricted to the upper part of this uppermost Tithonian zone. The same appears to be true of its Sverdrup Basin equivalent, which occurs immediately below the basal Berriasian *Buchia okensis* s.s. and *Praetollia* (*Pseudocraspedites*) aff. *anglicus* fauna in the continuous Buchanan Lake section and is replaced by a different association of *Buchia* that lacks *B.* n. sp. aff. *okensis* and *B. okensis* s.s. already in the *Taimyroceras? canadensis* beds of the *Buchia terebratuloides* s.l. Zone (Fig. 7). Nor is this characteristic association of *Buchia* forms known to occur in any collection from the older beds of that zone according to my studies.

Also, the zonal contemporaneity of these three ammonite faunules is indicated by the recorded association of *Praetollia* (*Praetollia*) *fedorovi* (i.e., of *Praetollia* cf. *antiqua* of Surlyk, 1973, p. 91, Pl. 1, Fig. 2) with *P.* (*P.*) *maynci* in East Greenland.

Casey (1973, p. 246), Casey *et al.* (1977, p. 30) and Saks *et al.* (1979, p. 99, Table 2) did not document their refusal to admit Jeletzky's (1973) generic assignment of his *Praetollia antiqua*. No comparably large representatives of *P. maynci* Spath 1952 were described and figured by these workers that demonstrate the assumed persistence of the characteristically sharp-topped, fine and very closely spaced ribbing habit of the inner whorls onto the advanced whorls and the adult living chamber. The only figured, sufficiently large whorl assigned to *P. maynci* (e.g., Shulgina, in Saks *et al.*, 1972, Pl. VI, Fig. 1) is neither conspecific nor congeneric with it. As noted by Casey (1973, p. 248), this ammonite appears to be a *Ronkinites*, the inner whorls of which it strongly resembles.

This rather dogmatic attitude is refuted by the data available to the only recent worker, other than myself, who had access to an abundant material of *Praetollia* (Surlyk, 1973; Surlyk *et al.*, 1973). Surlyk (1973, Pl. 1, Fig. 2) figures and assigns correctly to *P.* cf. *antiqua* a moderately-sized ammonite, in which the typical *P. maynci*-like ornament of the early part of the whorl grades into a considerably coarser and more widely spaced *P. antiqua*-like ornament on its oral quarter. This terminal ornament is also similar to the adult ornament of "*Borealites*" *fedorovi* Klimova 1969. Furthermore, Surlyk *et al.* (1973, Pl. 4, Fig. 3) figure a comparably large shell of *P. maynci* that exhibits a similar adoral coarsening and wider spacing of primaries and secondaries. The adoral one-third of this whorl also becomes morphologically similar to the advanced whorls of *P. antiqua* and "*B.*" *fedorovi*. Finally, some of the largest known shells in the original material of *P. maynci* (e.g., its var. *contigua;* see Spath, 1952, Pl. I, Fig. 1a; Pl. III, Fig. 1;, Pl. IV, Fig. 2) exhibit a similar adoral coarsening and wider spacing of primaries and secondaries. So far as one can judge, the width of the umbilicus is rather variable in the type material of *P. maynci*, so that the most widely umbilicate *P. m.* var. *aberrans* (see Spath, 1952, Pl. III, Fig. 7; this paper, Pl. V, Fig. 3) does not seem to differ much from *P. antiqua and "B." fedorovi* in this respect also. Therefore, all presently known distinctions between *P. maynci* and *P. antiqua*, and hence between *Praetollia* Spath 1952 and *Borealites* Kilmova 1969, are extremely trivial. This and the occurrence of *P. maynci* and *P. antiqua* in the same beds in East Greenland (Surlyk, 1973, p. 91, 92) and, apparently, in the Sverdrup Basin (above and Fig. 7), indicates that these two genera are synonymous on the generic and subgeneric levels.

On the specific level, *Praetollia antiqua* Jeletzky 1973 appears to be indistinguishable from the previously published *Borealites fedorovi* Klimova 1969 in every taxonomically significant aspect of their morphology, including the adult size. Therefore, the writer has suppressed *Praetollia antiqua* in favor of *Praetollia* (*Praetollia*) *fedorovi*.

Age and Correlation. P. (*P.*) *fedorovi* (=*P.* cf. *antiqua* of Surlyk, 1973, Pl. 1, Fig. 2) coexisted with P. (*P.*) *maynci* and *Hectoroceras kochi* in East Greenland. These beds were placed later into the *Hectoroceras kochi* Zone by Surlyk (1978, p. 31; 1979, p. 196), who has defined it by the first appearance of that ammonite. However, these beds are older than the North Siberian *Hectoroceras kochi* Zone that excludes beds where *Hectoroceras* appears for the first time and coexists with *P.* (*P.*) *maynci* and *Chetaites sibiricus* (e.g., Bassov *et al.* in Saks *et al.*, 1972, p. 18, 20, 36, 50, 51, Fig. 6; Casey *et al.*, 1977, p. 31, Pl. II,

Figs. 8, 9). These beds are expressly included in the *Chetaites sibiricus* Zone by most Soviet workers. These East Greenland beds have also yielded the *Buchia terebratuloides-B. unschensis* fauna (Surlyk and Zakharov, 1982, p. 749, Tables 2, 3) that, generally speaking, does not range above the *Chetaites sibiricus* Zone (Zakharov, 1979, 1981). These data attest to the correspondence of these East Greenland beds to the upper part of the *Chetaites sibiricus* Zone and the *Praetollia fedorovi* beds of the Sverdrup Basin, including the *Praetollia anglicus* and *Praetollia maynci* var. *aberrans* biofacies of the latter (Fig. 9). Hence these two Eurasian biochronological units are also latest Tithonian deposits.

The next younger beds of this continuous East Greenland section contain *Hectoroceras kochi* associated with *Praetollia* (*Pseudocraspedites*) aff. *anglicus* and typical *B. okensis* var. *canadiana* (Surlyk, 1973, p. 91, 92; Pl, 1, Figs. 1, 3). These beds were also included in the *Hectoroceras kochi* Zone by Surlyk (1978, p. 31; 1979, p. 106). Unlike the underlying *P. maynci* and *P. fedorovi*-bearing beds, they are correlative with the North Siberian *Hectoroceras kochi* Zone. The presence of *P.* aff. *anglicus* and *B. okensis* var. *canadiana* permits a direct correlation of these East Greenland beds with the basal Berriasian *Buchia okensis* s.s.-*Pseudocraspedites* aff. *anglicus* beds of the Sverdrup Basin. The reported association of these East Greenland taxa with abundant representatives of *B. unschensis, B. terebratuloides* and *B. fischeriana* (Surlyk and Zakharov, 1982, p. 749) is unusual as similar *Buchia* associations are so far only known from the uppermost part of the *B. terebratuloides* s.l. Zone in North America. However, we may be dealing with equivalents of the basal beds of the Canadian *Buchia okensis* s.s. and *Pseudocraspedites* aff. *anglicus* Zone where one can expect the presence of similarly mixed *Buchia* associations (i.e., faunal "overlap beds"). In the Sverdrup Basin such mixed faunas are expected in the unfossiliferous 8.5 m interval that separates the GSC locs. 52488 and 52490 in the Buchanan Lake section.

The correlation of the Canadian *Praetollia fedorovi* beds with the beds in Subpolar Urals that have yielded its type material (Klimova, 1969) is complicated by the absence of *P. fedorovi* in their most complete, apparently continuous section situated on Yani-Manya River (Gol'bert *et al.*, 1972, p. 13-16; Casey *et al.*, 1977, p. 22; Gol'bert and Klimova, 1979, p. 36). The *Chetaites sibiricus* Zone (i.e., Bed 1) of that section has only yielded *Chetaites sibiricus* (Casey *et al.*, 1977, Pl. I, Fig. 3), "*Praetollia*" sp. (Casey *et al.*, 1977, Pl. II, Fig. 3) and an unfigured "*Buchia volgensis*" (Casey, *et al.*, p. 22). The ribbing habit of this "*Praetollia*" sp. resembles that of comparable whorls of *Praetollia fedorovi* (e.g., Klimova, 1969, Pl. I, Fig. 2; Jeletzky, 1973, Pl. 5, Figs. 1b, 1d). However, the reported weakening of its secondaries on the venter (Casey *et al.*, 1977, p. 31) indicates a juvenile *Taimyroceras*. This bed is, therefore, likely to represent the basal part of the *Chetaites sibiricus* Zone where neither *Hectoroceras kochi* nor *P. maynci* are known to occur. Because of this and the considerations presented below the next younger bed of the Yani-Manya section, with only *Hectoroceras tollijense* and "*Borealites*" sp. indet. and which was correlated with the *Hectoroceras kochi* Zone by Gol'bert *et al.* (1972, p. 16-18) and Gol'bert and Kilmova (1979, p. 3), may well be older and coeval with the *Praetollia fedorovi* beds of the Sverdrup Basin and East Greenland.

All known material of *P. fedorovi* was found in the Yatria River section where the gritty and pebbly sandstones assigned to the *Hectoroceras kochi* Zone proper overlie disconformably older Upper Tithonian (i.e., older Upper Volgian) beds with *Craspedites* (*Taimyroceras*) cf. *taimyrensis* (Klimova 1969, p. 128, 129; Gol'bert *et al.*, 1972, p. 19, 31, 34). This richly fossiliferous but thin (about 1.2m) Bed 1 has yielded (Gol'bert and Klimova, 1972, p. 37): "*Hectoroceras tollijense* (Nik.), *Borealites fedorovi* Klim., *B. radialis* Klim., *B. explicatus* Klim., *B. mirus* Klim." Specifically identifiable buchias are absent according to these workers and Zakharov (1979, p. 126). I disagree with the dating of these *P. fedorovi*-bearing beds and correlate them with the upper part of the *Chetaites sibiricus* Zone and the Canadian *Praetollia fedorovi* beds. My reasons are: 1) The genus *Hectoroceras* and the species *Hectoroceras kochi* appear already in the upper part of the *Chetaites sibiricus* Zone in Northern Siberia while *P.* (*P.*) *fedorovi* is confined to the equivalent beds in East Greenland (again with *Hectoroceras*) and the Sverdrup Basin. Therefore, the presence of *Hectoroceras* is only diagnostic of the upper part of the *Chetaites sibiricus* and the *Hectoroceras kochi* Zones combined. 2) The generic nature and age significance of "*Hectoroceras*" *tollijense* are highly questionable and Sazonova (1977, p. 90) is possibly right in rejecting its assignment to *Shulginites* Casey 1973 and in treating it instead as a latest Volgian *Garniericeras* (compare Jeletzky, 1979, p. 6, footnote). 3) Even if Casey (1973, p. 240, 243) should be right in treating the "*Hectoroceras tollijense*" of Klimova (in Saks *et al.*, 1972, p. 202-204, Pl. XL, Figs. 1-4) from Bed 1 as true *Hectoroceras* closely related to but not conspecific with *H. kochi*. this would not indicate that this bed represents the *Hectoroceras kochi* Zone proper and is basal Berriasian for reasons presented under (1). 4) Bed 2 of the Yatria section, which apparently overlies gradationally its "*H. tollijense*"-and *P. fedorovi*-bearing Bed 1 (Gol'bert and Klimova, 1979, p. 37), contains *Praetollia* (*Praetollia*?) *surітiformis* (Klimova 1972) that is confined to the basal Berriasian *Buchia okensis* s.s. and *Pseudocraspedites* aff. *anglicus* Zone in the Sverdrup Basin (see next section).

In England the *Praetollia fedorovi* beds and the equivalent (?upper only) part of the *Chetaites sibiricus* Zone appear to be represented by a regional hiatus that separates the regional "*Runctonia*" *runctoni* Zone from the overlying *Hectoroceras kochi* Zone (Fig. 9). The allegedly basal Cretaceous "*Runctonia*" *runctoni* (Casey 1973, p. 210, 220, Table 2) is a *Taimyroceras* (Jeletzky 1979, p. 6, footnote). This subgenus appears to be largely or ?entirely restricted to beds older than the *Chetaites sibiricus* Zone and its equivalents (*Taimyroceras pulcher* of Casey, Messezhnikov, and Shulgina 1977 may or may not be an exception). Furthermore, the next older regional zone of *Craspedites lamplughi* appears to be coeval *not* with the North Siberian *Chetaites chetae* Zone but with the next older *Craspedites taimyrense* Zone and its equivalents (see p. 231 and Fig. 9).

The assignment of *Runctonia* to *Praetollia* proposed by Casey *et al.*, (1977, p. 31), is inadmissible because of the marked weakening to interruption of secondary ribs on its venter. This feature is diagnostic of the subgenus *Taimyroceras* but totally unknown in *Praetollia*. Therefore, *Runctonia* is a junior synonym of *Taimyroceras*, (Jeletzky, 1979, p. 6, footnote). The same is true of the "*Subcraspedites* (*Volgidiscus*)" *pulcher* of Casey *et al.*, (1977, p. 28, 29, Pl. 1, Fig. 2, Text-fig. 2). This ammonite, found in the Jurassic-Cretaceous boundary beds of the Subpolar Urals, is also a *Taimyroceras* because its secondaries are completely interrupted on the venter and its external suture is very simple and *Craspedites*-like (only 3 clearly defined auxiliary lobes). This ammonite is an older Late Volgian (?Chetae Zone) type.

The dating of the *Praetollia fedorovi* beds and their equivalents necessitates the abandonment of the previously proposed (Jeletzky, 1973, p. 50, Fig. 3) correlation of the *Praetollia antiqua* and *Buchia terebratuloides* s.l. fauna of the Buchanan

Lake section with: "part or all of the *Chetaites chetae* Zone of Northern Siberia. ... ". The well documented correspondence of the next older *Taimyroceras? canadensis* beds to the *Taimyroceras taimyrensis* Zone of Northern Siberia and the *Craspedites nodiger* Zone of Central Russia (Fig. 9) indicates instead that no diagnostic ammonites of the North Siberian *Chetaites chetae* fauna have been found in the Sverdrup Basin yet. Because of an apparent absence of hiatuses in the Jurassic part of the Deer Bay Shale, its beds containing the *Taimyroceras? canadensis* and the *Praetollia fedorovi* faunas apparently are separated by an interval of unfossiliferous (no ammonites) shale corresponding to the North Siberian *Chetaites chetae* Zone (Fig. 9).

Buchia okensis s.s. and *Pseudocraspedites* aff. *anglicus* Zone

General Remarks. The bulk of the data provided in the earlier description of the "*Buchia okensis* and *Craspedites* (*Subcraspedites*) aff. *suprasubditus* Zone" of Arctic Canada (Jeletzky, 1973, p. 52-54, 56, Figs. 2, 3) remain valid. Furthermore, most of the diagnostic buchias and ammonites of this zone have been figured adequately and described briefly already (Jeletzky, 1964, p. 30, 32, 34; Pl. I, Fig. 1; Pl. II, Figs. 1, 2, 4, 5; Pl. III, Figs. 1, 2; 1973, Pl. 6, Fig. 1; 1979, p. 3-6). This zone is widespread in the Sverdrup Basin (e.g., in the Buchanan Lake section, on Mackenzie King Island and at GSC loc. C-4764 in Mokka Fiord, Axel Heiberg Island and in Richardson Mountains.

Taxonomic Status of Diagnostic Ammonites. The new zonal designation proposed here reflects the new craspeditid taxonomy discussed above (p. 230, 231). The ammonites previously designated as either "*Tollia* (*Craspedites*)" or "*Craspedites* (*Subcraspedites*)" aff. *suprasubditus* (Jeletzky, 1964, Pl. II, Fig. 1; Pl. III, Fig. 2; 1971a, p. 4, Fig. 2; 1973, Pl. 6, Fig. 1), "*Tollia* (*Subcraspedites*)" aff. *spasskensis* (Jeletzky, 1964, Pl. II, Fig. 4), "*Subcraspedites* (*Subcraspedites*)" *anglicus* Shulgina 1972 and "*S.* (*S.*) *plicomphalus*" Shulgina 1972 *non* Sowerby 1823 were recently assigned to the new genus *Pseudocraspedites* by Casey, Messezhnikov and Shulgina (1977, p. 30, 31). In spite of some reservations about the validity of *Pseudocraspedites*, it will be used here as a subgenus of *Praetollia* s.l. for the craspeditid ammonites of the *Buchia okensis* s.s. Zone of the Sverdrup Basin and Northern Siberia. These pseudocraspeditids from the Sverdrup Basin shall be formally named in another paper. *Pseudocraspedites* is considered as a subgenus of *Praetollia* s.l. because of their very close similarity (except in advanced growth stages) combined with a temporal overlap (e.g., *Praetollia* (*Pseudocraspedites*) *anglicus* is latest Tithonian on Mackenzie King Island) and a complete vertical intergradation in the Buchanan Lake section. The external suture of the two subgenera is identical in all essential details.

Typical representatives of *Praetollia* s.s. and *P.* (*Pseudocraspedites*) differ in the shape of their advanced whorls and the width of the umbilicus. *Praetollia* s.s. is characterized by a more compressed oval to egg-shaped whorl cross-section with a high to moderately arched venter and an involute to moderately involute umbilicus. These features persist to the end of its adult living chamber. *P.* (*Pseudocraspedites*) has similar inner whorls in shape and ornament. However, its adult penultimate and ultimate whorls become slightly compressed to markedly depressed, rounded-triangular or broadly hoofshaped. The umbilicus is, as a rule, moderately evolute. *P.* (*Pseudocraspedites*) has a considerably larger adult shell and its more or less elevated primaries usually are transformed into prominent, broadly rounded, *Subcraspedites*-like bullae on the adult ultimate and, sometimes, adult penultimate whorl. Finally, *P.* (*Pseudocraspedites*) is very feebly sculptured to smooth (except for the umbilical bullae) on the ultimate and, sometimes, penultimate whorl. *Praetollia* s.s., in contrast, retains most of its progressively coarsening ribbing habit till the adult oral mouth border.

"*Tollia* (*Subcraspedites*)" aff. *hoeli* of Jeletzky (1964, Pl. III, Figs. 1, 3) is now assigned tentatively to *Praetollia* s.s. and synonymized with *Borealites*? *suritiformis* Klimova 1972 (in Gol'bert *et al.*, 1972, p. 152, 153, Pl. I, Figs. 1-Iv). This *Surites*-like form is characterized by the persistence of sculpture and of the relatively compressed oval whorl cross-section onto the ultimate whorl. It retains dichotomous rib bundles through most of its ontogeny but does not develop the *Subcraspedites*-like, round-topped umbilical bullae on the ultimate whorl. It combines these *Praetollia* s.s.-like features and its characteristic suture with a uniquely large size (i.e., for *Praetollia* s.s.) and an exceptionally large umbilicus of 40 to 45 per cent (i.e., for *Praetollia s.l.). This Praetollia* (*Praetollia*?) *suritiformis* (Klimova 1972) is restricted to the *Buchia okensis* s.s.-*Pseudocraspedites* aff. *anglicus* Zone in the Sverdrup Basin.

The citation of "*S.* aff. *kochi*" (Wiedmann, 1968, Table 4) from an unspecified locality in Canada is an error. I have never seen either *Hectoroceras* or any other keeled craspeditid from the Jurassic-Cretaceous boundary beds of either western or Arctic Canada, and believe that no such record exists in the Canadian and American literature. Dr. J. Wiedmann also confirmed this *lapsus calami*. Judging by the generic abbreviation "*S.*" and the placement of the fossil in the early Berriasian graph, *Subcraspedites* aff. *hoeli* was meant.

Age and Correlation. The identity of all *Buchia okensis* s.s. forms of this Sverdrup Basin zone with those from western British Columbia and northern Eurasia (Zakharov, 1981, p. 116-221, 226, 227, Pls. XXXI-XXXV), indicates the contemporaneity of this Canadian Arctic zone with the other two equivalents. Its gradational superposition on the *B. terebratuloides* s.l. Zone in the Sverdrup Basin and western British Columbia confirms this conclusion. These data attest to the basal Berriasian age (i.e., Grandis and basal Occitanica Zones) of the *B. okensis* s.s. - *Pseudocraspedites* aff. *anglicus* Zone of the Sverdrup Basin and its equivalents elsewhere in the Boreal Realm (Figs. 9, 10).

Within the Boreal Realm, the *Buchia okensis* s.s. - *Pseudocraspedites* aff. *anglicus* Zone was previously correlated tentatively with the *Chetaites sibiricus* and *Hectoroceras kochi* Zones of northern Siberia (Jeletzky, 1973, p. 56, Fig. 3). However, its gradational superposition on *Praetollia fedorovi* beds corresponding to the upper part of the *Chetaites sibiricus* Zone, and other arguments presented in connection with the correlation and dating of these beds, indicate that the *Buchia okensis* s.s.-*Pseudocraspedites* aff. *anglicus* Zone is correlative with the restricted North Siberian *Hectoroceras kochi* Zone only (Fig. 9). This correlation is confirmed by the occurrence of *Praetollia* (*Praetollia*?) *suritiformis* in this Canadian zone and in the inferred equivalents of the restricted *Hectoroceras kochi* Zone in the Subpolar Urals.

The *Buchia okensis* s.s.-*Pseudocraspedites* aff. *anglicus* Zone is overlain by the unfossiliferous beds of the Deer Bay Formation that are believed to include the rest of the Berriasian because of the Valanginian ammonites found in its still higher beds (Jeletzky, 1979).

TIME RANGE OF TYPE RIASANIAN BEDS AND THEIR CORRELATION WITH THE BERRIASIAN OF MANGYSHLAK, NORTH CAUCASUS AND CRIMEA

The Type Riasanian and the Position of the Tithonian-Berriasian Boundary in the Central Russian Basin

Recent investigations of Casey *et al.*, (1977) and Messezhnikov *et al.* (1979) contradict the widespread assumption (Fig. 1) that the type Riasanian beds are coeval with the *Tirnovella occitanica* and *Fauriella boissieri* Zones only, and that they are separated from the underlying Upper Volgian beds by a hiatus that corresponds to the basal Berriasian *Pseudosubplanites grandis* Zone and the topmost Tithonian part of the *Paraulacosphinctes transitorius* Zone.

These workers discovered *Hectoroceras kochi* in some sections of the Riasanian beds. This predominantly basal Cretaceous species (Figs. 9, 10) occurs at the very base of some sections (e.g., Kostino Village), but in some other sections (e.g., Kuzminskoye Village) somewhat above the base of the Riasanain beds. The basal beds here contain *Riasanites* and "*Euthymiceras*" ex gr. *transfigurabilis* in association with *Garniericeras subclypeiforme,* allied *Garniericeras, ?Craspedites* ex gr. *mosquensis* (small and poorly preserved shells) and numerous *Buchia fischeriana.* [Following the suggestion of Luppov *et al.* (1979, p. 160) I do not believe that the Riasanian "*Euthymiceras*" ex gr. *transigurabilis* is either congeneric with or closely allied to the true *Euthymiceras* ex gr. *euthymi* (see below)]. Messezhnikov *et al.* (1982, p. 225, 227) do not list any other buchias from this basal bed. Zakharov (1981, p. 225, 227), however, has identified rare *Buchia okensis* in it, while stressing that *B. unschensis* s.s. is absent here but common in the overlying bed with *Hectoroceras kochi.*

The allegedly Tethyan berriasellids Riasanites Spath 1924 and "*Euthymiceras*" ex gr. *transfigurabilis* (Bogoslowsky) were found to range down into the basal *Garniericeras* spp. and *?Craspedites* ex gr. *mosquensis*-bearing bed of the Riasanian beds in the Kuzminskoye profiles (Figs. 10, 11). This increases significantly their previous known time range in the middle Riasanian beds (i.e., from the base of *Surites tzikwinianus* Zone to the base of *Surites spasskensis*-bearing beds; see Jeletzky, 1971a, 1973, p. 58, 60, 62, Fig. 3; Casey, 1973, p. 216-226, Table 2).

Only the uppermost Volgian *Buchia unschensis* (Pavlov) s.s. was found in the Riasanian bed containing *Hectoroceras kochi* by Messezhnikov *et al.* (1979, p. 76). All other typically Berriasian *Buchia* species (apparently including the holotype of *Buchia okensis* figured by Pavlov, 1907, Pl. I, Fig. 11 and its other examples cited by various workers) were found to be confined to the overlying Riasanian beds, as indicated by their recorded association with *Surites spasskensis* (Pozharisskaya, 1971, p. 118). Messezhnikov *et al.* (1979, p. 77, 78) do not think that the distribution of diagnostic ammonites and buchias in the type Riasanian beds was influenced by their redeposition from underlying beds.

These new, startling data have been interpreted by Casey *et al.* (1977, p. 18, Table 2), Messezhnikov *et al.* (1979, p. 80) and

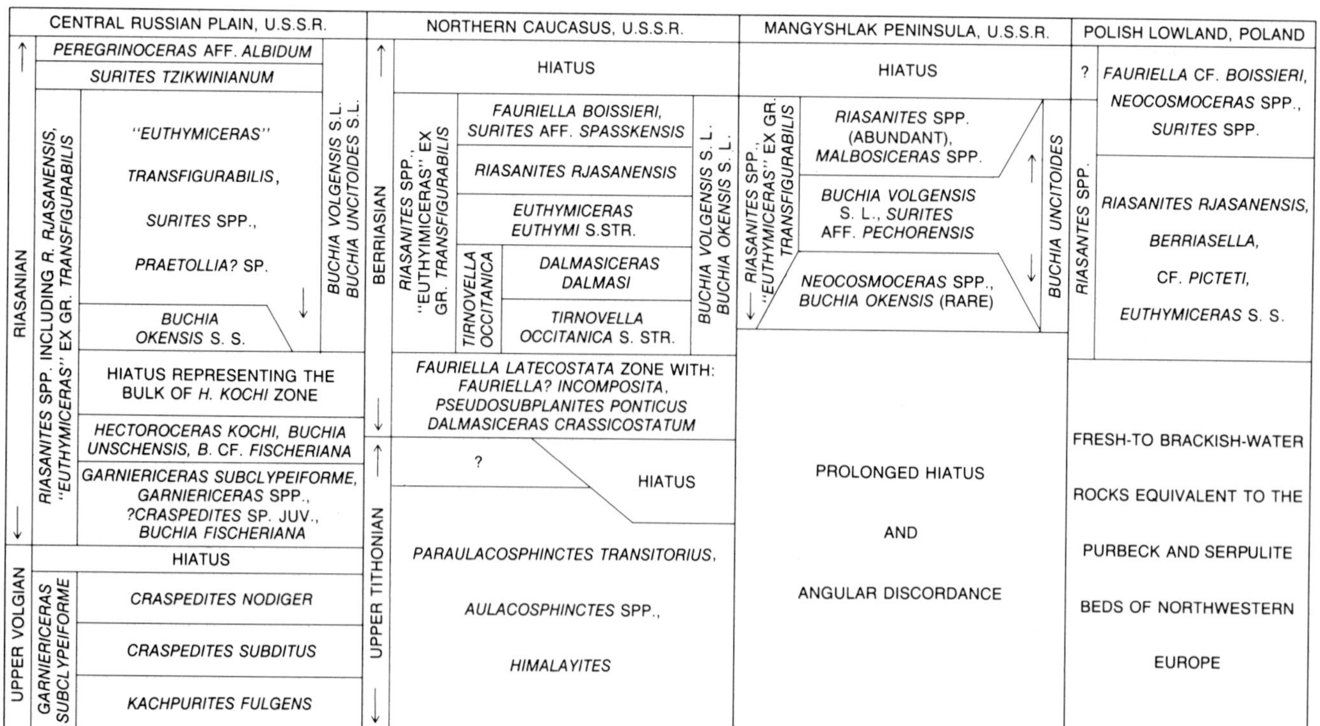

Figure 11. Proposed Correlation of Jurassic-Cretaceous Boundary beds of Central Russian Plain with those of southern part of U.S.S.R. and Central Poland.

Saks *et al.* (1979, p. 98, Table 2) in the sense that some of the sections of the Riasanian beds on the Oka River include the basal Cretaceous beds corresponding to the *Chetaites sibiricus, Praetollia maynci, "Praetollia" (recte Taimyroceras) runctoni* and *Subcraspedites antiqua (recte Praetollia fedorovi)* beds of other Boreal regions. Casey *et al.* (1977) also infer the presence of a hiatus, separating the *Riasanites*- and *Garniericeras*-bearing basal Riasanian beds from the underlying *Craspedites nodiger* Zone and corresponding to most of the *Chetaites chetae* Zone of northern Siberia.

The stratigraphic conclusions of Casey *et al.* (1977) and Messezhnikov *et al.* (1979), including the presence of a hiatus between the basal Riasanian bed at the Kuzminskoye Village and the Nodiger Zone, are accepted; but the dating of these events by them and Saks *et al.* (1979) is not. As demonstrated earlier (Figs. 5, 10), the *Chetaites sibiricus, Taimyroceras runctoni, Praetollia maynci* and *Praetollia fedorovi* beds are not basal Berriasian (i.e. Grandis Zone), but topmost Tithonian (i.e., *Buchia terebratuloides* s.l. or *Durangites* Zone). Furthermore, *Buchia fischeriana* is not a basal Berriasian but a Late Tithonian index fossil. The identification of rare *B. okensis* in this *B. fischeriana*-bearing bed must refer to the *B.* n. sp. aff. *okensis*, which is synonymized with true *B. okensis* s.s. by Zakharov (1981, p. 116-121, 226, 227). *Buchia* n. sp. aff. *okensis* may be associated with the latest representatives of *B. fischeriana* in the boundary beds separating the *B. fischeriana* and *B. terebratuloides* s.l. Zones, and in the lower part of the latter zone, in western British Columbia. This evidence of *Buchia* agrees with that provided by the ammonite fauna of this bed which has entirely Late, but not the latest, Volgian (and hence Late, but not the latest, Tithonian) affinities. The genus *Garniericeras* and the species *G. clypeiforme* are strictly Late Volgian taxa which are not known to range above the Nodiger Zone in Central Russia (Shulgina, in Saks *et al.*, 1969, p. 160; Gerassimov, 1969, p. 95-96). Furthermore, *G. margaritae* Shulgina 1969, a close ally of *G. clypeiforme*, was found to range from the *Craspedites okensis* Zone to the *Craspedites taimyrense* Zone inclusive in Northern Siberia (Shulgina, in Saks *et al.*, 1969, p. 161; Fig. 9 of this paper). The genus *Craspedites* has exactly the same time range as *Garniericeras*. It is not known to range either into the latest Jurassic *Chetaites sibiricus* Zone of Northern Siberia or in the equivalent *Praetollia fedorovi* Beds of western Siberia and the Sverdrup Basin. The basal Riasanian bed of the Kuzminskoye Village sections is, therefore, an obvious correlative of the *Chetaites chetae* Zone of Northern Siberia and of the lower (?basal) part of the *Buchia terebratuloides* Zone in western British Columbia and northwestern California. Because of the inferred presence of a hiatus between this bed and the underlying Nodiger Zone (Casey *et al.*, 1977, p. 18, Table 2), the former is correlated with the upper part of the *C. chetae* Zone only (Fig. 9). This conclusion agrees with the fact that *Hectoroceras kochi*, that occurs in the immediately overlying bed of this profile, is associated with numerous *B. unschensis* s.s. and rare *B.* cf. *fischeriana* but not with *B. okensis* s.s. As shown earlier (Figs. 9-11) *B. unschensis* s.s. is an almost exclusively Late Volgian (i.e., Late Tithonian) species. It is restricted to the *Buchia terebratuloides* s.l. Zone throughout the Boreal Realm and the North Pacific Province of western North America, except for East Greenland where it ranges up into the basal beds of *Buchia okensis* s.s. Zone. Zakharov's (1981, p. 225) attempt to explain away this type Riasanian faunal association by postulating a diachronic southward migration of *B. unschensis* s.s. in unconvincing, because this species was found to hold its proper latest Tithonian level (i.e., the upper part of *B. terebratuloides* s.s. Zone in association with *Substeueroceras* and *Parodontoceras*) in the southernmost points of its penetration into the North Pacific Faunal Province. Because of the abundance of *B. unschensis* s.s., combined with the absence of *B. okensis* s.s., this *Hectoroceras kochi*-bearing bed of the type Riasanian beds cannot represent the *H. kochi* Zone proper. It must correspond, instead, to that uppermost? part of the underlying *Chetaites sibiricus* Zone where its name fossil is associated with *B. unschensis* s.s. and *H. kochi*, while *B. okensis* s.s. is still absent (Zakharov, 1981, p. 225, Table 3).

The next younger Riasanian level of the Kuzminskoye Village profiles, with *Surites* spp., rare *Buchia okensis* s.s. (?late forms only; compare Pavlow, 1907, Pl. I, Fig. 10; Gerassimov, 1955, Pl. XIII, figs. 9, 10; Pozharisskaya, 1971, Pl. XXVII, Figs. 2, 3, 4; Pl. XXVIII, Figs. 1, 2), numerous *B. volgensis* and numerous *B. jasikovi* (i.e., *B. uncitoides* s.l. of Jeletzky, 1965, and in this paper), is definitely younger than the *B. okensis* s.s. Zone proper. It is correlative, instead, with the lower part of *B. jasikovi* Zone of Zakharov (1981, p. 227, 228, Table 3) and the lower part of the *B. uncitoides* s.l. Zone of western British Columbia (Figs. 5, 10). In these beds rare to fairly common late forms of *B. okensis* s.s. coexist either with *B. uncitoides* s.l. alone or with *B. uncitoides* s.l. and *B. volgensis*. This dating of the second and third levels from base of the Kuzminskoye Village profiles indicates the absence of most or all of the restricted *B. okensis* s.s.-*H. kochi* Zone. This earliest Berriasian zone appears to be represented by a hiatus separating these beds (Figs. 10, 11).

This re-appraisal of the age of the two basal beds of the type Riasanian on the Oka River demonstrates that the genus *Riasanites* Spath 1924 and the *"Euthymiceras"* ex gr. *transfigurabilis* (Bogoslovsky) are not restricted to equivalents of the Mid- to Late Berriasian (i.e., Occitanica and Boissieri Zones) of southwestern Europe as is believed by nearly all recent workers (Figs. 1, 2). Instead, these ammonites range down into the equivalents of the North Siberian *Chetaites chetae* and *C. sibiricus* Zones that correspond to the upper part of *Buchia terebratuloides* s.l. Zone of Western and Arctic North America and, hence, to the upper part of the *Paraulacosphinctes transitorius* Zone (i.e., the regional *Durangites* Zone) of the European Tethys. This conclusion brings us at least halfway back to the view of oldtime specialists (e.g., Pavlow, 1897, 1907, corr. table opp. p. 84; Spath, 1947, p. 54, 55, 56; Arkell 1946, 1956, p. 492, 493) who considered the *Riasanites rjasanensis* Zone to be entirely latest Jurassic (i.e., Transitorius time). This conclusion calls for a new approach to the correlation of the *Riasanites* Zone of the type Riasanian beds with the long known occurrences of *Riasanites* and *"Euthymiceras"* ex gr. *transfigurabilis* on the Mangyshlak Peninsula and in the Crimean-Caucasian region of the European Tethys, which shall be attempted in the following sections.

Time Ranges of Boreal Ammonites and Buchias in the Marginal Tethyan Basins of Southern U.S.S.R.

Mangyshlak Peninsula. The recent paper of Luppov *et al.* (1979, in Russian) is an excellent summary of the macroinvertebrate paleontology and biochronology of the Berriasian and Valanginian rocks of the Mangyshlak Peninsula; it forms the basis of this critical review. However, it does not add much to the earlier paper of Louppov *et al.* (1975) that was published in French.

According to Luppov *et al.* (1979, p. 161, my translation from Russian), "Reliable finds of *Riasanites* are mostly restricted to the upper part of the Berriasian – the regional zone of *Riasanites* and *Pygurus rostratus*. However, rare specimens allied to the species *R. subrjasanensis* Nik. were found in the lower regional zone."

This latter regional zone, the *Neocosmoceras* and *Septaliphora semenovi* Zone, is correlated with the Occitanica Zone of the type Berriasian because *Neocosmoceras* does not range below that zone in southeastern France (Le Hégarat, 1973). The older Berriasian (i.e., Grandis Zone), perhaps including the basal part of the Occitanica Zone, and the Tithonian rocks are unknown and presumed to be absent in the Mangyshlak Region. This is indicated by the fact that according to Luppov *et al.*, (1979, p. 161, my translation from Russian). "The lowermost part of the Mangyshlak Berriasian, its basal zone, at the level only 1-2 m above the base of the Cretaceous section, already contains *Neocosmoceras, Subalpinites* and, slightly higher, *Euthymiceras* sp. nov. ex gr. *euthymi* Pict. These forms are absent in the lower part of the stage in the southeastern France and also in the Crimea." These equivalents of the Occitanica Zone overlie unconformably various pre-Tithonian Jurassic rocks. Their basal conglomerate includes derived fossils ranging from the Callovian to Kimmeridgian (Luppov *et al.*, 1979, p. 159).

The time range of *"Euthymiceras"* ex gr. *transfigurabilis* is even more restricted in the Mangyshlak sections. According to Luppov *et al.* (1979, p. 160, my translation from Russian), *"Euthymiceras"* (quotation marks of Luppov *et al.*; my remark) *transfigurabilis* Bogosl. and forms closely allied to it occur together with the *Riasanites* in the upper regional zone. The time range of *Surites kozakovianus* Bogoslovsky and of a new *Surites* species closely allied to *S. pechorensis* in the Berriasian section of Mangyshlak remains uncertain because all their representatives studied are derived from old, poorly localized collections. However, the third form – a new species closely allied to *S. pechorensis* – was found associated with ammonites of the species group *"E. transfigurabilis"* in the middle regional zone (Luppov *et al.*, 1979, p. 161)."

As pointed out by Luppov *et al.* (1979, p. 162, my translation from Russian), "Buchias have the greatest value for an interregional correlation. Following forms occur in the Berriasian: *B. volgensis* Lah., *B. okensis* Pavl., *B. uncitoides* Pavl., *B.* ex gr. lahuseni (Pavl.), and *B.* ex gr. *terebratuloides* Lah. According to the latest study of Zakharov (1977), the species *Buchia okensis* and *B. uncitoides* are for all practical purposes the zonal Berriasian species in all regions of the globe where buchias occur. In Mangyshlak, the species *B. okensis* (Pl. II, Fig. 4) is present in small numbers in the lower beds while *B. uncitoides* (Pl. II, Fig. 5) occurs throughout the Berriasian section and characterizes it. This permits the conclusion that the lower *Buchia*-Zone of *B. okensis* is either absent in Mangyshlak or is represented by its upper part only. This agrees well with the conclusions based on ammonites. According to its buchias, the part of the Berriasian present in Mangyshlak corresponds to the upper zone – that of *B. uncitoides*. *B. volgensis* (Pl. II, Fig. 3) is an even more important species in the Berriasian of Mangyshlak. In spite of being present throughout its section, the mass occurrences of its shells are restricted to the middle part of the stage. This permitted the differentiation of the regional zone with *B. volgensis.*"

I agree with these conclusions. It must be added, however, that the specimen of *B. okensis* (Pavlow) figured by Luppov *et al.* (1979, p. 162, Pl. II, Fig. 4) is a typical late form that is largely confined to the upper part of its zone and the lower part of the overlying, Late Berriasian *Buchia uncitoides* Zone. The *B. volgensis* figured by Luppov *et al.* (1979, Pl. II, Fig. 3) is also a late form, transitional to *B. keyserlingi.* Such forms of *B. volgensis* s.l. closely approach its late forms occuring in the Late Riasanian (i.e., the so called Spasskensis Zone of authors) and assigned to *B. merklini* by Pozharisskaya (1971, p. 120, 121, Pl. XXIX, Fig. 6, 6a). The Early Berriasian representatives of *B. volgensis* differ from these Late Berriasian forms in various combinations of the following morphological features: less regular concentric sculpture, lesser thickness of the shell combined with a longer, greater downward curvature and a distinct right-hand curvature of the beak of the left valve (e.g., Surlyk, 1978, Pl. 1, Fig. 2; Pl. 2, Fig. 3; Pl. 4, Fig. 8). The form assigned to *B. lahuseni* (Luppov *et al.*, 1979, Pl. II, Fig. 2) does not belong to this latest Jurassic species but is a variant of *B. volgensis* s.l. No figures of *B.* ex gr. *terebratuloides* are available and the true nature of this form cannot be recognized. However, I consider the presence of true *B. terebratuloides* s.l. in the Mid-to Late Berriasian to be most unlikely.

These affinities of the *Buchia* fauna of the Mangyshlak Berriasian indicate correlation of its lower zone containing late forms of *B. okensis* s.s. in association with late forms of *B. volgensis* and *B. uncitoides* with the *Buchia uncitoides* s.l. Zone of northern Siberia (i.e., *B. jasikovi* Zone of Zakharov, 1981, p. 227, 228, Table 3) and western British Columbia. These were shown above to be Mid- to early Late Berriasian (Figs. 5, 10). Furthermore, this lower zone is obviously coeval with that bed of the type Riasanian sequence in which *Riasanites* spp. and *"Euthymiceras"* ex gr. *transfigurabilis* are associated with *Surites* spp., late forms of *B. okensis* s.s. and *B. volgensis*, and *B. uncitoides* (Figs. 9-11). The equivalents of the underlying beds of the Kuzminskoye Village profiles (i.e., *Hectoroceras kochi-Buchia unschensis* s.s. and *Garniericeras subclypeiforme-Buchia fischeriana* beds; see Fig. 10) are not represented in the Mangyshlak profiles.

This review strongly suggests that *Riasanites* first appears in the Mangyshlak sections in the beds that are considerably younger than those in which this genus appears in the type Riasanian sections. However, this conclusion must remain tentative so long as the Mangyshlak sections alone are considered. The presence of a regional hiatus in these sections corresponding to the Tithonian, the Grandis Zone and, possibly, the lower part of the Occitanica Zone, leaves open the possibility that *Riasanites* was originally present in some of these beds prior to their destruction throughout the region. However, the absence of *Riasanites* in the equivalents of the Grandis Zone and older rocks in the northern Caucasus and Crimea indicates that this was not the case.

The Riasanian *"Euthymiceras"* ex gr. *transfigurabilis* definitely appear much later in the Mangyshlak sections than in the type Riasanian beds in the Central Russian Basin.

It is important to point out that the first appearance of *Riasanites* in the Mangyshlak sections appears to coincide approximately with that of other Boreal immigrants, such as *Buchia* and, probably, *Surites* s.l.

The above data invalidate Luppov *et al.'s* (1979, p. 161, my translation from Russian) conclusion that "Secondly, ammonites belonging to the genus *Riasanites*, which are known on the Russian Platform from the very base of the Ryazanian Horizon, appear in large numbers only in the upper part of the Mangyshlak section – in the upper regional zone. Insofar as the *Riasanites* occur, in addition to Mangyshlak, also in Northern Caucasus and have evidently evolved out of the Mediterranean forms (*sic*; writer's remark), there are no reasons to assume their later appearance in Mangyshlak in comparison with the Moscow area. If so, the lower beds of the Mangyshlakian Berriasian are older than the Ryazanian Horizon proper."

Northern Caucasus. As already pointed out by Grigorieva (1938, p. 87-88) for the northwestern Caucasus, "*Berriasella (Riasanites)*" *rjasanensis* var. *maikopensis* Grigorieva and "*Protacanthodiscus (Euthymiceras)*" *transfigurabilis* (Bogoslovsky) are either associated with such diagnostic Late Berriasian species as *Protacanthodiscus (Pomeliceras) breveti* (Pomel) and *P. (Malbosiceras) malbosi* (Pictet) or occur in beds overlying those containing these species. These old data have been confirmed recently by Sakharov (1975, 1976, 1979).

The following review is based on the more detailed paper of Sakharov (1976), published in Russian, because it lists *Buchia* and *Surites* in addition to other ammonites, which alone are listed in the other papers of Sakharov (1975, 1979).

According to Sakharov (1976, p. 41) the earliest representatives of "*Euthymiceras*" ex gr. *transfigurabilis* and *Riasanites* occur in the bed 3 representing the *Dalmasiceras dalmasi* Subzone of his standard section. This subzone comprises the upper part of the Mid-Berriasian *Tirnovella occitanica* Zone. These ammonites were only found in the Uruch River section of this subzone where they are recorded as "*E.*" cf. *transfigurabilis* and *Riasanites* sp. indet. Elsewhere in the area the first representatives of these Riasanian ammonites make their first appearance in the beds 5 to 7 of the standard section which overlie beds 2 and 3 containing (combined list): *Tirnovella occitanica, T.* cf. *occitanica, T.* cf. *nicolovi, Dalmasiceras* aff. *dalmasi, D.* aff. *crassicostatum, Neocomites* cf. vasseuri and *Buchia volgensis.* Beds 2 and 3 represent the *Tirnovella occitanica* s.s. and *Dalmasiceras dalmasi* Subzones of the *Tirnovella occitanica* Zone. *Buchia volgensis* occurs in Bed 2 but is not recorded from the underlying oldest Berriasian Bed 1 of the area, which contains (Sakharov, 1976, p. 41). "*Fauriella Latecostata. F.*(?) *incomposita, Pseudosubplanites ponticus, P. combesi, Malbosiceras paramimounum, M. malbosi, Dalmasiceras crassicostatum.* This fauna, which is absent in Mangyshlak area, is designated as *Fauriella latecostata* Zone. It corresponds to the *Pseudosubplanites grandis* Zone of the type Berriasian."

Except for the previously mentioned isolated occurrence of riasanitid ammonites in the Bed 3 which was apparently questioned later (Sakharov, 1979), they are restricted to beds 5 to 9 of Sakharov's (1976, p. 40) standard section.

"*Euthymiceras*" ex gr. *transfigurabilis* are restricted to beds 5-6. However, Sakharov treats the underlying Bed 4 without ammonites, as part of this *Euthymiceras euthymi* zone. Bed 5 has yielded: *Euthymiceras euthymi, E.* aff. *euthymi,* "*E.*" *transfigurabilis,* "*E.*" *transcaspius, Neocosmoceras* aff. *rerollei* and *Himalayites* cf. *kasbensis.* Bed 6 yielded: *Euthymiceras euthymi, E.* aff. *euthymi,* "*E.*" *transfigurabilis,* "*E.*" *transcaspius, Neocosmoceras* aff. *rerollei, N.* cf. *sayni, Himalayites breveti, H.* cf. *kasbensis* and *Riasanites* sp. Sakharov (1976, p. 40) also lists *Buchia volgensis* (Lahusen) and *B. surensis* (Pavlow). The absence of figures precludes a decision whether this *B. volgensis* is an early or a late form. However, its association with a form listed as *B. surensis* suggests strongly that Bed 6 corresponds to the middle to upper part of the Riasanian beds not older than Bed 4 of Messezhnikov *et al.* (1979, p. 78). Judging by the description and figures provided by Pavlow (1907, p. 30, Pl. II, figs. 23-25) and Pozharisskaya (1971, p. 119, Pl. XXVIII, Fig. 3; Pl. XXIX, Fig. 1), *B. surensis* is one of the late forms of the *Buchia volgensis* species group restricted to the *Surites* s.l.-bearing part of the Riasanian beds and the basal Valanginian Undulatoplicatilis Zone.

Except for the previously mentioned occurrences of *Riasanites,* this genus is restricted to Beds 7 to 9 of the standard section (Sakharov, 1976, p. 40) designated as the *Riasanites rjasanensis* Zone.

The following ammonites and buchias are listed from Bed 7: *Riasanites rjasanensis, R. r. maikopensis, R. swistowianus, Riasanites* spp., "*Euthymiceras*" *transfigurabilis, Blanfordiceras* aff. *wallichi, Himalayites breveti, Negreliceras negreli, Tirnovella* cf. *subalpina, Buchia volgensis, B. okensis, B. trigonoides* and *B. merklini.*

The presence of *B. okensis* and *B. trigonoides* in this list seems to indicate Early instead of Late Riasanian affinities of the *Buchia* fauna. However, the presence of *B. merklini* Pozharisskaya (1971, p. 120, 121, Pl. XXIX, Figs. 6, 6a) indicates the middle to upper part of the Riasanian beds not older than Bed 4 of Messezhnikov *et al.* (1979). This form, which was described from specimens found in the Spasskensis Zone (Pozharisskaya, 1971), is a late form of the *Buchia volgensis* species group closely related to and possibly synonymous with *B. surensis* (Pavlow, 1907) (compare Pozharisskaya, 1971, Pl. XXVIII, Fig. 3 and Pl. XXIX, Figs. 6, 6a with Pavlow, 1907, Pl. II, Fig. 25). This strongly suggests, in turn, that the "*B. trigonoides*" listed by Sakharov (1976) belongs to one of the forms from the Upper Riasanian and basal Valanginian (Undulatoplicatilis Zone) figured by Pavlow (1907, Pl. IV, Fig. 25) under that name and by Gerassimov (1955, Pl. XIII, Figs. 8-10) under *B. spasskensis.* Many of these buchias (e.g., Gerassimov, 1955, Pl. XIII, Figs. 9, 10) either closely resemble the smallish late forms of *B. okensis* or must be transferred to the latter forms (Jeletzky, 1965, p. 24). Such a form could have been identified easily as *B. okensis* by Sakharov. I feel certain that these forms have nothing to do with true "*Aucella*" *trigonoides* Lahusen (1888, p. 36, 37, Pl. II, Figs. 21-24), which was erroneously recorded to occur principally in association with "*Hoplites rjasanensis*", but actually is but a triagular shaped variant of the almost exclusively Late Volgian *Buchia fischeriana* (d'Orbigny). In conclusion, typical representatives of *B. okensis* (Pavlow) closely resembling its holotype (Pavlow, 1907, Pl. I, Fig. II) also range up into the lower Upper Riasanian beds and their equivalents in North America (Jeletzky, 1965, p. 23, 24; Pl. X, Fig. 2 and earlier in this paper).

Bed 8 of the Sakharov's (1976, p. 40) section has only yielded rare *Riasanites* sp. while Bed 9 contains *Riasanites rjasanensis, R. subrjasanensis,* and *R.* aff. *subrjasanensis* in association with numerous *Fauriella boissieri.* Finally, the topmost Berriasian Bed 10, which is disconformably overlain by Valanginian beds, yielded no ammonites, except for *Ptychophylloceras calypso* associated with numerous pelecypods and *Toxaster granosus.*

Sakharov (1976, p. 42; my translation from Russian) evaluates the age of his beds 7-10 as follows: "Deposits of the *Euthymiceras euthymi* Zone (*i.e.* beds 4-6; my remark) are overlain by argillaceous limy rocks of *Riasanites rjasanensis* Zone (beds 7-10) in which one can clearly recognize two different complexes of the ammonites. The lower complex characterizes the Subzone of *Riasanites rjasanensis s. str.* while the upper characterizes the subzone of *Fauriella boissieri.*

"The two subzones of the *Riasanites rjasanensis* Zone have in common the presence of ammonites belonging to the genus *Riasanites.*" He also notes the presence of a solitary representative of *Surites* ex gr. *spasskensis* in the *Fauriella boissieri* Subzone.

The following correlation of the regional zones of *Euthymiceras euthymi* and *Riasanites rjasanensis* with the Berriasian Zones of southeastern France, Crimea and Mangyshlak is proposed in a later, summary paper (Sakharov, 1979, p. 185, my translation from Russian):. "Because of the presence of conspecific ammonites of the genus

Euthymiceras, the Euthymiceras euthymi Zone is correlative with the zones of *Euthymiceras euthymi-Dalmasiceras dalmasi* and *Berriasella privasensis-Spiticeras spitiense* of the Crimea and then with the *Malbosiceras paramimounum* of the Berriasian in southeastern France. In Mangyshlak the beds with *Neocosmoceras* and *Septaliphoria semenovi* correspond to this part of our section. The deposits of the lower part of the stage are eroded away on Russian Platform. The correlation of the lower part of the *Riasanites rjasanensis* Zone (Subzone of *Riasanites rjasanensis s.str.* of Northern Caucasus) with the zone of *Riasanites rjasanensis* of the Russian platform does not present any difficulties (*sic;* writer's remark). Because of the presence of diagnostic species in common, the *Fauriella boissieri* Subzone of Northern Caucasus is equally definitively correlative with the *Fauriella boissieri* Zone of the Crimea, Bulgaria and Southeastern France (Drustchits and Vakhrameev, 1976; Sakharov, 1976a, Le Hégarat, 1973)."

There is no doubt that the *Riasanites* and *"Euthymiceras"* ex gr. *transfigurabilis* of the North Caucasian sections are restricted to the Berriasian beds that are considerably younger than those in which these ammonites make their first appearance in the type Riasanian sections. Furthermore, the evidence of the North Caucasian sections is conclusive (unlike that of the Mangyshlak sections) where the total range of these riasanitids is concerned because of the presence of an indubitable equivalent of the Grandis Zone underneath the lowermost bed containing them. The above cited, contrary conclusion of Sacharov (1979, p. 185) is just as untenable as that of Luppov *et al.* (1979).

Crimea. Most of the recent descriptions of the Berriasian stratigraphy and paleontology of Crimea are rather generalized (e.g., Drustchits and Vakhrameev, 1976, p. 196, 198, 199; Gorbatchik, Drustchits and Yanov, 1975; Drustchits, 1960). They do not permit evaluation of the local time ranges of *Riasanites* and *"Euthymiceras"* ex gr. *transfigurabilis* in terms of the standard Tethyan zones characteristic of the Mediterranean Province and easily recognizable in Crimea. However, their generalized range charts of diagnostic Tithonian and Berriasian ammonites indicate that *Riasanites,* at least, is restricted to the Late and late Middle Berriasian zones of *Berriasella privasensis-Spiticeras spitiense* and *Euthymiceras euthymi-Dalmasiceras dalmasi.* Khimiashvili (1976, p. 118) indicates a similar time range of *Riasanites* in Crimea in his generalized range chart of diagnostic ammonites present in different regions of the Mediterranean Province. The above charts do not indicate *"Euthymiceras"* ex gr. *transfigurabilis* in Crimea, except Drustchits (1963, p. 11) who, according to Le Hégarat (1973, p. 305), lists (?) *"E." transfigurabilis* (Bogoslovsky) and *Riasanites rjasanensis* (Lahusen).

These scant data, suggesting the rarity of riasanitids in the Crimea and the restriction of *Riasanites* to its late Mid-and Late Berriasian, were confirmed by more recent research of Kvantaliani and Lyssenko (1979a, 1979b). A detailed description of the principal Crimean sections provided by these workers (1979b, p. 629-632, Corr. Table) indicate the total absence of riasanitids in the *Pseudosubplanites grandis-Berriasella jacobi* Zone and the local equivalents of the Occitanica Zone. Furthermore, no riasanitids were found in the *Euthymiceras euthymi* Zone that is correlative with the Paramimounum Zone of the basal Upper Berriasian. The first riasanitids, which Kvantaliani and Lyssenko (1979a) assign to the new genus *Tauricoceras,* appear in the next younger beds. These beds, designated as *T. crassicostatum* Zone, are coeval with the *Riasanites rjasanensis* Zone of Northern Caucasus and the Picteti Subzone of southwestern Europe. Kvantaliani and Lyssenko (1979b, p. 632; my translation from Russian) state:

"Because of the presence of *Malbosiceras malbosi* (Pict.) in the bed 1, this part of the profile is correlated with the *M. malbosi* subzone of the Lower Berriasian. The bed 2 is assigned to the zone of *Spiticeras spitiense* because of the prevalent numbers of *Spiticeras (S.) spitiense* (Blanf). (as it was proposed previously (1)). The bed 3 is assigned to the zone of *D. dalmasi* because of the presence of *Dalmasiceras punctatum.* The bed 4 is assigned to the zone of *E. euthymi* because of the appearance of representatives of the genus *Euthymiceras,* which occur stratigraphically above the *D. dalmasi* zone in all sections studied by us. Previously the two last mentioned zones were treated as a single zone of *D. dalmasi – E. euthymi.* Numerous ammonites belonging to the new genus *Tauricoceras* (5) appear above the *E. euthymi* zone in this, as well as in other profiles. The riasanitids occur rarely somewhat higher where they are represented by scattered fragments."

Riasanites (including the very closely related *Tauricoceras*) is restricted in Crimea to beds younger than those in which it makes its first appearance in the Northern Caucasus and Mangyshlak. All other generically undetermined riasanitids appear to be similarly restricted.

General Biochronological Conclusions

The use of the independently dated *Buchia* Zones of the Jurassic-Cretaceous boundary beds of western and Arctic Canada has revealed that the type Riasanian beds on the Oka River include much older beds than hitherto believed. Their basal bed, characterized by *Garniericeras clypeiforme* and *Buchia fischeriana,* is correlative with the upper part of the *Chetaites chetae* Zone of Northern Siberia and the upper part of the *Paraulacosphinctes transitorius* Zone of the Tethyan Realm (Figs. 10, 11). *Riasanites* and *"Euthymiceras"* ex gr. *transfigurabilis* were found to range to the very base of the type Riasanian beds. Therefore, they appear much earlier in the Boreal Central Russian Basin than in the Mangyshlakian, North Caucasian and Crimean Basins of southern U.S.S.R. that form part of the Tethyan Realm. As indicated by *Buchia* species associated with the earliest riasanitid ammonities known in these basins, these ammonites are restricted to the uppermost part of their known time range in the type Riasanian sections (Fig. 11). This amply documented strong discrepancy of time ranges of riasanitid ammonites invalidates the universally held idea about their Tethyan origin and northward migration from the Crimean-Caucasian geosynclinal seas into the Central Russian Boreal Basin. There is every reason to infer instead their northern origin and southward migration from the Central Russian Boreal Basin. This invalidates all arguments advanced by Casey (1962, 1963, 1973), Enay (1972, 1973) and Zeiss (1968, p. 49, 139, footnote; 1974, 1977, 1979, p. 18, Corr. Table, and in Westermann, 1982, p. 142, Corr. Table) in favour of the equivalence of the bulk of the Upper Volgian stage with either the Lower (Grandis Zone) or Lower and Middle (Grandis and Occitanica Zones) Berriasian beds of southeastern France.

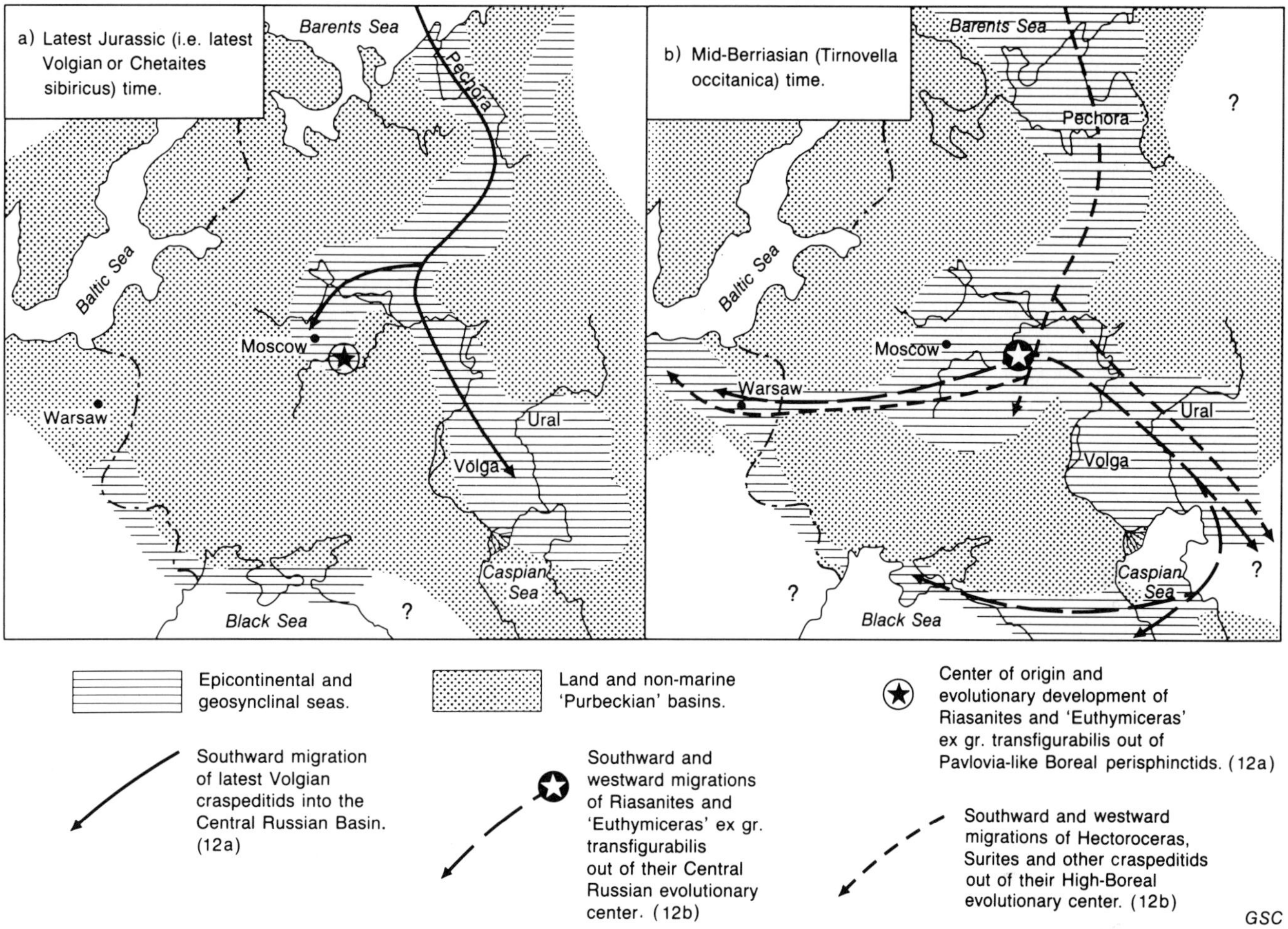

Figure 12. Paleobiogeography of *Riasanities* and *'Euthymiceras'* ex gr. *transfigurabilis* in latest Jurassic (i.e. latest Volgian or *Chetaites sibiricus*) time (Fig. 12a) and earliest Cretaceous (i.e. mid-Berriasian or *Tirnovella occitanica*) time (Fig. 12b).

These data also vindicate (i.e., except for the age of the *Chetaites sibiricus* Zone and its equivalents; Figs. 9, 10) the ideas of recent Soviet workers (Fig. 1) comprehensively discussed and documented by Saks *et al.* (1972). The recent discreditation (e.g., by Casey *et al.*, 1977; Messezhnikov *et al.*, 1979) of the key argument of these workers and of Jeletzky (1965, 1968, 1971a, 1973, 1974, 1979) postulating a prolonged hiatus that separates the basal Riasanian beds from the topmost Volgian beds, is made irrelevant by the pronounced discrepancy of time ranges of the riasanitid ammonites (Fig. 11). The contradiction between the arguments of Casey, Enay and Zeiss and the independent evidence of Tethyan ammonites with Late Tithonian affinities discovered in the Upper Volgian rocks of Northern Siberia (Shulgina, 1967), East Greenland (Surlyk, 1978) and other Arctic basins, is also resolved thereby. Shulgina (1967) and Saks *et al.* (1972) were entirely correct in their appraisal of the biochronological significance of these North Siberian *"Virgatosphinctes"* spp., *Aulacosphinctes tripartitus* Shulgina 1967, *Berriasella*? *borealis* Shulgina 1967 and *Lemencia*? aff. *richteri* (Oppel). Regardless of whether or not future researchers confirm the original identifications of Shulgina (1967) or segregate these ammonites into new perisphinctid genera (e.g., *Praechetaites* Sazonova and Sazonov 1979b), these ammonites are typical Late Tithonian taxa, closely allied to the Tethyan genera into which they were originally placed.

LATE VOLGIAN AND RIASANIAN PALEOBIOGEOGRAPHY AND PALEOGEOGRAPHY OF EASTERN EUROPE

The earlier appearance of the riasanitid ammonites in the type Riasanian beds, as compared with their first appearance in the Mangyshlakian, North Caucasian and Crimean sections, indicates that these ammonites had migrated southward into the marginal Tethyan basins of southern U.S.S.R. from this Boreal habitat, not *vice versa*. There is additional biotic evidence for this southward migration in that the first appearance of these Riasanian ammonites in Mangyshlak and Northern Caucasus (but apparently not in Crimea) appears to coincide approximately with that of other Boreal immigrants, such as various *Buchia* forms and (?)*Surites* spp. Furthermore, the numbers of these Riasanian ammonites, buchias and (?)*Surites* spp. increase progressively upward in these Mangyshlakian and North Caucasian sections following their initially sparce to very rare occurrences there.

Yet another line of reasoning indicating the Boreal origin and a southward migration of *Riasanites* and *"Euthymiceras"* ex gr. *transfigurabilis* is provided by the latest Volgian and Riasanian (or latest Tithonian and Berriasian) paleogeography of Eastern Europe. As indicated in Figure 12a compiled from numerous Soviet sources (e.g., Sazonova, 1971, 1977; Saks *et al.*, 1972, 1973; Makridin, Migachova and Sterlin, in Krimgolts *et al.*, 1972), the earliest Riasanian (=late Late Tithonian) basin of Central Russia with the oldest known *Riasanites* and *"E."* ex gr. *transfigurabilis* was closed in the south and southeast, around the present Caspian Sea. Only nonmarine to lagoonal rocks of that time are known from the southern part of the Russian Platform and within the southern and eastern parts of the Caspian Depression. Furthermore, most of the latest Jurassic was a time of nondeposition and erosion in the Mangyshlak region and in the greater northeastern part of the northern Caucasus. Finally, the latest Tithonian marine faunas of the Crimean and Caucasian geosynclinal seas totally lack Boreal elements. Conversely, the Central Russian Basin of that time is characterized by a total absence of Tethyan faunal elements (*Riasanites* and *"E."* ex gr. *transfigurabilis* being discredited).

In contrast, this Central Russian Boreal Basin must have had an open marine connection with the high Boreal basins of the European and Siberian Arctic *via* the Pechora Seaway. This northern seaway is documented by the presence of the uppermost Jurassic marine rocks within the Pechora Basin and farther north. Furthermore, its existence is indicated by repeated penetration of the latest Volgian craspeditids, perisphinctids and buchias characteristic of these high Arctic basins into the restricted Central Russian Basin. Therefore, *Riasanites* and *"E."* ex gr. *transfigurabilis,* which were confined to the Central Russian Basin at that time, must have either evolved out of some older Boreal perisphinctids (e.g., pavlovids) inside of that basin, or they must have migrated into it from the north out of the high Boreal basins (Fig. 12a). The writer prefers the former hypothesis but this problem is beyond the scope of this paper. This latest Jurassic paleogeography persisted right through the Early Berriasian.

The paleogeography of Eastern Europe must have changed rather drastically in the Mid-Berriasian (i.e., early Occitanica time) when the Mangyshlak area and the greater part of the Northern Caucasus apparently were flooded for the first time since earlier Late Jurassic. This coalescence of the epicontinental Boreal sea of Central Russia with the Tethyan geosynclinal seas of Southern U.S.S.R., is attested by the appearance of *Riasanites, "Euthymiceras"* ex gr. *transfigurabilis,* buchias and (?)*Surites* in the Mangyshlak area and in the Northern Caucasus. These inferred paleogeographic changes are shown in Figure 12b.

The sudden appearance of *Riasanites* (including *Tauricoceras* Kvantaliani and Lyssenko, 1979) and apparently *"E."* ex gr. *transfigurabilis* in the later part of the *Fauriella boissieri* Zone of Crimea and their obvious abundance there (Kvantaliani and Lyssenko, 1979b) are believed to reflect their abrupt westward migration out of the Pechora-Mangyshlak Seaway, after they had been established in Mangyshlak and/or North Caucasus.

The appearance of *Riasanites, "Euthymiceras"* ex gr. *transfigurabilis* and *Surites* spp. in the Upper Berriasian of Central Poland (Marek and Raczynska, 1973; Dembovska and Marek, 1979) suggests strongly that these Central Russian forms migrated thereto from the Central Russian Basin *via* the newly opened Mid-Berriasian generation of the Brest Seaway (i.e., across the Welt of Polessie; see Fig. 12b).

The alternative, hitherto preferred hypothesis of migration of these ammonites into the Central Polish Basin from the south *via* the Lvov Seaway (Sazonova, 1971, 1977, p. 30, 31, Figs. 7, 8; Marek and Raczynska, 1979, p. 449, Fig. 1) does not appear feasible because of the previously documented Boreal origin of *Riasanites* and *"Euthymiceras"* ex gr. *transfigurabilis,* combined with their very late appearance in the Crimean Basin (Kvantaliani and Lyssenko, 1979b, p. 632, Corr. Table). Furthermore, *Surites* are totally unknown in the Crimean Basin and rare to very rare in the Mangyshlak and Northern Caucasus. The existence of the Brest Seaway in the Mid-to Late Berriasian is, therefore, required by the presence of numerous *Riasanites, "E."* ex gr. *tranfigurabilis* and *Surites* spp. (all either conspecific or very closely allied to the Central Russian species) in the Central Polish Basin. However, the simultaneous existence of the Lvov-Black Sea Seaway is indicated by an abundance of true Tethyan berriasellids in the Mid- to Late Berriasian basin of Central Poland (Marek and Raczynska, 1973, p. 373, 375, 376, Table 1; 1979, p. 448, 449, Table 1). Unlike the Central Russian riasanitids and *Surites,* these true berriasellids could only have reached the Central Polish Basin from the southeast.

LATEST JURASSIC AND EARLIEST CRETACEOUS ZOOGEOGRAPHY AND THE HYPOTHESIS OF ALLOCHTONOUS TERRANES OF WESTERN NORTH AMERICA

The exceptionally strong development of latitudinal, climatic paleobiotic belts in the latest Jurassic and earliest Cretaceous (p. 182-183, Fig. 3) permits the test of the validity of the hypothesis of "allochtonous (or suspect) terranes" currently fashionable with North American Cordilleran geologists (e.g., Jones *et al.*, 1977; Coney *et al.*, 1980; Yorath and Chase, 1981; Monger *et al.*, 1982) and geophysicists (e.g., Packer and Stone, 1974; Irving and Yole, 1972; Monger and Irving, 1980; Monger and Yole, 1980) for Late Tithonian and younger times. These terranes comprising British Columbia, Alaska and other Cordilleran regions (e.g., the Wrangellia, Alexander, Bridge River and Stikinia terranes) are said to have experienced huge northward displacements (commonly from the northern subequatorial or even from the low southern latitudes) in the Late Jurassic and later times. The hypothesis is based solely on measurements of magnetic paleolatitudes of these terranes.

The Late Tithonian and Berriasian marine macroinvertebrate faunas are nearly uniform across the "allochtonous" Terrane II of Monger *et al.* (1982, p. 72, Fig. 1), between western Vancouver Island (Wrangellia) and the Tyaughton Lake-Manning Park areas of the Tyaughton Trough (Bridge River terrane). These faunas (see Jeletzky, in Jeletzky and Tipper, 1968, p. 80-83; 1970, p. 649-654, Table XI-8; 1971b, p. 35, 36, and above) consist everywhere of the same *Buchia* species with mostly minor admixtures of the same belemnites, pelecypods, gastropods and brachiopods. Their ammonites are more varied but almost exclusively (with one exception) Tethyan in their affinities. Among buchias, all endemic species (e.g., *B.* cf. *blanfordiana, B.* n. sp. aff. *okensis, B. piochii* s.s., *B. okensis* s.s., *B. uncitoides spasskensoides* and *B. tolmatschowi*) that had arisen and evolved in the Tyaughton and Insular Troughs, are ubiquitous. Furthermore, these Late Tithonian and Berriasian faunas are extremely closely allied to the contemporary faunas of the intracratonic Peace River Basin, that is separated from Terrane II by the allegedly allochtonous Terrane I. Except for the *Buchia tolmatschowi* fauna, all Late Tithonian and Berriasian *Buchia* faunas of the Tyaughton and Insular Troughs also occur in the uppermost Fernie Group and the Monteith Formation of the Peace River Basin (see Jeletzky, 1968a, p. 4-7; 1970, Table XI-8; 1971b, p. 35, 36; numerous unpublished fossil reports and earlier in this paper; Frebold and Tipper, 1970, p. 14-17, Table I). Furthermore, all other marine macroinvertebrates of these Peace River faunas, including their Tethyan ammonites, are nearly identical with those of the contemporary faunas of western British Columbia. The distinctive presence of fairly numerous, high Boreal craspeditid and dorsoplanitid ammonites in some of these faunas (e.g., Jeletzky, 1964, Pl. I, Figs. 2, 6, 7; 1968a, p. 4-7; 1970, Pl. XI-8; Frebold and Tipper, p. 14-17, Table I) allies them to the contemporary high Boreal faunas of the intracratonic basins of Arctic Canada.

All Late Tithonian and Berriasian marine macroinvertebrate faunas of British Columbia are very closely allied to those of the Porcupine Plain-Richardson Mountains Trough and the Sverdrup Basin (see Jeletzky, 1964, Pl. I, figs. 3, 8; Pl. II, Figs. 2, 3; Pl. IV, Figs. 2-4; Table I; 1968a, p. 4-8; 1971a, p. 3, Fig. 2; 1973, p. 47-62, Fig. 3; 1975b, text and Figs. 1, 3-7 and p. 192-217, Fig. 10 of this paper). Most of the buchias and belemnites of these high Boreal, intracratonic basins of Arctic Canada are conspecific with their counterparts in the intracratonic and geosynclinal basins of British Columbia (Figs. 5, 10). The close affinities of these faunas are further stressed by far reaching overlaps of some of their most diagnostic elements. A number of buchias native to the Tyaughton and Insular Troughs (e.g., *B.* cf. *blanfordiana, B. piochii* s.s., *B. uncitoides* s.l., *B. tolmatschowi*) and most of their Tethyan ammonites either become progressively more scarce northward and then disappear within these Arctic intracratonic basins or only become relatively (and progressively) more scarce within them. Conversely, a number of characteristically high Boreal fossils (e.g., *Buchia unschensis* s.s., *B. terebratuloides obliqua, B. fischeriana, Arctotis* spp., *Meleagrinella* spp., *McLearnia* spp. and all high Boreal ammonites) that abound in these basins (especially in the more northerly Sverdrup Basin) either become progressively more scarce southward and then disappear or only progressively become more scarce in that direction.

The Late Tithonian and Berriasian marine macroinvertebrate faunas of western British Columbia are just as closely allied with their counterparts in southeastern Alaska as they are with the contemporary faunas of the Peace River Basin (see Jones, 1973; Imlay, 1980, p. 66, 67; 1981, p. 17-23, Fig. 12). Furthermore, the close affinities and overlapping relationships of these faunas of southeastern Alaska with their high Boreal North Alaskan counterparts parallel exactly the relationships, described above, between the East Pacific faunas of the geosynclinal and intracratonic basins of British Columbia on the one hand and the intracratonic basins of Arctic Canada on the other.

All these paleozoogeographic data indicate that all latest Jurassic and earliest Cretaceous marine basins of western and northeastern British Columbia, including those of allegedly allochtonous Terranes I and II of Monger, Price and Tempelman-Kluit (1982, p. 72, Fig. 1), were situated at approximately the same mid-northern paleolatitude, just south of the southern limit of the Boreal Realm as defined earlier (Fig. 3). The same must have been true of all of southeastern Alaska.

The character of the Late Tithonian and Berriasian faunas of the Insular and Tyaughton Troughs and the Peace Rive Basin militates against the assumption that any of them could have been transported northward even from the paleolatitude of northwestern California, either at that time or later. The contemporary North Pacific faunas of that region already differ markedly in either the total lack or extreme rarity of a number of buchias characteristic of the British Columbian basins. Furthermore, they differ in a considerably greater frequency of occurrence and diversity of their Tethyan ammonites and the appearance of some other southern molluskan types absent in the British Columbian basins. I discussed previously that the utterly different character of the Late Tithonian and Berriasian marine faunas and facies prevalent still further south in the Caribbean province (Fig. 3) rules out that any parts of these basins have been transported from there since the Late Tithonian.

Because of the typically shelf-like character of all their nearly identical latest Jurassic and basal Cretaceous marine faunas (especially belemnites), the Terranes I and II and the Peace River Basin must have been situated geographically close to each other in terms of their paleolongitudes. Furthermore, this character of marine faunas indicate the absence of any deep oceanic basins between these three megatectonic blocks of western Canada at least since the Late Tithonian. Therefore, Terranes I and II must have been accreted to the North American Craton by that time.

The Tyaughton and Insular Troughs of Terrane II must have been connected with the Peace River Basin by a submeridional epicontinental seaway that was sufficiently wide and shallow to allow an entirely free exchange of their shelf-type faunas right across the allegedly allochtonous Terrane I. This seaway was originally inferred by Jeletzky and Tipper (1968, p. 83, 84) and named the Vanderhoof Strait by Jeletzky (1970, p. 654, Fig. XI-2).

The Arctic intracratonic basins of Canada must have been situated close to the north of the North Pacific basins and approximately at their present latitudes and longitudes (Fig. 4). The geographically intermediate position of the Porcupine Plain-Richardson Mountains Trough is indicated by the transitional character of its faunas. This suggests, in turn, the presence of an easily accessible, epicontinental (because of an exclusively shelf-like character of all faunas concerned) seaway connecting that trough with the intracratonic and geosynclinal basins of British Columbia; it was named the Dawson Strait by Jeletzky, 1970, p. 654, Fig. XI-2). All of western and Arctic Canada, including Terranes I and II of British Columbia and South Yukon and Alaska, formed a closely knit tectonic ensemble accreted to the North American Craton at least by the Late Tithonian. All component tectonic blocks of this ensemble must have been situated approximately at their present day latitudes and longitudes (i.e., within the limits of precision of the paleozoogeographic method defined below) and were not separated by any deep oceanic basins since that time (Jeletzky and Tipper, 1968, p. 80-83; Jeletzky, 1970, p. 654, Fig. XI-2; and this paper Fig. 4).

The late Tithonian and Berriasian faunas of western and Arctic North America are almost as closely allied to the contemporary shallow marine faunas of Northeast Siberia as they are among themselves. This indicates a latitudinally and longitudinally adjacent paleogeographic positions of these parts of the two continents and the existence of a broad belt of marine shelf connecting them beneath the Bering Strait at least since the Late Tithonian. This inferred Bering Shelf must have flanked from the south the ancient Arctic (or Hyperborean) Platform (Jeletzky, 1975b, p. 2, 3, 49). Therefore, the North American Plate extended into eastern Siberia, at least up to the Yuzhnyi Anui Fold Belt as proposed by most of the recent Soviet workers (e.g., Smirnov and Marchenko, 1976) and some recent North American workers (e.g., Jeletzky, 1975b; Churkin and Trexler, 1981).

The paleozoogeographic method employed here is "coarse grained" in the sense that the paleogeographic positions of the marine basins concerned can only be recognized approximately. Furthermore, it cannot be used during the periods of far-reaching levelling of the climatic conditions (e.g., Hettangian or Sinemurian). My research of the post-Sinemurian Jurassic and Cretaceous marine faunas indicates that the climatologically caused latitudinal changes of their composition become obvious only when north-south distances of 500 to 1,000 km are involved. Jones (1973) climatic zonation of Early Cretaceous marine faunas of the Pacific slope of North America has yielded similar results, and so did Tipper's (1982) attempt to climatically zone the Pliensbachian ammonite faunas of that region and to demonstrate their major north-south displacements. Even when the paleozoogeographic evidence for or against north-south displacements of tectonic blocks involves distances approaching 1,000 km, one has to be cautious as such displacement of faunal belts may also be caused by paleoceanic currents, connecting seaways, or impenetrable land barriers (compare Arkell, 1956, p. 606; Jeletzky, 1971b, p. 77, 78). Tipper's (1982) attribution of up to 1,000 km north-south displacements for the Pliensbachian ammonite assemblages within the Cordilleran Belt of North America to transcurrent faulting may be valid. However, these differences could also have been either strongly influenced or caused by the presence of land barriers separating the individual Pliensbachian basins and/or by the either southward or northward flowing oceanic currents within them. Therefore, my criticisms of the hypothesis of allochtonous (or suspect) terranes are directed *only and alone* against assumptions of north-south movements exceeding 1,000 km. There the paleozoogeographic evidence becomes decisive, as such displacements would inevitably cause the appearance of high Tethyan faunas in the midst of the utterly different North Pacific or high Boreal faunas that are alone present in British Columbia, Alaska and northeastern Siberia.

The latest Jurassic and earliest Cretaceous paleogeography and tectonic setting of the Cordilleran Belt of North America, adhered to by myself (Figs. 3, 4), is incompatible with many of the measurements of magnetic paleolatitudes made in the Jurassic and later rocks of its individual tectonic blocks. Consequently, all such measurements conflicting with the paleozoogeographic evidence presented above are considered spurious and caused by an overprinting and/or disorganization of original inclinations by the extremely violent intrusive and tectonic history of the region. Well documented instances of such falsifications of paleomagnetic records are rapidly multiplying outside of the region (e.g., Schmidt and Embleton, 1981; Orlenok and Chebanenko, 1982, p. 9, 10, Fig. 2; Khudoley, 1981, p. 174-177, Fig. 5) and some of these workers infer that paleomagnetic measurements made in strongly dislocated and/or metamorphosed rocks are generally unreliable. Such evidence is also available for the Cordilleran Belt of North America. For example, mutually exclusive Jurassic paleomagnetic data were obtained in adjacent stations in Alaska (Packer and Stone, 1974, p. 991). Some of them indicate absence of major displacements of the rocks concerned as compared with their present position. Diametrically opposed results were also obtained by Symons (1971) and Irving, Monger and Yole (1980) for the Early Jurassic Topley Intrusions of the Stikinia and the so called Island Intrusions (Mid-Jurassic) of the Wrangellia. Symons' (1971) rejection of either major displacement or rotation of these intrusions since their injection and cooling agrees perfectly with the paleozoogeographic data and so is obviously more trustworthy than the opposite conclusions of Irving, Monger and Yole (1980). It is also supported by the obvious continuity of the so called Island (=Coast) Instrusions of the Wrangellia with the earlier phases of the Coast Intrusions of the mainland of British Columbia stressed recently by Jeletzky (1976, p. 63-65; 1978, p. 28, 29).

ACKNOWLEDGEMENTS

The growth of ideas presented in this paper was greatly facilitated by discussion with many colleagues both within and outside of the Geological Survey of Canada and by the study of numerous museum collections in North America, Western Europe and England during the last 30 years or so. Sincere thanks are expressed to all the colleagues, too numerous to mention by name, in Canada, the United States of America, Great Britain, Western Europe and the Soviet Union who have furthered the writer's research one way or another. Mrs. K. Vincent prepared all of the text figures from drafts provided by the writer, edited the manuscript and ably supervised the photography of all fossils published in this paper. Ms. J. White photographed all fossils reproduced in it.

REFERENCES

Allemann, F., Grün, W., and Wiedmann, J., 1975, The Berriasian of Caravaca (Prov. of Murcia) in the Subbetic Zone of Spain and Its Importance for Defining this Stage and the Jurassic-Cretaceous Boundary: Colloque Sur la Limite Jurassique-Crétacé, Lyon, Neuchâtel, septembre 1973: *in* Mémoires du Bureau de Recherches Géologiques et Minières, France, no. 86, p. 14-22.

Anderson, F.M., 1945, Knoxville Series in the California Mesozoic: Geological Society of America Bulletin, v. 56, p. 909-1014.

Arkell, W.J., 1946, Standard of the European Jurassic: Geological Society of America Bulletin, v. 57, p. 1-34.

———, 1956, Jurassic Geology of the World: Oliver and Boyd, Edinburgh and London, 806 p.

Arkell, W.J., Furnish, W.M., Kummel, B., Miller, A.K., Moore,

R.C., Schindewolf, O.H., Silvester-Bradley, P.C., and Wright, C.W., 1957, Mollusca 4, Cephalopoda, Ammonoidea: Treatise on Invertebrate Paleontology, R.C. Moore, ed.; Part L: Geological Society of America and University of Kansas Press. p. L1-L490.

Barthel, K.W., Cediel, F., Geyer, O.F., and Remane, J., 1966, Der subbetische Jura von Gehegin (Provinz Murcia, Spanien): Mitteilungen der Bayerischen Staatssammlung für Paläontologie und historische Geologie, Heft 6, p. 167-211.

Biro-Bogoczky, L., 1980, Estudio Sobre el Limite Entre el Titoniano y el Neocomiano en la Formation Lo Valdes, Provincia de Santiago (33°50′ Lat. Sur) Chile, Principalemente Sobre la Base de Ammonoideos: Actas II Congreso Argentino de Paleontologia y Bioestratigrafia y I Congreso Latinoamericano de Paleontologia, Buenos Aires, 1978, t. V., P. 137-152.

Bodylevsky, V.J., 1936, The Upper Volgian Fauna of Novaya Zemlya: Trudy Arkticheskogo Nauchno-Issledovatel'skogo Instituta, v. 49, p. 113-136.

Bodenbender, G., 1892, Sobre el Terreno Jurasico y Cretaceo en los Andes Argentinos, Entre el Rio Diamante y Rio Limay: Boletin de Academia Nacional de Ciencias de Cordoba (Argentina), v. XIII, p. 1-44.

Bonet, F., 1956, Zonificacion Microfaunistica de Las Calizas Cretacicas del Este de Mexico: Asociacion Mexicana Géologos Petroleros, Boletin, v. 8, p. 389-488.

Burckhardt, C., 1912, Faunas Jurassiques et Crétaciques de San Pedro del Gallo: Instituto de Geologia México, Boletin 29, 264 p.

Busnardo, R., Le Hégarat, G., et Magne, J., 1965, Le stratotype du Berriasien; Colloque sur le Crétacé Inférieur (Lyon, Sept. 1963): *in* Mémoires du Bureau des Recherches Géologiques et Minières, France, no. 34, p. 5-33.

Churkin, M., Jr., and Trexler, J.H., Jr., 1981, Continental Plates and Accreted Oceanic Terranes in the Arctic: *in* The Ocean Basins and Margins, v. 5. The Arctic Ocean: Plenum Press, New York and London, p. 1-20.

Callomon, J.H., 1961, The Jurassic System in East Greenland: Geology of the Arctic: Proceedings, 1st International Symposium on Arctic Geology, v. I, p. 258-268.

Cantú Chapa, A., 1963, Étude biostratigraphique des Ammonites du centre et de l'est du México (Jurassique Superieur et Crétacé): Société Géologique de France, Mémoire 99, n. ser., v. 42, pt. 4, 102 p.

———, 1967, El limite Jurassico-Cretácico en Mazatepec, Puebla; *in:* Estratigrafia del Jurassico de Mazatepec, Puebla (Mexico); Instituto Mexicano del Petroléo. Technologia de la Exploración. Seccion Geologia, Monografia I, p. 3-24.

———, 1968, Sobre una association *Proniceras-Durangites-"Hildoglochiceras"* del Noreste de Mexico: Instituto Mexicano del Petroléo, Seccion Geologia, Monografia 2, p. 19-26.

———, 1971, La serie Huasteca (Jurassico medio-superior) del Centro Este de Mexico: Instituto Mexicano del Petroléo, Revista, v. 3(2), p. 17-40.

———, 1982, The Jurassic-Cretaceous Boundary in the Subsurface of Eastern Mexico: Journal of Petroleum Geology, v. 4, p. 311-318.

Casey, R., 1962, The Ammonites of the Spilsby Sandstone and the Jurassic-Cretaceous Boundary: Geological Society of London Proceedings, Session 1961-62, nos. 1593-1602, p. 95-100.

———, 1963, The Dawn of the Cretaceous Period in Britain: South-East Union of Scientific Societies Bulletin, v. 117, p. 1-15.

———, 1973, The Ammonite Succession at the Jurassic-Cretaceous Boundary in Eastern England: The Boreal Lower Cretaceous: Geological Journal, Special Issue No. 5, p. 193-266.

Casey, R., Messezhnikov, M.S., and Shulgina, N.I., 1977, Comparison of the Jurassic-Cretaceous boundary beds of England, Russian Platform, Subpolar Ural and Siberia: Izvestiya Akademii Nauk SSSR, Seriya Geologicheskaya, No. 7, p. 14-33 (in Russian).

Coates, J.A., 1974, Geology of the Manning Park Area, British Columbia: Geological Survey of Canada Bulletin 238, 177 p.

Coney, P.J., Jones, D.L., and Monger, H.W.H., 1980, Cordilleran Suspect Terranes: Nature, v. 288, p. 329-333.

Crickmay, C.H., 1930a, Fossils from Harrison Lake Area, British Columbia: Contributions to Canadian Paleontology, Natural History Museum, Canada, Bulletin 63, p. 33-60.

———, 1930b, The Structural Connection Between the Coast Range of British Columbia and the Cascade Range of Washington: Geological Magazine, v. LXVII, p. 482-491.

———, 1932, A New Jurassic Ammonite from the Coast Range of California: American Midland Naturalist, v. 13, p. 1-7.

———, 1962, Gross Stratigraphy of Harrison Lake Area, British Columbia: Article 8, p. 1-12: (Published by author, available at Evelyn LeMille Books, Calgary.)

Dembowska, Ya, and Marek, S., 1979, The Jurassic-Cretaceous Boundary in the Polish Lowland: Akademia Nauk SSSR, Sibirskoye Otdelenie, Institut Geologii i Geofiziki; Verkhniaya Yura i yee Granitsa s Melovoi Sistemoi, "Nauka" Press, Novossibirsk, p. 40-43 (in Russian).

Drustchits, V.V., 1960, Lower Cretaceous Deposits of the Crimea: *in* Drustchits, V.V. and Kudriavtsev, M.P., (eds.): Atlas Nizhnemelovoi Fauny Severnogo Kavkaza i Kryma: Gosudarstvennoye Nauchno-Issledovatel'skoye Izdatel'stvo Neftianoi i Gorno-Toplivnoi Literatury, Moscow, p. 53-74 (in Russian).

———, 1963, Subdivisions of the Lower Cretaceous of the Crimea and Caucasus: *in* Biostratigraphy and Ammonites of the Lower Cretaceous of Crimea and Caucasus: Moscow State University Press (in Russian).

Drustchits, V.V. and Vakhrameev, V.A., 1976, The Jurassic-Cretaceous Boundary: *in* Granitsy Geologicheskich System: "Nauka" Press, Moscow, p. 185-224 (in Russian).

Enay, R., 1964, L'étage tithonique: Colloque du Jurassique, Luxembourg, 1962: Volume des Comptes Rendus et Mémoires, Luxemburg, p. 355-379.

———, 1972, Paleobiogéographie des Ammonites du Jurassique Terminal (tithonique/volgien/portlandien s.l.) et Mobilité Continentale: Géobios, v. 5, p. 355-407.

———, 1973, Upper Jurassic (Tithonian) Ammonites: *in* Hallam, A., ed., Atlas of Palaeobiogeography: Elsevier Scientific Publishing Company, Amsterdam, p. 297-307.

Enay, R., and Géyssant, J.R., 1975, Faunes Tithoniques des Chaines Bétiques (Espagne méridionale); Colloque sur la limite Jurassique-Crétacé, Lyon-Neuchâtel, septembre 1973: *in* Mémoires du Bureau de Recherches Géologiques et Minières, France, no. 86, p. 39-55.

Enay, R., Martin, C., Monod, O., and Thieuloy, J.P., 1971, Jurassique Supérieur à Ammonites (Kimmeridgien-Tithonique) dans l'autochtone du Taurus de Beysehir (Turque méridionale): Annales, Institute Geologici Publici Hungarici, v. LIV (2), p. 397-422.

Fortier, Y.O., Blackadar, R.G., Glenister, B.F., Greiner, H.R., McLaren, D.J., McMillan, N.J., Norris A.W., Roots, E.G., Souther, J.G., Thorsteinsson, R., and Tozer, E.T., 1963, Geology of the North-Central Part of the Arctic Archipelago, Northwest Territories (Operation Franklin): Geological Survey of Canada Memoir 320, 671 p.

Frebold, H., 1957, The Jurassic Fernie Group in the Canadian Rocky

Mountains and Foothills: Geological Survey of Canada Memoir 287, 197 p.

———, 1961, the Jurassic Faunas of the Canadian Arctic, Middle and Upper Jurassic Ammonites: Geological Survey of Canada Bulletin 74, 43 p.

Frebold, H., and Tipper, H.W., 1970, Status of the Jurassic in the Canadian Cordillera of British Columbia, Alberta, and Southern Yukon: Canadian Journal of Earth Sciences, v. 7, p. 1-21.

Gerassimov, P.A., 1955, Index Fossils of the Mesozoicum of Central Regions of the European Part of USSR, Part I: Lamellibranchiate, Gastropod and Scaphopod Mollusks and Brachiopods of the Jurassic Deposits: Geologicheskoye Upravlenie Tsentral'nykh Rayonov Ministerstva Geologii i Okhrany Nedr, GNTI Press, Moscow, 276 p. (in Russian).

———, 1969, Upper Substage of the Volgian Stage of the Central Part of Russian Platform: Akademiya Nauk SSSR, Ministerstvo Geologii SSSR, Mezhduvedomstvennyi Stratigraficheskii Komitet, "Nauka" Press, Moscow, 144 p. (in Russian).

———, 1971, About Berriasian and Lower Valanginian of Russian Platform: Doklady Akademii Nauk SSSR, v. 198(5), p. 1156-1157 (in Russian).

Gerrassimov, P.A., and Michailov, N.P., 1966, Volgian Stage and the Unified Stratigraphic Scale of the Upper Part of the Jurassic System: Izvestiia Akademii Nauk SSSR, Serija geologicheskaya, no. 2, p. 118-138.

Gerassimov, P.A., Migacheva, E.E., Naidin, D.P., and Sterlin, B.P., 1962, Jurassic and Cretaceous Deposits of Russian Platform: Moscow State University Press, 195 p. (in Russian).

Gerth, H., 1926, Beiträge zur Paläontologie and Stratigraphie des Lias, Doggers, Tithons und der Unterkreide in der Kordilleren im Süden der Provinz Mendoza (Argentinien). III. Die Fauna des Neokom in der Argentinischen Kordillere: Geologische Rundschau, Bd. 17A (G. Steinmann's Festschrift), p. 463-494 (also published as La Fauna Neocomiana de la Cordillera Argentina en la parte meridional de la provincia de Mendoza: Actas, Academia Nacional Ciencias Republica Argentina, v. IX, 1ª entrega, 1925).

Gol'bert, A.V., Klimova, I.G., and Saks, V.N., 1972, The Standard Section of the Neocomian for West Siberia in the Subpolar Transuralian Region: Ministerstvo Geologii SSSR, Sibirskii Nauchno-Issledovatel'skii Institut Geologii, Geofiziki i Mineral'nogo Syria, "Nauka" Press, Novossibirsk, 184 p. (in Russian).

Gol'bert, A.V., and Klimova, I.G., 1979, Jurassic-Cretaceous Boundary Beds and the Marine Lower Cretaceous in the Neocomian Standard Section for Western Siberia (in exposures adjacent to the subpolar Urals): Akademiya Nauk SSSR, Sibirskoye Otdelenie, Institut Geologii i Geofiziki, Verkhnaya Yura i Granitsa yee s Melovoi Systemoi, "Nauka" Press, Novossibirsk, p. 35-40 (in Russian).

Gorbachik, T.N., Drustchits, V.V., and Yanin, B.T., 1975, Lower Cretaceous Deposits of the Watershed Bel'bek-Al'ma (Crimea): Vestnik Moskovskogo Gossudarstvennogo Universiteta, No. 6, p. 19-31 (in Russian).

Grigorieva, O.K., 1938, A Lower Valanginian Ammonite Fauna from the Basin of the Belaya River on the Northern Slope of the Caucasus (Maikop area): *in* Azov-Black Sea Geological Trust, Materials on the Geology and Economic Minerals, Collection no. 1. Rostov on Don, p. 83-123.

Hakansson, E., Birkelund, T., Piasecki, S., and Zakharov, V., 1981, Jurassic-Cretaceous Boundary Strata of the Extreme Arctic (Peary Land, North Greenland): Geological Society of Denmark Bulletin, v. 30, p. 11-42.

Hallam, A., 1969, Faunal Realms and Facies in the Jurassic: Palaeontology, v. 12, p. 1-18.

Hölder, H., 1979, Jurassic: *in* Robison, R.A. and Teichert, C., eds., Treatise on Invertebrate Paleontology, Part A: Introduction, Fossilization (Taphonomy), Biogeography and Biostratigraphy: p. A390-A417.

Hoedemaeker, Ph. J., 1979, The Jurassic-Cretaceous Boundary Near Miravetes (Caravaca, SE Spain); Arguments for its Position at the Base of the Occitanica-Zone: Cuadernos de Geologia, Universidad de Granada, v. 10, p. 235-247.

———, 1982, Ammonite Biostratigraphy of the Uppermost Tithonian, Berriasian, and Lower Valanginian along the Rio Argos (Caravaca, SE Spain): Rijksmuseum van Geologie en Mineralogie, Leiden, Scripta Geologica, no. 65, 81 p.

Housa, V., 1965, La Klippe de Stramberk: *in* Réunion Extraordinaire de la Societé Géologique de France: Carpates Tchécoslovaques, Bulletin, Societé Géologique de France, v. 7, p. 1074-1077.

———, 1975, Geology and Paleontology of the Stramberg Limestone (Upper Tithonian) and the associated Lower Cretaceous beds: Colloque sur la limite Jurassique-Crétacé, Lyon, Neuchâtel, septembre 1973: *in* Mémoires, Bureau de Recherches Géologiques et Minières, France, no. 86, p. 342-349.

Imlay, R.W., 1939, Upper Jurassic Ammonites from Mexico: Geological Society of America Bulletin, v. 59, p. 1-78.

———, 1942, Late Jurassic Fossils from Cuba and their Economic Significance: Geological Society of America Bulletin, v. 53, p. 1417-1478.

———, 1943, Jurassic Formations of Gulf Region: American Association of Petroleum Geologists Bulletin, v. 27, p. 1407-1538.

———, 1952, Correlation of the Jurassic Formations of North America, Exclusive of Canada: Geological Society of America Bulletin, v. 63, p. 953-992.

———, 1955, Characteristic Jurassic Mollusks from Northern Alaska: United States Geological Survey Professional Paper 274-D, p. 69-96.

———, 1959, Succession and Speciation of the Pelecypod *Aucella:* United States Geological Survey Professional Paper 314-G, p. 155-169.

———, 1965, Jurassic Marine Faunal Differentiation in North America: Journal of Paleontology, v. 39, p. 1023-1038.

———, 1980, Jurassic Paleobiogeography of the Conterminus United States in its Continental Setting: United States Geological Survey Professional Paper 1062, 134 p.

Imlay, R.W., and Jones, D.L., 1970, Ammonites from the *Buchia* Zones in Northwestern California and Southwestern Oregon: United States Geological Survey Professional Paper 647-B, 59 p.

Irving, E., and Yole, R.W., 1972, Paleomagnetism and the Kinematic History of Mafic and Ultramafic Rocks in Fold Mountain Belt: *in* Irving, E., ed., The Ancient Oceanic Lithosphere; Canada Department of Energy, Mines and Resources, Earth Physics Branch, v. 42, p. 87-95.

Irving, E., Monger, J.W.H., and Yole, R.W., 1980, New Paleomagnetic Evidence for Displaced Terranes in British Columbia: *in* Strangeway, D.W. ed., The Continental Crust and its Mineral Deposits: Geological Association of Canada Special Paper 20, p. 441-456.

Jeletzky, J.A., 1950, Stratigraphy of the West Coast of Vancouver Island between Kyuquot Sound and Esperanza Inlet, British Columbia: Geological Survey of Canada Paper 50-37, 52 p.

———, 1964, Illustrations of Canadian Fossils: Lower Cretaceous Marine Index Fossils of the Sedimentary Basins of Western and Arctic Canada: Geological Survey of Canada Paper 64-11, 101 p.

______, 1965, Late Upper Jurassic and Early Lower Cretaceous Fossil Zones of the Canadian Western Cordillera, British Columbia: Geological Survey of Canada Bulletin 103, 70 p.

______, 1966, Upper Volgian (latest Jurassic) Ammonite and Buchias of Arctic Canada: Geological Survey of Canada Bulletin 128, 51 p.

______, 1967, Stratigraphy and Paleontology of Lower Cretaceous and Upper Jurassic Rocks of Taseko Lakes (92-O) and Pemberton (92-J) Map-Areas: Geological Survey of Canada Paper 67-1A, p. 65-68.

______, 1968a, Macrofossil Zones of the Marine Cretaceous of the Western Interior of Canada and their Correlation with the Zones and Stages of Europe and the Eastern Interior of the United States: Geological Survey of Canada Paper 67-72, 66 p.

______, 1968b, Stratigraphy and Paleontology of Lower Cretaceous and Upper Jurassic Rocks of Northeastern Corner of Mount Waddington Map-Area, British Columbia (92N): Geological Survey of Canada Paper 68-1, Part A, p. 103-106.

______, 1970, Cretaceous Macrofossils: Biochronology Chapter, Pt. V: R.W. Douglas, ed., Geological Survey of Canada, Centennial Edition of the Economic Geology Series No. 1, p. 649-662.

______, 1971a, Biochronology of Jurassic-Cretaceous Transition Beds in Canada: Geological Survey of Canada Paper 71-16, 8 p.

______, 1971b, Marine Cretaceous Biotic Provinces and Paleogeography of Western and Arctic Canada; Illustrated by a Detailed Study of Ammonites: Geological Survey of Canada Paper 70-22, 92 p.

______, 1971c, Cretaceous and Jurassic Stratigraphy of Some Areas of Southwestern British Columbia: Geological Survey of Canada Paper 71-1A, no. 122, p. 221-227.

______, 1973, Biochronology of the Marine Boreal Latest Jurassic, Berriasian and Valanginian in Canada: *in* Casey, R. and Rawson, P.F., eds., The Boreal Lower Cretaceous: Geological Journal, Special Issue No. 5, p. 41-80.

______, 1974, Jurassic-Cretaceous Transition Beds in Canada: Colloque du Jurassique, Luxembourg, 1967: *in* Mémoires du Bureau de Recherches Géologiques et Minières, France, no. 75, p. 701-707.

______, 1975a, Intervention: Discussions Sur la Position de la Limite Jurassique-Crétacé: Colloque sur la Limite Jurassique-Crétacé, Lyon, Neuchâtel, septembre 1973: *in* Mémoires du Bureau de Recherches Géologiques et Minières, France, no. 86, p. 383.

______, 1975b, Jurassic and Lower Cretaceous Paleogeography and Depositional Tectonics of Porcupine Plateau, Adjacent Areas of Northern Yukon and Those of Mackenzie District: Geological Survey of Canada Paper 74-16, 52 p.

______, 1976, Mesozoic and ?Tertiary Rocks of Quatsino Sound, Vancouver Island, British Columbia: Geological Survey of Canada Bulletin 242, 243 p.

______, 1978, Causes of Cretaceous Oscillations of Sea Level in Western and Arctic Canada and Some General Geotectonic Implications: Geological Survey of Canada Paper 77-18, 44 p.

______, 1979, Eurasian Craspeditid Genera *Temnoptychites* and *Tollia* in the Lower Valanginian of Sverdrup Basin, District of Franklin: With Comments on Taxonomy and Nomenclature of Craspeditidae: Geological Survey of Canada Bulletin 299, 89 p.

______, 1982, Jurassic-Cretaceous Boundary Beds of Western and Arctic Canada and the Tithonian-Berriasian Boundary in the Boreal Realm: Journal of Paleontology, v. 56, March, 1982, Part II of III, Supplement to no. 2, North American Paleontological Convention III, Abstracts of Papers, p. 13, 14.

Jeletzky, J.A. and Tipper, H.W., 1968, Upper Jurassic and Cretaceous Rocks of Taseko Lakes Map-Area and their Bearing on the Geological History of Southwestern British Columbia: Geological Survey of Canada Paper 67-54, 218 p.

Jones, D.L., 1973, Structural Elements and Biostratigraphical Framework of Lower Cretaceous Rocks in Southern Alaska: *in* R. Casey and P.F. Rawson, eds. The Boreal Lower Cretaceous: Geological Journal, Special Issue No. 5, p. 1-18.

Jones, D.L., Bailey, E.H., and Imlay, R.W., 1969, Structural and Stratigraphic Significance of the *Buchia* Zones in the Colyear Springs – Paskenta Area, California: United States Geological Survey Professional Paper 647-A, 24 p.

Jones, D.L., Silberling, N.J., and Hillhouse, J., 1977, Wrangellia: A Displaced Terrane in Northwestern North America: Canadian Journal of Earth Sciences, v. 14, p. 2565-2577.

Kaufmann, E.G., 1973, Cretaceous Bivalvia: *in* Hallam, A., ed., Atlas of Palaeobiogeography: Elsevier Scientific Publishing Company, Amsterdam, p. 353-383.

Khain, V.E., 1981, Pacific Ocean and the Pacific Mobile Belt: Vestnik Moskovskogo Gosudarstvennogo Instituta, Geology, no. 1, p. 3-13.

Khimiashvili, N.G., 1976, Tithonian and Berriasian Ammonoidea of the Caucasus: Akademiya Nauk Gruzinskoi SSR, Institut Paleobiologii, "Metziereba" Press, Tbilisi, 180 p. (in Russian).

Khudoley, K.M., 1959, Representatives of the Genus *Virgatosphinctes* from Late Jurassic Deposits of Southern Sikhote-Alyn: Leningrad, Vsessoyuznii Nauchno-Issledovatel'skii Geologicheskii Institut, Informatsionnyi Sbornik no. 24, p. 117-126 (in Russian).

______, 1963, Late Jurassic Ammonites of Southern Sikhote-Alyn: Leningrad, Vsessoyuznii Nauchno-Issledovatel'skii Geologicheskii Institut, Trudy, novaya seriiya, v. 81, p. 61-73 (in Russian).

______, 1970, Tithonian Marine Zoogeographical Provinces of the Pacific Ocean: Akademiya Nauk SSSR, Doklady, v. 194, p. 667-669 (in Russian).

______, 1974, Circum-Pacific Mesozoic Ammonoid Distribution: Relation to Hypotheses of Continental Drift, Polar Wandering, and Earth Expansion: *in* Kahle, C.F., ed., Plate Tectonics-Assessments and Reassessments. American Association of Petroleum Geologists Memoir 23, p. 295-330.

______, 1981, Criticism of the Hypotheses of Continental Drift, Plate Tectonics and Meandering of Geographic Poles: Akademiya Nauk SSSR, Trudy XXI sessii Vsessoyuznogo Paleontologicheskojo Obstchestva: Paleontologiya, Paleobiogeografiya i Mobilism: Magadan Publishing House, p. 166-179 (in Russian).

Klimova, I.G., 1969, About the Early Berriasian of Western Siberia: Geology and Geophysics (Novossibirsk), no. 4, p. 128-132 (in Russian).

Konovalov, V.P., 1977, The Genus *Buchia* in the Pacific Province: *in* Evolutsiya Organicheskogo Mira Tikhookeanskogo Poyassa: Khronologicheskie paleobiogeograficheskie rubezhy: Akademiya Nauk SSSR, Dal'nevostochnyi Nauchnyi Tsentr, Biologo-Pochvennyi Institut, Vladivostok, p. 139-155 (in Russian).

Krantz, F., 1926, Die Ammoniten des Mittel – und Obertithons: Geologische Rundschau, Sonderband 17A (G. Steinmann's Festschrift), p. 428-462.

Krimgol'ts, G.Y., ed., 1972, Jurassic System (of U.S.S.R.): Akademiya Nauk SSSR, Ministerstvo Geologii, Stratigrafiia SSSR, v. X, 524 p. (in Russian).

Kvantaliani, I.V., and Lyssenko, N.I., 1979a, New Berriasian Genus *Tauricoceras*: Soobstcheniia Akademii Nauk Gruzinskoi SSR, Tbilisi, v. 93, p. 629-632 (in Russian).

______, 1979b, On the Zonal Subdivision of the Crimean Berriasian:

Soobstcheniia Akademii Nauk Gruzinskoi SSR, Tbilisi, v. 94, p. 629-632 (in Russian).

Lahusen, J., 1888, Ueber die Russischen Aucellen: Mémoire, Comité Géologique, St. Petersburg, v. VIII(1), 45 p.

Leanza, A.F., 1945, Ammonites del Jurasico Superior y del Cretaceo Inferior de la Sierra Azul, en la Parte Meridional de la Provincia de Mendoza: Anales Museo de la Plata, Nueva Serie, Paleontologia, Seccion A; Paleozoologia, 6, Moluscos, no. 1, 100 p.

———, 1947, Upper Limit of the Jurassic System: Geological Society of America Bulletin, v. 58, p. 833-842.

Leanza, H.A., 1975, *Himalayites andinus* n. sp. (Ammonitina) del Tithoniano Superior de Neuquén, Argentina: Actas del I Congreso Paleontologia Argentino de Biostratigrafia: Tucoman, t. 1, p. 581-588.

———, 1980, The Lower and Middle Tithonian Ammonite Fauna from Cerro Lotena province of Neuquen, Argentina: Zitteliana (Munich), v. 5, p. 3-45.

———, 1981a, The Jurassic-Cretaceous Boundary Beds in West-Central Argentina and their Ammonite Zones: Neues Jahrbuch fur Geologie und Palaontologie, Abhandlungen, v. 161, p. 62-92.

———, 1981b, Faunas de Ammonites del Jurasico Superior y del Cretacico Inferior de America del Sur, Con Especial Consideracion de la Argentina: Comité Sudamericana del Jurasico y Cretacico: Cuencas sedimentarias del Jurasico y Cretacico de America del Sur, v. 2, p. 559-597.

Leanza, H.A. and Hugo, C.A., 1977, Succession de Ammonites y Edad de la Formacion Vaca Muerta y Sincronicas Entre los Paralelos 35° y 40° l.s., Cuenca Neuquina-Mendocina: Asociacion Geologica Argentina, Revista, v. XXXII, p. 248-264.

Le Hégarat, G., 1971, Signification Stratigraphique de L'horizon de Stramberg (Tchécoslovaquie): Comptes Rendus, Académie des Sciences (Paris), v. 273, Séries D., no. 15, p. 1254-1257.

———, 1973, Le Berriasien du sud-est de la France: Documents des Laboratoires de Géologie de la Faculté des Sciences de Lyon no. 43, (fasc. 1-2 for 1971), p. 1-576.

Lisson, C.I., 1907, Contribucion a la Geologia de Lima y Sus Alrededores. 2. Observaciones paleontologicas: Libreria é Imprenta Gil, Lima, p. 27-64.

Louppov, N.P., Bogdanova, T.N., and Lobatchova, S.V., 1975, Le Berriasien de Manguychlak Comme un Lien Entre le Berriasien du Domaine Méditerranién et Celui du Domaine Boreal (thèses): Colloque sur la limite Jurassique-Crétacé, Lyon, Neuchâtel, septembre 1973: *in* Mémoires du Bureau de Recherches Géologiques et Minières, France, no. 86, p. 129-134.

———, 1979, Paleontological Basis of Correlation of the Berriasian and Valanginian Rocks of Mangyshlak with Those of Southeastern France, Northern Part of the Federal Republic of Germany and Russian Platform: Akademiya Nauk SSSR, Sibirskoye Otdelenie, Institut Geologii i Geofiziki, Verkhniaya Yura i Granitsa yee s Melovoi Systemoi, p. 159-168 (in Russian).

Luppov, N.P., Eristavi, M.S., and Drustchits, V.V., 1958, Superfamily Berriaselliacea: *in* Luppov, N.P. and Drustchits, V.V., eds, Ammonoidea (Ceratites and Ammonites), Dibranchiata Volume: Principles of Paleontology, Mollusca-Cephalopoda II: Moscow, GNTI-Press, p. 96-104 (in Russian).

Marek, S., 1967, The Infravalanginian of Kujawy: Institut Geologiczny, Bulletin 200: Z Badan Stratigraficzno-Paleontologicznych w Polsce, v. II, Warszawa, Wydawnictwa Geologiczne, p. 133-236, (Polish with English summary).

Marek, S. and Raczynska, A., 1973, The Stratigraphy and Paleogeography of the Lower Cretaceous Deposits of the Polish Lowland Area: Casey, R. and Rawson, P.F., eds., The Boreal Lower Cretaceous: Geological Journal, Special Issue, no. 5, p. 369-386.

———, 1979, Paläogeographie der Unterkreide des Nordpolnischen Beckens: Aspekte der Kreide Europas: International Union of Geological Sciences, Series A, no. 6, Stuttgart, p. 447-462.

Matsumoto, T., 1978, Japan and adjacent areas (Chapter 4): *in* Moullade, M. and Nairn, A.E.M., eds., The Phanerozoic Geology of the World, Vol. II: The Mesozoic, A: Elsevier Scientific Publishing Company, Amsterdam, p. 79-144.

Mazenot, G., 1939, Les Palaeohoplitidae Tithoniques et Berriasiens du Sud-Est de la France: Société géologique de France, n.s., Mémoire 41, 303 p.

Messezhnikov, V.A., Zakharov, V.A., Shulgina, N.I. and Alekseev, S.N., 1979, The Stratigraphy of the Riasanian Horizon on Oka River: Akademiya Nauk SSSR, Sibirskoye Otdelenie, Institut Geologii i Geofiziki: Verkhnaya Yura i Granitsa yee s Melovoi Sistemoi, "Nauka" Press, Novossibirsk, p. 71-81 (in Russian).

Monger, J.W.H., and Irving, E., 1980, Northward Displacement of North-Central British Columbia: Nature, v. 285, p. 289-294.

Monger, J.W.H., Price, R.A., and Tempelman-Kluit, D.J., 1982, Tectonic Accretion and the Origin of the Two Major Metamorphic and Plutonic Welts in the Canadian Cordillera: Geology, v. 10, p. 70-75.

Nikitin, S., 1884, Allgemeine Geologische Karte von Russland, Blatt 56, Jaroslawl, etc.: Mémoires, Comité Géologique, v. I, 153 p.

Oloriz, F.F., and Tavera, J.M., 1979a, El Tithonico: *in* El Jurasico de las Zonas Extremas de las Cordilleras Beticas. III. Zona Subbetica: II Col. Estrat. y Paleogeogr: Jurasico Espana, Granada, p. 151-169.

———, 1979b, III. 11. El Tithonico en la zona Subbetica: Cuadernas de Geologia (Madrid) Universidad de Granada, no. 10, p. 489-508.

———, 1981, Stratigraphische Position der Kalke von Stramberg (CSSR) – Überarbeitung der Jüngsten Hypothesen: Neues Jahrbuch für Geologie und Paläontologie, Monatshefte, Heft 1, p. 41-49.

Orlenok, V.V., and Chebanenko, I.I., 1982, On the Insufficiency of Presently Available Results of Marine Geophysical Investigations for the Support of the Spredding Hypothesis: Akademiya Nauk Ukrainskoi SSR, Institut Geologicheskich Nauk, Tectonika i Stratigrafiya, No. 22, Kiev, "Naukova Dumka" Press, p. 3-11.

Packer, D.R., and Stone, D.B., 1974, Paleomagnetism of Jurassic Rocks from Southern Alaska and the Tectonic Implications: Canadian Journal of Earth Sciences, v. 11, p. 976-997.

Paraketsov, K.V., 1980, The Problem of a Zonal Subdivision of the Upper Jurassic and Lower Cretaceous Rocks of Northeastern U.S.S.R. Using Buchias: Biostratigrafiya i Korrelatsiya mesozoiskich otlozhenii severo-vostoka SSSR, Magadan, p. 91-106 (in Russian).

Pavlov, A.P., 1897, Voyage Géologique par la Volga de Kazan à Tzaritsyn: Guide Excursion, VII Congrès Géologique International, v. XX, 40 p.

———, 1907, Enchaînement des Aucelles et Aucellines du Crétacé Russe: Nouveaux Mémoires, Société Impériale Naturalistes de Moscou, v. 17, 93 p.

Pozharisskaya, G.F., 1971, Berriasian and Lower Valanginian Aucellas of Russian Platform: Ministerstvo Geologii SSSR, Vsessoyuznyi Nauchno-Issledovatel'skii Geologorazvedochnyi Neftianoi Institut (VNIGNI), Trudy, No. 110. Berrias Russkoi Platformy (Stratigrafiya, fauna ammonitov i aucell, Sazonova, N.T., ed.), p. 111-141 (in Russian).

Remane, J., 1964, Untersuchungen zur Systematik und Stratigraphie der Calpionellen in der Jura-Kreide-Grenzschichten des Vocontischen Troges: Palaeontographica, Abteilung A., v. 123, Stuttgart, p. 1-57.

Roddick, J.A., 1965, Vancouver North, Coquitlam, and Pitt Lake Map-Areas, British Columbia: With Special Emphasis on the Evolution of the Plutonic Rocks: Geological Survey of Canada Memoir 335, 276 p.

Sakharov, A.S., 1975, Reference Sections of the NE Caucasus Berriasian; Colloque sur la limite Jurassique-Crétacé, Lyon, Neuchâtel, septembre 1973: *in* Mémoires du Bureau de Recherches Géologiques et Minières, France, no. 86, p. 68-76.

______, 1976, The Berriasian Standard Section of Northeastern Caucasus: Izvestiya Akademii Nauk SSSR, seriya geologicheskaya no. 1, p. 38-46 (in Russian).

______, 1979, Stratigraphy of Berriasian Deposits of Northern Caucasus: Akademiya Nauk SSSR, Sibirskoye Otdelenie, Institut Geologii i Geofiziki, Verkhnaya Yura i Granitsa yee s Melovoi Sistemoi, "Nauka" Press, Novossibirsk, p. 181-186 (in Russian).

Saks, V.N., and Shulgina, N.I., 1974, Basic Problems of the Upper Volgian, Berriasian and Valanginian Stratigraphy in the Boreal Zone: Acta Geologica Polonica, v. 24, p. 543-560.

Saks, V.N., Messezhnikov, M.S., and Shulgina, N.I., 1979, Stratigraphy of Jurassic-Cretaceous Boundary Beds in the Boreal Belt: Akademiya Nauk SSSR, Sibirskoye Otdelenie, Institut Geologii i Geofiziki, Verkhnaya Yura i Granitsa yee s Melovoi Systemoi. "Nauka" Press, Novossibirsk, p. 93-102, (in Russian).

Saks, V.N., Ronkina, Z.Z., Shulgina, N.I., Bassov, V.A. and Bondarenko, N.M., 1963, Stratigraphy of Jurassic and Cretaceous Systems of Northern U.S.S.R.: Izdatel'stvo Akademii Nauk SSSR, Moskva-Leningrad, 227 p. (in Russian).

Saks, V.N., Ronkina, Z.Z., Bassov, V.A., Zakharov, V.A., Messezhnikov, M.S., Shulgina, N.I., Yudovnii, E.G., Pospelova, G.A., Anuchin, A.V. and Larionova, G.Y., 1969, Standard Section of Upper Jurassic Deposits of the Kheta River Basin (Khatanga Depression): "Nauka" Press, Leningrad, 208 p. (in Russian).

Saks, V.N., Shulgina, N.I., Bassov, V.A., Burdykina, M.D., Gol'bert, A.V., Dagis, A.S., Zakharov, V.A., Ivanova, E.F., Klimova, I.G., Messezhnikov, M.S., Nal'niaeva, T.I., Pokhialainen, V.P., Romanova, E.Ye., Ronkina, Z.Z., Sazonova, I.G., Trushkova, L.Ya., Yudovnii, E.G., 1972, Jurassic-Cretaceous Boundary and Berriasian Stage in Boreal Realm: "Nauka" Press, Novossibirsk, 371 p. (in Russian) (also available in English translation by the Israel Program for Scientific Translation, Jerusalem, 1975, from the United States Department of Commerce).

Saks, V.N., Bassov, V.A., Dagis, A.A. Dagis, A.S., Ivanova, E.F., Meledina, S.V., Messezhnikov, M.S., Nalniaeva, T.I., Zakharov, V.A. and Shulgina, N.I., 1973, Paleozoogeography of Boreal-Realm Seas in Jurassic and Neocomian: Pitcher, M.G., ed., Arctic Geology, American Association of Petroleum Geologists Memoir 19: p. 219-229.

Sapunov, I., 1977, Ammonite Stratigraphy of the Upper Jurassic in Bulgaria, IV, Tithonian: Substages, Zones and Subzones: Geologia Balcanica, v. 7(2), p. 43-64.

Sato, T., 1962, Études Biostratigraphiques des Ammonites du Jurassique du Japon: Mémoires de la Société Géologique de France, n.s. no. 94, 122 p.

Sazonova, I.G., 1971, Berriasian and Lower Valanginian Ammonites of the Russian Platform: *in* Berrias Russkoi Platformy (Stratigrafiya, Fauna Ammonitov i Autsell): Vsesoyuznyi Nauchno-Issledovatel'skii Geologorazvedochnyi Neftianoi Institut (VNIGNI), Trudy, no. 110, p. 1-110 (in Russian).

______, 1977, Ammonites of Jurassic-Cretaceous Boundary Beds of the Russian Platform: Vsessoyuznii Nauchno-Issledovatel'skii Geologorazvedochnii Neftianoi Institut (VNIGNI), Trudy, no. 195, 128 p. (in Russian).

Sazonova, I.G., and Sazonov, N.T., 1979a, The Problem of Recognition of the Upper Stage of the Jurassic System and the Lower Stage of the Cretaceous System in the East-European Platform: Akademiya Nauk SSSR, Sibirskoye Otdelenie, Institut Geologii i Geofiziki, Verkhnaya Yura i Granitsa yee s Melovoi Systemoi, "Nauka" Press, Novossibirsk, p. 86-93 (in Russian).

______, 1979b, The Jurassic-Cretaceous Boundary in the East-European Platform: Aspekte der Kreide Europas, International Union of Geological Sciences, Series A, no. 6, Stuttgart, p. 487-494.

Schmidt, P.W., and Embleton, B.J.J., 1981, Magnetic Overprinting in Southeastern Australia and the Thermal History of its Rifted Margin: Journal of Geophysical Research, B 86, no. 5, p. 3998-4008.

Shulgina, N.I., 1967, Tithonian Ammonites of Northern Siberia: Akademiya Nauk SSSR, Sibirskoye Otdelenie, Institut Geologii i Geofiziki: Problemy Paleontologicheskogo Obosnovaniya detal'noi stratigrafii mesozoya Sibiri i Dal'nego Vostoka: "Nauka" Press, Leningrad, p. 131-177 (in Russian).

Smirnov, A.M., and Marchenko, A.F., 1976, The Basement Complex of Geosynclyne of Kamchatka (About the Problem of Structural Connections Between Northeastern Asia and North America): Akademiya Nauk SSSR, Dal'nevostochnyi Geologicheskii Institut. Geologicheskie i paleontologicheskie aspekty razvitiia Dal'nego Vostoka, Vladivostok, p. 3-19 (in Russian).

Sokolov, D.N., 1926, Mesozoic Fossils from Bol'shezemel'skaya Tundra and Kashpur: Travaux, Musée Géologique de l'Académie des Sciences, U.R.S.S., v. III, p. 15-62 (in Russian).

Spath, L.F., 1947, Additional Observations on the Invertebrates (Chiefly Ammonites) of the Jurassic and Cretaceous of East Greenland: I, The *Hectoroceras* Fauna of S.W. Jameson Land: Meddelelser om Gronland, v. 132, no. 3, 70 p.

______, 1950, A New Tithonian Ammonoid Fauna from Kurdistan, Northern Iraq: Bulletin, British Museum of Natural History, Geology, v. 1, no. 4, p. 95-137.

Stevens, G.R., 1965, The Jurassic and Cretaceous Belemnites of New Zealand and a Review of the Jurassic and Cretaceous Belemnites of the Indo-Pacific Region: New Zealand Geological Survey, Palaeontological Bulletin no. 36, 283 p.

______, 1967, Upper Jurassic Fossils from Ellsworth Land, West Antarctica, and Notes on Upper Jurassic Biogeography of the South Pacific Region: New Zealand Journal of Geology and Geophysics, v. 10, p. 345-393.

______, 1968, The Jurassic System in New Zealand: New Zealand Geological Survey, Department of Scientific and Industrial Research, 21 p.

Surlyk, F., 1973, The Jurassic-Cretaceous Boundary in Jameson Land, East Greenland: Casey, R., and Rawson, P.F., eds., The Boreal Lower Cretaceous: Geological Journal, Special Issue no. 5, p. 81-100.

______, 1977, Stratigraphy, Tectonics and Palaeogeography of the Jurassic Sediments of the Areas North of Kongs Oscars Fjord, East Greenland: Gronlands Geologiske Undersogelse Bulletin 123, 56 p.

______, 1978, Submarine Fan Sedimentation along Fault Scarps on Tilted Fault Blocks (Jurassic-Cretaceous Boundary, East Greenland): Gronlands Geologiske Undersogelse Bulletin 128, 108 p.

______, 1979, Jurassic-Cretaceous Boundary Beds of the Wollaston Forland Area, East Greenland: Akademiya Nauk SSSR, Sibirskoye Otdelenie, Institut Geologii i Geofiziki, Verkhnaya Yura i Granitsa yee s Melovoi Sistemoi: "Nauka" Press, Novossibirsk, p. 105-106 (in Russian).

Surlyk, F., and Zakharov, V.A., 1982, Buchiid Bivalves from the

Upper Jurassic and Lower Cretaceous of East Greenland: Paleontology, v. 25, p. 727-753.

Surlyk, F., Callomon, J.H., Bromley, R.G., and Birkelund, T., 1973, Stratigraphy of the Jurassic-Lower Cretaceous Sediments of Jameson Land and Scoresby Land, East Greenland: Gronlands Geologiske Undersogelse Bulletin 105, 76 p.

Sutherland Brown, A., 1966, Tectonic History of the Insular Belt of British Columbia: Canadian Institute of Mining and Metallurgy, Special Volume No. 8, p. 83-100.

———, 1968, Geology of the Queen Charlotte Islands, British Columbia: British Columbia Department of Mines, Bulletin 54, 226 p.

Symons, D.T.A., 1971, Paleomagnetism of Jurassic Intrusions, British Columbia: Geological Survey of Canada Paper 70-63, 17 p.

Takanashi, H., 1973, The Isokusa Formation and its Late Upper Jurassic and Early Lower Cretaceous Ammonite Fauna: Tohoku University Science Reports, 2 series (Geology), Special Volume no. 6, Prof. K. Hatai, Memorial Volume, p. 319-336.

Tipper, H.W., 1969, Mesozoic and Cenozoic Geology of the Northeast Part of Mount Waddington Map-Area (92-N), Coast District, British Columbia: Geological Survey of Canada Paper 68-33, 103 p.

———, 1982, Offset of an Upper Pliensbachian Geographic Zonation in the North American Cordillera by Transcurrent Movement: Canadian Journal of Earth Sciences, v. 18, p. 1788-1792.

Tipper, H.W., and Richards, T.A., 1976, Jurassic Stratigraphy and History of North-Central British Columbia: Geological Survey of Canada Bulletin 270, 73 p.

Tozer, E.T., and Thorsteinsson, R., 1964, Western Queen Elizabeth Islands, Arctic Archipelago: Geological Survey of Canada Memoir 332, 242 p.

Vakhrameev, V.A., Ill'ina, V.I., Kalacheva, E.D., Krimgolts, G.Ya, Meledina, S.V., Messezhnikov, M.S., Repin, Yu.S., and Fokina, N.I., 1982, Zones of the Jurassic System in U.S.S.R.: Academy of Sciences of the U.S.S.R., Ministry of Geology of the U.S.S.R., Interdepartmental Stratigraphic Committee of the USSR, Transactions, v. 10, 192 p. (in Russian).

Verma, H.M., and Westermann, G.E.G., 1973, The Tithonian (Jurassic) Ammonite Fauna of Sierra Catorce, San Luis Potosi, Mexico: Bulletins of American Paleontology, v. 63, no. 277, 320 p.

Waagen, W., 1864, Der Jura in Franken, Schwaben und der Schweiz. Preisschrift der Universitaet Muenchen.

Warren, P.S., and Stelck, C.R., 1958, the Nikanassin-Luscar Hiatus in the Canadian Rockies: Transactions, Royal Society of Canada, Series III, Section IV, v. LII, p. 55-62.

Westermann, G.E.G., (leader) *et al.*, 1982, Circumpacific Jurassic Research Group (I.G.C.P. Project No. 171), Report No. 1, Hamilton, Ontario, 162 p., Appendix A (Limited distribution).

Wiedmann, J., 1968, Das Problem Stratigraphischer Grenzziehung und die Jura/Kreide Grenze: Eclogae Geologicae Helvetiae, v. 61(2), p. 321-386.

———, 1980, Paläogeographie und Stratigraphie im Grenzbereich Jura/Kreide Sudamerikas: Münsterische Forschungen in Geologie und Paläontologie, no. 51, p. 27-61.

Wilson, D.G., 1976, Studies of Mesozoic Stratigraphy, Tanquary Fiord to Yelverton Pass, Northern Ellesmere Island, District of Franklin: Geological Survey of Canada Paper 76-1A, no. 95, p. 449-451.

Yefimova, A.F., Kinassov, V.P., Paraketsov, K.V., Polubot'ko, I.V., Repin, Yu.S. and Dagis, A.S., 1968, Field Atlas of the Jurassic Fauna and Flora of Northeastern U.S.S.R.: Ministerstvo Geologii RSFSR, Severo-vostochnoe Ordena Trudovogo Krasnogo Znameni Geologicheskoye Upravlenie, Magadan Publishing House, 381 p.

Yegoyan, V.L., 1975, Tithonian-Berriasian Boundary is the Boundary of Jurassic and Cretaceous Systems: Colloque sur la Limite Jurassique-Crétacé, Lyon, Neuchâtel, septembre 1973: *in* Mémoires du Bureau de Recherches Géologiques et Minières, France, no. 86, p. 363-369.

Yorat, C.J. and Chase, R.L., 1981, Tectonic History of the Queen Charlotte Islands and Adjacent Areas – A Model: Canadian Journal of Earth Sciences, v. 18, p. 1717-1739.

Zakharov, V.A., 1979, Zonal Subdivision of the Boreal Upper Jurassic and Neocomian Deposits on the Basis of Buchias: Akademiya Nauk SSSR, Sibirskoye otdelenie, Institut Geologii i Geofiziki. Verkhnaya Yura i Granitsa yee s Melovoi Systemoi, "Nauka" Press, Novossibirsk, p. 122-130 (in Russian).

———, 1981, Buchiids and the Biostratigraphy of the Boreal Upper Jurassic and the Neocomian: Akademiya Nauk SSSR, Sibirskoye Otdelenie, Institut Geologii i Geofisiki, Trudy no. 458; "Nauka" Press, Moscow, 271 p. (in Russian).

Zakharov, V.A., Surluk, F., and Dalland, A., 1981, Upper Jurassic-Lower Cretaceous *Buchia* from Andoy, Northern Norway: Norsk Geologisk Tidsskrift, v. 61, p. 261-269.

Zeiss, A., 1968, Untersuchungen zur Paläontologie der Cephalopoden des Unter-Tithon der Südlichen Frankenalb: Bayerische Akademie der Wissenschaften, Mathemathisch-Naturwissenschaftliche Klasse, Abhandlungen, Neue Folge, Heft 132, 190 p.

———, 1974, Berechtigung und Gliederung der Tithon-Stufe und Ihre Stellung im Oberen Jura: Colloque sur la Limite Jurassique-Crétacé, Luxembourg, 1967, Paris: *in* Mémoires du Bureau de Récherches Géologiques et Minières, France, no. 75. p. 283-291.

———, 1977, Some Ammonites of the Klentnice Beds (Upper Tithonian) and Remarks on Correlation of the Uppermost Jurassic: Acta Geologica Polonica, v. 27, p. 269-386.

———, 1979, The Problem of Correlation in the Upper Jurassic and Some Considerations about the Jurassic-Cretaceous Boundary: Akademiya Nauk SSSR, Sibirskoye Otdelenie, Institut Geologii i Geofiziki, Verkhniaya Yura i Granitsa yee s Melovoi Systemoi: "Nauka" Press, Novossibirsk, p. 14-27 (in Russian).

Zittel, K.A. von, 1868, Die Cephalopoden der Stramberger Schichten: Museum (K.) Bayerischen Staates: Paläontologische Mitteilungen, v. 2, p. 33-118.

Comments by A.ZEISS (Erlangen, F.R.G.)

I have just published a paper in German language dealing with the global correlation of marine beds around the Jurassic/Cretaceous boundary (with a correlation table, Zeiss 1983). As my results differ considerably from those of Jeletzky I feel obliged to provide the reader with critical comments on three of the major subjects, at least.

1) *Correlation at the Base: The B. cf. blanfordiana Zone and its Equivalents in Europe.* Jeletzky claims that the age of this zone should now be considered as 'late Lower' Tithonian (=late Middle Tithonian) and 'late Lower' Volgian (=late Middle Volgian), respectively. This opinion is based on a revision of an ammonite earlier determined tentatively as *'Substeueroceras stantoni'* (Jeletzky, 1965) and now assigned to *Pectinatites (Paraberriasella).* The first determination of this small ammonite, which has not been refigured despite of additional preparation work, was not as bad as Jeletzky seems to believe

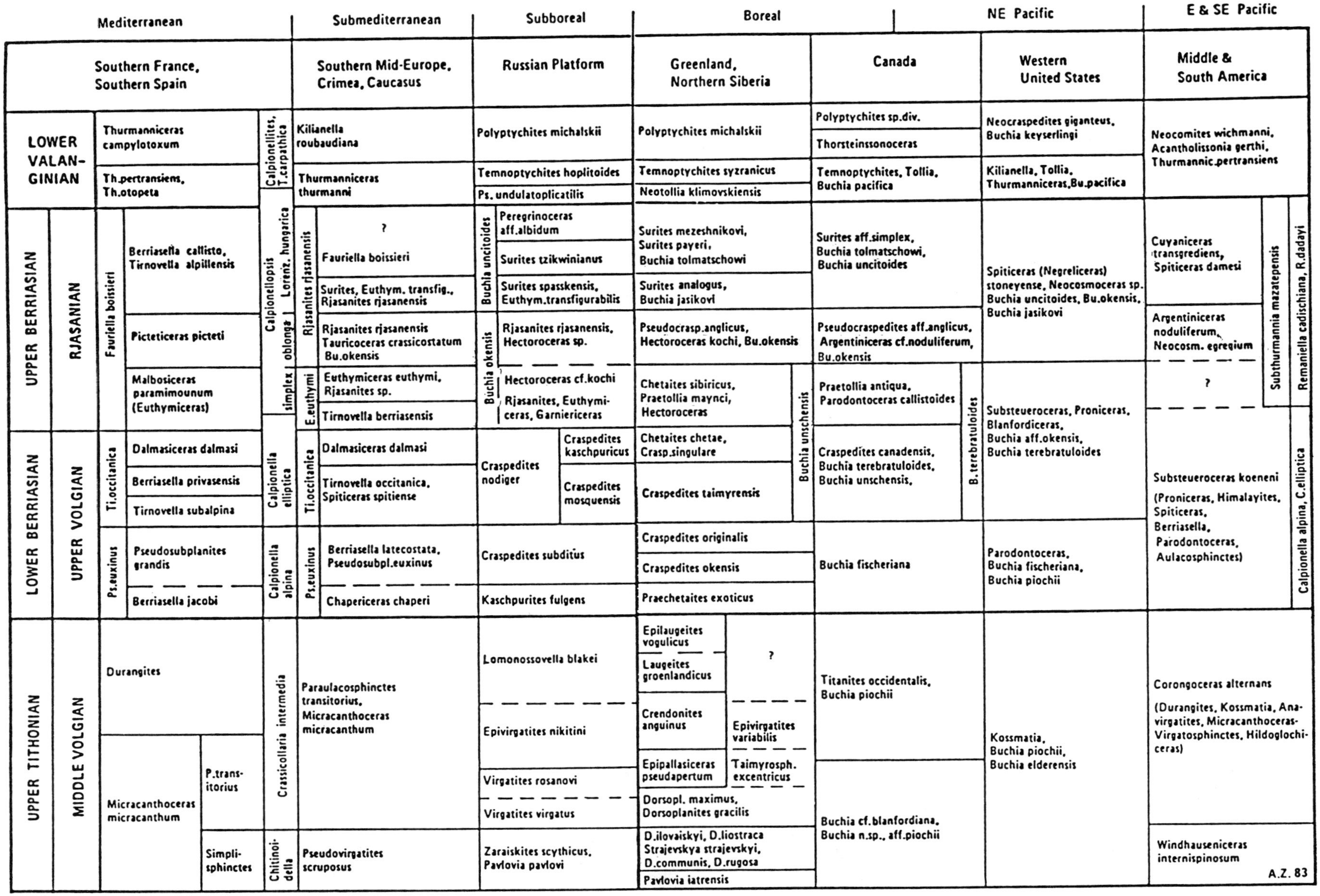

Figure 1. Correlation chart of uppermost Jurassic and lowermost Cretaceous stages and zones.

now. It is rather similar to the holotype (cf. Imlay and Jones, 1970); the absence of an external furrow or a weakening of the ribs suggests that this specimen is a forerunner of *S. stantoni* in the lower part of the Upper Tithonian. I made similar observations in the lower Upper Tithonian of Lower Austria where forms very close to the younger *P. transitorius* exist, but still without an external furrow, and which are thus considered as forerunners of *P. transitorius*. The specimen in question has nothing to do with *Paraberriasella* (for a good diagnosis see Donze and Enay, 1961, p. 149), a genus which has the ventral furrow limited to young stages and double bifurcations at the end of the body chamber. *Paraberriasella* has nothing to do with *Pectinatites*, as Jeletzky assumes. Donze & Enay (1961) write: "Nous ne pensons à des lines phyletiques quelconques entre ces deux genres". The ammonite originally determined as "resembling *Gymnodiscoceras*" is now compared with *Substreblites*; but the specimen in question is too badly preserved to allow a decision to which genus it might belong. Thus there is no need to change the correlation of the following zones: *Buchia* cf. *blanfordiana*/*B. piochii* and *B. elderensis* (with *Kossmatia*)/ *Windhauseniceras internispinosum*/ *Corongoceras alternans* and *Durangites*, which were correctly placed by Imlay and Jones (1970) and Imlay (1980). All these zones belong to the Upper Tithonian.

Furthermore, Jeletzky correlates the late Middle Volgian *Epivirgatites* zones of the USSR with the upper Middle Tithonian *M. ponti* zone. This is wrong according to all research work on the boundary beds Middle/ Upper Tithonian (Uppermost Lower/ Middle Volgian) in eastern central Europe (Kutek and Zeiss, 1975; Zeiss, 1977); hence the *Epivirgatites* zones correspond to the Upper Tithonian. These wrong datings and correlations by Jeletzky cause further mis-correlations (see below).

2) *The Age of the Substeueroceras Beds.* Jeletzky alleges that the *Substeueroceras* beds of America are coeval with the upper part of the *transitorius* zone (*Durangites* zone) of Europe. He explains the low frequency of *Durangites* in the *Substeueroceras* beds of Mexico with a cooling of the sea in these regions, while in Spain temperatures continued to be warm enough for *Durangites*. But there is no reason for such a 'cooling', Mexico and Spain lying approximately on the same paleolatitude. Thus, the *Durangites*/*Kossmatia* assemblage of Mexico and the *Durangites* zone of Southern Europe are of nearly the same age. There is no need to unite this assemblage with that of *Substeueroceras*. As Verma and Westermann (1973) have shown, both are well separated in general. The same is valid for the separation of the *Corongoceras* and *Substeueroceras* beds of South America (e.g. Birö, 1981). The younger age of the *Substeueroceras* beds is demonstrated by Cantú Chapa (1968), too, who records that *Substeueroceras* beds in Mexico contain calpionellids. The mentioned calpionellid species are characteristic for the Lower Berriasian (Zones B and C) of Mediterranean Europe. This means that the main part of the *Substeueroceras* beds has to be correlated with the Lower Berriasian of Europe, and not with the Upper Tithonian. Only at the base there may be some faunal overlapping, as Trejo (1975, 1980) indicates that questionable *Substeueroceras* and *Parodontoceras* occur already together with *Crassicollaria massutiniana* in the uppermost subzone (A3) of the *Crassicollaria*/ *Durangites* zone. Also some observations of Cantú Chapa (1982) and Olóriz and Tavera (1983) seem to give hints for such an overlapping. Further research work is necessary to clear up the transitional character of the uppermost *Durangites* zone in more detail and precision. Higher up in the sections studied by Cantú Chapa (1968) presumed equivalents of the *Argentiniceras* and *Spiticeras* zones have yielded a calpionellid assemblage approximately of zone D of Remane, clearly indicating an Upper, not a Lower Berriasian age. An open problem is the correlation of the base of the *Argentiniceras* zone of Argentina. According to Cantú Chapa (1968) the base of this zone would be coeval with that of the *Subthurmannia mazatepensis* zone, but due to correlation with *Buchia*-bearing *Substeueroceras* and *Argentiniceras* beds, it would lie somewhat higher. The intervall in question is documented in Argentina by non-fossiliferous beds (see Spath 1952, p. 29-31), while Birö (1981) from Chile mentions *P. callistoides* in the *Argentiniceras* zone. Thus, there may be a transitional faunal interval, but further research work is needed to clarify the problem. It is extremely important to undertake comparative studies on the exact vertical ranges of the calpionellid and ammonite species concerned.

Summarizing all evidence it must be stressed that the *Substeueroceras* beds are younger than supposed by Jeletzky; consequently also the *Buchia terebratuloides*, *B. fischeriana* and *B. unschensis* zones belong into the lower Berriasian, as they contain *Substeueroceras* and *Parodontoceras* in the western part of North America. The superjacent *B. okensis* zone is of Upper Berriasian Age (*Boissieri* Zone). The three *Buchia* species first mentioned are characteristic for the Upper Volgian, too, which has yielded now also calpionellids of zones B and C of Remane (see Sazonova & Sazonov 1983), thus indicating once more its correspondence to the Lower Berriasian. The opinion of Jeletzky that the alleged *Virgatosphinctes* of the Upper Volgian of the USSR indicate Upper Tithonian, cannot be maintained. These ammonites are unrelated to true *Virgatosphinctes*; their inner whorls have a Boreal habitus of ribbing and coiling (see also Zeiss, 1968). They belong to a special boreal lineage *(Praechetaites)* which cannot be used as evidence for Upper Tithonian vs. Lower Berriasian.

3) *Stratigraphic Position of the Lower Rjasanian Beds.* Jeletzky's claim of an Upper Tithonian age for the basal Rjasanian beds is contradicted by my correlation of the Upper Volgian with the Lower Berriasian of southern Europe and the *Substeueroceras*/ *Pardodontoceras* beds of America. The basal Rjasan beds are characterized by *Buchia fischeriana, unschensis, Euthymiceras, Rjasanites* and *Hectoroceras*. According to Surlyk and Zakharov (1982) the Upper Volgian *B. fischeriana* extends into the basal Rjasanian *Ch. sibiricus* zone, where *B. unschensis* still occurs together with/ or above *B. terebratuloides*. This zone is coeval with the *B.* aff. *okensis* zone in North America. As *Substeueroceras* in California occurs in the middle part of this zone indicating at least a *Privasensis*-age, the upper part may correspond to the *Dalmasi-*/ *Euthymi*-subzones of Europe. The two oldest Rjasanian beds with *Buchia fischeriana* and *B. unschensis* probably correspond to the uppermost *B.*

unschensis/*B. terebratuloides*/*B.* aff. *okensis* zones in North America. However, a lowermost *B. okensis* age cannot be completely excluded as in East Greenland the first three species occur together with *B. okensis* in the *Okensis*-Zone as does *B.* aff. *okensis* in B.C. In the Caucasus and adjacent regions the first appearance of *Neocosmoceras* (and *Euthymiceras*) is above the *Dalmasi*/*Paramimounum*-boundary, i.e. in the *Euthymi*-subzone. This corresponds rather well to the appearance of *Neocosmoceras* and *Negreliceras* in the *Okensis*-/*Unicitoides*-Zone of North America. In the Caucasus together with those ammonites *B. volgensis* has been reported as well as *B. surensis*, which has been found also in the *Chetaites*/*B. terebratuloides* beds of the northeastern Soviet Union (Parakezov and Parakezova 1979) Therefore this horizon may be coeval with the *Sibiricus*-Zone. A revision of the buchias from the Caucasus region is necessary before a final correlation can be drawn.

I am fully aware that many correlations at the Jurassic/Cretaceous boundary can be traced only tentatively at present. To solve the difficult problems of correlation it is necessary to undertake

1) more detailed work on the exact vertical ranges and quantitative distribution of ammonite species, and
2) comparative research work on all fossil groups concerned.

The discussion of many other important and interesting subjects of the paper of Jeletzky has been postponed for want of space. They will be discussed elsewhere.

REFERENCES NOT MENTIONED PREVIOUSLY

Cantu Chapa, A., 1968, El limite Jurasico-Cretacico en Mazatepec, Puebla (Mexico): Ingeneria Petrolera, v. 8, p. 5-26.

______, 1982, The Jurassic-Cretaceous boundary in the Subsurface of Eastern Mexico: Journal of Petroleum Geology, v. 4, p. 311-318.

Donze, P., and Enay, R., 1961, Les Céphalopodes du Tithonique inférieur de la Saint-Concors pres Chambery (Savoie): Traveau de Laboratoire Géologique de Lyon, N.S., v. 7, 236 p.

Kutek, J., and Zeiss, A., 1975, Tithonian-Volgian Ammonites from Brzostowka near Tomazow Mazowiecki, Central Poland: Memoir Bureau de Researches Géologiques et Minieres, v. 86, p. 123-128.

Olóriz, F., and Tavera, J.M., 1983, Correlation of the Tithonian in Central Sector of the Betic Cordilleras (Spain) in the Light of Recent Studies: Bulletin de Academie Polska de Sciences, Série Sciences de la Terre, v. 30, p. 145-156.

Parakezov, K.V., and Parakezowa, G.I., 1979, Volzhskie otlozhenija Severo-Vostoka SSSR i ich biostratigrafitsheskaja charakteristika. In: Verchn. Jura i graniza ee s melovoj sist., p. 81-85.

Sazonova, J.G., and Sazonov, N.T., 1983, The Berriasian of the European realm: Zitteliana, v. 10, p. 439-446.

Surlyk, F., and Zakharov, V., 1982, Buchiid bivalves from the Upper Jurassic and Lower Cretaceous of East Greenland: Palaeontology, v. 25, p. 727-753.

Trejo, M., 1975, Los Tintinnidos Mesozoicos de Mexico: Bulletin de Bureau Reserches Géologique et Minieres, v. 75, p. 95-104.

______, 1980, Distribution Estratigrafica de los Tintinidos Mesozoicos Mexicanos: Revista del Instituto Mexicana Petrolifero, v. 12, p. 4-13.

Zeiss, A., 1983, Zur Frage der Äquivalenz der Stufen Tithon/Berrias/Portland in Eurasien und Amerika. Ein Beitrag zur Klärung der Weltweiten Korrelation der Jura-/Kreide Grenzschichten im Marinen Bereich: Zitteliana, v. 10, p. 427-438.

REPLY by J.A. Jeletzky

I disagree with all major criticisms made by Zeiss. The reasons are as follows.

1) Correlation at the Base of the Upper Tithonian: The B. cf. blanfordiana Zone and Its European Equivalents. Zeiss's claim that my late early Tithonian (=late early Volgian) dating of this zone is "based on a revision of an ammonite earlier tentatively determined as '*Substeueroceras stantoni*' (Jeletzky, 1965) and now assigned to '*Pectinatites (Paraberriasella)*' ... " is a misunderstanding. The late lower Volgian dating of this zone (p. 185-189) is based principally on the presence of diagnostic *Buchia russiensis* (Pavlow 1907) sensu Zakharov 1981. Furthermore, the zone occurs in a well established sequence of Upper Jurassic *Buchia* zones of western and Arctic Canada that can be readily correlated with the standard lower and upper Volgian zones of northern U.S.S.R. Zakharov (1981, p. 95, 223, 224) has shown that a number of Canadian and Californian buchias previously assigned to the younger *B. piochii* s. l. are synonyms of the Eurasian *B. russiensis* s. l. He stresses that *B. russiensis* s. l. is essentially restricted to the middle and upper parts of the "Middle Volgian substage" in northern U.S.S.R. wherever the age of its beds is reliably established by the associated Boreal ammonites (i.e., to the upper *Buchia mosquensis* Zone). The validity of all previous claims of the occurence of *B. russiensis* s. l. in the basal Upper Volgian and equivalent lower Upper Tithonian beds in the U.S.S.R., European Arctic and North America is queried. My research, and that of Surlyk and Zakharov (1982, p. 747, 748, Table 2) in East Greenland, confirm the above conclusions. Since *B.* cf. *blanfordiana* is restricted to the upper part of the intercontinental *Buchia russiensis* s. l. Zone in western North America, its regional zone there is an upper Lower Volgian deposit. Its late early Tithonian age is demonstrated by the conformable and gradational superposition of the *Buchia piochii* s.s. Zone that contains the oldest late Tithonian ammonites known in western Canada (i.e., *Proniceras* sp. indet. juven. and *Notostephanus*? sp. indet.; p. 189) and California (p. 191-192). This demonstrates that: a) the *Buchia* cf. *blanfordiana* Zone is older than the *Durangites-Kossmatia-Corongoceras* complex of Mexico and the *Windhauseniceras internispinosum* Zone of western South America; b) the *Kossmatia* occurs only in late Early Tithonian rocks in California but ranges through part or all of the early Tithonian and all of the late Tithonian in Mexico; and c) the invalidity of Zeiss's correlation of the Upper Tithonian and Upper Volgian rocks of Central and Eastern Europe.

The true nature of the above mentioned ammonite found in the *Buchia* cf. *blanfordiana* Zone is rather unimportant. However, I insist that it is quite unlike *Substeueroceras* and is much more comparable with *Pectinatites (Paraberriasella)* ex gr. *blondeti*.

2) The Age of Substeueroceras Beds. This is the focal point of our disagreement about correlations of the Tethyan and Boreal facies of the Jurassic-Cretaceous boundary beds.

a) I reject Zeiss's assertion of an essential contemporaneity of "the *Durangites-Kossmatia* assemblage of Mexico with the *Durangites* Zone of Southern Europe". He provides no new specific evidence contradicting the overwhelming ammonite and *Buchia* evidence for the early late Tithonian age of this complex, the equivalent western South American *Corongoceras alternans* and *Windhauseniceras internispinosum* Zones, and the equivalent *Buchia piochii* s.s.-*B. fischeriana* Zone of California; nor against my correlation (p. 191-193 and 197-203) of these regional American zones with the European *Micracanthoceras microcanthum* Zone. The same applies to the abundant evidence demonstrating the late Late Tithonian age of the next younger *Berriasella-Substeueroceras* complex of Mexico, the equivalent *Substeueroceras koeneni* Zone of South America, and the equivalent *Buchia terebratuloides* s. l. Zone of California, British Columbia and the Sverdrup Basin. Furthermore, *Durangites heilprini*, several species of *Kossmatia*, and *Corongoceras cordobai* are associated with all diagnostic *Berriasella* species of the *Berriasella-Substeueroceras* complex and several of its *Substeueroceras* species at the fossil locality 25 of Verma and Westermann (1973, p. 135) while *Kossmatia bifurcata* occurs in place associated with *Substeueroceras koeneni tabulatum* at their locality 19c (Verma and Westermann, 1973, p. 124, 125). This demonstrates that the two complexes are ill-defined and no more than subzones of the same broad faunal zone. The ascendance of *Durangites, Kossmatia*, and *Corongoceras* into the beds characterized by the *Berriasella-Substeueroceras* fauna is proof of the late Late Tithonian age of that fauna.

b) Zeiss rejects my suggestion that the gradual late Tithonian cooling of the Mexican basins influenced the relative rarity of *Durangites* in the *Berriasella-Substeueroceras* complex there because these basins are situated at the same lattitude with southern Spain where *Durangites* is abundant in the uppermost Tithonian beds. However, my conclusion is strongly supported by the increasingly strong high-Boreal affinities of the Tithonian buchias and belemnites of the adjacent marine basins of California and western British Columbia. These affinities reached their peak in the latest Tithonian *Buchia terebratuloides* s.l. Zone that is correlative with the *Berriasella-Substeueroceras* complex of Mexico. The disappearance of *Kossmatia* from California in the latest early Tithonian discussed in (1) and its survival in the Mexican basins is one argument suggesting this gradual Tithonian cooling. The slightly later replacement of the endemic North Pacific lineage of *Buchia piochii* s.s.-*Buchia* ex gr. *lahuseni* (inferred to be "warm-loving" because of superabundance in California and British Columbia contrasted with rarity or absence in the Canadian Boreal basins further north) first by the high-Boreal *Buchia fischeriana* and then by the high-Boreal *Buchia terebratuloides* s.l. faunas is another such argument. Finally, the Boreal affinities of the Californian belemnite faunas reach their peak at the same latest Tithonian time (Anderson, 1945; Saks *et al.*, 1973, p. 221). The following early Berriasian warming trend is documented by an almost total absence of *B. okensis* s.s. and *B.* n. sp. aff. *okensis* in these rocks of California. This contrasts with the abundance of the latest Tithonian forms of *B.* n. sp. aff. *okensis* there and with the continuing abundance of *B. okensis* s.s. and *B.* n.sp. aff. *okensis* farther north in the early Berriasian rocks of western British Columbia and Arctic Canada. These buchias are replaced in the Californian equivalents of *Buchia okensis* s.s. Zone by much smaller and much finer and much more closely ribbed *B. uncitoides* and *B. uncitoides jasikovi*. These homeomorphs of *B. piochii* s.s. are much less common in the equivalent rocks (i.e., *Buchia okensis* s.s. Zone) in western British Columbia and are still more rare yet farther north on the mainland of western Canada. These relationships clearly indicate the "warm-loving" nature of the above subspecies of *B. uncitoides* s.l. and the "cold-loving" nature of *B. okensis* s.s. and *B.* n.sp. aff. *okensis*. The northward migration of the latter in the early Berriasian in comparison with the latest Tithonian must reflect an increase of the water temperatures in the early Berriasian basins of California.

This Tithonian cooling trend in the North Pacific basins of western North America suggests a gradually increasing influence of cold current(s) that must have been flowing southward out of high Boreal basins of North America and Northeast Siberia *via* the already extant Bering Strait (Saks and Nalniaeva, 1979, p. 75; Jeletzky, 1971b, p. 77, 78). This was followed apparently by a period of a gradual waning of these currents that lasted at least through the Berriasian, beginning with the *Buchia okensis* s.s. time. This, and the other data (p. 243), invalidate Zeiss's claim that the Mexican Late Tithonian basins must have had the same water temperature as their Spanish counterparts (well protected from the Boreal influences by the Purbeckian landmass!) simply because they were situated at the same latitude.

c) Zeiss's opinion that the calpionellid fauna of the Mexican *Substeueroceras* beds is correlative with "Zone B of the mediterranean area of Europe and, perhaps, to some degree also Zone C." is invalidated by the evidence of their ammonite fauna presented in my paper and in (a). This fauna is late Late Tithonian and was always recognized as such by North and South American micropaleontologists (p. 197) and by most or ?all Mexican micropaleontologists (e.g., Bonet, 1956; Trejo, 1960, 1975; Salinas in Cantu, Chapa, 1980, p. 180). As already recognized by these micropaleontologists, the calpionellids also contradict Zeiss. The paper of Cantu Chapa utilized by him merely lists *Calpionella alpina* (Lorenz), *C. elliptica* (Cadisch), *Nannoconus steinmanni* (Kamptner), *Tintinnopsella longa* (Colomb), *Globochaeta alpina* (Lombard) and *Fibrosphaera* sp. without stating their relative abundance in the samples, morphology, etc. These microfossil lists do not permit any choice between the calpionellid zones A and B the recognition of which depends on the relative abundance of these and other species in the samples and their specific, chronologically restricted features, rather than on their presence and absence (e.g., Remane, 1964, p. 37-46, Beilage I and in Le Hégarat and Remane, 1968, p. 43-46, Table VII; p. 54, Table X). Furthermore, these identifications of *C. alpina, C. elliptica* and *T. longa* are obviously derived from the older papers of Bonet (1956) and Trejo (1960), which were compiled before the introduction of modern calpionellid taxonomy and biostratigraphy by Remane (1964). A revision of the original and/or additional material from these localities by a tintinnid specialist is, therefore, required. So far as I know, this was done only by Trejo (1975) who provides some of the critical information.

Zeiss does not say which of the Mexican calpionellid zones or subzones of Trejo (1975) he considers to be equivalent to the

Zone B and, in part, to the Zone C of the Mediterranean standard. However, only Trejo's (1975, p. 98) *Crassicollaria massutiniana-Crassicollaria parvula* Subzone of the *Crassicollaria* Zone is relevant to this discussion. It alone has yielded "*Substeueroceras?, Hildoglochiceras* aff. *grossicostatus*, cf. *Parodontoceras* sp., and cf. *Micracantoceras sp. ...* " and was assigned to the upper Tithonian. The immediately overlying *Calpionella* Zone, or more precisely *Calpionella alpina* Zone, did not yield any elements of the *Substeueroceras* fauna. Trejo (1975, p. 98) correlates the above subzone with the upper part of the Zone A of the *Crassicollaria* Zone or with the middle and upper parts of Remane's (1964) Subzone of *C. brevis-C. massutiniana* of the Mediterranean standard. This is indicated by the prevalence of *C. massutiniana* and *C. parvula* in this Mexican subzone combined with the rarity of *C. elliptica* and *Tintinnopsella carpathica* (an early form). Furthermore, this subzone contains very scarce representatives of the diagnostic *C. intermedia* and a solitary specimen of the diagnostic *C. brevis*. All these forms are diagnostic of the so called *brevis-massutiniana* association of Remane (1964, p. 47, Fig. 18), the validity of which was confirmed by all subsequent work in the western Mediterranean region (e.g., Remane in Le Hégarat and Remane, 1968, p. 30, etc.; Remane in Barthel *et al.*, 1966, p. 202, Table 3; Allemann et al., 1975; Hoedemaeker, 1982, p. 9, 10). This fauna was found everywhere directly below another calpionellid fauna assignable to the Zone B. Therefore, the calpionellid fauna of the *Substeueroceras* beds of Mexico fully supports their traditional assignment to the upper Upper Tithonian (i.e., the regional *Durangites* Zone or the upper part of the Transitorius Zone) based on their ammonite fauna. All my inter-realm correlations based on this age of the *Substeueroceras* beds and their equivalents throughout western North and South America stand.

d) The demonstration of the late Tithonian age of the *Substeueroceras* ex gr. *koeneni*-bearing rocks of North and South America confirmed above militates against a Berriasian age of the "alleged *Virgatosphinctes* of the Upper Volgian of the USSR ... " claimed by Zeiss. It is no longer critical for me whether these assuredly Early to Late Tithonian ammonites are true *Virgatosphinctes* (or closely allied genera derived from that Tethyan stock), as I suppose, or its homeomorphs derived from some Boreal perisphinctid stock.

3) The Stratigraphic Position of Lower Riasanian Beds. The refutal of Zeiss's correlation of the *Substeueroceras* beds and their equivalents presented in (2) also invalidates all his arguments for an early Berriasian age of these basal beds containing *Buchia unschensis* s.s. fauna above and *Buchia fischeriana* s.l. fauna below (p. 235-237; Figs. 10, 11). However, the association of numerous *B. unschensis* s.s. and *B. fischeriana* s.l. with *B. okensis* s.s. in East Greenland (Surlyk and Zakharov, 1982) lacks the stratigraphic significance assumed by Zeiss. This association is an example of a mixed fauna of *Buchia terebratuloides* s.l. and *Buchia okensis* s.s. Zones that occurs in their boundary beds. Because of the presence of only two specimens of *B. okensis* s.s. (Surlyk and Zakharov, 1982, p. 740), this fauna is correlative with faunas of the uppermost part of *Buchia terebratuloides* s.l. Zone in California, British Columbia and the Sverdrup Basin containing rare to very rare *B. okensis* s.s. Contrary to my previous opinion (p. 233-234), the presence of *Hectoroceras kochi* and *Praetollia (Pseudocraspedites)* aff. *anglicus* do not contradict this conclusion as these two ammonites are now known to make their first appearance in the North Siberian *Chetaites sibiricus* Zone. Such solitary to very rare occurrences of index species either closely above or closely below their fossil zones are a normal phenomenon that does not detract from their zonal value. This phenomenon is particularly common in *Buchia* species which commonly exhibit continuous "streams" of gradually changing forms (so called *plexi*) that connect several of their successive paleontological species and preclude their sharp morphological differentiation near the boundaries of their zones.

Zeiss's criticism of my conclusion that *Riasanites* and "*Euthymiceras*" ex gr. *transfigurabilis* appeared about two and a half *Buchia* Zones and several ammonite zones later (see Fig. 11) in Mangyshlak, Caucasus and Crimea than they did in the Riasanian Beds is again based *solely* on his already refuted (under 2) correlation of the *Substeueroceras* beds with the early and early mid-Berriasian. Therefore, my contrary conclusions, the inevitably following Boreal origin of these ammonites, and their southward migration are valid.

REFERENCES NOT MENTIONED PREVIOUSLY

Cantú Chapa, A., 1980, El Limite Jurasico-Cretacico en Mexico: Asociacion Paleontologica Argentina, Actas Segundo Congreso Argentino de Paleontologia y Bioestratigrafia y Primer Congreso Latinoamericano de Paleontologia, t. V. Simposio Limite Jurasico-Cretacico, p. 177-184.

Le Hégarat, G. and Remane, J., 1968, Tithonique Supérieur et Berriasien de la Bordure Cévenole. Corrélation des Ammonites et des Calpionelles: Geobios, Lyon, No. 1, p. 7-70.

Saks, V.N. and Nalniaeva, T.N., 1979, The Peculiarities of Distribution of the Boreal Belemnoidea: Akademia Nauk SSSR, Sibirskoye Otdelenie, Trudy Instituta Geologii i Geofiziki, No. 411: Usloviia Sustchestvovaniya Mesozoiskich Morskich Faun, p. 9-22 (in Russian).

Trejo, H., 1960, La familia Nannoconidae su Alcance Estratigrafico en America: Boletin de la Asociacion Mexicana de Geologos Petroleos, v. XII, p. 259-315.

Jurassic-Cretaceous Biochronology and Paleogeography of North America,
edited by G.E.G. Westermann,
Geological Association of Canada Special Paper 27, 1984

MID-CRETACEOUS MOLLUSCAN BIOSTRATIGRAPHY AND PALEOGEOGRAPHY OF SOUTHWESTERN PART OF WESTERN INTERIOR, UNITED STATES

William A. Cobban
United States Geological Survey, Federal Center, Denver, Colorado 80225
Stephen C. Hook
Getty Oil Company, Exploration and Production Research Center, Houston, Texas 77215-0070

ABSTRACT

The area of this report comprises part of New Mexico, Arizona, Colorado, and Utah, and centres around the Four Corners. Rocks treated range in age from middle Cenomanian to late Turonian and include at least 17 ammonite zones.

The Western Interior epeiric sea transgressed into the area from the east. A conspicuous embayment, the Seboyeta Bay, formed in central New Mexico early in middle Cenomanian time. By the end of the middle Cenomanian, the Seboyeta embayment had encroached into eastern Arizona. In the late Cenomanian, the sea rapidly transgressed westward over the eastern half of Arizona and much of Utah. Maximum transgression occurred about the end of this stage. By the middle Turonian, the sea retreated northeastward from much of Arizona, but the shoreline in Utah remained nearly stationary. During the late Turonian, the sea was transgressive at first and regressive later.

A few of the regressive near-shore sandstone beds have extensive bivalve and gastropod faunas. Shaly deposits formed farther offshore have sparser faunas usually dominated by inoceramids and thin-shelled bivalves. Ammonite faunas are characterized by Tethyan forms in some areas and Temperate ones in others.

RÉSUMÉ

La région traitée dans ce rapport comprend des parties du Nouveau-Mexique, l'Arizona, le Colorado, le Utah et des centres du voisinage des "Four Corners". L'age des roches traiteés s'étend du Cénomannien moyen au Turonnien avancé et comprennent au moins 17 zones d'ammonites.

La transgression de la mer de l'Intérieur Ouest débuta du côté est de cette région. Une baie importante, la Baie de Seboyeta, se forma au centre du Nouveau-Mexique au début du Cénomannien moyen. Par la fin du Cénomannien moyen, la Baie de Seboyeta atteigna l'est de l'Arizona. Durant le Cénomannien avancé, la mer a subit une transgression rapide vers l'ouest et noya la moitié est de l'Arizona ainsi que la plupart du Utah. Cette évènement marqua la transgression marine maximale. Au Turonnien moyen, la mer effectua une régression nord-est de la plupart de l'Arizona. Cependant, la côte du l'Utah demeura quaisiment stationnaire. Au Turonnien avancé, la mer a subit une transgression en premier suivie d'une régression plus tard.

Quelques strates sableuses d'avant-côte formeés durant la régression (i.e., "regressive nearshore sandstone beds") renferment des faunes importantes de bivalves et gastéropodes. Des dépots de "shale" sont retrouvés plus aux large et contiennent des faunes moins importantes ordinairement domineés par des inocéramides et des bivalves au coquillage mince. Les faunes d'ammonites sont caractérisées par des formes Téthyienne dans certains endroits et des formes tempérées dans d'autres.

INTRODUCTION

Rocks and fossils reported in this paper range in age from middle Cenomanian through most of the late Turonian. The area investigated includes much of the Four Corners States, the western two-thirds of New Mexico, the northeastern half of Arizona, much of the southeastern half of Utah and much of the southwestern half of Colorado.

Faunal data are based on collections made by many members of the United States Geological Survey and the New Mexico Bureau of Mines and Mineral Resources, as well as by students and other individuals. Guide fossils to many of the zones are lacking in some areas, however, and many collections contain specimens that may be diagnostic species but cannot be identified with certainty owing to poor preservation or fragmentary condition. At best, this paper is a preliminary report giving our interpretations of changes of transgressions and regressions and their effects on molluscan distribution.

ROCKS INVESTIGATED

Cenomanian-Turonian rocks in the area investigated are assigned to many formations and members (Fig. 1). In general, the names Dakota Sandstone and overlying Mancos Shale have been applied in all four states, but in the eastern more open seaway area, other names are used in place of Mancos. Likewise, along the southwest margin of the seaway, other names are used locally.

New Mexico

In east-central New Mexico, the standard Great Plains nomenclature has long been applied: Dakota Sandstone (oldest), Graneros Shale, Greenhorn Limestone, and Carlile Shale (youngest). In recent years, a Thatcher Limestone Member has been recognized in the Graneros Shale (personal observations); the Greenhorn Limestone has been divided into Lincoln Limestone Member (oldest), Hartland Shale Member, and Bridge Creek Limestone Member (youngest); and the Carlile Shale has been divided into Fairport Member (oldest), Blue Hill Member, Codell Sandstone Member, Juana Lopez Member, and an upper unnamed shaly member (e.g., Hook and Cobban, 1980, Fig. 4).

In west-central New Mexico, tongues of sandstone largely of a western source complicate the stratigraphy and necessitate a different nomenclature. The Dakota Sandstone is intertongued extensively with the Mancos Shale, and the sequence is, from oldest to youngest, Oak Canyon Member (of the Dakota), Cubero Tongue (of the Dakota), Clay Mesa Tongue (of the Mancos), Paguate Tongue (of the Dakota), Whitewater Arroyo Tongue (of the Mancos), and Twowells Tongue (of the Dakota) (Landis *et al.*, 1973). Where the Oak Canyon, Cubero, Clay Mesa, and Paguate cannot be differentiated, the sequence consists of the main body of the Dakota overlain by the Whitewater Arroyo Tongue of the Mancos, and that in turn, by the Twowells Tongue of the Dakota (Landis *et al.*, 1973, Fig. 2); or an unnamed shale tongue of Mancos may separate the Paguate Tongue from the main body of the Dakota (Hook *et al.*, 1980,

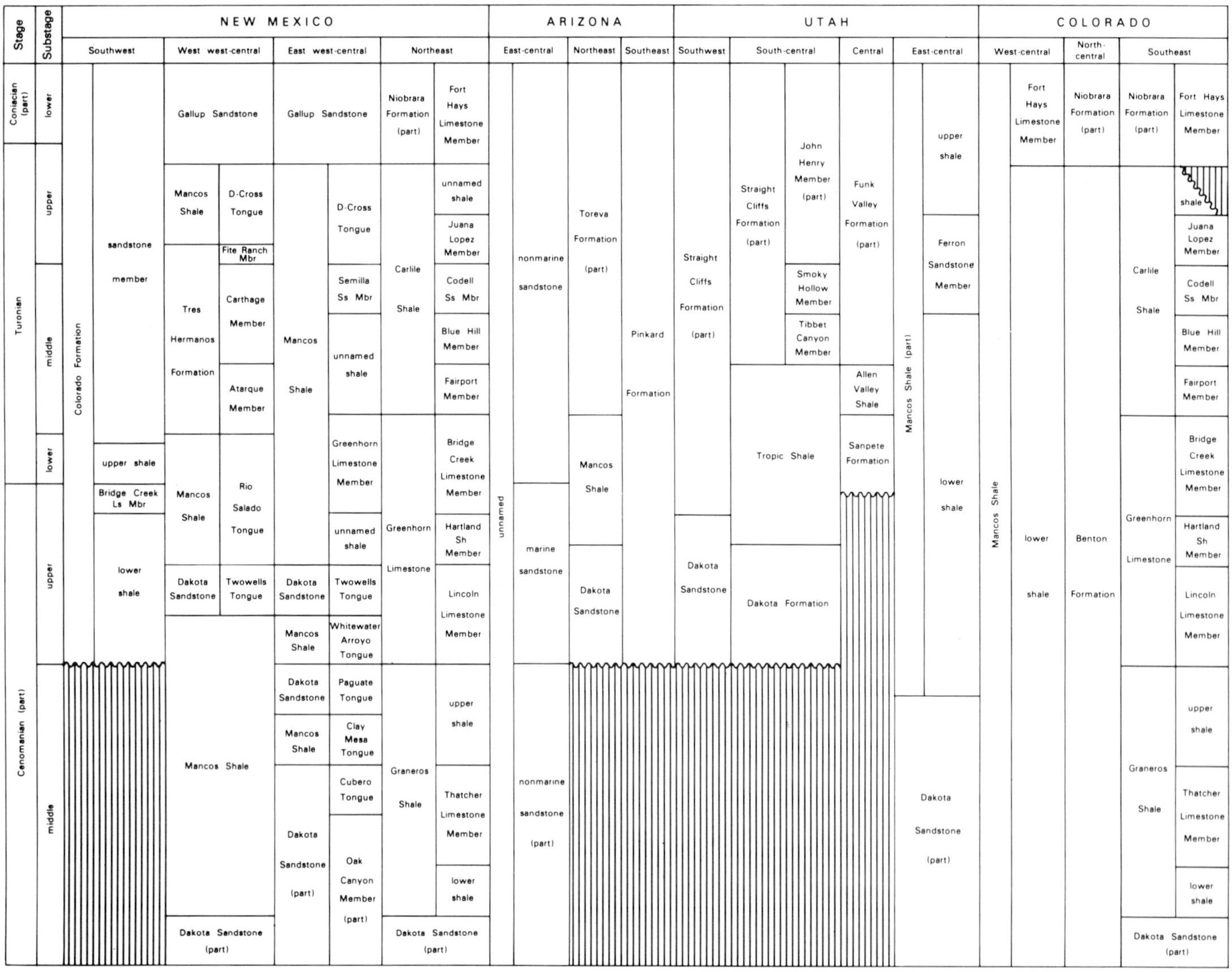

Figure 1. Chart showing correlation of middle Cenomanian to early Coniacian rocks in parts of New Mexico, Arizona, Utah, and Colorado.

Fig. 2); or the Twowells Tongue may be separated from a basal tongue of the Dakota by a thick lower unnamed tongue of Mancos Shale (Hook and Cobban, 1979, Fig. 5).

In much of west-central New Mexico, the Twowells Tongue is overlain by the Rio Salado Tongue of Mancos Shale and that in turn is overlain by the Tres Hermanos Formation and the D-Cross Tongue of the Mancos (Hook *et al.*, 1983). In this part of New Mexico calcareous beds, mainly equivalent to the Bridge Creek Limestone Member of the Greenhorn Limestone farther east, have been referred to as the Greenhorn Limestone (e.g., Molenaar, 1973, Figs. 2, 11). Farther north in the San Juan Basin, these calcareous beds have been referred to as a "limestone unit" in the Mancos Shale (O'Sullivan *et al.*, 1972, p. E-18-E19, Fig. 10) or as the Greenhorn Limestone Member (Peterson and Kirk, 1977, Fig. 2).

Nomenclature of middle Cenomanian-Turonian rocks in southern New Mexico varies considerably owing to major east-west facies changes and the long distances between outcrops. Texas names have been applied in the San Andres Mountains (Eagle Ford Formation by Kottlowski *et al.*, 1956) and in the El Paso area (Boquillas Formation by Strain, 1968). Farther west, Colorado Shale or Colorado Formation is usually applied to these rocks (e.g., Darton, 1916, p. 44; Spencer and Paige, 1935, p. 30; Jones *et al.*, 1967, p. 35; Cunningham, 1974; Elston, 1960; Morrison, 1965).

Arizona

The largest area of mid-Cretaceous rocks in Arizona crops out on Black Mesa in the northeastern part of the State, where Cenomanian-Turonian rocks are included in the Dakota Sandstone, Mancos Shale, and Toreva Formation (Repenning and Page, 1956). Southeast of the Black Mesa along the Arizona-New Mexico boundary, rocks of this age have been assigned to the Dakota Sandstone, Mancos Shale, and Mesaverde Formation (O'Brien, 1956; Young, 1957). South of Black Mesa along the Mogollon Rim in the Show Low area, unnamed marine mid-Cretaceous rocks crop out (Finnell, 1966; McKay, 1972). Much farther south in the south-central part of the State, marine rocks of mid-Cretaceous age have been recorded but not formally named (Campbell, 1904, p. 245; Ross, 1925, p. 25-28). In the Clifton-Morenci area in southeastern Arizona, marine sandstone and shale of late Cenomanian age have been named the Pinkard Formation (Lindgren, 1905a, p. 105; 1905b, p. 73.)

Utah

Mid-Cretaceous rocks crop out extensively in much of Utah. In the southern part in the Kaiparowits basin, marine rocks of late Cenomanian-middle Turonian age are included in the Dakota Formation, Tropic Shale, and lower part of the Straight Cliffs Formation (Peterson and Waldrop, 1965, Fig. 3; Peterson and Kirk, 1977, Fig. 4). Farther west, the Tropic Shale becomes sandy and disappears, and the sequence is then Dakota Formation overlain by Straight Cliffs Formation (Averitt and Threet, 1973). In the Henry Basin in south-central Utah, the sequence is Dakota Formation overlain by Mancos Shale; the latter is divided into several members of which the lower two are Tununk Shale Member (upper Cenomanian-middle Turonian) and Ferron Sandstone Member (middle and upper Turonian) (Peterson and Kirk, 1977, Fig. 4). Farther northwest, a different nomenclature was proposed by Spieker (1946) for mid-Cretaceous rocks in the Sanpete Valley in central Utah where the lower two units, Sanpete Formation and overlying Allen Valley Shale, are of Turonian age. Along the Book Cliffs in east-central Utah the sequence is Dakota Sandstone and Mancos Shale – the latter having a Ferron Sandstone Member that separates a lower unnamed shale from an upper one (Fisher *et al.*, 1960, Pl. 10).

Colorado

Middle Cenomanian-upper Turonian rocks in eastern Colorado are included in the Graneros Shale, Greenhorn Limestone, and Carlile Shale. The sequence is best exposed and documented in the Pueblo area (Cobban and Scott, 1972). The Graneros consists of the Thatcher Limestone Member separating a lower unnamed shale member barren of molluscan fossils from an upper unnamed shale member containing molluscan fossils. The Greenhorn consists of, from oldest to youngest, Lincoln Limestone Member, Hartland Shale Member, and Bridge Creek Limestone Member. Four members usually make up the Carlile Shale; these are, from oldest to youngest, Fairport Member, Blue Hill Member, Codell Sandstone Member, and Juana Lopez Member (Scott, 1969, p. 41-49). Locally, a thin unnamed calcareous shale unit may overlie the Juana Lopez Member.

In the central part of Colorado, Cenomanian-Turonian rocks are usually included in the Benton Formation (e.g., Stark *et al.*, 1949, p. 52-53). Rocks of this age farther west in Colorado are the Dakota Sandstone and Mancos Shale. The latter may have a Ferron Sandstone Member (e.g., Cashion, 1973) or a Juana Lopez Member (e.g., Haynes *et al.*, 1972), or locally, a calcareous Greenhorn Limestone equivalent (e.g., Steven *et al.*, 1974).

AMMONITE SEQUENCE

At least 17 ammonite zones of middle Cenomanian-late Turonian age can be recognized in the Four Corners States. Ten of the better known and more widely-distributed range zones are used in this report to provide time intervals short enough for showing step-by-step changes in transgressions and regressions of the epeiric sea. These zones and their generally accepted ages are:

Ages	Ammonite Range Zones
Late Turonian	*Scaphites whitfieldi*
Middle Turonian	*Prionocyclus hyatti* *Collignoniceras woollgari*
Early Turonian	*Mammites nodosoides*
Late Cenomanian	*Sciponoceras gracile* *Metoicoceras mosbyense* *Calycoceras canitaurinum*
Middle Cenomanian	*Plesiacanthoceras* aff. *wyomingense* *Acanthoceras alvaradoense* *Conlinoceras tarrantense*

BIOSTRATIGRAPHY AND PALEOGEOGRAPHY

Range Zone of *Conlinoceras Tarrantense*

The holotype of this species, *Metacalycoceras*? *tarrantense* Adkins (1928, p. 241, Pl. 28, Fig. 3; Pl. 29, Fig. 1) came from the basal beds of the Eagle Ford Group in Texas. The species has been found in the Oak Canyon Member and Cubero Tongue of the Dakota Sandstone in west-central New Mexico (Cobban, 1977, p. 22, Pl. 3, Fig. 9; Pl. 4, Figs. 1-3, 6). A closely allied form, *C. gilberti* (Cobban and Scott, 1972, p. 61, Pl. 1; Pl. 2, Figs. 5-9, 13-18; Pl. 3, Figs. 5-7, 11; Text-figs. 23, 24) is present in the Thatcher Limestone Member of the Graneros Shale in southeastern Colorado.

During the time of *C. tarrantense,* the Cretaceous epeiric sea transgressed from the east across the eastern two-thirds of Colorado as well as across the northeastern part of New Mexico (Fig. 2). A conspicuous embayment, Seboyeta Bay, extended into west-central New Mexico where the Oak Canyon Member and Cubero Tongue of the Dakota Sandstone were deposited (Hook *et al.,* 1980, p. 44). The western edge of Seboyeta Bay was probably a little east of Gallup; diagnostic marine fossils have not been found west of Crownpoint. The southern limits of the embayment were probably at Socorro inasmuch as *C. tarrantense* occurs only 30 cm above the top of the Dakota Sandstone 40 km north of the former coal-mining town of Carthage, whereas the top of the Dakota is of post-*tarrantense* age at Carthage. Northern limits of Seboyeta Bay are poorly known, but the presence of *Acanthoceras amphibolum* Morrow of post-*tarrantense* age in the Dakota Sandstone near Tierra Amarilla suggests that the shoreline was farther east, possibly near Taos.

The western edge of the seaway in Colorado has not been determined. *Calycoceras tarrantense* has been found in sandy beds near the top of the Dakota Sandstone at one locality in the Delta area. The coal beds in the Dakota Sandstone in western Colorado were probably formed in swamps that were marginal to the *tarrantense* sea.

Molluscan fossils are abundant in the western or shoreward part of Seboyeta Bay especially in the area east and northeast of Grants, New Mexico, where the lower part of the Cubero Tongue of the Dakota and dark brown weathered ferruginous sandstone concretions in the Oak Canyon Member contain a diversified assemblage of bivalves, gastropods, and ammonites (Cobban, 1977, Table 2, Pls. 1-5). Bivalves diagnostic of the Oak Canyon Member and Cubero Tongue include *Inoceramus eulessanus* Stephenson, *Exogyra columbella* Meek, *Plicatula arenaria* Meek, and *Camptonectes symmetricus* Herrick and Johnson. The gastropod *Arrhoges modesta* (Cragin)? is found in great abundance (Cobban, 1977, Pl. 2, Fig. 4). Five genera of ammonites are present, and *Conlinoceras tarrantense* is the most common species.

In southeastern Colorado and northeastern New Mexico, a single bed of hard, ferruginous, silty limestone usually makes up the Thatcher Limestone Member of the Graneros Shale (Cobban and Scott, 1972, Figs. 4-6). Fossils are generally crushed and fragmented. Locally, ammonites may occur abundantly as a single species (Cobban and Scott, 1972, Fig. 7). *Conlinoceras gilberti* Cobban and Scott is the dominant form in most collections. Among the bivalves, *Inoceramus eulessanus* Stephenson and *Plicatula arenaria* Meek are the most common species. A variety of gastropods is present in places.

Range Zone of *Acanthoceras Alvaradoense*

The ammonite species selected as a guide to this zone was described by Moreman (1942, p. 205, Pl. 32, Fig. 6; Text-fig. *2I, 2T*) from the Eagle Ford Group of Texas. *Acanthoceras alvaradoense,* of middle Cenomanian age, has been found as far north as the Black Hills area in eastern Wyoming.

The presence of a well-preserved and diversified *A. alvaradoense* fauna at the base of the Boquillas Formation in New Mexico just northwest of El Paso, Texas (Strain, 1976, p. 82) resting disconformably on lower Cenomanian rocks, reveals a considerable southward transgression of the epeiric sea. Inasmuch as the basal part of the Colorado Formation in the Cooke Range north of Deming, New Mexico contains the *Calycoceras canitaurinum* (Haas) fauna of post-*alvaradoense* age, the western limits of the *alvaradoense* sea were somewhere between there and El Paso (Fig. 2).

In the northern part of New Mexico in the Chama-Tierra Amarilla area, the oldest fossils in the Dakota Sandstone represent the range zone of *Acanthoceras amphibolum* Morrow of post-*alvaradoense* age suggesting that the *alvaradoense* shoreline was somewhere to the east. *Acanthoceras alvaradoense* is present in the Clay Mesa Tongue of the Mancos Shale at several localities southeast, south, and southwest of Grants, which reveals that Seboyeta Bay was still in existence but much broadened.

Records of *A. alvaradoense* in the eastern part of Colorado are lacking, but marine rocks of this age should be present in the post-Thatcher part of the Graneros Shale at Pueblo. The shoreline was probably not far southwest of the Delta-Montrose area in southwestern Colorado, where *A. alvaradoense* has been collected from the top of the Dakota Sandstone.

In contrast to the rich ammonite fauna at El Paso, Texas, rocks of *alvaradoense* age are poorly fossiliferous in New Mexico and Colorado. Only nine taxa have been recorded from the Clay Mesa Tongue of the Mancos Shale (Cobban, 1977, Table 1). *Inoceramus arvanus* Stephenson may be restricted to this tongue, but most of the other species range up into the Paguate Tongue of the Dakota. No gastropods are known from rocks of *alvaradoense* age in New Mexico and Colorado and only three genera of ammonites have been recorded (Cobban, 1977, Table 1).

Range Zone of *Plesiacanthoceras* aff. *Wyomingense*

Plesiacanthoceras wyomingense was described by Reagan (1924, p. 181, Pl. 19, Figs. 1, 2) as *Metoicoceras wyomingensis.* The holotype came from the Frontier Formation in east-central Wyoming where *P. wyomingense* forms a distinct zone above that of *Acanthoceras amphibolum* Morrow. In western New Mexico, a species related to *P. wyomingense* seems to lie in the zone of *A. amphibolum.*

The presence of *P.* aff. *wyomingense* at Mesa Redonda west of St. Johns, Arizona (Kirkland, 1982) reveals a westward expansion of the Seboyeta embayment into eastern Arizona at

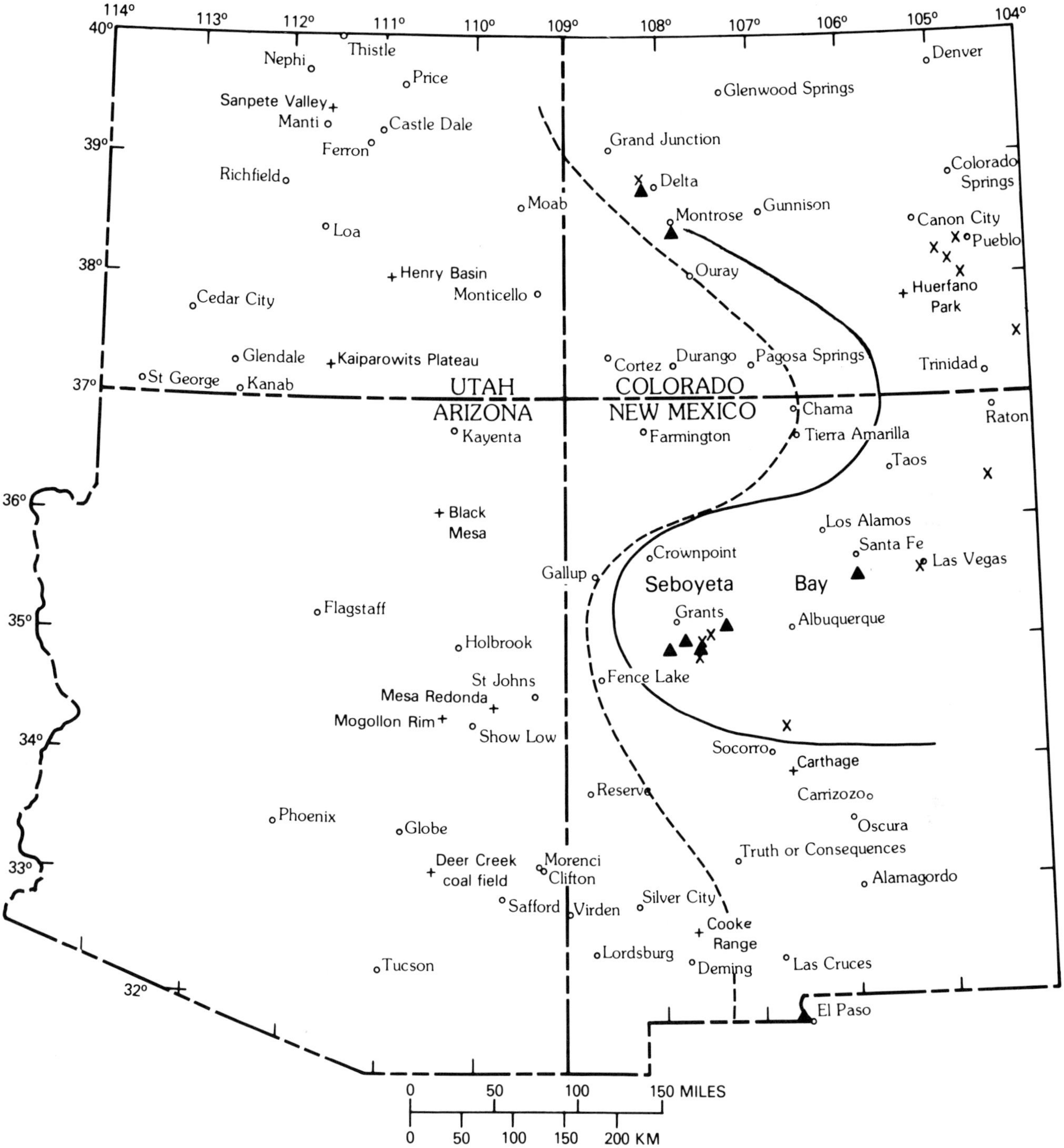

Figure 2. Map of parts of the Four Corners States showing localities (X) where *Conlinoceras tarrantense* (Adkins) has been found. Solid line shows the general position of the shoreline. Solid triangles (▲) mark localities where *Acanthoceras alvaradoense* Moreman has been found, and the general position of the shoreline of that time is marked by short dashes.

the close of middle Cenomanian time (Fig. 3). In southwestern New Mexico, the shoreline remained east of the Cooke Range near Deming. The recent discovery (by G.R. Scott, 1981, pers. commun.) of *Acanthoceras amphibolum* in the upper part of the Dakota Sandstone in the Chama-Tierra Amarilla area in northern New Mexico suggests that the shoreline was not far to the west in that area. The northwest trending shoreline in southwestern Colorado was probably much like that of *alvaradoense* time. However, an encroachment of the sea into eastern Utah is suggested by the possible presence of the marker bentonite bed in the basal part of the Mancos Shale along the Utah boundary area west of Grand Junction, Colorado. The marker bentonite bed (or X-bentonite bed) (Cobban and Scott, 1972, p. 4-28, Fig. 10) is widely distributed over the Great Plains and in areas farther west in Wyoming, Colorado, and New Mexico; it lies in the range zone of *Acanthoceras amphibolum*.

Fossil molluscs are abundant and varied in the Seboyeta embayment, where they occur in shallow-water sandstone such as the Paguate Tongue of the Dakota (Cobban, 1977, Table 1). Among the more abundant bivalves are *Idonearca blandpiedi* Stephenson, *Pinna petrina* White, *Phelopteria*? cf. *aquilerae* (Bose), *Inoceramus rutherfordi* Warren, *Exogyra trigeri* Coquand, *E. levis* Stephenson, *Ostrea beloiti* Logan, and *Aphrodina* cf. *munda* (Stephenson). A large gastropod fauna is

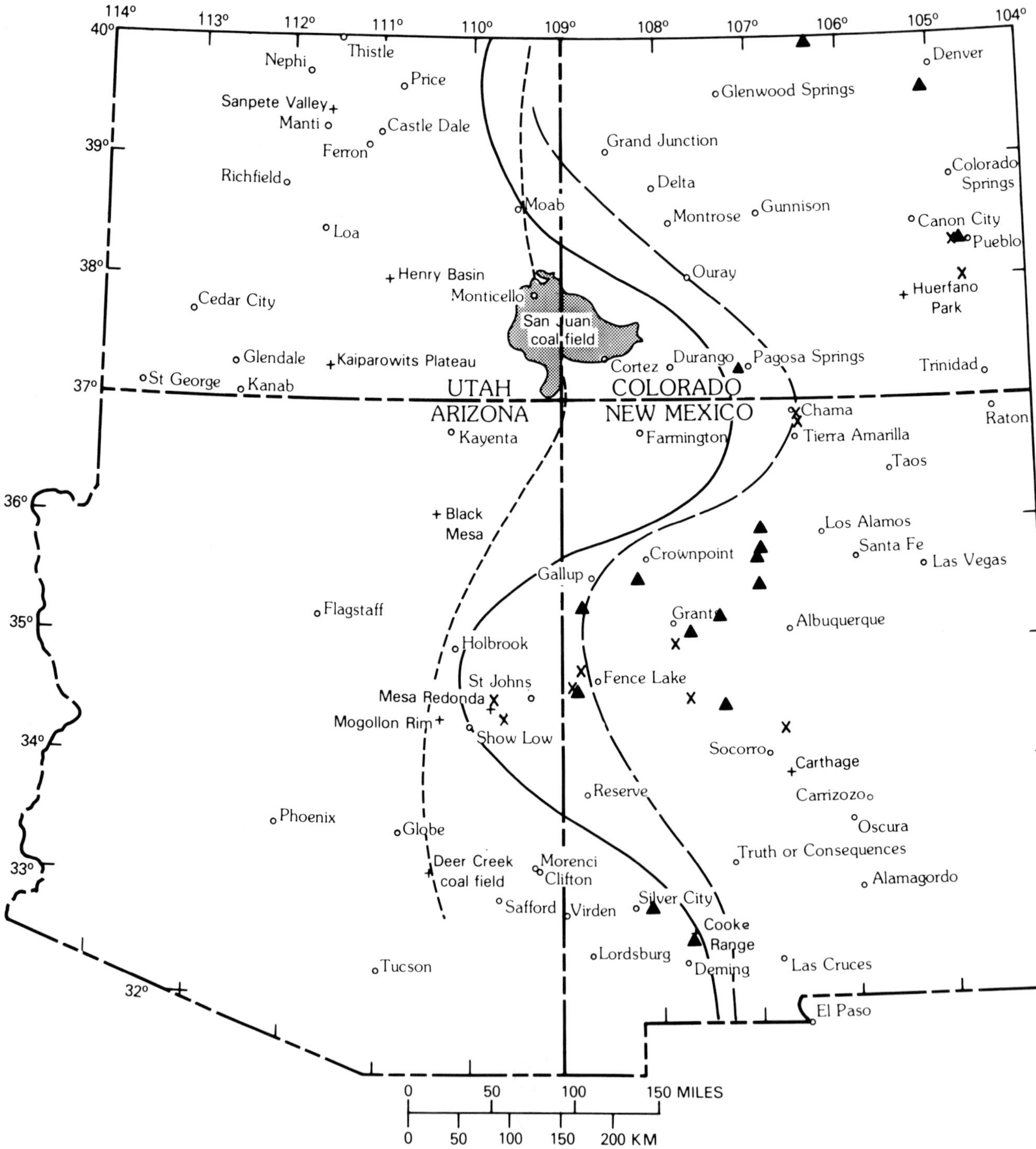

Figure 3. Map of parts of the Four Corners States showing localities (X) where *Plesiacanthoceras* aff. *wyomingense* (Reagan) or *Acanthoceras amphibolum* Morrow has been found. Solid line shows the general position of the shoreline. Solid triangles (▲) mark localities where *Calycoceras canitaurinum* (Haas) has been found, and the general position of the shoreline of that time is marked by short dashes. The approximate position of the shoreline during the time of *Acanthoceras alvaradonese* Moreman is shown by the long dashes.

present as well as nine genera of ammonites. The abundance of *Exogyra* (3 species) and *Pycnodonte* indicates a Tethyan influence as far north as Seboyeta Bay.

Range Zone of *Calycoceras Canitaurinum*

The holotype of *Calycoceras canitaurinum* (Haas, 1949, p. 9, Pls. 1-3; Pl. 4, Figs. 1, 2, 4; Text figs. 1-4) came from the basal part of the Cody Shale in northern Wyoming. This early-Late Cenomanian species, originally assigned to *Mantelliceras,* is widely distributed in the Western Interior from southwestern New Mexico to north-central Montana. The range zone of this species is the same as that of *Dunveganoceras pondi* Haas, which has been a guide to one of the Cretaceous ammonite zones in the Western Interior (Cobban, 1951a, Fig. 2; Cobban and Reeside, 1952, p. 1017). In the area of the present report, *D. pondi* has not been found, whereas *C. canitaurinum* is widely distributed. Accordingly, *C. canitaurinum* is used as the guide fossil to the oldest of the late Cenomanian zones in the southern part of the Western Interior.

During the time of *C. canitaurinum,* the epeiric sea advanced westward across the southwestern and northwestern corners of New Mexico to form a broad embayment into eastern Arizona (Fig. 3). At the Four Corners, the oldest molluscs found are from the upper part of the Dakota Sandstone and represent the range zone of *Metoicoceras mosbyense* Cobban of post-*canitaurinum* age. The *canitaurinum* shoreline may have been here or a little to the east. Western limits of the sea in eastern Utah or possibly in western Colorado are unknown. *Calycoceras canitaurinum* was observed in nonsandy shale about 6 or 7 m above the base of the Mancos Shale at Pagosa Springs, Colorado, and the shoreline should be some distance to the west. Similar shale is present in the lower part of the Mancos Shale above the marker bentonite bed along the Utah boundary west of Grand Junction, Colorado, which suggests that the shore must have been farther west in eastern Utah. Coal beds in the Deer Creek coal field in southeastern Arizona and some of the coal beds in the San Juan coal field in southeastern Utah may have formed marginal to the *canitaurinum* sea.

Molluscan fossils are fairly abundant in the broad Seboyeta embayment, but elsewhere they are sparse. Those in the embayment are in the Whitewater Arroyo Tongue of the Mancos Shale and in the overlying Twowells Tongue of the Dakota Sandstone. Forty-three taxa have been listed from these units (Cobban, 1977, Table 3). Common bivalves include *Idonearca depressa* White, *Pinna petrina* White, *Inoceramus prefragilis* Stephenson, *Plicatula* cf. *ferryi* Coquand, *Pycnodonte* aff. *kellumi* (Jones), *Exogyra trigeri Coquand, and E. levis* Stephenson. Cerithid gastropods are fairly abundant locally. At one locality in west-central New Mexico, a nerineid gastropod was collected. Nerineids are typical of the warmer water of the Tethyan sea. Ammonites are not common, and only five genera have been found.

Range Zone of *Metoicoceras Mosbyense*

The guide fossil was described by Cobban (1953, p. 48, Pl. 6, Figs. 1-14; Pl. 7, Figs. 1-3) from the Mosby Sandstone Member of the Greenhorn Formation of central Montana.

Metoicoceras mosbyense is widely distributed in western New Mexico, eastern Arizona, southeastern Utah, and southwestern Colorado, and reveals a major westward advance of the epeiric sea in mid-late Cenomanian time (Fig. 4). The species occurs in shallow-water sandstone (Twowells Tongue) in westcentral and northwestern New Mexico and the adjoining part of Arizona. In southwestern New Mexico, *M. mosbyense* is found in an argillaceous fine-grained sandstone unit in the Colorado Formation in the Silver City-Virden area, but to the east in the Cooke Range, and to the northwest in the Clifton-Morenci area of Arizona, the species is found in dark grey clay shale. On the north and west sides of Black Mesa in northeastern Arizona and in the Kaiparowits Plateau area in southern Utah, *M. mosbyense* occurs in the uppermost sandstone unit of the Dakota Formation. The trend of the shoreline during this time was probably northwestward in Arizona and northeastward in Utah.

Molluscan fossils collected from an unnamed sandstone unit in the Deer Creek coal field, about 100 km west of Morenci, Arizona, were determined by T.W. Stanton (in Ross, 1925, p. 27) to be *Exogyra* aff. *laeviuscula* Roemer, *Trigonarca* cf. *depressa* (White), *Callista* (*Dosiniopsis*?) n. sp., *Glauconia coalvillensis* (Meek), and *Turritella* sp. The *Exogyra* is probably *E. levis* Stephenson, the *Trigonarca* is probably *Idonearca depressa* Meek, and the *Callista* may be the *Aphrodina* sp. of Cobban (1977, Table 3), all of which suggest the fauna of the Whitewater Arroyo Tongue of the Mancos Shale and Twowells Tongue of the Dakota Sandstone of areas farther northeast in New Mexico. The farthest advance of the late Cenomanian sea into southern Arizona may have occurred during the range zone of *Metoicoceras mosbyense.* In southern Utah some of the coal beds in the Dakota Formation in the San Juan and Kaiparowits coal fields may be also of this age.

Fossils molluscs are locally abundant and varied, especially along the Arizona-New Mexico boundary near St. Johns, where pinnids, exogyras, oysters, pectinids, and other bivalves are well preserved. Gastropods, however, are neither common nor varied, and ammonites are restricted to two genera. An arcuate belt of brackish-water rocks containing abundant *Flemingostrea prudentia* (White) parallels the shoreline in the Black Mesa-Kaiparowits Plateau area (Fig. 4).

Range Zone of *Sciponoceras Gracile*

Sciponoceras gracile, of late late Cenomanian age, was described as *Baculites gracilis* Shumard (1860, p. 596). Shumard's specimens came from the Eagle Ford Group of Texas. The species is known from many localities in the Western Interior from southwestern New Mexico to north-central Montana. In the area of our report, *S. gracile* occurs in the basal part of the Bridge Creek Limestone Member of the Greenhorn Limestone in eastern Colorado (Cobban and Scott, 1972, p. 47, Pl. 17, Figs. 9-29; Text-fig. 18), in the Greenhorn Limestone Member of the Mancos Shale in northwestern New Mexico, in the Bridge Creek Limestone Member of the Colorado Formation in southwestern New Mexico (Hook and Cobban, 1981, Fig. 3), in the basal part of the Mancos Shale in northeastern Arizona (Repenning and Page, 1956, p. 267), and in the lower part of the Tropic Shale in southern Utah (Stanton, 1893, p. 35; Gregory and Moore, 1931, p. 100).

The presence of *S. gracile* at Glendale in southwestern Utah and farther northwest at a locality 8 km east of Cedar City (Richardson, 1927, p. 470, 473) indicates a continued westward advance of the late Cenomanian sea (Fig. 3). Coal beds in the Harmony coal field southwest of Cedar City may have formed during this time. From the Cedar City area, the shoreline trended northeastward toward the southwestern corner of Wyoming. Two important control points along the way are near Ferron and Thistle, Utah. At a locality 25 km south of Ferron, the ammonite *Euomphaloceras* (*Kanabiceras*) *septem-*

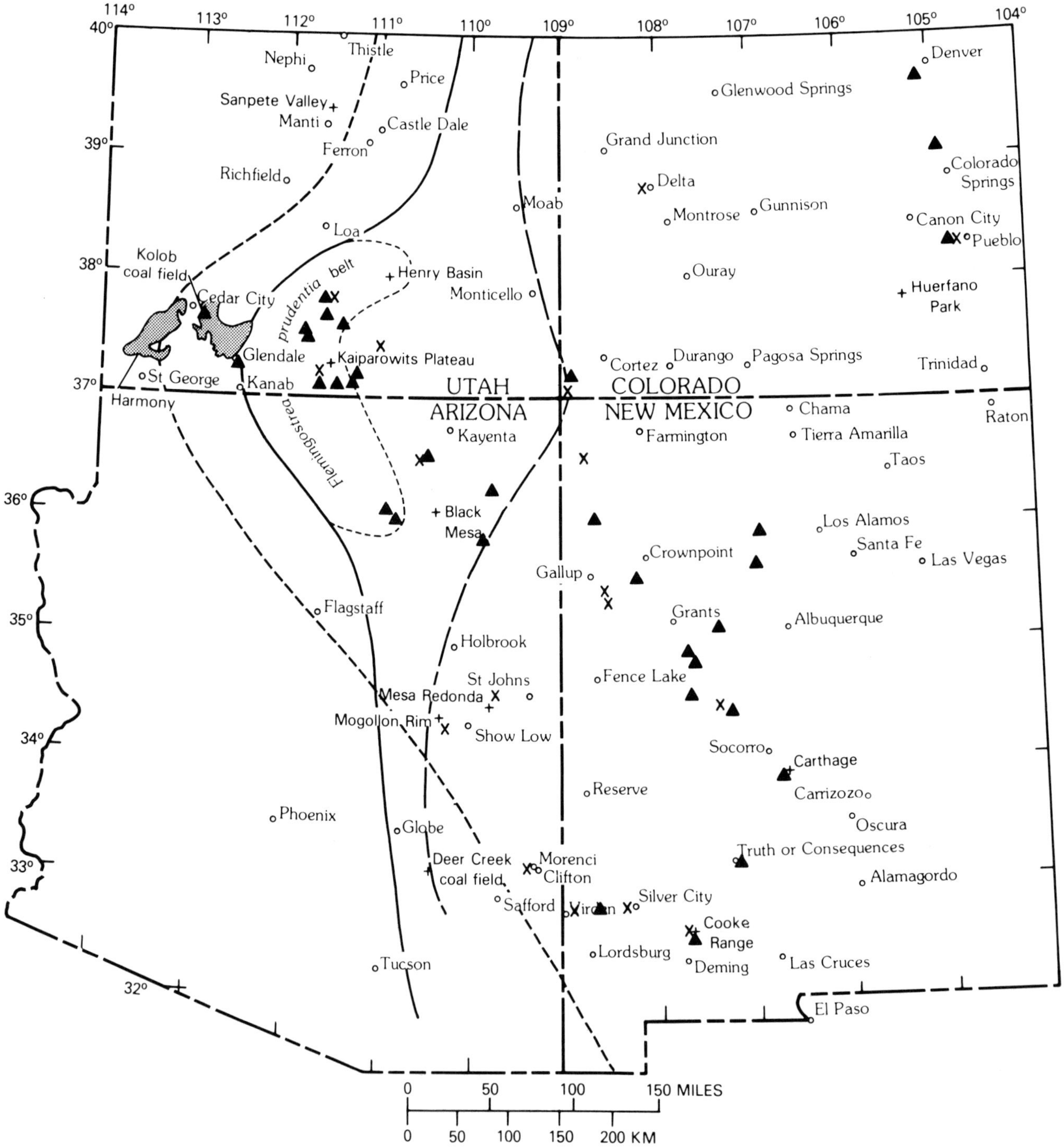

Figure 4. Map of parts of the Four Corners States showing localities (X) where *Metoicoceras mosbyense* Cobban has been found. Solid line shows the general position of the shoreline. Solid triangles (▲) mark localities where *Sciponoceras gracile* (Shumard) has been found, and the general position of the shoreline of that time is marked by short dashes. The approximate position of the shoreline during the time of *Calycoceras canitaurinum* (Haas) is shown by the long dashes.

seriatum (Cragin), restricted to the range zone of *S. gracile*, has been found in sandy beds at the base of the Mancos Shale (Cobban, 1976, p. 120, Pl. 1, Figs. 5, 6). Near Thistle, the oldest marine Cretaceous fossils are in the base of the Sanpete Formation and represent the range zone of *Mammites nodosoides* of post-*gracile* age.

The western limits of the late late-Cenomanian sea are unknown in Arizona. An ammonite, *Metoicoceras geslinianum* (d'Orbigny) of *gracile* age, was discovered recently in silty limestone concretions in a sandstone unit 10 km west of Show Low in the central-eastern part of the State (J.I. Kirkland, pers. commun., 1981). Cretaceous rocks in this area are unnamed and consist largely of fine- to coarse-grained feldspathic sandstone (Finnell, 1966; McKay, 1972). The shoreline was probably not far to the southwest, and a general northeastward retreat of the sea from the Deer Creek coal field area is suggested.

Long ago, Stanton (1893, p. 35) drew attention to the large and varied molluscan fauna in rocks of *gracile* age in southern Utah, and lengthy lists of fossils have been given by later geologists (e.g., Gregory, 1951, p. 36-37). *Pycnodonte newberryi* (Stanton) is the most abundant bivalve and outcrops are locally paved with this form (Lawyer, 1972, Pl. 2, Fig. 4; Rigby *et al.*, 1974, Fig. 40; Hook and Cobban, 1977, Fig. 3). Common bivalve genera found in limestone concretions in the Tropic Shale include *Inoceramus, Exogyra, Pycnodonte, Lima, Psilomya, Camptonectes, Pleuriocardia, Lucina,* and *Corbula.* Sohl (1967, p. 17) listed the gastropods *Turritella, Cerithiella, Perissoptera, Pseudomaura, Cerithioderma*?, *Mesorhytis, Paladmete,* and *Ringicula,* and noted that these represented a variety of feeding types (deposit feeders, plant browsers, filter feeders, carnivores). A variety of ammonites is also present in these concretions and includes *Sciponoceras, Allocrioceras, Worthoceras, Scaphites, Placenticeras, Euomphaloceras* (*Kanabiceras*), *Metoicoceras,* and at least one undescribed genus. In the Black Mesa area in northeast Arizona, *Calycoceras, Pseudocalycoceras,* and *Moremanoceras* are also present. Molluscs are much less varied in southeastern Colorado and northeastern New Mexico, where the rocks are represented by limestone and limy shale rather than by limestone concretions in calcareous shale. Gastropods are sparse, and only a few genera of bivalves are present (Cobban and Scott, 1972, p. 24). Ammonites, however, are abundant and varied including some genera not recorded from Utah. Molluscs, other than *Pycnodonte newberryi* (Stanton), are not common or varied in southwest New Mexico (Hook and Cobban, 1981, Fig. 3).

Range Zone of *Mammites Nodosoides*

Mammites nodosoides (Schlüter) has long been accepted as a guide to the lower Turonian (de Grossouvre, 1901, Table 35). A lectotype from the lower Turonian of Czechoslovakia has been selected recently (Wright and Kennedy, 1981, p. 76, Fig. 23).

In the area of our report, *M. nodosoides* has been recorded from southeastern Colorado (Cobban and Scott, 1972), northern New Mexico (Pillmore and Eicher, 1976), west-central New Mexico (Cobban and Hook, 1979), eastern Arizona (Miller and Breed, 1963), and southern Utah (Lawrence, 1965; Lawrence and Stokes, 1965; Lawrence *et al.*, 1966) (Fig. 5). The eastern Arizona record was based on an ammonite now assigned to *Plesiacanthoceras* aff. *wyomingense* (Reagan).

Problems arise concerning fossils identified as *M. nodosoides* in the Western Interior. An ammonite described by Morrow (1935, p. 467, Pl. 51, Fig. 2; Pl. 52, Fig. 2a-c; Text-fig. 2) as *Mammites wingi* is easily confused with *M. nodosoides* unless specimens are well preserved or show their septal sutures.

Mammites nodosoides ranges through most of the upper half of the Bridge Creek Limestone Member of the Greenhorn Limestone in southeastern Colorado. The species has been found also in the Greenhorn Limestone Member of the Mancos Shale in northwestern New Mexico and in the equivalent calcareous shale unit in the Mancos in the west-central part. Limestone concretions near the top of the Mancos Shale in the Fence Lake area and near Truth or Consequences also contain *M. nodosoides.* Farther southeast, near El Paso, Texas, the species has been found in limestone concretions in a shale unit like that of the Mancos in the Truth or Consequences area but were included in the Boquillas Formation. Paralleling this northwest-trending shale belt is a near-shore sandstone belt that also contains *M. nodosoides* (e.g., in the Cooke Range).

In southern Utah, *M. nodosoides* has been recorded from the middle of the Tropic Shale as far west as the Glendale area (Lawrence and Stokes, 1965). Just east of Cedar City, about 60 km northwest of Glendale, a shaly unit in the dominantly sandstone sequence contains the bivalve *Mytiloides* of the *mytiloides-duplicostatus* group which suggests a *M. nodosoides* age. Stanton (1893, p. 37) noted the occurrence of marine and brackish-water fossils 30 km west of Cedar City. Inasmuch as the entire Cretaceous sequence is nonmarine 70 km southwest of Cedar City (C.M. Molenaar, pers. commun., 1981), the *nodosoides* shoreline was between these localities. The early Turonian seems to have been the time of maximum transgression in southern Utah.

Much farther northeast in Utah near Thistle, *Mammites* has been found in nearshore sandstone near the base of the Sanpete Formation. A northeast trend of the shoreline between the Cedar City area and Thistle is apparent.

Aside from the bivalve *Mytiloides mytiloides* (Mantell), fossil molluscs are scarce in the large area of carbonate deposition that covered Colorado, much of New Mexico, and parts of Arizona and Utah. The belt of noncalcareous shale bordering the west edge of the carbonate area contains many fossiliferous limestone concretions especially in western New Mexico. Here the bivalves *Phelopteria gastrodes* (Meek), *Plicatula ferryi* Coquand, *Pleuriocardia pauperculum* (Meek), and *Veniella mortoni* Meek and Hayden are abundant as well as thin-shelled oysters and the gastropods *Gyrodes conradi* Meek, *Turritella whitei* Stanton, and *Pyropsis coloradoensis* Stanton. Nine genera of ammonites are present including the Tethyan forms *Neoptychites cephalotus* (Courtiller) and *Fagesia superstes* (Kossmat). Large and varied bivalve assemblages are locally present in the near-shore sandstone beds, such as in the Sanpete Formation in central Utah.

Range Zone of *Collignoniceras Woollgari*

The ammonite for this zone was described from the Middle Chalk of England (Mantell, 1822, p. 197, Pl. 21, Fig. 16; Pl. 22,

Fig. 7). A middle Turonian age is usually accepted. The species has been found at many localities in the Four Corners States (Fig. 5). In places, an early and a late form of the species can be distinguished (Cobban and Hook, 1979) but in the present report, no separation is made.

In the eastern part of the area, *Collignoniceras woollgari* occurs rarely in limestone beds at or near the top of the Bridge Creek Limestone Member of the Greenhorn Limestone and as impressions in the overlying Fairport Member of the Carlile Shale. Farther west in the area of the San Juan Basin, Black

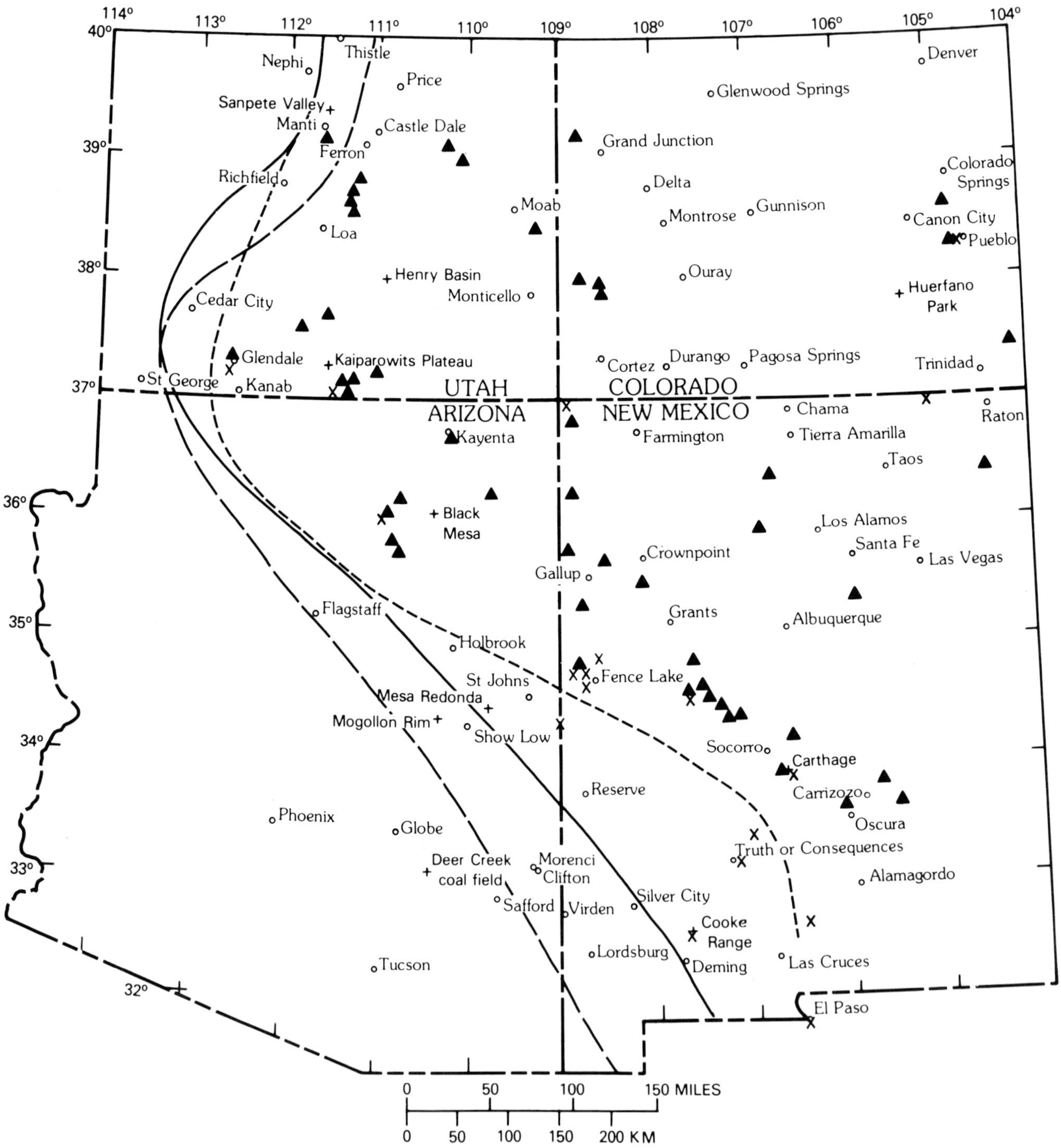

Figure 5. Map of parts of the Four Corners States showing localities (X) where *Mammites nodosoides* (Schlüter) has been found. Solid line shows the general position of the shoreline. Solid triangles (▲) mark localities where *Collignoniceras woollgari* (Mantell) has been found, and the general position of the shoreline of that time is marked by short dashes. The approximate position of the shoreline during the time of *Sciponoceras gracile* (Shumard) is shown by the long dashes.

Mesa, Kaiparowits Plateau, and Henry basin, *C. woollgari* is abundant as impressions in shale or as small internal molds in thin olive-grey siltstone lenses in the Mancos and Tropic Shales and in the Tununk Member of the Mancos Shale. In west-central New Mexico, the species is found in sandstone concretions in the basal regressive sandstone unit (Atarque Member) of the Tres Hermanos Formation as well as in limestone concretions in the upper part of the underlying Rio Salado Tongue of Mancos Shale.

The distribution of *C. woollgari* in New Mexico and the presence of this species in the basal regressive sandstone unit of the Tres Hermanos reveals a general regression of the epeiric sea in a northeastward direction in Arizona and New Mexico. The Salt Lake coal field in the Fence Lake area may have formed marginal to the shoreline during this time.

In southwest Utah, the shoreline seems to have remained nearly where it was in *nodosoides* time. The *Prionotropis* sp., recorded by Richardson (1927, p. 473) in the Glendale area, may be *C. woollgari*. If this interpretation is correct, the *woollgari* shoreline was somewhere between Glendale and Cedar City. A westward transgression of the sea in central Utah is suggested by the Allen Valley Shale, a 180 to 240 m thick unit of chiefly grey marine shale that overlies the nearshore sandstones of the lower Turonian Sanpete Formation.

Aside from *Mytiloides* of the *mytiloides-hercynicus* groups, small thin-shelled oysters, and *C. woollgari*, molluscan fossils are scarce in the calcareous shaly beds in eastern Colorado and northeastern New Mexico. In areas of noncalcareous shale deposition farther west and southwest, as many as nine genera of ammonites are present, but there is very little else (Cobban and Hook, 1979, p. 11). A much larger fauna of bivalves and gastropods and a few ammonites are present in nearshore sandstones in west-central New Mexico (basal regressive sandstone unit of Tres Hermanos Formation), northeastern Arizona (Toreva Formation, J.I. Kirkland, pers. commun., 1981), and western Colorado (unnamed sandy unit in Mancos Shale west of Grand Junction). Nineteen genera of bivalves, seven genera of gastropods, and six genera of ammonites have been recorded from the basal part of the Tres Hermanos Formation in west-central New Mexico (Cobban and Hook, 1979, p. 11). Bivalves usually found in these sandstone units include *Pinna petrina* White, *Pleuriocardia* aff. *pauperculum* (Meek), and *Veniella mortoni* Meek and Hayden. Common gastropods are *Pugnellus (Gymnarus) fusiformis* Meek, *Gyrodes conradi* Meek, *G.* aff. *depressa* Meek, *Pyropsis coloradoensis* Stanton, and *Carota* cf. *dalli* (Stanton). Ammonite diversity is low in the northern part of the area but greatly increases southward in New Mexico where some Tethyan forms (e.g., *Hoplitoides* ex gr. *koeneni-wohltmanni, Spathites rioensis* Powell, and *Neoptychites*? sp.) have been recorded (Cobban and Hook, 1979, p. 11).

Range Zone of *Prionocyclus Hyatti*

Prionocyclus hyatti (Stanton) was based on specimens from the Codell Sandstone Member of the Carlile Shale in Huerfano Park, Colorado (Stanton, 1893, p. 176, Pl. 42, Figs. 5-8). The species is transitional from *Collignoniceras* to *Prionocyclus*. A late middle Turonian age is usually assigned.

Prionocyclus hyatti has been found at many localities in the Four Corners States (Fig. 6). Specimens occur in concretions in the Blue Hill Member and Codell Sandstone Member of the Carlile shale in southeastern Colorado and northeastern New Mexico, in the Semilla Sandstone Member of the Mancos Shale in west-central New Mexico, and in the Tres Hermanos Formation farther south. The species also occurs in silty beds in the upper part of the Tununk Member of the Mancos in southern Utah and in the upper part of the Mancos Shale beneath the Ferron Sandstone Member much farther north near Price. Probably part of the Ferron Member and its coal beds in the Emery coal field were formed during this time.

At Carthage in south-central New Mexico, the middle nonmarine part of the Tres Hermanos Formation (Carthage Member) is of *hyatti* age inasmuch as *Collignoniceras woollgari* has been collected from the basal regressive sandstone unit, and *Prionocyclus macombi* Meek of post-*hyatti* age has been found in the top of the member. *Prionocyclus hyatti*, however, has been found in a nearshore sandstone bed in the Tres Hermanos Formation near Oscura about 75 km southeast of Carthage and the shoreline was somewhere between these localities.

During the time of *P. hyatti*, the sea regressed northeastward in New Mexico. Possibly some of the coal beds in the Salt Lake coal field along the Arizona border were formed during this time. The shoreline probably trended northwest across the northeastern part of Arizona. *Prionocyclus hyatti* has not been identified for certain in the Black Mesa area, although *Inoceramus howelli* White of this age has been found on the north side of the mesa.

In Utah an eastward regression of the sea is indicated by a shift in fossil localities eastward compared to those of *woollgari* time and by the presence of *P. hyatti* in the lower regressive marine sandstone part of the Ferron Sandstone Member of the Mancos Shale as well as in the underlying transition beds to the Mancos Shale. Some of the coal beds in the Emery and Henry Mountains coal fields may have been formed late in *hyatti* time.

Over much of the area of the Four Corners States, molluscan fossils are rather sparse in rocks of *hyatti* age. Two exceptions are the Codell Sandstone Member of the Carlile Shale in Huerfano Park in south-central Colorado and the Semilla Sandstone Member of the Mancos Shale in the Rio Puerco valley northwest of Albuquerque, New Mexico. The beds now named the Codell Sandstone Member in Huerfano Park were referred to by Stanton (1893, p. 28) as the Pugnellus sandstone for the abundance of the gastropod *Pugnellus fusiformis* Stanton. Stanton (1893, p. 29) listed some 23 kinds of bivalves and 17 forms of gastropods from this unit. The Semilla Sandstone Member does not have as varied a fauna, but at least 20 kinds of bivalves and a few species of gastropods were listed from the type section (Dane *et al.*, 1968, p. F17-F19).

Bivalves that are useful guides to the range zone of *Prionocyclus hyatti* include *Inoceramus howelli* White, *I. flaccidus* White, *Exogyra suborbiculata* (Lamarck), *Ostrea malachitensis* Stanton, and *Lopha bellaplicata novamexicana* Kauffman. The Tethyan ammonites *Coilopoceras springeri* Hyatt, *Herrickiceras costatum* (Herrick and Johnson), and *Spathites puercoensis* (Herrick and Johnson) are found in this zone in New Mexico.

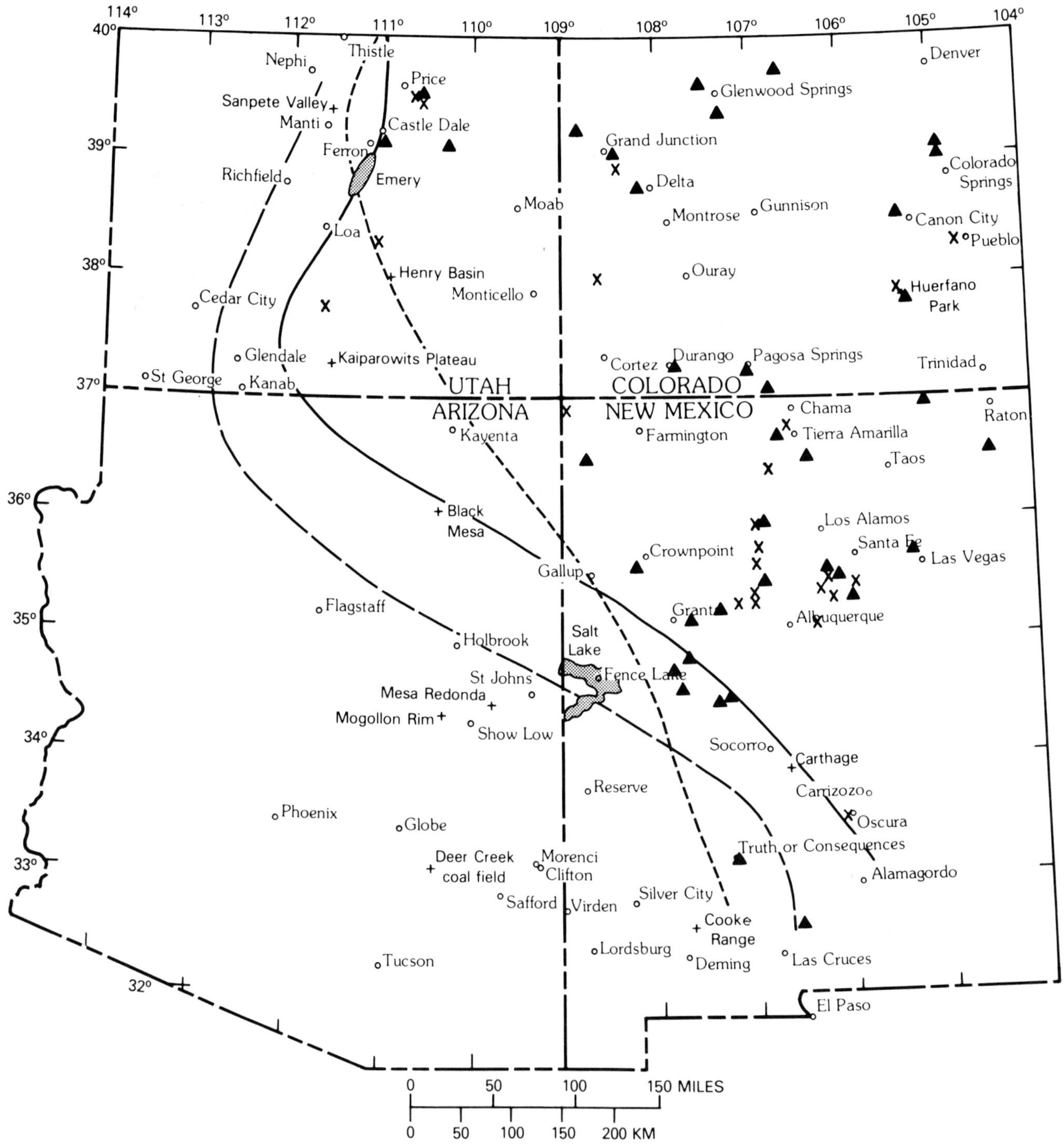

Figure 6. Map of parts of the Four Corners States showing localities (X) where *Prionocyclus hyatti* (Stanton) has been found. Solid line shows the general position of the shoreline. Solid triangles (▲) mark localities where *Scaphites whitfieldi* Cobban has been found, and the general position of the shoreline of that time is marked by short dashes. The approximate position of the shoreline during the time of *Collignoniceras woollgari* (Mantell) is shown by the long dashes.

Range Zone of *Scaphites Whitfieldi*

Cobban (1951b, p. 24, Pl. 4, Figs. 30-40; Pl. 5, Figs. 1-4) described the guide fossil to this late Turonian zone from the Turner Sandy Member of the Carlile Shale of the Black Hills. Many localities are known for this species in Colorado, Utah, and New Mexico (Fig. 6).

A southwestward transgression of the sea in western New Mexico occurred during the time of *Scaphites whitfieldi* when the D-Cross Tongue of the Mancos Shale was deposited over the Tres Hermanos Formation. *Scaphites whitfieldi* has been collected from nearshore sandstone in the Truth or Consequences area. Inasmuch as rocks of this age are nonmarine in the

Cooke Range near Deming, a shore position between these localities is indicated. Molenaar (1973, Fig. 7) pointed out that the strike of the shoreline in western New Mexico during this time was north-northwest. This trend probably continued across the northeastern corner of Arizona and on to central Utah where scaphites closely related to *S. whitfieldi* have been found near Ferron just above the Ferron Sandstone Member of the Mancos Shale (Cobban, 1976, p. 124, Pl. 1, Fig. 4). From central Utah, the shoreline curved northeastward and entered Wyoming at some point east of its southwestern corner. Over much of the western two-thirds of Colorado and the north half of New Mexico, the uppermost part of the Juana Lopez Member of the Carlile and of the Mancos Shales was deposited during this time.

Molluscan fossils had little diversity during the time of *S. whitfieldi. Inoceramus perplexus* Whitfield, *Lopha lugubris* (Conrad), *Baculites yokoyamai* Tokunaga and Shimizu, *Scaphites whitfieldi* Cobban, and *Prionocyclus novimexicanus* (Marcou) are the usual forms. A rarity in the Western Interior is the ammonite *Reesideites* cf. *minimus* (Hayasaka and Fukada).

REFERENCES

Adkins, W.S., 1928, Handbook of Texas Cretaceous Fossils: Texas University Bulletin 2838, 385 p.

Averitt, Paul, and Threet, R.L., 1973, Geologic Map of the Cedar City Quadrangle, Iron County, Utah: United States Geological Survey Quadrangle Map GQ-1120, Scale 1:24,000.

Campbell, M.R., 1904, The Deer Creek Coal Field, Arizona: United States Geological Survey Bulletin 225, p. 240-258.

Cashion, W.B., 1973, Geologic and Structure Map of the Grand Junction Quadrangle, Colorado and Utah: United States Geological Survey Miscellaneous Investigations Series Map I-736, Scale 1:250,000 (reprinted 1979).

Cobban, W.A., 1951a, Colorado Shale of Central and Northwestern Montana and Equivalent Rocks of Black Hills: American Association of Petroleum Geologists Bulletin, v. 35, p. 2170-2198.

______, 1951b, Scaphitoid Cephalopods of the Colorado Group: United States Geological Survey Professional Paper 239, 42 p.

______, 1953, Cenomanian Ammonite Fauna from the Mosby Sandstone of Central Montana: United States Geological Survey Professional Paper 243-D, p. 45-55.

______, 1976, Ammonite Record from the Mancos Shale of the Castle Valley-Price-Woodside Area, East-Central Utah: Brigham Young University, Geology Studies, v. 22, p. 117-126.

______, 1977, Characteristic Marine Molluscan Fossils from the Dakota Sandstone and Intertongued Mancos Shale, West-Central New Mexico: United States Geological Survey Professional Paper 1009, 30 p.

Cobban, W.A., and Hook, S.C., 1979, *Collignoniceras woollgari woollgari* (Mantell) Ammonite Fauna from Upper Cretaceous of Western Interior, United States: New Mexico Bureau of Mines and Mineral Resources Memoir 37, 51 p.

Cobban, W.A., and Reeside, J.B., Jr., 1952, Correlation of the Cretaceous Formations of the Western Interior of the United States: Geological Society of America Bulletin, v. 63, p. 1011-1043.

Cobban, W.A., and Scott, G.R., 1972, Stratigraphy and Ammonite Fauna of the Graneros Shale and Greenhorn Limestone Near Pueblo, Colorado: United States Geological Survey Professional Paper 645, 108 p.

Cunningham, J.E., 1974, Geologic Map and Sections of Silver City Quadrangle, New Mexico: New Mexico Bureau Mines and Mineral Resources, Geologic Map 30, Scale 1:24,000.

Dane, C.H., Kauffman, E.G., and Cobban, W.A., 1968, Semilla Sandstone, A New Member of the Mancos Shale in the Southeastern Part of the San Juan Basin, New Mexico: United States Geological Survey Bulletin 1254-F, 21 p.

Darton, N.H., 1916, Geology and Underground Water of Luna County, New Mexico: United States Geological Survey Bulletin 618, 188 p.

Elston, W.E., 1960, Reconnaissance Geologic Map of Virden Thirty-Minute Quadrange (N. Mex.): New Mexico Institute Mining Technology Geological Map 15, Scale 1:126,720.

Finnell, T.L., 1966, Geologic Map of the Chediske Peak Quadrangle, Navaho County, Arizona: United States Geological Survey Quadrangle Map GQ-544, Scale 1:62,500.

Fisher, D.J., Erdmann, C.E., and Reeside, J.B., Jr., 1960, Cretaceous and Tertiary Formations of the Book Cliffs, Carbon, Emery, and Grand Counties, Utah, and Garfield and Mesa Counties, Colorado: United States Geological Survey Professional Paper 332, 80 p.

Gregory, H.E., 1951, The Geology and Geography of the Paunsaugunt Region, Utah: United States Geological Survey Professional Paper 226, 116 p.

Gregory, H.E., and Moore, R.C., 1931, The Kaiparowits Region, a Geographic and Geologic Reconnaissance of Parts of Utah and Arizona: United States Geological Survey Professional Paper 164, 161 p.

Grossouvre, Albert de, 1901, Recherches Sur la Craie Supérieur, Part 1, No. 2, Stratigraphie Générale: Mémoires pour Servir a L'explication La Carte Geologique Détaillée de la France, p. 561-1013.

Haas, Otto, 1949, Acanthoceratid Ammonoidea from Near Greybull, Wyoming: American Museum of Natural History Bulletin, v. 93, art. 1, 39 p.

Haynes, D.D., Vogel, J.D., and Wyant, D.G., 1972, Geology, Structure and Uranium Deposits of the Cortez Quadrangle, Colorado and Utah: United States Geological Survey Miscellaneous Investigations Series Map I-629, Scale 1:250,000.

Hook, S.C., and Cobban, W.A., 1977, *Pycnodonte newberryi* (Stanton) – Common Guide Fossil in Upper Cretaceous of New Mexico: in Kottlowski, F.E., *et al.*: New Mexico Bureau of Mines and Mineral Resources Annual Report 1976-77, p. 48-54.

______, 1979, *Prionocyclus novimexicanus* (Marcou) – Common Upper Cretaceous guide fossil in New Mexico: *in* Kottlowski, F.E., *et al.*: New Mexico Bureau of Mines and Mineral Resources Annual Report 1977-78, p. 35-42.

______, 1980, Some Guide Fossils in Upper Cretaceous Juana Lopez Member of Mancos and Carlile Shales, New Mexico: in Kottlowski, F.E., *et al.*: New Mexico Bureau of Mines and Mineral Resources Annual Report 1978-79, p. 38-49.

______, 1981, Late Greenhorn (Mid-Cretaceous) Discontinuity Surfaces, Southwest New Mexico: New Mexico Bureau of Mines and Mineral Resources Circular 180, p. 5-21.

Hook, S.C., Cobban, W.A., and Landis, E.R., 1980, Extension of the Intertongued Dakota Sandstone-Mancos Shale Terminology into the Southern Zuni Basin: New Mexico Geology, v. 2, p. 42-44, 46.

Hook, S.C., Molenaar, C.M., and Cobban, W.A., 1983, Stratigraphy and Revision of Nomenclature of Upper Cenomanian to Turonian (Upper Cretaceous) Rocks of West-Central New Mexico: New Mexico Bureau of Mines and Mineral Resources Circular 185, p. 7-28.

Jones, W.R., Hernon, R.M., and Moore, S.L., 1967, General Geology of Santa Rita Quadrangle, Grant County, New Mexico: United Geological Survey Professional Paper 555, 144 p.

Kirkland, J.I., 1982, Reassessment of the Age of the Late Cretaceous Section at Mesa Redonda, Apache County, Arizona: Journal of Paleontology, v. 56, p. 547-550.

Kottlowski, F.E., Flower, R.H., Thompson, M.L., and Foster, R.W., 1956, Stratigraphic Studies of the San Andres Mountains, New Mexico: New Mexico Bureau of Mines and Mineral Resources Memoir 1, 132 p.

Landis, E.R., Dane, C.H., and Cobban, W.A., 1973, Stratigraphic Terminology of the Dakota Sandstone and Mancos Shale, West-Central New Mexico: United States Geological Survey Bulletin 1372-J, 44 p.

Lawrence, J.C., 1965, Stratigraphy of the Dakota and Tropic Formations of Cretaceous Age in Southern Utah: in Geology and Resources of South-Central Utah – Resources for Power: Utah Geological Society Guidebook to Geology of Utah, no. 19, p. 71-91.

Lawrence, J.C., and Stokes, W.L., 1965, *Mammites* Zone in the Tropic Shale of Late Cretaceous Age in Southern Utah [abst.]: Utah Academy Science, Arts, and Letters, Proceedings, v. 42, pt. 1, p. 168-169.

Lawrence, J.C., Herman, George, and Elison, J.H., 1966, North American Distribution of the Turonian (Late Cretaceous) Ammonite *Mammites Nodosoides* (Schlotheim): in Abstracts for 1965: Geological Society of America Special Paper 87, p. 293.

Lawyer, G.R., 1972, Sedimentary Features and Paleoenvironment of the Dakota Sandstone (Early Upper Cretaceous) Near Hanksville, Utah: Brigham Young University Geology Studies, v. 19, p. 89-120.

Lindgren, Waldemar, 1905a, Description of the Clifton Quadrangle [Ariz.]: United States Geological Survey Geologic Atlas, Folio 129, 13 p.

______, 1905b, The Copper Deposits of the Clifton-Morenci District, Arizona: United States Geological Survey Professional Paper 43, 375 p.

Mantell, Gideon, 1822, The Fossils of the South Downs, or Illustrations of the Geology of Sussex: London, Lupton Relfe, 327 p.

McKay, E.J., 1972, Geologic Map of the Show Low Quadrangle, Navajo County, Arizona: United States Geological Survey Quadrangle Map GQ-973, Scale 1:62,500.

Miller, H.W., Jr., and Breed, W.J., 1963, Lower Turonian (Cretaceous) Ammonite from Mesa Redonda, Arizona: Plateau, v. 35, p. 123-130.

Molenaar, C.M., 1973, Sedimentary Facies and Correlation of the Gallup Sandstone and Associated Formations, Northwestern New Mexico: in Cretaceous and Tertiary Rocks of the Southern Colorado Plateau: Four Corners Geological Society Memoir, p. 85-110.

Moreman, W.L., 1942, Paleontology of the Eagle Ford Group of North and Central Texas: Journal of Paleontology, v. 16, p. 192-220.

Morrison, R.B., 1965, Geologic Map of the Duncan and Canador Peak Quadrangles, Arizona and New Mexico: United States Geological Survey Miscellaneous Geologic Investigations Map I-442, Scale 1:48,000, Separate Text.

Morrow, A.L., 1935, Cephalopods from the Upper Cretaceous of Kansas: Journal of Paleontology, v. 9, p. 463-473.

O'Brien, B.R., 1956, Geology of Cienega Amarilla Area, Catron County, New Mexico, and Apache County, Arizona: Unpub. M.A. Thesis, University Texas, Austin.

O'Sullivan, R.B., Repenning, C.A., Beaumont, E.C., and Page, H.C., 1972, Stratigraphy of the Cretaceous Rocks and the Tertiary Ojo Alamo Sandstone, Navajo and Hopi Indian Reservations, Arizona, New Mexico, and Utah: United States Geological Survey Professional Paper 521-E, 65 p.

Peterson, Fred, and Kirk, A.R., 1977, Correlation of the Cretaceous Rocks in the San Juan, Black Mesa, Kaiparowits and Henry Basins, Southern Colorado Plateau: in New Mexico Geological Society Guidebook, 28th Field Conference, San Juan Basin III: p. 167-178.

Peterson, Fred, and Waldrop, H.A., 1965, Jurassic and Cretaceous Stratigraphy of South-Central Kaiparowits Plateau, Utah: in Geology and Resources of South-Central Utah – Resources for Power: Utah Geological Society and Intermountain Association of Petroleum Geologists Guidebook to the Geology of Utah, No. 19, p. 47-69.

Pillmore, C.L., and Eicher, D.L., 1976, Lower Part of the Marine Cretaceous at Gold Creek, Vermejo Park, New Mexico: New Mexico Geological Society Guidebook 27th Field Conference, p. 171-176.

Reagan, A.B., 1924, Cretacic Mollusca of Pacific Slope: Pan-American Geologist, v. 41, p. 179-190.

Repenning, C.A., and Page, H.G., 1956, Late Cretaceous stratigraphy of Black Mesa, Navajo and Hopi Indian Reservations, Arizona: American Association of Petroleum Geologists Bulletin, v. 40, p. 255-294.

Richardson, G.B., 1927, The Upper Cretaceous Section in the Colob Plateau, Southwest Utah: Washington Academy of Science Journal, v. 17, p. 464-475.

Rigby, J.K., Hintze, L.F., and Welsh, S.L., 1974, Geologic Guide to the Northwestern Colorado Plateau. Pt. 1, Thistle to Green River, Utah, via. U.S. Highway 50-6; Pt. 2, Green River to Salina, via Interstate 70; Pt. 3, Salina to Thistle, via U.S. Highway 89: Brigham Young University Geology Studies, v. 21, pt. 2, Studies for Students No. 9, 119 p.

Ross, C.P., 1925, Geology and Ore deposits of the Aravaipa and Stanley Mining Districts, Graham County, Arizona: United States Geological Survey Bulletin 763, 120 p.

Scott, G.R., 1969, General and Engineering Geology of the Northern Part of Pueblo, Colorado: United States Geological Survey Bulletin 1262, 131 p.

Shumard, B.J., 1860, Descriptions of New Cretaceous fossils from Texas: Academy of Science of St. Louis, Transactions, v. 1, p. 590-610.

Sohl, N.F., 1967, Upper Cretaceous Gastropod Assemblages of the Western Interior of the United States: in A Symposium on Paleoenvironments of the Cretaceous Seaway in the Western Interior: Geological Society America, Rocky Mountain Section, 20th Annual Meeting: Golden, Colorado, Colorado School of Mines, p. 1-37.

Spencer, A.C., and Paige, Sidney, 1935, Geology of the Santa Rita Mining Area, New Mexico: United States Geological Survey Bulletin 859, 78 p.

Spieker, E.M., 1946, Late Mesozoic and Early Cenozoic History of Central Utah: United States Geological Survey Professional Paper 205-D, p. 117-161.

Stanton, T.W., 1893, The Colorado Formation and Its Invertebrate Fauna: United States Geological Survey Bulletin 106, 288 p.

Stark, J.T., Johnson, J.H., Behre, C.H., Jr., Powers, W.E., Howland, A.L., Gould, D.B., and others, 1949, Geology and Origin of South Park, Colorado: Geological Society of America Memoir 33, 188 p.

Steven, T.A., Lipman, P.W., Hail, W.J., Jr., Barker, Fred, and

Luedke, R.G., 1974, Geologic Map of the Durango Quadrangle, Southwestern Colorado: United States Geological Survey Miscellaneous Investigations Map I-764, Scale 1:250,000.

Strain, W.S., 1968, Cerro de Muleros (Cerro de Cristo Rey): in Delaware Basin Exploration, 1968 Guidebook: West Texas Geological Society Publication 68-55, p. 82.

______, 1976, New Formation Names in the Cretaceous at Cerro de Cristo Rey, Dona Ana County, New Mexico: in Lovejoy, E.M.P., Geology of Cerro de Cristo Rey Uplift, Chihuahua and New Mexico: New Mexico Bureau of Mines and Mineral Resources Memoir 31, p. 77-82.

Wright, C.W., and Kennedy, W.J., 1981, The Ammonoidea of the Plenus Marls and the Middle Chalk: Palaeontographical Society (Monograph), 148 p.

Young, Keith, 1957, Cretaceous Ammonites from Eastern Apache County, Arizona: Journal of Paleontology, v. 31, p. 1167-1174.

Jurassic-Cretaceous Biochronology and Paleogeography of North America,
edited by G.E.G. Westermann,
Geological Association of Canada Special Paper 27, 1984

PALEOBIOGEOGRAPHY AND EVOLUTIONARY RESPONSE DYNAMIC IN THE CRETACEOUS WESTERN INTERIOR SEAWAY OF NORTH AMERICA

Erle G. Kauffman
Department of Geological Sciences, University of Colorado, Boulder, CO 80309

ABSTRACT

Comparison of modern biogeographic divisions with Phanerozoic counterparts reveals numerous methodological and conceptual discrepancies that impair the study of the dynamics and evolution of paleobiogeographic units. By applying modern analytical criteria to the definition of global Cretaceous, mollusc-based, biogeographic divisions, a predictable distribution of these large-scale ecological units is obtained. Cretaceous units were fewer, broader, and had more diffuse ecotonal boundaries than modern analogs. Evidence for Cold Temperate Cretaceous divisions and polar glaciation is lacking. Cretaceous paleobiogeographic divisions reflect broad climatic gradients in a generally warm and equable world. Three paleobiogeographic divisions characterized the Cretaceous Western Interior Seaway of North America during most of its history: a Cool Temperate Northern Interior Subprovince; a Mild Temperate Central Interior Subprovince; and a Warm Temperate Southern Interior Subprovince. To the south lay the Subtropical Gulf and Atlantic Coast Subprovince, and the Tropical Caribbean Province. Stratigraphic data reveal five third-order eustatic cycles which inundated the interior of North America during the Cretaceous. In each, the distribution of paleobiogeographic units was dramatically altered within a 1 to 2 Ma interval around peak transgression and eustatic highstand. These peaks were associated with abrupt paleoceanographic changes involving warming and amelioration of climates, normalization of salinity, and rapid fluctuations between well oxygenated waters and anoxic events. During these 1 to 2 Ma intervals, abrupt northward migration of Subtropical organisms as much as 1500 miles into west-central Canada, and of Tropical reef biotas into the southern United States, depict rapid northward incursions of Subtropical and Warm Temperate biogeographic units. These were followed rapidly by southward emigration to normal distributions. Major extinction events are commonly associated with these incursions, as are periods of rapid evolution among marine taxa. The major centers of endemism and the most rapid rates of evolution in North American Cretaceous molluscs occurred within paleobiogeographic ecotones.

RÉSUMÉ

La comparison des divisions biogéographiques modernes avec des contreparties Phanérozoique révèle de nombreuses méthodologies et conceptualisations inconstantes qui nuisent à l'étude de la dynamique et l'évolution des unités paléobiogeographiques. En applicant des critères analytiques modernes pour définir des divisions biogéographiques du Crétacé globale basé sur des mollusques; on obtient une distribution prévisible de ces unités écologiques de grande échele. Les unités du Crétacé sont moins nombreuses, plus larges et comprennent des limites d'écotone plus diffuses que des analogues modernes. Rien indique une glaciation polaire et des divisions d'un Crétacé Froid. Les divisions paléobiogéographiques du Crétacé réflètent des gradients climatiques larges dans un monde générallement chaud et calme. Trois divisions paléobiogéographiques caractérisent le Passage Marin de l'Ouest Interieur d'Amérique du Nord durant la plupart du Crétacé: une Sous-province Tempéreé-froide de L'intérieur Nord; une Sous-province Tempéreé-douce de L'intérieur central; et une Sous-province Tempéreé-chaude de L'intérieur Sud. Au sud de cette région on retrouve le Golfe Sub-tropical, la Sous-province de la Côte Atlantique et la Province Tropicale des Antilles. Les donneés stratigraphiques révèlent cinq cycles eustatiques du troisième ordre qui ont innondé l'intérieur de l'Amérique du Nord durant le Crétacé. Pour chaque cycle, la distribution des unités paléobiogéographiques a été dramatiquement altéreé à l'intérieur d'un intervalle de un à deux Ma durant une transgression marine maximale. les pics de transgression étaient associés avec des changements paléocéanographiques soudains invoquant; un réchauffement et une amélioration climatique, une normalisation de la salinité et des fluctuations rapides entre des eaux bien oxygénées et des évènements anoxiques. Durant ces intervalles de un à des Ma on note des migrations nordiques importantes de la part des organismes sub-tropicaux – jusqu'à 1500 milles à l'intérieur de l'ouest-central Canadien; et d'autre part des biotopes des récifs tropicaux pénétrant dans le sud des Etats-Unis. Ces phénomènes se traduisent par des incursions rapides vers le nord des unités biogéographiques sub-tropical et tempéreé-chaude. Celles-ci ont été suivit par une émigration sud aux distributions normales. Des évènements d'extinction majeures ainsi que des périodes d'évolution rapides entre les taxons marins sont fréquemment associés avec ces incursions. Les centres endémiques majeurs et les taux d'évolution les plus rapides au niveau des mollusques du Crétacé de l'Amérique du Nord sont observés à l'intérieur des écotones paléobiogéographiques.

INTRODUCTION

The widely held concept that ancient paleobiogeographic units – provinces, regions, and realms – were static to slowly shifting, unusually broad divisions of the global biota is not defensible from detailed regional studies (e.g., Pleistocene to Recent biogeographic changes), nor from theory generated by plate tectonics, paleoceanography, paleoclimatology, and modern ecology. The continuous restructuring of continents and ocean basins has produced major barriers and pathways to biotic distribution within a few million years; global sealevel fluctuations reaching hundreds of meters have alternately drowned and exposed large portions of the Earth's continents. Such eustatic changes have greatly altered habitat distribution, migratory pathways, and global climate within 10 Ma intervals – the scale of third-order tectonoeustatic cycles (Vail *et al.*, 1977). These earthbound forces, and extraterrestrial influences on climatic change, should have produced profound short-term shifts in the distribution of the world biota within each geologic period.

Further, biogeographic divisions of the global biota are large-scale ecological units (Kauffman and Scott, 1976), composed of interacting and at least loosely structured communities; as such they should be sensitive to, migrate with, and change niche size relative to shifts in major environmental factors through time. In particular, regional shifts in the large-scale ecotones that mark competitive, ecologically disrupted boundaries between biogeographic units should be recognizable in the fossil record. A prediction can be made that paleobiogeographic units of all ages were highly dynamic, as they were during Pleistocene to Recent glacial and interglacial cycles.

Why, then, the seemingly great discrepancy between the majority of ancient paleobiogeographic reconstructions and the modern distribution of biotas? For example, the majority of papers in Hallam (1973) and Hughes (1973) portray a Phanerozoic world partitioned into a relatively few, broadly distributed biogeographic units, usually "realms" and "provinces" applied inconsistently by various authors to divisions of approximately the same magnitude, and imply that these units were relatively stable over 10 to 35 million year intervals (one-third of a geologic period). Compare these concepts with the Pleistocene to Recent biogeographic divisions of the global marine ecosystem (e.g., Ekman, 1967; Coomans, 1962; Hall, 1964; Addicott, 1966, 1969; Hazel, 1970; Culver and Buzas, 1981a, 1981b; 1982a, 1982b, among many papers on the North American margins). Was the ancient oceanographic-biotic system really that distinct from the modern systems? Theoretically, the only real biogeographic distinctions between the geologically atypical, regionally extensive, glacial intervals (including the Recent) and the more prevalent warm, equable, Phanerozoic climates characterized by higher sea-level stand, should be related to: (a) plate arrangements, affecting distribution of both landmasses and marine systems; and (b) the generally broader Phanerozoic thermal gradient during non-glacial intervals. The latter predictably would result in warmer, more extensive oceans, lacking cold polar zones, and thus broader and somewhat fewer latitudinally distributed, thermally regulated, climatic zones (e.g., restricted to Cool Temperate-Tropical climatic zones of today's oceans). Even these predictable differences do not explain the great discrepancy between, for example, the number of paleobiogeographic divisions plotted for Ordovician trilobites (Whittington, 1973), Silurian brachiopods (Boucot and Johnson, 1973), or Cretaceous cephalopods (Stevens, 1973; Matsumoto, 1973) and the number of modern biogeographic divisions or predictions made for Phanerozoic biotas based on Holocene biogeographic distributions.

Kauffman (1973) addressed many of these problems and identified three important discrepancies between modern and ancient biogeographic analyses: (1) lack of consistency in the definition of paleobiogeographic divisions among Phanerozoic analyses, and between the Phanerozoic and Recent biotas. The criteria used for definition of modern biogeographic units are not consistently applied to analyses of Phanerozoic biotas (*e.g.*, degree of internal endemism, and percent of species crossover at biogeographic boundaries). Even the well defined terminology of modern biogeographic units is inconsistently applied to Phanerozoic distributions (realms, regions, provinces, subprovinces; see Ekman, 1967; and guidelines for quantification by Coomans, 1962; Hazel, 1970; Kauffman, 1973; and Culver and Buzas, 1981a, 1981b; 1982a, 1982b). Without consistency in the definition and classification of modern and ancient biogeographic units, their dynamic changes in space and time cannot be documented.

(2) It is common practice among paleontologists to define paleobiogeographic units based on only one phylum, class, or order of fossil organisms, and more commonly, on only selected taxa (usually genera) within that group. Obviously, emphasis on cosmopolitan eurytopic taxa (e.g., many cephalopods) will give quite a different biogeographic picture than plots for the same time period based on ecologically restricted, stenotopic taxa (e.g., the reef biota). Multitaxial biogeographic plots, such as those that can now be constructed for the marine biotas of the North Atlantic margins (ref. cit.), or at least plots based on all genera or species within a major class or phylum (e.g., Hazel, 1970; Kauffman, 1973; Culver and Buzas, 1981a, 1981b, 1982a, 1982b), are mandatory if paleontologists hope to integrate modern and ancient biogeography, and study the dynamics of paleobiogeographic change.

(3) The stratigraphic resolution of most paleobiogeographic plots is insufficient to show the dynamic changes that predictably should characterize biotic distribution through time. The predominant method of authors in Hallam (1973) and Hughes (1973), for example, is to plot the distribution of taxa within

intervals representing one-third of a geologic period, i.e., 10 to 35 million year accumulations of data. Yet many factors of the global environment that directly affect regional patterns of biotic distribution have a smaller periodicity; e.g., third-order tectonoeustatic fluctuations which emplace and remove extensive epicontinental seas from the craton average 5 to 10 Ma in duration; major biogeographic changes within such cycles may be abrupt. Studies in the Cretaceous of North America indicate that latitudinal shifts in the northern boundaries of single paleobiogeographic provinces of over 1500 miles have taken place within a 2 Ma interval, dated radiometrically (e.g., Late Cenomanian-Early Turonian). Many large-scale climatic cycles have a shorter periodicity, and even the 21,000, 42,000 and 100,000 year Milankovitch-like cycles of the Mesozoic and Cenozoic can be shown to have caused major latitudinal changes in biogeographic distribution during certain intervals. Tectonic and other paleogeographic controls on regional migration and distribution of organisms can certainly act within 10 Ma intervals.

In a stage by stage global paleobiogeographic analysis of all Cretaceous bivalve genera employing measurements of percent endemism in defining and classifying biogeographic units, Kauffman (1973) demonstrated that, even with data from a single major class of Mollusca, it was possible to divide the Cretaceous world into a predictably large number of discrete paleobiogeographic units and to plot their dynamic changes through time within 5 Ma or shorter intervals. In this study, three realms, three regions, seven provinces, and 12 subprovinces of the Cretaceous, representing Cool Temperate to Tropical climatic zones, were identified and their changes plotted through time relative to plate tectonic and tectonoeustatic controls. Based on modern patterns of biogeographic distributions for Mollusca (e.g., Hall, 1964), these were predictable numbers and size-ranges for biogeographic units in the Cretaceous, with its warmer, more equable climates and broader temperature gradients which lacked cold climatic zones around the poles. The origins, spread, decline and extinction of Cretaceous paleobiogeographic units, as defined on percent endemism within each unit through time, were found to be primarily related to plate tectonic control on ocean history, to dispersal and migration of marine organisms, to relative isolation of large gene pools across expanding or contracting oceans, and, thus, to relative development of endemic biotas. These biogeographic patterns were secondarily modified by major tectonoeustatic changes linked to the history of sea-floor spreading. Whereas this study established a new methodology for paleobiogeographic analysis, with important links to modern analyses, it only documented the broad patterns of change for Bivalvia in Cretaceous paleobiogeographic units at 5 Ma intervals, on the average, and identified major causes for change.

This paper explores, in detail, some of the ideas laid down in my 1973 paper, within the great Western Interior Cretaceous epicontinental seaway of North America and the adjacent Gulf Coast and Caribbean basins. Emphasis is placed on interpretation of the dynamics of paleobiogeographic change and on factors controlling major shifts in biotic distribution within the basin. Further, the effects of large scale biogeographic changes, and their causes, on patterns of evolution are documented for several Cretaceous molluscan groups. These analyses have been made possible because of the immense stratigraphic and paleobiologic data-set for this basin available from the works of many students and colleagues in the University of Colorado Cretaceous Basin Analysis Research Group. The high-resolution stratigraphic standard that we have collectively developed for the Western Interior Cretaceous basin is unparalleled, and makes possible the dissection of biogeographic and evolutionary patterns within the basin in the context of a very detailed and accurate matrix of geologic time.

GEOLOGY AND OCEANOGRAPHY OF THE WESTERN INTERIOR CRETACEOUS SEAWAY

Introduction

At peak flooding, the Cretaceous seaway of Western Interior North America extended from Arctic Canada and Alaska south to the Gulf of Mexico, with probable intermittent connections to the Hudson's Bay region (Fig. 1; Williams and Stelck, 1975; Kauffman, 1977a), approximately 3000 miles (4800 km). Shoreline deposits of Minnesota (Bergquist, 1944; Cobban and Merewether, 1983) and central Utah (Ryer, 1977)

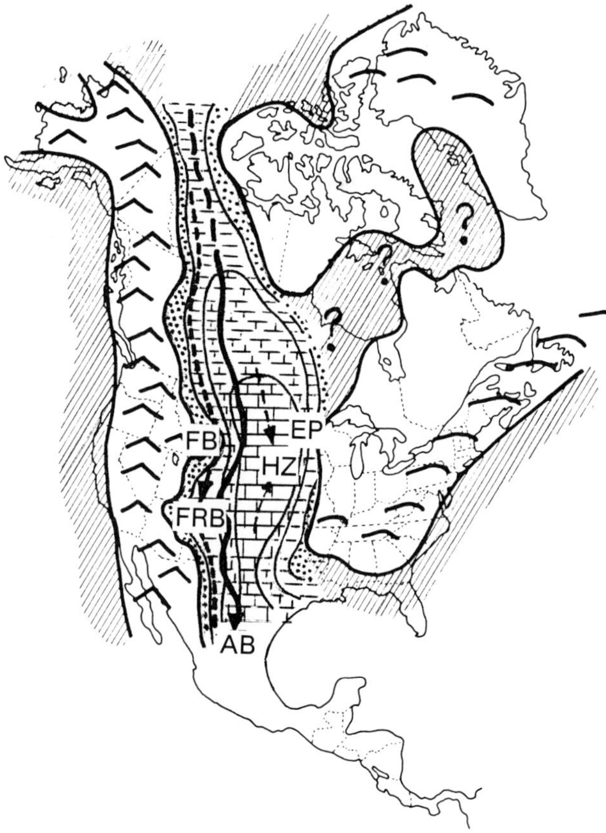

Figure 1. Extent of the Western Interior Cretaceous Seaway, and major facies distribution patterns, within each of the four major tectonic zones of the basin during peak Lower Turonian eustatic rise and transgression. Modified from Kauffman (1969), and Williams and Stelck (1975, Fig. 5). Key: *AB*-Axial Basin; *EP*-Eastern Stable Platform; *FB*-Foreland Basin; *FRB*-Forebulge zone; *HZ*-Hinge Zone.

suggest a maximum width of nearly 1000 miles (1620 km). The size and shape of the seaway changed in response to tectonoeustatic, tectonic and sedimentologic processes (e.g., delta-building). The major elongate trough of the seaway occupied a tectonically generated foreland basin (Price, 1973; Jordan, 1981). To the west lay the great Cordilleran thrust belt, associated with numerous intrusive bodies, and with volcanic centers localized in Idaho-Montana and New Mexico-Arizona. To the east lay the broad, planed, stable cratonic platform. The sedimentary basin was asymmetrical both in terms of subsidence and sedimentary thickness, and in terms of paleobathymetry (Fig. 2; Kauffman, 1977a). In an east-west transect, the Western Interior Cretaceous Basin was divisible into four tectono-sedimentologic "provinces" (Fig. 2; Kauffman, 1977a): (1,2) A pair of western to west-central, rapidly subsiding troughs, separated by a forebulge zone, with a steep western slope and moderately sloping eastern edge: these are (1) the tectonically generated western foreland basin containing thick coarse clastics (Jordan, 1981), and; (2) an axial basin, the zone of maximum subsidence and the deepest part of the seaway at maximum flooding; (3) an east-central, temporally shifting, tectonic hinge zone between the foreland basin and stable platform, characterized by episodic block uplift or isostatic flexure, intermittent basin-swell topography, and rapidly changing fine-grained marine facies cut by complex regional disconformities; and (4) a stable eastern platform zone characterized only by broad subsidence-uplift episodes, widespread uniform, fine-grained clastic and carbonate facies, and shallow, mainly photic water depths.

The seaway was initially flooded from the north end in the Aptian Stage; during the Aptian-early Late Albian stages an extensive northern arm of the seaway (Fig. 3; McGookey *et al.*, 1972; Williams and Stelck, 1975), and a minor southern arm encroached slowly and irregularly into the basin, meeting in middle Late Albian time in central and southern Colorado (the Kiowa-Skull Creek eustatic cycle; Kauffman, 1969, 1977a). After a brief, partial, Late Albian retreat, the two arms of the seaway once again joined in latest Albian-earliest Cenomanian time and remained as a continuous marine system between the proto-Gulf of Mexico and the Circum-Boreal Seaway for nearly 35 Ma, draining from the interior for the last time in the Middle Maastrichtian ("Fox Hills time"; Waage, 1968) but with a remnant arm persisting into the Middle Paleocene (Cannonball Sea). During latest Albian-Middle Maastrichtian time, however, five major third-order tectonoeustatic cycles and numerous important fourth-order transgressive-regressive cycles altered the shape, bathymetry, and sedimentational patterns of the basin (Kauffman, 1969, 1977a); all third-order and some fourth-order cycles can be linked, through precise biostratigraphic-geochronologic correlations, to global tectonoeustatic sea-level changes (Kauffman, 1977a, 1979; Hancock and Kauffman, 1979; Vail *et al.*, 1977). Peak rises of sea-level recording third-order cycles in the Western Interior basin occurred in the middle Late Albian (Kiowa-Skull Creek Cycle), the middle Early Turonian (Greenhorn Cycle), the Lower Coniacian-Late Santonian (Niobrara Cycle, containing four discrete fourth-order cycles), the late Early Campanian (Claggett Cycle), and the middle Late Campanian stages (Bearpaw cycle) (Kauffman, 1977a). Fourth-order cycles thought to be related to smaller sealevel changes, or stillstand events, are expressed as smaller

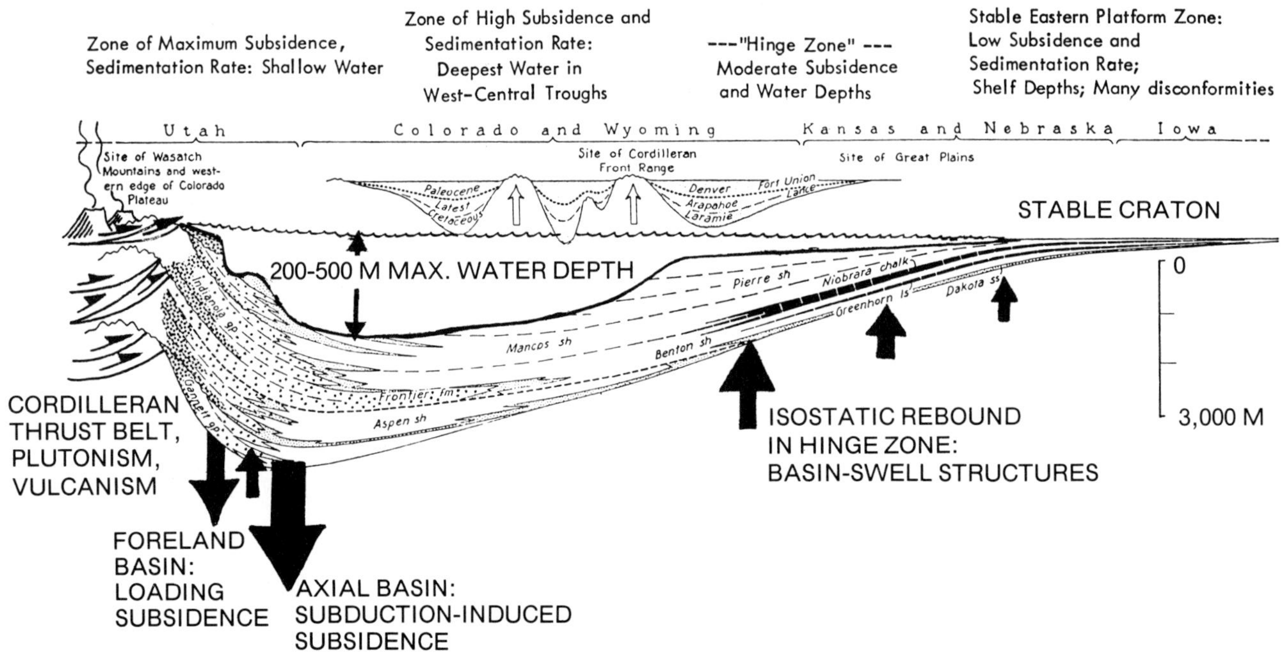

Figure 2. Generalized structural and stratigraphic cross-section through the Western Interior Cretaceous Basin (modified from Kauffman, 1977a), modeled after a peak transgressive phase and eustatic high-stand, showing the major structural divisions of the basin, sediment thicknesses and facies characteristic of each division, and relative paleobathymetry. Position, direction, and size of arrows indicate relative thrusting, subsidence, and uplift at the western margin and within the basin during periods of eustatic rise.

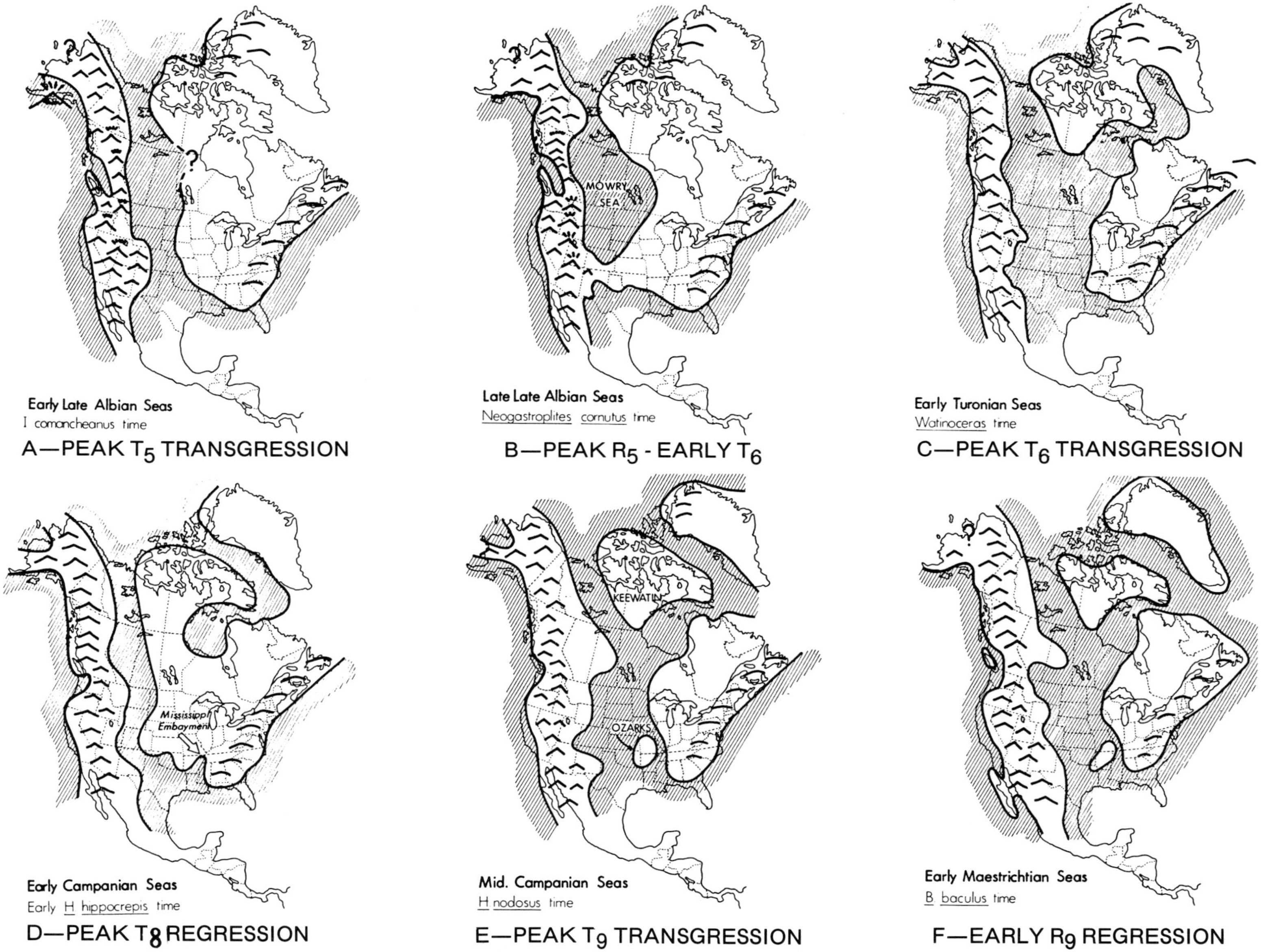

Figure 3. Generalized maps of North America showing extent of Western Interior Cretaceous Seaway during major third-order eustatic fluctuations (transgressions and regressions, utilizing Kauffman's 1977a nomenclature), in temporal progression from A – F (oldest to youngest). Maps from Williams and Stelck, 1975, slightly modified.

regional strandline fluctuations, progradational events, and/or regional disconformities. These sealevel fluctuations, coupled with related tectonic modification of the foreland basin, exercised the principal control on climate and biogeographic distribution in the Western Interior Seaway through time.

Tectonoeustatic sealevel changes within the Western Interior Basin are expressed as regionally correlative migrations of strandlines (e.g., Weimer, 1959, 1960; Gill and Cobban, 1973) and basinal sedimentary packages which, when fully preserved, consist of an upward-fining and/or more calcareous facies-suite overlain by a lithologically very similar, but reversed, upward-coarsening facies-suite (Figs. 5, 6, 9; Hattin, 1964; Kauffman, 1967, 1969, 1977a). Collectively, the strandline migrational surfaces (disconformities) and the sediment-suites comprise the marine cyclothem for each fluctuation of sealevel. Internally, each cyclothem bears lithologic, paleobiologic, and geochemical evidence of major changes in climate and currents, water stratification, temperature, and chemistry of the epicontinental sea which had a profound effect on the marine biota and its patterns of distribution through time.

Detailed interpretation of the eustatic, tectonic, sedimentologic, and ecologic-paleobiogeographic evolution of the Western Interior Cretaceous Seaway and the correlation of these events regionally to establish the magnitude of causal phenomena, demand a complex and refined system of high-resolution stratigraphy with world-wide application. We are fortunate that the Cretaceous System of correlation established for the Western Interior Seaway is perhaps the most refined anywhere in the world, combining isochronous surfaces of over 600 layers of volcanic ash and other "geologically instantaneous" events with a rapidly expanding bank of geochronologic data (> 50 K-Ar dates/40 Ma), an emerging and highly precise stable-isotope chemostratigraphy, and a biostratigraphic system of interval and assemblage biozones with resolution averaging 0.25-0.33 Ma/zone for most of the Cretaceous sequence.

Tectonic Framework

The Western Interior epicontinental sea occupied an Andean-type foreland basin, the origin and structural history of which was controlled by: (a) plate-tectonic activity, timing and intensity along the subduction zone of the Pacific coast of North America; (b) the timing and magnitude of regional compressional (thrust) tectonics in the western Cordilleran orogenic belt; and (c) the structural integrity of the crust in front of the Cordilleran thrust belt. All of these factors worked in concert to shape the Western Interior Basin, the history of epicontinental marine incursions, and biogeographic evolution of the seaway.

Figure 2 shows the major tectonic divisions of the basin (after Kauffman, 1977a) during the Cretaceous, from west to east: (a) the tectonically active, topographically elevated western belt of thrusting, intrusion, and vulcanism, which episodically provided immense amounts of terrigenous clastic sediments to the rapidly subsiding western trough of the seaway. Price (1973) and Jordan (1981) demonstrated that this trough was formed by crustal loading through eastward pileup of thrust plates, and accumulation of thick, predominantly coarse-terrigenous clastic sequences east of the thrust belt. Coastal-plain and predominantly shallow-water marine sequences characterized the western margin of the epicontinental sea, indicating that subsidence responded in proportion to tectonic plus sedimentologic loading of the crust. (b) The west-central trough along the basin-axis in which, regionally, the thickest accumulation of marine sediments developed, generally medium to fine-grained clastics (McGookey et al., 1972, Figs. 31, 32, 35, 43). This was the deepest part of the basin, based on slope analysis and bathymetric indicators among the biota, during sealevel rise and highstand. Rapid subsidence of the basin along its axis, with subsidence outstripping sediment loading, is thus indicated; this center of subsidence was outboard of predicted basinal development due to crustal and resultant sediment loading (Jordan, 1981), and thus a distinct, more remote mechanism for mid-basin subsidence must be called upon, possibly, some kind of a "drag effect" related to rapid, shallow plate-subduction originating at the Pacific margin of North America. Separation of the thick sedimentary sequences of the foreland trough from those of the basinal axis by a zone of thinner sedimentary sequences suggests some kind of foreland-bulge between them. Sedimentation patterns indicate that subsidence throughout these depositional basins was intermittent and separated by times of basin filling.

(c) The broad, migrating, east-central tectonic hinge zone (Fig. 2) bounded on the west by a moderately inclined slope descending into the axial basin, along which rapid facies change took place from thin sequences of eastern platform carbonates and clays cut by numerous regional disconformities to thick westward sequences of clays, silts and sands lacking major disconformities (Fig. 2; and McCookey *et al.*, 1972, Fig. 43). Subsidence was moderate and episodic west of the hinge zone, and occurred sporadically at very low levels on the eastern platform. The hinge zone was a region of tectonic adjustment to pulses of subsidence and loading in the basins to the west. It was characterized by periodic development of block uplifts separated by smaller basins. Analyses of stratigraphic thicknesses across this hinge (from Hattin, 1962, 1964, 1965, 1975, 1982, for Kansas; Kauffman, 1969, 1977b; Kauffman *et al.*, 1969; Scott, 1969; and Cobban and Scott, 1972 for Colorado), shows thinning of certain stratigraphic intervals, but not others, across basement blocks of the hinge zone; this dates their movement as correlative with episodes of thrusting to the west and with times of eustatic sealevel rise (Fig. 4).

(d) The tectonically stable platform occupied the eastern one-third of the seaway. During maximum transgression the seaway inundated the stable craton and the Transcontinental Arch. Subsidence was low and episodic, affecting broad areas, and sedimentation was characterized by slow accumulation of pelagic carbonates and fine-grained terrigenous clastics in low-energy but predominantly photic marine environments. Coarser clastics (sand) were thin and mature, and restricted to very nearshore facies.

Many aspects ofthe tectonic history in the Western Interior Basin may have affected the geographic distributional patterns of Cretaceous marine organisms. (a) Rates, timing, and magnitude of subsidence events controlled water depth, water mass distribution, and the current systems involved in dispersal and the development of habitats for marine organisms. Active subsidence and deep water occurred only during eustatic rise of sealevel; basin filling, reduced subsidence, and a predominance of shallow water habitats, characterize regression and eustatic fall. Varying patterns of subsidence along the basinal axis determined north-south continuity of depth-zones and water-masses. (b) The size and shape of the basin and relative constriction of its apertures (Williams and Stelck, 1975, Text-figs. 3-6) influencing current and tidal circulation, were controlled by sealevel changes, crustal shortening, block-uplift of large parts of the basin, and/or emplacement of tectonic blocks in basinal apertures. (c) Rainfall and evaporation levels in the Western Interior seaway controlled surface salinity and were affected by episodic development of extensive mountain ranges along the western side of the basin through thrusting and uplift. These elevated ranges caused extensive updrafts, air turbulence, and greater precipitation than to the east. Increased runoff from these ranges intermittently produced brackish surface-layers on the seaway, restricting planktonic dispersal. Intermittent reduction of those ranges would have greatly altered climatic patterns toward drier phases. Similarly, episodes of massive volcanic eruptions potentially altered the atmosphere of the Western Interior in terms of cloud-cover, heat-penetration, and thus annual temperature and rainfall – critical factors in the Cretaceous history of biogeographic dispersal as they are in modern marine environments (Kauffman, 1975). Over 600 volcanic eruptions are recorded in middle and Upper Cretaceous marine

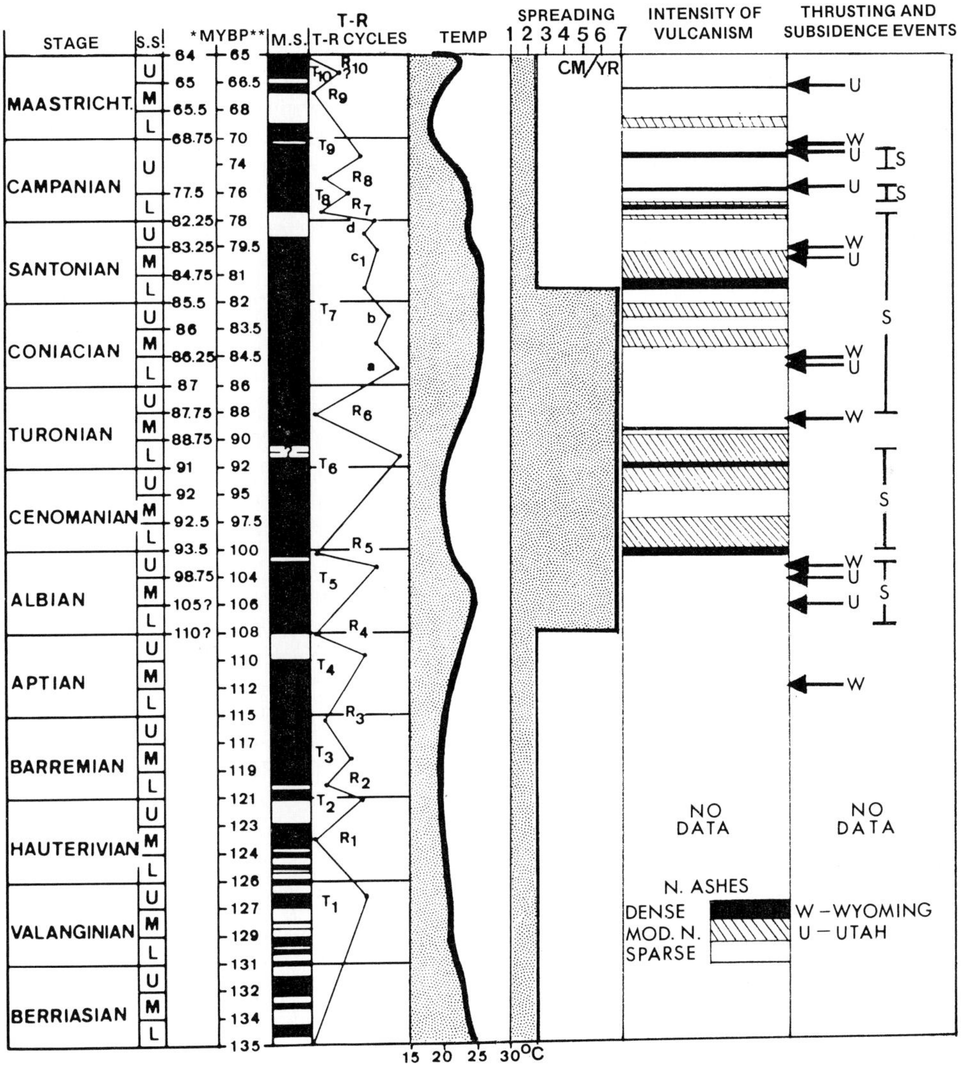

Figure 4. Geochronologic and magnetostratigraphic dating of global sealevel changes; third-order transgressive-regressive (T-R) epicontinental cycles (from Kauffman, 1977a); average marine temperature fluctuations (averaged from data in Kauffman, 1977a); Atlantic and Pacific half-spreading rates averaged (Larson and Pitman, 1972; Sheridan *et al.*, 1982); episodes of intense (black) and moderately intense (striped) volcanism, measured by number and magnitude of volcanic ash events (partially from Kauffman *et al.*, 1976); initiation of major thrusting (arrows) in Utah (U) and Wyoming (W) (data from Villien and Kligfield, pers. commun., 1984, and Wiltschko and Dorr, 1983, respectively); and times of most active basin subsidence (S), determined stratigraphically, in the Western Interior Seaway. Note correlation of active tectonism, subsidence, and volcanism with periods of active plate spreading and global sealevel rise, as contrasted to periods of relative tectonic quiescence during sealevel fall. Generalized time scales from Kauffman (Western Interior Seaway, left column, 1977a) and Van Hinte (global; right column, 1976).

strata of the Western Interior by 0.1 cm-3 m thick ashes and bentonites. (d) Episodes of thrusting and uplift are marked by massive influx of terrigenous clastics, and with them greater turbidity, varying chemistry, and fluctuating water-depths affecting biotic migration over broad areas.

From these observations a synthetic theory for the dynamics of the Western Interior foreland basin can be developed and applied to the interpretation of Cretaceous paleobiogeographic patterns, as follows. Acceleration of rates of seafloor spreading in the Cretaceous was associated with development of topographically positive regions on the ocean floor, causing global tectonoeustatic sealevel rise up to 300 metres and widespread, synchronous, epicontinental transgression. Active spreading also correlated with increase in subduction rates along the Pacific boundary of North America. This activated compressional deformation in less stable, inboard portions of the craton, resulting in the development of an asymmetrical foreland basin bounded on the west by linear fold and thrust belts, intrusive igneous bodies, and active vulcanism. Coincident subsidence, eustatic drowning and greatly increased sedimentation in the Western Interior epicontinental basin was predictable. Water-depth in this seaway depended upon rates of eustatic rise and basin subsidence balanced against rates of sedimentation. High-resolution correlation of events in the Western Interior Cretaceous Basin clearly supports this theory (Fig. 4). Periods of active Cretaceous plate spreading (Rona, 1973, Pitman, 1978) bracket both the interval of major thrusting, uplift, and volcanism in the Western Interior Basin, and the interval of greatest eustatic drowning, subsidence, and marine sedimentation (Kauffman *et al.*, 1979b, Fig. 2). All major thrusting events (dated by large-scale movement on thrusts and resultant sedimentary events), and the majority of episodes of intense volcanism (measured by numbers, spacing, thickness and distribution of volcanic ashes) are precisely correlative to episodes of marine transgression and eustatic rise (Fig. 4). Episodes of rapid basinal subsidence preserving the thickest coarse clastic sequences in the foreland trough, and continuous transgressive sediment packages recording maximum water depth in the axial basin, are all associated with sealevel rise and transgression. Contemporaneous subsidence in the central basin exceeded sedimentation, protecting fining-upward transgressive sequences from erosion and creating the deepest water along the basinal axis; consequently clastic wedges emanating from the western thrust belt during sealevel rise do not cross the basinal axis (see McGookey *et al.*, 1972, Fig. 43). The relationship between regional tectonic activity, active volcanism, extensive basin subsidence, rapid sedimentation, and sealevel rise holds for all five of the major third-order tectonoeustatic pulses in the Western Interior Cretaceous (Fig. 4).

The second part of the theory predicts that during regressive phases of each third-order eustatic cycle, slowing or cessation of seafloor spreading and subduction resulted in low levels of epicontinental tectonism – thrusting, folding, subsidence, intrusion and vulcanism – initiating destruction of Cordilleran ranges through erosion. Falling sealevel, epicontinental regression, and slowing of tectonically generated basinal subsidence resulted in filling of the epicontinental basin with progressively coarser sediment; basinal subsidence during regression was primarily related to sediment-loading. The history of the Western Interior Basin supports this theory. No major thrusting or intrusive episodes, and less than 25 percent of the volcanic ash deposits are associated with regional marine regression and eustatic fall. Rates of sedimentation near the thurst belts generally decline, but the central part of the basin fills with sediment to the point where large scale progradational events produce sand bodies that cross the former sites of the foreland and axial basins without reflecting any topographic depression during regression (Fig. 2; and McGookey *et al.*, 1972, Fig. 43).

This tectonic theory predicts, therefore, that: (a) immigration of taxa into the Western Interior Basin from both northern Cool Temperate and southern Warm Temperate to Subtropical paleobiogeographic sources, (b) establishment of broad biogeographic ecotones, and (c) rapid shifts in paleobiogeographic boundaries within the basin, would be possible mainly during transgression and eustatic highstand. During these intervals the size and depth of the seaway, and the diversity and quality of paleoenvironments would be favorable for development of diverse biotas, high productivity, and establishment of migratory pathways at several levels within the marine water column. This favored widespread migration of permanent plankton, nekton, and of the planktotrophic larvae of many benthic marine organisms. Paleobiogeographic data presented subsequently wholly supports this interpretation.

Conversely, eustatic lowering produced regressive phases of the Western Interior Seaway, constriction of basinal apertures, basin filling, decrease in basinal size, restriction and reduction of both depth-controlled and spatially-controlled niches for marine organisms. These greatly diminished opportunities for immigration and/or spread of biotas from adjacent biogeographic units, especially those of deep- and warm-water environments. Regressive phases of basinal development would thus favor expansion of endemic shallow-water taxa with Temperate zone affinities, and with preferences for clastic substrates. Regional retreat of southern paleobiogeographic assemblages and increased endemism during eustatic fall are predictable and observed within the basin.

Patterns of Sedimentation

Patterns of sedimentation in the Western Interior Seaway, depicting the evolution of pelagic and benthic environments and niche diversity, were closely linked to both regional tectonic and tectonoeustatic controls (Kauffman, 1977a). At peak Cretaceous flooding (e.g., Early Turonian; Coniacian-Santonian) all major marine facies were developed in the basin; niche diversity was high and many habitats were extensively developed – a favorable setting for immigration and proliferation of biotas having diverse paleobiogeographic origins. Kauffman (1967, 1969, 1977a) described and traced 12 major marine facies through five third-order tectonoeustatic cyclothems of the Western Interior Basin, and showed them to contain diverse, benthic, marine paleocommunities of both northern and southern origin. Detailed biostratigraphic and geochronologic correlation of transgressive and regressive peaks of Western Interior cyclothems to similar peaks in Western Europe and other Cretaceous areas of the world by Hancock and Kauffman (1979) and Kauffman (1979) proves their tectonoeustatic origin.

Kauffman (1967, 1977a) suggested that the basic sedimentation model for a single tectonoeustatic cycle in rapidly subsiding basinal settings of the Western Interior Seaway was that of a symmetrical cyclothem (Fig. 5). This model contains 12 major marine and marginal-marine lithofacies arranged in an upward-fining sequence, from nearshore sands to offshore pelagic carbonates, during transgression (eustatic rise); essentially the same facies-suite occurs in reverse order (upward coarsening) during regression and eustatic fall. The model proposes diachronous transgressive and regressive disconformities bounding each cyclothem shoreward, with gradual shoreward loss of pelagic carbonates and shales but lateral retention of the basic symmetry of the cyclothem. In this eustatic model points of maximum transgression of the eastern and western strandline facies correlate precisely with the point of maximum development of pelagic carbonates in the upward-fining basinal sequence in both third- and fourth-order cycles.

Several modifications of this model have been suggested by subsequent work, in each tectonic zone of the basin. (1) Along the western margin of the seaway and in the rapidly subsiding foreland basin the cyclothems are preserved in thick sequences of coarse, continental, marginal-marine, and nearshore marine terrigenous sediments cut by multiple disconformities mainly developed across the top of progradational units. Thin, fossiliferous marine shaley units represent times of peak-transgression and eustatic rise (Ryer, 1977). Third-order cyclothems of this region are broadly symmetrical, but are frequently interrupted by major strandline reversals, representing upward-coarsening progradational events (Ryer, 1977). These strand migrations comprise asymmetrical fourth-order cyclothems with marked transgressive disconformities, little or no preserved transgressive sediment, and thick regressive sequences. Fourth-order cycles

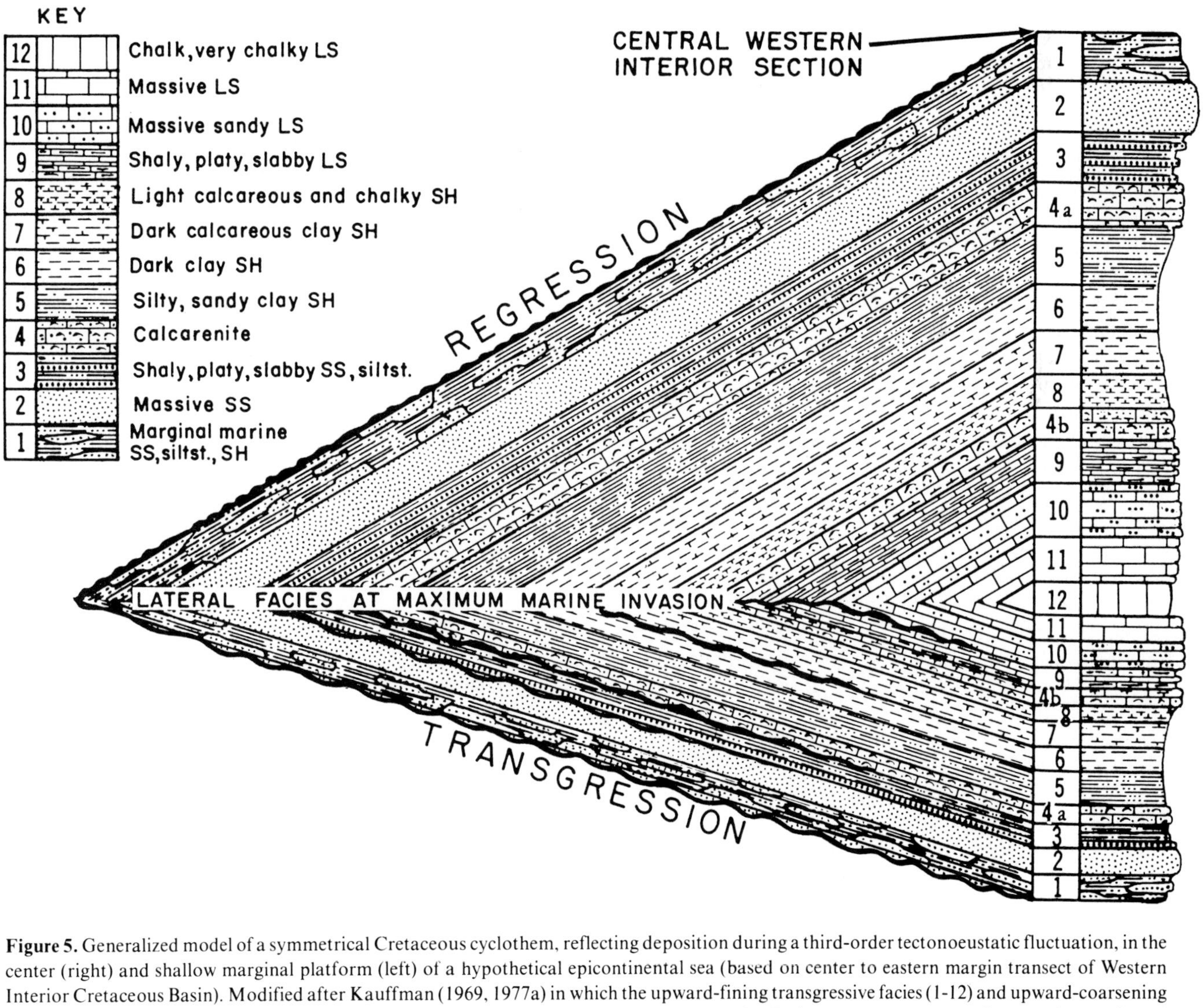

Figure 5. Generalized model of a symmetrical Cretaceous cyclothem, reflecting deposition during a third-order tectonoeustatic fluctuation, in the center (right) and shallow marginal platform (left) of a hypothetical epicontinental sea (based on center to eastern margin transect of Western Interior Cretaceous Basin). Modified after Kauffman (1969, 1977a) in which the upward-fining transgressive facies (1-12) and upward-coarsening regressive facies (12-1), are defined in detail as preserved in the axial part of the Western Interior Basin. Note shoreward-basinward diachronism of facies during transgression-regression, bounding disconformities, and concentration of intra-cycle disconformities on transgressive hemicycle.

are most extensively developed on third-order regressive (eustatic) hemicycles. Some fourth-order cyclothems also appear to be linked to eustatic fluctuations or still-stand events (e.g. progradational sequences); others were caused by regional tectonic, subsidence, and sedimentation events.

(2) In the deep, west-central, axial part of the basin the thickest and most diverse sequence of marine sediments is preserved (McGookey *et al.*, 1972, Figs. 30-35, 43) and characterized by dark silty clay and calcareous clay-shales, intercalated with thinner distal (basinal) extensions of pelagic carbonate units from the eastern shelf during peak transgression, and with basinal extensions of prograding sand-dominated facies from the west during regression. These cyclothems are basically symmetrical with upward-fining (and/or more calcareous) facies during eustatic rise and transgression, and with upward-coarsening sequences during regression and eustatic fall. The cyclothems become somewhat asymmetrical on the eastern flank of the basin where regional disconformities of small duration cut mainly the transgressive sequences, increasing in magnitude eastward over the hinge zone and onto the stable platform. It is from Cenomanian through Santonian rocks in this region that Kauffman (1967, 1977a) developed the cyclothemic model for the Western Interior Cretaceous Basin (Fig. 5). The symmetry of the cyclothems in the axial part of the Western Interior Basin is strong evidence for rapid subsidence during transgression which exceeded tectonic and sediment loading, dropping transgressive sequences rapidly into deep-water zones before they were extensively eroded. However, even the most symmetrical mid-basin cyclothems have significantly different coarse-clastic sequences bounding them. Transgressive sandstones and calcarenites track the transgressive disconformity in a narrow belt of nearshore facies, with few progradational sequences; none of these progradational sequences cross the axis of the depositional basin, and most do not even reach the basinal axis, confirming that this was the deepest part of the seaway during eustatic rise (see McGookey *et al.*, 1972, Fig. 43). Regressive coarse-clastics in the cyclothems, however, are characteristically more abundant and occur as widespread progradational facies which extend across the depositional axes of the basin without significant change, in some cases reaching the hinge zone and eastern platform. This confirms basinal filling during regression and eustatic fall, with subsidence barely keeping up with sedimentation (e.g., McGookey *et al.*, 1972, Fig. 43; Gill and Cobban, 1973).

(3) The east-central tectonic hinge zone of the Western Interior Basin is characterized by extensive facies change and rapid eastward thinning of dominantly fine-grained marine sedimentary sequences (Fig. 2; McGookey *et al.*, 1972, Fig. 43). Sedimentary units comprising transgressive sequences are variable in thickness across the hinge as a result of active basement flexure and block uplift during eustatic rise and, thus, numerous disconformities are developed which cut the sequences on the transgressive hemicyclothems and at peak regression. Thus, the cyclothems are sub-symmetrical, with transgressive sequences significantly shortened by these disconformities but with little change in facies sequences (Kauffman, 1967, 1977a).

(4) The eastern platform zone encompasses one-third of the Western Interior Basin (Fig. 1), and is characterized by thin sedimentary sequences dominated by clayey, silty, and calcareous shales, and by pelagic carbonates deposited in shallow (photic) waters. Thin mature sands abound in the nearshore facies, suggesting limited clastic sources on the broad flat eastern coastal plain. Cyclothems are subsymmetrical near the hinge zone, with thin discontinuous transgressive sequences cut extensively by regional disconformities, bypass surfaces, and condensation zones; third-order cyclothems become asymmetrical shoreward (Hattin, 1975) with nearly complete regressive sequences overlying extensive transgressive (ravinement) disconformities preserving little or no transgressive sediments other than lag deposits. The extensive development of pelagic carbonates composed of diverse calcereous microplankton in photic shelf environments seems unusual. This probably reflects exceptional warmth and clarity of the shallow eastern platform seas during the Cretaceous, and also suggests different ecological ranges for certain Cretaceous calcereous plankton than documented for their modern counterparts. Extensive immigration of Warm Temperate to Subtropical calcereous plankton into the seaway, including the larvae of many benthic organisms, appears to have taken place rapidly across the shallow eastern platform during peak rise of sealevel.

Variations in the development and/or preservation of Cretaceous cyclothems in the basin reflect secondary tectonic, sedimentologic, and oceanographic influences on the expression of global eustatic events, not predominant regional control on development of these cycles. Paleobathymetric analyses and facies distribution within these Cretaceous cyclothems indicate that deep-water migration routes for northern and southern biotas developed in the central and west-central parts of the basin only during eustatic sealevel rise and transgression. Extensive development of pelagic carbonates suggests that peak transgression was characterized over parts of the basin by the immigration of warm water masses, and by increases in salinity and biotic productivity. Changes from fine-grained clastics to pelagic carbonates signals the point of change in water mass and predicts a shift from Temperate to Subtropical climates and marine environments of the upper water column. Favorable benthic habitats expanded during transgression; transgressive events were times of maximum opportunity for immigration and exchange of biotas within the seaway, especially from southern Warm Temperate to Tropical paleobiogeographic units. Conversely, shallowing, cooling, and decrease in salinity characterize regressive facies and eustatic fall; these conditions favored limited colonization by shallow-water Temperate and endemic marine biotas. Southward shift of warm-water paleobiogeographic units is predictable from facies analysis during marine regression.

Paleoceanography of the Western Interior Seaway

Tectonic and sedimentologic patterns within the Western Interior Seaway also reflect major oceanographic fluctuations which exercised primary control on the paleobiogeographic distribution of marine organisms. The most obvious of these are tectonoeustatic changes of sealevel, which caused broad transgressive-regressive migrations of the strand-line, major changes in shape, size, and continuity of the seaway, and alternating episodes of deepening and shoaling. Eustatic effects were greatly enhanced by coincident tectonic activity which exercised strong controls on the orientation, physiography, and paleobathymetry of the seaway. Five major third-order tectonoeustatic cycles (Vail *et al.*, 1977), averaging about 9 Ma in duration, and more numerous fourth-order cycles depicting relative sealevel changes and eustatic stillstand events, characterize the paleoceanographic history of the Western Interior Seaway (Fig. 6). All third-order and some fourth-order cycles have transgressive and regressive peaks which are correlative worldwide within one or two regional biostratigraphic zones of 0.2 to 0.5 Ma duration, proving their tectonoeustatic origin (Kauffman, 1977a, 1979; Hancock and Kauffman, 1979). Widely correlative transgressive peaks in the middle Late Albian (Kiowa-Skull Creek Cyclothem), middle Early Turonian (Greenhorn Cyclothem), Lower Coniacian through Late Santonian (Niobrara Cyclothem; divided into four fourth-order eustatic pulses); late Early Campanian (Clagget Cyclothem), and middle Late Campanian (Bearpaw Cyclothem) are expressed in the centre of the basin by maxima in upward-fining clastic sequences and/or deposition of pelagic carbonates, and marginally as maximum shoreward migrations of the strand line which overprint virtually all regional tectonic, sedimentation, and subsidence phenomena. Hancock and Kauffman (1979, Fig. 5) and others propose maximum middle and Late Cretaceous elevation of sealevel during peak rise in Euramerica as about 300 m.

These large tectonoeustatic fluctuations regulated the paleoceanography of the Western Interior Seaway. During each eustatic rise the following sequence of changes can be generalized from sedimentologic, geochemical, and paleobiologic data. (1) With initial transgression, limited interconnection of separate northern Cool Temperate and southern Warm Temperate to Subtropical water masses, allowed immigration and extensive mixing of northern and southern marine biotas, and established broad biogeographic ecotones documented by the geographic extent of mixing of characteristic northern and southern taxa (Sohl, 1967, 1971; Kauffman, 1967, 1973, 1977a, 1979). Northern taxa migrated as far south as Texas (Stephenson, 1952), and Subtropical taxa as far north as central Saskatchewan (Douglas, 1972; Caldwell and Evans, 1963). The Tropical reef line seems to have shifted north into Texas with each major eustatic rise. (2) Enlargement and progressive deepening of the epicontinental seaway, reflecting eustatic rise, accelerated subsidence and deepening along the basinal axis. Major northern and southern marine currents moved into the Western Interior Basin (Kauffman, 1975), overriding constrictions at the northern and southern apertures of the seaway (narrow necks and shallow sills) that earlier had prevented free circulation of tides, currents, and immigration of many biotic constituents from the Gulf of Mexico and Circumboreal Sea. (3) With widespread flooding and open-marine circulation in the basin around peak transgression, a regional maritime climate developed with warming temperatures, rising surface salinity, and declining seasonality. These changes were accompanied by extensive northern migration of warm waters, and rapid northward shifts of Warm Temperate to Subtropical biotas of the Caribbean, Gulf of Mexico, and southern Western Interior, reflecting rapid, short-term (1 to 3 Ma) changes in watermass distribution, currents, temperature and salinity which occur near the eustatic highstand for each cycle. (4) Coincident draining and shoaling of the Western Interior Basin occurred during eustatic fall and regression (average 4 to 4.5 Ma duration). The size of the basin was reduced, the maritime climate disrupted, and basinal circulation became restricted as barriers to incursion of large currents from the north and south were re-established. With these changes, cooling, greater seasonality, increased rainfall, and decreased surface salinity in the seaway predictably forced warmer water, more normal marine biotas southward toward the Gulf of Mexico and Caribbean, leaving behind more eurytopic Cool- to Mild-Temperate biotas.

This oceanographic model is supported by new geochemical and paleobiological data. Two powerful geochemical tools have been applied recently to the interpretation of water chemistry, climate and temperature in the Cretaceous seaway of North America: (a) state isotope geochemistry, primarily involving the use of carbon ($\delta^{13}C$) and oxygen ($\delta^{18}0$) isotopes to sense temperature and/or salinity levels through time, and (b), along with $\delta^{13}C$ analysis, the use of organic carbon (C_{org}) values to sense original environments of deposition – the preservation of organic carbon is directly related to dissolved-oxygen levels within the water column and benthic substrate. C_{org} level is also a rough measure of biological productivity. Temperature, salinity, and oxygen levels have strong controls on the dispersal of marine organisms, ecosystem structure, and thus on the characteristics of paleobiogeographic units.

Scholle and Arthur (1980) and Pratt (1983; and pers. commun., 1983) generated detailed stable isotope curves for the Western Interior Cretaceous based on $\delta^{18}0$ and $\delta^{13}C$ analysis of carbonate rock fragments and calcite prisms from inoceramid bivalve shells. These curves span the Greenhorn and Niobrara Cyclothems with data largely drawn from the central and eastern part of the seaway. These data (Fig. 6) demonstrate fluctuations in aquatic chemistry of the Western Interior seaway that profoundly affected the dispersal of marine organisms (Fig. 6).

Analyses of oxygen and carbon isotopes show unusually negative values for both $\delta^{13}C_{org}$ and $\delta^{18}0$ through most of the transgressive-regressive history of the seaway, compared to modern marine systems. Negative $\delta^{13}C_{org}$ values range between -25 and -27 (°/oo *vs.* PDB) during all but peak-transgressive phases (eustatic highstand) of basin history. Peak-transgressive values climb slightly to -23 (°/oo *vs.* PDB). $\delta^{18}0$ values range from -7 to -11 during most of the basin's marine history, climbing to -3 to -5 (°/oo *vs.* PDB) at peak-transgression and eustatic highstand. Values around -3 $\delta^{18}0$ are considered to represent near-normal marine (saline) conditions for epicontinental Cretaceous seaways (M.A. Arthur, pers. commun., 1983). Interpreted as temperature, $\delta^{18}0$ fluctuations indicate temperature range and projected values which are much too severe for the biota associated with the samples, so that salinity fluctuations are called upon to explain these values.

Interpreted as a signal of salinity fluctuation, these values suggest slightly to moderately brackish conditions during most of the marine history of the seaway, especially in the upper part of the water-column. A widespread brackish cap probably developed here as a result of internal freshwater runoff which

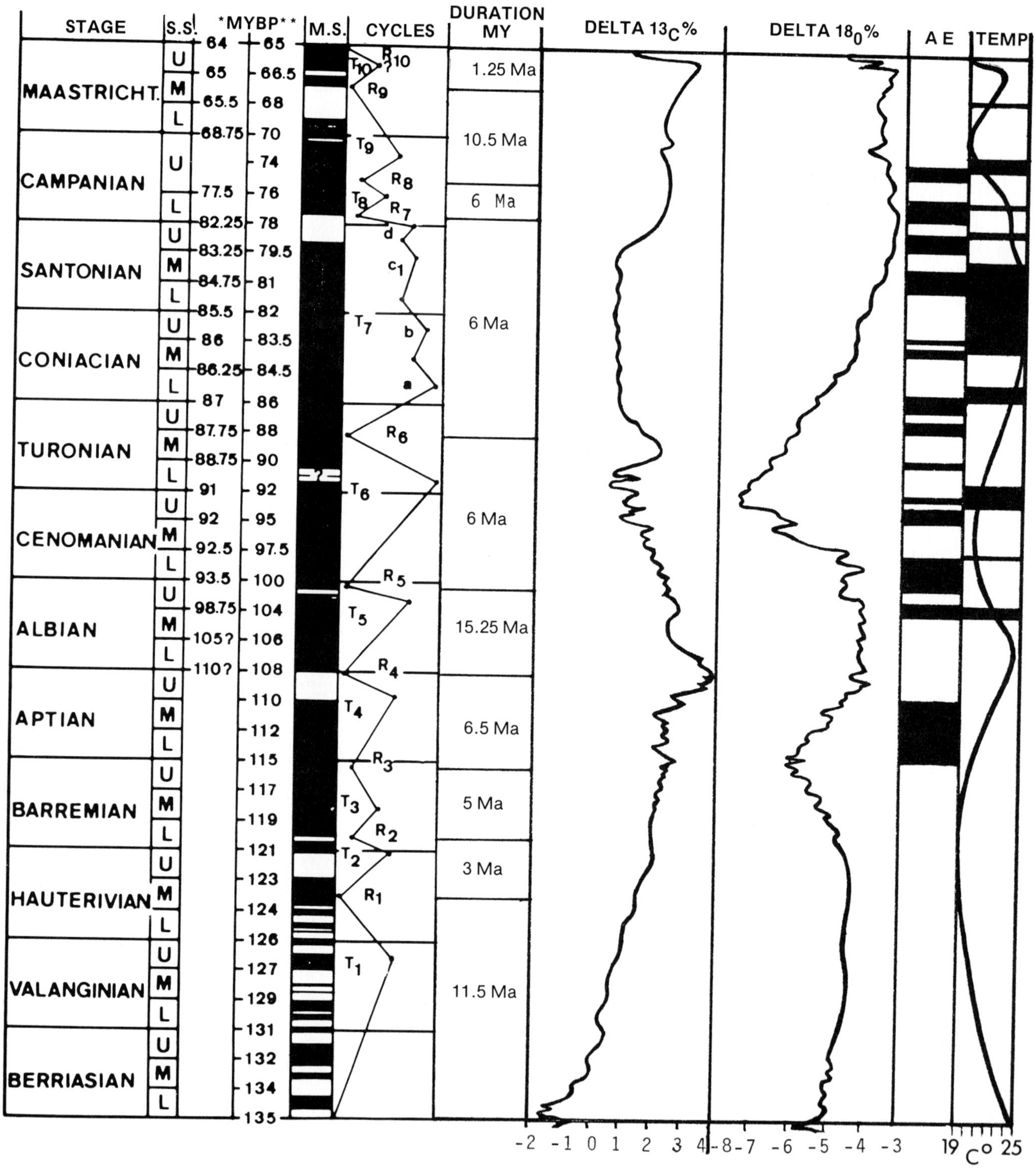

Figure 6. Global third-order tectonoeustatic cycles (T_1-R_1) plotted against magnetostratigraphic and radiometric standards (from Kauffman, 1977a; see explanation, Fig. 4 herein). T_5-T_{10} widely inundated the Western Interior Cretaceous Basin. Note radiometric times of peak transgressive and regressive events, cycle durations, and major paleoceanographic fluctuations as defined by geochemical profiles ($\delta^{13}C_{arb}$, $\delta^{18}O_{org}$). Geochemistry generalized from Scholle and Arthur (1980); Pratt (1983). Right hand columns show anoxic events (AE) and rapid warm-water incursions (TEMP) as black bands, marine temperature curve generalized from data in Kauffman (1977a).

was not offset by tidal and current exchange of normal seawater into the basin. $\delta^{18}O$ isotopes indicate that normal-marine salinity was developed only near eustatic highstand and peak transgression. Consequently, immigration of plankton, nekton, and the planktonic larvae of many normal marine benthic organisms from northern Cool Temperate and southern Warm Temperate to Subtropical subprovinces was greatly restricted during much of the basin's marine history by brackish water in the photic part of the water column. Near peak transgression, isotopic data (Fig. 6) suggest that shifts to normal salinity were relatively rapid, and lasted less than 2 Ma for the Greenhorn and Niobrara Cycles. The rapid normalization of salinity near peak-transgression in the Western Interior Seaway might reflect eustatic rise reaching some threshold point at which warm southern waters were deep enough to overstep barriers to circulation in the southern end of the seaway and thus to flood rapidly into the interior. Oceanographically widespread immigration of normal-marine biotas into the Western Interior Seaway were therefore short lived, and primarily associated with sealevel highstand.

Organic-carbon analyses for the same interval (all data from Colorado and Kansas) demonstrate further restrictions to biotic migration during transgressive episodes which are associated with probable marine oxygen deficiencies. C_{org} curves show relatively high values (2 to 7% by weight) of organic carbon at several intervals during transgression of the Greenhorn and Niobrara cycles. These are commonly linked to episodes of reduced surface salinity determined isotopically; establishment of a subsaline cap on the surface of the Western Interior Seaway resulted in density stratification of the water-column, greatly reducing vertical exchange of oxygen from the surface to the bottom. With reduction or elimination of existing deep-water oxygen through metabolic processes and aerobic bacterial decay, the benthic and lower to middle parts of the water-column became stagnant, greatly restricting immigration and habitation by all but the most adaptive organisms. Reduced oxygen in the seaway is projected for intervals early in transgression, during middle transgression, near peak transgression, and during early regression of the eustatic cycles (Fig. 6). Oxygenation of the seaway to normal levels occurred mainly near peak transgression in association with normalization of surface salinity.

Finally, measurements of marine paleotemperatures can be interpreted from biological evidence and, broadly, from facies (e.g., pelagic carbonates vs. clay shales). These interpretations assume that Cretaceous organisms lived generally within the same thermal regime as their analogs in modern biogeographic provinces (see Hall, 1964). Based on these concepts, Kauffman (1973) provided lists of Cretaceous bivalve genera which characterized each major climatic zone; previously Sohl (1967, 1971) had done this for gastropods. Following these works, this study assumes that Cretaceous Tropical climatic zones were characterized by biogeographic units containing reefs or equivalent frameworks; and other taxa characteristic of modern Tropical carbonate platforms. Subtropical zones were characterized by similar taxa, but without formation of significant biotic frameworks. Warm Temperate environments contained diverse, large, thick-shelled molluscs, but lacked reef-builders and other taxa specialized for habitats of carbonate platforms. Faunas of the northern Gulf of Mexico are comparable. Mild Temperate faunas are interpreted as being moderately diverse assemblages of medium-sized, shelled invertebrates similar to those of the modern Virginian Province. Cool Temperate biotas are characterized by medium to small shelled organisms with moderate diversity. Specific taxa lists are being developed for these divisions.

Based on these comparisons, Cretaceous faunas from the interior of North America indicate that Cool Temperate to Warm Temperate climatic zones prevailed in the Western Interior seaway during most of its history. However, major northward migrations of Warm-Temperate and Subtropical biotas occurred near peak-transgression in all third-order and some fourth-order cyclothems.

Cretaceous Paleobiogeography of the Western Interior Seaway

Modern concepts of Cretaceous paleobiogeography for North America have only developed over the past two decades. Many early paleontologists recognized the unique aspect of faunas of the Western Interior Cretaceous, but none attempted to explain this in paleobiogeographic terms until Sohl (1967, 1971), Jeletzky (1971), Kauffman (1973), and Scott (1977) established the presence of two to three major paleobiogeographic divisions of the seaway, representing average expressions of broad climatic zones (Fig. 7). From north to south, these were: (a) a Cool Temperate biota derived from the northern Circum-Boreal Sea through southward invasion of the "middle" Cretaceous arm of the seaway (Fig. 3; Williams and Stelck, 1975, Text-figs. 2, 4); characteristic taxa of this biota are given by Jeletzky (1971; cephalopods), Sohl (1967, 1971; gastropods), and Kauffman (1973; bivalves). During average climatic conditions for the Western Interior (Temperate Zone-dominated) the northern paleobiogeographic division extended from the Cretaceous Circum-Boreal Sea to the southern borders of Montana, Idaho, and North Dakota (Fig. 7). (b) a Mild Temperate biota found between northern Wyoming – South Dakota, and north-central Utah, northern Colorado, and Kansas (Fig. 7); this division also encompasses most of a broad zone of mixing (ecotone) of northern Cool Temperate and southern Warm Temperate biotas (see Sohl, 1967, Fig. 7). The unique elements of this subprovince are mainly rapidly evolving endemic taxa; high levels of endemism were possibly caused by widespread ecological distruption of northern and southern ecological units (communities) in the broad paleobiogeographic ecotone. (c) a Warm Temperate southern biota derived from the Gulf and Atlantic Coast Subprovince (Kauffman, 1973) during northward invasion of the southern arm of the Western Interior Seaway (Fig. 3). This biota normally occurred between southeastern Utah, northeastern Nevada, northern Arizona, New Mexico, northern Texas, central to southern Kansas, and northwestern Oklahoma (Fig. 7). Typical biotas are described by Scott (1970, 1977).

Paleobiographic concepts and nomenclature have been applied variably to these divisions. Most early authors considered

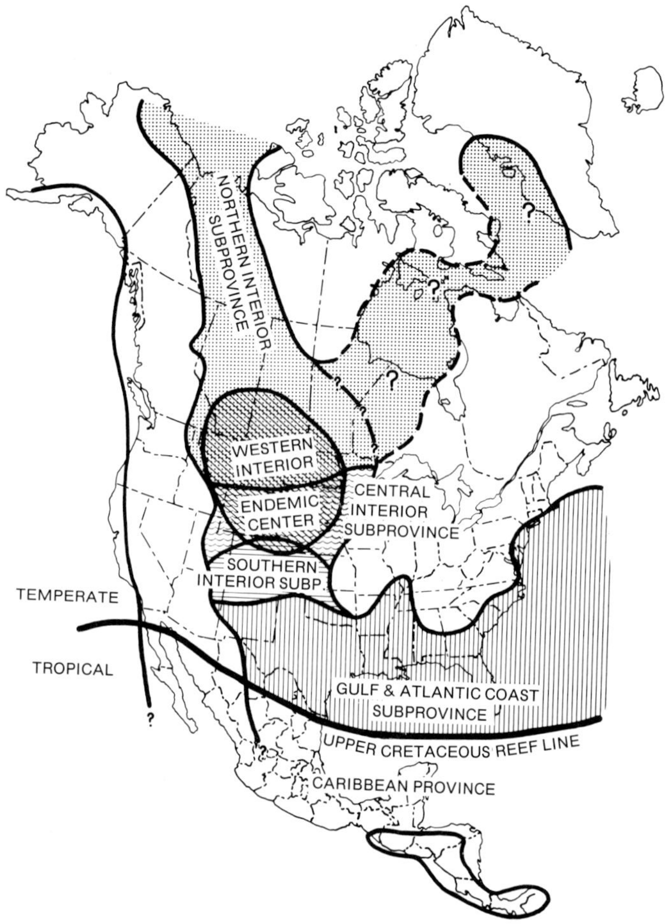

Figure 7. Generalized map of the Western Interior Cretaceous Seaway of North America, showing average distribution of paleobiogeographic units (Subprovinces, Provinces) during early to middle transgression and middle to late regression associated with all third-order tectonoeustatic cycles. Map also shows average extent of paleobiogeographic ecotone (endemic center), within which most rapid evolution took place. Subprovinces and provinces defined on percent molluscan endemism (10 to 25% and 25 to 50%, respectively). From north to south, subprovinces reflect Cool Temperate, Mild Temperate, Warm Temperate, and Subtropical climatic zones, as defined in modern American coastal settings by Hall (1964) and others.

the entire Western Interior Cretaceous biota as all or part of the "North American Province". Jeletzky (1971) recognized "northern and southern divisions" of this biota and Sohl (1971) recognized "northern" and "southern" divisions of the "Western Interior Region". Kauffman (1973), based on measures of endemism for Cretaceous bivalves and comparison of these values to modern biogeographic concepts, determined that bivalve endemism within the Western Interior was only 5 to 10% and not deserving of provincial rank. Kauffman thus named a "Western Interior Endemic Center", extending from the central into the northern part of the basin, and recognized the southern interior biota as a "northern arm of the Gulf and Atlantic Coast Subprovince"; the North American Province was expanded to include the Atlantic and Gulf Coast biota, as well as the Western Interior biota. Subsequent work has demonstrated that the bivalves are biogeographically somewhat atypical of this seaway, and that Western Interior endemism is higher for other groups of invertebrates. This allows a more refined paleobiogeographic division of the seaway. Scott (1977) recognized a Southern Western Interior Province for Warm Temperate biotas of Kansas, Oklahoma, New Mexico, northeastern Arizona, and northern Texas (Fig. 7); calculations of endemism percentages suggest that this division is of subprovincial rank, utilizing the scheme of Kauffman (1973).

In this paper and in Coates *et al.* (1984), a new scheme of paleobiogeographic division is employed, one which is consistent with modern biogeographic concepts and their ancient equivalents (Kauffman, 1973). Figure 7 shows these divisions as averaged for the Western Interior Seaway during most of its marine history but exclusive of peak transgression. Following Jeletzky (1971) with modification, the northern Cool Temperate biota of the seaway is termed the Northern Interior Subprovince; the Mild Temperate biota of Colorado, northern New Mexico, and Kansas is termed the Central Interior Subprovince (new name); and following Scott (1977), the southern Warm Temperate biota is called the Southern Interior Subprovince. All subprovinces, along with the Gulf and Atlantic Coast Subprovince of Kauffman (1973), are placed within the North American Province of Kauffman (1973). Percentages of endemism for genera and species groups range between 10 and 25 per cent for all subprovinces recognized above.

A modified use "North American Endemic Center" (Kauffman, 1973) is applied to the main zone of paleobiogeographic overlap (ecotone) between the Cool Temperate Northern Interior biotas, and the Warm Temperate Southern Interior biotas (Fig. 7; Sohl, 1967, Fig. 7). This zone incorporates most areas of endemism within the Interior Seaway and spans the northernmost edge of the Southern Interior Subprovince, most of the Central Interior Subprovince, and the southern third of the Northern Interior Subprovince. Endemism is particularly evident in ammonite groups like the Gastroplitidae (e.g., Reeside and Cobban, 1960), Scaphitidae (e.g., Cobban, 1951a, 1969), and Baculitidae (e.g., Cobban, 1951b, 1958, 1962a, 1962b); and among bivalves of the Inoceramidae (e.g., the Albian-Cenomanian *"Inoceramus" dunveganensis, "I." athabaskensis*, and *"I." moberliensis* lineages; Kauffman, 1977c). In utilizing the term "endemic center" in this evolutionary sense, I remove it from the hierarchy of paleobiogeographic classification of Kauffman (1973).

South of the Western Interior Seaway and connected to it through a somewhat narrow aperture (Fig. 7) lay the Cretaceous Gulf of Mexico and Caribbean Sea, at the northern edge of, and within the Tropical Tethyan Realm, respectively (Fig. 7; Kauffman, 1973). During times of normal distribution of paleobiogeographic units in the Western Interior Seaway (Fig. 7), the Gulf and Atlantic Coast Subprovince contained a mixture of southern Warm Temperate and Subtropical taxa; the biota had great diversity, abundance of medium to large-size, shelly species, and many reef and carbonate platform-associated taxa but without the formation of large organic frameworks. Detailed analysis of this region will probably allow eventual subdivision of the Gulf Coast biotas into a northern group dominated by Warm Temperate organisms related to those of the Southern Interior Subprovince, and a southern group dominated by Subtropical organisms related more to those of Tethys.

The Tethyan biota is represented by the Caribbean Province and dominated by Tropical mollusc, echinoderm, and larger foraminifer assemblages, as well as by reef-like buildups of corals, calcareous algae, rudistid bivalves, and associated taxa.

An understanding of these paleobiographic divisions is necessary for the interpretation of dynamic paleobiogeographic changes in the Western Interior. Episodically, these changes involved rapid immigration and emmigration of Subtropical and Tropical taxa through the southern half of the Seaway. Figure 7 is an average expression of paleobiogeographic distribution patterns within the Western Interior Seaway, not a static situation. It provides, however, a useful base for a study of paleobiogeographic dynamics within the basin.

Dynamic Changes in Paleobiogeographic Units of the Western Interior Seaway

The average Upper Cretaceous paleobiogeographic distribution of marine biotas, reflecting climatic zones in North America (Fig. 7) has been determined by careful analyses of the distribution of many marine groups. This has led to identification of characteristic taxa for each subprovince which allow it to be mapped in time and space, and to the definition of broad ecotonal boundaries (e.g., Sohl, 1967) between subprovinces. These boundaries can be mapped through time in the same way that Gill and Cobban (1973) have mapped Cretaceous strand lines. With these data in hand, it is now possible to examine the hypothesis that paleobiogeographic units like those of the Western Interior epicontinental seaway are dynamic – shifting continuously in response to changes in tectonoeustacy, paleoceanography and regional geology.

The distributions of North American paleobiogeographic units and their boundaries have been plotted from regional biotic data throughout the 34 Ma Albian-Maastrichtian history of the Western Interior Seaway. Data have then been lumped into substage groupings, averaging 2.0 Ma in duration in order to map and compare the paleobiological data through the paleobiogeographic history of the Western Interior Cretaceous Seaway (Figs. 8-13). Paleobiogeographic plots at this resolution reveal dynamic shifts over short time intervals in the North American subprovinces, and especially in the distribution of the Southern Interior and Gulf and Atlantic Coast Subprovinces during maximum transgression and eustatic highstand.

In summary, prior to long-term connection of the southern and northern arms of the seaway in the earliest Cenomanian (Fig. 3; Williams and Stelck, 1975), Cool Temperate biotas, which occupied the northern arm, were only slightly differentiated from those of the Circum-Boreal Seaway, where they had originated. Only at maximum Late Albian penetration into central Colorado (Fig. 3; Williams and Stelck, 1975, Text-Fig. 4) did significant endemism develop within the northern arm among gastroplitid ammonites and inoceramid bivalves (Reeside and Cobban, 1960, Kennedy and Cobban, 1976). Similarly, during the Albian, the more restricted southern arm of the seaway (Fig. 3; Williams and Stelck, 1975, Text-Figs. 2, 4) contained Warm Temperate to Subtropical biotas which were less diverse but very similar to those of the Gulf and Atlantic Subprovince. The two biotas overlapped briefly during a short middle Late Albian connection of these seaways. Without significant endemism in either arm of the seaway, distinct paleobiogeographic units could not be defined and these two biotas are grouped with the Circum-Boreal and Gulf and Atlantic Subprovinces, respectively, in plotting Albian paleobiogeography. Further, Kauffman noted (1973, Fig. 4) that rates of pre-Cenomanian seafloor spreading and opening of the Atlantic were not sufficient to achieve widespread genetic isolation of the North American biota from that of Western Europe; distances across the Atlantic were still within reach of westward larval dispersal for many taxa. As a consequence, North American endemism in general was restricted, and paleobiogeographic differentiation occurred at very low levels during the Albian. Endemism at subprovincial levels did not occur in the Gulf and Atlantic Coast or the Western Interior until the Cenomanian.

(2) Following Early Cenomanian connection of the northern and southern arms of the Western Interior Seaway, and due to extensive genetic isolation among North American marine biotas as the Atlantic Ocean basin opened to distances which exceeded larval dispersal ranges for many benthic taxa from European marine ecosystems – discrete North Temperate and Tethyan Realm provinces and subprovinces became rapidly differentiated in North America (Kauffman, 1973, Fig. 4). Figure 7 defines the average expression of these new paleobiogeographic units during Cenomanian-Maastrichtian time, a pattern that was generally persistent, with boundary migrations rarely exceeding 200 miles (320 km), through most of the transgressive and regressive phases of each eustatic cycle over 35 Ma. This relative biogeographic stability was, in part, probably due to the ecological restriction of the biota during these inververvals; diversity was relatively low due to subnormal surface salinities and less than normal oxygen in the middle and lower water column. Ecological generalists, many with cosmopolitan and intercontinental distribution, and low-diversity, endemic biotas occupied these stressed environments, making precise tracing of paleobiogeographic fluctuations difficult; these physically accommodated ecosystems, dominated by eurytopic generalists, would be expected to show low levels of paleobiogeographic response to broadly and slowly changing environments of long-term transgression and regression.

(3) Detailed paleobiogeographic mapping does reveal exceptionally rapid, broad shifts in North American subprovinces for each cyclothem just prior to, during, and just after peak rise of sealevel and epicontinental transgression. These are especially evident during the extensive Early Turonian and Coniacian through Santonian transgressions (eustatic highstands). In each cycle the rate and magnitude of paleobiogeographic change is correlative with the relative magnitude and duration of the third-order eustatic fluctuation. These abrupt biogeographic changes are fully predictable from our basin models utilizing tectonic, sedimentologic, biologic and geochemical evidence.

SUMMARY

Several important geologic and oceanographic events converged during peak rise of sealevel of create conditions favorable to widespread rapid shifts in the distribution of paleobiogeographic units. Maximum elevation of sealevel accompanied maximum subsidence of the Western Interior Basin due to tectonic and sediment loading, modified probably by shallow plate subduction. Subsidence outstripped sedimentation, so that the deepest water of the basin history was developed during peak transgression and, with it, maximum opportunities for development of strong, multilayered current systems down the western and central part of the basin. These conditions favored immigration of biotas from surrounding ecosystems through mobility of adult plankton and nekton, or by dispersal of the planktonic larvae of benthic organisms in the upper water column. Both rise of sealevel and deepening of the basin were progressive however, so that rapid immigration of taxa near peak-rise of sealevel is not predictable from these factors alone. Rapid biogeographic changes are predictable, however, from geochemical and biological data which suggest that brackish, oxygen-deficient waters within Cool to Warm Temperate climatic belts dominated the seaway during much of the transgressive and regressive history of the cyclothems. These conditions favored immigration of more generalized northern faunal elements into the basin; this was countered, however, by broad warming associated with increasing maritime climates during transgression. Deleterious chemical environments of the seaway further discouraged widespread immigration of normal-marine taxa from northern and southern biogeographic units. Marine larvae, especially, are very sensitive to abnormal water-chemistry and temperature (Kauffman, 1975; Scheltema, 1977). Further, the possible development of at least a partial sill at the constricted southern end of the seaway, holding back major warm currents during sealevel rise until some depth-threshold was reached, cannot be ruled out as a contributing factor to retention of a semi-stable Temperate biogeography during most of transgressive history in the seaway (Fig. 7).

However, sedimentologic, geochemical, and paleobiologic data clearly show rapid, major changes in the watermass of the Western Interior Seaway within 1 to 2 Ma of peak-eustatic rise during the middle Late Albian, Late Cenomanian to Early Turonian, during transgressive pulses of the Coniacian-Santonian, during the late Early Campanian, and during the middle Late Campanian. These changes are variably characterized by onset of pelagic (foraminiferal-coccolith) carbonate deposition in dominantly non-calcareous sequences; by isotopic changes indicating normalization of salinity and extensive oxygenation throughout the water column; by episodic decrease in organic-carbon content of sediments associated with periods of benthic oxygenation; by physical evidence of current-mixing in deep-water sediments, and by the migration and introduction of numerous taxa characteristic of the Gulf and Atlantic Coast Subprovince (Warm Temperate to Subtropical) and, to a lesser degree, of the Tropical Caribbean Province. These rapid physical and chemical shifts are indicated in data presented on Figures 4-6. The extent of northward migrations of warm water taxa from the Gulf and Caribbean biotas are shown on Figures 8-13, with Figure 7 being the average expression of paleobiogeographic distributions between sealevel peaks. Radiometric dating of strata containing evidence for large, rapid shifts in paleobiogeographic distributions indicate that these changes took place in less than a million years, usually within a half million years. Events associated with biogeographic migrations depicted in Figures 8-13 during eposides of peak-rise of sealevel in the Greenhorn and Niobrara Cycles is drawn from the more extensive work of Coates, Kauffman, and Sohl (in press).

The Middle Late Albian Incursion. The first trough connection between the southern and northern arms of the Western Interior seaway occurred in the middle Late Albian (zones of *"Inoceramus" comancheanus*; *"I." bellvuensis*: Kauffman, 1975; Kauffman *et al.*, 1976) with the initial junction in south-central Colorado. This connection followed a long period of progressive incursion of the northern and southern arms of the seaway (Aptian-Middle Albian; 15 Ma); during this interval Cool Temperate Circum-Boreal biotas migrated as far south as northern Colorado, and Warm Temperate to Subtropical biotas migrated as far north as southern Colorado, Oklahoma, and central Kansas. Excepting for important development of endemic ammonites and bivalves in the distal (southern) reaches of the northern arm the Cool Temperate biota was a diminished version of the Circum-Boreal Canadian-Alaska biota. The distal reaches of the southern arm of the seaway had its strongest affiliations to the Warm Temperate-Subtropical biota of the Texas Gulf Coast, and only low levels of endemism.

In contrast to the paucity of southern warm water elements that immigrated into the Western Interior seaway during Aptian-Middle Albian time, at the southern aperture of the seaway this interval was marked by significant paleobiogeographic change. Warm Temperate Aptian biotas were replaced first by Subtropical assemblages and ultimately by Tropical reef-associated biotas in northern Mexico, central Texas, southern New Mexico, and southern Arizona during the Albian. Stoyanow (1949) documented cosmopolitan and Warm Temperate biotas in southern Arizona in the Early Aptian (*Deshaysites* Assemblage), Subtropical marginal Tethyan forms in the Middle and Upper Aptian (*Kazanskyella* through *Hypacanthoplites* assemblages), and fully Tropical Tethyan biotas, including rudist-coral reefs during the Albian (Scott and Brenckle, 1977). This sequence depicts a significant northward migration of the Temperate-Subtropical and Subtropical-Tropical biogeographic boundaries across Mexico and Texas during the Aptian-Albian, associated with rising sealevel. Compare the Albian reef line (Fig. 8) with the average Upper Cretaceous reef line (Fig. 7) 650 to 800 km to the south. Maximum northward Lower Cretaceous migration of these boundaries culminated in the development of rudist-dominated frameworks throughout central Texas by the early Middle Albian (e.g., Perkins, 1974, Fig. 2) and their proliferation into the Upper Albian of Texas, southern Arizona, and southern New Mexico. Young (1972) also noted episodic northward migrations of "cosmopolitan" Tethyan ammonites into central Texas during the Albian, and related this to breaching of the "barrier reef" (bank) at the southern edge of the Comanche shelf. Despite progressive northward migration of Tethyan and Marginal Tethyan biotas into Texas during the Aptian-Albian, none entered the southern end of the Western Interior basin before the Late Albian.

By the time warm Tethyan waters first overstepped the res-

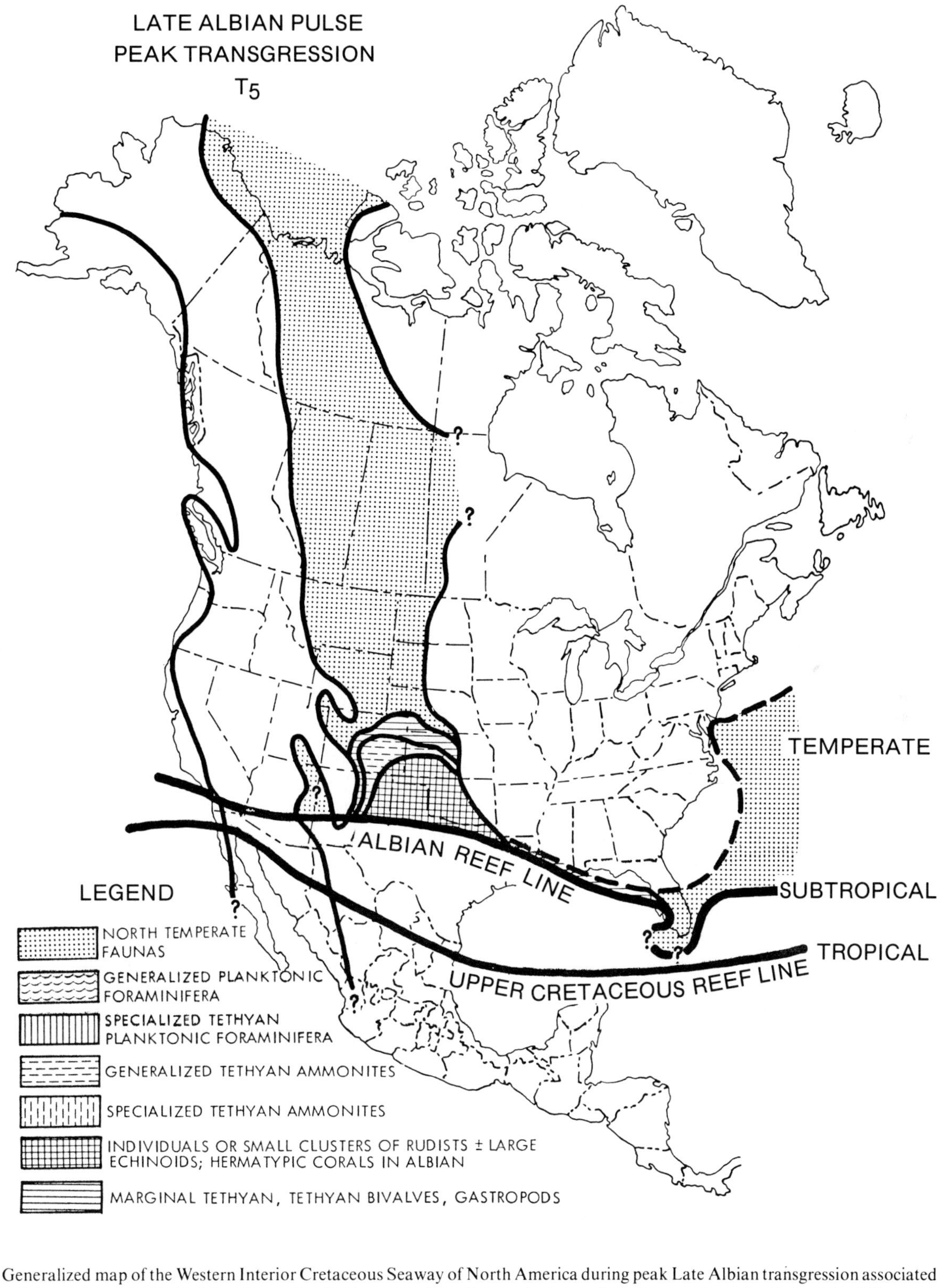

Figure 8. Generalized map of the Western Interior Cretaceous Seaway of North America during peak Late Albian transgression associated with the Kiowa-Skull Creek Eustatic Cycle. Map shows distribution of various Subtropical organisms (patterns keyed to lower left of diagram) defining northward shift of Temperate-Subtropical paleobiogeographic units during brief incursion of warm normal marine waters into the seaway at eustatic high-stand. Note in particular brief northward shift of Albian Tropical reef line from normal Upper Cretaceous position (heavy dark line). Compare with more normal paleobiogeographic distributions in Figure 7.

tricted southern aperture of the Western Interior Seaway, therefore, a major buildup of Tethyan biotas had already occurred just to the south, and the Tethyan reef boundary had attained its maximum northerly position for the entire Cretaceous (Fig. 8). The stage was thus set for a rapid northward Tethyan incursion into the Western Interior seaway once the Late Albian eustatic rise widened and then breached the restricted southern aperture of the basin. Early Late Albian immigration of Subtropical biotas mainly from the southeast took less than a half million years to reach southern Kansas, northwestern Oklahoma, and New Mexico.

Twenhofel (1924) and Scott (1970, 1977) documented the following sequence of events. (a) Rapid marine transgression into the southern part of the Western Interior Basin during the early Late Albian; marginal Tethyan influence was initially strong. The "Champion Draw Bed" (Cragin, 1895; Twenhofel, 1924, Scott, 1970, 1977: upper *Venezoliceras kiowanum* Zone) of southern Kansas contains the highest diversity (36 species) and the strongest marginal-Tethyan faunal affinities (60 to 65% similarity with the Texas Gulf Coast) for the entire Late Albian transgression. The fauna includes small isolated, hermatypic coral colonies *(Astrocoenia)* but no reefs. Other Subtropical elements include *Venezoliceras*, bivalves like *Texagryphaea, Ceratostreon*, large *Trachycardium*, and *Cucullaea*, trochid and granulated turritellid gastropods, and the Tethyan echinoid *Salenia*. There is little endemism in this biota. (b) Continued middle Late Albian transgression (zone of *Adkinsites bravoensis*) extended the sea into southern Colorado and central Kansas. Partial closing of the southeastern aperture of the seaway by a major deltaic complex, shifted the principal source of warm southern waters and biotas to the Chihuahua Trough. Marginal-Tethyan influence remained high, involving many of the same molluscan genera that characterized the Champion Draw biota, additional Subtropical bivalves like *Roudaria* and *Neithea fredericksburgenesis* (Cragin), generalized planktonic foraminifera, and an expanded Tethyan ammonite assemblage including *Adkinsites, Venezoliceras*, and *Manuaniceras*. But collectively, the middle Late Albian fauna had less similarity (50%) to the Marginal Tethyan faunas of Texas than before, and lacked hermatypic corals; decrease in Tethyan influence was coincident with increasing percentages of normally Warm Temperate taxa, and significantly higher levels of endemism, marking origination of the Southern Interior Subprovince (Scott, 1977). Marginal Tethyan biotas extended further north during the middle Late Albian than earlier in the transgression. (c) During the latter part of the middle Late Albian (*"Inoceramus" comancheanus* Zone) the two arms of the Western Interior Seaway first joined in central Colorado. Mixing and competition among the Marginal Tethyan – Warm Temperate southern biota and the Cool Temperate northern biota ensued. An initially sharp biogeographic boundary gave way to a broad zone of faunal mixing (ecotone) between Colorado and Texas. Marginal Tethyan bivalves like *Texigryphaea* and large *Trachycardium* reached south-central Colorado while northern inoceramid lineages reached north Texas. Overall Tethyan influence declined sharply and diversity decreased. Strong eastern (Kansas: 24%) and western (New Mexico-Colorado: 9%) endemic centers (Scott, 1977, p. 167) developed. The Marginal Tethyan – Temperate boundary retreated southward into New Mexico during the Late Albian, leaving behind isolated Subtropical faunal elements. (d) During the late Late Albian (*"Inoceramus" bellvuensis* Zone) the Western Interior seaway attained its maximum Lower Cretaceous development at peak-transgression; this was closely followed by basinal filling, shoaling, and rapid Late Albian regression. Whereas a few Marginal Tethyan molluscs reached their maximum penetration into north-central Colorado at this time (Fig. 8), the fauna of the southern seaway was predominantly Warm Temperate. The Warm Temperate – Marginal Tethyan boundary retreated into Texas, beyond the aperture of the seaway, and the Cretaceous reef-line retreated into northern Mexico (Fig. 7 *vs.* Fig. 8) during the Late Albian regression (T_6; Fig. 4). The Late Albian incursion of Subtropical biotas into the Western Interior lasted approximately 0.5 to 1.0 Ma (Fig. 4), with maximum northward migration of the Tropical Caribbean Province-Subtropical Atlantic Coast-Southern Interior Subprovince boundary of about 1100 km.

The Late Cenomanian-Early Turonian Incursion. One of the most dramatic incursions of Tethyan and Marginal Tethyan biotas into the Western Interior Seaway occurred near and at the peak of the great Cenomanian Transgression (T_6; Fig. 4; Hancock and Kauffman, 1979), in a 2.5 Ma interval spanning latest Cenomanian and Early Turonian time. (a) This transgression began during the latest Albian (*Neogastroplites* zones) with eustatic rise, re-advance of the northern (Cool to Mild Temperate) arm of the seaway, and buildup of Subtropical Tethyan influence northward across the Texas Comanche Shelf without breaching the southern aperture of the Interior Seaway. By latest Albian, the northern arm of the seaway reached southern Colorado (Fig. 3), bearing a discrete Northern Interior Subprovince biota rich in endemic ammonite and bivalve species. Equivalent strata in Texas record northward transgression across the Comanche Shelf, and a change from cosmopolitan molluscs of the *Mortoniceras wintoni* assemblage to Tropical-Subtropical Tethyan assemblages (including many endemics) of the *Drakeoceras gabrielense* Zone (Young, 1972); extensive platform carbonates developed concurrently.

(b) During the Early Cenomanian the Western Interior Temperate biota was still largely isolated from that of Gulf Coast shelf seas by a shallow clastic sedimentary sill (Woodbine Formation and equivalents) across the southern aperture. The Early Cenomanian biota of the Western Interior was predominantly Cool to Mild Temperate, with high endemism among the dominant inoceramid bivalves and ammonites. The Northern and Central Interior Subprovinces were broadly differentiated, with a diffuse ecotonal boundary. In Texas, however, Young (1972) documented increasing Early Cenomanian endemism and Tethyan influence among ammonite faunas of the Comanche Shelf. Thus, Early Cenomanian eustatic rise coincidently caused buildup of Tropical to Subtropical Tethyan biotas in the Gulf Coast and initiated widespread invasion of the Interior Seaway, mainly from the north. But the restricted southern aperture of the basin remained as an effective barrier to northward immigration of warm-water taxa, as did the slightly brackish, oxygen-restricted chemistry of the interior seaway at this time.

(c) By Middle Cenomanian time the barrier at the southern aperture of the Western Interior basin was breached with continued eustatic rise. A broad zone of intermixing of northern and southern Temperate biotas was quickly established in the southern one-third of the Western Interior and Texas (Stephenson, 1952); rare Subtropical elements (rudists, specialized planktonic foraminifera, etc.), reached southern Colorado. Gradual northward expansion of Warm Temperate and cosmopolitan taxa and short-lived incursions of generalized, eurythermal, planktonic foraminiferans (Eicher, 1977, Fig. 1) suggest increasing water temperatures and slow northward expansion of the Southern Interior Subprovince. Still, Temperate zone biotas predominated throughout the seaway during the Middle Cenomanian.

(d) Warm-water cosmopolitan and marginal-Tethyan (Subtropical) taxa first immigrated into the southern half of the Western Interior Seaway during the Late Cenomanian, beginning with a few specialized (stenothermal), planktonic foraminiferans and calcareous nannoplankton, which then diversified gradually through the substage forming widespread pelagic carbonates (Eicher, 1977, fig. 1; Kauffman, 1984). The gradual buildup of Tethyan influence in the southern part of the Western Interior seaway was abruptly altered through a remarkable series of environmental and biological events during the Late Cenomanian (*Dunveganoceras pondi, D. albertense*, and *Sciponoceras gracile* zones) and Early Turonian (Kauffman and Scholle, 1977; Scholle and Kauffman, 1977; Kauffman, 1984). These gave rise to an extensive, short-term incursion of Tethyan waters which carried the Warm Temperate-Subtropical paleobiogeographic boundary nearly 1300 miles (2080 km) to the north (Fig. 9) and back again in 2.5 to 3 Ma, as follows.

(e) Near peak-transgression (lower *S. gracile* Zone) stable isotope, sedimentologic, and biologic data indicate initiation of a sharp rise in water temperature and salinity levels throughout the central and southern part of the seaway. This marks abrupt flooding of Gulf and Caribbean watermasses into the Western Interior as a result of rapid eustatic rise and/or sudden breaching of barriers in the southern aperture of the seaway in less than 0.5 Ma. This generally correlates with a marked Late Cenomanian decline in the effectiveness of the Stuart City "reef" as a barrier to northward ammonite migration (Young, 1972). (f) This temperature/salinity change produced a dramatic increase in marine diversity throughout the southern half of the Western Interior basin. Koch (1978, 1980) reported 170 macrofaunal taxa in the *S. gracile* Zone, a six-fold increase over diversity recorded from immediately underlying strata (Kauffman, 1984). Eicher (1977, Fig. 1) recorded an increase from 0 to 30 benthic calcareous foraminiferal species, and from 10 to 15 planktonic species, across this marked environmental boundary. This rapid increase in diversity resulted from: increased rates of evolution due to temperature and salinity stress (Kauffman, 1978); widespread mixing of southern and northern biotas leading to temporary niche partitioning and exploitation of new habitats; widespread endemism among Warm Temperate elements; and immigration and ecological accommodation of many cosmopolitan and some marginal Tethyan taxa, including warm-water radiolitid rudists, the large echinoid *Holaster*, Subtropical ammonites *(Desmoceras, Pseudouhligella, Anisoceras)*, large *Trachycardium* and gastropods like *Nerinea* and *Cassiope* (Sohl, 1967, 1971). The now diffuse Subtropical (Gulf and Atlantic Coast Subprovince) – Warm Temperate (Southern Interior Subprovince) boundary migrated northward into New Mexico and southern Utah, and the Warm Temperate – Mild Temperate (Central Interior Subprovince) boundary as far north as central Montana and southern Alberta, by the end of the Cenomanian.

(g) A major extinction event occurred just below and at the Cenomanian – Turonian boundary (Kauffman, 1984), largely involving the loss of endemic, Warm Temperate and cosmopolitan elements of the *Sciponoceras gracile* Zone macrofauna (Koch, 1978) and the calcareous benthic microfauna (Eicher, 1977, Fig. 1). This was associated with a significant shift in oxygen isotope values indicating a major drop in salinity just prior to peak-transgression (Pratt, 1983, and pers. commun.) and severe stratification of the water-column leading to an oxygen crisis in the seaway, reflecting a global anoxic event (The Bonarelli Event). Sediments immediately overlying the extinction level intermittently contain high levels of C_{org} and only rare, eurytopic latest Cenomanian molluscs. (h) Reoxygenation of the seaway accompanying still rising temperature and salinity levels within the first 0.5 Ma of the Turonian produced abrupt northward migration of the Subtropical biota. The Subtropical Gulf and Atlantic Coast Subprovince – Southern Interior Subprovince (Warm Temperate) boundary reached into southern Canada, Montana, and North Dakota as defined by the northernmost extension of keeled, planktonic foraminiferans and relatively diverse, marginal-Tethyan macrofaunas, predominantly molluscs like *Cassiope* (Sohl, 1967, 1971), radiolitid rudists, *Holaster* (echinoid) and varied Tethyan ammonites *(Fagesia, Vascoceras, Ampakabites, Choffaticeras*, and *Nigericeras)*. This maximum incursion of Tethyan biotas during the Middle Cretaceous (Fig. 9) coincided with peak transgression (T_6; Fig. 4), peak eustatic rise, and maximum marine warming (zones of *Watinoceras coloradoense* and *Mammites nodosoides*). (i) This Subtropical biota diversified during the middle and late Early Turonian (one Ma). Kauffman (1984) recorded a rapid change from 2 to 13 to 28 molluscan taxa, and Eicher (1977, Fig. 1) recorded an increase from 15 to 20 planktonic foraminifera species through the lower three inoceramid zones of the Early Turonian in Colorado. Warm to Mild Temperate biotas spread well up into central Alberta with correlative retreat of the Northern Interior Subprovince; (j) Subtropical immigrants in the Western Interior biota gradually decreased in numbers, diversity, and biogeographic range during the latest Early and early Middle Turonian (*Mytiloides labiatus* s.s. Zone; *Collignoniceras woollgari* Zone), 2.5 to 3 Ma after their abrupt immigration. This decline was associated with initial eustatic fall, early regression (R_6; Fig. 4), and lowering of epicontinental marine temperatures and salinity in the seaway. Kauffman (1984) recorded an 86 per cent decrease in macrofossil diversity and Eicher (1977) documented a similar decline in planktonic foraminiferal diversity in the basinal center during the early Middle Turonian. The Subtropical – Warm Temperate paleobiogeographic boundary retreated southward nearly 1,000 miles (1,600 km) during this 1.5 Ma period, and Cool and Mild Temperate biotas became re-established to the

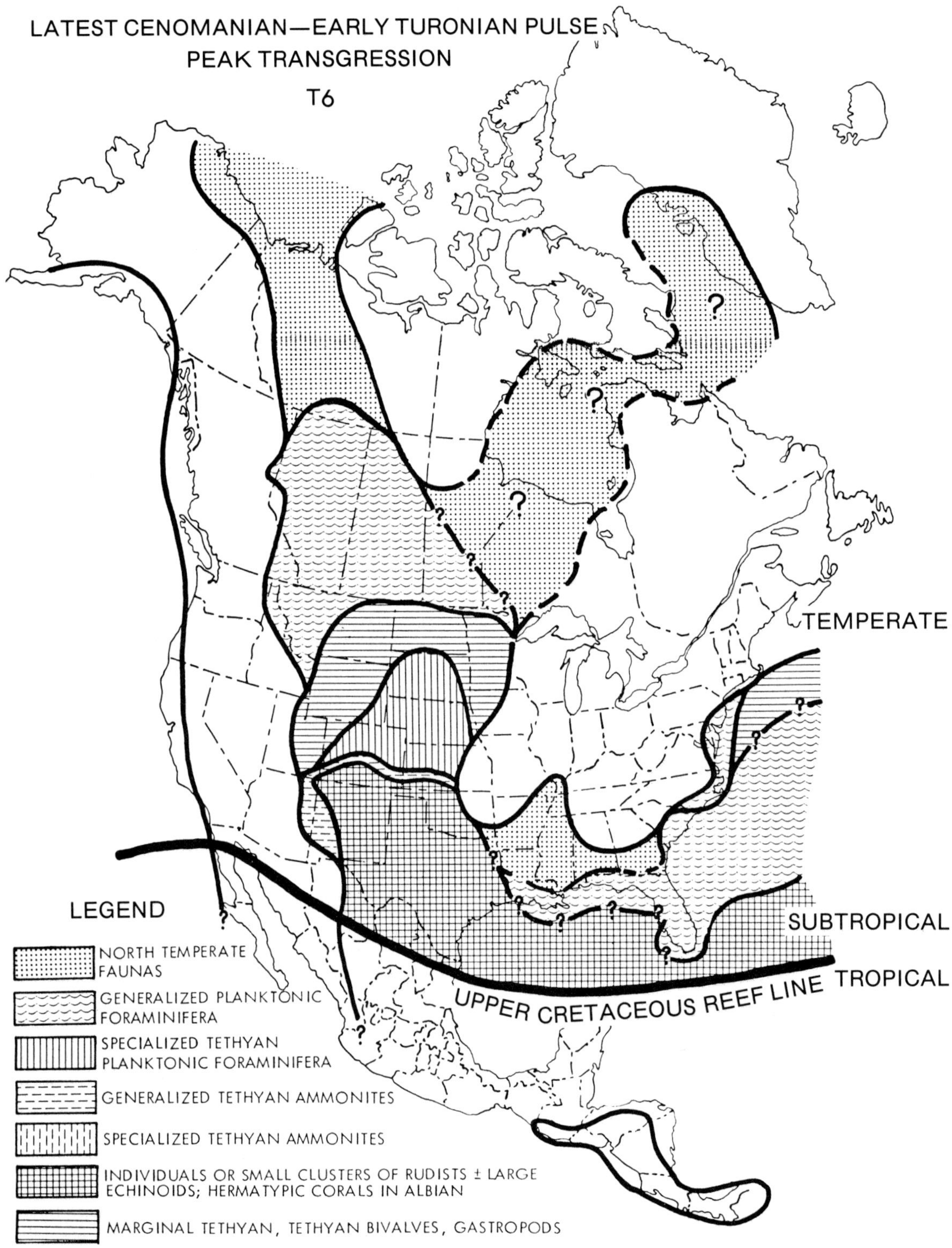

Figure 9. Generalized map of Western Interior Cretaceous Seaway of North America during peak Early Turonian transgression associated with the Greenhorn Eustatic Cycle. Map shows distribution of various Subtropical organisms (patterns keyed to lower left of diagram), defining extensive, short-term, northward migration of Warm Temperate and Subtropical paleobiogeographic units during rapid incursion of warm, normal marine waters into the seaway at eustatic highstand. Compare with more normal paleobiogeographic distributions in Figure 7.

south, attaining their "normal" biogeographic position (Fig. 7) by the Middle Turonian.

The Coniacian-Santonian Tethyan Incursion. The second-most extensive incursion of Subtropical biotas into the Western Interior Seaway occurred during the Early Coniacian to Late Santonian Niobrara eustatic cycle (Fig. 10), a time of sealevel highstand and global warming (Fig. 4). Four separate incursions were each linked to a small fourth-order transgressive pulse (T_{7a}, T_{7b}, T_{7c}, T_{7d}; Fig. 4); minor regressions separate these pulses. The Lower Coniacian (T_{7a}) and Middle Santonian (T_{7c}) transgressions were significantly more extensive than the others. The Coniacian-Santonian eustatic cycle consists in the central and eastern part of the seaway of abundant pelagic chalks or limestones bearing numerous Subtropical biotic elements during transgressive peaks; these alternate with calcareous shale or finely laminated, pelagic marls yielding a cosmopolitan warm-water biota during regression. The entire pelagic carbonate interval in the southern and central part of the basin contains diverse planktonic foraminiferans, including keeled stenothermal forms and calcareous nannoplankton; Kent (1967) provides partial documentation. The following sequence of events characterize this incursion, and are abstracted from Coates *et al.* (1984).

(a) During rapid (0.5 Ma) initial eustatic rise of sealevel associated with the Niobrara marine cycle early and middle transgressive sediments of the Western Interior and northern Gulf Coast consisted of sands, calcarenites, and clays lacking significant pelagic carbonate and bearing Temperate endemic and cosmopolitan biotas of moderately low diversity. The basin was somewhat subsaline and oxygen-restricted throughout.

(b) During the latest Turonian (zone of *Prionocyclus quadratus*) and earliest Coniacian (middle T_{7a} transgression; Fig. 4) pelagic carbonates containing low diversity assemblages of Warm Temperate to Subtropical microplankton (Kent, 1967) became abruptly widespread (<250 Ka) in the central and southern part of the seaway. These, and moderately diverse warm-water macrofaunas, document northward expansion of the Southern Interior Subprovince to at least southern Wyoming. Scholle and Kauffman (1977) noted a significant change in $\delta^{18}O$ and $\delta^{13}C$ isotopic values associated with the initial spread of these uppermost Turonian – lowest Coniacian carbonates suggesting an abrupt increase in water temperature and/or normalization of salinity. Coincidently, abundance and diversity of warm-water biotas increased sharply. Macrofaunal composition changed rapidly from wholly Warm Temperate to predominantly warm-water cosmopolitan assemblages, but initially lacked Subtropical elements. The similarities between this diversification event and the one immediately following the Late Cenomanian change in temperature/salinity values are striking.

(c) A major extinction event followed, abruptly eliminating most Warm Temperate and many cosmopolitan marine taxa and marking the Turonian – Coniacian boundary (basal Niobrara Formation and equivalents). Isotopic and biologic data indicate stable temperatures and near-normal salinity levels across the boundary. But Lower Coniacian limestones enriched in pyrite, marcasite, and limonite and organic-rich marls contain only sparse eurytopic molluscs and foraminifers (Kent, 1967; Kauffman, 1984) recording widespread, short-term (0.1 Ma) oxygen depletion in the Western Interior Basin, one which has a global imprint. This anoxic event, coupled with ecological stress imposed by relatively rapid temperature/salinity increase during the latest Turonian and earliest Coniacian was probably responsible for the extinction event.

(d) Isotopic and biologic data (Scholle and Kauffman, 1977) suggest a second, smaller, increase in temperature and/or salinity within the Western Interior seaway at the first peak of Niobrara transgression (T_{7a}). This coincides with abrupt appearance (<0.25 Ma) and diversification of the Lower Coniacian biota (*Inoceramus erectus* and *I. deformis* zones), including southern Interior Warm Temperate groups, generalized warm-water cosmopolitan stocks, and Subtropical biotas from the Tethyan margin (e.g. the rudist *Durania* in small clusters, Tethyan ammonites like *Forresteria, Barroisiceras* and *Peroniceras*, keeled planktonic foraminifers to Wyoming and South Dakota, and generalized planktonics into southern Canada; Fig. 10). This marks rapid northward migration of the diffuse Gulf and Atlantic Coast Subprovince – Southern Interior Subprovince boundary approximately 600 miles (1,000 km) to central Wyoming and South Dakota (Fig. 10). Simultaneously, the Southern Interior – Central Interior Subprovince boundary migrated nearly 700 miles (1,100 km) northward into southern Canada. No core-Tethyan faunas entered the Western Interior Basin.

(e) Subsequent fourth-order fluctuations in sealevel during Coniacian-Santonian time produced small-scale changes in the distributions of these Subtropical and Warm Temperate Western Interior biotas. Gradual decline in marginal-Tethyan biotic influence and slow southward retreat of the Subtropical – Warm Temperate boundary began in the late Early Coiacian and culminated in the Middle Coniacian, coincident with a fourth-order regression (R_{7a}) and a small but abrupt decline in salinity and water temperature. Concurrently, molluscan and calcareous microplanktonic diversity decreased in the northern and central parts of the basin (Kent, 1967; Kauffman, 1984); cosmopolitan ecological generalists replaced specialized Subtropical elements of the biota during the late Early Coniacian. By Middle Coniacian time (lower *Scaphites ventricosus* Zone) the Subtropical-Warm Temperate biogeographic boundary retreated approximately 400 miles (650 km) south to Colorado.

(f) A second Coniacian-Santonian incursion of warm-water biotas occurred during the late Middle to Late Coniacian fourth-order transgressive peak (T_{7b}; Zone of *Scaphites ventricosus* and *Magadiceramus subquadratus*). It is represented near the basinal center as far north as Wyoming by widespread, pelagic chalks and limestones containing diverse calcareous foraminifer and nannoplankton with Tethyan affinities (Kent, 1967). Sparse isotopic data across this limestone suggest a small increase in temperature and/or salinity. This interpretation is confirmed by the reappearance of a Marginal Tethyan macrofauna up to northern Colorado and southern Wyoming, including sparse radiolitid rudists (*Durania*) and the ammonite *Peroniceras*. Northward re-advance of the Subtropical Gulf Coast – Warm Temperate, Southern Interior paleobiogeographic

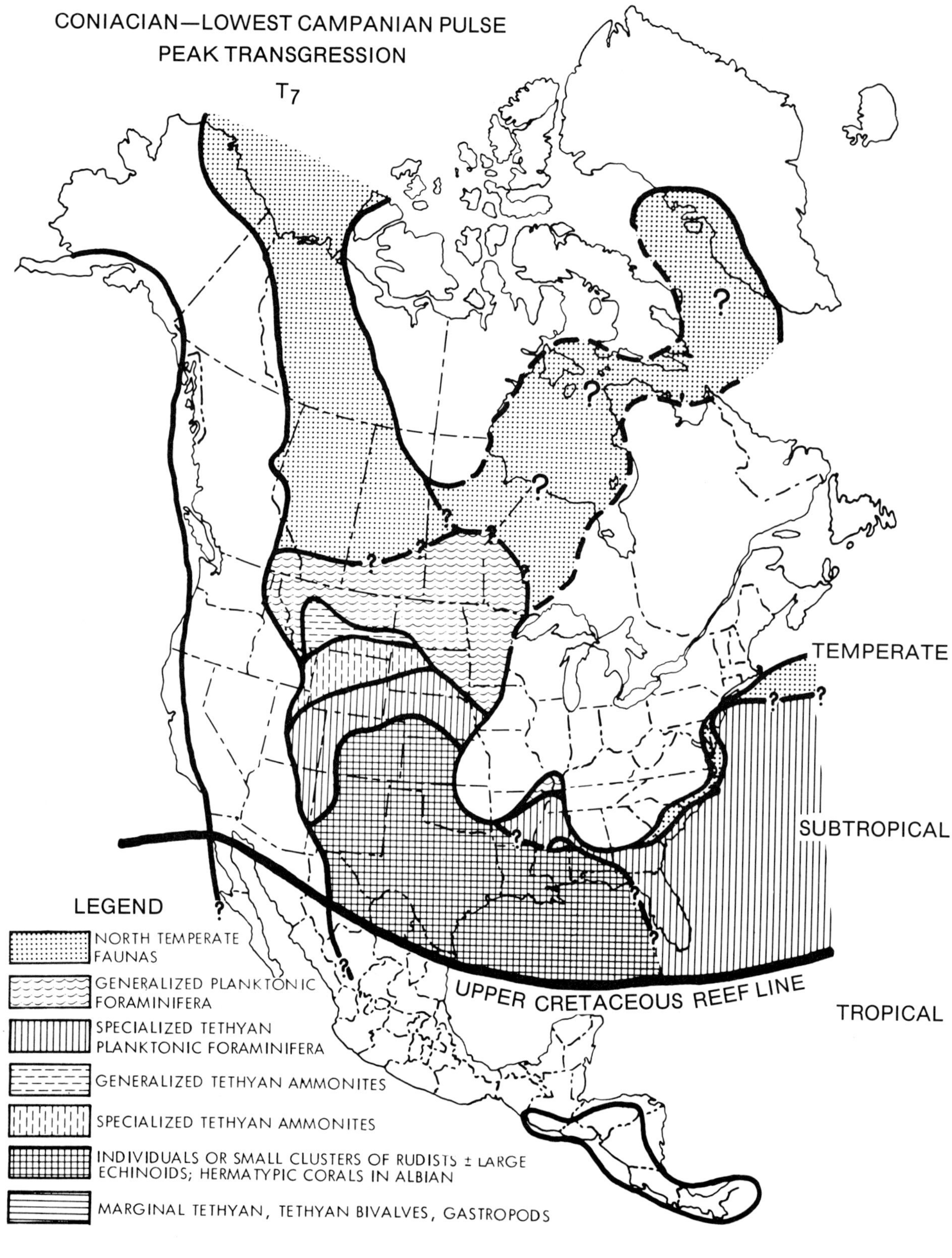

Figure 10. Generalized map of Western Interior Cretaceous Seaway of North America during peak Coniacian-Late Santonian fourth-order transgressions within third-order Niobrara Eustatic Cycle. Map shows distribution of various Subtropical organisms (patterns keyed to lower left of diagram), defining extensive, short-term northward migration of Warm Temperate and Subtropical paleobiogeographic units into the seaway during rapid incursion of warm, normal marine waters at eustatic highstand. Compare with more normal paleobiogeographic distributions in Figure 7.

boundary to at least the Early Coniacian (Fig. 10) is indicated by these distributions.

(g) A second small regressive pulse (R_{7b}) occurred during the latest Coniacian and Early Santonian (zones of *Scaphites depressus, Clioscaphites saxitonianus,* and *C. vermiformis*); numbers and diversity decreased among specialized calcareous microplankton while a few typical, Subtropical elements persisted up to southern Wyoming, including some specialized planktonic foraminifers, rare *Durania,* and ammonites of the Family Texanitidae (*Texanites, Protexanites*). A slight southward retreat of the Subtropical-Warm Temperate climatic/biogeographic boundary into south-central Colorado is indicated.

(h) The third major Tethyan biotic incursion (T_{7c}; Fig. 4), associated with one of the two most extensive fourth-order transgressions of the Niobrara cycle, is represented by late Middle Santonian chalks (*Clioscaphites choteauensis* Zone) in the central and southern Interior seaway. These contain diverse, warm water, calcareous microplankton up to south-central Wyoming (Kent, 1967; Frerichs *et al.,* 1975), large rudists (*Durania*) to Colorado and Kansas, and texanitid ammonites into Wyoming. Generalized calcareous microplankton reached central Canada (Fig. 10); the Subtropical-Warm Temperate biogeographic boundary again shifted 500 to 1000 miles (800 to 1600 km) northward in 0.5 Ma or less.

(i) A minor one-Ma regressive pulse (R_{7c}; Fig. 4) occurred during the Late Santonian (zones of *Desmoscaphites erdmanni* and *D. bassleri*) and is represented by widespread return to deposits of shaly chalk and calcareous shale. The calcareous microbiota, rudist bivalves, and Tethyan ammonites all declined in numbers as the Subtropical – Warm Temperate biogeographic boundary retreated into southern Wyoming, northern Colorado, South Dakota and Nebraska. However, generalized planktonic foraminifera remained as far north as southern Canada, indicating little expansion of the Cool Temperate biota during this regression.

(j) A final northward migration of Subtropical biotas during the Coniacian – Santonian transgressive event occurred during the latest Santonian and earliest Campanian (Zone of *Scaphites hippocrepis*). It is represented by northward spread of pelagic chalks and limestones, and by a small increase in planktonic foraminiferal numbers and diversity including an increase in diversity of keeled species. Gulf Coast ammonites (*Scaphites hippocrepis – S. leei* plexus; Cobban, 1969) reached northeastern Alberta and rudists (*Durania*) immigrated as far as southern Colorado. The Subtropical – Warm Temperate boundary reached southern Wyoming. This pulse was notably less extensive than previous T_7 incursions.

(k) An abrupt (0.5 Ma) decline in numbers and diversity of calcareous microbiotas, and rapid southward emigration of specialized, Subtropical macrofaunas and keeled planktonic foraminifera, occurred during the latest Santonian – Early Campanian regression (R_{7d}; *Haresiceras placentiforme* Zone). The Subtropical Gulf Coast-Warm Temperate Southern Interior Subprovince boundary shifted southward about 600 miles (1,000 km) into Texas in less than one million years. Cool Temperate biotas of the Northern Interior Subprovince expanded once again into the northern United States to a position similar to that in Figure 7 by middle Early Campanian.

Thus, the Coniacian – Santonian transgressive peak (T_7; Fig. 4) in the Western Interior seaway was characterized by fluctuating levels of Subtropical Tethyan biotic influence throughout its 5 Ma history; each fluctuation was linked to small fourth-order transgressive-regressive episodes. Alternating northward migrations of Subtropical elements to southern Wyoming, and their subsequent retreat to southern Colorado, mark these fluctuations of the Gulf and Atlantic Coast – Southern Interior Subprovince boundaries. Figure 4 suggests that each Subtropical incursion lasted 0.5 Ma or less; southward retreat of the Marginal Tethyan boundary was a slower process. Marginal-Tethyan influence was strongest between the Early Coniacian and the Middle Santonian, and gradually weakened during the Late Santonian and Early Campanian, with retreat of the Subtropical – Warm Temperate paleobiogeographic boundary into Texas associated with falling sealevel.

The Early Late Campanian Incursion. A moderately extensive and rapid (2.5 Ma) marine regression (R_{7d}; Fig. 4) followed the Coniacian – Santonian Subtropical incursion, causing emigration of Marginal Tethyan and Warm Temperate biotas to the Texas Coastal Plain. This regression terminated during the middle Early Campanian (Fig. 4). A number of smaller transgressions followed during Campanian – Maastrichtian time (T_8 – T_{10}; Fig. 4), reflecting additional eustatic fluctuations, but within the Western Interior Basin none were of the magnitude of the previous events. Nevertheless, each ensuing transgression (eustatic rise) in the Western Interior Basin was characterized by rapid northward immigration of Subtropical and Warm Temperate biotas and the paleobiogeographic units which they occupied.

A small incursion of Subtropical faunal elements occurred during the early part of the Late Campanian (Fig. 11), near the transgressive peak of the Claggett Cyclothem (T_8, Fig. 4). Transgression and the first Subtropical biotic incursions began in the late Early Campanian (*Baculites obtusus* Zone) and peaked in the early Late Campanian (*B. mclearni, B. asperiformis,* and *B.* sp. smooth zones). Subtropical immigrants at peak-transgression included the *Trachyscaphites spiniger* ammonite lineage in Wyoming, scattered rudistid bivalves ("*Icthyosarcolites*" *coralloidea* Hall and Meek) in central Montana, and generalized planktonic foraminifer, in central Wyoming (Mello, 1971). Stenothermal, keeled, planktonic foraminifer occurred in abundance only near the southern aperture of the seaway. At maximum incursion (early Late Campanian), the Subtropical-Warm Temperate boundary lay across northern Wyoming, southern Montana and central South Dakota (Fig. 11) – the average northern range of Subtropical ammonites, rudists, and planktonic foraminifera. Sohl (1967, 1971) noted that the Subtropical gastropod genera *Trachytriton* and *Desmieria* have a similar Campanian distribution. The approximate duration of this Tethyan pulse was 1.25 Ma (Fig. 4). A prominent regression followed (R_8; Fig. 4) causing southward retreat of many but not all Subtropical elements. Mello (1971) noted abundant planktonic foraminifer in central Wyoming during the Claggett regression (*B. gilberti – B. scotti* zones); rare Subtropical gastropods and rudists remained in the southern and central part of the seaway during this regression.

The Middle Late Campanian Tethyan Incursion. The last major transgression in the history of the Western Interior seaway began and peaked in the middle Late Campanian (*Didymoceras nebrascense* through *Baculites compressus* zones; Bear-

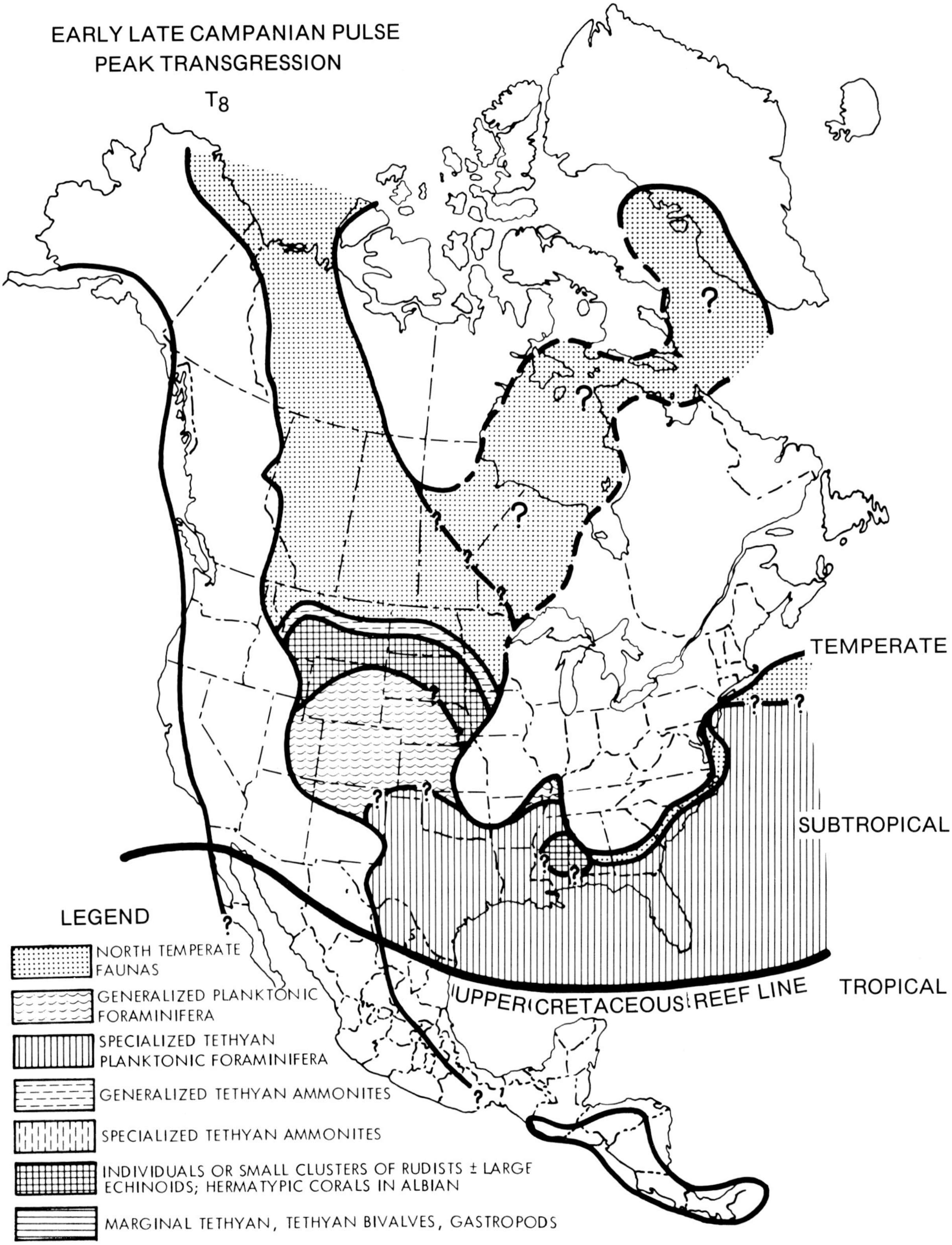

Figure 11. Generalized map of Western Interior Cretaceous Seaway of North America during peak early Late Campanian transgression associated with the Claggett Eustatic Cycle. Map shows distribution of various Subtropical organisms (patterns keyed to lower left of diagram), defining extensive, short-term northward migration of Warm Temperate and Subtropical paleobiogeographic units during rapid incursion of warm, normal marine waters into the seaway at eustatic highstand. Compare with more normal paleobiogeographic distributions in Figure 7.

paw Cyclothem; Kauffman, 1977a). A major incursion of subtropical biotas accompanied the transgression, reaching central Canada (Fig. 12), and included rudistid bivalves ("*Icthyosarcolites*"), Subtropical gastropods (*Trachytriton, Desmieria;* Sohl; 1967, 1971), and generalized planktonic foraminifer (Mello, 1971) during and just following peak-transgression (*B. compressus* Zone). The bulk of the faunal evidence suggests that the Subtropical (Gulf and Atlantic Coast) – Warm Temperate (Southern Interior) boundary extended across northern Wyoming. A major regression followed (R_9; Fig. 4), and all Subtropical biotas disappeared from the Western Interior during the latest Campanian to Early Maastrichtian with southward expansion of Cool to Mild Temperate biotas into the United States. The middle Late Campanian incursion lasted one Ma; the main pulse was less than 0.5 Ma long (Fig. 4).

The Maastrichtian Tethyan Incursion. The Early and Middle Maastrichtian marks the major M_9 regression (Fig. 4) of epicontinental seas from the Western Interior Cretaceous Basin. This regression was interrupted by a minor transgressive reversal and brief northward spread of warm-water biotas during the late Early to Middle Maastrichtian (*B. clinolobatus* Zone). Mello (1969) reported an assemblage of mostly generalized, planktonic foraminifera in South Dakota, coincident with the peak of this minor transgression. Speden (1970) reported sparse rudists ("*Icthyosarcolites*") in the Lower Fox Hills Formation along with the warm-water bivalve *Parapholas.* The northward extent of these more typical Subtropical elements is shown in Figure 13. The evidence does not suggest a major displacement of the Subtropical – Warm Temperate Boundary at this time; possibly it reached northern New Mexico and southern Colorado. The duration of this incursion was about 0.25 Ma.

Summary

Five major and five minor northward incursions of southern Warm Temperate and Subtropical organisms from the Gulf and Atlantic Coast Subprovince far into the Western Interior Seaway of North America were predominantly associated with short intervals around peak-transgressions, reflecting eustatic highstands. Isotopic, sedimentologic, and paleobiologic data suggest that relatively abrupt increases in temperature and salinity, and re-establishment of vertical circulation, immediately preceded and/or accompanied these incursions, but that somewhat more gradual declines in temperature and salinity terminated them. This in turn suggests that warm Subtropical surface-waters rapidly flooded northward into the epicontinental basin after a prolonged period of eustatic rise which finally reached some critical threshold of depth, overcoming circulational barriers formed in the constricted, partially silled, southern aperture of the Western Interior Seaway. Prior to and just after such incursions, geochemical and paleobiological data strongly suggest that the Western Interior Seaway was slightly brackish, broadly oxygen-restricted, and wholly within Temperate climatic belts.

Evidence for rapid northward incursions of Subtropical organisms is mainly found in: widespread pelagic carbonates, reflecting floods of calcareous microplankton (including stenothermal, keeled, planktonic foraminifera) and calcareous benthonic foraminifera; isolated specimens or small clusters of rudist bivalves; diverse Subtropical gastropods and bivalves; larger irregular echinoids; rare hermatypic colonial corals; and among Tethyan ammonites. These Subtropical elements, and numerous warm-water, cosmopolitan taxa appeared, immigrated northward for considerable distances, and then emigrated from the basin within short periods of time (0.25 to 2.5 Ma intervals). Regressions greatly restricted or eliminated Subtropical biotas from the Western Interior Seaway, although a few more adaptive taxa remained during lesser regressive pulses of the Coniacian and Santonian. The paleobiogeographic fluctuations of the Central Interior – Southern Interior Subprovince boundary, and the Southern Interior – Gulf and Atlantic Coast Subprovince boundary, correlate with migrations of the Subtropical and Warm Temperate climatic zones from the Gulf Coastal Plain into the Western Interior Basin and back over distances as great as 1,300 miles (2080 km).

In general, there seems to be a patterned biotic response to these incursions of subtropical Tethyan biotas, and similarity in the biotas of each incursion – i.e., one to two genera of rudists, one to five genera of Subtropical ammonites, a moderate number of other molluscan genera, one to two hermatypic coral and larger echinoid genera, floods of planktonic and benthonic, calcareous microbiotas, and only moderately complex community structure. Many communities were composed equally of Subtropical, cosmopolitan and eurytopic Temperate organisms during these incursions. The duration of each incursion was too short, and Tropical environmental influence too marginal to allow development of complex Tethyan ecosystems in the Western Interior. There also seems to have been a relatively consistent, north to south, geographic distribution pattern of warm-water faunal elements during these incursions. This probably reflects variations in thermal tolerance among Warm Temperate to Subtropical organisms versus the nature of the thermal gradient in the Western Interior Seaway. Thus, generalized (eurythermal) planktonic foraminifera and nannoplankton usually extended farthest to the north, miscellaneous Subtropical bivalves and gastropods next-farthest, ammonites of Tethyan affinities next, specialized (keeled) planktonic foraminifera next, and hermatypic corals, radiolitid rudists, and larger echinoids had the least northerly range during any incursion. Exceptions exist as apparent ecological tolerances shifted in groups through time (e.g., rudists) or due to gaps in biogeographic documentation.

In many cases these rapid paleobiogeographic incursions were associated with other major changes in the biota, such as major extinctions, breakdown of ecological structure, increases in evolutionary rates, and high levels of endemism.

EVOLUTIONARY RESPONSE TO PATTERNS OF PALEOBIOGEOGRAPHIC CHANGE

Kauffman, in his analyses of evolutionary rates for Western Interior Cretaceous Mollusca (1972, 1977b, 1978), noted that in virtually all lineages studied, rates of evolution were primarily controlled by the ecological features of each group, such as: trophic strategy; habitat needs; adaptive breadth (degree of

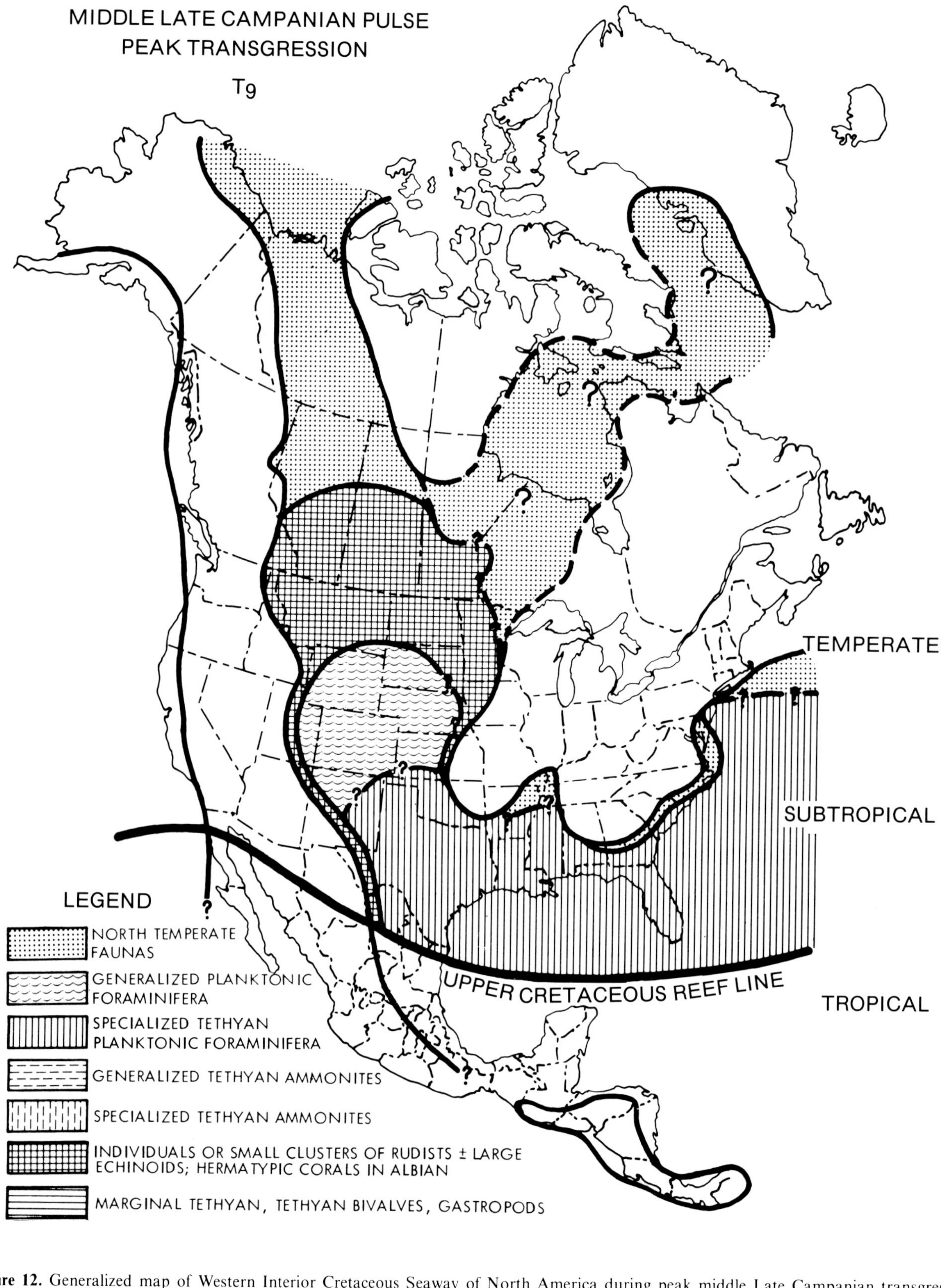

Figure 12. Generalized map of Western Interior Cretaceous Seaway of North America during peak middle Late Campanian transgression associated with the Bearpaw Eustatic cycle. Map shows distribution of various Subtropical organisms (patterns keyed to lower left of diagram), defining extensive, short-term migration of Warm Temperate and Subtropical paleobiogeographic units during rapid incursion of warm, normal marine waters into the seaway at eustatic highstand. Compare with more normal paleobiogeographic distributions in Figure 7.

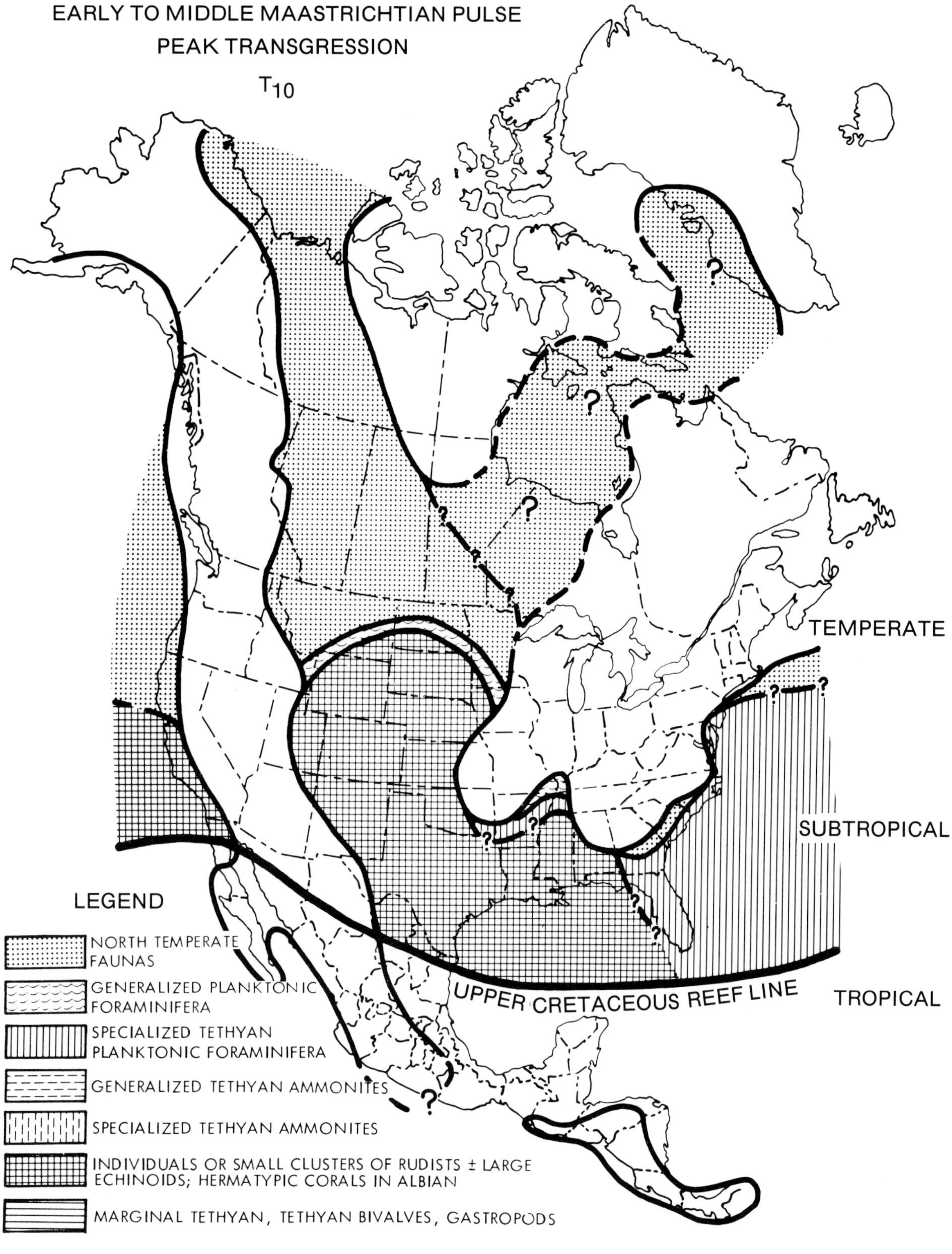

Figure 13. Generalized map of Western Interior Cretaceous Seaway of North America during a small Middle Maastrichtian eustatically-generated transgression associated with deposition of the Fox Hills Sandstone. Map shows distribution of various Subtropical organisms (patterns keyed to lower left of diagram), defining extensive, short-term migration of Warm Temperate and Subtropical paleobiogeographic units during rapid incursion of warm, normal marine waters into the seaway at small eustatic highstand. Compare with more normal paleobiogeographic distributions in Figure 7.

eurytopy *vs.* stenotopy); degree of habitat protection from unpredictable perturbations; and by the rate and magnitude of biological stress placed on these lineages through time by shifting physical, chemical, and biological factors of the environment. These initial analyses showed that virtually all episodes of rapid evolution recorded for different Cretaceous lineages were associated with peak eustatic rise in sealevel (peak-transgression) and throughout regression. These were times of rapid environmental changes followed by longer deterioration of marine habitats through loss or great restriction of prime niches; diminished ecospace; declining salinity, temperature and, sporadically, oxygen in the seaway; and by onset of highly turbid shallow-water conditions with increasing competition among surviving taxa for resources and habitats. Thus, in general, stenotopic, shallow-marine organisms, especially epibenthic and neritic taxa, were most affected by major fluctuations in sealevel, temperature, salinity, oxygen, turbidity, trophic resources, habitat size, and niche diversity. They were first to respond to eustatically related changes during peak-transgression and early regression. More eurytopic taxa did not undergo rapid evolution and extinction until high stress levels had been reached, as during the most rapid oceanographic changes associated with peak-transgression, or during the last phases of a major eustatic fall and widespread epicontinental regression. Some highly adapted taxa (e.g., intertidal and brackish water forms), and others with highly protected niches (deep burrowing or deep water taxa), showed little change in association with sealevel fluctuations.

Equally important was the documentation of major changes in evolutionary rates *within* lineages (Kauffman, 1977d, 1978) in direct relation to the rate and intensity of environmental change causing biological stress. With these relationships broadly documented, we can now turn our attention to paleobiogeographic impact on rates and patterns of evolution within the Western Interior Seaway.

Two major kinds of paleobiogeographic controls on Cretaceous evolutionary history can be envisioned in the seaway, one short-term and one longer-term. (1) The effect, during peak eustatic rise of sealevel, of major stress placed on Western Interior biotas due to rapid changes in oceanographic factors (e.g., temperature, salinity and oxygen). This was amplified by ecological disruption of the normal Western Interior biota through abrupt northward displacement of Temperate climatic belts and by the broad competitive boundaries established between Mild to Cool Temperate resident biotas and rapidly immigrating Warm Temperate to Subtropical biotas of the Gulf and Atlantic Coast Subprovince and the Caribbean Province. (2) Longer-term disruption of Western Interior ecosystems at major boundaries of paleobiogeographic units, such as the broad ecotonal zones of intermixing between discrete biotas of the Northern, Central, and Southern Interior Subprovinces, and between these Temperate Zone biotas and the Subtropical biota of the Gulf and Atlantic Coast Subprovince of the North American Province (Fig. 7; Kauffman, 1973). These two effects are difficult to separate in some cases.

Several tests have been made (Kauffman, 1972, 1977d, 1978) in an attempt to relate evolutionary rates to sealevel changes among Cretaceous molluscs of the Western Interior and Gulf Coast (Fig. 14 summarizes data); these analyses utilize number of species arising within lineages per million years as a measure of evolutionary rates (Kauffman, 1978, Fig. 1). Whereas these data clearly show that the highest evolutionary rates for many groups occurred during middle to late regressive stages of eustatic fall (e.g., Kauffman, 1972, Figs. 1-5; 1977d, Figs. 5-7; Figs. 3-5) associated with rapidly deteriorating epicontinental marine environments and loss or restriction of many prime habitats (i.e., high stress conditions), the second highest peaks of evolution in these data occur just before or during peak transgression, associated with rapid environmental and paleobiogeographic fluctuations. For example (Fig. 14), Kauffman has shown (1972, 1977d, 1978) that the middle to peak-transgressive phases of the Cenomanian eustatic rise (T_6; Fig. 4) are times of especially rapid speciation among many lineages of Bivalvia (*Nucula*, Tellinidae, *Yoldia, Cymbophora, Callistina, Inoceramus, Mytiloides, Camptonectes* and *Pycnodonte*), as well as among certain lineages of scaphitid ammonites. Kauffman has further shown unusually rapid to punctuated evolution in morphologic features of the *Inoceramus pictus* lineage during the latest Cenomanian (1977d, Fig. 4). These bursts of evolution coincide with the interval of greatest oceanographic and paleobiogeographic change during eustatic rise, as determined from geochemical, sedimentologic and biologic data. Important but less spectacular evolutionary peaks were also documented near the peak Coniacian-Santonian rise of sealevel (T_7; Fig. 4) for the bivalve lineages of *Protodonax* (intertidal), *Cremnoceramus*, and the ammonite genera *Baculites, Scaphites, Pteroscaphites*, and *Clioscaphites*. Peak Campanian transgressions show increasing evolutionary rates among the bivalves lineages of *Thyasira, Protodonax, Endocostea, Platyceramus* and *Pycnodonte* (Kauffman, 1978, Figs. 3-5) and the ammonites of the *Trachyscaphites* and *Hoploscaphites* lineages (Kauffman, 1977d, Fig. 6 modified to show sealevel highstand in Santonian and late Early Campanian).

Similar analyses of evolutionary rates among Cretaceous Mollusca of the Texas Gulf Coast (Kauffman, 1972, Figs. 1-4) strongly support observations from the Western Interior that abrupt paleobiogeographic and environmental changes around eustatic highstand and peak-transgression result in accelerated speciation rates. Analyses of 18 lineages of marine gastropods and 11 lineages of bivalves from the Gulf Coast show that the highest evolutionary rates for both classes and virtually all tested lineages occur at and just after peak-transgression of the late Early Campanian eustatic rise (T_8; Fig. 4); a strong secondary peak in bivalve evolution occurs early in the Late Campanian to lower Maastrichtian regression (R_9; Fig. 4), in association with a small fourth-order Early Maastrichtian transgressive peak. The stronger evolutionary effect of these maximum transgressions in the Gulf Coast may represent larger-scale thermal changes at the transgressive peak because of closer proximity to the migrating Subtropical-Warm Temperate climatic boundary. All new data that have been generated on lineage evolution in the Gulf and Western Interior biotas support these early observations.

I conclude from these extensive data that abrupt changes in marine environments (rapid fluctuations in temperature, oxygen, and salinity), and in the position of paleobiogeographic boundaries as they rapidly migrated northward across the Western Interior Seaway at peak eustatic rise, have severely stressed marine biotas of these areas, driving evolutionary rates upward within many lineages. It is difficult to sort out the physico-chemical stresses from those inherent in biological competition set up by immigration and establishment of ecologically disrupted ecotones, or from migrations of paleobiogeographic units over hundreds to more than 1000 miles (1600 km) during one to three million-year intervals. Nevertheless, the distribution of the data showing rapid evolution clustered tightly around transgressive peaks and limited to the known range of paleobiogeographic mixing (Sohl, 1967, 1971), or ecotone, strongly suggest at least a partial control on evolutionary rates due to biological stress related to rapid shifts in paleobiogeographic units.

A second biogeographic control on evolutionary rates and patterns among Cretaceous biotas is related to the establishment of broad paleobiogeographic ecotones at the junctions of Western Interior and Coastal Plan subprovinces. Though dynamic in the sense of broad and frequent north-south migrations, these mixing zones of Temperate and Subtropical biotas persisted for the 35 Ma history of the connection between northern and southern arms of the seaway. Unlike modern biogeographic ecotones, which extend for only tens of miles on either side of provincial boundaries, Cretaceous ecotones of the Western Interior were broad and graded over 500 to 1000 miles (800 to 1600 km, e.g., Colorado to Montana in the Campanian; Sohl, 1967, 1971) during much of their history, completely encompassing the Central Interior Subprovince. During major paleobiogeographic shifts, as near peak eustatic rise and transgression, the entire ecotone moved hundreds of miles north or south, and expanded or contracted in size. These perturbations should have had a profound effect on evolutionary history of Cretaceous marine biotas, as should the very existence of large biogeographic ecotones in the seaway.

Biogeographic units are expanded ecological units composed

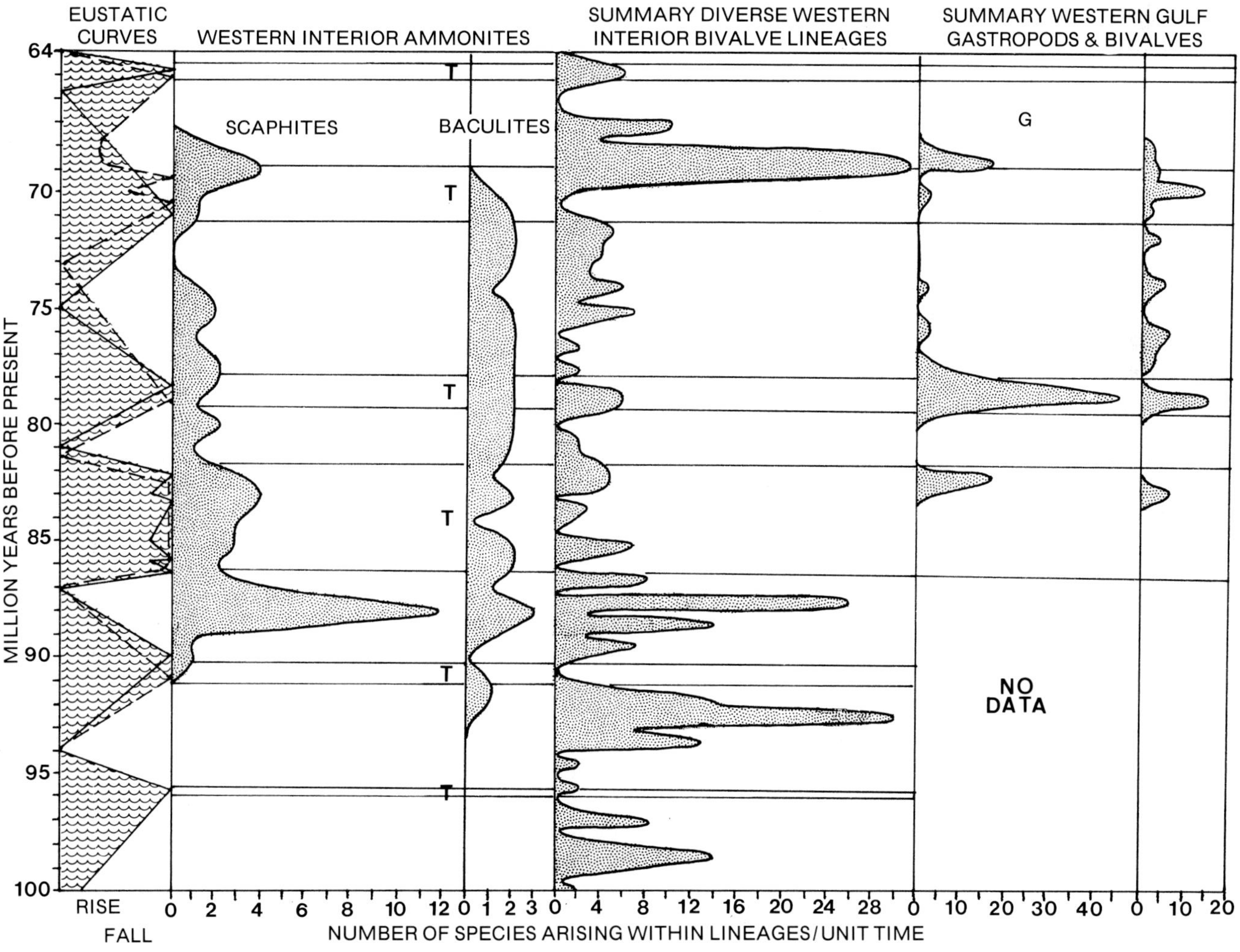

Figure 14. Summary diagram of molluscan evolutionary rates, as defined by number of species arising within single lineages per million-year interval, plotted against third-order Cretaceous tectonoeustatic cycles of the Western Interior Seaway (solid line) and western Gulf Coastal Plain (dashed line) (data from Kauffman, 1972, 1978, 1979). Note high correlation of episodes of rapid evolution with: (a) high-stress conditions during middle and late phases of eustatic fall and epicontinental regression, and (b) with rapid thermal, salinity, and dissolved oxygen fluctuations associated with warm water incursions during late transgressions and near eustatic highstands.

of numerous, at least broadly structured communities. For this reason one could predict that both community and biogeographic ecotones are zones of biological competition between similarly adapted taxa from discrete bounding communities and are thus biologically stressed and ecologically disrupted; these conditions favor increased endemism and more rapid evolution. Data exist from modern marine ecotones to support this contention. Newman (1979) and Spivey (1981) have demonstrated among living barnacles of the Pacific Coast and Gulf of Mexico that biogeographic ecotones with strong environmental gradients are characterized by extensive faunal mixing, lower than normal population size for many taxa (reflecting competition for the same space and resources), the highest levels of endemism for the group, and the presence of refugia species in open competition with species that had previously displaced them in well-structured ecosystems. Consequently, diveristy is higher than normal for barnacles in biogeographic ecotones.

These limited observations suggst a hypothesis which might be applied to understanding the evolutionary history of Cretaceous biotas in the Western Interior of North America, i.e.: Ecotones are zones of disruption of ecosystems because of the broad competition between similarly adapted taxa from different communities or biogeographic units in the same space – the zone of biogeographic overlap. They comprise a stressful environment, yet offer opportunities for new and unique adaptations because of the incomplete and inefficient use of ecospace made possible by establishment of a broad competitive boundary. This situation fosters the origination and rapid evolution of endemic taxa specifically adapted to ecotone niches. On the limited scale of modern ecotonal boundary zones this may result in a small number of endemics (ecotonal species) with restricted population size and geographic range; but during the Phanerozoic when climatic gradients, biogeographic boundary-zones, and thus ecotones were much broader and stabilized over longer periods, these ecotones may have acted as major centers of evolution among more widely spread endemic stocks. If so, they are extremely important to evolutionary theory, and may partially explain the broad variations in evolutionary rates and patterns observed for the biotas of the Western Interior Cretaceous Seaway (Kauffman, 1962, 1977d, 1978) and elsewhere. This hypothesis is testable with the extensive evolutionary data and high-resolution stratigraphy available for the Western Interior Cretaceous Seaway.

As a basic test of this hypothesis the principal biogeographic ranges of most Western Interior endemic molluscs (data from numerous sources) have been plotted against the average distribution of Temperate biogeographic subprovinces in the Western Interior (Fig. 7); the areal extent of this center of Western Interior endemism overlaps both major paleobiogeographic boundary zones in the center of the basin and extends north well into Canada, generally supporting the hypothesis that the boundary zones are a major center of endemic evolution.

The northern extension of this endemic center beyond the Mild Temperate-Cool Temperate biogeographic boundary seems puzzling at first, but the Campanian data of Sohl (1967; Fig. 7) clarify this in showing that the main zone of biogeographic overlap of gastropod taxa partially defining the broad

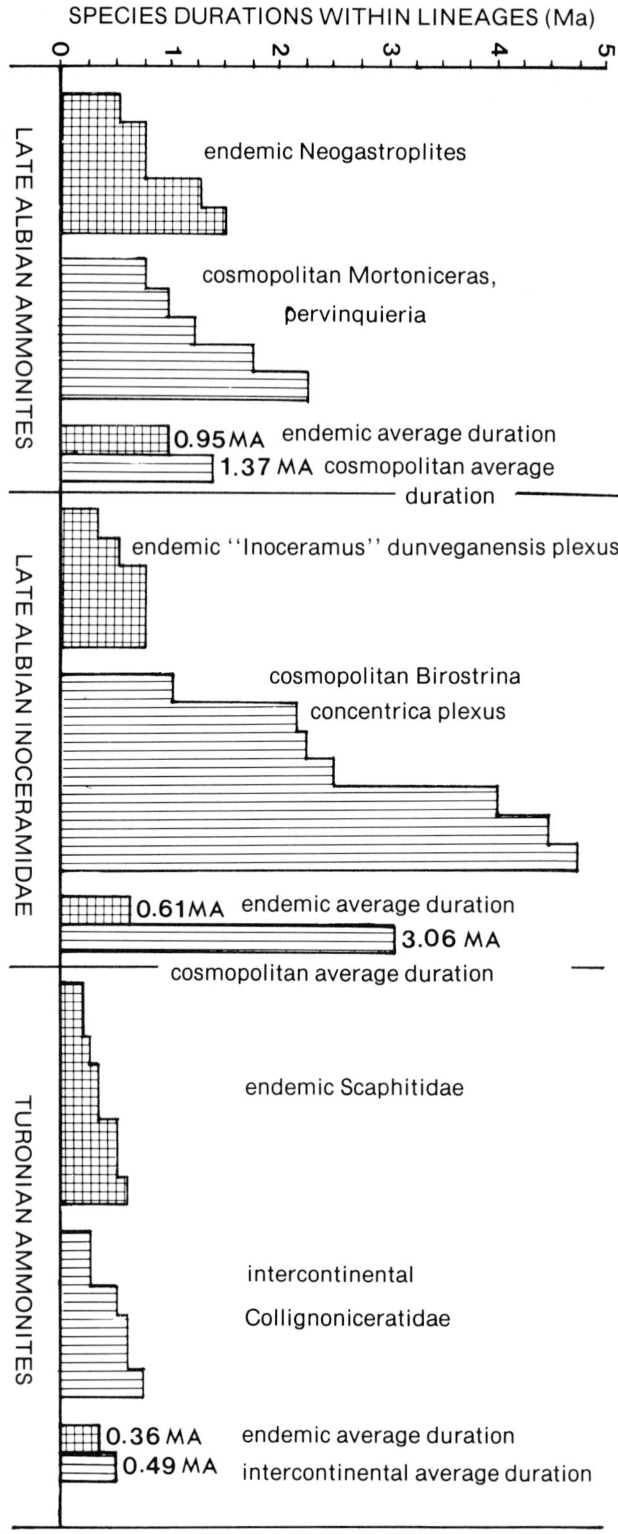

Figure 15. Comparative evolutionary rates between intercontinental or cosmopolitan molluscs and those of the Western Interior paleobiogeographic ecotones, as measured by average species durations within lineages of ammonites and inoceramid bivalves. Note higher rates of evolution for endemic ecotonal lineages arising at the competitive, biologically stressed boundaries between paleobiogeographic subprovinces.

ecotone between the Northern Interior (Cool Temperate) Subprovince and the Southern Interior (Warm Temperate) Subprovince, actually extends from central Colorado well into Montana and, probably, southern Alberta. My own unpublished data from Cretaceous Bivalvia supports Sohl's observations completely. This northern offset of the ecotone from the Central Interior-Northern Interior Subprovince boundary probably reflects the dominance of southern warm-water surface currents over deeper, northern cool-water counter-currents in the seaway. This hypothesis is further supported by the relatively greater magnitude of northward Subtropical Tethyan biotic incursions during peak-transgression, as opposed to the more limited southward migration of Cool Temperate organisms between transgressive peaks. Thus, the general biogeographic distributional patterns and close spatial correlation through time of both endemic lineages and the Western Interior ecotones support the hypothesis that these zones of biogeographic overlap and ecosystem disruption fostered evolution of unique endemic taxa over a broad area in the Cretaceous, and possibly at other times during the Phanerozoic.

A second test of the hypothesis can be made by plotting the evolutionary rates of endemic taxa within these ecotones against those of contemporaneous taxa with broader Temperate Zone to cosmopolitan biogeographic distribution. Lineages of ammonites and inoceramid bivalves, representing the best studied molluscs for the Western Interior Seaway with a broad range of biogeographic distributional patterns, were selected for testing. Figure 15 shows the results of this analysis; in every case, evolutionary rates of endemic lineages arising within the biogeographic ecotone were more rapid than those of contemporaneous and ecologically similar lineages with broader paleobiogeographic ranges. Future analyses will focus on the timing and biogeographic migration patterns of evolution in these endemic centers.

CONCLUSIONS

Analyses of Cretaceous paleobiogeographic distribution patterns in the Western Interior Seaway of North America, utilizing modern concepts of biogeography and similar analytical levels of endemism to define biogeographic units, reveals the presence of three subprovinces, representing (north to south) Cool Temperate, Mild Temperate and Warm Temperate climatic zones. These subprovinces were definable during most of the history of the seaway. These subprovinces equate in general biotic character to modern Atlantic Coastal provinces between Nova Scotia and the Carolinas plus the northern Gulf of Mexico. Their greater biogeographic spread, compared to modern analogs, reflects a broader and warmer climatic gradient during the Cretaceous, lacking cold polar zones. Whereas an average distributional pattern for these subprovinces can be plotted (Fig. 7), reflecting conditions during most early to mid-transgressive and mid- to late-regressive eustatic episodes, these subprovinces were highly dynamic in space and time, especially during peak-transgression (eustatic highstand).

Each transgressive peak for third-order tectonoeustatic cycles in the Western Interior is characterized by a short (0.5-2.5 Ma) interval during which southern Warm Temperate and Subtropical taxa, biogeographic units, and ecotonal boundary zones migrated rapidly far to the north in the seaway, normally reaching the Canadian border or southern Alberta and Saskatchewan. Similar but lesser incursions of warm-water taxa are associated with some fourth-order transgressive peaks as well, suggesting possible eustatic control on these 1.5 Ma long events. The rapidity with which these migrations took place, displacing Temperate Zone subprovinces and their ecotonal boundaries northward, suggests some threshold effect at the southern end of the Western Interior Basin. Possibly, a structural and sedimentologic barrier or sill existed at the southern end of the seaway and was rapidly overstepped at some critical point in sealevel rise, allowing warm-water masses to flood quickly into the Interior. During these peak-transgressive intervals, a four-fold biogeographic division of the Interior seaway was possible, therefore, with the Gulf and Atlantic Coast Subprovince (Subtropical to very Warm temperate) migrating across the central and northern part of the United States interior, and Temperate subprovinces being displaced northward. Retreat of these warm-water biotas during sealevel fall was somewhat slower, but still took less than 1.5 Ma in most cases.

These migrations had a profound effect on the evolution of lineages and ecosystems in the Western Interior Cretaceous. In particular, the development and rapid migration of broad paleobiogeographic ecotones at subprovince boundaries, characterized by extensive overlap of southern warm-water and northern cool-water taxa and disruption of ecological structure, provided a biologically and physico-chemically stressed environment in which the origin and rapid evolution of endemic molluscan taxa were favored. Tests of this hypothesis are wholly positive. Plots of the biogeographic range of the ecotones *vs.* the principal regions where endemic taxa have arisen in the Western Interior Seaway show almost complete overlap (Fig. 7). Evolutionary rates of endemic taxa arising in these ecotonal zones are slightly to significantly greater than those of contemporaneous, co-occurring taxa with similar ecological requirements but broader biogeographic distribution. From these analyses, I conclude that paleobiogeographic ecotones were dynamic, in space and time; they migrated broadly and rapidly, were ecologically disrupted for long intervals, and thus presented unique opportunities for the origination and rapid evolution of initially endemic taxa. Biogeographic ecotones probably have been important centers for evolution throughout the Phanerozoic and are deserving of careful study in consideration of evolutionary theory.

REFERENCES

Addicott, W.O., 1966, Late Pleistocene Marine Paleoecology and Zoogeography in Central California: United States Geological Survey Professional Paper 523-C, p. C1-C21.

———, 1969, Tertiary Climatic Change in the Marginal Northeastern Pacific Ocean: Science, v. 165, p. 583-586.

Bergquist, H.R., 1944, Cretaceous of the Mesabi Iron Range, Minnesota: Journal of Paleontology, v. 18, p. 1-30.

Boucot, A.J., and Johnson, J.G., 1973, Silurian Brachiopods: *in* Hallam, A., ed., Atlas of Paleobiogeography: Amsterdam, London, New York, Elsevier Scientific Publishing Company, p. 59-65.

Caldwell, W.G.E., and Evans, J.K., 1963, A Cretaceous Rudist from

Canada and a Redescription of the Holotype of *Ichthyosarcolites coraloidea* (Hall and Meek): Journal of Paleontology, v. 37, p. 615-620.

Coates, A.G., Kauffman, E.G., and Sohl, N.F., 1984, Tethyan Paleographic Incursions into North Temperate Seas: Cretaceous Research, (in press).

Cobban, W.A., 1951a, Scaphitoid Cephalopods of the Colorado Group: United States Geological Survey Professional Paper 239, 42 p.

———, 1951b, New Species of *Baculites* from the Upper Cretaceous of Montana and South Dakota: Journal of Paleontology, v. 25, p. 817-821.

———, 1958, Two New Species of *Baculites* from the Western Interior Region: Journal of Paleontology, v. 32, p. 660-665.

———, 1962a, New *Baculites* from the Bearpaw Shale and Equivalent Rocks of the Western Interior: Journal of Paleontology, v. 36, p. 126-135.

———, 1962b, *Baculites* from the Lower Part of the Pierre Shale and Equivalent rocks in the Western Interior: Journal of Paleontology, v. 36, p. 704-718.

———, 1969, The Late Cretaceous Ammonites *Scaphites leei* Reeside and *Scaphites hippocrepis* (DeKay) in the Western Interior of the United States: United States Geological Survey Professional Paper 618, p. 1-29.

Cobban, W.A., and Merewether, E.A., 1983, Stratigraphy and Paleontology of Mid-Cretaceous Rocks in Minnesota and Contiguous Areas: United States Geological Survey Professional Paper 1253, p. 1-52.

Cobban, W.A., and Scott, G.R., 1972, Stratigraphy and Ammonite Fauna and the Graneros Shale and Greenhorn Limestone Near Pueblo, Colorado: United States Geological Survey Professional Paper 645, 108 p.

Coomans, H.E., 1962, The Marine Mollusk Fauna of the Virginian Area as a Basis for Defining Zoogeographical Provinces: Beaufortia, v. 9, no. 98, p. 83-104.

Cragin, F.W., 1895, The Mentor Beds: A Central Kansas Terrane of the Comanchean Series: American Geologist, v. 16, p. 162-165.

Culver, S.J. and Buzas, M.A., 1981a, Distribution of Recent Benthic Foraminifera in the Gulf of Mexico: Smithsonian Contributions to Marine Science, no. 8, 898 p.

———, 1981b, Foraminifera; Distribution of Provinces in the Gulf of Mexico: Nature, v. 290, p. 328.

———, 1982a, Recent Benthic Foraminiferal Provinces between Newfoundland and Yucatan: Geological Society of America Bulletin, v. 93, p. 269-277.

———, 1982b, Distribution of Recent Benthic Foraminifera in the Caribbean Region: Smithsonian Contributions to Marine Science, no. 14, 382 p.

Douglas, R.G., 1972, Paleozoogeography of Late Cretaceous Planktonic Foraminifera in North America: Journal of Foraminiferal Research, v. 2, no. 1, p. 14-34.

Eicher, D.L., 1977, Stratigraphic Distribution of Cretaceous Foraminifera, Rock Canyon Anticline; *in* Kauffman, E.G., ed., Cretaceous Facies, Faunas, and Paleoenvironments Across the Western Interior Basin: Mountain Geologist, Rocky Mountain Association of Geologists, v. 14, no. 3, 4, p. 153-154.

Ekman, S., 1967, Zoogeography of the Sea: London, Sidgwick and Jackson, 2nd Edition, 417 p.

Frerichs, W.E., Atherton, C.C., and Shive, P.N., 1975, Planktonic Foraminifera from the Chalky Members of the Niobrara Formation: Journal of Foraminiferal Research, v. 7, p. 149-156.

Gill, J.R., and Cobban, W.A., 1973, Stratigraphy and Geologic History of the Montana Group and Equivalent Rocks, Montana, Wyoming, and North and South Dakota: United States Geological Survey Professional Paper 776, 37 p.

Hall, C.A., 1964, Shallow Water Marine Climates and Molluscan Provinces: Ecology, v. 45, no. 2, p. 226-234.

Hallam, A., ed., 1973, Atlas of Palaeobiogeography: Amsterdam, London, New York, Elsevier Scientific Publishing Company, 531 p.

Hancock, J.M., and Kauffman, E.G., 1979, The Great Transgressions of the Late Cretaceous: Journal of the Geological Society of London, v. 136, pt. 2, p. 175-186.

Hattin, D.E., 1962, Stratigraphy of the Carlile Shale (Upper Cretaceous) in Kansas: Kansas Geological Survey Bulletin 156, 155 p.

———, 1964, Cyclic Sedimentation in the Colorado Group of West-Central Kansas: *in* Merriam, D.F., eds., Symposium on Cyclic Sedimentation: Kansas Geological Survey Bulletin 169, p. 205-217.

———, 1965, Stratigraphy of the Graneros Shale (Upper Cretaceous) in Central Kansas: Kansas Geological Survey Bulletin 178, 83 p.

———, 1975, Stratigraphy and Depositional Environment of Greenhorn Limestone (Upper Cretaceous) of Kansas: Kansas Geological Survey Bulletin 209, 128 p.

———, 1982, Stratigraphy and Depositional Environment of Smoky Hill Chalk Member, Niobrara Chalk (Upper Cretaceous) of the Type Area, Western Kansas: Kansas Geological Survey Bulletin 225, 108 p.

Hazel, J.E., 1970, Atlantic Continental Shelf and Slope of the United States; Ostracode Zoogeography in the Southern Nova Scotian and Northern Virginian Faunal Provinces: United States Geological Survey Professional Paper 529-E, p. 1-21.

Hughes, N.F., eds., 1973, Organisms and Continents through Time: Palaeontological Association Special Papers in Palaeontology, no. 12, Systematics Association Publication no. 9, 334 p.

Jeletzky, J.A., 1971, Marine Cretaceous Biotic Provinces and Paleogeography of Western and Arctic Canada: Illustrated by a Detailed Study of Ammonites: Geological Survey Canada Paper 70-22, p. 1-92.

Jordan, T., 1981, Thrust Loads and Foreland Basin Evolution, Cretaceous, Western United States: American Association of Petroleum Geologists Bulletin, v. 65, no. 12, p. 2506-2520.

Kauffman, E.G., 1967, Coloradoan Macroinvertebrate Assemblages, Central Western Interior United States: *in* Kauffman, E.G., and Kent, H.C., Co-organizers, Paleoenvironments of the Cretaceous Seaway in the Western Interior: Colorado School of Mines, Special Publication, p. 67-143.

———, 1969, Cretaceous Marine Cycles of the Western Interior: Mountain Geologist, Rocky Mountain Association of Geologists, v. 6, no. 4, p. 227-245.

———, 1972, Evolutionary Rates and Patterns of North American Mollusca: Proceedings of the XXI International Geological Congress, Montreal, 1972, sec. 7, p. 174-189.

———, 1973, Cretaceous Bivalvia: *in* Hallam, A., ed., Atlas of Palaeobiogeography: Amsterdam, London and New York, Elsevier Publishing Company, p. 353-383.

———, 1975, Dispersal and Biostratigraphic Potential of Cretaceous Benthonic Bivalvia in the Western Interior: *in* Caldwell, W.G.E., ed., The Cretaceous System in the Western Interior of North America: Geological Association of Canada Special Paper 13, p. 163-194.

———, 1977a, Geological and Biological Overview: Western Interior Cretaceous Basin: *in* Kauffman, E.G., ed., Cretaceous Facies, Faunas, and Paleoenvironments across the Western Interior Basin: Mountain Geologist, Rocky Mountain Association of Geologists, v. 14, nos. 3, 4, p. 75-99.

______, 1977b, Upper Cretaceous Cyclothems, Biotas, and Environments, Rock Canyon Anticline, Pueblo, Colorado: *in* Kauffman, E.G., ed., Cretaceous Facies, Faunas, and Paleoenvironments across the Western Interior Basin: Mountain Geologist, Rocky Mountain Association of Geologists, v. 14, nos. 3, 4, p. 129-152.

______, 1977c, Illustrated Guide to Biostratigraphically Important Cretaceous Macrofossils, Western Interior Basin, U.S.A.: *in* Kauffman, E.G., ed., Cretaceous Facies, Faunas, and Paleoenvironments across the Western Interior Basin: Mountain Geologists, Rocky Mountain Association of Geologists, v. 14, nos., 3, 4, p. 225-274.

______, 1977d, Evolutionary Rates and Biostratigraphy: *in* Kauffman, E.G., and Hazel, J.E., eds., Concepts and Methods of Biostratigraphy: Stroudsburg, PA, Dowden, Hutchinson, and Ross, Inc., p. 109-141.

______, 1978, Evolutionary Rates and Patterns among Cretacous Bivalvia: Philosophical Transactions of the Royal Society of London, Ser. B., no. 284, p. 277-304.

______, 1979, Cretaceous: *in* Robison, R.A., and Teichert, C., ed., Treatise on Invertebrate Paleontology: Pt. A, Introduction: Fossilization (Taphonomy), Biogeography and Biostratigraphy: Geological Society of America, University of Kansas Press, p. A418-A487.

______, 1984, The Fabric of Cretaceous Marine Extinctions: *in* Berggren, W., and Van Couvering, J., eds., Catastrophies and Earth History: The New Uniformitarianism: Princeton University Press, p. 151-246.

Kauffman, E.G., and Scholle, P.A., 1977, Abrupt Biotic and Environmental Changes during Peak Cretaceous Transgressions in Euramerica: North American Paleontological Convention II, Abstracts of Papers, 1977; Journal of Paleontology, v. 51, supplement to no. 2, p. 16.

Kauffman, E.G., and Scott, R.W., 1976, Basic Concepts of Community Ecology and Paleoecology: *in* Scott, R.W., and West, R.R., eds., Structure and Classification of Paleocommunities: Stroudsburg, PA, Dowden, Hutchinson, and Ross, Inc., p. 1-28.

Kauffman, E.G., Cobban, W.A., and Eicher, D.L., 1976, Albian through Lower Coniacian Strata and Biostratigraphy, Western Interior United States: *in* Reyment, R.A., and Thomel, G., eds., Mid-Cretaceous Events: Annales du Museum d'Histoire naturelle de Nice, T. 4, p. xxiii-1 – xxiii-56.

Kauffman, E.G., Powell, J.D., and Hattin, D.E., 1969, Cenomanian-Turonian Facies across the Raton Basin: Mountain Geologist, Rocky Mountain Association of Geologists, v. 6, p. 93-118.

Kennedy, W.J., and Cobban, W.A., 1976, Aspects of Ammonite Biology, Biogeography, and Biostratigraphy: Palaeontological Association, Special Papers in Palaeontology, no. 17, p. 1-94.

Kent, H.C., 1967, Paleoenvironments during Coniacian time in Colorado: *in* Kauffman, E.G. and Kent, H.C., Co-organizers, Paleoenvironments of the Cretaceous Seaway in the Western Interior: Colorado School of Mines, Special Publication, p. 203-236.

Koch, C.F., 1978, Bias in the Published Fossil Record: Paleobiology, v. 4, no. 3, p. 367-372.

______, 1980, Bivalve Species Duration, Areal Extent and Population Size in a Cretaceous Sea: Paleobiology, v. 6, no. 2, p. 184-192.

Larson, R.L., and Pitman, W.C., III, 1972, World-side Correlation of Mesozoic Magnetic Anomalies, and its Implications: Geological Society of America Bulletin, v. 83, p. 3645-36662.

Matsumoto, T., 1973, Late Cretaceous Ammonoidea: *in* Hallam, A., ed., Atlas of Palaeobiogeography: Amsterdam, London and New York, Elsevier Scientific Publishing Company, p. 421-429.

McGookey, D.P., Haun, J.D., Hale, L.A., Goodell, H.G., McCubbin, D.G., Weimer, R.J., and Wulf, G.R., 1972, Cretaceous System: *in* Geologic Atlas of the Rocky Mountain Region: Rocky Mountain Association of Geologists, p. 190-228.

Mello, J.F., 1969, Foraminifera and Stratigraphy of the Upper Part of the Pierre Shale and Lower Part of the Fox Hills Sandstone (Cretaceous) North-Central South Dakota: United States Geological Survey Professional Paper 611, 121 p.

______, 1971, Foraminifera from the Pierre Shale (Upper Cretaceous) at Red Bird, Wyoming: United States Geological Survey Professional Paper 393-C, p. C1-C54.

Newman, W.A., 1979, California Transition Zone: Significance of Short Range Endemics: *in* Gray, J., and Boucot, A.J., eds., Historical Biogeography, Plate Tectonics and the Changing Environment: Corvallis, Oregon State University Press, p. 399-416.

Perkins, B.F., 1974, Paleoecology of a Rudist Reef Complex in the Comanche Cretaceous Glen Rose Limestone of Central Texas: *in* Perkins, B.E., ed., Aspects of Trinity Division Geology: Geoscience and Man, v. 8, School of Geoscience, Louisiana State University, p. 131-174.

Pitman III, W.C., 1978, Relationship Between Eustacy and Stratigraphic Sequences of Passive Margins: Geological Society of America Bulletin, v. 89, p. 1389-1403.

Pratt, L.M., 1983, Isotopic Studies of Organic Matter and Carbonate in Mid-Cretaceous Strata near Pueblo, Colorado: *in* Kauffman, E.G., ed., Depositional Environments and Paleoclimates of the Greenhorn Tectonoeustatic Cycle, Rock Canyon Anticline, Pueblo, Colorado: Field Excursion Guidebook, Penrose Conference on Cretaceous Paleoclimates, October 3-7, 1983, p. 77-98.

Price, R.A., 1973, Large-Scale Gravitational Flow of Supracrustal Rocks, Southern Canadian Rockies: *in* DeJong, K.A., and Scholten, R., eds., Gravity and Tectonics: New York, J. Wiley Publishing, p. 491-502.

Reeside, J.B., and Cobban, W.A., 1960, Studies of the Mowry Shale (Cretaceous) and Contemporary Formations in the United States and Canada: United States Geological Survey Professional Paper 335, p. 1-126.

Rona, P.A., 1973, Relations Between Rates of Sediment Accumulation on Continental Shelves, Sea-Floor Spreading, and Eustacy Inferred from the Central North Atlantic: Geological Society of America Bulletin, v. 84, p. 2851-2872.

Ryer, T.A., 1977, Coalville and Rockport Areas, Utah: *in* Kauffman, E.G., ed., Cretaceous Facies, Faunas, and Paleoenvironments across the Western Interior Basin: Mountain Geologist, Rocky Mountain Association of Geologists, v. 14, nos. 3, 4, p. 105-128.

Scheltema, R.S., 1977, Dispersal of Marine Invertebrate Organisms; Paleobiogeographic and Biostratigraphic implications: *in* Kauffman, E.G. and Hazel, J.E., eds., Concepts and Methods of Biostratigraphy: Stroudsburg, PA, Dowden, Hutchinson and Ross, Inc., p. 73-108.

Sheridan, R.E., *et al.*, 1982, Early History of the Atlantic Ocean and Gas Hydrates on the Blake Outer Ridge: Results of Deep Sea Drilling Project Leg 76: Geological Society of America Bulletin, v. 93, p. 876-885.

Scholle, P.A., and Arthur, M.A., 1980, Carbon Isotopic Fluctuations in Cretaceous Pelagic Limestones: Potential Stratigraphic and Petroleum Exploration Tool: American Association of Petroleum Geologists Bulletin, v. 64, p. 67-87.

Scholle, P., and Kauffman, E.G., 1977, Paleoecological Implications of Stable Isotope Data from Upper Cretaceous Limestones and Fossils from the U.S. Western Interior: North American Paleontology Convention II, Abstracts of Papers, 1977; Journal of Paleontology, v. 51, supplement to no. 2, p. 24-25.

Scott, G.R., 1969, General and Engineering Geology of the Northern Part of Pueblo, Colorado: United States Geological Survey Bulletin 1261, 131 p.

Scott, R.W., 1970, Paleoecology and Paleontology of the Lower Cretaceous Kiowa Formation, Kansas: University of Kansas Press, Paleontological Contributions, Article 52 (Cretaceous 1), 94 p.

———, 1977, Early Cretaceous Environments and Paleocommunities in the Southern Western Interior: *in* Kauffman, E.G., ed., Cretateous Facies, Faunas, and Paleoenvironments across the Western Interior Basin: Mountain Geologist, Rocky Mountain Association of Geologists, v. 14, nos. 3, 4, p. 155-168.

Scott, R.W., and Brenckle, P.L., 1977, Biotic Zonation of a Lower Cretaceous Coral-Algal-Rudist Reef, Arizona: Proceedings of the 3rd International Coral Reef Symposium, University of Miami, p. 183-189.

Sohl, N.F., 1967, Upper Cretaceous Gastropod Assemblages of the Western Interior of the United States: *in* Kauffman, E.G. and Kent, H.C., Co-organizers, Paleoenvironments of the Cretaceous Seaway: Colorado School of Mines, Special Publication, p. 1-37.

Sohl, N.F., 1971, North American Cretaceous Biotic Provinces Delineated by Gastropods: Allen Press, Lawrence, KS, North America Paleontological Convention, Proceedings, Chicago, 1969, v. 2, pt. 1, p. 1610-1638.

Speden, I.G., 1970, The Type Fox Hills Formation, Cretaceous (Maestrichtian), South Dakota, Pt. 2, Systematics of the Bivalvia: Bulletin of the Peabody Museum of Natural History, Yale University, v. 33, 222 p.

Spivey, H., 1981, Origins, Distributions, and Zoogeographical Affinities of the Cirrepedia in the Gulf of Mexico: Journal of Biogeography, v. 8, p. 153-176.

Stephenson, L.W., 1952, Larger Invertebrate Fossils of the Woodbine Formation (Cenomanian) of Texas: United States Geological Survey Professional Paper 242, p. 1-211.

Stevens, G.R., 1973, Cretaceous Belemnites: *in* Hallam, A., ed., Atlas of Palaeobiogeography: Amsterdam, London and New York, Elsevier Scientific Publishing Company, p. 385-401.

Stoyanow, A., 1949, Lower Cretaceous Stratigraphy in Southeastern Arizona: Geological Society of America Memoir 28, 169 p.

Trexler, D.W., 1967, Nannoplankton (Coccoliths) as Possible Indicators of Paleocurrents for the Cretaceous Sea of the Western Interior: *in* Kauffman, E.G. and Kent, H.C., Co-organizers, Paleoenvironments of the Cretaceous Seaway in the Western Interior: Colorado School of Mines, Special Publication, p. 197-201.

Twenhofel, R., 1924, The geology and Invertebrate Paleontology of the Comanchean and "Dakota" Formations of Kansas: Kansas Geological Survey Bulletin 9, 128 p.

Vail, P.R., Mitchum, R.M., and Thompson III, S., 1977, Relative Changes of Sealevel from Coastal Onlap: Memoirs of the American Association of Petroleum Geologists, no. 26, p. 63-87.

Van Hinte, J.E., 1976, A Cretaceous Time Scale: American Association of Petroleum Geologists Bulletin, v. 60, p. 498-516.

Waage, K.M., 1968, The Type Fox Hills Formation, Cretaceous (Maestrichtian), South Dakota: Part 1, Stratigraphy and Paleoenvironments: Peabody Museum of Natural History, Bulletin 27, p. 1-157.

Weimer, R.J., 1959, Upper Cretaceous Stratigraphy, Colorado: *in* Haun, J.D., and Weimer, R.J., eds., Symposium on Cretaceous Rocks of Colorado and Adjacent Areas: Rocky Mountain Association of Geologists, Guidebook, 11th Field Conference, Washakie, Sand Wash, and Piceance Basins, p. 9-16.

———, 1960, Upper Cretaceous Stratigraphy, Rocky Mountain Area: American Association of Petroleum Geologists Bulletin, v. 4, p. 1-20.

Whittington, H.B., 1973, Ordovician Trilobites: *in* Hallam, A., ed., Atlas of Palaeobiogeography: Amsterdam, London and New York, Elsevier Scientific Publishing Company, p. 13-18.

Williams, G.D., and Stelck, C.R., 1975, Speculations on the Cretaceous Palaeogeography of North America: *in* Caldwell, W.G.E., eds., The Cretaceous System in the Western Interior of North America: Geological Association of Canada Special Paper 13, p. 1-20.

Wiltschko, D.V. and Dorr, J.A., Jr., 1983, Timing of Deformation in Overthrust Belt and Foreland of Idaho, Wyoming, and Utah: Bulletin of the American Association of Petroleum Geologists, v. 67, no. 8, p. 1304-1322.

Young, K., 1972, Cretaceous Paleogeography: Implications of Endemic Ammonite Faunas: Bureau of Economic Geology, University of Texas, Geological Circular 72-2, p. 1-13.

Jurassic-Cretaceous Biochronology and Paleogeography of North America,
edited by G.E.G. Westermann,
Geological Association of Canada Special Paper 27, 1984

SUMMARY OF SYMPOSIUM PAPERS ON THE JURASSIC-CRETACEOUS BIOCHRONOLOGY AND PALEOGEOGRAPHY OF NORTH AMERICA

G.E.G. Westermann
Department of Geology, McMaster University, Hamilton, Ontario L8S 4M1

INTRODUCTION

The diversity in subject matter and scope of articles in this symposium volume reflect the variety of research carried out at present on Jurassic and Cretaceous events in North America. Much of the research is of the type done by the two honorees, Ralph W. Imlay and George W. Jeletzky.

This summary will attempt to clarify the interrelations among such different fields of endeavour as biochronology, paleobiogeography, paleogeology and plate tectonics. Clearly, no continent-wide and certainly no global understanding is possible without strict time frames in the sequence of events in the development of the earth's surface, hydrosphere, atmosphere and biosphere; "4-dimensional" biomes, i.e., changing biogeographic provinces, can be better understood and interpreted in the light of plate tectonic theory, which they may support or limit in particular cases.

This review follows the papers as sequenced geographically, i.e., clockwise from the eastern trailing edge of the subcontinent and the opening North Atlantic Ocean, to the more complicated Gulf region and Mexico, to the largely active Pacific leading edge, the Arctic, and finally, to the cratonic interior. This leads to a mixture of research methods applied and taxa used but this obviously cannot be helped.

Imlay, using the fixist, conservative model, gave a splendid overview of the paleogeography, biogeography, biostratigraphy and chronostratigraphy of the entire subcontinent. He developed this synthesis almost entirely on his own during several decades (see Kauffman's address), culminating in his monumental work of 1980. Some details, however, were outdated by the time of publication as the manuscript was submitted in 1976; the maps and biostratigraphic charts provided in this volume are updated. Imlay carried out abundant regional and supraregional chronostratigraphic correlations with admirable success (see Callomon's views discussed below) even without defining proper chronostratigraphic zones. He treated all data as factually as possible and evidently considered an independent North American zonal system superfluous within the framework of the European standard stages and zones. Imlay's series of paleogeographic maps illustrate better than any brief review the major transgressive cycles and outlines of the Jurassic seas, especially in the continental interior.

EAST COAST AND NORTH AMERICAN BASIN

The opening history of the central Atlantic (southern North Atlantic), as is now considered well established by most geologists adhering to plate tectonic theory, began south of the present Azores-Gibralter Ridge, i.e., the North American basin, and later extended south- and northward. Triassic extension resulted in crustal thinning by deep listric faults marked by basaltic intrusions, while the consequent subsidence and marine deposition began with halite in the Late Triassic (Ascoli *et al.*, this volume). Continental separation and the generation of the ocean floor between the present Azores-Gibraltar ridge and Greater Antilles started in the late Mid-Jurassic (165-170 Ma, DNAG time scale; Kent and Gradstein *in* Palmer, 1983) (Fig. 1), while separation between the then subsided Grand Banks and Portugal occurred only in the earliest Cretaceous (ca. 130-135 Ma, DNAG scale), and the opening between Labrador and Ireland occurred in the mid-Cretaceous (ca. 100-110 Ma, DNAG scale; Hallam, 1981). The Grand Banks of Newfoundland and the Lusitanian Basin of central Portugal are good subjects for a pre-drift comparison of their sedimentary and biotic histories, particularly since the Portuguese Jurassic is well known and both basins of deposition were in close proximity during the Jurassic.

The Proto-Atlantic (Hispanic Corridor)

Exton and Gradstein (this volume) compared, in detail, the Triassic- Early Jurassic rock sequence and the Early Jurassic fossil succession of both the Grand Banks and the Lusitanian Basin, and found they were, at least lithostratigraphically, closely similar. Based on diverse microfossil dates (microflora, Foraminifera, and Ostracoda) in the Grand Banks subsurface, over 3050 m of predominantly marine sediments are believed to be present. All of the standard European Jurassic stages have

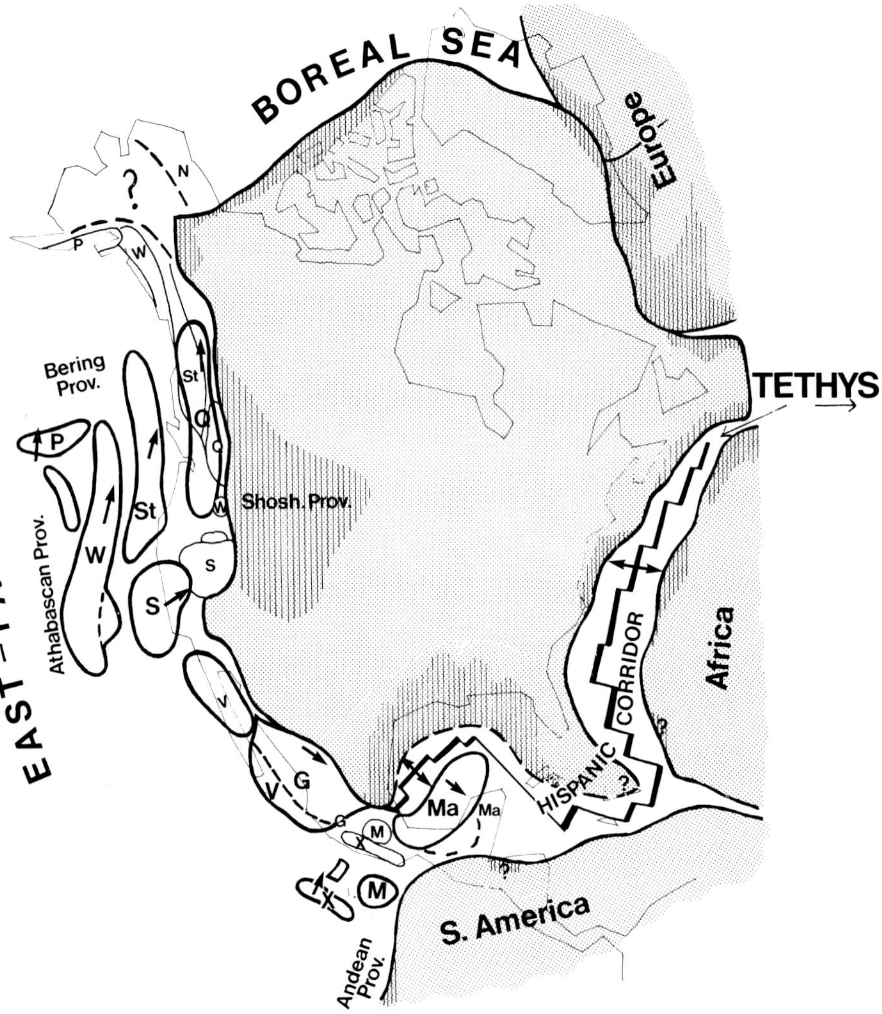

Figure 1. Schematic and simplified reconstruction of North America in the Early and Middle Jurassic, with maximally allowable displacements of allochthonous terranes and Atlantic opening; cratons in grey. Terrane abbreviations: G, Goodness; M, Mixteca; Ma, Maya; N, North Slope; P, Peninsular; Q, Quesnellia; S, Sonoma; ST, Stikinia, V, Vizcaino; W, Wrangellia; X, Xolapa.

been recognized, and the authors believe that the Grand Banks Jurassic sequence is among the stratigraphically most complete in North America. In central Portugal, and probably the Grand Banks according to correlation, sedimentation began in (? Middle-) Late Triassic to earliest Jurassic grabens and fault-bounded basins. Triassic redbeds and Hettangian evaporites lie unconformably on Precambrian and Paleozoic basement. These sediments are overlain by thick-bedded dolostones of the Sinemurian-early Pliensbachian followed by limestones, marls and shales extending well into the Middle Jurassic and now reflecting normal marine conditions. The Coimbre Dolostones of Portugal have their counterpart in the Murre carbonate of the Grand Banks, while the superjacent Lusitanian limestones and pelites are represented by the Whale Unit. Even a calcareous, probably regressive, facies at the Pliensbachian-Toarcian boundary within an argillaceous series is common between both basins. Immediately southward, on the Scotian Shelf and in southern Portugal, however, predominant calcareous deposition continues throughout this time, marking the western extension of the West-Tethys carbonate platform.

Exton and Gradstein concluded that, within the limits of resolution of Grand Banks biostratigraphy, lithofacies changes in the Grand Banks basins are synchronous with those of the Lusitanian basins. Resolution, however, is low since biostratigraphy for the entire Lusitanian Lower Jurassic is divided into only three ostracod-foraminiferal zones. The first, the *Frondicularia terquemi* Assemblage Zone (Upper Sinemurian-basal Toarcian) is incomplete with its base being a facies boundary (elsewhere in Europe it includes the basal Jurassic). It contains two subzones: the *Brizelina liasica* Acme Subzone which is correlated with the upper Margaritatus Standard Zone, Upper Pliensbachian (although the range zone of *B. liasica* is said to be Ibex through Margaritatus Zones), and the *Ogmoconcha* cf. *ambo* Range Subzone which is coeval to the Upper Spinatum through Lower Tenuicostatum Zones, Upper Pliensbachian and Lower Toarcian. The second, the *Bairdiacyoris* sp. Assemblage Zone, is correlated with the Tenuicostatum Zone through the Variabilis Zone, late Early and Middle Toarcian. The third, the *Lenticulina dorbignyi* Assemblage Zone, extends from the Early Toarcian Thoarcense Zone through the entire Aalenian.

Thus, each microfaunal zone corresponds to three to six ammonite standard zones or approximately one entire stage, and microfauna is virtually absent in the Hettangian and most of the Sinemurian.

On the Grand Banks, only two Early Jurassic microfaunal zones can be distinguished. The *Involutina liasica* "Interval Zone" contains numerous species of the Lusitanian *F. terquemi* Zone with which it is, therefore, correlated. Only *I. liasica* and one other Grand Banks foraminiferal species have not been found in Portugal, while Portuguese ostracod species appear to be absent, possibly due to poor sampling from the Grand Banks. Higher up in the Grand Banks sequence follow, as in Portugal, the *Bairdiaecypris* sp. and *L. dorbignyi* Zones. All in all, as many as 20 to 22 species are common between both regions.

The Early Jurassic foraminiferal assemblages of the Grand Banks and Portugal, however, belonged to cosmopolitan faunas which occupied epicontinental seas before the fragmentation of Laurasia during the later Jurassic. The adjacent margin of the West-Tethys carbonate platform had a different, geographically more restricted fauna of larger foraminiferas.

The biogeographic history of the earliest (oceanic) North Atlantic and its ancestral (cratonic) proto-Atlantic, i.e., the Hispanic Corridor, can also be inferred from the relative affinities between the eastern Pacific and Mediterranean molluscan faunas (see below under Mexican ammonites) (Fig. 2). There is a remarkable similarity between the curves for the Hispanic Corridor "openness" and for the eustatic sea level of Hallam (1978, 1981 slightly modified) but not to the curve of Vail and Todd (1981).

The Jurassic-Cretaceous Boundary

The chronostratigraphy of the North American Basin is now well established around the important Jurassic-Cretaceous boundary. At present, this time horizon can usually be placed with 3 m precision within the almost 3 km long well cores. Ascoli, Poag and Remane have established a multitaxial microfossil zonation for the Tithonian to Valanginian, from the Baltimore Canyon Trough in the South to the East Newfoundland Basin in the North. This zonation was based on 28 offshore wells drilled by different companies and achieved by first extending the existing Canadian East coast zonation to the New England Shelf with four successive calcareous/arenaceous foraminiferal-ostracod-calpionellid zones, and then correlating, using calpionellid zonation, with the Mediterranean ammonite standard zones. The Mediterranean calpionellid zones are recognized for the first time in five wells throughout most of the

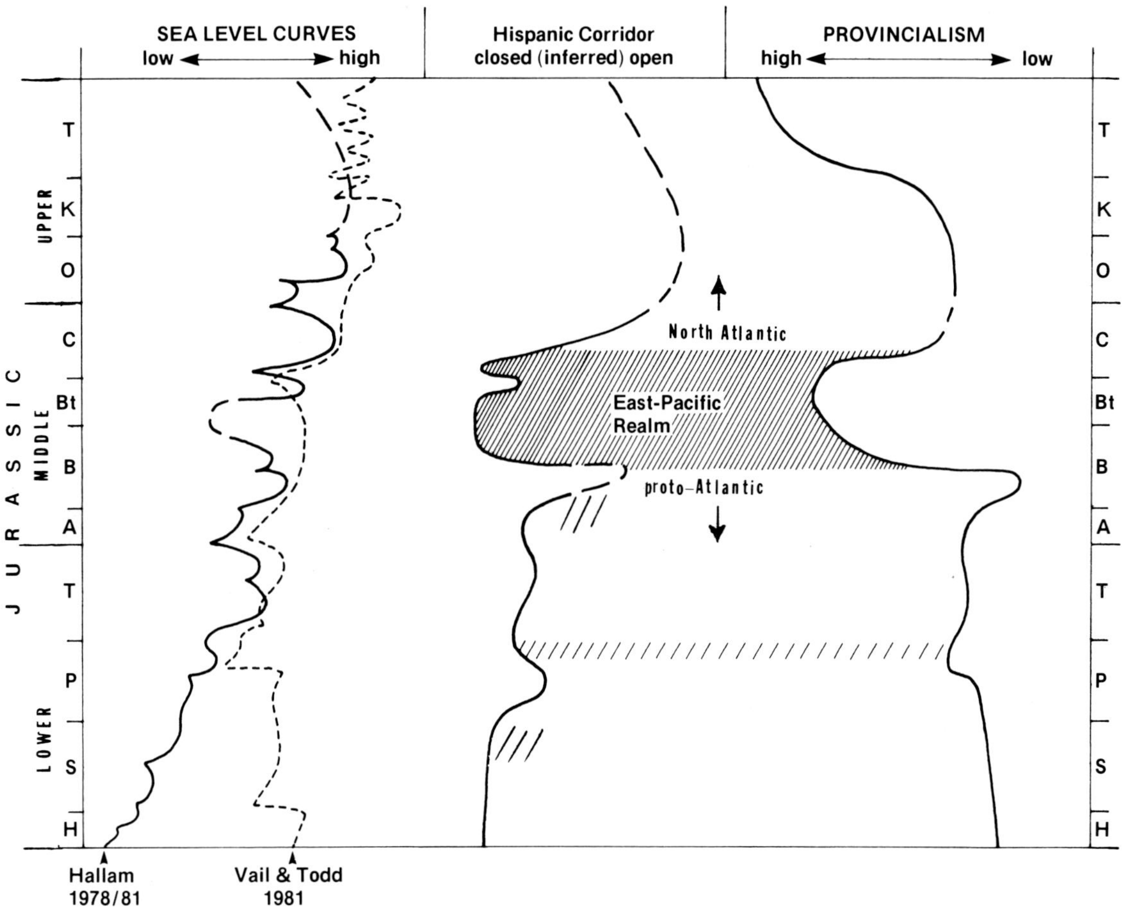

Figure 2. The Jurassic evolution of the Hispanic Corridor, i.e., the cratonic "proto-Atlantic" and oceanic North Atlantic, as inferred from the molluscan affinities between the eastern Pacific and the Mediterranean. This curve resembles the eustatic sea level curve of Hallam and, except for the earliest Jurassic, the tentative curve for global ammonoid provincialism.

North American Basin and used for the calibration of the consistently present foraminifer-ostracod zones. Calpionellids were apparently more stenotopic than the foraminifers and ostracods. As in the Lower Jurassic, the foraminiferal fauna is very similar to that in Iberia (Portugal, Spain) and Northwest Africa.

A littoral to inner sublittoral biota dominated by arenaceous foraminifers and ostracods, can be distinguished from a middle to outer sublittoral and neritic biota with mainly calcareous foraminifers and calpionellids (Ascoli *et al.*, this volume). The calpinellids were clearly planktonic as indicated by their typical occurrence in micritic and pelagic limestones, their rare association with benthos, and their frequent association with abundant radiolarians, coccoliths and nannocanids. There appears to be, however, an interesting zoogeographic dependence of habitat. While in the Mediterranean Tethys, calpionellids have usually been found in outer sublittoral and abyssal sediments, the East Newfoundland and Scotia Basin occurrences indicate only the middle to outer shelf, up to at least 50 km landward of the shelf edge.

MEXICO AND THE GULF

A convergent island arc system existed since the Triassic along the Pacific margin of North America and trench sedimentation extended along the Mexican coast. By late Cretaceous time, volcanic rocks and intrusives accumulated as far south as Panama, unless these originated in an eastern Pacific arc and were later accreted to the craton. Eastern Mexico, as part of the Gulf of Mexico Basin, lay behind this "trough island arc system" of Scott (this volume), i.e., the behind-the-arc basin of Longoria (this volume) which was partly closed by eastward thrusting at the end of the Jurassic. Partly enclosing the basin were the Florida-Bahama Platform in the northeast and the Yucatan Platform in the southeast.

Eastern Mexico

The Jurassic formations and events of Eastern Mexico are the key to the complex history of the Gulf of Mexico. The consensus of interpretation holds that the Gulf of Mexico is a Jurassic rift basin closely connected with the history of the central (southern North) Atlantic. The Gulf was structurally extended in the southeast by a left-lateral transform fault, i.e., the Sonora-Mojave and Torreon-Monterrey megashears, extending perhaps from the northwestern corner of Mexico all the way to the southern part of the Gulf (Fig. 1, arrow north of G). All of "Pacific" and southern Mexico, including the Yucatan Peninsula which originally lay within the "Gulf", would have moved some 1000 km southeastward (see Longoria; Scott; Taylor *et al.*, all in this volume). The exact position of this megashear or zone of megashears, however, remains cryptic (compare Taylor *et al.*, Fig. 21, this volume, after Campa and Coney, 1982, with Longoria, this volume, Fig. 5) owing to subsequent extensive tectonic and volcanic activity.

The initiation of the Gulf opening is evidenced by the thick sequence in rift valleys of Permo-Triassic continental deposits and volcanics, followed by Early Jurassic marine clastics (ca. 2000 m). In the Huayacocotla Mountains (Sierra Madre Oriental) deposition occurred presumably in a NW-SE graben structure (aulacogen of Schmidt-Effing, 1980) representing Tectonic Event I of Longoria. The Early Jurassic Tethyan faunas of east-central Mexico, however, tend to be cosmopolitan so that flooding from either the "proto-Atlantic" (Western Tethys) or the Pacific cannot be clearly distinguished on faunistic grounds. The early Middle Jurassic was again a time of extremely low sea-level and continental redbed deposits, i.e., Bathymetric Event II of Longoria which he related to the opening of the Gulf.

During the widespread flooding of the widening Gulf in the Middle Jurassic, evaporites were deposited, although the age of the evaporites, including massive halite, cannot be determined accurately (Imlay, this volume).

In east-central Mexico, the mid-Jurassic transgression occurred in the (present) Huayacocotla Mountains area, during the late Bathonian or Callovian. Tethyan ammonites of about Middle Callovian age (*Reineckia* s.s.) occur in the shales of the Santiago Formation, i.e., not much above the basal sandstone (Tepexic Formation) from which the Andean, ?latest Bathonian *Neuqueniceras* association has been recorded in rare unsubstantiated findings (Westermann, *in* Taylor *et al.*, this volume). The progressive regional transgression from the central Atlantic toward the Mexican interior and the persistent deepening of the sea, determined stratigraphically as well as microfaunally, continued through the Late Jurassic; at the same time, active sea floor spreading occurred in the Gulf and an island-arc system developed on the Pacific side of east-central Mexico (Longoria, this volume). While the earliest *wide-spread* marine sediments and Tethyan biota are clearly of Oxfordian age (as also described by Alencaster and Scott, both this volume), the first West-Tethyan assemblage in Mexico, as in the Andes, is the mid-Callovian *Reineckia* s.s. association which was mentioned above. It appears, therefore, that the beginning of mid-Atlantic rifting, dated as Mid-Callovian by deep-sea drill cores in the Blake Bahama basin (Sheriden and Gradstein, 1981) established the first normal salinity connection across the Gulf (see Scott). Alternatively, this *Reineckeia* fauna could perhaps have come by sea across Mexico, as Imlay and probably Alencaster believed. According to Westermann *et al.*, the pre-Callovian proto-Atlantic seaway (Hispanic Corridor) connecting West-Tethys with the eastern Pacific may have by-passed the Gulf, although this seems unlikely (see Longoria). The Late Jurassic sea reached from the Atlantic to the Pacific coast of Mexico as is evident from the molluscan faunas, including the boreal bivalve *Buchia* of northern Mexico (see contributions in this volume by Alencaster, Buitron, and Imlay).

The sea began to recede in the Early Cretaceous. This is interpreted by Longoria (this volume) as a consequence of compressional tectonism, subduction on the Pacific side and temporary cessations of sea-floor spreading in the Gulf. Shallow-water platforms, originally present mainly around the Yucatan, became extensive, except during a major Albian-Cenomanian subsidence. These platforms are noteworthy for their rudist reefs which Alencaster and Scott described. The Albian-Cenomanian deepening of the sea may be related to renewed sea-floor spreading in the Gulf (Longoria, this volume).

PACIFIC (CORDILLERAN) MEXICO

As much as eighty per cent of the southern part of the North American Cordillera of Mexico is made up of *suspect terranes* (Campa and Coney, 1983, earlier) resembling the Cordillera of Canada and the United States (see Taylor *et al.*, this volume, Fig. 21). The eastern terranes had already accreted to the craton by Jurassic time, but were cut during the Jurassic by the Mojave-Sonora and Torreón-Monterrey Megashears/Discontinuities. According to Campa and Coney, western and southern Mexico is made up largely of several distinct Jurassic-Cretaceous submarine volcanic arc terranes, mostly with unknown basements, that accreted against the North American craton during the Late Cretaceous-Early Tertiary. Much of the critical evidence, however, is hidden under thick Tertiary volcanics, e.g., the Trans-Mexico Volcanic Axis reaching from the Pacific to the Gulf. Paleogeography under the hypothesis of allochthonous terranes is, of course, diametrically opposed to the conventional reconstructions of Imlay and Alencaster (both this volume), whether or not the theory of (mega-) plate tectonics is accepted. Yet the relative and absolute positions of the supposed terranes during the Jurassic remains, so far, unknown. A good example is the Mixteca terrane of the Sierra Madre Occidental in southern Mexico which has been analyzed faunistically by Westermann (*in* Taylor *et al.*, and Westermann *et al.*, both this volume). This is the only terrane in Mexico with well documented Late Bajocian-Early Callovian ammonite faunas. The faunas consist of different mixtures of Andean and Mediterranean species together with endemic elements (Fig. 3). The biogeographic analysis places this terrane during mid-Jurassic time in the vicinity of both the proto-Atlantic seaway (Hispanic Corridor), connecting to the West-Tethys, and the Central Andes (southern Peru). The abundant phylloceratids in some Mixteca faunas also support an oceanic rather than cratonic environment, although continental beds are also present. Ammonite biogeography, therefore, tends to support the hypothesis of an allochthonous Mixteca terrane (Fig. 1), rather than the conservative view of a marine Oaxaca Embayment penetrating far inland, with a somewhat restricted northwesterly opening, as illustrated by Imlay and Alencaster.

Beginning with the Oxfordian, the ammonite faunas are clearly Tethyan and cosmopolitan throughout Mexico, reflecting the largely open oceanic connection across Mexico from the Pacific to the Gulf and West-Tethys (Alencaster, this volume). Early and Mid-Cretaceous trans-Mexican seas continue the Late Jurassic transgressive trend with land masses prevailing along the southern rim and bordered by platforms along the Gulf.

In the Neocomian-Aptian sequence, Alencaster distinguished a variety of depositional environments based on lithology and fauna. Thin-bedded limestones, commonly with chert and interbedded shales and containing ammonites, developed in the basins; platform facies with calcareous and carbonaceous shales

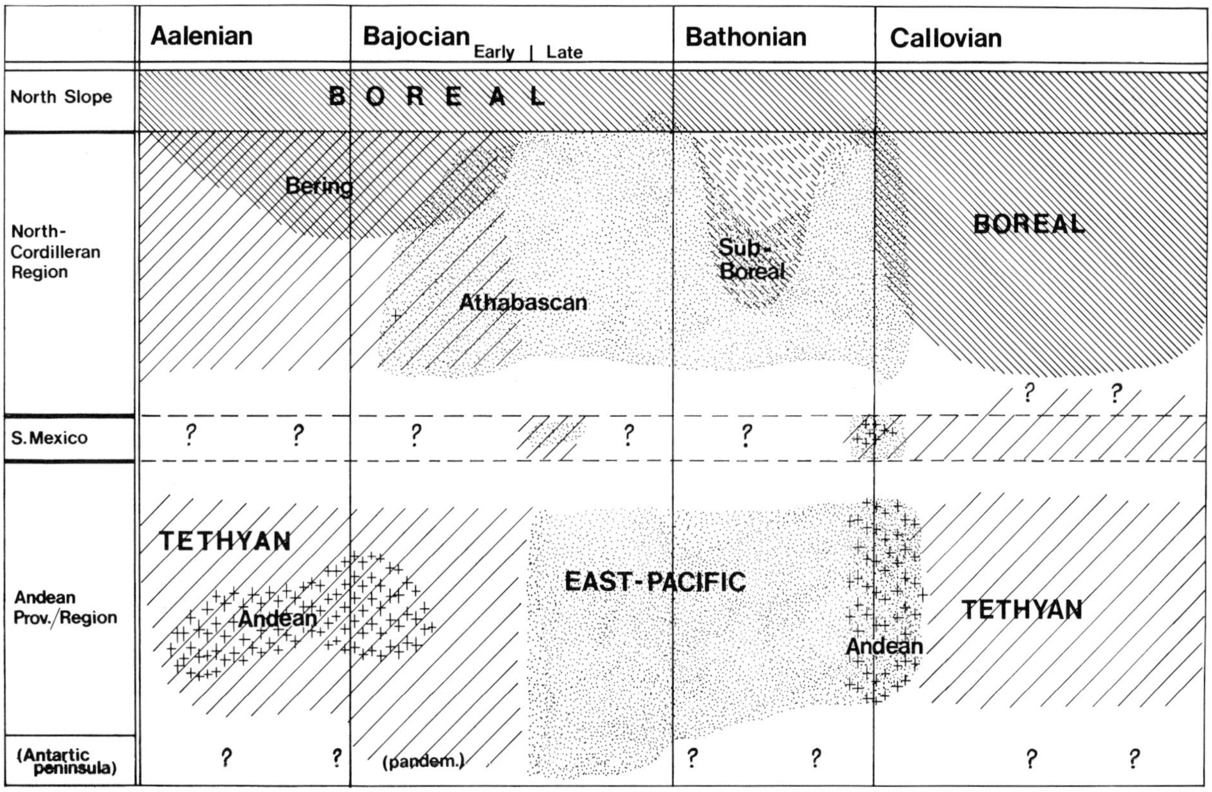

Figure 3. Biogeographic evolution of the eastern Pacific ammonoid realms, regions, and provinces during the critical time of the Middle Jurassic. For northeastern Pacific biogeographic units see Figure 4. Note the Tethyan/East-Pacific realms ecotone (overlap) in the Mixteca terrane of southern Mexico, which has predominantly Andean faunas at this time.

or nearshore clastics, contain coal beds, oysters, corals, gastropods, rudists, other bivalves, and sometimes ammonites. Abundant reef development occurred particularly during the Albian-Cenomanian maximal extent of the shallow seas. The Turonian regression continued throughout the remainder of the Cretaceous, with the Pacific margin becoming a continuous land mass and with shallow platforms receding toward the present Gulf except for the stable shallow-marine Yucatan Platform. The invertebrate faunas now belong to the Caribbean Subprovince of the Mediterranean Province (Tethyan Realm).

CORDILLERA OF CANADA AND UNITED STATES

Allochthonous Terranes and Ammonites

The Canadian part of the Cordillera is clearly divided into a western belt with abundant plutonic rocks and volcanics and an eastern belt characterized by thick sedimentary sequences. The dividing line lies roughly along the Rocky Mountain Trench in northern British Columbia and the Kootenay Arc in the south. The western belt is an excellent example of a classical "eugeosyncline" and the eastern belt of a "miogeosyncline" (Tipper, this volume). In the last 15 years radically new concepts involving both latitudinal and longitudinal large-scale displacements of numerous structural blocks (*terranes*) within the western belt have been developed (e.g., Coney *et al.*, 1980) and are followed by Tipper and by Taylor *et al.* (both this volume). The western belt consists of many fault-bounded terranes that are "suspect" of being allochthonous with respect to the North American craton and, in some cases (Fig. 1), may be exotic having come from far offshore or even from across the Pacific being accreted against the westward drifting craton. Accretion was usually followed by a northward displacement along right-lateral transform faults (megashears). The evidence for latitudinal (N-S) shifts of terranes comes from structural, paleomagnetic and, sometimes, paleontologic studies; documentation of longitudinal (E-W) shift, however, depends largely on paleontology alone, e.g., the Asiatic fusilininaceans recorded from the Permian of British Columbia (Monger and Ross, 1971).

In the Lower and Middle Jurassic, several principal tectonostratigraphic terranes can be distinguished in the Canadian Cordillera, each having internally consistent rock and fossil successions (Fig. 1). The principal three western terranes (Wrangellia, Stikinia and Quesnellia) are allochthonous and "eugeosynclinal", while the Rocky Mountains and Liard Troughs areas are autochthonous and in their relative original position on the continent. The rock units of these displaced terranes each can be grouped into four episodes of deposition separated by unconformities. The hiatuses appear to be relatively coeval among the terranes, separating depositional sequences differing only in detail and comprising volcanics, black shales and coarse clastics. The fourth episode involves a widespread marine transgression culminating in the Albian when all terranes were firmly locked together. Amalgamation of the terranes, however, probably occurred much earlier and all, with the possible exception of Wrangellia, appear to have been in roughly their present position with respect to one another by the Late Jurassic.

The disrupted Wrangellia terrane reaches northward to southern Alaska where the Peninsular terrane continues along the Pacific margin, and southward possibly into northeastern Oregon (Fig. 1). Other terranes important in the United States during the Jurassic are the North Slope terrane of Alaska, John Day of Oregon, Sonomia of Nevada, Eastern Klamath Mountains and Northern Sierra terranes along the Oregon-California border, and the Franciscan, Calaveras and Baja terranes in California (Taylor *et al.*, this volume). Most of them appear to be less well known than their Canadian counterparts because of less favourable outcrop conditions. Their history is probably similar, however, with accretion to the craton being completed during the Jurassic followed by right-lateral displacement along strike-slip faults.

The principle of biogeographic analysis of terrane displacement is based on dissimilarities between coeval fossil assemblages where the dissimilarities cannot be explained by present geographic location or facies. For example, different faunal or floral provinces/subprovinces are juxtaposed today, or two terranes belonging to the same province are in radically different latitudes today. This procedure, therefore, is based on precise biozonation for time-correlation and detailed taxonomy for biotic province recognition. Faunal analyses can, in favourable cases, provide indications that are beyond the reach of paleomagnetic studies; e.g., they can show whether a terrane at a given time was in the northern or southern hemisphere (which commonly cannot be done paleomagnetically) or provide longitudinal (E-W) constraints.

The northeastern Pacific ammonite stratigraphy and biogeography for the Jurassic have been refined in the papers by Taylor *et al.*, and Callomon. Three new ammonite provinces are distinguished (Fig. 4): 1) the Bering Province along the north Pacific margin and north slopes of eastern Siberia and northwestern North America during the Aalenian and Early Bajocian; 2) the Athabascan Province for most of the allochthonous North Cordilleran Region (North American Cordillera) from (north) Alaska to Nevada mainly during the early Jurassic; while 3) the Shoshonean Province is distinguished for the cratonic Western Interior and "Fernie Basin" (Rocky Mountains-Liard Troughs) seas during late Middle Jurassic. The first belongs to the Boreal Realm, the others to the East-Pacific Realm. It is important to realize, however, that through time, the boundaries of biogeographic units may change markedly, that the units may change rank, and that taxa may thus apparently shift between provinces or realms. Consequently, biogeography may set limits to tectonic hypotheses and be supportive of one pre-existing hypothesis rather than of another, but on their own biogeographic data rarely suffice to reconstruct paleopositions of terranes.

For the Cordilleran Jurassic, as discussed by Taylor and coauthors, the earliest ammonite distributions (Sinemurian, Pliensbachian) are also the least ambiguous indicators of movements of terranes relative to the craton. The Early Jurassic data are compatible with, and perhaps supportive for, northward displacement of at least 500 km for Quesnellia, 1800 km for Stikinia and 2500 km for the Alaskan Peninsula and Wrangellia terranes. Similar displacements may also be suggested by the Bajocian (mid-Jurassic) faunas for the latter three terranes. To the south, post-Pliensbachian movement of about 1000 km of the John Day terrane and several hundred kilometres for Oregonian Wrangellia and Sonomia are also suggested (Fig. 1).

While most of the data derived from Middle and Late Jurassic biogeography are ambivalent and also compatible with stationary positions of the terranes, the data do falsify the hypothesis of Jurassic positions of Wrangellia and the Alaskan Peninsula terrane in the *southern* hemisphere or as *far* off-shore.

Biostratigraphy and Chronostratigraphy

The construction of a supra-regional biozonation requires, as a minimum, the taxonomic definition of the local fossil assemblages, followed by their arrangement in the proper sequence or sequences. This second step becomes more difficult as provinciality and endemism increase; both reached maxima in the later half of the North American Jurassic and the earliest Cretaceous. The third and most advanced step is the global

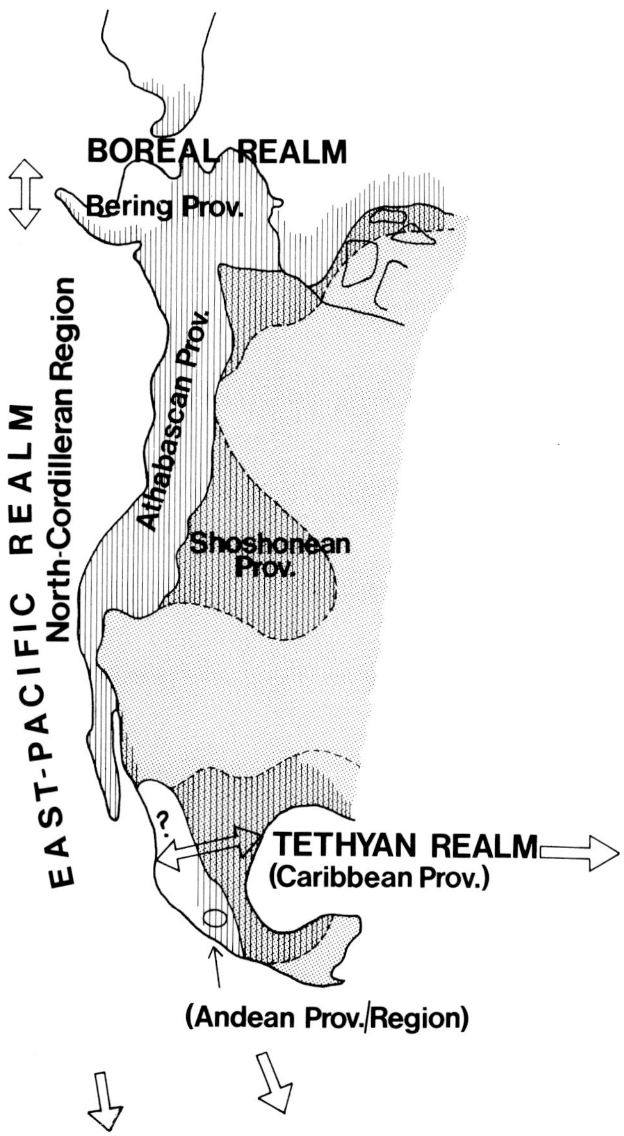

Figure 4. Jurassic ammonite realms, regions and provinces of North America; craton in grey and seas on the present continent hatched. Single-head arrows indicate the extent of biogeographic units, double-headed arrows indicate their shifting boundaries.

time-correlation of these zones, usually for historic reasons, with a standard zonation in Europe.

The upper half of the Jurassic has been the subject of thorough biostratigraphic, taxonomic, biogeographic and chronostratigraphic analysis by Callomon (this volume). He distinguished some eighty Late Bajocian to Tithonian ammonite faunas based on a critical evaluation of the extensive existing literature and aided by his personal knowledge of many of the outcrops, particularly in northern Canada and eastern Greenland. The western North American faunas, representing factual data rather than derived faunal zones, belong to four parallel time-ordered sequences which he attempted to correlate among themselves and with the standard successions in eastern Greenland and Europe. The separate sequences are based on geographical areas with distinct faunas, i.e., biogeographic provinces (Fig. 4). The epicratonic Western Interior now has been named the Shoshonean province and the mobile Cordillera the Athabascan Province (Taylor *et al.*). Both belong to the North Cordilleran Region which, in turn, is part of the East-Pacific Realm best developed during the Late Bajocian to Early Callovian (Westermann, 1981; Taylor *et al.,* this volume).

Callomon's more informal use of a "Pacific Realm" shows up the difficulties in defining the boundaries of biogeographic units. A very few ammonite genera are ambi-Pacific (e.g., the Early Bajocian *Pseudotoites*) and occur in the western Pacific together with a host of Tethyan and pandemic genera. One could, therefore, consider these parts of the western Pacific as overlap areas between the Tethyan Realm and a Pacific Realm. During the Late Bajocian to Early Callovian development of the East-Pacific Realm proper, however, only single specimens of Eastern Pacific genera (except for the curious *Neuqueniceras,* see Westermann, 1981) have been found within the rich Tethyan assemblages of Australasia (i.e., *Xenocephalites*) and Japan (i.e., a dubious *Lilloettia*). It seems prudent, therefore, to treat single specimens as (necrotic) "drifters" and to define realms by the distributions of dominant taxa.

The chronostratigraphic revision has led to the following changes of former stage correlations as exemplified by Imlay (1980; this volume). 1) All *Cranocephalites* faunas/zones present in south Alaska and the Canadian Arctic are dated as Late Bajocian; a supporting argument presented for lowering the first appearance of *Cranocephalites* (i.e., *C. borealis*) is the homeomorphy of this species with the ancestral Early Bajocian *Chondroceras.* 2) The *Epizigazagiceras* fauna of British Columbia and the *"Cranocephalites" costidensus* fauna of south Alaska, both containing *Parareineckeia,* are also lowered, from the Lower Bathonian into the Upper Bajocian, because of the associated *Leptosphinctes* [= *"Cobbanites"*]; yet *Leptosphinctes* need not here be restricted to the Upper Bajocian as in Europe, nor is its morphologic distinction from Bathonian perisphinctids as clear as supposed by Callomon. 3) The assemblages, e.g., the *Kepplerites maclearni* fauna, of the Shoshonean Province, Western Interior, formerly dated as Early Callovian, are now placed in the Late (Upper) Bathonian. 4) The supposedly Early Bathonian perisphinctid *Duashnoceras floresi* from southern Mexico is being reassigned to the late Stephanoceratinae of early Late Bajocian age, a taxononic and chronostratigraphic position recently confirmed by the field work of Westermann (*in* Taylor *et al.,* this volume). 5) Ten

Cardioceras faunas are distinguished in the Lower and Middle Oxfordian of the Western Interior, contrasting with only two previous distinguished faunas (Imlay, 1980, Fig. 15). 6) *"Titanites" occidentalis* Frebold, the largest known American ammonite (147 cm) and the only alleged Tithonian ammonite of the Rocky Mountain Trough (Fernie Basin), cannot be assigned to that genus nor to any known subfamily so that its age within the Kimmeridgian-Tithonian interval is also unknown.

WESTERN AND ARCTIC CANADA

Jurassic/Cretaceous Boundary Definition

Of great importance for time-correlations is the definition of the boundaries of the systems/periods, and the global contemporaneity of these boundaries. The exact position of the Jurassic-Cretaceous boundary in the different faunal realms and provinces remains controversial because ammonite provincialism and endemism are at an all-time high during this critical time.

Jeletzky (this volume) examined in great detail the Jurassic-Cretaceous boundary beds of western and Arctic Canada and the problem of the Tithonian-Berriasian stages in the Boreal Realm using the documentation he assembled during a life-time of work. The problems are indeed complex. Chronostratigraphic definitions and correlations have to be combined with detailed taxonomic analysis on a global scale, and the pertinent perisphinctacean ammonites and the bivalve *Buchia* are both "taxonomic nightmares." Dispute, therefore, comes as no surprise as illustrated by Zeiss' opposing position in his *Comments* (see below).

Significantly, *Buchia* species, due to their large geographic ranges across the boundaries of ammonite provinces and realms, have become essential chronostratigraphic guides while ammonite species define the standard zones. Jeletzky placed the Jurassic-Cretaceous boundary in British Columbia and the Canadian Arctic at the contact between the *Buchia terebratuloides* (s.l.) and *B. okensis* (s.s.) Zones. In Mexico, these bivalves are associated with Tethyan ammonites, i.e., those of the latest Tithonian *Substeueroceras-Berriasella* Complex and the lower Berriasian *Subthurmannia tenochi* Zone. These faunas were, in turn, considered coeval with the South American *Substeueroceras koeneni* and *Argentiniceras noduliformis* Zones respectively, and the latter zones finally correlated with the European *Paraulacosphintes transitorius* and *Pseudosubplanites grandis* Zones (base of Berriasian) of the southwestern European standard sequence. This long chain of global correlations uses, by definition, the conventional system boundary between the Tithonian and Berriasian stages as defined by the ammonite standard zones listed above, rather than the higher boundary supported, e.g., by Wiedmann (1975).

In the Canadian Arctic, the topmost Tithonian *Buchia terebratuloides* Zone has yielded ammonites, including *Praetollia federovi* and *P. maynci,* that have been used as indexes for allegedly Berriasian zones in Northern Siberia and East Greenland, but that strongly indicate to Jeletzky their Jurassic age. Furthermore, the basal Cretaceous *Buchia okensis* (s.s.) Zone is considered coeval with at least the middle and upper parts of the northern European *Hectoroceras kochi* Zone, while it is represented by a hiatus in the type Riasanian sections of central Russia.

Jeletzky considers the base of the Riasanian stage of Central Russia to be coeval with the upper *Chetaites chetae* Zone of North Siberia and places both in the Upper Tithonian rather than the Berriasian. The *Riasanites* and *"Euthymiceras"* ammonite species from the lower Riasanian, however, are said to make their appearance in southern Russia and Poland only well up in the Berriasian indicating their "slow" migration and Boreal origin. According to Jeletzky, these species are useless for age-correlations such as between the Upper Volgian and the allegedly coeval Lower (and Middle) Berriasian.

In the *Comments,* A. Zeiss (chairman, International Subcommission of Jurassic Stratigraphy) disagrees strongly with many of Jeletzky's taxonomic identifications and time-correlations. He insists instead that the Upper Volgian is coeval with the Lower Berriasian and, by extension, part of the Cretaceous. This includes the important *Buchia terebratuloides* Zone and even the earlier *Buchia fisheriana* Zone of western and Arctic Canada. In his *Reply,* Jeletzky refutes the arguments of his adversary point by point.

THE WESTERN CRATON OF CANADA AND UNITED STATES

Jurassic

The North American craton, as outlined on the maps of Taylor *et al.* (this volume), experienced repeated marine transgressions which are illustrated by a series of maps by Imlay (this volume). During the first half of the Jurassic the cratonic seas, represented by strata precisely dated by ammonites, are confined largely to marginal embayments of the Pacific in what is now eastern British Columbia and southwestern Alberta and to Arctic flooding of the north Yukon and the northwest corner of the Northwestern Territories. These embayments were probably sporadic and shifting. The Late Bajocian Western Interior sea transgressed southward and eastward into the present states of Idaho, Montana, western Wyoming and northern Utah, reaching its maximum extend (Fig. 4) in the Bathonian when the sea reached southwestern Manitoba. If Callomon's (this volume) age revision of the allegedly Early Callovian ammonite faunas as Late Bathonian is correct, the Western Interior sea withdrew completely from the craton at the end of the Bathonian (compare Taylor *et al.,* Fig. 15 with Imlay, Fig. 3). The sea advanced again by the Late Callovian and essentially covered the same basin in the Oxfordian. The sea was more stable on the northwestern corner of the craton in eastern Alaska, the northern Yukon, and a small part of the Northwest Territories. Again the entire western part of the North American craton was above sea level at the end of the Jurassic.

During the late Mid-Jurassic, as described by Taylor *et al.* and analysed by Callomon, the southern parts of the Western Interior sea contained ammonite faunas with a high proportion of endemics defining the new Shoshonean Province of the East Pacific Realm. Boreal (and Sub-Boreal) ammonite faunas that lived together with pandemics on the Canadian part of the craton belonged parly to a typically northern Pacific and Nearctic fauna now designated as belonging to the Bering Province. Tethyan ammonite faunas on the craton were limited to the Early Pliensbachian and Oxfordian of southwestern Alberta and southeastern British Columbia. (Figs. 2, 4).

Cretaceous

The well known coalescence of the Arctic and Caribbean marine embayments forming a continuous seaway for the remainder of the Cretaceous occurred during the mid-Cretaceous. The southern part of this seaway has been the subject of biostratigraphic and paleogeographic analysis by Cobban and Hook, while the molluscan biota in relation to the paleoceanography were examined in the paper of Kauffman (both this volume).

The Cenomanian-Turonian interval in the states of Utah, Colorado, Arizona and New Mexico along the western coastline of this cratonic sea permitted the zone-to-zone tracing of the transgressive-regressive cycles (Cobban and Hook). A rapid regional transgression in the Cenomanian is followed by more localized changes in sea-level and a partial regression at the end of the Turonian. A few of the regressive near-shore sandstone beds have yielded extensive bivalve and gastropod faunas, while the more off-shore shales have sparse faunas usually dominated by inoceramids and other thin-shelled bivalves. Within this relatively small region the ammonites are characterized by warm Tethyan forms in some ares and temperate forms in others.

The epicontinental seaway which flooded the interior foreland basin of North America underwent major tectonic, sedimentologic, and eustatic changes that determined the distribution of marine facies and biotas. Kauffman distinguishes five major and numerous minor tectoneustatic events which are defined by facies and strandline changes; these are identifiable throughout the basin and commonly overprint local tectonic/sedimentologic events. Eustatic rise was characteristically accompanied by active tectonism, volcanism, and massive tectonic and sediment loading along the western basin margin; basin subsidence rates were high, exceeding crustal loading, and thus preserved complete transgressive sediment suites. These geologic events had a marked influence on the biotope. The rises in sea level resulted in the introduction and expansion of marine habitats and the amelioration of climates. Furthermore, biogeographic boundaries gradually shifted north until near peak transgression, when isotopic and biotic data demonstrate abrupt temperature and/or salinity increases within the basin and rapid spread of subtropical organisms (warm water subprovinces) up to 1000 miles north into the basin centre. During the eustatic rises, Kauffman argued, the concentrations of anoxic events, intervals of sediment starvation, deep basinal environments and currents, and abrupt paleoceanographic changes broadly modified the composition and geography of marine ecosystems. Eustatic fall, on the other hand, was characterized by reduced tectonism, volcanism, and subsidence. This resulted in the filling and shallowing of the basin and a great reduction in marine habitats. Temperature and salinity levels gradually dropped so that a slightly brackish sea developed. As a consequence, biogeographic units gradually retreated southward and diversity decreased markedly within ecological units. It comes, therefore, as no surprise that evolutionary patterns reflect these major environmental events; Kauffman suggested that major extinctions can be correlated with both peak eustatic rise and fall events.

Environmental and biotic events ("event stratigraphy") in the Cretaceous Western Interior Basin have been dated with the exceptional resolution of approximately 100,000 years (0.1 Ma), a secular resolving power of 1,000.

REFERENCES

Campa, M.F. and Coney, P.J., 1983, Tectono-stratigraphic Terranes and Mineral Resource Distributions in Mexico: Canadian Journal of Earth Sciences, v. 10, p. 240-252.

Coney, P.L., Jones, D.L., and Monger, J.W.L., 1980, Cordilleran Suspect Terranes: Nature, v. 28, p. 329-333.

Hallam, A., 1978, Mesozoic Geology and the Opening of the North Atlantic: Journal of Geology, v. 79, p. 129-157.

———, 1981, A Revised Sea-level Curve for the Early Jurassic: Journal of the Geological Society of London, v. 138, p. 735-743.

Imlay, R., 1980, Jurassic Paleobiogeography of the Conterminuous United States in its Continental Setting: United States Geological Survey Professional Paper 1062, p. 134.

Monger, J.W.H., and Ross, C.A., 1971, Distribution of Fusulinaceans in the Western Canadian Cordillera: Canadian Journal of Earth Sciences, v. 8, p. 259-278.

Schmidt-Effing, R., 1979, The Huayacocotla Aulacogen in Mexico (Lower Jurassic) and the Origin of the Gulf of Mexico: *in* The Early Opening of the Central North Atlantic: Luisiana Geological Society, p. 75-86.

Sheridan, R.E., and Gradstein, F.M., 1981, Early History of the Atlantic Ocean and Gas Hydrates in the Blake Outer Ridge, Results of the Deep Sea Drilling Project: Episodes, v. 1981, no. 2, p. 16-22.

Vail, P.R., and Todd, R.G., 1981, Northern North Sea Jurassic Unconformities, Chronostratigraphy and Sea-level Changes from Seismic Stratigraphy: *in* Illing, L.V. and Hobson, G.S., ed., Petroleum Geology of the Continental Shelf of Northwest Europe: Heydon, London, p. 216-235.

Westermann, G.E.G., 1981 (preprint 1980), Ammonite Biochronology and Biogeography of the Circum-Pacific Middle Jurassic: *in* House, M.R., and Senior, J.R., eds., The Ammonidea: Systematics Association Special Volume 18, p. 459-498.

Wiedmann, J., 1975, The Jurassic-Cretaceous Boundary as one of the Mesozoic System Boundaries: *in* Colloque sur la Limite Jurassique-Crétacé, Lyon, Neuchatel 1973: Bureau de Recherches Géologiques et Minières (Luxembourg) Mémoir 86, p. 355-362.

The Geological Association of Canada Special Papers

20 **The Continental Crust and Its Mineral Deposits: A Volume in Honour of J. Tuzo Wilson**
Edited by D.W. Strangway, 1980

21 **Cretaceous Rocks and Their Foraminifera in the Manitoba Escarpment**
By D.H. McNeil and W.G.E. Caldwell, 1981

22 **The Buchans Orebodies: Fifty Years of Geology and Mining**
Edited by E.A. Swanson, D.F. Strong and J.G. Thurlow, 1981

23 **Sedimentation and Tectonics in Aluvial Basins**
Edited by A.D. Miall, 1981

24 **Major Structural Zones and Faults of the Northern Appalachians**
Edited by P. St-Julien and J. Béland, 1982

25 **Precambrian Sulphide Deposits: H.S. Robinson Memorial Volume**
Edited by R.W. Hutchison, C.D. Spence and J.M. Franklin, 1982

26 **Glacial Lake Agassiz**
Edited by J.T. Teller and Lee Clayton, 1983

27 **Jurassic-Cretaceous Biochronology and Paleogeography of North America**
Edited by G.E.G. Westermann

Orders and requests for information should be sent to
Geological Association of Canada Publications
Business and Economic Service Ltd.
111 Peter Street, Suite 509
Toronto, Ontario M5V 2H1